Thawing Permafrost

J. van Huissteden

Thawing Permafrost

Permafrost Carbon in a Warming Arctic

Springer

J. van Huissteden
Faculty of Science
Vrije Universiteit
Amsterdam, The Netherlands

ISBN 978-3-030-31378-4 ISBN 978-3-030-31379-1 (eBook)
https://doi.org/10.1007/978-3-030-31379-1

This Springer imprint is published by the registered company Springer Nature Switzerland AG.
The registered company address is: Gewerbestrasse 11, 6330 Cham, Switzerland

Preface

That there is something wrong with the climate, and permafrost has something to do with it, is a vague public notion. For the media, permafrost is now and then a source of alarming messages from a strange and faraway world in the north. There are indeed reasons for serious concerns, and even alert, as some recent scientific publications show. However, alert or not, a better understanding of the permafrost world and its environmental processes and ecosystems is urgently needed. The 'permafrost carbon feedback' is an uncertain force in the climate system. Permafrost thaw is not just a change of soil temperature. If permafrost thaws, changes occur from a landscape scale down to the scale of tiny interstices in soils where greenhouse-gas-producing microbes live. Entire ecosystems may change profoundly and abruptly.

The permafrost carbon feedback is the subject of this book. Permafrost soils are a large reservoir of frozen organic matter – the amount of carbon is nearly twice the CO_2-carbon in the atmosphere. It has been compared to a freezer that is switched off: its organic contents are exposed to rapid decomposition. The release of carbon from thawing permafrost is a form of feedback in the climate system that enhances human-induced global warming. Warming of Arctic soils results in a perturbation of the carbon cycle of permafrost surface soils, and deeper subsoil. The result is an enhanced release of greenhouse gases. This increases the societal and economic challenge to reduce greenhouse gas emissions and mitigate climate change.

However, matters are not as simple as that. How fast will the decomposition be? What greenhouse gases are emitted – the well-known carbon dioxide, the stronger greenhouse gas methane or the fairly unknown but far more potent nitrous oxide? What are the transient and long-term effects? To what extent are ecosystems resilient to large future environmental changes, and can they compensate for carbon losses by increased CO_2 storage? To answer these questions, research efforts on the effects of climate change on permafrost soils and ecosystems have hugely intensified the last decennia. This includes the efforts of scientists in experimenting to understand permafrost and its ecosystems, and in monitoring the remotest permafrost areas, under often challenging conditions. This requires significant innovations in instrumentation and research methods and has spurred much interdisciplinary work in the natural sciences.

Meanwhile, anthropogenic global warming has become a scientifically recognised fact, although it is still the subject of political–ideological clashes. However, Arctic temperature records speak for themselves, and those who live in the Arctic or visit Arctic areas frequently have hands-on experience with recent changes. Based on my own observations in the Siberian tundra, the scientific literature on changes in other permafrost areas and the projections of future permafrost, I realized that I was studying a world that is changing rapidly. At the same time, the complexity of the permafrost part of the Earth system became more and more apparent.

When I was asked to write a book on the permafrost carbon cycle by Springer Verlag, there was some hesitation on my part, given the wide range of topics that have to be covered. A meaningful overview of this subject is quite a task, given the large amount of research papers that have been published, in a large range of specialist and non-specialist journals, from geophysics to biology. It is a moving target, as shown by the number of new papers that are appearing each month that present new data on rapidly changing permafrost or integrate new fields of science in the subject, such as machine learning or metagenomics.

However, from a quick scan of existing textbooks, it is learned that there were not many texts that integrate physical and ecosystem aspects of the carbon cycle of permafrost. Moreover, the recent developments in the science of the permafrost carbon cycle and the rapid changes that occur in permafrost regions justify a textbook on the recent scientific developments in this field. Therefore, the focus in this book lies on recent research of the last two decades and on recently reported phenomena that show the increasing impact of climate change on permafrost, such as abrupt thaw. The subject matter of the book is restricted to northern hemisphere high latitudes. This is not to dismiss the importance of Antarctic permafrost or the permafrost on the Tibetan Plateau (the 'Third Pole'), but the largest masses of permafrost carbon occur in the north of Eurasia and North America.

This book is intended for an audience that has a background in natural sciences but is relatively new to the subject of permafrost. It could serve as an introductory and reference work for scientists embarking on permafrost carbon research (at MSc or PhD level) or for science writers active in communication of science to the public. It is not intended specifically as a course textbook, although the text undoubtedly can be used for course building, and parts of the text can serve as reference material. With this book, I hope to have contributed to a better use of recently gained knowledge on permafrost and permafrost ecosystems and further the advancement of this knowledge.

Amsterdam, The Netherlands J. van Huissteden

Acknowledgements

Many people have contributed directly and indirectly to this book. First, I want to thank Han Dolman, Jorien Vonk, Jef Vandenberghe and Juul Limpens, who reviewed all or some chapters of this book. Jef Vandenberghe was my promotor in 1990, and he introduced me to paleo-permafrost during my MSc thesis research. I also want to thank many colleagues in the permafrost research community who permitted me to use graphs, diagrams and photos from their own publications and who kindly provided me with high-quality copies of this material.

Without my own research experience in Siberia, this book could not have been written. This research would have been impossible without the help of the scientists of the Siberian Branch of the Russian Academy of Science institutes, with whom I collaborated in Siberia; the Institute for Biological Problems of the Cryolithozone (IBPC) in Yakutsk; and the Institute for Water and Environmental Problems in Barnaul. My special thanks go out to Prof. Trofim Maximov, Dr. Alexander Kononov, Sergey Karsanaev and Roman Petrov for accompanying me and my colleagues on field trips to the north of Yakutia and to Prof. Alexander Puzanov and Dr. Nelley Kovalevskaya for their cooperation with INSAR work in Yamal.

Tatyana Strioekova and her crew of the Fisheries and Nature Inspection in Chokurdagh are thanked for their logistic help and hospitality. Their efforts have made the establishment of a research station in the Siberian tundra a success. I also thank the many colleagues at my own department, Wageningen and Zürich University, who accompanied me during field work and contributed to the research work at the station: Frans-Jan Parmentier, Cinzia Berrittella, Yanjiao Mi, Artem Budishchev, Josh Dean, Kanayim Teshebaeva, Ron Lootens, Monique Heijmans, Daan Blok, Bingxi Li, Peng Wang, Gabriela Schaepman and her team. Over time, many other foreign colleagues visited the Kytalyk site and other sites near Chokurdagh. My special thanks are for Lutz Schirrmeister, for helping me out with a broken generator. Furthermore, I owe much to colleagues whom I met at the regularly held workshops where we discussed research work in Eastern Siberia and at permafrost conferences: Prof. Atsuko Sugimoto, Prof. Takeshi Ohta, Prof. Oleg Anisimov, Prof Hans Joosten.

The Dutch governmental science funding organisation NWO and the European Union Framework Programmes (in particular with the TCOS and PAGE21 programmes) have provided the research funding that is needed to do permafrost research in the high north of the Eurasian continent.

Last but not least, I want to thank my wife, Jet Rootlieb, who had to endure an all-too-busy husband during the writing of this book, at a time when other people start to enjoy their retirement. My daughter Corine van Huissteden checked the text, with special attention to terminology that needs explanation for non-specialists.

About the Book

The first four chapters have an introductory character, providing the more basic knowledge on permafrost and soil processes, without referring much to future changes of permafrost. In Chaps. 5, 6, 7, and 8, the emphasis is laid on future changes and how these will affect the carbon cycle. Chaps. 5, 6, and 7 focus on observations and experimental results and Chap. 8 on models.

- Chapter 1 is an introductory chapter giving a basic overview of permafrost and its environment, in the context of past, present and future climate.
- Chapter 2 treats the energy fluxes from the atmosphere to the soil and vice versa in more detail, in particular how these are modified by vegetation and snow cover.
- Chapter 3 explores the role of ice in permafrost: how permafrost accumulates large amounts of ice and the role of ground ice when permafrost thaws. In particular, the geomorphological and hydrological effects of permafrost thaw are included here.
- Chapter 4 is the carbon cycle chapter. It describes how carbon accumulates in permafrost soils, how it is preserved and what determines the rate of decomposition of this carbon. It includes also the ecosystem processes that are responsible for methane emission in permafrost wetlands.
- Chapter 5 treats physical changes of permafrost soils that are expected to influence the carbon cycle of permafrost: increase of temperature, active layer thickness, subsidence, erosion and hydrological change. It discusses several field experiments that may shed light on future changes.
- Chapter 6 includes the changes in vegetation by shifts in climate, how these may affect the carbon cycle and permafrost carbon stores, and permafrost temperature. Included are the effects of permafrost thaw on vegetation and a section on the carbon cycle effects of other human disturbances of vulnerable permafrost environments.
- Chapter 7 focusses on a specific subject: deep permafrost sources of methane as opposed to the superficial wetland ecosystem source.

- Chapter 8 includes models in the discussion. For projecting future permafrost changes, models are essential. However, these model projections also strongly depend on how permafrost ecosystem processes are represented in models. The chapter discusses how permafrost carbon cycle processes are represented in models and discusses improvements to the modelling of permafrost.

Contents

About the Author

I received my training as a physical geographer in the 1970s and 1980s, when human-induced climate change began to become an issue but was not yet at the forefront of public attention. My thesis research on ice age fluvial deposits in the Netherlands (published in 1990 at VU University, Vrije Universiteit in Amsterdam, Netherlands) was my first intensive experience with permafrost because of the ubiquitous soil and sedimentary structures testifying to the presence of Pleistocene permafrost. These sedimentary structures also testify of permafrost degradation, for instance the formation of thaw lakes. Since 1992, I have been affiliated with VU University as an assistant and associate professor. After my thesis research, I have spent about a decade researching soil conservation in temperate peatlands, including carbon flux measurements and model development. However, permafrost kept my interest, and I grabbed with both hands the opportunity to start research on permafrost carbon emissions in Northeastern Siberia.

Thanks to my Russian colleagues of the RAS Institute for Biological Problems of the Cryolithozone (IBPC), a new research station had been established a few years before in the tundra of the Indigirka Lowlands. Siberia contains the largest permafrost area in the world, and only a handful stations existed in this area, with state-of-the art carbon flux measurements. Despite all logistical challenges, the Kytalyk research station has provided energy balance and carbon flux data and has grown in importance. Research at the station has provided input for many research papers. The input from biologists, from the IBPC and from Wageningen and Zürich Universities has been very important here and helped me to develop a broader understanding of ecosystem processes, besides the physical side of the permafrost system. My research work on paleo-permafrost and 14 years of permafrost research in Siberia has resulted in 53 research papers, as first author or contributor.

Chapter 1
Introduction

Abstract The permafrost carbon feedback is the potential release of greenhouse gases from thawing permafrost that enhances global warming. Research on permafrost – perennially frozen ground – therefore has shifted its focus from physical frozen ground processes towards the carbon cycle of permafrost soils. In this chapter, a general introduction is given to basic concepts with respect to permafrost. Next, an overview of environmental conditions of High Latitude permafrost regions describes their variation of present climate, vegetation, ecosystems, soil carbon, geomorphology and geology since the Last Glacial. The projections for future climate in permafrost regions are discussed, based on the IPCC climate model projections. The uncertainty in these climate model projections results mainly from future greenhouse gas emission scenarios, determined by the success or failure of human societies to curb greenhouse gas emissions. The permafrost carbon feedback has important social consequences: the stronger this feedback is, the larger the human effort to moderate climate change needs to be.

For many people, permafrost is something of desolate barren lands in faraway northern regions. However, permafrost stretches far south in the North-American and Eurasian continents; large areas of permafrost are covered with forests, it is not uncommon to see agricultural activities in the southern part of the present-day permafrost zone, and cities have been built on permafrost. Nearly one quarter of the land surface of the northern hemisphere has permafrost soils. During glacial periods, permafrost reached present-day temperate latitudes.

The potential emission to the atmosphere of the greenhouse gases carbon dioxide and methane from permafrost soils is the prime reason for the strong increase of scientific interest in permafrost. It is a feedback in the climate system that enhances human-induced global warming as shown by model projections (e.g. Schuur et al. 2008, 2015), but is subject to large uncertainties (IPCC 2013). Besides the frozen permafrost soil itself, the ecosystems and their reaction to climate change play an important role.

J. van Huissteden, *Thawing Permafrost*,
https://doi.org/10.1007/978-3-030-31379-1_1

This chapter serves as a general introduction to world of permafrost for those that are not yet familiar with it, and as a general introduction to the relation of permafrost with climate, and the climate change that is expected to deeply affect permafrost.

1.1 What Is Permafrost and Where Does it Occur?

Although the word *permafrost* suggests something that is permanently frozen, permafrost needs neither be permanent nor frozen. The official definition of, endorsed by the International Permafrost Association, is: Ground (soil or rock) that remains at, or below 0 °C for at least 2 years (AGCR 1988; French 2018). This does not mean that the ground is *frozen,* (that is, contains ice); water in the pores of the soil or rock may be too salty to freeze. Such material (with temperature below zero, but lacking ice) is called cryotic. Second, the '*at least two years*' in the definition means that permafrost does not need to be truly permanent. The time limit of 2 years in the definition is meant to distinguish permafrost from soils that have below-zero temperatures only seasonally. Permafrost is a significant part of what is known as the cryosphere – the part of Earth's surface where water occurs as ice, including lake and sea ice, snow cover, glaciers and ice caps.

Permafrost occurs in cold climate regions in the north and south of the Earth, and at high elevations, e.g. the Alps in Europe or the Tibet Plateau. Figure 1.1 shows a map of Northern Hemisphere permafrost distribution. The large permafrost areas are located on the northern American and Asian continents. In particular in the Eurasian continent, permafrost stretches far south. Besides the extensive permafrost areas in the north, also the Tibetan plateau contains a large area of permafrost because of its high elevation. Below as much as 23.9% of the ice-free land area of the Northern Hemisphere permafrost may occur (Brown et al. 2002; Zhang et al. 2008). Another estimate of potential permafrost area, based on mean annual air temperatures, indicates that 19–25% of the Northern Hemisphere land area may contain permafrost (Gruber 2012). Permafrost also occurs on the sea floor of the Arctic Ocean, for reasons explained below. Going southward on the map of Fig. 1.1, we see successive zones of continuous, discontinuous, sporadic and isolated permafrost. Therefore, the permafrost percentages above do not indicate the actual area of frozen soil; in large parts of permafrost realm, permafrost is only present at favourable locations, determined by vegetation and soil conditions (see below).

At the soil surface of permafrost soils, the top layer – called the active layer – thaws in summer and refreezes in winter each year. This active layer may have a thickness ranging between a few decimetres to more than a metre. The base of the active layer is often referred to as permafrost table – by analogy to water table. The active layer is generally thinnest in cold climates and thickest at the southern limits of the permafrost zone. However, it is not only the temperature above the permafrost that determines the temperature regime in the soil, but also surface conditions like vegetation cover, topography, winter snow cover and soil properties including the soil water content. These will be discussed extensively in the next chapters. Climate warming will cause an increase of the active layer thickness, affecting the carbon and nutrient cycle of permafrost soils. Active layer thickness (abbreviated ALT throughout this book) is an important permafrost state variable.

Below a certain depth – in the order of 10–20m – the daily and seasonal temperature fluctuations at the surface do not penetrate. This is the isothermal zone of the permafrost. Figure 1.2 shows a schematized vertical temperature profile of permafrost. In the isothermal zone, the permafrost temperature is determined only by the geothermal heat gradient (the temperature gradient caused by to heat flow from the Earth's interior), the yearly average temperature at the surface, and by long-term climatic fluctuations (see Chap. 2; Romanovsky et al. 2002). Note that in Fig. 1.2, the base of the permafrost is the depth at which the temperature reaches 0 °C, not

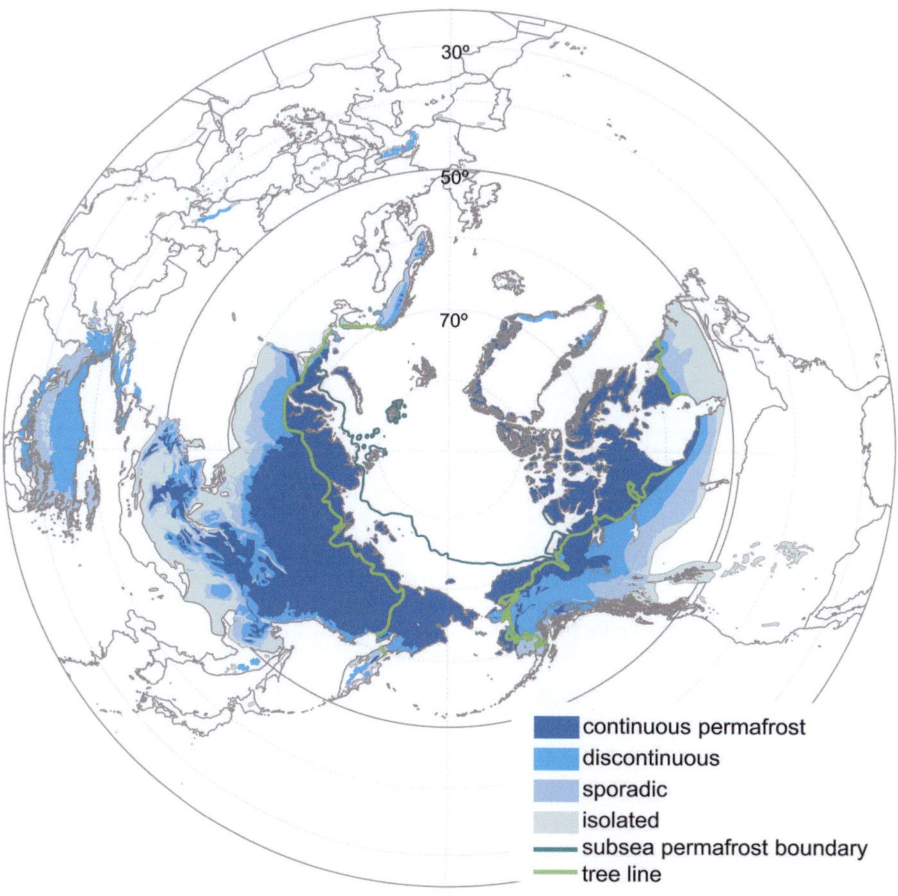

Fig. 1.1 Distribution of permafrost on the Northern Hemisphere and Antarctica (Brown et al. 2002; International Permafrost Association; based on GIS data from National Snow and Ice Data Center, accessed July 2019, http://nsidc.org/data/ggd318)

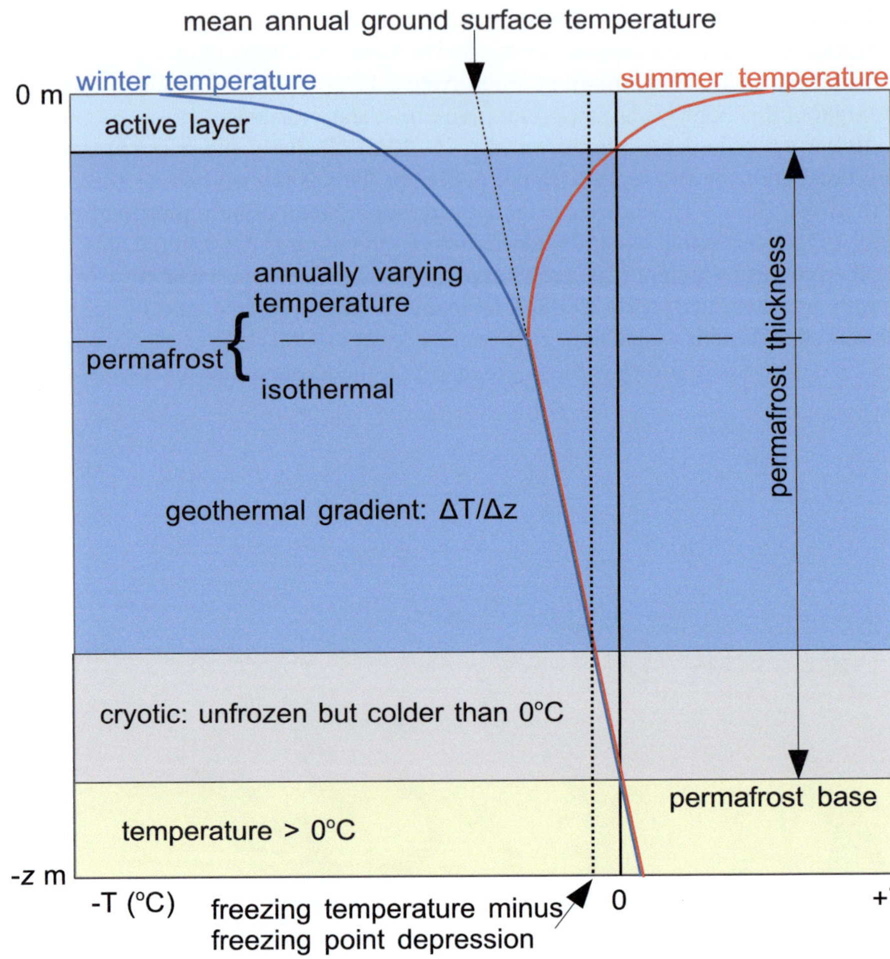

Fig. 1.2 Schematic vertical temperature profile through permafrost, from base to top. The blue line is a typical winter temperature profile, the red line a summer temperature profile. The thin dashed line represents the geothermal temperature gradient; the thick dashed vertical line the freezing temperature corrected for freezing point depression after French 2018)

the depth at which the deepest ice is found, in line with the definition of permafrost. The groundwater may contain salts, which cause freezing point depression (decrease of the temperature at which ice forms). This results in a cryotic layer at the base of the permafrost.

The lower boundary of the permafrost is determined by the local geothermal heat gradient, thermal properties of subsurface rocks and the average surface temperature (Fig. 1.2). The depth of this boundary ranges from more than a kilometre in High Arctic areas to a few tens of metres (Williams and Smith 1991). Extreme depths of up to 1470 m have been found under the Anabar plateau in northern

Siberia (Tumel 2002; Serreze and Barry 2014). Large permafrost depths of hundreds of metres are associated with areas of the Arctic that have not been covered with ice caps during the Pleistocene, such as lowlands in northern and eastern Siberia. The thickness of permafrost can be estimated from:

$$T_z = T_s + z\frac{\Delta T}{\Delta z} \tag{1.1}$$

Here $\Delta T/\Delta z$ is the geothermal temperature gradient (the average increase of temperature with depth), T_s the average temperature at the ground surface, and T_z the temperature at depth z (French 2018). However, thermal conductivity of rocks and the geothermal heat flux show considerable spatial and depth variations which influence the permafrost thickness, therefore Eq. 1.1 is only a generalised approximation.

Permafrost at large depths reacts slowly to climate change, resulting in relict permafrost at depth in areas where the current climate and land surface conditions do not support stable permafrost (Delisle 1998; Osterkamp and Romanovsky 1999; Kitover et al. 2012, 2013). Relicts of glacial age permafrost in the shape of a 93 m thick permafrost layer have been found at depths below 350 m in Poland in an area of low geothermal gradient (Šafanda et al. 2004; Szewczyk and Nawrocki 2011). In the discontinuous permafrost zone of Alaska, 740 thousand years old relict permafrost ice, indicating formerly continuous permafrost, has been found close to the surface (Froese et al. 2008).

Within the permafrost, rock or sediment bodies may exist that are not frozen, or have a temperature above 0 °C. Such a feature is called a talik (Fig. 1.3); taliks are important for groundwater flow in permafrost. Here again, the distinction between unfrozen (not containing ice) and temperature can be made. Taliks may be unfrozen but have a below-zero temperature, in which case they are called cryotic taliks; noncryotic taliks have a temperature above 0 °C (ACGR 1988). Taliks with tempera-

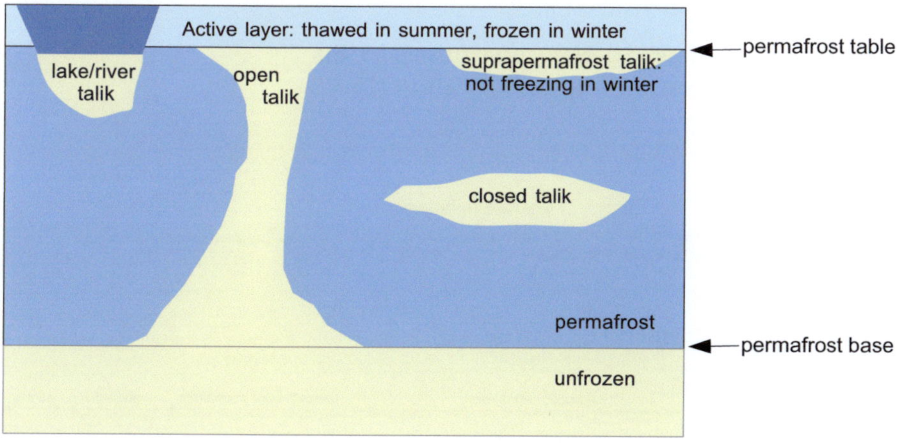

Fig. 1.3 Types of unfrozen bodies (taliks) within permafrost

tures above zero may result from local permafrost thaw as a result of climate warming or changes of the ground surface which add heat to the permafrost. Taliks develop below water bodies because of their thermal properties (Chap. 2). Quite often, talik development and eventual refreezing of taliks is related to changes in river courses or lakes. If a talik occurs close to the surface, with only in winter frozen soil above it, it is called a supra-permafrost talik. Non-cryotic taliks within the permafrost may be the result of past permafrost thaw, after which refreezing of permafrost from above has isolated the unfrozen layer from the active layer. Taliks further can be classified as open or closed. Open taliks connect the active layer with unfrozen material below the permafrost; closed taliks are isolated unfrozen layers entirely enclosed by frozen material. Often, they develop by refreezing of open taliks or lake taliks.

The geographical limits of permafrost are determined in general by the climate. Where mean annual air temperatures are below 0 °C, permafrost may occur. However, this limit is quite diffuse. The local occurrence of permafrost is governed by the soil heat balance, which in turn is influenced by microclimate, vegetation and soil thermal properties (Brown et al. 2002). For instance, at the southern limits of permafrost, peat soils tend to contain more permafrost than mineral soils because of the thermal properties of peat. The soil heat balance and soil thermal properties are discussed in Chap. 2. Relict permafrost, which exists because thaw did not progress deep enough to make it disappear, is common at the southern permafrost boundary. The examples of relict ice age permafrost from Poland cited above are an extreme case; relict permafrost due to smaller amplitude surface climate change than glacial-interglacial cycles is common.

Broad geographical zones are distinguished in permafrost, based on spatial continuity (Fig. 1.4, Zhang et al. 2008):

- Continuous permafrost: 90% or more of the land surface underlain by permafrost (46.9%);
- Discontinuous permafrost: 50–90% permafrost, with locally non-permafrost areas (19.2%;
- Sporadic permafrost: 10–50% permafrost (17.1%);
- Isolated permafrost: 10% or less permafrost (16.8%).

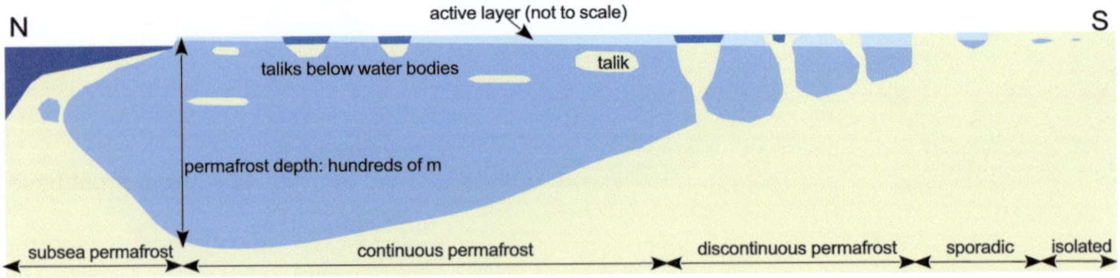

Fig. 1.4 North-south vertical cross section through continental Arctic permafrost, indicating zonation of permafrost based on spatial continuity of permafrost presence

The numbers in brackets are the percentage of total permafrost area covered by the respective permafrost classes. Continuous permafrost, occupying nearly half of the total permafrost region, occurs where the mean annual air temperature (MAAT) is several degrees below zero; in North America its southern boundary follows broadly the −6 to −8 °C MAAT isotherm (French 2018). The range of actual permafrost percentages in the permafrost classes is very broad and based on very large scale mapping. A more precise approach is given by Gruber (2012) based on probabilities of permafrost occurrence attached to MAAT.

Besides permafrost on land, subsea permafrost is found on the Arctic Ocean shelf (Figs.1.1 and 1.4). During the ice ages, a part of the now flooded seafloor was a wide, unglaciated, lowland area, where the cold glacial climate resulted in growth of thick permafrost. During the Holocene, these areas were flooded by the sea. In the Laptev sea north of Siberia continuous subsea permafrost is found down to seafloor depths of 65 m, discontinuous permafrost occurs down to depths of 100–120 m (Romanovskii et al. 2004). Because of the continuous above-zero sea water bottom temperatures, this subsea permafrost is a relict and will ultimately disappear.

Despite its size, Greenland contains relatively small areas of permafrost. Most of Greenland is covered by an ice cap. Beneath glaciers and ice caps, often liquid water exists. Because of the pressure of the ice beneath a glacier or ice cap, the melting point of ice is lower. Furthermore, frictional heat generated by glacier ice movement, the insulating effect of ice, and geothermal heat flow contribute to melting of ice at the base of ice caps and glaciers. Therefore, the limit of glaciers is taken on maps also as the limit of permafrost. However, parts of the Greenland ice cap, both interior parts and northern fringes, are frozen to the bottom (MacGregor et al. 2016). If permafrost is present below a glacier or ice cap, the ice is cold-based; below warm-based ice there is no permafrost (Sugden and John 1976).

Although the presence of ice is not a necessary condition of permafrost, it occurs abundantly in permafrost soils, and is intimately linked to the carbon cycle of permafrost. A very important feature is the occurrence of excess ice. The normal porosity of loose sediments or soil material is in the order of 30 to 40% of the volume. Water expands when it freezes by 9%. However, soils and sediments in permafrost

Fig. 1.5 Ground ice in ice-rich Pleistocene deposits in northeastern Siberia. (Photo by the author)

areas may contain much more ice than results from expansion of freezing alone; 'soils' with locally over 100 weight % of ice occur. If a frozen soil column with excess ice thaws, it will settle, with a column of water remaining above the water-saturated soil material. The volume of this supernatant water relative to the total volume is a measure for the amount of excess ice (French 2018). Accumulation of excess ice in the soil results in uplift of the soil surface, known as frost heave. There are several processes that cause accumulation of ice in soils over time, such as ice segregation forming ice lenses, and the formation of subsoil ice bodies like ice wedges and injection ice. This will be discussed in Chap. 3. The total mass of ice in permafrost soils is known as ground ice (Fig. 1.5). When permafrost contains frozen soil material and the soil temperature is raised above the freezing point, we speak of thawing permafrost, not melting permafrost. It is only the permafrost ice that melts.

Permafrost terminology is often derived from languages spoken in the areas where permafrost occurs. We have seen already the word 'talik', which is derived from a Russian word for thaw, таять (tayat'), several other terms derived from local languages will follow.

1.2 Research on Permafrost: A Shifting Focus from Ice to Carbon

Scientific research on permafrost covers many aspects of the natural sciences. The changes occurring in permafrost areas, and their increasing population also have spurred interest from the social sciences. The earliest scientific studies at the end of the nineteenth century and the early twentieth century were of an explorative nature, often by geologists. For instance, a classical work is the doctoral thesis of De Leffingwell (1919) on the Alaskan North Slope, who was the first to study the effects of intense frost action on tundra soils. Next, large scale exploration of the Arctic for fossil fuels and military activities in the previous century initiated geotechnical engineering studies, to develop stable buildings and infrastructure on permafrost.

Soil scientists (e.g. Tedrow 1977) have studied permafrost soils from the perspective of increasing agricultural activities at the southern fringes of the Arctic, and to understand soil forming processes under extreme climatic conditions. Soil forming processes are a range of chemical, physical and biological processes that contribute to the development of a vertical profile of varying soil composition (organic matter, minerals, precipitates like iron oxides and salts). The rates of these processes also depend on climate. In particular around round major population centres in Subarctic regions there is agricultural activity, including growing of crops.

Biologists have long been interested in adaptations of ecosystems and species to the extreme Arctic climate conditions. This research was stimulated by the increasing interest in the High North after World War II, and by the '60s of the twentieth century a large number of scientific studies on permafrost soils and ecosystems appeared in print. In particular the Arctic research sites in the North American LTER (Long Term Ecological Research) network have contributed much to understanding of the human-induced changes in ecological processes, energy and carbon flows in Arctic terrestrial ecosystems (e.g. Oechel et al. 1993; Hobbie et al. 1995).

Another reason to study Arctic land ecosystems is their analogy with Pleistocene deposits from cold glacial climates. The theory of ice ages advanced in 1837 by the Swiss-American naturalist Louis Agassiz, also spurred the interest of geologists in High Latitudes. Soon it was realized, that not only glaciers advanced during the cold climate episodes of the Pleistocene, but also cold climate zones bordering the glaciers moved southwards. This is known as the periglacial zone: unglaciated land, with a cold climate, influencing soils and deposition of sediments (French 2018). Traces of the presence of permafrost and other features in sediment and landforms that are characteristic for cold environments, are found in sedimentary deposits in now temperate areas. Comparing these traces with present-day permafrost leads to a better understanding of the paleoclimate during glacials, although such comparisons are not always straightforward (e.g. Vandenberghe and Pissart 1993; Isarin et al. 1998; Van Huissteden et al. 2003).

Note that the term 'periglacial' is not synonymous with permafrost. It is used specifically in the research fields of geology and geomorphology. Periglacial processes include cold-climate processes of weathering of rocks, transport by mass wasting processes and formation of particular landforms, all occurring in cold climates, but permafrost presence is not necessary.

In the last decades of the twentieth century it became clear that the release of greenhouse gases in the atmosphere would lead to rapid climate change in the Arctic, and also climate scientists started to study the significance of permafrost (Schuur et al. 2008). Mean annual air temperature increase in the Arctic may amount several degrees in the coming century, and it is quite clear that this warming process is well underway (see below). Not only temperature, but also precipitation will increase; both air temperature and increase of snow cover thickness in winter will enhance permafrost thaw. A warmer climate also leads to loss of glacier ice and sea ice cover, which in turn feeds back on permafrost (e.g. Parmentier et al. 2015).

The most recent climate projections of the IPCC AR5 reports show that the expected changes are drastic compared to changes that we have seen in recent decades. Temperature and precipitation changes are expected to be highest in winter, winter warming may amount up to 10 °C depending on the greenhouse gas concentration pathway. Likewise, the changes in permafrost landscapes may be drastic. Future climate change is discussed in more detail below.

This has changed the focus of permafrost research towards the carbon cycle of permafrost (e.g. Zimov et al. 2006; Schuur et al. 2008). The main greenhouse gases that will be released are carbon-containing greenhouse gases, CO_2 and CH_4. Locally also the potent non-carbon greenhouse gas nitrous oxide, N_2O is released (Repo et al. 2009). These gases are produced by decomposition of organic matter in soils by soil fauna and bacteria (Chap. 4). The commonly used term 'greenhouse gas' is a somewhat unfortunate term, since a the physical process of warming in a greenhouse is different from the radiative warming effects of these gases. However, in this book the commonly used terminology will be followed.

Permafrost soils store nearly twice as much carbon in the form of organic matter, than there is carbon in the atmosphere as carbon-containing greenhouse gases (Tarnocai et al. 2009; Hugelius et al. 2014). Changes in this huge soil carbon reservoir evidently will affect fluxes of carbon greenhouse gases to the atmosphere and ocean. This has become known as the permafrost carbon feedback, a positive feedback to climate change, driven by the increasing release of greenhouse gases from permafrost soils when climate warms, further enhancing climate warming.

1.3 The Permafrost Carbon Feedback

The permafrost carbon feedback is the main subject of this book. Since the northern hemisphere has by far the largest land area with carbon-rich permafrost soils, the emphasis in this book lies on northern hemisphere permafrost.

From the climate scientists' viewpoint, the research questions to be addressed have become: how much carbon will be released by climate warming, how fast, and in which way – as CO_2 or CH_4? The effect of these gases on climate differs. The direct radiative effect of atmospheric CO_2 on the climate is less than that of CH_4, but the lifetime of CH_4 is much shorter because of its oxidation processes in the atmosphere and soils. Comparing and adding the climate warming effect of greenhouse gases is therefore not simple (Chap. 4). As rules of thumb, metrics have been developed; the most well known is the Global Warming Potential (GWP), which expresses the warming effect of a pulse emission of 1 kg of a greenhouse gas to that of 1 kg of CO_2, with respect to a certain time frame. On a timescale of 20 years, the GWP of CH_4 is 86–84, on a timescale of 100 years 34–28, depending on whether climate change – carbon feedbacks are included or not (IPCC 2013). The GWP of N_2O is 268 (20 years) and 298 (100 years). The release of these greenhouse gases from permafrost will amplify global warming, but the uncertainty to which extent is large; this depends on the emitted quantities and the ratio of CO_2 to CH_4 (Schuur et al. 2008, 2015; see below).

From first biochemical principles, the rate of bacterial decomposition of organic matter in soils will increase approximately exponentially when the soil temperature increases (given sufficient substrate availability). However, in a soil environment, and in particular in permafrost soils, matters are more complex (Chap. 4). In the case of permafrost soils, decomposition cannot be isolated from other aspects of permafrost: permafrost soil freeze-thaw processes, geomorphology, ecosystems and water cycle – all affect temperature change, substrate, soil aeration and moisture conditions, and therefore also soil biota.

Important aspects of the vulnerability of soil organic carbon in permafrost soils to large scale emission of carbon are the excess ice content in combination with carbon content and the rate at which this soil carbon can decompose: its lability (Schuur et al. 2008). In particular the widespread frozen, ice-rich peat soils in Subarctic areas are sensitive to thaw. Another type of ice-rich deposits of loess and slope-wash that are frequently discussed in literature are yedoma deposits in northern Siberia and Alaska (see below). Their carbon content is much lower than that of peat, yet higher than that of most mineral soils. However, these deposits are widespread and vulnerable to thaw due to their high excess ice content (Zimov et al. 2006; Schirrmeister et al. 2013; Murton et al. 2015).

The emission of CH_4 from permafrost has become the most well-known symptom of the permafrost carbon feedback in the media, since it is a stronger greenhouse gas than CO_2, and its release as bubbles in lake bottom seeps, and recently even craters, is more spectacular than the diffuse and invisible increase in CO_2 (and CH_4) from soil organic matter decomposition. Nevertheless, CO_2 emissions will be a considerable, or even the largest part of the permafrost carbon feedback (Schuur et al. 2015; Schädel et al. 2016). Three main groups of sources of CO_2 and CH_4 can be distinguished, of which the first two are in particular also important CO_2 sources (Van Huissteden and Dolman 2012):

1. Ecosystem emissions, representing emissions from the abundant wetlands on permafrost soils. Here, CH_4 emission is a property of the ecosystem irrespective of climate change. Changes (positive or negative) in these emissions will be the result of complex interactions between vegetation, soil ecosystem and hydrology, climate and permafrost thaw (e.g. O'Connor et al. 2010; Van Huissteden and Dolman 2012).
2. Decomposition of old organic matter in permafrost soils and sediments, liberated by permafrost thaw (e.g. Walter et al. 2006; O'Connor et al. 2010). This depends on hydrological changes and geomorphological processes triggered by permafrost thaw.
3. Liberation of CH_4 from deeper reservoirs in the permafrost, e.g. gas pockets in taliks, or methane hydrates at or near the base of deep permafrost. Methane hydrates consist of CH_4 enclosed in an icy structure of water molecules, which is stable under high pressure and low temperatures, but may become unstable if temperature at depth rises. Reservoirs at large depth in or below permafrost are less likely to be released on a large scale soon, but gas from shallower reservoirs on land and in subsea permafrost may be released already (O'Connor et al. 2010; Walter Anthony et al. 2012; Chap. 7).

A short note on units is appropriate here. Permafrost soils contain a huge mass of carbon. Nevertheless, this carbon mass is often expressed in scientific publications in grammes of carbon: Petagrammes (Pg, 10^{15} gram), Gigatonnes (10^9 tonne = 1 Pg) is another commonly used unit. Furthermore, it is the carbon component that is usually indicated in mass calculations and reported values. For instance, a mass of methane consists for 75% of its mass of carbon. The mass of CH_4 is then specified as Pg CH_4-C, the mass of its carbon.

1.4 Setting the Stage

This section goes into more detail on the climatic, biological and geologic conditions of permafrost areas, and serve as a general background reference for the following chapters. This demonstrates that there is no 'typical' permafrost environment, but spatial variability in environmental conditions is a basic feature, which affects the permafrost carbon cycle strongly.

1.4.1 Climate in Permafrost Areas

Our knowledge of climate in permafrost areas is restricted by the low density of the observation network. The number of weather stations in the Arctic is small and they are widely dispersed (Figs. 1.6 and 1.7). Satellite remote sensing and the use of data-constrained weather forecast models for filling in the gaps (reanalysis, Dee et al. 2011) have improved this situation, although these cannot replace surface data collection. Furthermore, there is intrinsic instrumental uncertainty of a major climate variable: precipitation and snowfall. Precipitation gauges tend to bias towards severely under-catching the amount of falling snow due to wind effects (Serreze and Barry 2014).

Intuitively, the occurrence of permafrost should broadly agree with the 0 °C mean annual *air* temperature (MAAT) isotherm (Fig. 1.2). The mean annual *ground surface* temperature (MAGST) usually differs from MAAT because of the presence of vegetation and presence of an insulating snow cover in winter (French 2018; see also Chap. 2). There is usually an offset between MAAT and MAGST which may

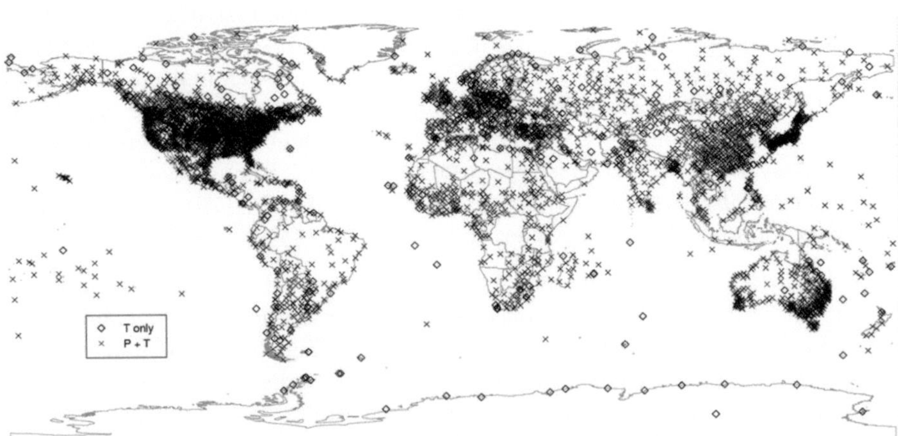

Fig. 1.6 Location of weather stations with at least 30 year monthly temperature or precipitation data. Diamonds: temperature data only; crosses: precipitation data. (From Peel et al. 2007, Fig. 2; published under Creative Commons License 4.0 https://creativecommons.org/licenses/by/4.0/)

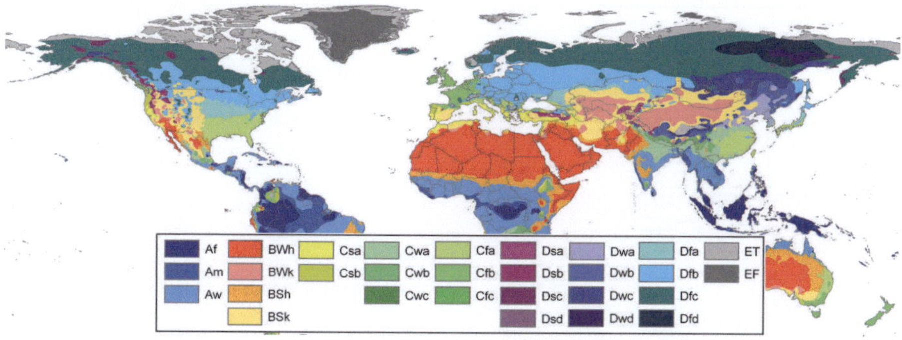

Fig. 1.7 Köppen-Geiger climate classification for the Northern Hemisphere. Letter codes in legend: A = tropical climates; B = arid climates; C = temperate climates; D = cold climates; E = Polar climates. Second letter for D climates: s = dry summer, w = dry winter, f = without dry season; third letter: a = hot summer, b = warm summer, c = cold summer, d = very cold winter. Second letter for E climates: T = Tundra, F = Frost climate. For full codes of A, B and C climates see Peel et al. (2007). (From Peel et al. 2016, Fig. 10 (modified, only Northern Hemisphere shown); published under Creative Commons License 4.0 https://creativecommons.org/licenses/by/4.0/)

amount up to 6 °C and results in a slightly warmer soil surface. Furthermore, the thermal properties (thermal conductivity and heat capacity) of the active layer differ between summer and winter because freezing and thawing results in a large change of these thermal properties (Fig. 1.7).

Next, temperature at the top op the permafrost (TTOP) below the active layer differs from the MAGST (Lachenbruch et al. 1988; Smith and Riseborough 2002; Fig. 1.8). Another – somewhat confusing – abbreviation is MAGT, which is also used for temperature at the top of the permafrost or that of the base of seasonal frost layer if permafrost is absent (Sazonova and Romanovsky 2003). The net effect is that the southern permafrost boundary follows approximately the −1 °C MAAT isotherm while the boundary between discontinuous and continuous permafrost varies between a MAAT of −6 to −8 °C (Brown 1970).

The carbon cycle in permafrost areas requires understanding of the climatic variables that affect ecosystems. First, the seasonal variability of the temperature is important for arctic ecosystems. For instance, the city of Yakutsk in eastern Siberia (62°N) has a MAAT of −8.8 °C, while Longyearbyen on Svalbard (78°N) experiences a MAAT of −7.5 °C. Despite the more modest MAAT, the vegetation near Longyearbyen is a low biomass tundra vegetation, often almost barren, while the surroundings of Yakutsk are covered with extensive larch forests. Both areas are located on continuous permafrost. In extremely continental Yakutsk, the mean July temperature is 19.5 °C and the growing season is long, with 5 months with temperature exceeding 5 °C, while in Svalbard the warmest month has an average temperature of 5.9 °C, with a growing season of barely 3 months. The reason for the small difference in yearly average is the contrasting winter temperature, which is considerably lower in Eastern Siberia than in Svalbard (Fig. 1.9).

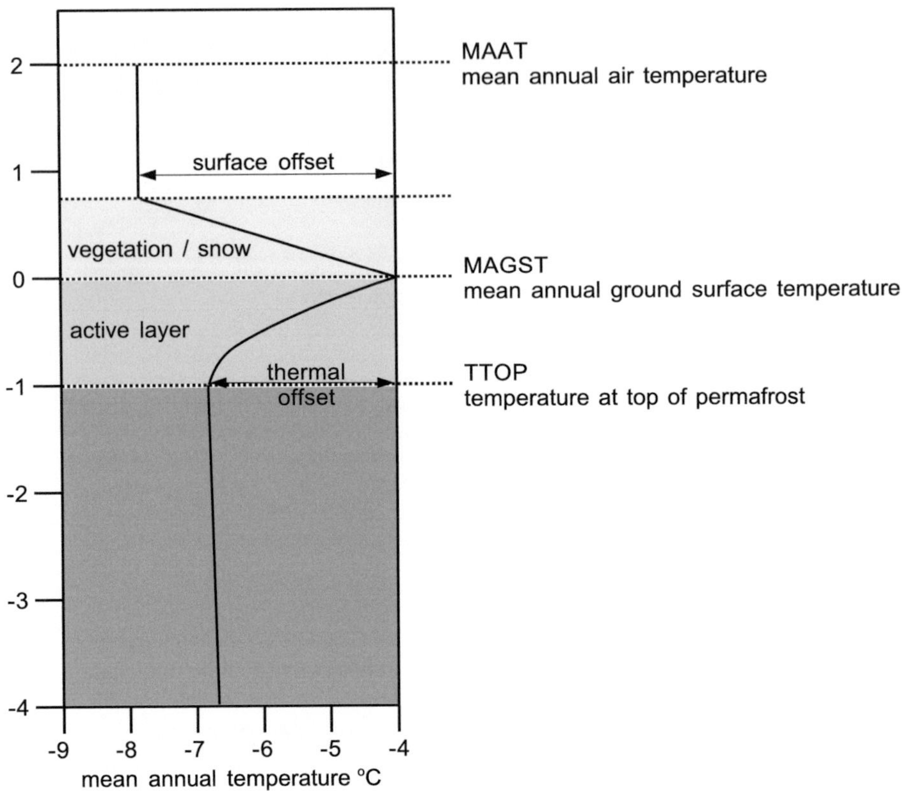

Fig. 1.8 Definition diagram with schematic temperature profile through the top of the permafrost, showing the relation between air temperature and permafrost temperature. (Based on Smith and Riseborough 2002)

Figure 1.9 shows the mean annual air temperatures and January and July means for the Arctic from 1985 till 2015, derived from ERA Interim Reanalysis data (a meteorological forecast model constrained by observation data). During July, only temperatures above the Greenland ice cap remain below zero. Temperatures along the Arctic Ocean coasts remain generally low in summer due to the cooling influence of the melting Arctic sea ice. The continental interiors of Eurasia and North America have July temperatures between 10 and 20 °C. In winter, a pattern develops where these continental interiors, together with the Greenland ice cap, show the lowest temperatures down to approximately −40 °C; these low temperature areas are linked by an area of low temperatures down to −30 °C over the frozen Arctic Ocean.

The unfrozen Pacific and Atlantic Oceans have a strong warming influence. The transport of heat from the Atlantic Ocean by the North Atlantic and Norwegian currents raises the Arctic winter temperatures (AMAP 2012; Serreze and Barry 2014). The moderating oceanic influence on winter temperatures is even visible far away

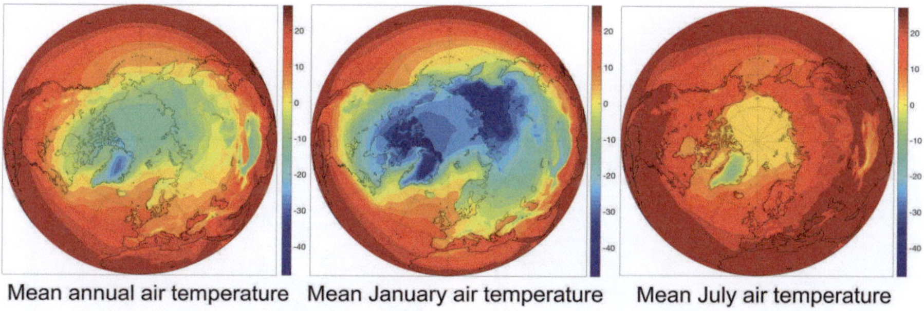

Fig. 1.9 Mean annual air temperature (MAAT), mean January air temperature and mean July air temperature in °C at 2 m height above the surface, averaged over 1985–2015. (Data from ECMWF ERA-Interim reanalysis; see Dee et al. 2011; https://www.ecmwf.int/en/forecasts/datasets/reanalysis-datasets/era-interim, accessed November 2017)

from the Pacific and Atlantic, along the frozen Arctic Ocean coasts of Siberia and Canada, where the Arctic Ocean reduces the winter extremes with respect to those on the continent.

The main Atlantic current that transports heat into the Arctic Ocean is the Norwegian current, which splits into two branches, one west of Svalbard and one into the Barents Sea. Outflow of cold water from the Arctic Ocean occurs along the East coast of Greenland and Baffin Bay between Greenland and the Canadian Archipelago. On the Pacific side, a smaller inflow of warm water enters via the Bering Strait. These currents also influence seawater salinity, which is important for sea ice formation. The Arctic Ocean, a relatively small ocean (ca. 1% of global ocean volume), receives 11% of global river runoff through the large north-flowing Eurasian and North American rivers (McClelland et al. 2012). This comparatively large freshwater influence facilitates sea ice formation (ACIA 2004).

Besides temperature, the duration and thickness of snow cover is the second important climatic variable for both vegetation and permafrost. Snow reflects solar radiation (typically 80–90% is reflected), and insulates vegetation and soil from extreme low temperatures in winter. Snow cover duration is longest in high northern latitudes, mountain ranges and the Tibetan plateau (Fig. 1.10). In Fig. 1.10 the 30 year average of February snow cover water equivalent is plotted for the northern hemisphere (in meters water equivalent). The water equivalent cannot be directly converted in thickness of the snow cover because of its highly variable density (multiplying snow water equivalent by factor of 3 gives a coarse approximation of snow cover thickness).

The snowpack density may vary between 100 and 800 kg m^{-3} depending on age (freshly fallen snow settles over time) and melting and refreezing events (changing snow into more dense ice). Throughout the season, loss through sublimation of snow occurs. The average density of a snowpack varies also with vegetation. Tundra snowpacks are generally thin, maximally 0.75 m deep, with a relatively high density of 380 kg m^{-3}. In forests, the snow cover tends to be thicker and lighter, 260 kg m^{-3} (Serreze and Barry 2014). Figure 1.10 shows large snowpack water equivalents at

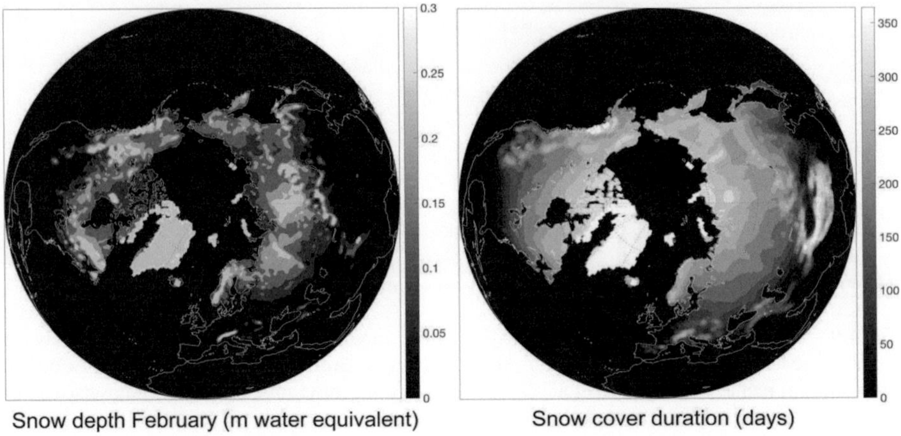

Fig. 1.10 Left: snow depth in February (in m water equivalent); the maxima over glaciers and ice caps may exceed 1 m. Right: average snow cover duration in days. Averages over 1985–2015. (Data source: see Fig. 1.9)

east coast margins in North America and Europe, and low snow water equivalents in continental interiors, as a result of continental scale precipitation distribution (see below).

Sublimation losses over the season may range from 15 to 50% (Liston and Sturm 2002). Wind redistribution of snow results in significant thickness variations. At wind speeds above 5 m s^{-1} snow is easily lifted from the surface and transported in suspension. On flat terrain irregular grooves and ridges are formed by wind erosion of snow, known as sastrugi, and low snow dunes in the order of a few decimetre height. Much thicker snow accumulations occur in the lee side of slopes and in valleys, while exposed plateaus and ridges may be blown free of snow (Serreze and Barry 2014; Liston and Sturm 2002). Thick snow accumulations decrease frost penetration in winter and are important sources of moisture in spring (e.g. Johansson et al. 2013).

Despite the winter snow cover, most of the Arctic is a comparatively dry world in terms of precipitation. The daily winter precipitation map (Fig. 1.11) shows low precipitation over the Arctic Ocean and adjacent continental landmasses. Areas bordering the open Pacific and Atlantic Oceans receive more precipitation, where moisture bringing low pressure areas (cyclones), driven by the westerly winds, can penetrate inland. The summer precipitation map of Fig. 1.11 shows a change of this pattern, with deeper continental penetration of cyclones from the Pacific and Atlantic. Continental interiors and areas near the Arctic Ocean generally have their precipitation maximum in summer, while the precipitation maximum for the margins of the continents near the Atlantic and Pacific Oceans is generally in winter (Serreze and Barry 2014). Only the Canadian High Arctic and the western Arctic Ocean remain relatively dry in summer.

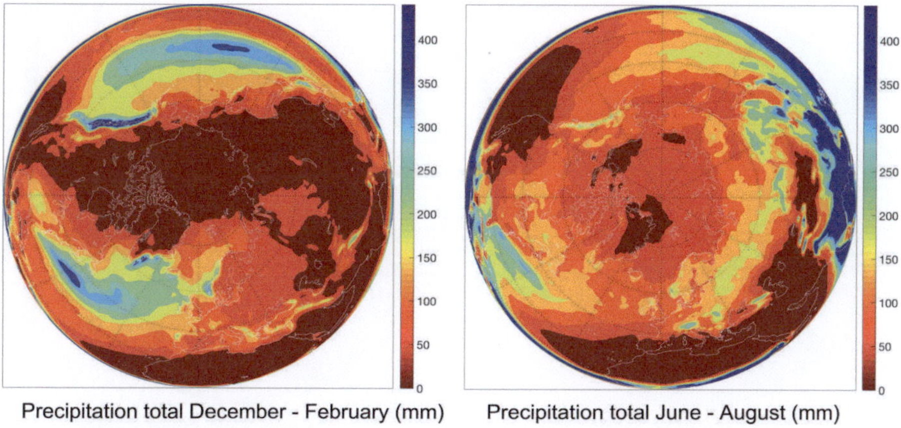

Fig. 1.11 Left: precipitation totals during winter months (December–January-February) in mm water. Right: precipitation totals during summer months (June–July-August). Averages over 1985–2015. (Data source: see Fig. 1.9)

Wind speeds are high over the Arctic Ocean (Serreze and Barry 2014). The wind regime affects convective heat transports, transporting warmer air masses in winter and cooler air masses in summer. In winter, cyclone activity over the Atlantic Ocean is stronger than in summer and penetrates from the Atlantic into the Barents and Kara seas north of Russia and western Siberia. This oceanic wind climate also affects the coastal regions along the Arctic Ocean. In the continental interiors where anticyclones develop in winter, winds tend to be light and skies clear, contributing to strong radiative heat losses at the earth surface.

Cloud cover may seem less important for permafrost, but in winter it affects the radiation balance, providing a source of thermal radiation as clouds trap thermal radiation and radiate it back to the surface. In summer, persistent cloud cover will reduce solar radiation reaching the surface and reflect more radiation back in space. When the effect on winter radiation dominates over that of the summer, clouds contribute to surface warming (Curry et al. 1996; Eugster et al. 2000). However, the exact sign of the net effect for the Arctic is uncertain, in particular because of the uncertainty in cloud observations and disagreement between models (Kay et al. 2016). Cloud cover also affects the primary productivity of ecosystems. Over the Arctic Ocean, low level stratus (layer-like) clouds occur frequently, notably in summer (Serreze and Barry 2014). This also affects the coastal regions (Fig. 1.12). Over the continental areas further south, convective cumulus clouds are more common in summer.

Persistent air pressure patterns develop in winter by radiative strong cooling of continental landmasses during the polar night. An anticyclone (high pressure area) develops over eastern Siberia and central Asia (Fig. 1.13). This favours the development of winter inversions (temperature increasing with height above the surface, rather than decreasing), resulting in absence of vertical exchange of air. This induces strong near-surface cooling, in particular in basins between mountain chains. An

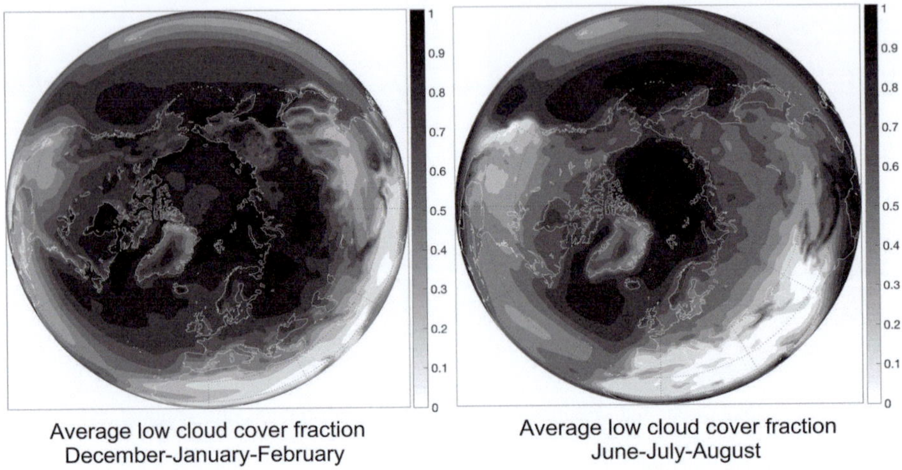

Fig. 1.12 Left: average low cloud cover fraction during winter months (December–January-February). Right: average low cloud cover fraction during summer months (June–July-August). Averages over 1985–2015. (Data source: see Fig. 1.9)

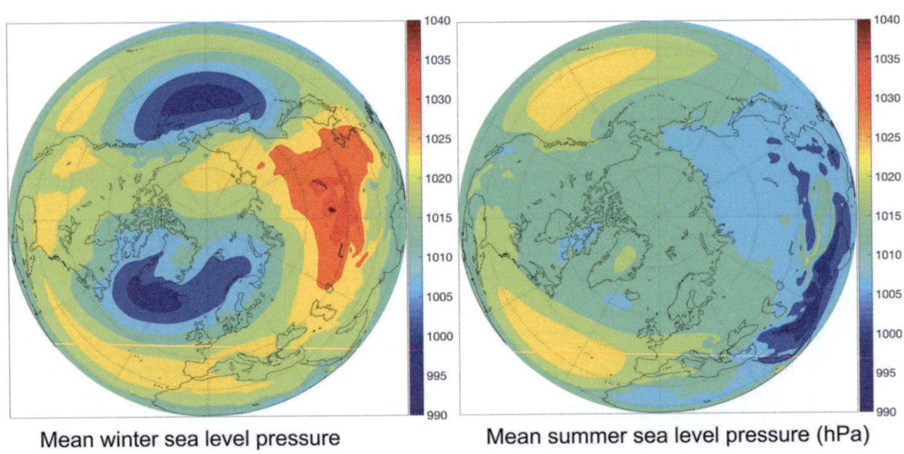

Fig. 1.13 Left: average mean sea level pressure (hectoPascal, hPa) during winter months (December–January-February). Right: average mean sea level pressure during summer months (June–July-August). Averages over 1985–2015. (Data source see Fig. 1.9)

example is the town of Verkhoyansk in the Yana river basin between two mountain ranges in northeastern Siberia. Here, winter temperature extremes below −60 °C are sometimes recorded (Serreze and Barry 2014).

Over the Pacific and Atlantic Oceans respectively, the Aleutian and Icelandic lows are found (Serreze and Barry 2014). Higher in the Arctic atmosphere, from the middle troposphere up into the stratosphere, a low pressure cell is present, containing a cold air mass (Waugh et al. 1997). In summer these pressure differences

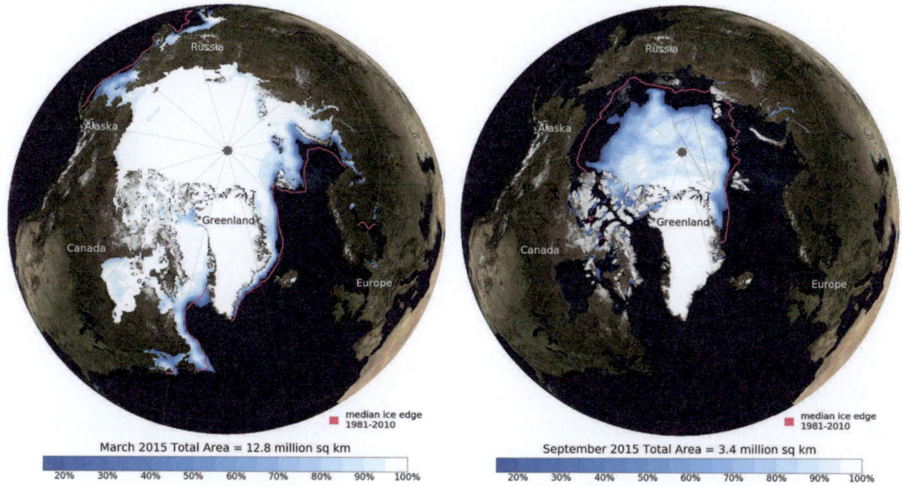

Fig. 1.14 Arctic ocean ice concentration (percentage of area covered with sea ice) at the end of winter (March 2015 average, left) and in early autumn (September 2015 average, right), derived from satellite remote sensing data. The arrow indicates the direction of transpolar drift. The pink lines give the 30 year average ice extent for 1981–2010 for comparison. Images courtesy of the National Snow and Ice Data Center, University of Colorado, Boulder

become less intense. However, a strong high pressure cell remains present above the Greenland ice cap, giving rise to strong winds transporting dense, cold air down the slope of the ice cap (katabatic winds).

The sea ice cover has a strong influence on the climate of the continental landmasses around the Arctic Ocean (Serreze and Barry 2014; Parmentier et al. 2013, 2015). The sea ice cover is largest at the end of the winter and has its maximum in March, and its minimum in early autumn in September (Fig. 1.14). Currently, the sea ice cover area is on a continuous decline (http://nsidc.org/arcticseaicenews/). The sea ice is under constant motion, driven by the Arctic Ocean currents.

The Transpolar Drift, driven by prevailing winds in summer and running from the Laptev Sea to Northern Greenland, drives ice away from the Siberian coast towards Canada and Greenland and out to the Atlantic Ocean along the Greenland east coast. Under present climatic conditions, the northern Siberian coast is ice-free by the end of summer (AMAP 2012, 2017; Fig. 1.14). The high reflection of sunlight by the sea ice cover contributes to atmospheric cooling during the summer, and the latent heat for melting of sea ice is for a large part extracted from the atmosphere. This results in summer cooling of nearby coastal landmasses. Moreover, these low temperatures contribute to low-level stratus cloud cover.

The Arctic climate has distinct natural variability over time: the Arctic Oscillation (AO). It is defined on the basis of atmospheric pressure variations over the Arctic (Serreze and Barry 2014), but has widespread effects on surface climate in the Arctic and beyond. In what is known as the 'positive phase', low pressure dominates over the Arctic, and the polar vortex, the low pressure area at higher levels in the

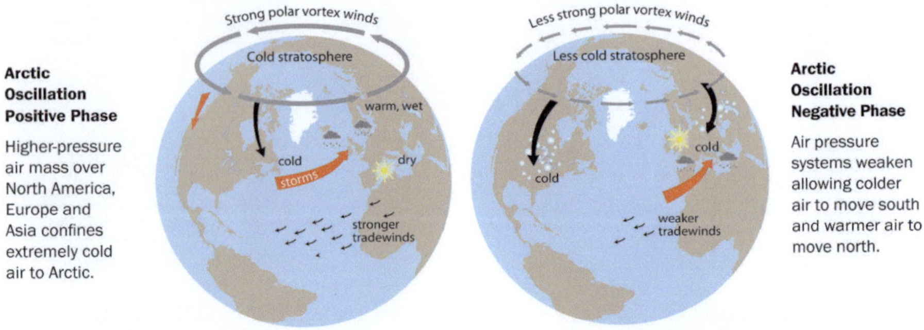

Fig. 1.15 The Arctic Oscillation and its effects on lower latitude wind patterns. (From AMAP (2012); https://www.amap.no/documents/download/1692/inline; accessed July 2019)

atmosphere over the North Pole is strong and stable, and the stratosphere is cold over the Arctic (Fig. 1.15). Mid-latitude pressure is high, which causes ocean storms to migrate farther north. This results in wetter and warmer conditions in the north-westerly coasts of the oceans (Scandinavia and Alaska), while the northeasterly coasts (Greenland, Northeast Asia) remain relatively cold. The influence reaches as far south as the Mediterranean and southeastern United States, which are drier. The jet stream (the high altitude zone of strong westerly winds) has a predominantly west-east orientation. In the negative phase, the air pressure over Arctic is relatively high. The jet stream has strong north – south meanders, driving cold air outbreaks over mid latitudes in winter. The weather patterns are opposite to those of the positive phase (Serreze and Barry 2014). The AO also influences sea ice cover; a negative phase of the AO decreases sea ice loss (Serreze and Barry 2014). However, recently the correlation of sea ice cover with AO phases have become weaker (Ogi et al. 2016).

The AO is related to the North Atlantic Oscillation (NAO), which is expressed by pressure variations of the Icelandic low and the Azores high. The AO and NAO have been seen as manifestations of the same weather pattern (Thompson and Wallace 1998) although there are reasons to doubt this long distance connection between Arctic and Atlantic climate (ACIA 2004). A specific periodicity of these phenomena appears to be absent. On the far eastern side of the Arctic Ocean, the climate is influenced by the Pacific Decadal Oscillation (PDO), which results in temperature and pressure variations of the northern Pacific Ocean. The physical drivers of these oscillations are still uncertain (Serreze and Barry 2014).

Fig. 1.16 From left to right: Sparsely vegetated permafrost soil near Isfjord Radio, Svalbard; Varied tundra near Kytalyk research station in the Indigirka lowlands, northeast Siberia with sedges, Dwarf Birch and peat mosses; boreal forest (Larch, Scots Pine, Birch) on permafrost at Spasskaya Pad research station, near Yakutsk, eastern Siberia. (Photos by the author)

Fig. 1.17 Thaw pond resulting from vegetation disturbance by frequent walking. (Photo by the author)

1.4.2 Vegetation in Permafrost Areas

For many, permafrost is almost synonymous with treeless tundra (Fig. 1.16). However, permafrost also occurs in continental sub-polar or subarctic climates with long, cold winters, but relatively warm summers in the continental interiors; these climates support tree growth. Large areas of permafrost lie south of the tree line (Fig. 1.1) and are covered by dense boreal forest. Tundra vegetation – ranging from lichens, grasses and mosses to shrub-dominated vegetation – mostly occurs in climates with a maritime influence, where summer temperatures are low as a result of influx of cold air from the Arctic Ocean. Almost barren areas are restricted to northern permafrost areas of northern Greenland and Canada and the Arctic Ocean islands as Svalbard.

The vegetation cover in permafrost interacts strongly with the permafrost soil heat exchange (e.g. Péwé and Brown 1973; McGuire et al. 2002). The vegetation is adapted to the cold air and soil climate, but it also influences its environment by influencing soil temperature (Chap. 2). In the Arctic summer, ice-rich permafrost is in thermal disequilibrium with the temperature at the soil surface. If the ice is exposed to these temperatures, it may easily thaw (Fig. 1.17). However, a vegetation cover of sufficient thickness insulates the soil surface from incoming radiation, and therefore will protect underlying ice-rich permafrost from thawing. In particular a peaty moss layer is an effective insulator. Disturbance of this protective vegetation cover, by natural processes as erosion or human-induced disturbance, will easily start thawing processes (e.g. Nauta et al. 2014).

Based on climate and biogeography, a broad subdivision of the Arctic land cover is the High Arctic and Low Arctic; in the High Arctic areas conditions are colder and more extreme than in the Low Arctic (Serreze and Barry 2014). Both temperature and moisture supply structure the Arctic vegetation. A more detailed subdivision is used for the Circum-Arctic Vegetation Map (CAVM, Walker et al. 2002, 2005) of mostly non-forest vegetations above the treeline. Five sub-zones have been distinguished on this map:

- Sub-zone A is often referred to as polar desert; it has a largely barren surface. Plant cover, biomass and net annual production in tonnes per hectare: <5% vascular plants, <40% lichens and mosses, <3 t ha^{-1} plant biomass, annual production <0.3 t ha^{-1} $year^{-1}$. If present, the vegetation consists of species of low stature: cyanobacteria, lichens, mosses, and scattered grasses and low forbs, usually with a compact cushion growth form (e.g. Fig. 1.16 left). The soil shows evidence of intense frost action, creating a polygon-like micro-relief; vegetation is determined by this micro-relief structure (Chap. 3). Shrubs are absent, wetlands with sedges, if existing, lack significant organic layers. Sub-zone A mainly occurs at high latitude coastal areas and islands, where mean July temperatures are as low as 2–3° C. The climate is usually windy with relatively small amounts of precipitation. Frost shattered rock cover and snowbanks are common in higher relief areas.
- Sub-zone B requires higher summer temperatures and has a more continuous, layered vegetation cover, consisting of a layer of mosses and a layer of herbs and prostrate dwarf shrub (shrubs with branches that lie on the ground surface). Plant cover, biomass and net annual production: 5–50% vascular plants, <60% lichens and mosses, 5–20 t ha^{-1} plant biomass, annual production 0.2–1.9 t ha^{-1} $year^{-1}$. Still many patches of bare ground occur between the short vegetation, which generally does not exceed 10 cm in height. Locally sedge wetlands may form.
- Sub-zone C and D represent the more typical tundra (e.g. Figure 1.16 mid), requiring higher summer temperatures up to 10 °C on average. Plant cover, biomass and net annual production for zone C: 5–50% vascular plants, 10–30 t ha^{-1} plant biomass, annual production 1.7–2.9 t ha^{-1} $year^{-1}$; zone D: 50–80% vascular plants, 30–60 t ha^{-1} plant biomass, annual production 2.7–3.9 t ha^{-1} $year^{-1}$. The moss and lichen cover is more continuous and vascular plants cover (mainly dwarf shrubs, grasses) is higher (up to 15 cm). Typically, Sub-zone D is similar to C, with a more continuous cover of vascular plants (including dwarf shrubs up to 40 cm height); in subzone C the vegetation is still open and patchy. The Dwarf Birch *Betula nana* is typical for dry sites. Wetlands have a dense vegetation consisting of sedges (*Carex* species) and the typical cotton grasses (*Eriophorum* species). *Sphagnum* (peat) mosses also appear as peat formers.

- Sub-zone E also includes higher shrubs up to 80 cm in height, which can make up a two- or three-storied vegetation with a moss layer, eventually a forbs layer, and a tall shrub layer (mostly Willow species). Summer temperatures are typically up to 12 °C. This sub-zone is close to the tree line. Plant cover, biomass and net annual production: 80–100% vascular plants, 50–100 t ha^{-1} plant biomass, annual production 3.3–4.3 t ha^{-1} $year^{-1}$.

A vegetation type that is also frequently mentioned is heath vegetation, denoting low shrub-dominated vegetation that is found on well-drained soils, mostly in sub-zones B, C and D. Typical plants of the Ericacea family in these zones are *Vaccinium, Ledum* and *Cassiope*. Another frequently mentioned vegetation type is tussock tundra; this is moist tundra where the Cotton Grass *Eriophorum vaginatum* forms closely spaced tussocks.

With higher and more extensive vegetation, also the vegetation biomass, the leaf area index (LAI, the amount of leaf area per square metre, m m^{-2}) and the productivity of the vegetation increases. Although this vegetation zonation appears to be a quite strict zonation with temperature, there may be considerable local variation. Mostly, this is related to topography. For instance, in the typical tundra of sub-zone C and D, wind-swept uplands and plateaus may occur where the vegetation attains the characteristics of sub-zone B and A, while in topographical lows along rivers tall shrub vegetation is found, with characteristics of sub-zone E. A similar zonation can be seen in mountain permafrost areas.

The shrub tundra grades into the tundra-taiga transition, where scattered trees become more and more common in the land cover. The northern limit of the boreal forest correlates with the position of the 10–13 °C July isotherm (Bonan et al. 1995). Four vegetation zones are commonly distinguished in the boreal zone: the forest – tundra transition zone, the open boreal woodland, the main boreal forest and the boreal – mixed forest (Oechel and Lawrence 1985). The southern boundary of the boreal forest is found where conifer-dominated forests are replaced by deciduous broadleaf forest, or steppe grassland, depending on the aridity of the climate.

On a continental scale, variations in the dominating tree species exist on this north – south zonation, related to precipitation and continentality of the climate (Baldocchi et al. 2000). The eastern Siberian forest experiences a colder and drier climate than that of the Canadian and Scandinavian boreal forests, and grows for a large part of its area on continuous and discontinuous permafrost (Bonan et al. 1995). Dominant tree species of the boreal forest are coniferous trees. These are the evergreen Spruce (*Picea*), Fir (*Abies*) and Pine (*Pinus*) and the deciduous Larch (*Larix*). Broadleaf deciduous trees and shrubs like Birch (*Betula*), Alder (*Alnus*) and Poplar (*Populus*) occur locally in early successional stages or in specific habitats like river floodplains. In the extreme continental climate of Siberia, Dahurian Larch (*Larix gmelinii*) dominates (Fig. 1.16 right). This tree with its shallow root system is well adapted to permafrost soils. It tolerates winter temperatures below 60 °C in the Verkhoyansk region in Yakutia. Permafrost may even be essential to the existence of the extensive Siberian Larch forests, as the permafrost soils provide water to the trees in dry summers by thawing of permafrost ice (Sugimoto et al. 2002;

Zhang et al. 2011). Other common species in this environment are Scots Pine (*Pinus sylvestris*) and Birch (*Betula sp.*).

1.4.3 Peatlands and Wetlands

The ubiquitous presence of wetlands is an important aspect of vegetation in permafrost regions. Wherever the topography is flat, poorly drained areas may be expected in any climate with a sufficient precipitation surplus. However, frozen soils limit drainage. Firstly, because subsurface drainage by groundwater flow is impeded by the soil ice, and secondly because micro-relief created by frost action may impede surface drainage (Chap. 3). This is essential for the carbon cycle of permafrost soils. Wetland soils accumulate soil organic matter because of slow decomposition of plant remains under waterlogged conditions, resulting in the formation of peat layers (Tarnocai and Stolbovoy 2006; Tarnocai et al. 2009). Charman (2002) is a source of information on peatland typology, morphology and processes.

Wetlands vary strongly in vegetation composition and rate of peat accumulation. Although for many people 'wetland' is synonymous with 'peatland', wetlands do not always accumulate peat; peat only accumulates when the rate of organic production exceeds the rate of decomposition. Strong temporal variation of the water table or hydro-chemical conditions may result in high decomposition rates, preventing peat accumulation. In a botanical sense, a peat forming vegetation may be present, but the resulting peat layer may be too thin to classify as it a peat soil or peat deposit in a soil science or geological sense. In most soil classifications a peat layer should be at least 40 cm thick and consist of a minimum percentage of organic material.

Globally, peatlands occupy about 4×10^6 km^2, the boreal and subarctic peatland area is estimated at 3.46×10^6 km^2, or about 87% of the world's peatlands (Vitt 2006). Many of these northern peatlands contain at least sporadic permafrost. In the boreal forest zone, about 24% is occupied by peatlands (Wieder et al. 2006). The long term carbon accumulation rate (LORCA, the rate at which carbon is stored in the soil by peat formation in time frames of hundreds or thousands of years), in the boreal zone amounts 13–20 g C m^{-2} year^{-1}. In subarctic and arctic peatlands it is lower, in the order of 1.2–16.5 g C m^{-2} year^{-1} (Kuhry and Turunen 2006 and references therein).

A major variable that strongly influences the botanical characteristics, peat accumulation rates and landform of peatlands is their water source (Vitt 2006). These water sources can be grossly subdivided in:

- Ombrogenous water – rain and snow melt water which is poor in nutrients and dissolved minerals. Ombrotrophic peatlands receiving only precipitation water generally have a vegetation dominated by *Sphagnum* mosses and are referred to as bogs or *mires*.
- Geogenous water – surface water or ground water which has been in contact with the soil and contains more dissolved minerals and nutrients.

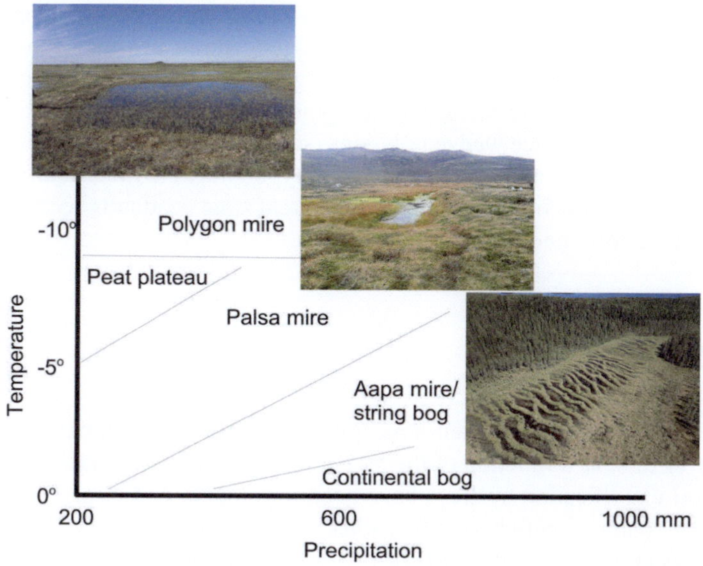

Fig. 1.18 Climate of Arctic and Boreal peatland types. Simplified after Vitt (2006). Photo polygon mire (Indigirka lowlands, Siberia) and palsa mire (Stordalen mire, Northern Sweden) by the author. For the polygon mire, see also Fig. 1.22. (Photo aapa/string mire source: Aapa mire, Sweden, Location: 62° 34′ 41.96″ N, 16° 40′ 56.14″ E, https://en.wikipedia.org/wiki/String_bog; licensed under Creative Commons Attribution-ShareAlike 4.0 International (CC BY-SA 4.0) license https://creativecommons.org/licenses/by/4.0/)

Minerotrophic peatlands with geogenous water have a vegetation where *Sphagnum* is less dominant or absent, are dominated by graminoid species (reeds, sedges and cottongrass) and brown mosses, and are known as fens. The geogenous waters can be subdivided again in three groups: stagnant water collected in depressions (topogenous), water flowing over, or through the soil, including seepage of ground water (soligenous) and flood water from rivers (limnogenous). The latter are generally more nutrient-rich. In continuous permafrost wetlands, seepage of groundwater is usually restricted to the active layer in summer (Woo 2012).

The second variable that determines peatland vegetation and form is climate, both temperature and precipitation (Vitt 2006). Bogs change from open vegetation with Sphagnum to an increasing tree cover, going from moist maritime climates to drier continental climates (Fig. 1.18). Going from south to north, *Sphagnum* bogs are replaced by aapa mires or string bogs, which have a distinct surface pattern of wet pools or lawns of sedges, alternating with drier strings of vegetation dominated by shrubs and *Sphagnum*. These in turn give way to palsa bogs and peat plateaus, in the sporadic and discontinuous permafrost zone. Palsas are elevated islands of ice-rich frozen peat, often with a dry lichen-rich surface, within minerotrophic fens. When these palsas coalesce in areas with more continuous permafrost they form extensive peat plateaus; palsas and peat plateaus are sensitive to abrupt thaw (Chap. 3).

Within the tundra zone, the wetland vegetation consists of graminoid species like sedges and cotton grass if the water source is geogenous, or moss-rich vegetation including brown mosses (true mosses with a generally reddish brown colour) and peat mosses (*Sphagnum*) in more ombrotrophic situations. Micro-relief, in particular the ridges created by ice wedge polygons (see next section), may impede surface drainage; this type of peatlands is known as polygon mires (Minke et al. 2007). The peats are generally much thinner than in the boreal zone peatlands because of lower carbon accumulation rates.

1.4.4 Soils

The soil is the interface between atmosphere, vegetation and rocks. Soil profiles develop over time as a function of the parent rock material, vegetation and climate. Various soil classification systems have been developed, and classifications differ between the countries with permafrost soils. A circumpolar soil atlas has been published (Jones et al. 2010; Fig. 1.19) based on the soil classification of the FAO soil map of the world, the World Reference Base for Soil Resources.

The first important feature of permafrost soils is their high content of organic matter in the shape of peat or other organic matter in mineral horizons, which drives the permafrost carbon feedback. The main reason is the very slow decomposition of organic matter, in particular when it becomes incorporated below the active layer in the permanently frozen soil. A northern circumpolar soil carbon database has been established to gather knowledge on the amounts of carbon stored in permafrost soils (Tarnocai et al. 2009; Hugelius et al. 2014). The latest estimate amounts 1300–1370 Gt C, with an uncertainty range of 930–1690 Gt C (Hugelius et al. 2014). For comparison: the upper end of the uncertainty range amounts to nearly twice the amount of CO_2 carbon in the atmosphere. See Chap. 4 for more on this subject.

A second aspect of many permafrost soils is the presence of a distinct micro-relief in many areas (Fig. 1.20), or patterned ground. Soil forming processes in permafrost soils are influenced by freeze and thaw, causing deformation of the soil horizons. These deformations are known as cryoturbation (Vandenberghe 1988; Walker et al. 2004; French 2018). These processes include: soil movements induced by frost heave (resulting from ice lens growth), processes causing sorting of soil material in coarse-grained and fine-grained fractions, semi-fluid movements of water-saturated fine-grained soil material during thaw or refreezing of the active layer, and contraction cracking by extremely low temperatures. Altogether, these processes cause distortion of soil horizons and redistribution of soil material. Cryoturbation includes translocation of organic material formed at the surface to deeper levels in the soil, a mechanism by which carbon is stored in permafrost soils (e.g. Walker et al. 2004; Bockheim 2007). See Chap. 3.

Typical patterned ground features are circles or small polygons (delineated by vegetation or cracks, Fig. 1.20a), and similar features that show sorting of stones and finer grained material (Fig. 1.20b). On slopes, these features elongate to

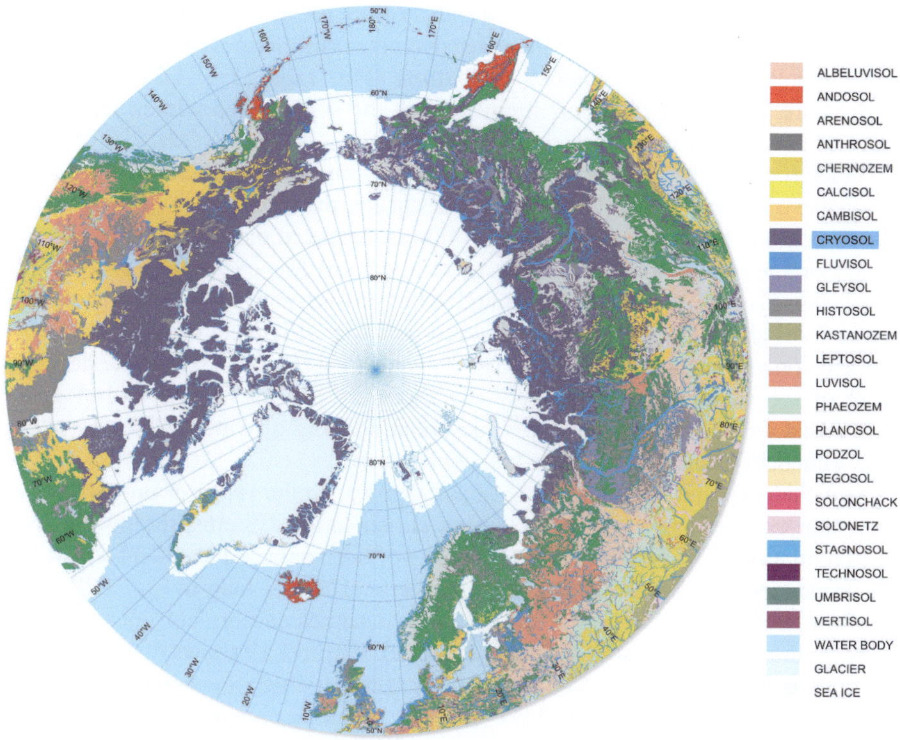

Fig. 1.19 Soil map of the northern circumpolar region. For a description of relevant legend units, see text. For an extensive legend see Jones et al. (2010). In the legend, the cryosols (typical for permafrost) are highlighted. From the Soil Atlas of the Northern Circumpolar Region (Jones et al. 2010), European Commission, Publications Office of the European Union, Luxembourg. Reuse permitted under Commission Decision of 12 december 2011 on the reuse of Commission documents (2011/833/EU)

Fig. 1.20 Patterned ground at the surface of permafrost soils. (**a**) non-sorted circles at a flat mountain top in northern Sweden; boundaries are lined with grasses with a thin snow cover. (**b**) sorted circles, Spitsbergen, Kap Linne. (**c**) Patterned ground at two spatial scales, large ice wedge polygons created by thermal contraction cracking (length of sides of the polygons 5–10 m) and small non-sorted circles within the polygons. (Photos by the author)

downslope directed lines. Patterned ground forms on at least two spatial scales (Fig. 1.21c). Larger features are the extensive networks of ice wedge polygons that occur ubiquitously on continuous permafrost (Fig. 1.20c, Fig. 1.21). These ice wedges are formed by contraction cracking of the soil in deep winter, when temperature may drop rapidly, creating contraction stresses (French 2018). Over time, the cracks become filled with ice by repeated cracking and infilling, creating a vertical wedge of ice in the soil (Chap. 3). At the surface, the result is a polygonal system of ridges along the wedges created by volume increase at the wedge location, with poorly drained pools in between; these are known as low-centered polygons. The ice wedges also may develop into a trough when erosion and thawing of ice removes ice and soil along the wedge. High-centred polygons (Fig. 1.21) develop by thawing, erosion and surface lowering above the wedge ice (French 2018). The development of ice wedges and their thaw processes are crucial for the hydrology of permafrost soils, since it results in widespread ponding of water, while ice wedge thaw enhances drainage (e.g. Liljedahl et al. 2016).

Soil classification is the grouping of soils with a similar range of properties into units that can be mapped. In soil science classification systems, the morphology (differentiation of the soil into different layers or horizons) and chemical and physical properties are used as distinguishing criteria. The soil map in Fig. 1.19 is the result of such a classification system. Soil class names end on ‘sol’, derived from Latin ‘solum’, meaning ‘soil’ (Jones et al. 2010). Figure 1.19 shows that most of the continuous permafrost area is dominated by cryosols, which are shaped by cryoturbation processes. Leptosols (weakly expressed, thin soils on bare rock) are common in mountainous areas. Gleysols are common; these are soils that are water saturated most of the time, and are usually in a chemically reduced state. Histosols are peat soils, accumulating organic matter under waterlogged conditions, and occur in the southern regions of the continuous permafrost zones of Canada and western Siberia. Podzols occur mainly in southern regions of the continuous permafrost zone, in particular in Siberia. Podzol soils are acid soils, with a clear, bleached horizon where iron, aluminium and organic matter have been leached out. These accumulate as oxides and amorphous organic material below the leached horizon, known as the podzol B horizon. These leaching processes require an active layer which is thick enough to allow the leaching process.

In the discontinuous permafrost zone, the podzols and histosols are also common, next to a wider variety of soils. Cambisols, young soils with minor soil development are abundant on de-glaciated landscapes in North America. Luvisols (more typical for temperate climates) have a clay-rich horizon below a topsoil which is depleted in clay because of downward clay translocation. Albeluvisols have a similar clay accumulation horizon, but it has an irregular, broken upper boundary, penetrated by cracks filled with bleached soil material from above. This type of soil is common in Eurasia. The crack formation is attributed to freeze-thaw processes (Jones et al. 2010).

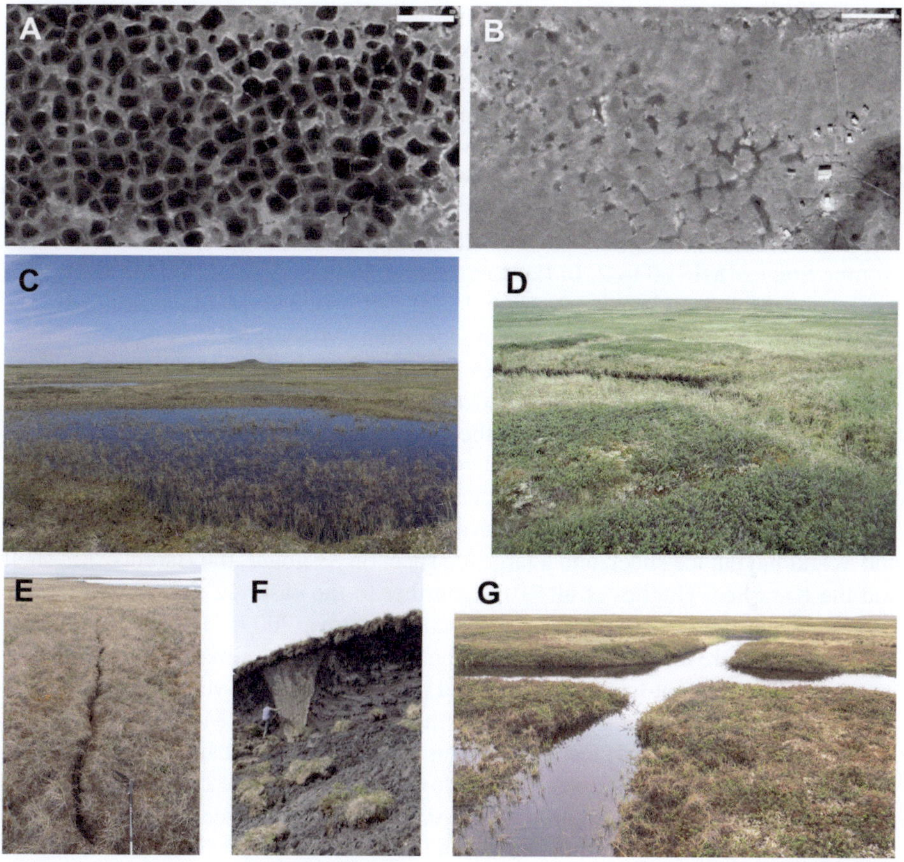

Fig. 1.21 (**a**) low-centered ice wedge polygons, Indigirka lowlands, eastern Siberia. The polygon centres are marked with ponds (dark areas). (**b**) high-centred polygons, same area; the ice wedges are marked with dark areas. Worldview panchromatic image, 0.4 m resolution July 102,015, scale bars 50 m, WorldView © 2019 MAXAR. (**c**) surface aspect of low-centred polygons in **a**. (**d**) Surface aspect of high-centred polygons in **b**. (**e**) fresh ice wedge crack. (**f**) ice wedge in cross section, note upturning of soil layers on the side of the wedge. (**g**) Thawing ice wedges; low polygon centre still intact. (Photos **c**, **d**, **e**, **g** by the author in Indigirka lowlands, northeast Siberia; photo **f** in northern Canada by Dr. L.M. Bröder)

1.4.5 Ice Age Permafrost

The repeated expansion and contraction of ice sheets and glaciers during the Quaternary period (the period in Earth history comprising the ice ages) has had pronounced influence on permafrost thickness, sediments and landforms in present permafrost areas, and has caused expansion and contraction of permafrost in the past (paleo-permafrost). The exact extent of Quaternary ice sheets and glaciers is the subject of debate, in particular for older glaciations and less well studied areas such as northern Siberia. Within each glacial, expansion and contraction phases occurred (stadials and interstadials) with different ice sheet configurations, and quite often only the maximum extent is known.

The last glacial cycle is known as the Eemian interglacial / Weichselian glacial (as it is known in Europe, or Sangamonian interglacial /Wisconsinan glacial (North America) and Mikulin interglacial/Valdai glacial (Russian plain). For glacials and interglacials, a time scale has been developed based on oxygen isotopes in deep sea sediments, which reflect the volume of ice-sheets (e.g. Bradley 2015). The Last Glacial correlates with the Pleistocene time stratigraphic units Marine Isotope Stages (MIS) 5 to 2. In Eastern Siberia, the last glacial cycle is split up in two separate cycles: Kazantsevo interglacial (MIS 5), Zyryan glaciation (MIS 4), Karginsky interstadial (MIS 3), Sartan glaciation (MIS 2); MIS 3 has been a relatively warm mild period with at times temperatures similar to those today (e.g. Murton et al. 2015). Figure 1.22 shows the inferred ice sheet distribution and paleo-permafrost distribution in the northern hemisphere (Vandenberghe et al. 2014).

Much of North America (Laurentide and Innuitan ice sheets, Margold et al. 2015) and northern Europe has been covered by ice sheets and glaciers (British Isles and Scandinavian ice sheet, and a large ice sheet on Svalbard, the Barentsz sea shelf and the Kara Sea, Hughes et al. 2016). Further to the east, ice sheet extent is less certain (Svendsen et al. 2004; Velichko and Faustova 2009; Hughes et al. 2016; Niessen et al. 2013). The current view is that the Siberian landmass was not covered by an ice sheet, except local mountain and plateau ice caps (Svendsen et al. 2004). This non-glaciated area extended towards the north of Alaska (Margold et al. 2015). During the earlier stages of the Weichselian, the European ice caps stretched further eastwards then during the Last Glacial Maximum (LGM) at ±21,000 years before present (BP) (Svendsen et al. 2004). At ±90,000 years BP the ice reached as far east as the Taymir peninsula and the Plutorana plateau in central Siberia. During the LGM the ice reached not further east than the Kara sea. However, Niessen et al. (2013) argue for an ice cap situated on the shelf north of the New Siberian Islands and Wrangel Island, based on seafloor morphology and sediments.

Large land areas with now temperate climates show traces of past permafrost. The exact timing of the maximum extent of permafrost during the Last Glacial is difficult to establish (see Vandenberghe et al. 2014 and Lindgren et al. 2016 for a discussion of uncertainties). It does not necessarily coincide with the maximum ice extent. Therefore, the maximum extent of permafrost during the Last Glacial is denoted as the Last Permafrost Maximum (LPM; Vandenberghe et al. 2014). Based on the map of Fig. 1.22 and other data sources, Lindgren et al. (2016) estimate the likely area of northern permafrost during the LPM between 32.7 and 35.3 million km^2. This includes 4.7 million km^2 of permafrost on exposed coastal sea shelves at the low sea levels of the LPM. This is 9.1 to 11.7 million km^2 larger than the current extent of permafrost on land (23.6 million km^2).

At the termination of the Last Glacial widespread permafrost thaw occurred (Walter et al. 2007). During this thaw period, probably 1000 Gt of carbon was lost from thawing permafrost soils to the atmosphere (Lindgren et al. 2018). This represented the permafrost carbon feedback of that time. During the Holocene, landscape

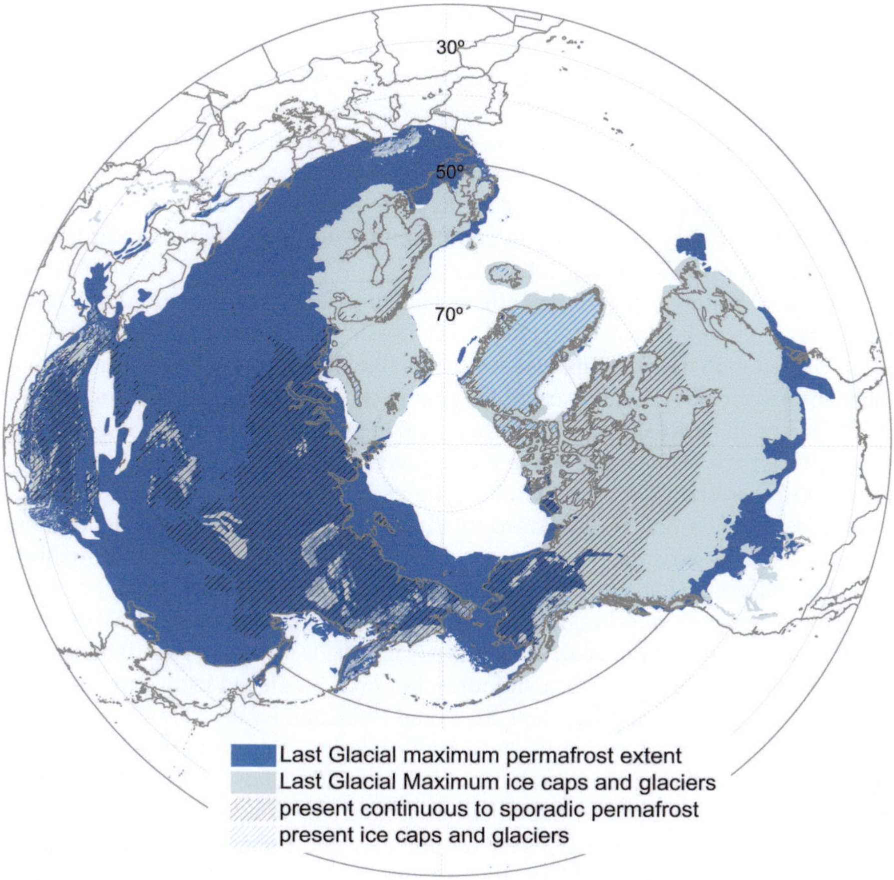

Fig. 1.22 Permafrost extent at approximately the time of the maximum extent of the ice sheets of the Last Glacial, the Last Glacial Maximum (LGM), or the Last Permafrost Maximum (LPM). Permafrost includes here continuous, discontinuous and sporadic permafrost on current land areas, and permafrost on emerged sea shelves. Shadings show the present extent of permafrost (continuous, discontinuous and sporadic) and ice caps/glaciers, derived from Fig. 1.1. Modified after Vandenberghe et al. (2014) and Lindgren et al. (2016); GIS data provided by Dr. Lindgren and National Snow and Ice Data Center as referenced in Fig. 1.1)

and climatic conditions for large scale peat accumulation became favourable in lowland areas left behind by thaw of the Laurentide and Scandinavian ice sheets. This resulted in the extensive peat deposits in northern boreal and temperate regions, described in Sect. 1.4.3. During the millennia of the Holocene, the permafrost carbon loss at the termination of the Last Glacial was more than compensated by carbon regain in peats and other soils.

The glacial history affects permafrost thickness. As mentioned in Sect. 1.1, below glacier ice permafrost is absent or thin, while in non-glacier covered areas such as northeastern Siberia, the permafrost depth reaches several hundreds of metres. In deglaciated areas, the permafrost starts to grow after deglaciation, provided the climate is sufficiently cold. For instance, in the Mackenzie Delta region, which has been at the border of the Laurentide ice sheet during the Wisconsinan glaciation, the permafrost depth may reach over 500 m in unglaciated areas, while in glaciated areas the depth declines to about 100 m (Burn and Kokelj 2009). Deep permafrost that exists now in non-glaciated areas, could be considered as relict with respect to the LPM permafrost, because its large thickness is the result of the past glacial climate, rather than being in equilibrium with the present warmer climate (Vandenberghe et al. 2014; Eq. 1.1).

The glacial history also strongly affects land forms, drainage systems and surface rock types and sediments. Compared to non-glaciated areas, glaciated areas are often covered with a suite of glacial sediments like till and glacial outwash, or show abundant bare rock surfaces created by glacial erosion. The glacial sediments are often piled up in steep-sloped hilly terrain, and in permafrost areas still may contain buried glacier ice. The slopes make these materials susceptible to erosion; such a landscape is termed paraglacial (Ballantyne et al. 2008). The drainage system in these areas is often poorly integrated, with numerous lakes in terrain depressions created by glacial erosion. The poor drainage of glacial terrain has promoted peat formation. Since the ice sheets also have dammed drainage systems of major rivers, extensive proglacial lake systems have existed, in particular in West Siberia and North America (Mangerud et al. 2004). These lake beds also have contributed to flat topography on which peatlands could form. For texts on glacial landforms and sediments see e.g. Bennett and Glasser (2011) and Benn and Evans (2014).

Non-glaciated permafrost areas have landforms resulting from weathering, erosion and sedimentation processes that are more similar to those under other climatic conditions, albeit that under cold climatic conditions the interaction with ground ice and physical weathering processes are more important (French 2018). In the unglaciated lowlands of eastern Siberia and northern Alaska, uninterrupted cold climate erosion and sedimentation processes throughout the Last Glacial and the Holocene have led to the deposition of the yedoma or Ice Complex deposits. These are thick successions of ice-rich loess (eolian dust deposits) and fluvially reworked sediments in a cold and dry steppe-tundra environment, and large volumes of ground ice (including ice wedges) that accumulated largely simultaneously with the sedimentation (Schirrmeister et al. 2011; Murton et al. 2015, 2017).

The Russian word едома originally means ‘hilly land’, and is used for areas with low hills and plateaus in northeast Siberia (Schirrmeister et al. 2013). In scientific literature it has become the name for the deposits that underly this topography in large areas. Although these sediments have modest amounts of organic matter compared to peats, this organic matter is a large carbon store because the large extent of Yedoma sediments (1.4 million km^2) and its vulnerability to thaw because of the high ice content (Zimov et al. 2006; Schirrmeister et al. 2011; Murton et al. 2015; Strauss et al. 2017).

The ice complex successions may be several tens of metres thick, and mostly date from the last glaciation (Schirrmeister et al. 2011) although similar deposits from earlier glacials have been reported. Murton et al. (2017) reports a thickness up to 80 m. The source of the fine-grained material in the Ice Complex deposits is assumed to be physical rock weathering and dry river beds. Murton et al. (2015, 2017) argue that the sediments are largely cold-climate loesses (windblown dust), while Schirrmeister et al. (2011) invoke a polygenetic depositional environment including loess, deposition by water in surficial meltwater runoff and colluvial deposition, varying from place to place. Murton et al. (2015) discuss various hypothesis for the sedimentary environment, and relate the yedoma loesses to the Pleistocene cold climate loess belts further westward in Europe. These have been deposited also largely under permafrost conditions, and are a rich source of information on paleo-permafrost during the Last Glacial (e.g. Vandenberghe et al. 2014). During deposition of these cold climate continental deposits, the vegetation was dominated by steppe vegetations, rich in forbs (Ashastina et al. 2018).

1.4.6 Geomorphology

Apart from the micro-topography created by frost action, ice-rich permafrost results also in larger scale geomorphological features. These are landforms created by growth of ground ice or by large scale thaw of ground ice; without being exhaustive on this subject, several striking features which affect the permafrost carbon cycle are introduced here. The formation mechanism of these landforms will be discussed in more detail in Chap. 3.

Palsas and pingos are features created by the accumulation of ground ice. Palsas and related forms, such as peat plateaus, and already have been mentioned in section are 1.4.3; they typically occur in peatlands in the discontinuous permafrost zone. The water which forms the ice is drawn from the surrounding soil. In northern Finland, where the name ‘palsa’ originates, they are typically 0.5–7 m in height and from 2 to 150 m in diameter (Seppälä 2011). Coalescing palsas form peat plateaus.

Pingos (Fig. 1.23) are conical hills found in the continuous permafrost zone. The word ‘pingo’ derives from Innuit language; in Siberia, they are known as ‘bulganniakh’, in Russian scientific literature also known as ‘hydrolaccoliths’. They can attain a larger height than palsas, heights up to 37 m have been recorded, although the majority of pingos is much lower (Grosse and Jones 2011). They result from local seepage of ground water through the permafrost. In cross section, they consist of a core of relatively pure ice, overlain by a mantle of soil. Growth of pingos may result in rupture of the overlying soil mantle at the top, after which the core of ice is exposed and thaw may result (Mackay 1998).

Ice-rich soils are preserved as long as the soil heat balance prevents thaw. However, this heat balance is easily disrupted, even in cold Arctic climates. For instance, disruption of a protective vegetation cover, which shades the soil from summer solar radiation may initiate thawing of excess ice, causing soil subsidence and the formation of ponds (Nauta et al. 2014). The thawing of ice-rich permafrost may have large effects on the land surface. The loss of volume by melting ice results in the formation of ponds and lakes (Hopkins 1949; Brouchkov et al. 2004; Jorgenson

Fig. 1.23 (**a**) Pingo on river floodplain; the pingo is approximately 20 m high. (**b**) thaw pond created by thawing of an ice lens in the subsurface. In the centre of the pool drowned dwarf birch branches protrude above the water; living dwarf birches on dry soil in the background. (**c**) thaw lake created by large scale thawing of ground ice, and active wave erosion on the banks. (**d**) Thaw slumps at the coast of Herschel Island, Northern Canada, photo by Dr. L. Bröder. (Photos **a**–**c** near Kytalyk Research Station, Indigirka lowlands, by the author)

et al. 2006; Grosse et al. 2013; Fig. 1.23). If a lake is large and deep enough it will not freeze to the bottom and it may expand by wave erosion, and sideways and downward heat transfer from the water to the permafrost (Hopkins and Kidd 1988; Plug and West 2009). In some permafrost areas, a dominant wind direction results in the formation of elongated lakes. Ultimately, thaw lakes are filled in with sediments or drained by erosion of thresholds which connect them with the river drainage system, or other lower-lying water bodies. (Mackay 1988; Boike et al. 2008; Van Huissteden et al. 2011; Grosse et al. 2013). Lowlands in permafrost areas that are underlain by fine-grained sedimentary deposits (including the yedoma deposits, but also peat, fluvial, lacustrine and marine deposits) are covered with thaw lakes and traces of ancient drained thaw lake basins (Fig. 1.24). Many of these are not the result of present thaw processes, but originate from the termination of the Last Glacial. The rapid transition towards the warmer Holocene climate has induced widespread lake thaw, which is evidenced by radiocarbon datings from thaw lake sediments throughout the Arctic and SubArctic (Walter et al. 2007). Overlapping thaw lake basins occur frequently, indicating repeated lake formation.

Soil which is supersaturated by water from thawing ice is susceptible to mass wasting processes ranging from gully formation (e.g. Boike et al. 2008) to landslides on sloping ground (e.g. Khomutov and Leibman 2014; French 2018). The

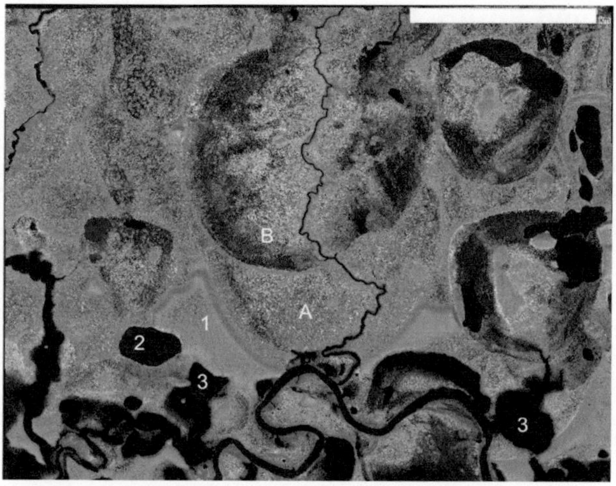

Fig. 1.24 Overlapping drained thaw lake basins, Indigirka Lowlands, Northeastern Siberia, near Kytalyk Research station. 1: Remnant of Pleistocene yedoma deposits; 2: actively expanding thaw lake of Fig. 1.24; 3: thaw lakes that have been connected to the river by relatively recent erosion; (**a**) oldest thaw lake bottom, drained at least 4000 years ago (Teltewskoi et al. 2016); (**b**) younger lake bottom, cutting into the older lake bottom. Scale bar 3 km. Wordlview panchromatic image, 0.4 m resolution July 102,015, WorldView © 2019 MAXAR

latter are known as cryogenic landslides, because of the role of thawing ground ice as a water source. The water drives a continuous flow of mud, emanating from headwalls with exposed ground ice that are sometimes tens of metres high and may remain active for many years (e.g. French 2018; Murton et al. 2017); Fig. 1.23. These processes liberate soil organic matter from permafrost soils and promote its subsequent decomposition, and transport and redeposition in rivers and sea (e.g. Walter et al. 2006; Schuur et al. 2008; Vonk et al. 2012; Nauta et al. 2014).

These processes of lake formation and thaw-driven erosion are often collected under the term thermokarst (e.g. 'thermokarst lake' instead of 'thaw lake'). However, the landforms created by thawing of ground ice are only very superficially similar to true 'karst' features, which are created by dissolution of limestone in non-permafrost environments, and therefore result from a quite different process than the phase transition of water. Moreover, in some publications quite sloppy usage of the term 'thermokarst' occurs, for instance by referring to 'a thermokarst', without proper reference to what kind process is underlying it, or what type of morphology – erosion, thaw subsidence, gully or mass wasting feature. I therefore prefer to abandon the term here, and resort to common, well-accepted geomorphological terminology: thaw lake formation, or other geomorphological terminology related to erosion and mass wasting. A better collective description for these phenomena is abrupt permafrost thaw (in short abrupt thaw) instead of thermokarst.

1.5 Recent and Future Climate Change

The effect of global warming as a result of greenhouse gas emissions is felt stronger in the Arctic than in temperate and tropical regions (IPCC 2013; Table 1.1, Fig. 1.25, 1.26, 1.27). The Arctic has warmed twice the global average over the past 50 years (Fig. 1.25). The late summer sea ice cover in the Arctic ocean is lower than it has ever been in the past 1450 years (Kinnard et al. 2011). There is extensive scientific literature on how warming of the Arctic climate is already affecting permafrost (AMAP 2017). For the future, climate warming will certainly have a widespread and deep effect on permafrost (e.g Romanovsky et al. 2010a). Even an immediate stop of greenhouse gas emissions would not stop warming of the climate, because

Table 1.1 Change of surface air temperature and precipitation over the Arctic according to the climate models presented in IPCC AR5, based on RCP scenario 4.5 only. The ranges for the temperature and precipitation are the ranges shown in model output maps of Fig. 1.27 and Fig. 1.28. The global mean for temperature and its likely range are derived from Table SPM.2 in the Summary for Policy Makers in IPCC AR5 (IPCC 2013)

Temperature	Summer ±2050	Winter ±2050	Summer ±2100	Winter ±2100
Arctic Ocean RCP 4.5	Up to 2 °C	4–9 °C	Up to 4 °C	5–11 °C
Arctic land RCP 4.5	1–3 °C	3–5 °C	2–4 °C	4–7 °C
Global mean RCP 4.5	1.4 (0.9–2.0)°C		1.8 (1.1–2.6)°C	
Precipitation				
Arctic Ocean RCP 4.5	0–20%	20–40%	0–30%	30–50%
Arctic land RCP 4.5	0–20%	10–30%	0–30%	20–40%

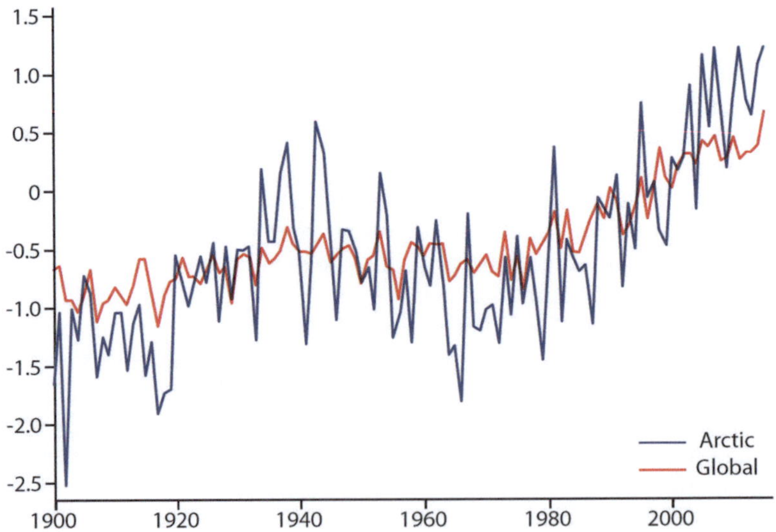

Fig. 1.25 Arctic (60–90°N) and global annual surface air temperature anomalies relative to the 1981–2010 mean value. (From AMAP 2017; revised from NOAA data)

of the long residence time of the most important greenhouse gas CO_2 in the atmosphere. Emission reductions can only influence the rate of already ongoing climate change – and therefore the rate at which permafrost landscapes will change in the coming centuries, and when these changes stabilize (McGuire et al. 2018). What the permafrost landscape looks like by the end of this century, is therefore strongly determined by political, economical and technological developments in human society today and the near future.

The changes in the Arctic are much stronger than the global mean temperature change because of the effect of polar amplification: warming reduces snow and sea ice cover, resulting in less solar radiation being reflected back in space, enhancing further warming (e.g. Serreze and Francis 2006; Serreze et al. 2009; AMAP 2017). It is a consistent feature in all climate models. It is also a distinctly seasonal feature: it is strongest in late autumn and early winter (November to December), because then the effect of less sea ice in the Arctic Ocean is strongest, shortly after the sea ice minimum in September (IPCC 2013 and references therein).

The projections of future climate change depend on future scenarios of the greenhouse gas emissions, in the latest IPCC AR5 report (IPCC 2013) these scenarios are the 'representative concentration pathways' (RCP). The RCP's are possible scenarios of future anthropogenic greenhouse gas emissions and the trajectory of resulting atmospheric greenhouse gas concentrations. These scenarios range from a future maximum followed by a decline of greenhouse gas emissions, that should be starting at about the time of writing of this book (scenario RCP 2.6), to a continued rise of emissions throughout this century (RCP 8.5). The numbers behind the abbreviation RCP correspond to the approximate amount of radiative heating of the climate expected from the scenario: RCP 8.5 means >8.5 W m^{-2} by the year 2100 (Moss et al. 2010). An often used in-between scenario is RCP 4.5, in which greenhouse gas emissions peak around 2060.

The climate model ensemble average of the IPCC AR5 shows that most of the Arctic climate warming occurs in winter (Table 1.1, Fig. 1.26). With respect to the reference period 1986–2005, the winter temperatures (December–February average) over the Arctic Ocean may have warmed by 4–9 °C around 2050, while the adjacent land surfaces in North America and Eurasia have warmed less, 3–5 °C. This assumes the moderate RCP 4.5 greenhouse gas emission scenario. There are regional differences – the numbers given here are the range of these differences on the maps of Fig. 1.26. At the end of this century, these projected values are 5–11 °C and 4–7 °C for the Arctic Ocean and land surface respectively. The Arctic Ocean warms more than the land surface in winter; most of the warming occurs near the Bering Strait and the Atlantic Ocean, in particular the Barentsz Sea. In summer (June–August), the land surface warming is more than that of the Arctic Ocean: 0–2 °C for the Arctic Ocean, while the North American and Eurasian land surfaces warm by 1–3 °C around 2050. For the end of the century, the projected warming for the Arctic Ocean is 0–4 °C in summer and for the land surface 2–4 °C.

However, the warming of the permafrost by the end of the century is strongly dependent on the RCP scenarios. The winter warming for the RCP 8.5 scenario over the Arctic land surface is expected to be 5–6 °C warmer than that for the RCP 4.5

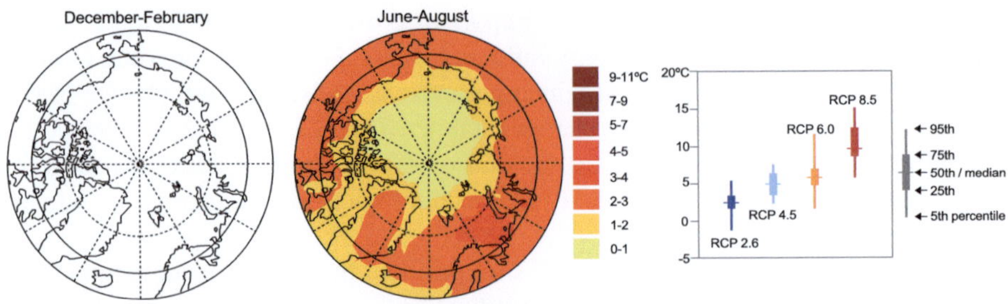

Fig. 1.26 Projected temperature changes at the end of the century (2081–2100) with respect to 1986–2005 in the RCP4.5 scenario, based on the 50% percentile of the CMIP5 climate model ensemble. Graph on right: modelled temperature change relative to 1986–2005, averaged over land areas in the Arctic (67.5°N to 90°N) in December to February. The fifth, 25th, 50th (median), 75th and 95th percentiles of the distribution of modelled 20-year mean changes are shown for the four RCP scenarios. Based on IPCC (2013), Annex I Fig. A1.8 and A1.9

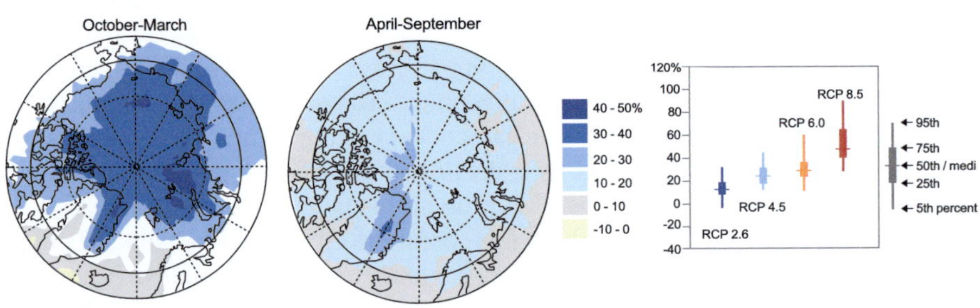

Fig. 1.27 Projected precipitation changes at the end of the century (2081–2100) with respect to 1986–2005 in the RCP4.5 scenario; the 50% percentile of the CMIP5 climate model ensemble is shown. Graph on right: Modelled precipitation change relative to 1986–2005, averaged over land areas in the Arctic (67.5°N to 90°N) in December to February. The fifth, 25th, 50th (median), 75th and 95th percentiles of the distribution of modelled 20-year mean changes are given for 2081–2100 for the four RCP scenarios. Based on IPCC (2013), Annex 1, Fig. A1.10 and A1.11

scenario. Parts of the Arctic land area may experience winter warming of over 15 °C if the warmest model projections become a reality (Fig. 1.26 right hand graph).

The precipitation also increases (Fig. 1.27). As a result of increased river runoff, freshwater storage in the Arctic Ocean has increased already; compared with the 1980–2000 average, the volume of freshwater in the upper layer of the Arctic Ocean has increased by 8000 cubic kilometres, or more than 11% (AMAP 2017). Winter precipitation (in mm water equivalent) in October–March is expected to increase over much of the Arctic Ocean by 20–40% in 2050 and by 30–50% at the end of the century. This is again based on the RCP 4.5 scenario. The winter precipitation increase is less over the land surface, 10–30% around 2050 and 20–40% at the end of the century, but it is in particular substantial in the drainage basins of the large East Siberian rivers (AMAP 2017) with substantial consequences for the hydrology of these drainage basins (Chap. 5). Also the April–September precipitation increases, although less than the winter precipitation: 0–20% over the Arctic Ocean and adjacent land masses around 2050, and 0–30% at the end of the century.

The strongest increase of precipitation occurs in autumn and winter. This precipitation is mostly fed by local (Arctic) evaporation, and to a lesser extent by moisture transport from lower latitudes. The retreat of sea ice in the Arctic Ocean, resulting in larger and more prolonged areas of open water in late summer and autumn, is an important moisture source for autumn precipitation. About 60% of the precipitation increase is originating from the Arctic itself, which is mostly the the autumn-winter precipitation, and 40% from distant moisture sources (Bintanja and Selten 2014). Increasing snow cover thickness may increase warming of the permafrost because of the insulating effect of snow (Chap. 2). However, with the decline of snow cover duration, also a general decline of snow cover thickness is expected over large areas; increase of snow cover thickness is most likely near the Arctic ocean. The year-to-year variability of snow cover duration is strong, and spatial variability over the Arctic varies with the data source – station or remote sensing data (AMAP 2017).

Sea ice loss in the Arctic Ocean warms land temperatures up to 1500 km inland, promoting temperature increase and thaw of permafrost (Lawrence et al. 2008). For the late autumn/early winter period, the warming is about a factor 4 stronger than the global average. Despite large year-to-year variations, there is a clear downward trend in sea ice extent. A record low minimum occurred in September 2012 and a record low maximum sea ice extent occurred in March 2016 (AMAP 2012, 2017). The decrease of the Arctic ice cover has been linked to increase of the surface and soil temperatures and increase in primary productivity of ecosystems (e.g. Parmentier et al. 2013, 2015), which in turn are important drivers of the permafrost carbon cycle.

1.6 The Uncertain Future of Permafrost

Clear evidence of recent permafrost changes comes from measurements of permafrost temperatures in boreholes in permafrost and systematic data collection on ALT (Figs. 1.28 and 1.29). Time series of systematically collected data are relatively short, because data collection is challenging in remote Arctic areas. Nevertheless, time series of active layer depth show an increase in most sites in Fig. 1.29. However, ALT is a somewhat less certain indicator for permafrost change than the deeper permafrost temperature. It may include a component of surface subsidence by loss of volume as ground ice thaws. Temperature observations at depth are a more certain indicator of warming permafrost. These also show statistically significant trends of rising temperature between 0.21 and 0.82 °C per decade at depths between 10 to 20 m (Osterkamp and Romanovsky 1999; Romanovsky et al. 2010a, 2010b). This indicates that warming of the permafrost penetrates down to larger depths. Sites without change, or a slight decrease of temperature do occur, in particular in areas

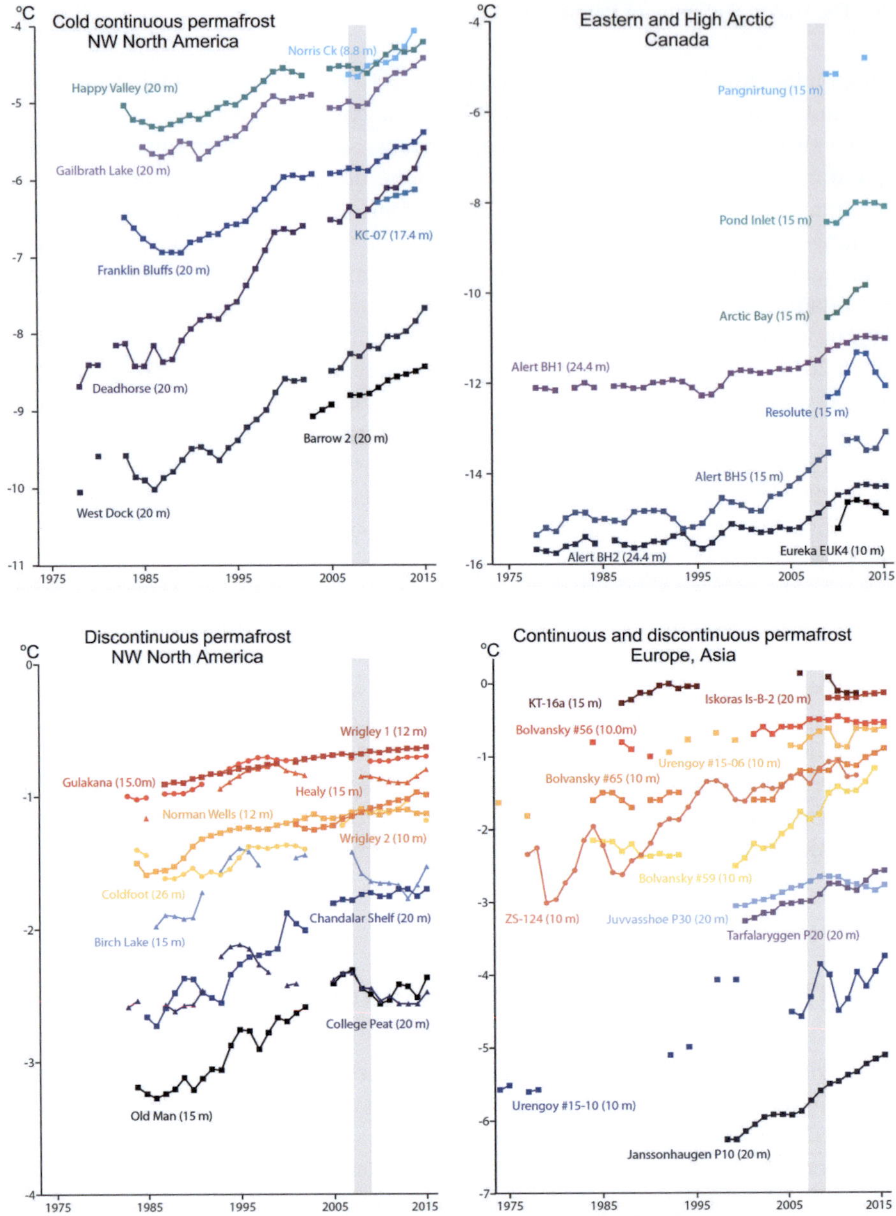

Fig. 1.28 Time series of annual mean ground temperature at depths of 9 to 26 m below the surface at selected measurement sites. Upper left: cold continuous permafrost of northwestern North America (Beaufort-Chukchi region), Upper right: discontinuous permafrost in Alaska and northwestern Canada. Lower left: cold continuous permafrost of eastern and High Arctic Canada (Baffin and Davis Strait). Lower right: continuous to discontinuous permafrost in Scandinavia, Svalbard, Russia/Siberia (Barentsz region). Temperatures are measured at or near the level of zero annual amplitude. The grey bar indicates the International Polar Year. Warm permafrost sites in red, cold permafrost sites in blue. From AMAP (2017). Data series are updated from Christiansen et al. (2010), Romanovsky et al. (2014, 2015) and Ednie and Smith (2015). Republished with permission of Arctic Monitoring and Assessment Programme

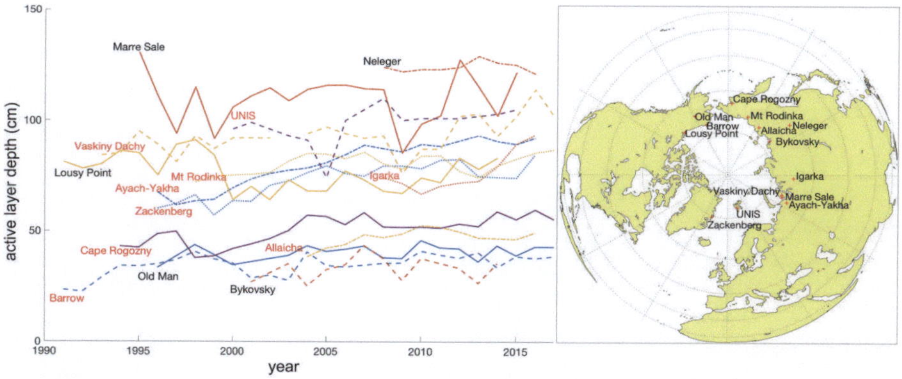

Fig. 1.29 Left: Time series of mean active layer depth at the end of the summer season at sites in permafrost regions. Site names printed in red indicate a significant trend ($p < 0.1$). Right: site location map. From CALM active layer database (https://www2.gwu.edu/~calm/); data provided by N. I. Shiklomanov (Barrow), V. Romanovsky (Old Man), D. Gilichinsky (Bykovsky), A. Abramov (Allaicha), A. Kotov/V. Razzhivin (Cape Rogozny), M. Lund (Zackenberg), Ole Humlum (UNIS), D. Kaverin (Ayach-Yakha), S. Davydov (Mt Rodinka), J. Chartrand (Lousy Point), M. Leibman (Vaskiny Daichy) F. Zepalov (Igarka), P. Konstantinov (Neleger), A. Vasiliev (Marre Sale)

with relatively warm, discontinuous permafrost. However, the overall trend is rising permafrost temperatures.

Temperature change is not the only effect of climate warming on permafrost. Geomorphological changes – formation of thaw ponds and lakes, and erosion of ice-rich permafrost – already has been mentioned above, and will be discussed extensively in Chap. 5. In particular there are reports of increasing pond formation throughout the Arctic (e.g. Jorgenson et al. 2006; Sannel and Kuhry 2011; Liljedahl et al. 2016) and reports of an increase of erosion features (Segal et al. 2016; Lewkowicz and Way 2019). Thaw lake expansion is uncertain. Two opposite climate driven processes occur – lake expansion by ground ice thaw (Grosse et al. 2013) and lake shrinkage by subsurface drainage and erosion (Yoshikawa and Hinzman 2003; Smith et al. 2005; Jones et al. 2011; Chap. 5). The formation and disappearance of lakes, ponds and wetlands is important for greenhouse gas fluxes of permafrost in a changing climate. It determines the fraction of the permafrost carbon stores that enter the atmosphere as CO_2 or CH_4 (Schädel et al. 2016; Fig. 1.30). Furthermore, vegetation change is expected to occur widely (e.g. Myers-Smith et al. 2011), which may compensate permafrost carbon losses partly. However, to what extent this may happen is debated, because an increasing role of fire and permafrost thaw is expected also (Chap. 6).

At what pace the permafrost land surface and carbon cycle will change as a result of global warming is still highly uncertain. First there is the uncertainty in the future greenhouse gas emission trajectories, next uncertainty in the climate change resulting from these emissions, as expressed in the spread between modelled climates (IPCC 2013). On top of that, there is large uncertainty in model predictions of future soil and permafrost temperatures and permafrost extent, depending on the ability of

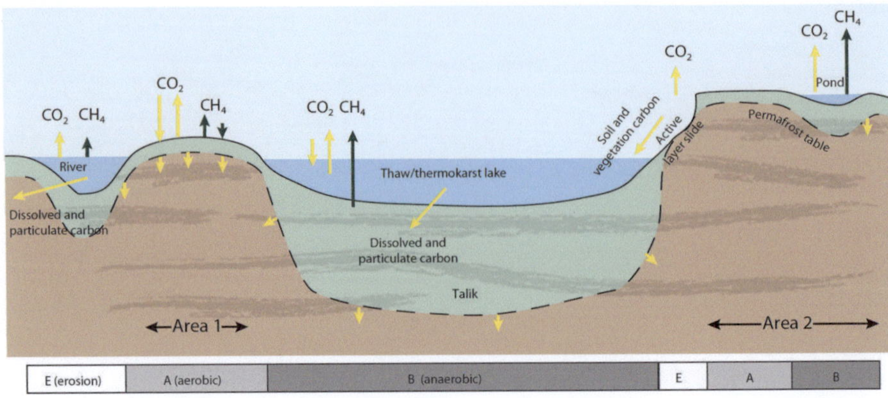

Fig. 1.30 Mechanisms by which carbon may be lost from Arctic soils in permafrost areas (excluding fire). Area 1 has a spatially homogenous increase in active layer thickness and Area 2 has spatially heterogeneous permafrost thaw with thaw pond and lake development, driven by differences in soil ice content (after Van Huissteden and Dolman 2012; from AMAP 2017). Republished with permission of Arctic Monitoring and Assessment Programme

models to represent permafrost physics and the carbon cycle (Slater and Lawrence 2013; McGuire et al. 2016, 2018; Chadburn et al. 2017; AMAP 2017). Most models still do not include the geomorphological and vegetation effects of permafrost thaw (Van Huissteden and Dolman 2012; Chap. 8). Another uncertainty in models is the effect of thicker snow cover and changing vegetation properties on permafrost temperature (Chadburn et al. 2017).

The modelled mean change of surface temperature over permafrost areas using the models of the IPCC AR5 report ranges between 1.5 and 10 °C by the end of the century, depending on future greenhouse gas emissions (Fig. 1.31). The modelled permafrost area by 2100 shows a very large uncertainty, due to differences in simulated surface climates among the models, and weaknesses in the representation of permafrost physics (Slater and Lawrence 2013). Averaged across all models and methods, the sensitivity of permafrost area to climate change is 1.67 x 10^6 km^2 °C^{-1}, but with a wide uncertainty margin (0.75–2.32x 10^6 km^2 °C^{-1}). In addition, disappearing sea ice in the Arctic Ocean affects permafrost thaw and the permafrost carbon cycle, in particular in Siberia (Vandenberghe et al. 2012; Parmentier et al. 2013).

Assuming the RCP4.5 emission scenario, near-surface permafrost is unlikely to be sustainable in the present-day discontinuous permafrost zone, where permafrost is the most vulnerable. With RCP8.5, permafrost existence will be reduced to the Canadian Archipelago, Russian Arctic coast, and east Siberian uplands at the end of the century. The permafrost area may be reduced from a present 15–16.5 million km^2 to less than 9.6 and 3.5 million km^2 in 2100 for respectively the RCP 4.5 and RCP 8.5 climates (Slater and Lawrence 2013). More recent model simulations indicate losses of permafrost between 3.2 and 4.7 million km^2 for the moderate RCP4.5 climate projections and between 5.7 and 16.1 million km^2 for the IPCC RCP8.5 climate projections by the year 2300, in less than three centuries (McGuire et al. 2018). These model estimates diverge widely due to structural differences of the

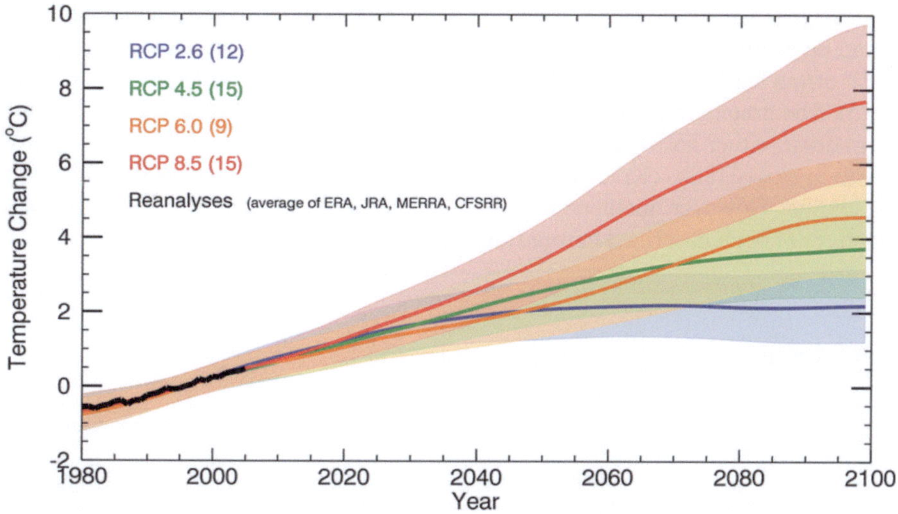

Fig. 1.31 Mean change in surface air temperature over the present-day permafrost area, for four representative concentration pathways (RCP's). The number of models contributing to each RCP estimate is given in brackets. The colour shaded area is one standard deviation across models. Republished with permission of American Meteorological Society from 'Diagnosing present and future permafrost from climate models' by Slater and Lawrence (2013), Journal of Climate, Vol. 26–15, 2013; permission conveyed through Copyright Clearance Center, Inc.

models (discussed in Chap. 8) and climate scenario uncertainty, but all point in the same direction: widespread permafrost thaw with large effects on ecosystems and the carbon balance.

The incidence of climatic extremes can be hugely important for permafrost; several examples will be discussed in Chaps. 5, 6 and 7. For instance, widespread ground ice thaw and erosion related to permafrost thaw have been attributed by various authors to a few extremely warm summers, as we shall see in the next chapters. Therefore, not only the change of the climate average is important, but also changes in the climate extremes.

This section also makes clear that strength of the permafrost carbon feedback depends on the RCP greenhouse gas emission scenarios. The strength of this feedback is highly relevant for the efforts to mitigate future climate change by reduction of greenhouse gas emissions from the various sectors of human society – industry, business, households, transport and agriculture. These efforts will need to be stronger, and for our societies and economies more far-reaching, if the strength of the permafrost carbon feedback is larger. It also indicates that mitigation of climate change can reduce permafrost and carbon loss, and climate change mitigation will become more and more an uphill struggle if it is not successful.

A perfect illustration of our incomplete knowledge of permafrost dynamics is the recent, puzzling discovery of craters, emitting methane, in continuous permafrost regions in western Siberia, in particular in the Yamalo-Nenets region. The craters reported so far are in the order of tens of metres cross section and reach depths of several tens of metres. The presence of ejecta, blocks of permafrost and tundra soil found up to hundred meter from one of the craters testify of an explosive process. Their origin has been linked to climate change (Leibman et al. 2014; Kizyakov et al. 2015). The craters may point to destabilization of deeper gas reservoirs within the permafrost, either gas hydrates or gas that becomes pressurized by cryogenic processes in closed taliks (Kraev et al. 2017). The debate on the origin of these features is ongoing. However, methane emission by formation of craters is a new process that has not been described before from permafrost areas.

References

ACIA (2004) Impacts of a warming Arctic: Arctic climate impact assessment. Cambridge University Press, Cambridge

AGCR (1988) Glossary of permafrost and related ground-ice terms. Associate Comittee on Geotechnical Research Techical Memorandum National Reseach Council Canada 142:157

AMAP (2012) Arctic climate issues 2011: changes in Arctic snow, water, ice and permafrost., SWIPA Overview report 2011

AMAP (2017) Snow, water, ice and permafrost in the Arctic (SWIPA) 2017. Arctic Monitoring and Assessment Programme (AMAP), Oslo, p 144

Ashastina K, Kuzmina S, Rudaya N, Troeva E, Schoch WH, Römermann C, Reinecke J, Otte V, Savvinov G, Wesche K (2018) Woodlands and steppes: Pleistocene vegetation in Yakutia's most continental part recorded in the Batagay permafrost sequence. Quat Sci Rev 196:38–61

Baldocchi D, Kelliher F, Black T, Jarvis P (2000) Climate and vegetation controls on boreal zone energy exchange. Glob Chang Biol 6(S1):69–83

Ballantyne C, Hall K, French H (2008) Paraglacial geomorphology: processes and paraglacial context. Geomorphology 95:1–2

Benn D, Evans D (2014) Glaciers and glaciation. Routledge, London, p 816

Bennett MM, Glasser NF (2011) Glacial geology: ice sheets and landforms. Wiley, Hoboken

Bintanja R, Selten F (2014) Future increases in Arctic precipitation linked to local evaporation and sea-ice retreat. Nature 509:479–482

Bockheim J (2007) Importance of cryoturbation in redistributing organic carbon in permafrost-affected soils. Soil Sci Soc Am J 71(4):1335–1342

Boike J, Wille C, Abnizova A (2008) Climatology and summer energy and water balance of polygonal tundra in the Lena River Delta, Siberia. J Geophys Res Biogeo 113(G3):G03025. https://doi.org/10.1029/2007JG00054

Bonan GB, Chapin FS, Thompson SL (1995) Boreal forest and tundra ecosystems as components of the climate system. Clim Chang 29(2):145–167

Bradley RS (2015) Paleoclimatology: reconstructing climates of the quaternary, vol 667, 3rd edn. Elsevier, Amsterdam

Brouchkov A, Fukuda M, Fedorov A, Konstantinov P, Iwahana G (2004) Thermokarst as a short-term permafrost disturbance, Central Yakutia. Permafr Periglac Process 15(1):81–87. https://doi.org/10.1002/ppp.473

Brown R (1970) Permafrost in Canada. University of Toronto Press, Toronto

Brown J, Ferrians O, Heginbottom JA, Melnikov E (2002) Circum-Arctic map of permafrost and ground-ice conditions, version 2. NSIDC: National Snow and Ice Data Center, Boulder

Burn C, Kokelj S (2009) The environment and permafrost of the Mackenzie Delta area. Permafr Periglac Process 20(2):83–105

Chadburn SE, Krinner G, Porada P, Bartsch A, Beer C, Belelli Marchesini L, Boike J, Ekici A, Elberling B, Friborg T, Hugelius G, Johansson M, Kuhry P, Kutzbach L, Langer M, Lund M,

Parmentier FJW, Peng S, Van Huissteden J, Wang T, Westermann S, Zhu D, Burke EJ (2017) Carbon stocks and fluxes in the high latitudes: using site-level data to evaluate Earth system models. Biogeosciences 14(22):5143–5169. https://doi.org/10.5194/bg-14-5143-2017
Charman D (2002) Peatlands and environmental change. Wiley, Chichester, p 301
Christiansen HH, Etzelmüller B, Isaksen K, Juliussen H, Farbrot H, Humlum O, Johansson M, Ingeman-Nielsen T, Kristensen L, Hjort J, Holmlund P, Sannel ABK, Sigsgaard C, Åkerman HJ, Foged N, Blikra LH, Pernosky MA, Ødegård R (2010) The thermal state of permafrost in the Nordic area during the international polar year. Permafr Periglac Process 21:156–181
Curry JA, Schramm JL, Rossow WB, Randall D (1996) Overview of Arctic cloud and radiation characteristics. J Clim 9(8):1731–1764
De Leffingwell E (1919) The Canning River region, northern Alaska. US Geol. Survey Professional Paper vol 109. US Government Printing Office, 251 p
Dee DP, Uppala S, Simmons A, Berrisford P, Poli P, Kobayashi S, Andrae U, Balmaseda M, Balsamo G, Bauer P (2011) The ERA-interim reanalysis: configuration and performance of the data assimilation system. Q J R Meteorol Soc 137(656):553–597
Delisle G (1998) Numerical simulation of permafrost growth and decay. J Quat Sci 13(4):325–333
Ednie M, Smith SL (2015) Permafrost temperature data 2008-2014 from community based monitoring sites in Nunavut, Geological Survey of Canada Open File, 7784
Eugster W, Rouse WR, Pielke RA Sr, Mcfadden JP, Baldocchi DD, Kittel TG, Chapin FS, Liston GE, Vidale PL, Vaganov E (2000) Land–atmosphere energy exchange in Arctic tundra and boreal forest: available data and feedbacks to climate. Glob Chang Biol 6(S1):84–115
French HM (2018) The periglacial environment, 4th edn. Wiley, Hoboken, p 515
Froese DG, Westgate JA, Reyes AV, Enkin RJ, Preece SJ (2008) Ancient permafrost and a future, warmer Arctic. Science 321(5896):1648–1648
Grosse G, Jones B (2011) Spatial distribution of pingos in northern Asia. Cryosphere 5(1):13–33
Grosse G, Jones B, Arp C (2013) Thermokarst Lakes, drainage, and drained basins. In: Shroder J, Giardino R, Harbor J (eds) Treatise on geomorphology, Glacial and Periglacial Geomorphology, vol 8. AcademicPress, SanDiego, pp 325–353. https://doi.org/10.1016/b978-0-12-374739-6.00216-5
Gruber S (2012) Derivation and analysis of a high-resolution estimate of global permafrost zonation. Cryosphere 6(1):221–233. https://doi.org/10.5194/tc-6-221-2012
Hobbie JE, Deegan LA, Peterson BJ, Rastetter EB, Shaver GR, Kling GW, O'Brien WJ, Chapin FT, Miller MC, Kipphut GW (1995) Long-term measurements at the arctic LTER site. In: Powell TM, Steele JH (eds) Ecological time series. Springer, Dordrecht, pp 391–409
Hopkins DM (1949) Thaw lakes and thaw sinks in the Imuruk Lake area, Seward Peninsula, Alaska. J Geol 57(2):119–131
Hopkins D, Kidd J (1988) Thaw lake sediments and sedimentary environments. In: Proceedings of the 5th international permafrost conference, 1998. pp 790–795
Hugelius G, Strauss J, Zubrzycki S, Harden JW, Schuur EAG, Ping CL, Schirrmeister L, Grosse G, Michaelson GJ, Koven CD, O'Donnell JA, Elberling B, Mishra U, Camill P, Yu Z, Palmtag J, Kuhry P (2014) Estimated stocks of circumpolar permafrost carbon with quantified uncertainty ranges and identified data gaps. Biogeosciences 11:6573–6593. https://doi.org/10.5194/bg-11-6573-2014
Hughes AL, Gyllencreutz R, Lohne ØS, Mangerud J, Svendsen JI (2016) The last Eurasian ice sheets–a chronological database and time-slice reconstruction, DATED-1. Boreas 45(1):1–45
IPCC (2013) Climate change 2013: the physical science basis. Contribution of Working Group I to the Fifth Assessment Report of the Intergovernmental Panel on Climate Change. Cambridge University Press. doi:https://doi.org/10.1017/CBO9781107415324
Isarin R, Huijzer A, Van Huissteden J (1998) Time-slice oriented multi-proxy database (MPDB) for palaeoclimate reconstruction. CAPS (Circumpolar Active-Layer Permafrost System) CD-ROM, Version 1
Johansson M, Callaghan TV, Bosiö J, Åkerman HJ, Jackowicz-Korczynski M, Christensen TR (2013) Rapid responses of permafrost and vegetation to experimentally increased snow cover in sub-arctic Sweden. Environ Res Lett 8(3):035025. https://doi.org/10.1088/1748-9326/8/3/035025

Jones A, Stolbovay V, Tarnocai C, Broll G, Spaargaren O, Montanarella L (2010) Soil atlas of the northern circumpolar region. European Commission, Publications office of the European Union, Luxembourg, p 144

Jones BM, Grosse G, Arp C, Jones M, Walter Anthony K, Romanovsky V (2011) Modern thermokarst lake dynamics in the continuous permafrost zone, northern Seward peninsula, Alaska. J Geophys Res Biogeosci 116(G2):G00M03. https://doi.org/10.1029/2011JG001666

Jorgenson MT, Shur YL, Pullman ER (2006) Abrupt increase in permafrost degradation in Arctic Alaska. Geophys Res Lett 33(2). https://doi.org/10.1029/2005gl024960

Kay JE, L'Ecuyer T, Chepfer H, Loeb N, Morrison A, Cesana G (2016) Recent advances in Arctic cloud and climate research. Curr Climate Change Rep 2(4):159–169

Khomutov A, Leibman M (2014) Assessment of landslide hazards in a typical tundra of central Yamal, Russia. In: Landslides in cold regions in the context of climate change. Springer, Cham, pp 271–290

Kinnard C, Zdanowicz CM, Fisher DA, Isaksson E, de Vernal A, Thompson LG (2011) Reconstructed changes in Arctic Sea ice over the past 1,450 years. Nature 479(7374):509–512

Kitover D, Renssen H, Van Balen R, Vandenberghe J (2012) Modeling permafrost response of the last glacial termination: first results of the VAMPER model. In: Tenth International Conference On Permafrost, 2012. pp 209–214

Kitover D, Balen R, Roche D, Vandenberghe J, Renssen H (2013) New estimates of permafrost evolution during the last 21 k years in Eurasia using numerical modelling. Permafr Periglac Process 24(4):286–303

Kizyakov A, Sonyushkin A, Leibman M, Zimin M, Khomutov A (2015) Geomorphological conditions of the gas-emission crater and its dynamics in Central Yamal. Earth's Cryosphere 19:15–25

Kraev G, Schulze E-D, Yurova A, Kholodov A, Chuvilin E, Rivkina E (2017) Cryogenic displacement and accumulation of biogenic methane in frozen soils. Atmos 8(6):105

Kuhry P, Turunen J (2006) The postglacial development of boreal and subarctic peatlands. In: Wieder R, Vitt DH (eds) Boreal peatland ecosystems. Springer, Heidelberg, pp 25–46

Lachenbruch AH, Cladouhos TT, Saltus R (1988) Permafrost temperature and the changing climate. In: Proceedings of the fifth international conference on permafrost. Tapir Publishers, Trondheim, pp 9–17

Lawrence DM, Slater AG, Tomas RA, Holland MM, Deser C (2008) Accelerated Arctic land warming and permafrost degradation during rapid sea ice loss. Geophys Res Lett 35(11):L11506. https://doi.org/10.1029/2008GL033985

Leibman MO, Kizyakov AI, Plekhanov AV, Streletskaya ID (2014) New permafrost feature–deep crater in Central Yamal (West Siberia, Russia) as a response to local climate fluctuations. Environ Sustain 4:68–80

Lewkowicz AG, Way RG (2019) Extremes of summer climate trigger thousands of thermokarst landslides in a high Arctic environment. Nat Commun 10(1):1329. https://doi.org/10.1038/s41467-019-09314-7

Liljedahl AK, Boike J, Daanen RP, Fedorov AN, Frost GV, Grosse G, Hinzman LD, Iijma Y, Jorgenson JC, Matveyeva N, Necsoiu M, Raynolds MK, Romanovsky VE, Schulla J, Tape KD, Walker DA, Wilson CJ, Yabuki H, Zona D (2016) Pan-Arctic ice-wedge degradation in warming permafrost and its influence on tundra hydrology. Nat Geosci 9(4):312–318. https://doi.org/10.1038/ngeo2674

Lindgren A, Hugelius G, Kuhry P, Christensen TR, Vandenberghe J (2016) Gis-based maps and area estimates of northern hemisphere permafrost extent during the last glacial maximum. Permafr Periglac Process 27(1):6–16

Lindgren A, Hugelius G, Kuhry P (2018) Extensive loss of past permafrost carbon but a net accumulation into present-day soils. Nature 560(7717):219

Liston GE, Sturm M (2002) Winter precipitation patterns in arctic Alaska determined from a blowing-snow model and snow-depth observations. J Hydrometeorol 3(6):646–659

MacGregor JA, Fahnestock MA, Catania GA, Aschwanden A, Clow GD, Colgan WT, Gogineni SP, Morlighem M, Nowicki SM, Paden JD (2016) A synthesis of the basal thermal state of the Greenland Ice Sheet. J Geophys Res Earth 121(7):1328–1350

Mackay JR (1988) Catastrophic lake drainage, Tuktoyaktuk peninsula area, District of Mackenzie. Curr Res Part D Geol Surv Canada 88:83–90

Mackay JR (1998) Pingo growth and collapse, Tuktoyaktuk Peninsula area, western Arctic coast, Canada: a long-term field study. Géog Phys Quatern 52(3):271–323

Mangerud J, Jakobsson M, Alexanderson H, Astakhov V, Clarke GK, Henriksen M, Hjort C, Krinner G, Lunkka J-P, Möller P (2004) Ice-dammed lakes and rerouting of the drainage of northern Eurasia during the last glaciation. Quat Sci Rev 23(11):1313–1332

Margold M, Stokes CR, Clark CD, Kleman J (2015) Ice streams in the Laurentide ice sheet: a new mapping inventory. J Maps 11(3):380–395

McClelland JW, Holmes R, Dunton K, Macdonald R (2012) The Arctic Ocean estuary. Estuar Coasts 35(2):353–368

McGuire A, Wirth C, Apps M, Beringer J, Clein J, Epstein H, Kicklighter D, Bhatti J, Chapin III F, De Groot B (2002) Environmental variation, vegetation distribution, carbon dynamics and water/energy exchange at high latitudes. J Veg Sci 13(3):301–314

McGuire AD, Koven C, Lawrence DM, Clein JS, Xia J, Beer C, Burke E, Chen G, Chen X, Delire C (2016) Variability in the sensitivity among model simulations of permafrost and carbon dynamics in the permafrost region between 1960 and 2009. Glob Biogeochem Cycles 30(7):1015–1037

McGuire AD, Lawrence DM, Koven C, Clein JS, Burke E, Chen G, Jafarov E, MacDougall AH, Marchenko S, Nicolsky D (2018) Dependence of the evolution of carbon dynamics in the northern permafrost region on the trajectory of climate change. Proc Natl Acad Sci 115(15):3882–3887

Minke M, Donner N, Karpov NS, de Klerk P, Joosten H (2007) Distribution, diversity, development and dynamics of polygon mires: examples from Northeast Yakutia (Siberia). Peatlands Int 1:36–40

Moss RH, Edmonds JA, Hibbard KA, Manning MR, Rose SK, Van Vuuren DP, Carter TR, Emori S, Kainuma M, Kram T (2010) The next generation of scenarios for climate change research and assessment. Nature 463(7282):747

Murton JB, Goslar T, Edwards ME, Bateman MD, Danilov PP, Savvinov GN, Gubin SV, Ghaleb B, Haile J, Kanevskiy M (2015) Palaeoenvironmental interpretation of Yedoma silt (ice complex) deposition as cold-climate loess, Duvanny Yar, Northeast Siberia. Permafr Periglac Process 26(3):208–288

Murton JB, Edwards ME, Lozhkin AV, Anderson PM, Savvinov GN, Bakulina N, Bondarenko OV, Cherepanova MV, Danilov PP, Boeskorov V (2017) Preliminary paleoenvironmental analysis of permafrost deposits at Batagaika megaslump, Yana Uplands, Northeast Siberia. Quat Res 87(2):314–330

Myers-Smith IH, Forbes BC, Wilmking M, Hallinger M, Lantz T, Blok D, Tape KD, Macias-Fauria M, Sass-Klaassen U, Lévesque E (2011) Shrub expansion in tundra ecosystems: dynamics, impacts and research priorities. Environ Res Lett 6(4):045509

Nauta AL, Heijmans MMPD, Blok D, Limpens J, Elberling B, Gallagher A, Li B, Petrov RE, Maximov TC, van Huissteden J, Berendse F (2014) Permafrost collapse after shrub removal shifts tundra ecosystem to a methane source. Nat Clim Chang 5(1):67–70. https://doi.org/10.1038/nclimate2446

Niessen F, Hong JK, Hegewald A, Matthiessen J, Stein R, Kim H, Kim S, Jensen L, Jokat W, Nam S-I (2013) Repeated Pleistocene glaciation of the East Siberian continental margin. Nat Geosci 6(10):842–846

O'Connor FM, Boucher O, Gedney N, Jones CD, Folberth GA, Coppell R, Friedlingstein P, Collins WJ, Chappellaz J, Ridley J, Johnson CE (2010) Possible role of wetlands, permafrost, and methane hydrates in the methane cycle under future climate change: a review. Rev Geophys 48(4). https://doi.org/10.1029/2010rg000326

Oechel WC, Lawrence WT (1985) Taiga. In: Chabot B, Mooney HA (eds) Physiological ecology of North American plant communities. Springer, pp 66–94

Oechel WC, Hastings SJ, Vourlitis G, Jenkins M, Riechers G, Grulke N (1993) Recent change of Arctic tundra ecosystems from a net carbon dioxide sink to a source. Nature 361(6412):520–523

Ogi M, Rysgaard S, Barber DG (2016) Importance of combined winter and summer Arctic Oscillation (AO) on September Sea ice extent. Environ Res Lett 11(3):034019

Osterkamp T, Romanovsky V (1999) Evidence for warming and thawing of discontinuous permafrost in Alaska. Permafr Periglac Process 10(1):17–37

Parmentier F-JW, Christensen TR, Sørensen LL, Rysgaard S, McGuire AD, Miller PA, Walker DA (2013) The impact of lower sea-ice extent on Arctic greenhouse-gas exchange. Nat Clim Chang 3(3):195–202. https://doi.org/10.1038/nclimate1784

Parmentier FJW, Zhang W, Mi Y, Zhu X, Huissteden J, Hayes DJ, Zhuang Q, Christensen TR, McGuire AD (2015) Rising methane emissions from northern wetlands associated with sea ice decline. Geophys Res Lett 42(17):7214–7222

Peel MC, Finlayson BL, McMahon TA (2007) Updated world map of the Köppen-Geiger climate classification. Hydrol Earth Syst Sci 11(5):1633–1644. https://doi.org/10.5194/hess-11-1633-2007

Péwé R, Brown T (1973) Distribution of permafrost in North America and its relationship to the environment: a review, 1963–1973. In: Permafrost: North American contribution [to the] second international conference. National Academies, Washington DC, p 71

Plug LW, West JJ (2009) Thaw lake expansion in a two-dimensional coupled model of heat transfer, thaw subsidence, and mass movement. J Geophys Res Earth Surface 114(F1):F01002. https://doi.org/10.1029/2006JF000740

Repo ME, Susiluoto S, Lind SE, Jokinen S, Elsakov V, Biasi C, Virtanen T, Martikainen PJ (2009) Large N_2O emissions from cryoturbated peat soil in tundra. Nat Geosci 2(3):189

Romanovskii N, Hubberten H-W, Gavrilov A, Tumskoy V, Kholodov A (2004) Permafrost of the east Siberian Arctic shelf and coastal lowlands. Quat Sci Rev 23(11):1359–1369

Romanovsky V, Burgess M, Smith S, Yoshikawa K, Brown J (2002) Permafrost temperature records: indicators of climate change. EOS Trans Am Geophys Union 83(50):589–594

Romanovsky VE, Smith SL, Christiansen HH (2010a) Permafrost thermal state in the polar Northern Hemisphere during the international polar year 2007–2009: a synthesis. Permafr Periglac Process 21(2):106–116

Romanovsky V, Drozdov D, Oberman N, Malkova G, Kholodov A, Marchenko S, Moskalenko N, Sergeev D, Ukraintseva N, Abramov A (2010b) Thermal state of permafrost in Russia. Permafr Periglac Process 21(2):136–155

Romanovsky VE, Kholodov AL, Smith SL, Christiansen HH, Shiklomanov NI, Drozdov DS, Malkova GV, Oberman NG, Marchenko SS (2014) Terrestrial permafrost. Bull Am Meteorol Soc 95:S139–S141

Romanovsky VE, Smith SL, Christiansen HH, Shiklomanov NI, Streletskiy DA, Drozdov DS, Malkova GV, Oberman NG, Kholodov AL, Marchenko SS (2015) Terrestrial permafrost. Bull Am Meteorol Soc 96:S139–S141

Šafanda J, Szewczyk J, Majorowicz J (2004) Geothermal evidence of very low glacial temperatures on a rim of the Fennoscandian ice sheet. Geophys Res Lett 31(7):L07211. https://doi.org/10.1029/2004GL019547

Sannel A, Kuhry P (2011) Warming-induced destabilization of peat plateau/thermokarst lake complexes. J Geophys Res Biogeo 116(G3):G03035. https://doi.org/10.1029/2010JG001635

Sazonova T, Romanovsky V (2003) A model for regional-scale estimation of temporal and spatial variability of active layer thickness and mean annual ground temperatures. Permafr Periglac Process 14(2):125–139

Schädel C, Bader MK-F, Schuur EA, Biasi C, Bracho R, Čapek P, De Baets S, Diáková K, Ernakovich J, Estop-Aragones C (2016) Potential carbon emissions dominated by carbon dioxide from thawed permafrost soils. Nat Clim Chang 6(10):950

Schirrmeister L, Grosse G, Wetterich S, Overduin PP, Strauss J, Schuur EAG, Hubberten H-W (2011) Fossil organic matter characteristics in permafrost deposits of the northeast Siberian Arctic. J Geophys Res 116:G00M02. https://doi.org/10.1029/2011jg001647

Schirrmeister L, Froese D, Tumskoy V, Grosse G, Wetterich S (2013) Yedoma: late Pleistocene ice-rich syngenetic permafrost of Beringia. In: Elias SAMC (ed) Encyclopedia of quaternary science, 2nd edn. Elsevier, Amsterdam, pp 542–552

Schuur EA, Bockheim J, Canadell JG, Euskirchen E, Field CB, Goryachkin SV, Hagemann S, Kuhry P, Lafleur PM, Lee H (2008) Vulnerability of permafrost carbon to climate change: implications for the global carbon cycle. Bioscience 58(8):701–714

Schuur EA, McGuire AD, Schadel C, Grosse G, Harden JW, Hayes DJ, Hugelius G, Koven CD, Kuhry P, Lawrence DM, Natali SM, Olefeldt D, Romanovsky VE, Schaefer K, Turetsky MR, Treat CC, Vonk JE (2015) Climate change and the permafrost carbon feedback. Nature 520(7546):171–179. https://doi.org/10.1038/nature14338

Segal RA, Lantz TC, Kokelj SV (2016) Acceleration of thaw slump activity in glaciated landscapes of the Western Canadian Arctic. Environ Res Lett 11(3):034025

Seppälä M (2011) Synthesis of studies of palsa formation underlining the importance of local environmental and physical characteristics. Quat Res 75(2):366–370

Serreze MC, Barry RG (2014) The Arctic climate system. Cambridge University Press, Cambridge

Serreze MC, Francis JA (2006) The Arctic amplification debate. Clim Chang 76(3–4):241–264

Serreze M, Barrett A, Stroeve J, Kindig D, Holland M (2009) The emergence of surface-based Arctic amplification. Cryosphere 3(1):11–19

Slater AG, Lawrence DM (2013) Diagnosing present and future permafrost from climate models. J Clim 26(15):5608–5623

Smith M, Riseborough D (2002) Climate and the limits of permafrost: a zonal analysis. Permafr Periglac Process 13(1):1–15

Smith LC, Sheng Y, MacDonald G, Hinzman L (2005) Disappearing arctic lakes. Science 308(5727):1429–1429

Strauss J, Schirrmeister L, Grosse G, Fortier D, Hugelius G, Knoblauch C, Romanovsky V, Schädel C, von Deimling TS, Schuur EA (2017) Deep Yedoma permafrost: a synthesis of depositional characteristics and carbon vulnerability. Earth Sci Rev 172:75–86

Sugden DE, John BS (1976) Glaciers and landscape. Edward Arnold, London, p 394

Sugimoto A, Yanagisawa N, Naito D, Fujita N, Maximov TC (2002) Importance of permafrost as a source of water for plants in east Siberian taiga. Ecol Res 17(4):493–503

Svendsen JI, Alexanderson H, Astakhov VI, Demidov I, Dowdeswell JA, Funder S, Gataullin V, Henriksen M, Hjort C, Houmark-Nielsen M (2004) Late quaternary ice sheet history of northern Eurasia. Quat Sci Rev 23(11):1229–1271

Szewczyk J, Nawrocki J (2011) Deep-seated relict permafrost in northeastern Poland. Boreas 40(3):385–388

Tarnocai C, Stolbovoy V (2006) Northern peatlands: their characteristics, development and sensitivity to climate change. Dev Earth Surface Process 9:17–51

Tarnocai C, Canadell JG, Schuur EAG, Kuhry P, Mazhitova G, Zimov S (2009) Soil organic carbon pools in the northern circumpolar permafrost region. Glob Biogeochem Cycles 23(2):GB2023. https://doi.org/10.1029/2008gb003327

Tedrow JC (1977) Soils of the polar landscapes. Rutgers University Press, New Brunswick. 638 p

Teltewskoi A, Beermann F, Beil I, Bobrov A, De Klerk P, Lorenz S, Lüder A, Michaelis D, Joosten H (2016) 4000 years of changing wetness in a permafrost polygon Peatland (Kytalyk, NE Siberia): a comparative high-resolution multi-proxy study. Permafr Periglac Process 27(1):76–95

Thompson DW, Wallace JM (1998) The Arctic oscillation signature in the wintertime geopotential height and temperature fields. Geophys Res Lett 25(9):1297–1300

Tumel N (2002) Permafrost. In: Shahgedanova M (ed) The physical geography of northern Eurasia, vol 3. Oxford University Press, Oxford, pp 149–168

Van Huissteden J, Dolman AJ (2012) Soil carbon in the Arctic and the permafrost carbon feedback. Curr Opin Environ Sustain 4(5):545–551. https://doi.org/10.1016/j.cosust.2012.09.008

Van Huissteden J, Vandenberghe J, Pollard D (2003) Palaeotemperature reconstructions of the European permafrost zone during marine oxygen isotope stage 3 compared with climate model results. J Quat Sci 18(5):453–464. https://doi.org/10.1002/jqs.766

Van Huissteden J, Berrittella C, Parmentier F, Mi Y, Maximov T, Dolman A (2011) Methane emissions from permafrost thaw lakes limited by lake drainage. Nat Clim Chang 1(2):119–123

Vandenberghe J (1988) Cryoturbations. In: Clark M (ed) Advances in Periglacial geomorphology. Wiley, New York, pp 179–198

Vandenberghe J, Pissart A (1993) Permafrost changes in Europe during the last glacial. Permafr Periglac Process 4(2):121–135

Vandenberghe J, Renssen H, Roche D, Goosse H, Velichko A, Gorbunov A, Levavasseur G (2012) Eurasian permafrost instability constrained by reduced sea-ice cover. Quat Sci Rev 34:16–23

Vandenberghe J, French HM, Gorbunov A, Marchenko S, Velichko AA, Jin H, Cui Z, Zhang T, Wan X (2014) The last permafrost maximum (LPM) map of the Northern Hemisphere: permafrost extent and mean annual air temperatures, 25–17 ka BP. Boreas 43(3):652–666

Velichko A, Faustova M (2009) The development of glaciations in late Pleistocene. Palaeoclimates and Palaeolandscapes of Extratropical space of northern hemisphere late Pleistocene. Holocene GEOS Press, Moscow, p 32e41. (in Russian)

Vitt DH (2006) Functional characteristics and indicators of boreal peatlands. In: Wieder RK, Vitt DH (eds) Boreal peatland ecosystems. Springer, Berlin, pp 9–24

Vonk JE, Sánchez-García L, Van Dongen B, Alling V, Kosmach D, Charkin A, Semiletov IP, Dudarev OV, Shakhova N, Roos P (2012) Activation of old carbon by erosion of coastal and subsea permafrost in Arctic Siberia. Nature 489(7414):137–140

Walker D, Gould W, Maier H, Raynolds M (2002) The circumpolar Arctic vegetation map: AVHRR-derived base maps, environmental controls, and integrated mapping procedures. Int J Remote Sens 23(21):4551–4570

Walker DA, Epstein HE, Gould WA, Kelley AM, Kade AN, Knudson JA, Krantz WB, Michaelson G, Peterson RA, Ping CL (2004) Frost-boil ecosystems: complex interactions between landforms, soils, vegetation and climate. Permafr Periglac Process 15(2):171–188

Walker DA, Raynolds MK, Daniëls FJ, Einarsson E, Elvebakk A, Gould WA, Katenin AE, Kholod SS, Markon CJ, Melnikov ES (2005) The circumpolar Arctic vegetation map. J Veg Sci 16(3):267–282

Walter Anthony KM, Anthony P, Grosse G, Chanton J (2012) Geologic methane seeps along boundaries of Arctic permafrost thaw and melting glaciers. Nat Geosci 5(6):419–426. https://doi.org/10.1038/ngeo1480

Walter KM, Zimov SA, Chanton JP, Verbyla D, Chapin FS 3rd (2006) Methane bubbling from Siberian thaw lakes as a positive feedback to climate warming. Nature 443(7107):71–75. https://doi.org/10.1038/nature05040

Walter K, Edwards M, Grosse G, Zimov S, Chapin F (2007) Thermokarst lakes as a source of atmospheric CH_4 during the last deglaciation. Science 318(5850):633–636

Waugh D, Plumb R, Elkins J, Fahey D, Boering K, Dutton G, Volk C, Keim E, Gao RS, Daube B (1997) Mixing of polar vortex air into middle latitudes as revealed by tracer-tracer scatterplots. J Geophys Res Atmos 102(D11):13119–13134

Wieder RK, Vitt DH, Benscoter BW (2006) Peatlands and the boreal forest. In: Wieder RK, Vitt DH (eds) Boreal peatland ecosystems. Springer, Berlin, pp 1–8

Williams PJ, Smith MW (1991) The frozen earth. Fundamentals of geocryology. Cambridge University Press, Cambridge. 306 p

Woo M-K (2012) Permafrost hydrology. Springer, Berlin, p 563

Yoshikawa K, Hinzman LD (2003) Shrinking thermokarst ponds and groundwater dynamics in discontinuous permafrost near council, Alaska. Permafr Periglac Process 14(2):151–160

Zhang T, Barry R, Knowles K, Heginbottom J, Brown J (2008) Statistics and characteristics of permafrost and ground-ice distribution in the northern hemisphere. Polar Geogr 31(1–2):47–68

Zhang N, Yasunari T, Ohta T (2011) Dynamics of the larch taiga–permafrost coupled system in Siberia under climate change. Environ Res Lett 6(2):024003

Zimov SA, Davydov SP, Zimova GM, Davydova AI, Schuur EAG, Dutta K, Chapin FS (2006) Permafrost carbon: stock and decomposability of a globally significant carbon pool. Geophys Res Lett 33(20). https://doi.org/10.1029/2006gl027484

Chapter 2
The Energy Balance of Permafrost Soils and Ecosystems

Abstract Thawing of permafrost ice requires energy transfer from the atmosphere and solar radiation to the permafrost soil. Snow and vegetation dominate the energy exchange at the surface of permafrost soils. The radiation balance (shortwave and longwave radiation) is the primary source of energy. Only a small part of this energy is available for warming the ground. Which part, depends strongly on the surface cover. A snow cover reflects solar radiation. A vegetation cover shades the soil surface and returns energy to the atmosphere as latent heat (vaporisation of water) and sensible heat (warming of the air). In this chapter, the basic principles of the exchange of energy at the surface are discussed: the radiation balance, the exchange of energy by turbulence- driven latent and sensible heat fluxes. The partitioning of these fluxes throughout the season is shown based on examples. This chapter shows why snow cover has a warming effect on permafrost and vegetation has a cooling effect. Next, the heat transfer into the soil is discussed. Here, the role of latent heat of freezing and thawing and the thermal properties of the soil are important, which vary strongly with soil material and soil water content. Together with the snow and vegetation characteristics, these thermal properties determine soil temperature and the active layer thickness.

Presence or absence of permafrost is determined by the heat balance of the soil, which does not depend on climate alone, but also on its surface cover, weather conditions and time of the day, soil properties and position in the terrain (Williams and Smith 1991). The energy input at the soil surface changes on long climatic time scales affecting permafrost thaw and refreeze, during the seasons by determining yearly variations in soil temperature and active layer thickness, with weather patterns that influence rate of thaw, soil wetness and soil thermal properties, and on a daily basis due to insolation.

Even subtle differences in soil and vegetation properties make a difference between thaw and freezing. This can be observed in temperate climates when a thin snow cover has fallen at near-zero air temperatures, in particular when previous days have been warmer. On bare or paved soil the snow readily thaws away, when heat stored in previous days in the subsoil is transferred the surface. On a lawn, or even on thin lines of moss between the pavement, it persists – protected by the insu-

J. van Huissteden, *Thawing Permafrost*,
https://doi.org/10.1007/978-3-030-31379-1_2

lating properties of the vegetation mat. Figure 1.20a shows a similar effect in the early fall on a mountain top with permafrost and vegetation-lined polygons in Northern Sweden.

The heat balance at the soil surface is determined by gains and losses of energy. These are the balance of:

1. The radiation balance, consisting of incoming and reflected solar radiation and incoming and outgoing thermal radiation of the earth surface (longwave radiation).
2. The latent heat exchange: evaporation or condensation of water.
3. The heat exchange by convection of air (sensible heat).
4. The ground heat flux, conduction of heat into or out of the ground, including latent heat for thawing of soil ice or freezing of soil water.
5. Localized and temporary heat balance elements, e.g. in the spring the thawing of snow and ponding of water at the surface.

The following discussion of the heat balance largely follows Williams and Smith (1991) and Bonan (2016), with examples of data from various permafrost locations. Figure 2.1 shows processes and state variables that affect the ground heat flux in permafrost soils.

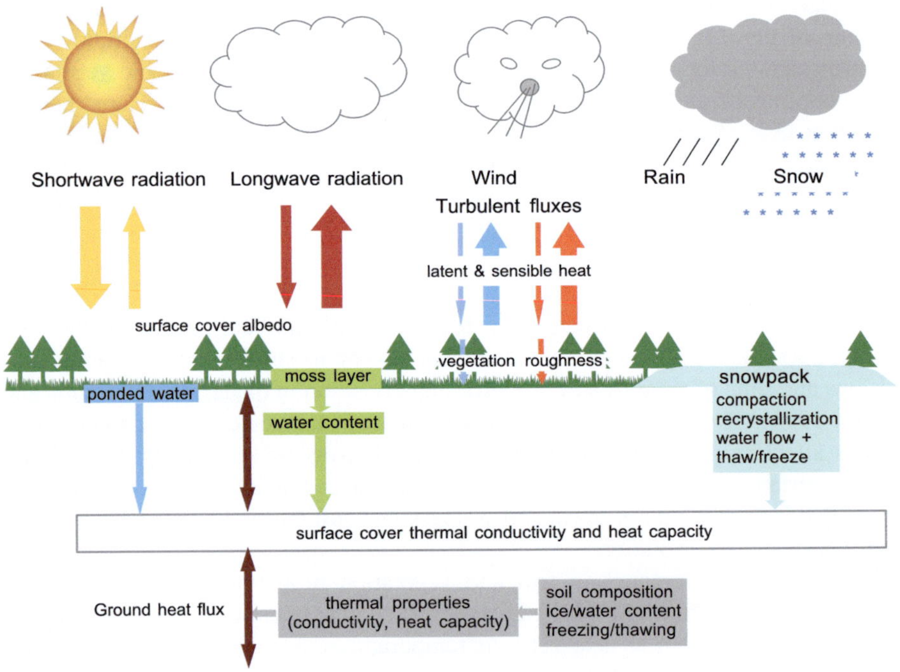

Fig. 2.1 Schematic overview of heat exchange processes to and from permafrost soil discussed in this chapter

2.1 The Radiation Balance

The main source of heat is incoming solar radiation (insolation), both shortwave radiation (the visible part of the electromagnetic spectrum, and 'near' infrared and ultraviolet), and longwave radiation. The wavelengths of shortwave radiation ranges between 1 and 5 μm. A part of the incoming solar radiation is coming directly from the sun: the direct shortwave radiation. Another part consists of diffuse radiation, which is solar radiation scattered by clouds and particles in the atmosphere. On cloudy days, the diffuse part dominates. The amount of solar energy from direct solar radiation per unit area (W m^{-2}) strongly depends on the incidence angle, determined by topography, latitude and season. The lower the sun angle in the sky, the larger the area over which the incoming radiation has to be distributed. A low sun angle also means more pronounced differences in the amount of solar radiation due to topography in high latitudes: south facing slopes catch more incoming solar radiation than slopes that face away from the sun (Fig. 2.2).

A part of the incoming solar radiation is reflected and is not available for heating up the surface. This reflected fraction is called the albedo of the surface. The albedo differs strongly for materials at the surface; differences that are highly relevant for the permafrost heat balance. Open water has the lowest albedo, less than a fraction of 0.1 is reflected, the rest is absorbed and heats the water. Moist soils, in particular if they contain a high percentage of organic matter, have an albedo less than 0.2. Dry soils and soils with less organic matter reflect more, with an albedo up to 0.4 for dry sand. Snow has the highest albedo; freshly fallen snow may have an albedo as high as 0.9. Older, and partly thawed snow has a lower albedo, down to 0.4. Ice has an albedo in the range of 0.3 to 0.4. Vegetation generally has an albedo below 0.2, although plant characteristics, such as the presence of reflective waxy leaf surfaces, and plant colour like the light colour of some lichens, can make large differences.

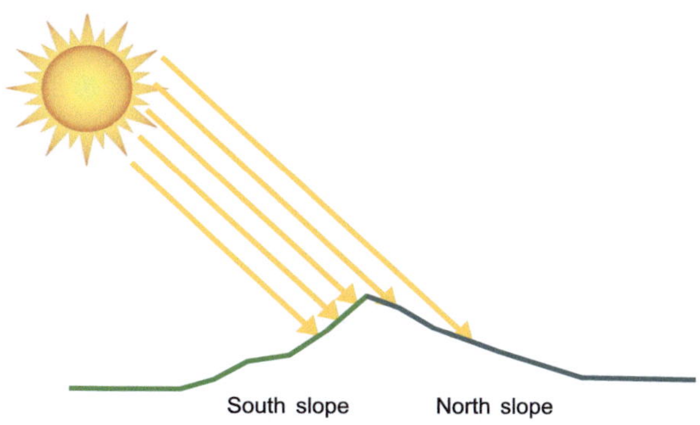

Fig. 2.2 Effect of sun angle and slope on incoming direct solar radiation

To summarize, the total shortwave radiation that is available for heating up the soil and vegetation surface is the not-reflected part of the sum of direct and diffuse incoming shortwave radiation:

$$R_{shortwave,total} = (1-a)\left(R_{in,shortwave,direct} + R_{in,shortwave,diffuse}\right) \tag{2.1}$$

in which a is the albedo. Besides shortwave radiation, also longwave (thermal infra-red) radiation has to be accounted for. Every object – including soils, plants, clouds – that has a temperature above the absolute zero emits radiation, proportionally to the fourth power of the temperature according to the Stefan-Boltzmann law:

$$R = \varepsilon\sigma T^4 \tag{2.2}$$

R is the amount of radiation per unit area (Wm^{-2}), T is the temperature in degrees Kelvin and σ is the Stefan-Boltzmann constant (5.67 10^{-8} Wm^{-2} K^{-4}). The constant ε is the emissivity factor. The Stefan-Boltzmann law assumes a perfect 'black body' which emits all radiation that would be produced according to the term σT^4 in Eq. 2.2 – which means an emissivity of 1. Natural objects have lower emissivity. Most natural materials at the earth surface have emissivities above 0.9 (e.g. water: 0.98, soil: 0.92–0.98); snow has a somewhat lower emissivity (0.8). The wavelength of the radiation is inversely related to temperature according to Wien's displacement law, which relates wavelength of emitted radiation to black body temperature. At normal temperatures at the Earth's surface, earth materials will hardly emit radiation in the visible part of the spectrum. The short wavelengths of solar radiation, and the wavelength of longwave radiation from the atmosphere and earth surface hardly overlap. To complete the radiation balance, the longwave radiation is added to Eq. 2.1:

$$R_{net} = (1-a)\left(R_{in,shortwave,direct} + R_{in,shortwave,diffuse}\right) + R_{in,longwave} + R_{out,longwave} \tag{2.3}$$

$R_{in,longwave}$ is the incoming longwave radiation, mostly radiation that is reflected by clouds or radiated by the greenhouse gases in the atmosphere. $R_{out,longwave}$ is the thermal radiation of the Earth's surface and vegetation (Eq. 2.2), going out into the atmosphere.

R_{net} is the sum of the radiation balance, the amount that is available for heating the surface. If R_{net} is negative, at night when there is little incoming shortwave radiation, the surface will cool because it emits longwave radiation, unless there are other sources of heat like convection of warm air. The nightly outgoing longwave radiation energy is always more than the incoming longwave radiation, in particular during nights with a clear sky, when the longwave radiation emitted by clouds is also minimal. During a clear winter night in temperate climates, the negative radiation balance may cool the surface below the freezing point. When this winter night takes months at high latitudes, it results in the intense freezing required for permafrost. This high-latitude winter cooling is hardly compensated in summer. Although the sun may remain above the horizon for several months, the incidence angle of the incoming radiation is low, reducing the amount of incoming shortwave radiation. Another factor that reduces incoming solar radiation in arctic summers is the persistence of low cloud cover in particular in maritime climates along the Arctic ocean seaboard. On the other hand, this cloud cover reduces longwave radiation heat losses in winter (Westermann et al. 2009; Serreze and Barry 2014).

2.2 Latent, Sensible and Conductive Heat Fluxes

The radiation balance is the primary source of energy at the surface, but its energy is partitioned in other energy fluxes. Two of these are driven by wind and are strongly affected by vegetation: the energy used for evaporation of water at the surface, and convective transport of heat – wind cooling or warming the surface. These energy fluxes take up most of the net radiation energy flux. The turbulence of the air flowing over the surface is an important parameter for these fluxes. The remainder is the conduction of heat into or out of the ground, which determines the permafrost temperature.

2.2.1 Partitioning of the Radiative Flux into Turbulent and Conductive Fluxes

The net incoming radiative energy is redistributed as latent, sensible and conductive heat fluxes, denoted by the letter Q below. The latent heat flux results from the evaporation of water, either by direct evaporation from wet surfaces (E), or by transpiration by plants (T). Both are taken together as 'evapotranspiration' (ET). If the combined water vapour flux ET per unit area is multiplied by the latent heat of evaporation of water λ, it represents a heat flux (Q_E, often written as λE), measured in W m^{-2} (λ = 2.477 10^6 J kg^{-1} at 10 °C, decreasing with higher temperature). Evaporation of 1 mm water at a temperature of 10 °C represents a heat flux of 28.7 W.m^{-2}. Usually this flux is directed away from the surface, but condensation (dew) results in a heat flux towards the surface. Q_E depends on:

- availability of water,
- the amount of solar energy available for evaporation,
- atmospheric conditions which help to carry water vapour away from the surface: low air humidity, wind, and turbulence of the air
- roughness of the surface, e.g. vegetation, which increases the area for exchange
- for the transpiration of plants the photosynthetic activity and plant physiological factors such as the conduction of water vapour through the respiration openings in the leaves, the stomata.

These factors will be discussed in the next section in more detail.

The second heat transfer process is the sensible heat flux Q_H. Sensible heat is the energy required to change the temperature of a substance when in contact with an object with a higher or lower temperature. It depends on the heat capacity of the substance (air in this case), which is the amount of energy that is required to change a unit mass one degree in temperature, without phase changes (unit J kg^{-1} K^{-1}). When cool air moves over a warm soil or vegetation surface, the soil and vegetation lose heat energy to the air, and the air warms up. Like the latent heat flux, Q_H can be positive, warming the surface (adding energy), or negative, or cooling the surface (removing energy). Similar to latent heat exchange, it depends on wind speed, turbulence and surface roughness, and is larger when temperature differences between surface and air are larger.

The third important heat flux – in particular for permafrost – is the ground heat flux Q_G, the amount of heat which is transferred into, or out of the ground by conduction. This heat conduction into the ground is determined by two thermal properties of the soil materials: the heat capacity and the thermal conductivity. The heat capacity is the amount of energy that is required to warm soil material with one degree. The thermal conductivity represents the speed at which the heat is transferred. The thermal conductivity is the quantity of heat that will flow through a unit area of the material within a certain time under the influence of a temperature gradient. If the temperature gradient is expressed in degrees Kelvin per meter ($K.m^{-1}$), the thermal conductivity is measured in Js^{-1} $m^{-2}(K\ m^{-1})^{-1}$ or W m^{-1} K^{-1}.

Insulating the wall of a house to reduce heating expenses is nothing else than reducing the thermal conductivity by adding materials that conduct heat poorly. Likewise, insulating materials on top of the soil reduce the ground heat flux. One such material that is ubiquitous in permafrost areas, is peat and mosses; another is a snow layer. In permafrost environments, the air temperature can be well above 20 °C, while at 50 cm depth the soil is still frozen. The process of heat flow into and out of the ground will be considered in more detail in the next sections.

Summing up all the energy fluxes we can draw up an energy balance of the surface, in which the radiation balance R_{net} is balanced by the turbulent and conductive heat fluxes:

$$R_{net} = Q_E + Q_H + Q_G \tag{2.4}$$

A storage term ΔS may be added to the right hand side of this equation, which describes the heat storage in the air in between the canopy of the vegetation cover and in biomass (wood, leaves). However, this term is negligible in low vegetation covers like tundra, and even in boreal forest very small (Tanaka et al. 2008; Ohta et al. 2008).

The ratio of Q_H over Q_E is known as the Bowen ratio, β. Other useful ratios are the fractions of net radiation that are used for evapotranspiration or the ground heat flux, Q_E/R_{net} and Q_G/R_{net}. The magnitude of these energy fluxes differs strongly between day and night (Fig. 2.3). During the day, R_{net} drives the other fluxes. During the night, R_{net} is much smaller than during the day and usually directed away from the surface (Fig. 2.3). Q_E and Q_H depend on weather conditions. Q_E is smaller during

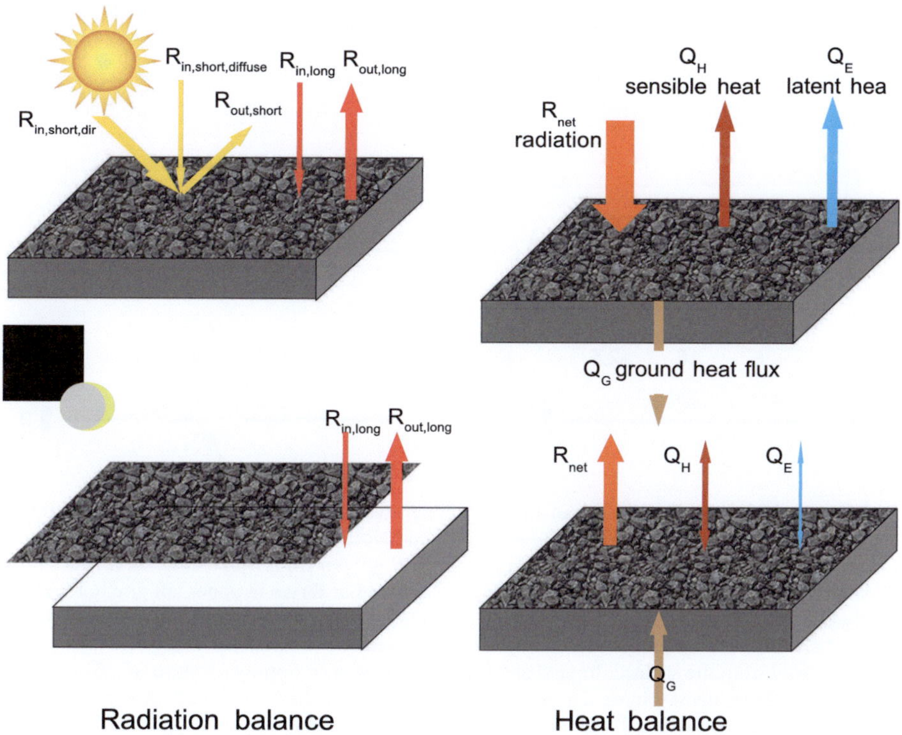

Fig. 2.3 Energy fluxes at the surface. Above: daytime fluxes; Below: nighttime fluxes. Left: radiation fluxes (R); in indicates incoming radiation, out outgoing radiation, short shortwave radiation; long longwave radiation, dir direct solar radiation, diffuse diffuse radiation. Right: Total energy balance. R_{net}: net radiation balance (sum of left side of figure); other flux symbols are explained in figure

the night because plant transpiration is absent in the dark, when photosynthesis is not possible and the stomata of the leaves are closed. Low windspeed and more stable atmosphere conditions during the night also contribute to lower Q_E and Q_H. Q_E may be directed towards the surface (manifested as dew formation).

During the night the soil surface tends to loose heat energy, cooling down the surface. The lost energy is partly compensated by the other heat fluxes, in particular the ground heat flux Q_G which changes sign during the night (Fig. 2.3). In tundra in the Lena delta, the summer heat balance was determined mainly by the the presence or absence of snow cover and the cloud cover (determining the incoming shortwave radiation), and the ground heat flux was determined by a strong thermal gradient in the frozen soil (Langer et al. 2011a). Although the incoming shortwave radiation was reduced by clouds, this was often combined with the advection (horizontal transport by wind) of warm air.

Q_E is generally the largest component of the energy balance, followed by the sensible heat flux Q_H. In dry conditions, the difference between Q_H and Q_E *is* small (β near 1), but in wet tundra environments with much open water Q_E will be the

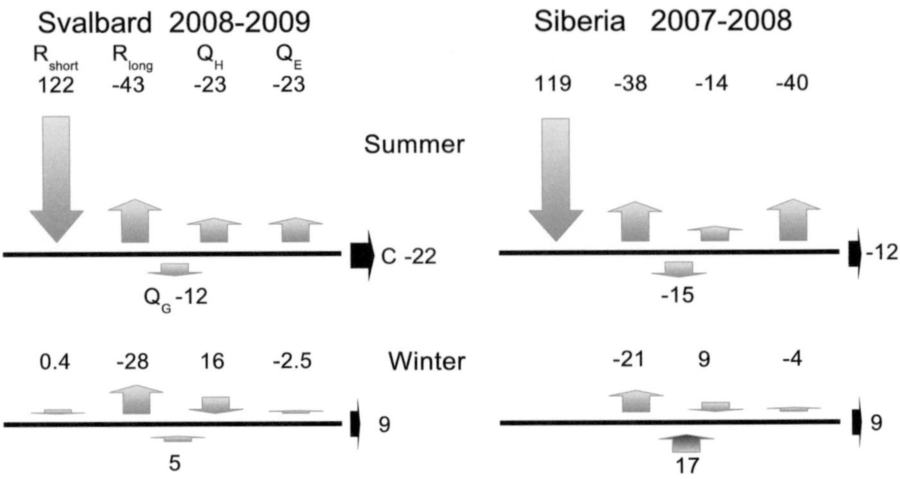

Fig. 2.4 Comparison of typical mean summer (upper two figures) and winter (lower two figures) energy fluxes (W m^{-2}) for Svalbard (Bayelva site) and Northern Siberian sites (Samoylov Island, Lena Delta). Fluxes are scaled relative to each other. Redrawn from Boike et al. 2012; data from Westermann et al. 2009 and Langer et al. 2011a, b. Legend of fluxes in upper left graph. R_{short}: net shortwave radiation; R_{long}: net longwave radiation; Q_H: sensible heat flux; Q_E: latent heat flux: C: energy balance closure term (see text). Sign of energy balance components: a negative sign means directed away from the surface. Republished with permission of Springer Nature AG, from Boike et al. (2012) 'Permafrost–physical aspects, carbon cycling, databases and uncertainties', In: Hüttl R., Schneider B., von Braun J. (eds) Recarbonization of the Biosphere. Springer, Dordrecht; permission conveyed through Copyright Clearance Center, Inc.

largest in summer (Boike et al. 2012; Fig. 2.4). Langer et al. (2011a) found that over dry tundra surfaces the summer Q_H and Q_E were approximately equal (36 ± 25 W m^{-2} and 28 ± 32 W m^{-2} respectively), while over wet tundra surfaces Q_E dominated (Q_H 1 ± 25 W m^{-2}, Q_E 64 ± 33 W m^{-2}). The large uncertainties reflect measurement uncertainty, and high spatial variation; various components of the heat balance are often measured on different spatial scales. Kasurinen et al. (2014) compared the evapotranspiration of 65 arctic and boreal research sites. In general, Q_E is higher for forest environments than for treeless ecosystems. Typical values for Q_E/R_{net} in arctic and boreal environments are 0.15–0.38.

Q_G, the component which determines the active layer depth, is in general the smallest component of the heat balance (Kasurinen et al. 2014), but in summer it is directed downward, and in tundra sites it may take up to 30% of the net radiation. Langer et al. (2011a) reports a value of Q_G/R_{net} of 0.2. These values are considerably higher than those at mid latitudes, where the ground heat flux is only a few percent of the heat balance and is often neglected. On the other hand, it is also the most uncertain component (Langer et al. 2011a). The persisting presence of ice in permafrost soils makes them a sink for heat in the summer; thawing of ice in the soil requires large amounts of heat energy (Boike et al. 2008, 2012; Langer et al. 2011a). At the same time the presence of an impervious frozen layer just below the surface impedes drainage and maintains high soil moisture levels, enhancing the latent heat flux.

In the winter heat balance, incoming solar radiation is very low or absent. The radiation balance is dominated by outgoing longwave radiation, cooling the surface (Langer et al. 2011a, b; Boike et al. 2012). Evaporation is much smaller and does not include the transpiration component from the vegetation which is dormant; it mostly consists of sublimation from snow and ice surfaces. The sensible heat flux Q_H is smaller than in summer and generally directed downward, when transport of warmer air over strongly cooled soil and snow surfaces warms the surface. The ground heat flux is directed upwards, transferring heat stored in summer in the soil to the surface again (Figs. 2.3 and 2.4).

2.2.2 Measurement Uncertainty

Q_E and Q_H are generally measured by micrometeorological methods, that are based on quantifying turbulent air movements in the planetary boundary layer, the layer of the atmosphere in which air flow is influenced by the surface properties of the Earth, where rapid vertical mixing occurs. A commonly used method is eddy covariance, in which the word ‘eddy’ derives from turbulent eddies (chaotically rotating three-dimensional whirls) that occur in any turbulent flow (Fig. 2.5). Turbulence is visible when smoke is released in turbulent air, and is shown by up- and downward and sideways movements of the smoke (eddies). Turbulence is felt as gusts and fluctuations of wind. Turbulence readily occurs at most wind speeds; only at very low wind speeds airflow may be expected to be laminar (air particles flowing in a straight line, without eddies).

Eddy covariance involves high frequency (10 Hz or more) measurements of wind speed in horizontal and vertical directions with a sonic anemometer to quantify the

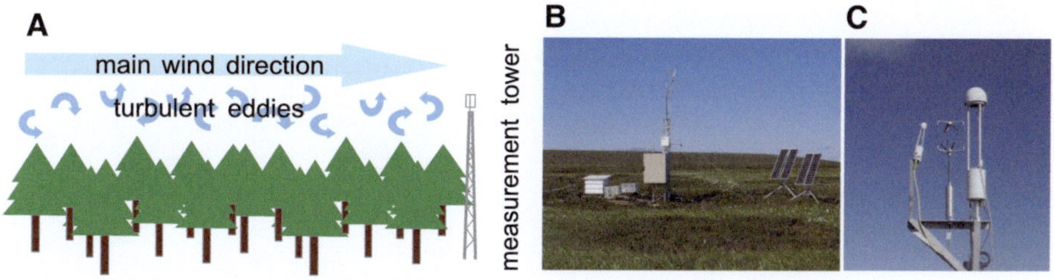

Fig. 2.5 Eddy covariance instrumentation. (**a**) The measurement principle is based on moving eddies in air flow which are captured at a high frequency (typically 10 Hz) sampling rate with a sonic anemometer (Burba 2013). (**b**) Eddy covariance instrumentation in Siberian tundra (Kytalyk research station); left: meteo-tower with radiation sensors on arms extending from the tower; middle: eddy covariance tower (3 m height); right: solar panels for power supply. C: detail of eddy covariance instrumentation; left: sensor for water vapour, air temperature and CO_2 concentration; middle: sonic anemometer; right: sensor for CH_4 concentration. Photos by the author

turbulent movements, and at the same rate measurements of the concentration of water vapour, and air temperature. When a turbulent eddy in the air flow passes the measurement instruments, it will result in vertical and horizontal fluctuations of wind speed, water vapour and temperature. Besides energy fluxes, also mass fluxes of CO_2 and CH_4 are measured in this way, by high frequency gas concentration measurements.

An extensive explanation of the theory behind eddy covariance is not feasible within the scope of this book; for that consult e.g. Foken (2008) or Aubinet et al. (2012). However, to understand the measurements and error sources therein, a short introduction of what is meant with the term 'covariance' here is necessary. The turbulent eddies create wind speed fluctuations in three dimensions. The wind speed vector at a certain moment can be decomposed into wind speeds in three directions in a Cartesian coordinate system; u and v are the decomposition of the wind vector in the horizontal plane (by convention in eastern respectively northern direction), and w the vertical component (Foken 2008). Turbulence makes the wind speed vary in the three directions. For calculations, this variable wind speed can be decomposed in an average speed with a superposed random fluctuating part, expressed as a deviation from the mean (Fig. 2.6a):

$$\begin{aligned} u &= \overline{u} + u' \\ v &= \overline{v} + v' \\ w &= \overline{w} + w' \end{aligned} \tag{2.5}$$

This is known as the Reynolds's decomposition; the terms with the bar are the averages, the terms with accents are the deviations from the average. The average of these deviations is zero. The Reynolds's decomposition is accompanied by rules for calculations with the averages. The rule for multiplication of two averages, e.g. the average of u and v is:

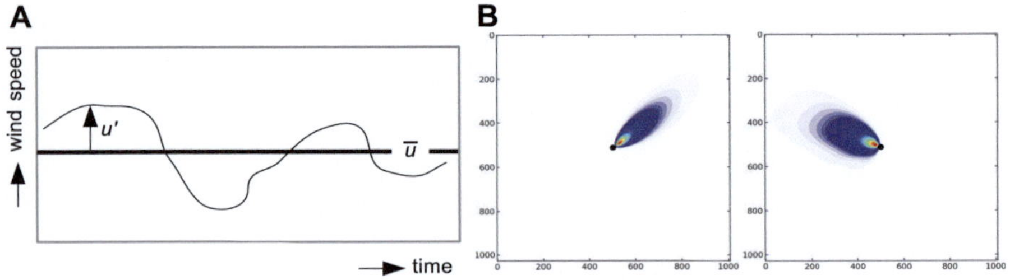

Fig. 2.6 (**a**) mean wind speed $\overline{u}$ and deviations from the mean u'. (**b**) Typical footprint areas of an eddy covariance tower for different wind directions and turbulence of a 3 m high tower located in the centre (black dot). Distances in m, the colours show the magnitude of the flux contribution from the area; color scale from red (high contribution) to bluegrey (low contribution). (**b**) courtesy of **a**. Budishchev

$$\overline{uw} = \overline{u}\,\overline{w} + \overline{u'w'} \tag{2.6}$$

Here, the last term is the same as covariance of the random fluctuations:

$$\overline{u'w'} = \frac{1}{N-1}\sum_{t=0}^{N}\left[\left(u_t - \overline{w_t}\right)\left(u_t - \overline{w_t}\right)\right] \tag{2.7}$$

Under conditions of steady state average wind speeds, an average horizontal wind without a consistent vertical component, and a homogeneous surface without large obstacles, the following basic equations can be used to derive fluxes of energy and substances from the variations of the vertical wind speed and concentration or temperature (Foken 2008):

$$\begin{aligned} u_*^2 &= -\overline{u'w'} \\ \frac{Q_H}{\rho c_p} &= \overline{T'w'} \\ \frac{Q_E}{\rho \lambda} &= \overline{q'w'} \\ \frac{Q_C}{\rho} &= \overline{c'w'} \end{aligned} \tag{2.8}$$

u_* is the friction velocity (m s^{-1}, see below), Q_E and Q_H are the latent and sensible heat fluxes, Q_C the flux of a trace gas, q the specific humidity of the air (kg kg^{-1}), c_p specific heat at constant pressure (J kg^{-1} K^{-1}), T temperature (K), ρ air density (kg m^{-3}), λ the heat of evaporation of water (J kg^{-1}), and c the concentration of a trace gas (e.g. CO_2, CH_4).

These equations state that the heat and substance fluxes can be computed from the covariance between the vertical windspeed and air temperature fluctuations (for Q_H) and vertical windspeed and specific humidity (mass fraction of water in air) for Q_E or gas concentration for Q_C (Foken 2008; Aubinet et al. 2012). On the raw fluxes computed from the covariances, corrections have to be applied for various error sources: a correction to align the coordinate system with the mean wind direction, correction for time lag between the wind speed and temperature, moisture, or concentration sensors, spectral corrections for missing parts of the frequency spectrum of turbulent eddies, density corrections (Aubinet et al. 2012). More extensive information on measurement principles of instruments can be found in Burba (2013), published by an instrument manufacturer.

An eddy covariance system measures fluxes from an upwind area, of which the size and location depends on the instrument height above the surface, wind direction and surface characteristics that affect turbulence, such as surface roughness (Aubinet et al. 2012). With typical instrumentation heights up to 5 m, most of the fluxes come from a roughly elliptical source area in the upwind direction with lengths in the order of 100 to 200 m. This area varies constantly with wind direction and turbulence of the air. To derive a statistical definition of the source area or footprint, footprint models are used (e.g. Kormann and Meixner 2001; Wilson 2015); see Fig. 2.6 for an example.

The radiation balance is measured with radiometers positioned above the ground surface. The ground heat flux is usually measured with heat flux plates and temperature sensors installed in the soil, which measure a vertical temperature profile (e.g. Van der Molen et al. 2007; Langer et al. 2011a, 2011b). Auxiliary measurements are made of the wind direction, vertical wind profile with anemometers at various heights, air temperature, moisture and air pressure. Soil data collection includes water table and soil moisture at various levels, soil temperature profiles and soil heat flux measurements. Vegetation data includes estimates of leaf area index (LAI), the total one-sided leaf area per unit of ground area ($m^2\ m^{-2}$) and biomass.

There is always measurement uncertainty in the fluxes, with as a result that the radiation balance and the sum of the other heat fluxes do not fully cancel out against each other – in other words, the balance is not fully closed. Therefore an energy balance closure term which includes these uncertainties is added in the energy balances. An important source of uncertainty is discrepancy between the ground surface areas that are covered by the various sensors. The eddy covariance footprint area is much larger than that of radiation sensors, which typically cover non-varying areas in the order of tens of m^2, and these are in turn much larger than the area used for the ground heat flux measurements, which is in the order of a square meter. Given the typical inhomogeneity of Arctic vegetation and soil, this incompatibility of sensor footprints is a source of error, but largely unavoidable in energy balance studies (Foken 2008; Langer et al. 2011ab).

Measurement of the heat fluxes puts high demands on the instrumentation, in particular in winter where temperatures can drop well below the normal operating temperatures of most electronics. Instruments such as anemometers become covered with hoar frost which disturbs air flow, and power supply of the instruments is critical in remote sites. For eddy covariance instruments in cold environments, the temperature of the instruments generated by their electronics has proven to result in anomalous fluxes, requiring additional corrections for older instrumentation (Burba et al. 2008; Burba 2013). Instrument failure may result in shorter and longer periods without data. In cases where data gaps are short, this is usually solved by filling the gaps with models that are driven by the data that are available (Aubinet et al. 2012), e.g. machine learning algorithms (Budishchev et al. subm).

Data correction and gap filling methods may differ among research groups, which may hamper data comparison to a certain extent; standardized online tools are available (e.g. https://www.bgc-jena.mpg.de/~MDIwork/eddyproc/). In the papers with published data cited below, these issues have been addressed in their methodology sections. In the data cited from these papers (e.g. Table 2.1), the differences due to methodology of data processing have not been assessed, and the data are presented as in the original papers. Where necessary, only unit conversions have been made to facilitate comparison.

Table 2.1 Summer heat exchange characteristics of vegetation on permafrost soils, varying from bare ground to forest

Vegetation	Barren (1)	Short tundra vegetation (2)	Wet tundra (3)	Wet tundra (4)	Wet tundra (4)	Tundra (5)	Low shrub (5)	Tall shrub (5)	Wood-land (5)	Forest (5)
Dominant species	Short grasses, mosses	*Cassiope*, *Salix*	*Carex*, moss, ponds, *Salix*	*Carex* mosses incl. *Sphagnum*	*Arctophila fulva*, *Carex*, mosses	*Betula nana*	*Betula glandulosa*	*Salix Betula gl.*	*Salix pulchra*	*Salix sp*, *Picea glauca*
Average / max. Shrub/tree height (m)	0	<0.3	<0.5	–	–	0.20–0.45	0.65–1.20	1.10–1.50	2.4–3.0	Trees: 7.3
Average / max shrub cover %	0	–	30	–	–	<1 / 5	29 / 100	65 / 100	75 / 100	10/01/75
LAI ($m^2 m^{-2}$) at peak biomass	<0.5	0.2–0.3	0.95	0.58	1.4	0.52	1.7	1.89	2.27	2.76
Albedo mid winter	0.8	0.8	0.8	–	–	0.85	0.85	0.85	0.6	0.18
Albedo snow melt period	0.65	0.8–0.2	0.8–0.2	–	–	0.82	0.78	0.68	0.49	0.18
Albedo summer	0.18	0.1–0.2	0.15–0.20	–		0.19	0.17	0.15	0.13	0.1
R_n ($W m^{-2}$) summer	79	262 ± 32[a]	81–104	300–500[a]	300–500[a]	99 ± 2.6	84 ± 4.4	80 ± 4.0	158 ± 5.7	109 ± 3.1
Bowen ratio (Q_H/Q_{LE}) summer	± 1	3.0 ± 0.4[a]	0.35–0.50	1.4[a]	1.4[a]	0.94	0.98	1.06	1.15	1.22
Ratio Q_H / net radiation summer	0.29	0.53 ± 0.04[a]	0.17–0.21	0.48 ± 0.12[a]	0.35 ± 0.14[a]	0.34	0.35	0.38	0.41	0.44
Ratio Q_{LE} / net radiation summer	0.29	0.20 ± 0.03[a]	0.49–0.42	0.35 ± 0.07[a]	0.29 ± 0.15[a]	0.36	0.35	0.36	0.36	0.37
Ratio Q_G / net radiation summer	0.15	0.22 ± 0.07[a]	0.20–0.19	0.12 ± 0.05[a]	0.16 ± 0.07[a]	0.12	0.12	0.08	0.1	0.05

(continued)

Table 2.1 (continued)

Vegetation	Barren (1)	Short tundra vegetation (2)	Wet tundra (3)	Wet tundra (4)	Wet tundra (4)	Tundra (5)	Low shrub (5)	Tall shrub (5)	Wood-land (5)	Forest (5)
Dominant species	Short grasses, mosses	*Cassiope, Salix*	*Carex*, moss, ponds, *Salix*	*Carex* mosses incl. *Sphagnum*	*Arctophila fulva, Carex*, mosses	*Betula nana*	*Betula glandulosa*	*Salix Betula gl.*	*Salix pulchra*	*Salix sp, Picea glauca*
Q_G (W m^{-2}) summer	12	53 ± 21[a]	15–20	–	–	12	11	6.4	15.8	5.5
R_n (W m^{-2}) winter	28	–	−23	–	–	–	–	–	–	–
Q_H winter	−16	–	−9	–	–	–	–	–	–	–
Q_{LE} winter	2.5	–	4	–	–	–	–	–	–	–
Q_G (W m^{-2}) winter	−5	–	−15.5	–	–	–	–	–	–	–
Roughness length (m)	0.01	–	0.001	–	–	0.04	0.08	0.18	0.74	1.6
Aerodynamic resistance r_a (s m^{-1})	–	91 ± 5[a]	–	63 ± 27[a]	62 ± 38[a]	43 ± 2.6	20.8 ± 1.6	24.3 ± 3.1	13.1 ± 1.2	11.2 ± 0.9
Bulk surface resistance r_c midday (s m^{-1})	–	342 ± 52[a]	50	46 ± 38[a]	46 ± 40[a]	175 ± 10.9	170 ± 35.9	172 ± 11.1	146 ± 14.7	169 ± 24.0
Nr observation summers	1	11	2	8	8	1	1	1	1	1
Energy balance closure	0.72	–	0.84	0.8	0.2	0.82	0.81	0.8	0.86	0.85

Data from (1) Westermann et al. (2009), (2) Lund et al. (2014), (3) Langer et al. (2011a, b); (4) Liljedahl et al. (2011) (5) Thompson et al. (2004), Sturm et al. (2005) and Beringer et al. (2005). The roughness lengths and resistances refer to latent heat

[a]summer midday values, otherwise seasonal/daily averages

2.3 Heat Balance of Vegetation Cover

The heat balance of vegetated soils is more complicated, since the heat fluxes also interact with vegetation, adding resistance to heat exchange between soil and atmosphere. For permafrost soils these modifications of the heat flux are important. An experiment with vegetation removal in the Siberian tundra (described in Chap. 6) illustrates the effect of vegetation on permafrost. Removal of a cover of dwarf birch shrubs (*Betula nana*) resulted in thawing of ground ice because the vegetation removal caused a change in the surface heat balance (Blok et al. 2010; Nauta et al. 2014); see also Fig. 1.17 for the effect of vegetation disturbance.

To understand the effect of a vegetation cover on the surface heat balance we need to consider the fluxes and the vegetation parameters that drive these fluxes in more detail. This requires some understanding of micro-scale meteorology: stability of the atmosphere, turbulent air flows and wind profiles. Below, only those aspects that contribute to an understanding of the vegetation effects on heat balance are included.

The exchange of sensible heat Q_H from an object to the surrounding air can be described as a flow of heat, proportional to the temperature difference between the object and the air, and inversely proportional to a resistance to heat transfer:

$$Q_H = -\rho c_p \frac{(T_a - T_s)}{r_H} \tag{2.9}$$

where T_a and T_s are the air and surface temperatures, ρ the density of air (kg m^{-3}, c_p the heat capacity of air (J kg^{-1} °C^{-1}), and r_H is the transfer resistance (in s m^{-1}). This is analogous to an electrical current through a resistor, where the current (in amperes) depend on the voltage and resistance of the resistor, according to Ohm's law. This transfer resistance depends on wind speed, surface roughness and turbulence of the airflow. The 'object' may be a single leaf in the vegetation canopy, but a similar reasoning holds for the canopy as a whole (Fig. 2.7). A similar resistance formulation can also be set up for latent heat. Any moist surface will lose heat when water evaporates. The latent heat of vaporization is 2,453,760 J kg^{-1} at 20 °C; it varies with temperature, for instance at 0 °C it is 2,500,800 J kg^{-1}. Over a snow surface, sublimation will occur; the latent heat of sublimation at 0 °C is slightly higher (2,834,100 J kg^{-1}) than the latent heat of evaporation but varies only slightly with temperature (Bonan 2016).

Evaporation ceases when the air is saturated with water. The amount of water in the air is expressed as a partial pressure of water vapour. The maximum amount of water that the air can hold – the saturation vapour pressure (e^*) – increases exponentially with higher air temperature (Fig. 2.8). Evaporation is related to the difference between the actual partial pressure of water vapour e_a and the saturation pressure: the vapour pressure deficit. A more well-known measure of the saturation of the air with water vapour is the relative humidity, which is the ratio between actual partial pressure of water vapour e_a and e_* expressed as a percentage. The resistance formula for latent heat flux can be formulated as follows (Bonan 2016):

$$\lambda E = Q_E = \frac{-\rho C_p}{\gamma} \frac{\left(e_a - e_T^*\right)}{r_S} \tag{2.10}$$

Here e_a is the actual vapour pressure, e_T^* the saturation vapour pressure at temperature T; $e_T^* - e_a$ is the vapour pressure deficit. $\lambda E = Q_E$ is the latent heat flux, ρ the density of air, C_p the heat capacity. The resistance term r_S has again the units s m^{-1}. The γ is the psychrometric constant, which depends on heat capacity, air pressure, and latent heat of vaporization: $\gamma = C_pP/0.622\lambda$. For evaporation over a snow surface the latent heat of evaporation is replaced by the latent heat of sublimation.

The exchange of sensible heat from a single leaf occurs on both sides of the leaf, and is driven by air flowing over the leaf. In the air flow a boundary layer is established just above the leaf surface (Bonan 2016). In this boundary layer, the air flow speed increases approximately parabolically from 0 m s^{-1} above the leaf surface to the normal wind speed at some distance above the surface (Fig. 2.7). This boundary layer is typically 1 to 10 mm thick, and results from frictional drag in the air flowing over the leaf surface. The thicker the boundary layer, the larger the resistance to heat flow. A thin boundary layer means that the leaf quickly exchanges heat with the surrounding air and adopts its temperature. The boundary layer resistance is related to leaf size and wind speed; a larger leaf has a higher resistance, a higher wind speed reduces the resistance.

Since sensible heat exchange occurs on both sides of the leaf, there are two resistances r_b for each side. These can be seen – again in analogy with electrical currents – as parallel resistances. Assuming both r_b's are equal, this leads according to

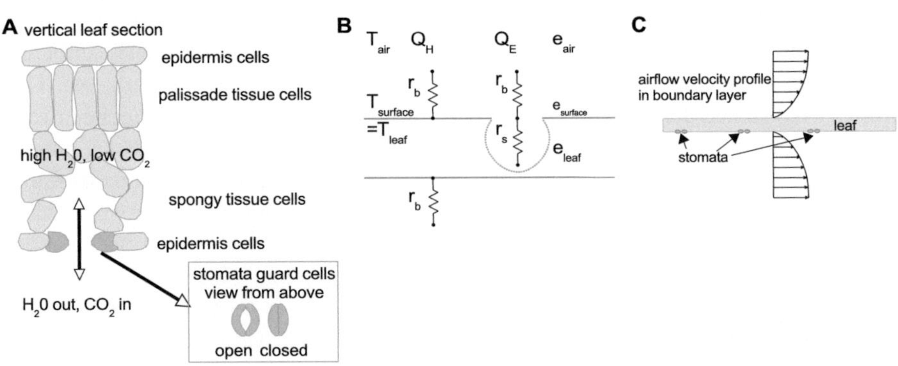

Fig. 2.7 Exchange of latent and sensible heat at the leaf level and resistance concept. (**a**) vertical cross section trough a leaf with stomata for water vapour and CO_2 exchange. (**b**) Resistance formulation schematic. Q_H: sensible heat flux (on both sides of leaf), r_b: resistance to heat exchange in the boundary layer in air above and below the leaf. T_{air}, $T_{surface}$: temperature of air and leaf surface. Q_E: latent heat exchange. Latent heat exchange includes two resistances: r_b for the boundary layer, and r_s for transfer of moisture through the leaf stomata. e_{air}, $e_{surface}$ and e_{leaf} are vapour pressures in air, at the leaf surface and inside the leaf. (**c**) Vertical air velocity profile in the boundary layer of a leaf. (After Bonan 2002)

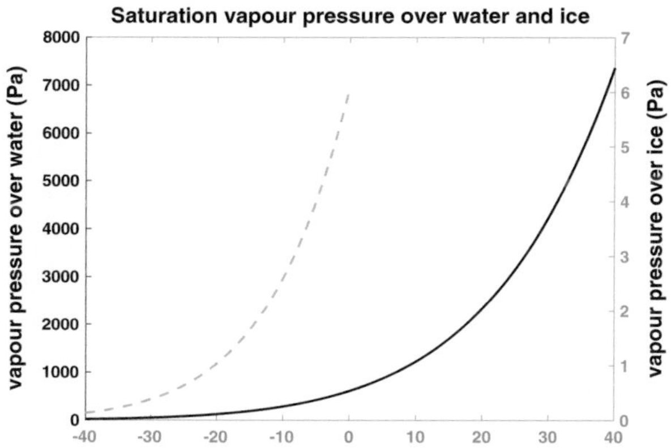

Fig. 2.8 Approximation of saturation vapour pressure over liquid water and ice after WMO (2008). Water: continuous line, left vertical axis; ice: grey dotted line, right vertical axis

the formula for parallel resistors ($R = r_1.r_2/(r_1 + r_2)$) to a combined resistance of $r_b/2$ for sensible heat exchange (Fig. 2.7).

For the transpiration part of evapotranspiration, water vapour has to be transferred from inside the leaf to the leaf surface, and next through the boundary layer (Fig. 2.7). Water vapour inside the leaf of higher plants is exchanged with the air by stomata. These are pores at the lower surface of leaves, which regulate the transport of gases, including CO_2 required for photosynthesis. Stomata open and close depending on environmental conditions. They open when high light levels increase photosynthesis rate and more CO_2 is required, they close under dry conditions when water vapour loss has to be prevented, and when temperature becomes too high or too low. Stomatal resistance may vary from 100 s m^{-1} when the pores are open, to greater than 5000 s m^{-1} when closed (Bonan 2016).

For vapour transport across the boundary layer, the same resistance r_b holds again. However, the stomatal resistance r_s is an additional resistance to heat flow, and adds to the boundary layer resistance. Since the stomata occur on one side the leaf only, we have to take the one-side boundary layer resistance for that. The total resistance for water vapour exchange is therefore $r_s + r_b$. Both sensible and latent heat transfer are highly important for sunlit leaves to maintain the temperature within limits. Without these heat exchange terms, a sunlit leaf will heat up strongly, since dissipation of heat by longwave radiation is too slow (Bonan 2016). The downside in high-latitude climates is, that plant leaves will cool very deeply in winter when the incoming solar radiation is absent, resulting in adaptations like the shedding of leaves and adaptation of leaf form.

The same principles that apply to the temperature and energy fluxes of a single leaf, also apply to that of the entire vegetation canopy. Different approaches for setting up a resistance network for exchange of energy with the vegetation canopy are

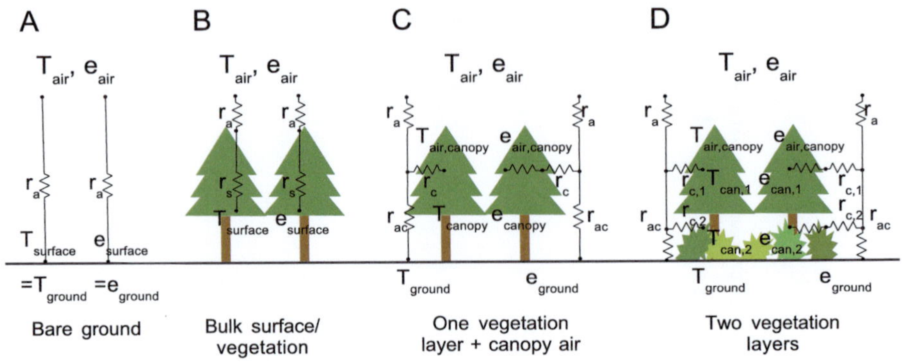

Fig. 2.9 Resistance networks for sensible and latent heat exchange, compared with the situation of bare ground. (**a**) bare ground; surface temperature and moisture assumed to be equal to/in equilibrium with ground surface. (**b**) simple bulk resistance model combining a surface resistance and aerodynamic resistance; temperature $T_{surface}$ and vapour pressure $e_{surface}$ are combined vegetation and ground surface temperature and vapour pressure. (**c**) Model including exchange within the vegetation canopy and air above it, and separate temperature/vapour pressure for ground surface, canopy and canopy air. (**d**) Similar to C for vegetation with two layers, with separate exchange with canopy air for both layers (redrawn from Bonan 2002). r_a: aerodynamic resistance; r_s: surface resistance; r_{ac}: resistance for aerodynamic processes within the canopy; r_c: resistance for exchange from vegetation to air within the canopy, including stomatal resistance for latent heat

shown in Fig. 2.9, compared with the approach for bare ground without vegetation. In the case of bare ground, we have only one resistance for exchange from surface to air for sensible and latent heat flux (Fig. 2.9a). The simplest bulk approach to vegetation includes two resistances in series, a surface resistance of exchange from the ground and vegetation surface to air combined, and an aerodynamic resistance for exchange with the atmosphere above (Fig. 2.9b). For latent heat, the first resistance includes the total stomatal resistance. A more complete model (Fig. 2.9c) includes heat exchange from the ground and vegetation separately with the air in the vegetation canopy, and the exchange of the canopy air with the atmosphere above it. As high (forest) vegetation usually has a multi-layered structure, this can be expanded with exchange for each vegetation layer (Fig. 2.9c). Important for the exchange energy of bare ground or a vegetation canopy with the atmosphere above it, is the turbulent mixing of the air, that occurs when there is wind blowing over the vegetated surface. This turbulent exchange of heat and water vapour is subject to *aerodynamic resistances, r_a* in Fig. 2.10a–d.

The one-resistor formulas for bare ground and sensible heat flux H (W m^{-2}) and latent heat fluxes E (kg m^{-2} s^{-1}, now expressed as exchange of water mass) are:

$$H = -\rho C_p \frac{(T_a - T_s)}{r_{ah}} \tag{2.11}$$

where T_a and T_s are air and surface temperatures, and r_{ah} the resistance for sensible heat (s m^{-1}), and

$$E = -\rho \frac{(q_a - q_s)}{r_{aw}} \quad (2.12)$$

where q_a and q_s are specific humidity (kg kg^{-1}) of the air and surface respectively, and r_{aw} the resistance for water transport (s m^{-1}). Similar resistance formulations can be drawn up in the case of the more complicated models in Fig. 2.9. Eq. 2.12 can be converted to heat exchange by multiplying both sides with the latent heat of vaporization λ and substituting $q = 0.622e/P$ (with P air pressure in Pa).

The rough ground and vegetation surfaces retard the air flow, by exerting a frictional drag. In the vegetation canopy and close above it, the wind speed is reduced. This results in a transfer of momentum (τ, mass × velocity per time unit, kg m^{-1} s^{-1}) from the air to the surface. In a vegetation canopy this is visible in the movement of leaves and branches. For this momentum flux a similar resistance formulation is used:

$$\tau = \rho \frac{(u_a - u_s)}{r_{am}} \quad (2.13)$$

Here u_a and u_s are horizontal wind strengths (m s^{-1}) in the in free air and immediately above the surface, r_{am} the resistance for momentum exchange. The result of this momentum transfer is a logarithmic increase of the horizontal mean wind speed with height above the surface (Fig. 2.10a).

This momentum transfer is created by turbulence of the air. The larger the height above the surface, the larger the wind speed and the turbulent eddies (Fig. 2.10a). The momentum flux is also known as the Reynolds shear stress.

These vertical turbulent movements of the air effectively enhance vertical transport of water vapour, sensible heat and other substances such as CO_2. Their importance is illustrated by calm nights without wind, when these vertical turbulent movements are absent. Water vapour accumulates then near the surface, as do gases

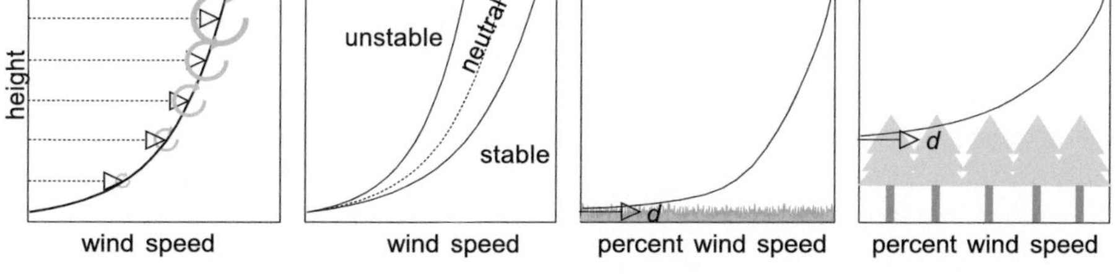

Fig. 2.10 (**a**) schematic logarithmic vertical wind speed profile with eddies (note that eddies are in reality not circular but irregular movements). (**b**) differences between vertical wind speed profiles in stable, neutral and unstable atmospheric conditions. (**c**, **d**) effect of vegetation height on the displacement height d of the wind profile (the height at which the wind speed profile extrapolates to zero, see arrow). Redrawn from Bonan (2002)

resulting from soil respiration such as CO_2 and CH_4. When cooling of the night air passes the dew-point, the accumulation of water vapour becomes visible as fog. As soon as the wind picks up, the accumulated vapours will be exchanged with the air at higher levels by vertical air movements: turbulent mixing. If this turbulent mixing is strong, the resistances in Eq. 2.11 and 2.12 are low, resulting in large fluxes.

Given that the resistances depend on wind speed and turbulence which vary with height, a relation between the resistances of momentum, heat and water vapour exchange, and the vertical wind profile is obvious. Without going into the physical and mathematical details (see e.g. Foken 2008 and Bonan 2016), the wind profile needs closer consideration. The ground and vegetation retard the motion of the air – a frictional drag. The average horizontal wind speed also increases with height above the surface, which results in a the approximately logarithmic wind profile. This shape of the wind profile is an expression of the shear stress exerted by the wind on the surface. It is furthermore influenced by the roughness of the vegetation at the surface and the stability of the air.

The shear stress can be expressed by a friction velocity, usually denoted by the symbol $u*$ which is a way to express shear stress exerted by the wind in units of velocity. It is related to τ in Eq. 2.13 and to the horizontal and vertical wind components by

$$u* = \sqrt{\frac{\tau}{\rho}} = \left[\left(\overline{u'v'} \right)^2 + \left(\overline{v'w'} \right)^2 \right]^{1/4} \tag{2.14}$$

where the terms $\overline{u'v'}$ and $\overline{v'w'}$ are the covariances between the horizontal (u, v) and vertical (w) components of the wind (Foken 2008). The larger these covariances, the larger the turbulent eddies in the wind are.

The role of the stability of the atmosphere is suppressing (stable atmosphere) or enhancing (unstable atmosphere) vertical movements. During the day, the surface is warmed by solar radiation, heating the surface air as well. Air parcels that are warmer than their surroundings will tend to move upwards, in particular when there is a steep temperature gradient at the surface, creating instability of the atmosphere. The instability enhances the vertical air movements created by turbulence, resulting in a more rapid exchange of energy and smaller wind speed gradients (Fig. 2.10b).

The higher the vegetation, the larger the roughness and the steeper the vertical wind profile (Fig. 2.10c, d). At some level above the ground, the horizontal wind speed should extrapolate to zero, due to the drag of surface roughness. This is called the displacement height or zero-plane displacement d. For bare ground, d is typically <0.05 m, in the case of a vegetation canopy, it is within the canopy – the taller the canopy, the larger d (Figure 2.10c, d). Generally, d is in the order of 2/3 to 3/4 of the average height of the vegetation.

The equation of the vertical wind profile is

$$u_z = \frac{u*}{k}\left[\ln\left(\frac{z-d}{z_{0m}}\right) - \Psi_m(\zeta)\right] \quad (2.15)$$

Here z is the hight above the surface (m), u∗ is the friction velocity of Eq. 2.14, *k* is the empirical Von Kármán constant (0.4), *d* the displacement height. The height z_{0m} (m) is a measure of the vegetation roughness, the roughness length. For instance, going from tundra to forest vegetation in Seward Peninsula, Alaska, the roughness length varies from 0.04 to 1.6 m, with canopy heights varying between 0.1 and 6.1 m (Beringer et al. 2005); for smooth surfaces, such as a snow cover, it may be smaller than 0.005 m. The function $\psi(\zeta)$ expresses the atmospheric stability conditions, it is zero under neutral conditions, and has negative values when the atmosphere is stable, positive values when unstable (Foken 2008). The level $z_{0m} + d$ is also known as the apparent sink of momentum, the level above the ground surface where all momentum of turbulence in the atmosphere appears to be lost.

The logarithmic wind profile applies only for heights above $z_{0m} + d$ and does not apply to wind profiles closer to the surface. In grassland, this is typically at a level of about 0.4 m, in tall forest this may be several metres (Fig. 2.10c, d). This does not mean that there is no air flow below this level within vegetation canopies; only the logarithmic wind profile does not hold here.

For sensible heat and moisture, similar logarithmic functions of the temperature and moisture difference with height can be derived (Bonan 2016):

$$T_z - T_s = \frac{-H}{k\rho c_p u*}\left[\ln\left(\frac{z-d}{z_{0h}}\right) - \Psi_h(\zeta)\right] \quad (2.16)$$

$$q_z - q_s = \frac{-E}{k\rho c_p u*}\left[\ln\left(\frac{z-d}{z_{0w}}\right) - \Psi_w(\zeta)\right] \quad (2.17)$$

Similar to momentum, the apparent height at which sensible heat and water vapour are exchanged with the atmosphere is not the ground surface or the top of the vegetation canopy, but a level within the canopy above the level at which the wind speed is zero, defined by $z_0 + d$. The surface values for temperature T_s and specific humidity q_s also refer to this level. The fluxes therefore appear to originate from a level within the canopy, determined by the wind gradient. The z_0 values are not the same for all all fluxes, those for momentum (z_{0m}), sensible heat (z_{0h}) and latent heat (z_{0w}) are different. Typically, the roughness length for sensible and latent heat exchange are 20% of that for momentum (Bonan 2016).

By substituting the temperature and moisture differences with height in the resistance eqs. 2.11 and 2.12, the aerodynamic resistance terms for sensible and latent heat flux can be related to the wind profile:

$$r_{ah} = \frac{1}{k^2 u_z}\left[\ln\left(\frac{z-d}{z_{0m}}\right) - \Psi_m(\zeta)\right]\left[\ln\left(\frac{z-d}{z_{0h}}\right) - \Psi_h(\zeta)\right] \tag{2.18}$$

$$r_{aw} = \frac{1}{k^2 u_z}\left[\ln\left(\frac{z-d}{z_{0m}}\right) - \Psi_m(\zeta)\right]\left[\ln\left(\frac{z-d}{z_{0w}}\right) - \Psi_w(\zeta)\right] \tag{2.19}$$

These equations give the aerodynamic resistances between a height z and the surface for sensible heat and latent heat exchange.

The meaning of Eq. 2.18 and 2.19 is that aerodynamic resistances are lower when the wind is stronger and atmosphere is unstable, resulting in a larger heat flux. Second, the resistances depend on canopy height, that determines the roughness length z_0 and displacement height d. Aerodynamic resistances for heat exchange of the vegetation canopy are high at lower vegetation height and decrease with higher vegetation; taller vegetation will have a larger surface area to exchange heat with the atmosphere. Tall vegetation canopies will therefore approach the temperature of the air more closely than short vegetation – these are better *coupled* to the the atmosphere than shorter vegetation. Short vegetation tends to have a higher temperature than the surrounding air in summer, as it exchanges heat gained by incoming solar radiation less effectively by turbulent fluxes. In a climate which has low temperatures during the summer, as occurs in tundra, this is advantageous (Grace et al. 1989; Bonan 2016).

So far, the aerodynamic resistances. However, within vegetation canopies, the *aerodynamic* resistance is not the only resistance to account for (Fig. 2.9). For the most simple approach with vegetation in Fig. 2.9b, a *surface resistance* is included which acts in series with the aerodynamic resistance. For latent heat, this includes the stomatal resistance and the resistance in the leaf boundary layer as in Fig. 2.7b, which is combined with resistance due to evaporation from the soil surface. Soil evaporation is regulated by aerodynamic processes within the canopy (Bonan 2016). Drier soils reduce evapotranspiration by increasing the bulk surface resistance to the latent heat flux (Liljedahl et al. 2011).

In the one-layer vegetation model of Fig. 2.9c, the total latent heat flux from the canopy is the sum of:

1. exchange between canopy air and the air above, governed by turbulent processes and the resistance of Eq. 2.19;
2. exchange between the ground surface and canopy air, governed by aerodynamic processes within the canopy, mostly below the displacement height;
3. exchange between canopy leaves and canopy air.

Analogous to Eq. 2.12, these latent heat fluxes can be formulated as follows. For exchange between canopy air and overlying air:

$$\lambda E_a = \frac{-\rho C_p}{\gamma} \frac{(e_{ac} - e_a)}{r_a} \tag{2.20}$$

Here r_a is the aerodynamic resistance of Eq. 2.19, e_{ac} the vapour pressure for canopy air. For exchange between ground surface and canopy air:

$$\lambda E_g = \frac{-\rho C_p}{\gamma} \frac{(e_g - e_{ac})}{r_{ac}} \tag{2.21}$$

Here r_{ac} is an aerodynamic resistance *within* the canopy, e_g is vapour pressure at the ground surface. For exchange between vegetation and canopy air:

$$\lambda E_v = \frac{-\rho C_p}{\gamma} \frac{(e_v - e_{ac})}{r_c} \tag{2.22}$$

Here e_v is vapour pressure of the vegetation, and r_c the canopy resistance. The canopy resistance combines the stomatal and boundary layer resistance (Fig. 2.9b) of all leaves. Similarly, the sensible heat exchange can be split up between an exchange between canopy air and air above the canopy, the exchange between the ground surface and canopy air (which is important for permafrost) and exchange between vegetation and canopy air (Fig. 2.9c).

Canopy conductance – the reciprocal of canopy resistance – relates to LAI, which can be used to estimate canopy resistance from leaf resistance (Bonan 2016):

- At lower LAI, canopy conductance increases linearly with LAI. Surface conductance is more important, because of a higher evaporation from the ground surface.
- At higher LAI (± 3 m^2 m^{-2}) the canopy conductance is roughly constant and related to leaf conductance. Forest has a LAI up to 6 m^2 m^{-2}. In a dense canopy with high LAI, less light penetrates to lower leaves, which close their stomata because of a reduction of photosynthesis. In dense canopies with high LAI direct evaporation from the soil becomes negligible.

The Penman-Monteith equation relates the evapotranspiration to the heat balance, and can be used to estimate evapotranspiration, or to estimate resistances if the heat balance has been measured (Bonan 2016):

$$\lambda E = \frac{\Delta(R_n - G) + \frac{\rho C_p}{r_a}(e^* - e_a)}{\Delta + \gamma \frac{r_c + r_a}{r_a}} = \frac{\Delta(R_n - G) + \rho C_p g_a (e^* - e_a)}{\Delta + \gamma (1 + g_a / g_c)} \tag{2.23}$$

Here, R_n is the net radiation, G the ground heat flux, r_c the canopy resistance, r_a the aerodynamic resistance, g_a and g_c the corresponding conductances (the reciprocal of resistance, *1/r*), γ the psychrometric constant, and Δ the slope of the saturation vapour pressure at air the ambient air temperature T_a ($\Delta = de* /dT_a$).

The theory in this section shows the dependency of the resistances to heat exchange on the vertical wind speed profile, turbulence and stability conditions of the atmosphere, as modified by the vegetation. It serves to understand the processes that influence heat balance at the soil surface in the presence of vegetation. Keep in mind, that the theoretical description in this section holds only for extensive, homogeneous surfaces. Areas with patchy vegetation with large height differences, and sloping terrain complicate airflow. Nevertheless, the model for sensible and latent heat transfer above is helpful to understand the heat exchange in vegetation canopies also in those cases.

2.4 Seasonality of the Surface Heat Balance Illustrated by Data

Data on heat fluxes in various Arctic vegetation types demonstrate the theory in the previous section. Vegetation cover characteristics influence the partitioning of the incoming net radiation among the other heat fluxes. The vegetation canopy height varies in high-latitude areas along with climatic and other ecological characteristics. This affects aerodynamical resistance to sensible and latent heat exchange. As a result, the heat exchange at the top of permafrost soils depends on vegetation characteristics, that seasonally change, for instance by shedding of leaves. In winter, the effect of snow cover has to be taken into account also. Low vegetation canopies are covered with snow for the largest part of the year, which changes the resistances to heat exchange. Moreover, it puts an insulating layer over the ground (see Sect. 2.5).

Variability of weather conditions throughout the season, and between years, has a large influence on the energy budget; this variability may mask the effects of ecosystem differences (Eugster et al. 2000). In particular the advection of maritime air masses onto adjacent coastal areas may have strong effects. Furthermore, seasonal changes in latent heat transfer may result from depletion of soil moisture reservoirs throughout the summer season (Westermann et al. 2009; Lund et al. 2014).

The amount of energy which is converted into heat for warming of permafrost may seem the most important topic in heat balance studies, but the latent heat is equally important as a component of the water balance. The water balance of permafrost soils also has a large effect on heat transfer and biogeochemical processes in these soils.

Below, the seasonal changes of the heat balance are discussed based on examples. As mentioned in Sect. 2.2.2, the energy balance is often not completely closed because of measurement uncertainty, with energy that is not accounted for in the balance. A closure term in the order of 10–20% of the incoming net radiation is not uncommon. In many cases, data are available for summer conditions only (e.g. Eugster et al. 2000; Beringer et al. 2005). In the discussion below, we focus on recent studies and studies that quantify year-round energy exchange or cover several years.

2.4.1 Summer

At high latitudes, the radiation balance in summer is dominated by the incoming shortwave solar radiation. Several factors influence the radiation balance:

- The albedo varies with vegetation cover; boreal forest has a lower albedo than shorter vegetation (Thompson et al. 2004; Table 2.1). Also, within tundra environments the albedo varies; dry tundra vegetation, in particular lichen-rich vegetation and shrub vegetation has a higher albedo than wet tundra with moist soil or standing water between the vegetation, and ponds (Blok et al. 2011; Juszak et al. 2014, 2016, 2017). This results in lower incoming shortwave radiation at the dry surface. These variations may be large enough to influence the energy balance (Eugster et al. 2000; Langer et al. 2011a). During the day and season, the albedo also varies with sun elevation and with cloud cover (Eugster et al. 2000; Juszak et al. 2016). On a small scale, the albedo of tundra vegetation is also influenced by the amount of dead standing biomass (leaves and branches); dark branches of shrubs may account for up to 71% of the absorption of the vegetation canopy (Juszak et al. 2014, 2016).
- Despite the nightly presence of the sun above the horizon above the Arctic Circle, the sun's angle at night is very low. Therefore, all summer heat flux measurements at Arctic sites display a clear diurnal pattern (Beringer et al. 2005; Westermann et al. 2009; Langer et al. 2011a).
- Besides the seasonal and daily variations, the shortwave radiation balance also varies with cloud cover, in particular in maritime cold climates. It may decrease as much as 95% under cloudy conditions. In Arctic coastal regions, weather patterns may include changes between onshore winds carrying low clouds and cold air, and offshore winds with more sunny conditions (Langer et al. 2011a; Liljedahl et al. 2011). Langer et al. (2011a) measured summer heat fluxes as high as 300 W m^{-2} under clear sky conditions in the Lena Delta, which was reduced to 20 W m^{-2} under overcast conditions.
- The incoming longwave radiation also varies strongly with cloud cover, with an opposite sign: it increases under overcast conditions. In the Lena Delta site, frequent overcast conditions increase the incoming longwave radiation by 35%, from about 280 W m^{-2} to 380 W m^{-2}. The outgoing longwave radiation varies with surface temperature. The net longwave radiation losses were as much as 100 W m^{-2} during clear sky conditions but almost zero under cloudy conditions (Langer et al. 2011a).

The turbulent heat balance components depend, next to weather variables on vegetation characteristics:

- Short tundra vegetation is aerodynamically smoother than taller forest vegetation, as is illustrated by the aerodynamical resistances for latent heat exchange in Table 2.1. In an experiment in Alaska (Beringer et al. 2005), the aerodynamical resistance of tundra vegetation to latent heat exchange varies from 43 s m^{-1} in tundra vegetation to 11 s m^{-1} in forest. Lund et al. (2014) report a value of 91 ± 5 s m^{-1} for the aerodynamic resistance of short Arctic tundra heath in Greenland. Wet tundra sedge/moss vegetations and standing water are aerodynamically smoother than drier vegetation types. In Alaskan tundra vegetations the amount of sensible heat transport in summer varies between 16% to 50% of the midday radiation balance (McFadden et al. 1998).
- The plant- and soil related surface resistances are calculated using the Penman-Monteith Eq. (2.23). It is generally smaller for the wet tundra sites. The canopy resistance for latent heat transfer may vary over two orders of magnitude, from below 10–100 s m^{-1} for tundra to more than 1000 s m^{-1} for Larch forest (Eugster et al. 2000). Within the tundra differences in heat exchange exist also. Wet tundra sedge vegetations and open water have a larger supply of moisture feeding the latent heat flux, than a drier dwarf shrub vegetation (Langer et al. 2011a).
- A specific feature of tundra and peatland bog vegetation is the abundance of mosses, which as non-vascular plants do not have the stomatal control of transpiration of vascular plants. Most mosses cannot extract water from the soil but rely on water storage within the moss layer (Elumeeva et al. 2011). This increases the surface resistance to latent energy exchange (Kasurinen et al. 2014); the surface resistance to water loss increases exponentially with decrease of the tissue water content (McFadden et al. 2003; Boike et al. 2008. Liljedahl et al. (2011) report an increase in surface resistance for latent heat exchange when the atmospheric demand for moisture is highest. Mosses therefore have a limited ability to supply moisture at low vapour pressure deficits. This is illustrated by an experiment by Blok et al. (2011) in the Siberian tundra, that shows that evaporation from the soil surface was increased by removal of the moss layer.

Unstable conditions enhance latent and sensible heat exchange. Above tundra, unstable conditions may develop during summer days with sufficient solar irradiation, while during the night neutral stratification develops. Only under very calm conditions, stable stratification develops with low sun angle during the night (Westermann et al. 2009; Langer et al. 2011a). The incoming solar energy during the day drives predominantly outgoing turbulent sensible and latent heat fluxes. The seasonal averaged sensible heat fluxes in Table 2.1 vary between 17% and 44% of the net radiation (highest in the Alaskan forest site). The latent heat fluxes vary between 29–49% (highest in the wet tundra of Northern Siberia). The ground heat flux is 5–20% of the radiation balance (lowest in the forest, highest in wet tundra).

In Table 2.1, latent and sensible heat fluxes are approximately equal for barren, tundra, shrub, and tall shrub vegetations, with Bowen ratios (Q_H/Q_E) near 1.0. Sites that report midday values show higher Bowen ratios (1.4–3.0), reflecting the larger sensible heat component at noon. On wet tundra sites, the Bowen ratio is lower because the abundant moisture availability induces a larger latent heat flux (however, see the effect of mosses mentioned above). The Bowen ratio may vary considerably throughout the season; at the barren site in Svalbard it varies between 0.25 and 2, depending on soil moisture. (Westermann et al. 2009). At the dry *Cassiope* heath in Greenland (Lund et al. 2014), the Bowen ratio is comparatively high. In a large database of energy flux partitioning over a wide range of Boreal to Arctic ecosystems (Eugster et al. 2000) the effect of soil moisture on the partitioning of fluxes between sensible and latent heat is clearly illustrated (Fig. 2.11).

In largely bare ground areas like those pictured in Fig. 1.21, the effect of the vegetation canopy on the latent and sensible heat fluxes is minimal. The sensible, latent and ground heat fluxes over sparse vegetation in Svalbard amount to about 40%, 40% and 21% of the radiation balance respectively (Westermann et al. 2009). This is seen for instance at High Arctic tundra in Northern Greenland (LAI 0.2–0.3), where the sensible heat and ground heat flux are large relative to latent heat (Lund et al. 2014). In Svalbard and northern Greenland, the summer latent heat flux is strongly coupled to soil moisture because of the minimal contribution by vegetation transpiration (Westermann et al. 2009). The soil evaporation is high in early summer, and decreases due to drying of the soil later in summer (Lund et al. 2014). In early summer, high soil water availability and evaporation is found in most permafrost ecosystems, because of an abundant supply of snow meltwater (Langer et al. 2011a).

Wet tundra has a higher latent heat flux and lower Bowen ratio than dry tundra (Langer et al. 2011a; Table 2.1; Fig. 2.12). The seasonally averaged latent heat flux and ground heat flux are considerably higher at the expense of the sensible heat flux; nearly half (42–49%) of the incoming radiative energy is balanced by the latent heat flux, the remainder is divided approximately equally between sensible heat flux (17–21%) and the ground heat flux (19–20%). The sensible heat flux may vary strongly, between 130 and −20 W m^{-2}. In wet soils, the conduction of heat into the ground is higher than in dry sites because of the effect of high moisture content on the thermal conductivity (Juszak et al. 2014; see Sect. 2.5). In wet tundra sites the ground heat flux may even restrict the energy available for evaporation (Langer et al. 2011a; Liljedahl et al. 2011).

In the woodland and forested sites of Beringer et al. (2005), the sensible heat flux increases, and the ground heat flux decreases with respect to the lower vegetations (Table 2.1). In general, the sensible heat flux of boreal forest tends to be higher than the latent heat flux, but in years with higher precipitation the latent heat flux may dominate (Matsumoto et al. 2008; Ohta et al. 2008).

The ground heat flux is crucial for thawing of the active layer of the permafrost in summer. However, only a relatively small part of the net radiation, in the order of 10% to 25% is spent on the ground heat flux for warming the soil; in forested sites it is even less, in the order of 5% (Table 2.1; Fig. 2.11; Matsumoto et al. 2008). Still, the soil is warmed up to depths in the order of 10 m or more. Most of the ground heat flux (about 60%) is used for thawing of ground ice, in particular in the early summer (Langer et al. 2011a, 2014). The presence of soil ice therefore creates a

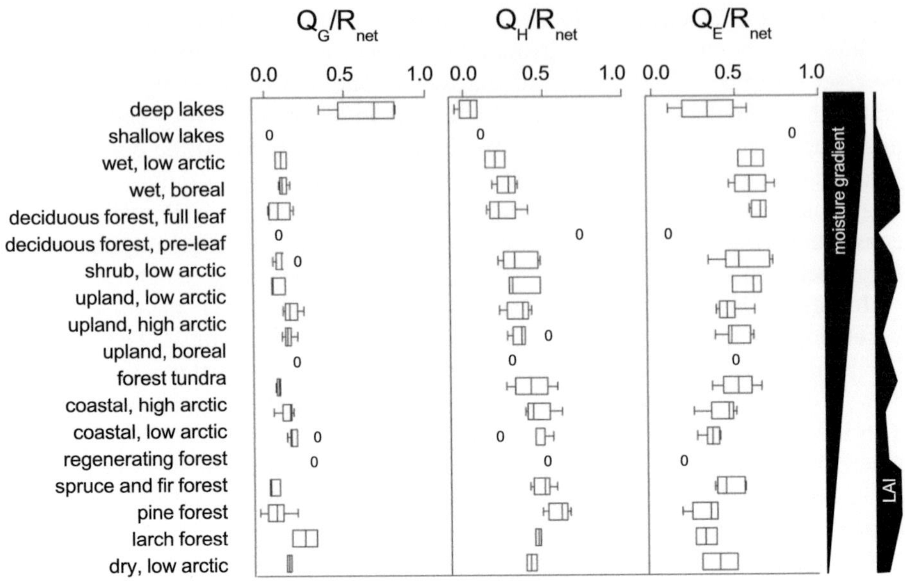

Fig. 2.11 Energy partitioning during summer (July) of a variety of northern ecosystems collected from a large number of publications, redrawn from Eugster et al. 2000. Values are sorted along the gradients of surface or soil moisture and leaf area index (LAI). The boxes of latent heat Q_E/R_{net}, sensible heat Q_H/R_{net} and Q_G/R_{net} show the interquartile range with the median value in the middle. Whiskers extend to the upper and lower values, and outliers are plotted individually as ellipses. Republished with permission of John Wiley and Sons Inc., from Eugster et al. 2000: Land–atmosphere energy exchange in Arctic tundra and boreal forest: available data and feedbacks to climate. Global change biology 6 (S1):84–115; permission conveyed through Copyright Clearance Center, Inc.

large heat sink in the soil in summer. When – by permafrost thaw – this heat sink becomes smaller, this may imply more energy for sensible heat and latent heat, which amplifies climate warming and creates a larger moisture source to the atmosphere (Eugster et al. 2000).

The spatial variation of the fluxes is rarely studied. The albedo varies with vegetation; likewise soil moisture and the presence of water at the surface varies also with the micro-topography discussed in the next chapter. The net effect may be an up to 75 W m^{-2} lower net radiation balance for the dry tundra, compared to the wet tundra (Langer et al. 2011a). Under clear-sky conditions during the night this situation is inverted, with higher outgoing longwave radiation, since the wet surface at night is warmer compared to dry tundra. Under cloudy conditions the spatial differences tend to vanish because of the smaller effect of albedo on the low incoming shortwave radiation. Juszak et al. (2014) compared the summer radiation and soil heat fluxes of dry dwarf shrub heath and wet sedge meadows in Siberian tundra. These vegetation types show a contrasting active layer thickness; the ALT of sedge meadows is approximately double that of dwarf shrub heath. They concluded that the deeper active layer below wet sedges is not an effect a of the canopy radiation

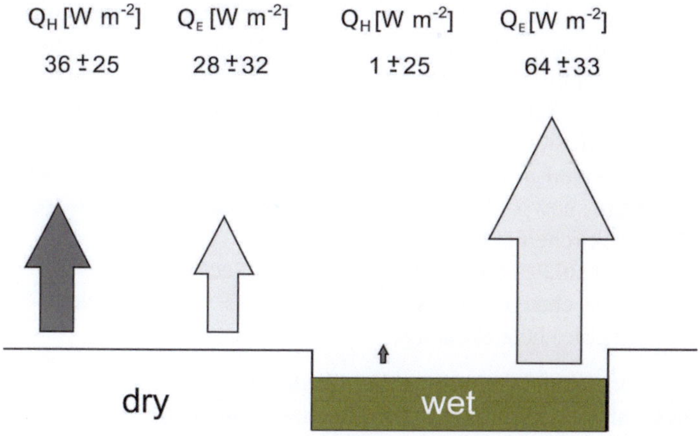

Fig. 2.12 Sensible (Q_H) and latent heat fluxes (Q_E) at wet and dry surfaces during summer 2008 in northern Siberian tundra. The values are estimated from the eddy covariance measurements of stationary and mobile eddy covariance systems. The areas of the arrows are scaled according to the heat flux values. (From: Langer et al. (2011a): The surface energy balance of a polygonal tundra site in northern Siberia – Part 1: Spring to fall, The Cryosphere 5, 151–171, DOI::https://doi.org/10.5194/tc-5-151-2011; Creative Commons License 3.0, https://creativecommons.org/licenses/by/3.0/)

budget, but that soil properties, such as a high soil albedo, high soil moisture and high thermal conductivity drive a higher ground heat flux.

Differences between wet and dry tundra surfaces also have a strong effect on the latent and sensible heat fluxes: over a wet surface, the latent heat flux dominates and the sensible heat flux may be negligible, resulting in low Bowen ratios (down to 0.02, Langer et al. 2011a). Over a pond surface, the sensible heat flux may be a fraction of the normal flux over a vegetated surface, while the latent heat flux is much higher (Fig. 2.12).

2.4.2 *Winter Cooling*

In the fall a snow pack starts to accumulate. The effect of topography and differences in vegetation cover on the snowpack thickness is large; topographical highs tend to be more exposed to wind and therefore may be blown free of snow, while lee sides of topographical obstacles, terrain hollows and taller vegetation collects snow (Sturm et al. 2001; Essery and Pomeroy 2004). In general, vegetation stabilizes the snow cover by reducing wind drift of snow, but interaction between snow cover and vegetation is more complicated than that. For instance, in Siberia tundra, large tussocks protruding above the snowpack in early winter create gaps in the snow layer, which leads to earlier active layer freezing (Kholodov et al. 2012). The main effects of the snowpack are:

- A large increase of the albedo, increasing the reflected solar radiation. The seasonal and year-to-year variability in the build-up of snow cover and its physical characteristics results in albedo changes. Older and thawing snow has a lower albedo than freshly fallen snow. The albedo effect is larger at low latitudes with mountain permafrost, than at higher latitudes, where in part of the winter no solar radiation is received at all. In the Arctic, the albedo effect is strongest in the spring when the snowpack is still present while the sun angle increases (Sturm et al. 2005; Euskirchen et al. 2007).
- The snowpack insulates the soil from heat loss (Sect. 2.5).
- The snow cover changes the surface roughness characteristics for turbulent (latent and sensible) heat exchange.

The vegetation strongly modifies the albedo of a snow-covered landscape. An almost entirely snow-covered tundra surface has a much larger albedo than a snow-covered forest (Bonan et al. 1992). A short tundra vegetation may become entirely covered by a snow pack with only a few branches of dwarf shrubs protruding above the snow. However, even modest differences in amount of branches protruding through the snow have an effect. At a completely snow-covered low shrub tundra, the albedo is 30% higher than at sites where shrubs protrude significantly above the snow (Sturm et al. 2005). Compared to low stature vegetation types like tundra (albedo about 0.85), woodland/forest sites have a much lower albedo of 0.18 (Table 2.1). In boreal conifer forest, the albedo is higher on non-deciduous trees (Pine) than on deciduous trees (e.g. Larch) (Matsumoto et al. 2008).

When the snow starts melting, subtle differences in shrub and tree cover cause increasing differences in albedo (Sturm et al. 2001, 2005). Flat sedge and dwarf shrub vegetations generally retain a high albedo longer than taller shrub vegetations and hummocky vegetations with tussocks (Loranty et al. 2011). Micro-relief elements, like vegetation tussocks and ice wedge polygon rims lose their snow cover rapidly and contribute strongly to albedo decrease (Langer et al. 2011a).

With respect to tundra, the effect of vegetation on albedo and the amount of solar shortwave radiation energy absorbed by the ecosystem increases, with factors of 1.06 for low shrubs, 1.22 for tall shrubs, 1.69 for woodland and 2.54 for forest at 65°N, over the entire year (Sturm et al. 2005). The effect is stronger at higher latitudes.

In the Arctic fall, the combination of an increasing albedo due to an increasing snow cover, short daytime and low sun angle strongly decreases the incoming shortwave radiation. Long-wave radiation starts to dominate the radiation balance. The net radiation balance becomes negative; the outgoing longwave (thermal) radiation of the surface is larger than incoming longwave and shortwave radiation, in particular under clear sky conditions. In one winter in the Lena Delta for example (Langer et al. 2011b), the incoming shortwave solar radiation amounted only 2 W m^{-2} in October and November; the total radiation balance was −20 W m^{-2}. A value of −28 W m^{-2} for the net longwave radiation was measured in High Arctic Svalbard (Westermann et al. 2009; Fig. 2.4). The atmospheric stratification in wintertime is dominantly neutral; stable stratification with negligible sensible and latent heat fluxes occur at clear sky conditions without wind (Westermann et al. 2009; Langer et al. 2011b).

The surface loss of longwave radiative energy is compensated by the ground heat flux and by the sensible heat flux (Fig. 2.4, Table 2.1). Heat stored in the soil in summer is transported back to the atmosphere. As seen above, the latent heat of thawing and freezing of water in the soils is a large heat sink; the freezing of soil water in winter releases a large amount of latent heat which compensates for the energy loss at the surface (heat transport in the soil is treated in more detail in Sect. 2.5).

The sensible heat flux also may become negative, supplying heat from the atmosphere to the surface (and cooling the air). In Western Svalbard, an oceanic coastal site with open sea presently persisting during most of the winter, the advection of warm air and rain falling on snow contributes to a sensible heat flux of 16 W m^{-2} (Westermann et al. 2009), while in the Lena Delta the sensible heat flux is comparatively low (5 W m^{-2}) and the ground heat flux supplies most heat (10–20 W m^{-2}) to compensate the radiative losses from the surface (Langer et al. 2011b). The latent heat flux is of minor importance, but not absent (Westermann et al. 2009), although the saturation vapour pressures are very low at low temperatures (Fig. 2.4).

However, the sensible heat flux is a fickle friend in winter, although it may be up to 43% of the net radiation. Occasionally it may still draw heat from the surface when the amount of incoming longwave radiation is high, and the temperature of the air above the snow is lower than that of the surface. In the maritime conditions of Svalbard, strong sensible and latent heat fluxes are sustained by strong winds. Sublimation from the snow surface may still drive a latent heat flux which cools the surface (Westermann et al. 2009). Therefore, which heat source to the surface is more important, sensible or ground heat flux, varies with weather patterns and climatic conditions.

In the course of the winter, the incoming longwave radiation from the atmosphere decreases with the cooling of the atmosphere. The amount of incoming longwave radiation during the polar night is strongly influenced by cloudiness (Langer et al. 2011b; Fig. 2.4 and 2.13). The outgoing longwave radiation also decreases with decreasing surface temperatures. For instance, incoming longwave radiation in Northern Siberia decreases from 300 W m^{-2} to 180 W m^{-2} with fluctuations in the order of 60 W m^{-2}. The outgoing longwave radiation decreased from 315 W m^{-2} to 200 W m^{-2}, corresponding to a surface temperature of −24 °C. The surface temperature followed roughly the fluctuations of the incoming longwave radiation, resulting in temperature fluctuations of up to 10 °C (Langer et al. 2011b).

The ground heat flux (which includes the heat flux through the snow cover) provides most of the energy to balance the radiation deficit, up to 81% of the winter net radiation (Fig. 2.13; Table 2.1). It may vary from year to year by a factor of two. The ground heat flux is largest under stable atmospheric conditions, when the turbulent fluxes are very small. The heat is for the largest part supplied by the latent heat of freezing soil water in fall and early winter. This latent heat supply varies spatially considerably, it is largest above freezing ponds (Langer et al. 2011b).

In the Arctic late winter and spring, the shortwave radiation energy source gradually returns, with a pronounced daily cycle. Heating of the surface around noon and

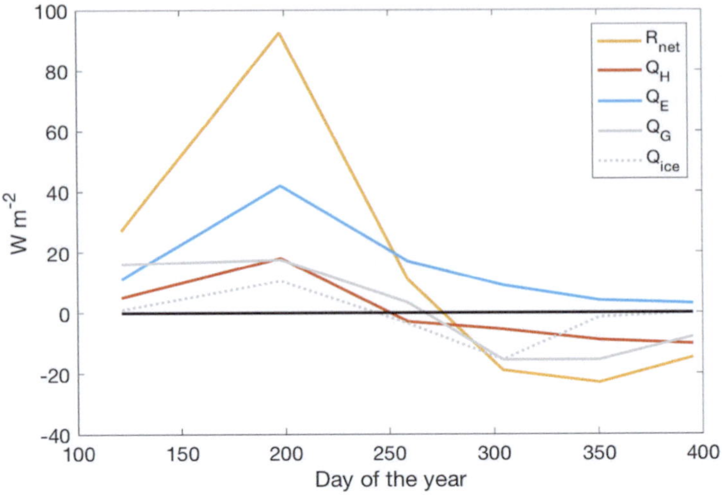

Fig. 2.13 Changing energy balance in the course of the year at the ground surface, at a tundra site in Siberia. Q_{ice} is the heat flux generated by thawing and freezing of ground ice. Other fluxes as in Fig. 2.3. After Langer et al. (2011a, b); data points have been put at midpoints of observation periods of Langer et al. (2011a, b) and averaged over two observation years if available

cooling during the night may induce unstable atmospheric conditions during the day and stable to neutral conditions during the night. The sensible heat flux increases in magnitude, while the ground heat flux loses the dominant position it has during the polar night, although cooling of the soil still continues during the period preceding snowmelt (Westermann et al. 2009). At the end of the winter, the sensible heat flux dominates the warming of the surface (Langer et al. 2011b).

In late spring the shortwave radiation starts to dominate the radiation balance. When the snow starts to melt, the latent heat energy consumed by the melting of the snowpack becomes a dominant component in the energy budget. By that time, also the ground heat flux has changed sign (Westermann et al. 2009; Langer et al. 2011b). Vegetation protruding above the snow cover has an important role in decreasing the albedo of the snow cover in spring; the albedo of a wet or old snow cover is also lower than that of fresh snow. This enhances the effect of increasing incoming shortwave radiation. The snow disappears first in places where the cover is thinnest. Micro-relief elements (tussocks, polygonal ridges) protrude above the snow cover (Langer et al. 2011a), further altering the radiation balance. In the tundra, relatively small differences in vegetation can generate radiation balance differences of up to 50 W m^{-2} (Loranty et al. 2011).

From these examples it is clear that permafrost soils cool most strongly during stable periods with clear-sky conditions during the polar night. Under these conditions the radiation balance is most negative, with minor incoming longwave radiation. The ground heat flux balances radiative energy losses. Since freezing of soil water is a source of latent heat in the soil, the soil water and ice content are important variables in the ground heat flux (see below). Of secondary importance is the sensible heat flux, in particular in periods with high wind speeds. Latent heat fluxes from surface to atmosphere are negligible in terms of energy, but still have an effect on the water balance: Langer et al. (2011b) estimate that sublimation of snow amounts to 60% of the snowpack water equivalent at the end of the winter.

2.4.3 Changes in the Heat Balance and Climate Change

Most observations of the energy balance contain only one or a few years of data. However, a 10 year long data set of Lund et al. (2014) of the Zackenberg site in Northeast Greenland shows significantly increasing latent and sensible heat fluxes in summer, and a decrease of the ground heat flux. Nevertheless, active layer thickness (ALT), has increased at a rate of 1.6 cm yr^{-1} at Zackenberg. The ratios of sensible, latent and ground heat flux over R_{net} correlate with ALT. In particular the latent heat flux became more important over the years, shown by a significantly decreasing Bowen ratio. Effects of Arctic greening (increased vegetation cover) were not seen during the observation period at this site. Therefore Lund et al. (2014) infer a relation with permafrost thaw; the increase of active layer depth results in less energy consumption for thawing of soil ice, resulting in more energy for warming of the atmosphere. This effect also implies a climate warming feedback: where the ALT decreases due to less intense ground freezing, it will result in more energy available for warming the atmosphere (Eugster et al. 2000).

Permafrost thaw causes changes in land surface characteristics (Chap. 3, 6). Thawing of excess ground ice can result in severe soil subsidence accompanied by ponding of water and rapid vegetation change (e.g. Jorgenson et al. 2006). These processes have also a pronounced effect on the energy balance. The energy balance of an intact peat plateau and a collapsed peat plateau was compared by Stiegler et al. (2016). At the collapse site, the soil was saturated with water, the albedo was slightly lower due to presence of bare, dark peat lumps, and the amount of *Sphagnum* mosses increased. At the dry peat plateau surface, the water content of the topsoil was about half of that of the collapse site. The surface and aerodynamic resistances for latent heat exchange of the collapse site are lower than those of the intact peat plateau (66 s m^{-1} vs. 66 s m^{-1} respectively 45 s m^{-1} vs. 73 s m^{-1}). This resulted in an increase of the latent heat flux at the collapse site (59 ± 39 to 64 ± 49 W m^{-2}) at the expense of the sensible heat exchange, resulting in a lower Bowen ratio (0.43 vs. 0.76). The ground heat flux at the collapse site increased from 7.2 ± 5.0 W m^{-2} to 8.6 ± 11.4 W m^{-2}, probably caused by the higher soil saturation of water which increased the thermal conductivity of the soil (Sect. 2.5).

Climate change also results in shifts in vegetation zones, which obviously results in changes of energy fluxes (Eugster et al. 2000; Myers-Smith et al. 2011). A lower winter/spring albedo of higher vegetation results in more radiative heat entering the vegetation cover in spring, contributing to more rapid snow melt and soil warming (Beringer et al. 2005). However, the lower aerodynamic resistance of higher vegetation results in higher turbulent fluxes, at the expense of the ground heat flux (Table 2.1). An additional effect is shading of the ground surface from incoming solar radiation by a larger mass of leaves and stems (Beringer et al. 2005; Eugster et al. 2000). Therefore, an increase of higher vegetation cover due to climate warming might retard warming of the soil. Conversely, removal of higher vegetation may result in increased ground heat flux in summer. Forest clear-cutting in Siberian taiga forest (Iwahana et al. 2005) and a shrub removal experiment in Siberian tundra by Blok et al. (2010) resulted an increase in ALT.

Another anthropogenic warming effect in the Arctic is deposition of light absorbing particles, including black carbon particles (soot) from combustion processes (AMAP 2015, 2017). In particular in spring, this decreases the albedo of the snow cover. The present black carbon deposition may have reduced snow cover duration for up to ten days compared to pre-industrial levels (Ménégoz et al. 2013), adding to the climate change induced warming of the permafrost. The effect of albedo reduction is strongest over the Russian Arctic, due to industrial processes including gas flaring in the oil and gas industry (AMAP 2015, 2017).

2.5 Ground Heat Flux

The ground heat flux is determined by thermal properties of the soil (heat capacity and thermal conductivity), which in turn relate to soil composition. Soil components that are relevant for the thermal properties vary strongly in space (including depth) and time. The variations over time are caused by variations in moisture content of the soil, and the phase changes of the soil water into ice and vice versa. When water freezes, its thermal conductivity increases fourfold, its heat capacity decreases, and latent heat is released. Also, spatial gradients of water content are common. Soil mineral composition and organic matter content vary less in time, but usually vary over depth, and in the horizontal plane. For simplicity, the ground heat flux is mostly treated here as a vertical flux only, but in particular lateral heterogeneities at the land surface (e.g. the formation of lakes and ponds) requires a three-dimensional treatment.

Temperature variations at the soil surface are largest, and decrease with depth. The depth to which daily surface temperature variations penetrate is in the order of decimetres. Variations related to weather patterns penetrate in the order of metres, seasonal variations penetrate down to depths of 10 meter or more. In summer, heat is transported into the ground, resulting in a vertical temperature profile with temperatures decreasing with depth. In winter this is reversed; temperatures increase with depth (Figs. 1.2 and 2.14). For the carbon cycle of permafrost, the soil temperatures are highly important, since process rates of decomposition of organic matter by bacteria depend on temperature.

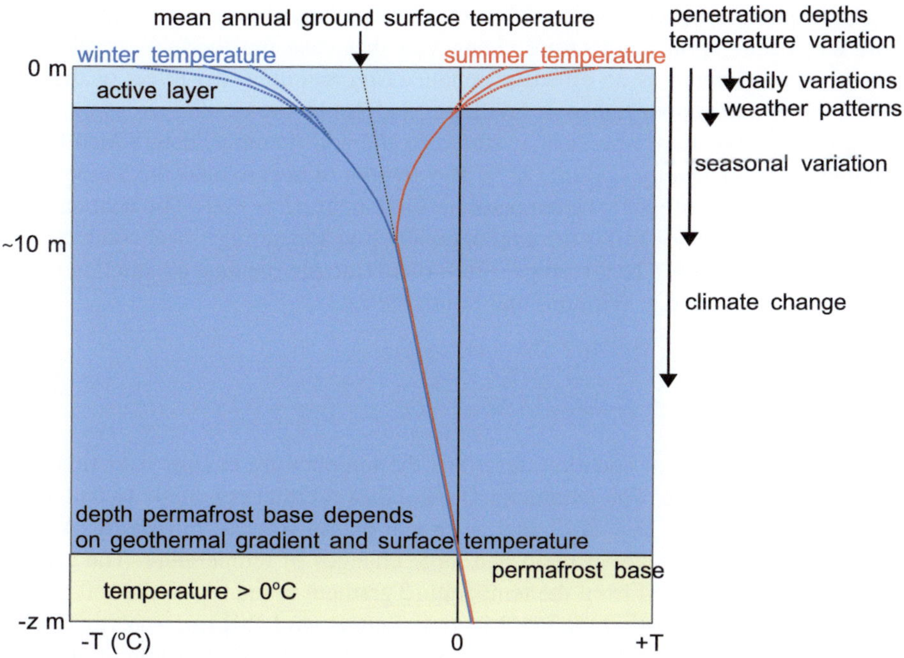

Fig. 2.14 Schematic vertical temperature profile with penetration depth scale for temperature variations at the surface. The blue line is a typical winter temperature profile, the red line a summer temperature profile. The dashed line represents the geothermal temperature gradient. Depths are approximate

2.5.1 Soil Profile Scale

First, the common case of conductive heat flow is considered here. The heat flux in the soil is largely determined by conduction, but also convective heat transfer by water flow or vapour transport may occur in soil pores (e.g. Hoekstra 1966; De Vries 1987). This is not included here for sake of simplicity. However, these processes are locally important, in particular in discontinuous permafrost (e.g. Williams and Smith 1991).

The heat flow is determined by the temperature gradient and the thermal conductivity of the soil material:

$$Q_G = -K\frac{dT}{dz} \tag{2.24}$$

where K is the thermal conductivity (W m^{-1} K^{-1}), the rate at which heat is transported in the soil over a unit thermal gradient. The lower the thermal conductivity, the slower heat is transported through the soil material. A low conductivity is characteristic for insulating materials. The larger the temperature gradient dT/dz, the larger the heat transport. In soils with both water and air, always some vapour transport will occur upon temperature change; in that case the thermal conductivity should be considered as an effective conductivity, including the effect of thermal conductivity and vapour transport (Williams and Smith 1991).

Next to conduction, also heat is stored in the soil material, determined by its volumetric heat capacity C (J m^{-3} K^{-1}). The amount of heat to raise the temperature of a given volume of soil by a temperature increment *dT* is *CdT*. The heat balance for a certain volume of soil is the sum of conduction and storage: heat conducted in = heat conducted out + heat storage. From this principle the heat conduction equation can be derived (e.g. Williams and Smith 1991):

$$\frac{\partial T}{\partial t} = D \frac{\partial^2 T}{\partial z^2} \tag{2.25}$$

This partial differential equation describes the temperature change over time and depth in a soil profile. The parameter *D* combines thermal conductivity and volumetric heat capacity: $D = K/C$, and is known as the thermal diffusivity. A high thermal diffusivity results in rapid and large changes of temperature. The ground heat flux can be derived from the temperature gradient at the top of the soil profile and the thermal conductivity in Eq. 2.24 (e.g. Langer et al. 2011a). Analytical solutions of Eq. 2.25 are available for a variety of initial and boundary conditions, and usually assume a homogeneous soil with time-invariant thermal properties. Since natural soils do not comply with these conditions, numerical simulations are usually applied, which include varying thermal properties.

Table 2.2 lists thermal properties of soils and their constituents, which show order of magnitude differences in thermal conductivity and diffusivity. Calculating the actual values of these properties for a given soil horizon requires knowledge of its mineral and organic constituents, porosity and water/ice content. The heat capacity of a mixture of minerals, organic matter, air, water and/or ice is a weighted average of the heat capacities of these components:

$$C_{soil} = \sum_{i=1}^{n} f_i C_i \tag{2.26}$$

(De Vries, 1966), in which f_i and C_i are respectively the volume fraction and volumetric heat capacities of the i-th constituent (minerals, organics, water, ice, air).

For conductivity this is less simple. First, the conducting materials behave analogous to parallel resistors in an electric circuit, letting more heat flow through the best conductor. Second, the interconnectedness of the material influences the conductivity – a solid conductor conducts better than one consisting of poorly interconnected grains, like sand grains. An additional problem is that thermal conductivity is difficult to measure experimentally. Several approaches have been advanced for calculating thermal conductivity. A model that includes organic matter, thawed and frozen soils is published by Balland and Arp (2005). In this method, first an estimate

Table 2.2 Thermal properties of soil materials

Material	Density kg m^{-3}	Mass heat capacity J kg^{-1} K^{-1}	Volumetric heat capacity J m^{-3} K^{-1}	Thermal conductivity W m^{-1} K^{-1}	Thermal diffusivity m^2 s^{-1}
Quartz	2660	800	2.128×10^6	8.80	4.135×10^{-6}
Clay minerals	2650	900	2.385×10^6	2.92	1.224×0^{-6}
Organic matter	1300	1920	2.496×10^6	0.25	0.100×10^{-6}
Water (0 °C)	1000	4180	4.180×10^6	0.57	0.136×10^{-6}
Ice (0 °C)	917	2100	1.926×10^6	2.24	1.163×10^{-6}
Air (0 °C, mean sea level pressure)	1.2	1010	1297	0.025	20.627×10^{-6}
Sand[1], P 0.4, 5% clay, 1% organic	1582	818	1.294×10^6	0.58	0.448×10^{-6}
Sand[1], P 0.4, θ 0.5	1782	1195	2.130×10^6	1.77	0.832×10^{-6}
Sand[1], P 0.4, water saturated	1982	1496	2.965×10^6	2.69	0.907×10^{-6}
Sand[1], P 0.4, water saturated + frozen	1948	1059	2.064×10^6	4.65	2.253×10^{-6}
Peat[1], P 0.8, 10% sand, 10% clay, dry	287	1726	0.495×10^6	0.03	0.059×10^{-6}
Peat[1], P 0.8, θ 0.5	687	3153	2.167×10^6	0.16	0.073×10^{-6}
Peat[1], P 0.8, water saturated	1087	3531	3.834×10^6	0.51	0.134×10^{-6}
Peat[1], P 0.8, water saturated + frozen	1021	1993	2.035×10^6	1.53	0.754×10^{-6}
Dry peat[2], P 0.8, θ 0.1			$0.9 \pm 0.5 \times 10^6$	0.14 ± 0.08	$0.16 \pm 0.01 \times 10^{-6}$
Frozen dry peat[2], P 0.8, θ 0.1			$0.7 \pm 0.3 \times 10^6$	0.46 ± 0.25	$0.68 \pm 0.19 \times 10^{-6}$
Wet peat[2] P 0.8, θ 0.7			$3.4 \pm 0.5 \times 10^6$	0.6 ± 0.17	$0.19 \pm 0.04 \times 10^{-6}$
Frozen wet peat[2], P 0.8, θ 0.7			$1.8 \pm 0.3 \times 10^6$	0.95 ± 0.23	$0.54 \pm 0.09 \times 10^{-6}$
Saturated peat[2], P 0.8, θ 0.8			$3.8 \pm 0.2 \times 10^6$	0.72 ± 0.08	$0.19 \pm 0.02 \times 10^{-6}$
Frozen saturated peat[2], P 0.8, θ 0.8			$2.0 \pm 0.05 \times 10^6$	1.92 ± 0.19	$0.96 \pm 0.09 \times 10^{-6}$
Air-dry mosses[3]				0.06	

(continued)

Table 2.2 (continued)

Material	Density kg m^{-3}	Mass heat capacity J kg^{-1} K^{-1}	Volumetric heat capacity J m^{-3} K^{-1}	Thermal conductivity W m^{-1} K^{-1}	Thermal diffusivity m^2 s^{-1}
Water-saturated mosses[3]				0.57	
snow[2,4]	190 ± 10		0.4 ± 0.04 × 10^6	0.05–0.22	0.54 ± 0.04 × 10^{-6}
ice[2]	920		1.9 × 10^6	2.3	1.2 × 10^{-6}

After Williams and Smith, 1989 and Balland and Arp 2005 with field data from Langer et al. 2011a, b

P: pore fraction of soil material; θ: water saturation fraction of pore fraction. 1: calculated from Eq. 2.30, 2.31 and 2.32, assuming $\theta = 1$ for saturated soil and $\theta = 0$ for dry soil. The volumetric heat capacity is obtained from the mass heat capacity by multiplying with the density. 2: field data from Langer et al. 2011a, b; estimated by calculating diffusivity from thermal gradient and heat capacity from soil constituents. 3: Data from Soudzilovskaya et al. (2013) on moss layers; 4: Saito et al. 2012

is made of the thermal conductivity K_{solid} of the soil (mineral and organic) solids, for fully water or ice saturated soils (K_{sat}) and for fully dry soil (K_{dry}), by determining the geometric mean of conductivities of the materials:

$$\begin{aligned} solids: K_{solid} &= K_{org}^{f_{org}} K_{quartz}^{f_{quartz}} K_{min}^{1-f_{org}-f_{quartz}} \\ unfrozen: K_{sat} &= K_{solid}^{1-f_{pores}} K_{water}^{f_{pores}} \\ (partly)\,frozen: K_{sat} &= K_{solid}^{1-f_{pores}} K_{ice}^{f_{pores}-f_{water}} K_{water}^{f_{water}} \\ dry\,soil: K_{dry} &= \frac{\left(aK_{solid} - K_{air}\right)\rho_b + K_{air}\rho_p}{\rho_p - (1-a)\rho_b} \end{aligned} \tag{2.27}$$

where *org* stands for organic matter, *quartz* for quartz and *min* for other minerals, e.g. clay. The *f*'s are the volume fractions of these materials within the total volume of the solids and the volume fraction of soil pores. The constant *a* has a value of 0.053, and ρ_b is the bulk density of the soil in g cm^{-3}, ρ_p the solid particle density (organic matter and minerals). Next, the soil conductivity is obtained from

$$K_{soil} = \left(K_{sat} - K_{dry}\right)K_e + K_{dry} \tag{2.28}$$

in which K_e depends on the water saturation fraction θ of the pores:

$$\begin{aligned} unfrozen: K_e &= \theta^{0.5\left(1+f_{org}-\alpha f_{sand}-f_{cf}\right)}\left[\left(\frac{1}{1+e^{-\beta\theta}}\right)^3 - \left(\frac{1-\theta}{2}\right)^3\right] \\ frozen: K_e &= \theta^{1+f_{org}} \end{aligned} \tag{2.29}$$

In the equation for unfrozen soil, α and β are adjustable parameters. Balland and Arp (2005) estimated these parameters by model fitting: $\alpha \approx 0.24 \pm 0.04$ and $\beta \approx 18.3 \pm 1.1$; f_{cf} is the volumetric fraction of coarse fragments, e.g. gravel. This model can also be used for assessing the thermal conductivity of other icy porous materials like a snowpack. Chadburn et al. (2015) show the importance of including organic matter and water content in soil thermal property estimation.

The water and ice content of the soils or sediments exert a large influence on the thermal properties and result in large changes when the soil thaws or freezes. Freezing causes a reduction of heat capacity but an increase in thermal conductivity. Peat soils, which usually have high porosities up to 90%, show an almost sixfold increase in diffusivity when it changes from dry peat to water-saturated, completely frozen peat.

Besides the thermal property changes during the freezing process, freezing and thawing results in the release or consumption of latent heat from the ice. This latent heat effect contributes strongly to the heat balance at the surface of permafrost soils by causing slow heating in spring and summer, and slow cooling in winter (Langer et al. 2011a, b). During freezing and thawing, this release of latent heat does not occur all at once at the crossing of the 0 °C, since not all soil water freezes at once, but over a certain temperature range. The presence of dissolved substances in the soil water, and electrically charged mineral surfaces (mostly clay minerals) depresses the freezing point, in particular in fine-grained soils. Clay soils still may contain some 10% of unfrozen water at temperatures below −5 °C (Williams and Smith 1991). The relation between unfrozen water content and temperature is known as the freezing characteristic or freezing curve, and varies strongly among soils, mainly with grain size (Fig. 2.15).

Based on Lunardini (1981), the change of stored heat within a temperature range T_1 to T_2 during freezing is the sum of the storage attributed to soil heat capacity and the storage of latent heat:

$$\Delta H_{soil} = \int_{T_1}^{T_2} \rho_w L_f d\theta_u \tag{2.30}$$

in which ρ_w is the density of water, L_f the latent heat of fusion of ice, and $d\theta_u$ the change in unfrozen water. The latent heat effect can be seen as an apparent heat capacity of the soil (Williams and Smith 1991). The release of latent heat warms the soil, and needs to be dissipated before further temperature change can occur. The apparent heat capacity is therefore larger than the heat capacity without phase change. Likewise, during thawing, the latent heat of thawing is a sink for heat, resulting in lower apparent heat capacity compared to no phase change.

According to Fig. 2.15, the unfrozen water content θ_u varies with temperature T during the freezing process, so does the apparent heat capacity, which is the sum of the normal volumetric heat capacity of the partly frozen soil $C_v(T)$ at temperature T, and the latent heat consumed by freezing of water:

$$C_{apparent} = C_v(T) + \rho_w L_f \left(d\theta_u / dT \right)_T \tag{2.31}$$

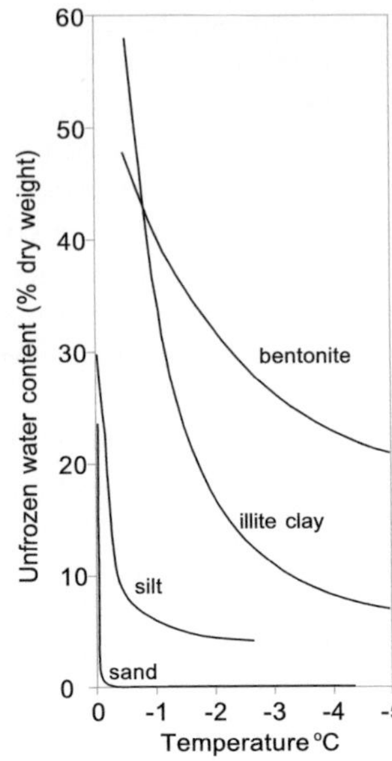

Fig. 2.15 Unfrozen water content at temperatures below 0 °C (freezing curve) in soils with various texture and mineralogy. Redrawn from Williams and Smith (1991), The Frozen Earth, Fig. 1.4; republished with permission of Cambridge University Press, permission conveyed through Copyright Clearance Center, Inc.

This apparent heat capacity is dependent on the slope of the freezing characteristic ($d\theta_u/dT$, Fig. 2.15). This slope is steepest just below zero degrees, resulting in a large change in apparent heat capacity, which may be two orders of magnitude higher than the above-zero heat capacity. At lower temperatures, and with most of the water frozen, the slope of the freezing characteristic decreases, in particular for coarse-grained soils. Figure 2.16 shows the change of unfrozen water content, thermal conductivity, apparent heat capacity, and thermal diffusivity D, for soils with different freezing characteristic below 0 °C. Since ice has a lower heat capacity than fluid water and a higher thermal conductivity, the thermal diffusivity of a soil should be higher in completely frozen state than in unfrozen state. However, during the freezing process the high apparent heat capacity due to latent heat effects results in much lower apparent thermal diffusivity, resulting in a slowdown of heat flow (Fig. 2.16d).

Laboratory measurements of the freezing curve are cumbersome (Williams and Smith 1991; Kurylyk and Watanabe 2013). Field measurements have been made with soil moisture TDR probes (Langer et al. 2011b). Various empirical models have been proposed for the freezing curve. The unfrozen water content is related to the thermodynamical equilibrium during phase change of water in soil pores, governed by the Clausius-Clapeyron relation (Kurylyk and Watanabe 2013; Chap. 3). The measurements are complicated by volume changes and movement of water during the freezing process. When ice forms during freezing, water is drawn to the freezing front (cryosuction). On the other hand, liquid water is retained in the pore space by capillary forces in larger soil pores and sorptive forces of electrically charged particles in clayey soils.

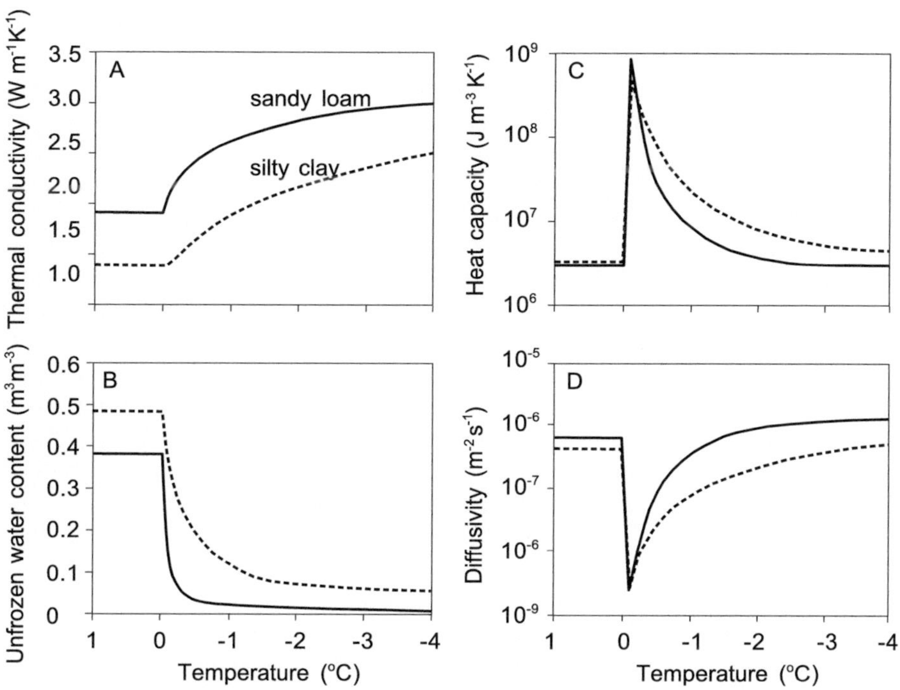

Fig. 2.16 (**a**) Change of thermal conductivity with temperature for two soils with different texture, during freezing. (**b**) Unfrozen water content (freezing characteristic) for the same soils. (**c**) Heat capacity change for these soils with apparent heat capacity below 0 °C. (**d**) Apparent thermal diffusivity for these soils. Redrawn from Williams and Smith (1991) The Frozen Earth, Figs. 4.1 and 4.2; republished with permission of Cambridge University Press, permission conveyed through Copyright Clearance Center, Inc.

Pore water retention in soils under unfrozen conditions is also driven by these capillary and sorptive forces. The water retention curve describes the amount of water retained in the soil under varying soil water potential (suction pressure). A relation of the freezing curve with water retention therefore has been employed in various models of the freezing curve. Kurylyk and Watanabe (2013) list several equations for the freezing curve; to be useful in calculating the apparent heat capacity, these equations should be differentiable. Soil water retention has often been related to specific surface area of soils, which is the total surface area of all particles in a unit mass of soil, which can be determined in soil laboratories. Williams and Smith (1991) give an empirical relation for the freezing characteristic based on Anderson and Tice (1972):

$$\ln \theta_u = 0.2618 + 0.5519 \ln S - 1.449 S^{-0.264} \ln T \tag{2.32}$$

Here θ_u is the unfrozen water content (g g^{-1}), S is the specific surface (m^2 g^{-1}), and T the temperature in degrees C below zero.

However, the assumption of one freezing characteristic is a rather simplistic approach. The behaviour of the unfrozen water content is subject to hysteresis at freezing and thawing (resulting in different curves for freezing and thawing). Moreover, soil structure changes by repeated freezing and thawing (Chap. 3), which results in variation of the freezing characteristic over time.

The freezing of soil water influences the ground heat flux for a long time in autumn and early winter (Romanovsky and Osterkamp 2000); it may take 2 months before the soil is completely frozen. The heat balance data from Langer et al. (2011a, b) show this effect clearly. The latent heat effect becomes more important with thicker active layer (Smith and Riseborough 2002). In summer the thawing of ground ice contributes to a strong heat sink in the ground. In northern Siberia, about 60% of the energy storage in the ground is attributed to the phase change of soil water; the remainder is used for soil warming down to a depth of 15 m (Langer et al. 2011a, b). When freezing of the soil starts, the temperature may stay at 0 °C for some time until most of the water is frozen (e.g. Romanovsky and Osterkamp 2000). This is known as the zero curtain, illustrated in Fig. 2.17.

The temperature profile and ground heat flux can be modelled numerically using the relations above (Williams and Smith 1991; Romanovsky and Osterkamp 2000; Mi et al. 2014). An important condition for correct modelling of the yearly

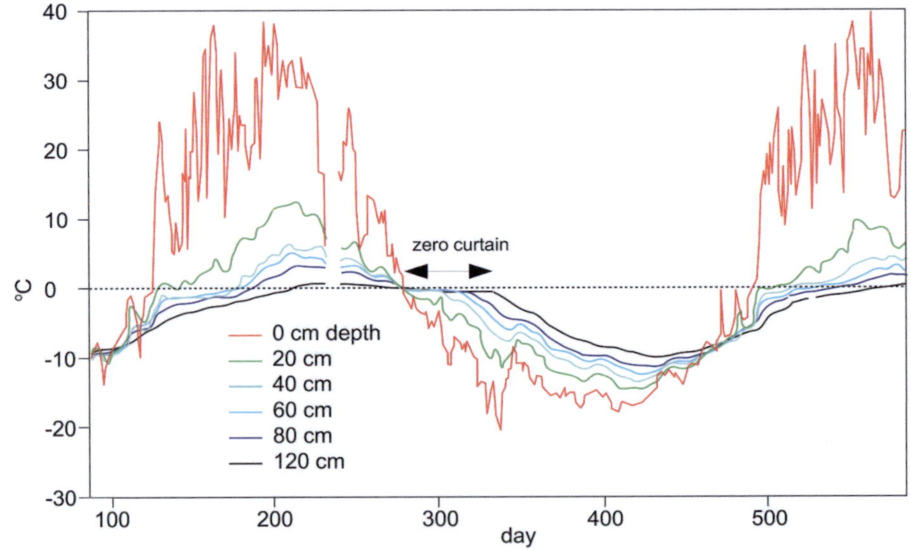

Fig. 2.17 Temperature time series of soil temperature at depths ranging from soil surface to 120 cm below surface in a soil profile in taiga forest at Spasskaya Pad research station near Yakutsk. The 'zero curtain' is shown by temperatures remaining stable at nearly 0 °C in the soil below 20 cm, where water content is higher than in the topsoil. Data provided by Prof. T.C. Maximov, Institute for Biological Problems of the Cryolithozone, Russian Academy of Sciences, Yakutsk

temperature cycle is, that the depth of the simulated soil profile should include deeper layers of permafrost down to at least the maximum depth of penetration of the yearly temperature cycle (Chadburn et al. 2015).

2.5.2 *The Effect of Ground Surface Conditions on Soil Temperature and Heat Flux*

A typical problem of moss-rich ecosystems is, where the ground surface is located exactly. In the case of peat soils with a moss cover, the distinction between the living mosses and lichens layer (bryophyte layer), and the organic soil may be quite difficult to make (Chadburn et al. 2015). In soil classifications, this is the distinction between living moss plants and dead material that has started to decompose. Although a living moss layer is part of the vegetation canopy, a dense moss layer is often taken as part of the soil for practical reasons. Correct computation of heat transfer through moss cover on the top of the soil is important, because of the insulating properties of mosses, in particular when the mosses are dry (Gornall et al. 2007; Chadburn et al. 2015; Soudzilovskaia et al. 2013). The mosses reduce the amplitude of soil temperature fluctuations during the growing season, but not the mean soil temperature (Soudzilovskaia et al. 2013). However, Porada et al. (2016) show a strong cooling effect of on average 2.7 °C of the moss layer in model experiments, and Matyshak et al. (2017) show that the presence of a thick organic horizon with an extremely low thermal diffusivity limits heat penetration to the top of the permafrost in summer.

For modelling the ground heat flux in organic soils, the thermal properties of the living moss layer can be taken as equal to those of the topsoil (Chadburn et al. 2015), but physical properties (density, composition) between the moss layer and topsoil usually differ. Experiments by Soudzilovskaia et al. (2013) show that the thermal conductivity and volumetric heat capacity of mosses were independent of the moss species and density of the mosses mat, and depended linearly on mat moisture content. They give useful regression relations for estimating the thermal properties of the bryophyte layer. Measured values of thermal conductivity are lower than those for peat, demonstrating the insulating properties of the moss layer (Table 2.2). Porada et al. (2016) modelled the thermal properties of the bryophyte layer with an approach similar to that of Balland and Arp (2005).

The temporary presence of ponded water or a snowpack also needs to be taken into account. In heat balance studies, the pond water surface and snow surface are taken as the reference surface for the heat balance (e.g. Langer et al. 2011a, b). Note that this is not the same as the ground surface as a soil scientist would define it. Standing water and snow have widely varying thermal properties, that influence the heat transfer from the air to the ground surface. Ponded water has a high heat capacity and a low albedo, transferring additional warmth to the soil surface in summer and having a considerable latent heat effect in autumn. The snowpack is a thermal insulator (Table 2.2), although the thermal properties of the snowpack vary with time.

Topsoil temperatures can be as much as 7 °C warmer than air temperatures below the snowpack, despite the cooling effect of a high albedo of the snow surface (e.g. Ekici et al. 2015). Experiments show that a thick snow cover has a profound influence on permafrost, increasing the ALT because the soil remains warmer in winter below the snow (e.g. Stieglitz et al. 2003; Hinkel and Hurd Jr 2006; Johansson et al. 2013). Variation in snow cover thickness can influence permafrost temperature as much as surface air temperature changes; this even holds for temperatures at depth in the permafrost (Stieglitz et al. 2003).

The snow cover is characterized by its thickness, snow water equivalent (SWE, the volume of water released on thawing), density, and microstructure. Freshly fallen snow has a very low thermal conductivity and a high albedo. Redistribution of snow by wind influences local snowpack thickness (Vionnet et al. 2012). Over time, various processes result in a denser snow cover and slightly lower albedo. The snow compacts under its own weight. Snowflakes are subject to metamorphic processes, changing the shapes of the flakes (Vionnet et al. 2012; Pinzer et al. 2012). These processes are caused by sublimation and refreezing of ice within the snowpack, driven by temperature gradients through the snowpack. Ice crystals sublimate, after which the water vapour from this sublimation refreezes on neighbouring grains (Yosida 1955). Up to 60% of the ice mass per day may be turned over in this way (Pinzer et al. 2012). A layer of depth hoar (loose, sugar-like grains) below the surface is formed by this metamorphosis. Further melting and refreezing of ice is driven by the surface heat balance (radiation, sensible and latent heat exchange or sublimation and surface hoar formation) and rain-on-snow events (Woo 2012). Water may percolate downward through the snowpack and may refreeze at depth. In particular extreme weather events like incursion of warm air masses and rain storms during winter contribute to changes of the snowpack (Bokhorst et al. 2016).

These processes result in variations of the thickness, density, albedo and thermal properties of the snowpack over time. Typical values for density of the snowpack are 191 ± 41 to 156 ± 52 kg m^3 and thermal conductivity 0.05–0.22 W m^{-1} K^{-1} (Saito et al. 2012; Lehning et al. 2002b). These values vary with depth in the snowpack. For correct modelling of soil temperatures and the ground heat flux, the snowpack and the bryophyte cover are crucial. Chadburn et al. (2015) have shown that calculated vertical temperature profiles significantly improve when the surface cover is included. In surface heat balance models, a bias in the simulation of the snowpack thickness translates into a bias in simulated surface temperatures (Ekici et al. 2015).

Several more or less detailed snow cover models that simulate snow cover thickness and thermal properties are available, and have been integrated in land surface and earth system models, e.g. Bartelt and Lehning (2002) and Lehning et al. (2002a, b); Vionnet et al. (2012); Saito et al. (2012); Dutra et al. (2010); Kuipers Munneke et al. (2011). The more detailed models usually are multi-layered (Dutra et al. 2012). However, both measurements and modelling of snowpack properties are still difficult, and less detailed models may give considerable deviations from observation data (Dutra et al. 2012; Pinzer et al. 2012; Saito et al. 2012).

2.5.3 Large Scale Approaches

The modelling of ground heat flux, soil temperature, and ALT, including parametrisation of snowpack development and spatially varying soil physical properties, is demanding with respect to data and parameter availability. For large scale simulations of ALT and soil temperatures, other approaches have been developed, to derive these parameters from generic meteorological and soil data. These are based on an approximation of the heat balance by thawing and freezing indices, the n-factors (Lunardini 1978, 1981). The freezing index or freezing degree-days is the sum of all daily air temperatures below 0 °C, the thawing index or thawing degree-days the sum of all daily temperatures above 0 °C.

A simplifying approach which has been widely used is the Stefan equation (Stefan 1891) for calculating ALT, in its simplest form (Kurylyk 2015):

$$\zeta = \sqrt{\frac{2kI_t}{L\omega\rho}} \tag{2.33}$$

Here ξ is the active layer depth (m), k is the thermal conductivity of the soil (W m^{-1} K^{-1}) in the thawed state, L the latent heat of freezing, ω is the ratio of the mass of pore ice over the mass of dry soil, and ρ is the dry bulk density of the soil. The Stefan equation assumes a linear temperature distribution in the thawing soil, no changes in thermal properties over time and depth, no heat transport below the thawing front, and the ground heat flux is used entirely used for thawing (Kurylyk 2015; Langer et al. 2011a). Several attempts have been made to adapt the Stefan equation for two- or multilayered soils; see Kurylyk (2015) for an assessment. Because of the simplifying assumptions, Stefan-type equations may result in systematic errors of the calculated ALT vs. observations. For Alaskan soils, systematic errors of up to 71% have been found (Romanovsky and Osterkamp 1997).

The TTOP (Temperature at Top Of Permafrost) model uses n-factors to estimate the mean annual temperature at the top of the permafrost, including temperature differences between the mean annual air temperature (MAAT), the mean annual ground surface temperature (MAGST) and the TTOP (Smith and Riseborough 1996, 2002):

$$\text{TTOP} = \text{MAAT} + \text{Surface Offset} + \text{Thermal Offset} \tag{2.34}$$

The Surface Offset is the temperature difference between MAGST and MAAT, resulting from the presence of vegetation and snow cover. The Thermal Offset is the temperature difference between the TTOP and MAGST. MAGST is usually several degrees warmer than the air temperature due to snow cover, while MAGST is in turn warmer than the TTOP due to the temperature gradient in the soil. TTOP is calculated from n-factors and thermal conductivities of the active layer:

$$TTOP = \frac{\frac{k_t}{k_f} n_t I_t - n_f I_f}{P} \tag{2.35}$$

Here k_t is the thermal conductivity of the active layer in a thawed state, k_f the same in frozen state, n_t is a scaling factor between summer air and ground surface temperatures which stands for the vegetation effect, n_f a scaling factor for winter air and ground surface temperatures which describes the snow cover effect. I_t and I_f are the thawing and freezing index. The n_f scaling depends on snow cover thickness and MAAT.

The TTOP model lumps a plethora of heat exchange processes in vegetation, snow and soil in a few empirically derived scaling factors. The latent heat effect of soil water freezing is not included in the model which limits its usefulness (Romanovsky and Osterkamp 2000). However, simplicity may be advantageous for large scale modelling, in cases where more complicated models suffer from high uncertainty in their parametrization. It has been applied to analyse environmental conditions that determine the permafrost limits in Canada (Smith and Riseborough 2002) and to assess discrepancies between paleoclimate models and paleopermafrost distribution in Europe (Van Huissteden et al. 2003).

Anisimov and Nelson (1997), Romanovsky and Osterkamp (1997), Sazonova and Romanovsky (2003) used an model based on Kudryavtsev et al. (1977), which is more detailed than the approaches above, and resulted in less error in computed ALT. This approach assumes simple sinusoidal functions for the seasonal temperature, determined by the MAAT and seasonal temperature range. The model needs information on average winter snow cover thickness, snow density and thermal conductivity, and thermal properties of the soil in frozen and thawed state.

2.6 Deeper Permafrost Temperature Profile and Lateral Heat Fluxes

Below the active layer, the seasonal fluctuations of the air temperature still penetrate, although being gradually dampened, down to the depth at which the the seasonal fluctuations are absent. This depth is known as the damping depth or depth of zero annual amplitude. At this depth, the temperature approximately equals the mean annual ground surface temperature. An analytical solution of the heat transport Eq. (2.25) illustrates this. If we assume a sinusoidal temperature fluctuation at the top of the soil and homogeneous thermal properties, the amplitude of the annual temperature fluctuation decreases exponentially with depth. The exponent is determined by thermal diffusivity D and the period length of the surface temperature fluctuation p:

$$A_z = A_0 \exp\left(\frac{z}{\sqrt{pD / \pi}}\right) \tag{2.36}$$

A_0 is here the amplitude of the temperature at the top of the soil, A_z the amplitude at depth z (negative downward from the surface), D the thermal diffusivity, and p the period of the temperature cycle at the top of the soil. If p is short (e.g. the daily cycle and effects of weather patterns), the damping depth is generally less less than a metre. The yearly cycle with a larger p penetrates much deeper, and may still be measurable at depths in the order of 10 to 15 m (Figs. 2.14 and 2.17). A higher thermal diffusivity results in a deeper penetration of the seasonal fluctuations. In continental climates with a large annual temperature amplitude, the seasonal fluctuations also penetrate deeper than in a maritime Arctic climate with a lower amplitude.

Under steady climate conditions at the surface, and homogenous thermal diffusivity with depth, the temperature below the damping depth should increase linearly with depth, following the *geothermal gradient*, the temperature gradient in the Earth's crust due to heat flow from the interior of the Earth (Eq. 1.1). This temperature gradient is approximately 1 °C per 30–60 m depth, but varies spatially with with geologic and tectonic make-up the crust. Therefore the temperature T_z at depth z ideally should follow the geothermal gradient (Fig. 2.14).

$$T_z = T_s + G.z \tag{2.37}$$

T_s is the long-term (climatic) average of the ground surface temperature, G is the geothermal temperature gradient. The geothermal gradient also offsets the temperature at zero annual amplitude slightly, with respect to the mean annual temperature at the surface. From the geothermal gradient an estimate of the total thickness of the permafrost can be derived (Eq. 1.1). This is usually not the depth down to which ice occurs, since dissolved salts in rocks and sediment depress the freezing point; furthermore, there is a slight freezing temperature decrease as a result of pressure at depth (−0.44 °C at 600 m depth (Osterkamp and Payne 1981). Within the depth range where ice may be expected, taliks frequently occur as a result of high solute concentrations of the pore water.

This vertical temperature profile is modified by climate cycles of longer duration. Multiyear periods of warming and cooling are reflected in the shape of the temperature profiles below the depth of annual variation (Fig. 2.18). Temperature data from boreholes in permafrost areas are a valuable archive for climate change in permafrost areas (Romanovsky et al. 2002; Biskaborn et al. 2015). Lachenbruch and Marshall (1986) and Osterkamp (2005) analysed deviations from the linear temperature profile in deep boreholes in Alaska, and found evidence of an overall increase in the permafrost surface temperature between 2 and 4 °C; also periods of relative cooling were reflected in the temperature profiles.

In the (unlikely) case of a stable climate after a change in surface temperature, the entire temperature profile would be shifted to left (cooling) or right (warming) over time (Fig. 2.18). This may take thousands of years, depending on the thermal diffusivity and depth of the permafrost (Lunardini 1981). Modelling of permafrost thickness in Alaska by Osterkamp and Gosink (1991) resulted in a 20,000 year lag between climate change and response of the thickness of the permafrost. Kitover et al. (2012) found a permafrost thickness change of 0.002–0.012 m yr.$^{-1}$ in model-

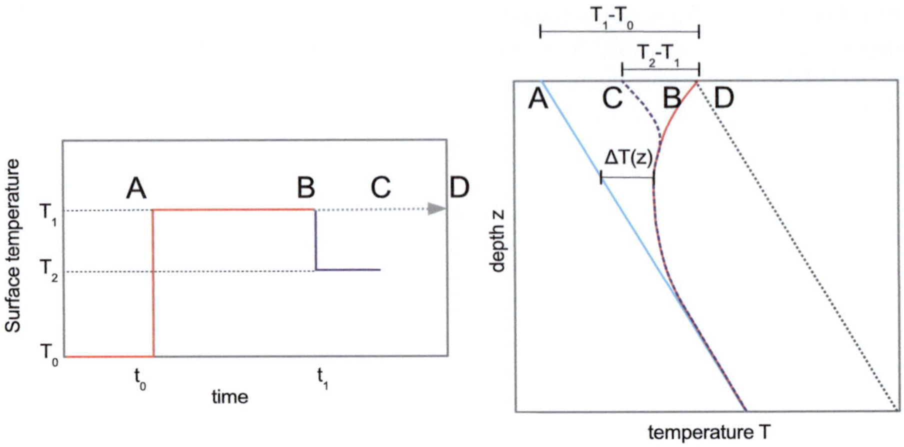

Fig. 2.18 Vertical temperature profile change by climate change. At A, the temperature profile (blue line) is at equilibrium with the surface temperature. After a step change in surface temperature T from T0 to T1, after some time the profile changes shape to the red line at B. If a subsequent climate cooling occurs (shift from T_1 to T_2), the profile would change over time in that of C (purple dotted line). In the ideal case that the profile would equilibrate with the new surface temperature T_1, again a linear profile would develop as at A, shifted to the right with respect to A (dotted grey line at D). Modified after Lachenbruch and Marshall (1986)

ling the effect of deglaciation near the limit of the North American Laurentide ice sheet. Their simulation of permafrost depth in Western and Central Europe (Kitover et al. 2013) at the Late Glacial Maximum ±20,000 years before present, resulted in a permafrost depth between 260–320 m and a thaw rate of 0.022–0.032 m $yr.^{-1}$ during deglaciation. In their model the permafrost disappeared between 11 and 9 kyr. Survival of relict ancient permafrost over millennia has been reported by Froese et al. (2008) in Alaska and by Szewczyk and Nawrocki (2011) from Poland. Therefore, during warming of the climate, permafrost may remain at depth for millennia. However, this permafrost is relict as soon as the top of the permafrost is detached by metres from the seasonally freezing topsoil, although it still will have a significant influence on deeper groundwater (Chap. 3).

2.7 Lakes and Other Water Bodies

So far, only vertical heat fluxes have been considered. However, lakes of various sizes and origin are abundant in the Arctic. Water bodies introduce lateral discontinuities in the surface heat exchange. The heat balance at the surface of water bodies differs strongly from that of neighbouring land areas: a low albedo and surface roughness in summer, more evaporation, a high heat capacity of the water and internal heat transport by convective and wind-driven currents.

Stratification of lake water masses reduces internal heat transport. During summer, lake surface waters warm by absorption of radiation, and cool by turbulent heat exchange with the atmosphere. The upper portion of lake water warms in summer more than deeper water, resulting in a layer of lighter, warm water with little exchange with deeper, colder water water. This is marked by a layer with a strong temperature gradient with depth, the thermocline (Spence et al. 2003). Only very shallow lakes may not develop stratification in summer (Oswald and Rouse 2004).

Many Arctic lakes are dimictic, which means that the lakes are stratified most of the year. Turbulent mixing of the water column is restricted to marked periods in spring and autumn. The stable stratification during summer is followed by cooling of the surface water in autumn, resulting in turbulent mixing of the lake water. Renewed stratification develops before freeze-up, when the lake surface water cools below the temperature of maximum density of water (for pure fresh water 4 °C). This causes a lighter, but now colder, water layer at the top, which allows the lake to freeze from the top (e.g. Oswald and Rouse 2004; Jammet et al. 2015).

The heat balance of lakes contains more terms than that for the land surface, since heat storage and lateral transport of heat has a more prominent role:

$$R_{net} = \Delta Q_S + Q_E + Q_H + Q_A + Q_B \tag{2.38}$$

ΔQ_S is the heat storage change in the water column, Q_A is the lateral heat transport by inflow and outflow of water, Q_B the heat transport through the bottom sediment, Q_E and Q_H the latent and sensible heat fluxes. The Q_A *term* can be neglected if the associated volumes are small with respect to the lake volume. The Q_B term is often neglected since it is usually <5% of the heat stored in the lakes (Oswald and Rouse, 2003). Nevertheless, Q_B is important for permafrost temperatures and the development of taliks below lakes (Boike et al. 2015). The storage term is calculated from the water heat capacity, water depth z and the average temperature change ΔT of the water column over time Δt:

$$\Delta Q_s = \rho_w C_w z \frac{\Delta \overline{T_w}}{\Delta t} \tag{2.39}$$

where ρ_w is the density of the water, and C_w its heat capacity (e.g. Oswald and Rouse, 2003).

Like on land, the latent and sensible heat fluxes and heat storage are driven by air-water temperature differences, the radiation balance and surface wind (Oswald and Rouse 2004). Absorption of radiation is enhanced by dissolved organic carbon (DOC) in lake water; in a lake with high DOC concentration, most of the incoming radiation was absorbed in the upper metre (Spence et al. 2003). In summer the latent heat flux is larger over lakes relative to the land surface (Eaton et al. 2001); a larger part of the incoming radiation is used for evaporation, and the Bowen ratio is lower than that of the land surface. This holds in particular for deeper lakes, while shallower lakes behave more like wetlands in this respect. Comparing the heat balance of lakes of various size and depth in northern Canada (including Great Slave Lake), Oswald and Rouse (2004) found that the largest evaporation over Great Slave Lake in Canada occurred with low incoming radiation, when cold and dry air masses moved over the lake, while for smaller lakes most evaporation occurs on days with a high net incoming radiation and warm, dry air.

Lake size and depth also determines the amount of heat stored in the water, since the deeper water column can store more energy. Stratification develops later in summer in large and deep lakes than in shallower lakes. The ice-free period and the summer evaporation is also larger for deep lakes. Storm events mix warm surface water to deeper levels (Oswald and Rouse 2004). The dimensionless Wedderburn number is a measure of the balance between wind mixing and stratification. A Wedderburn number of 1 is a threshold value at which the wind shear starts mixing water above and below the thermocline. For large Wedderburn numbers ($>>1$) the buoyancy forces promoting stratification are stronger than the wind stress, while for small values the wind stress is large enough to overcome stratification (Boike et al. 2015; Read et al. 2011).

For effects of lakes on permafrost it is crucial whether they freeze to the bottom or not (Arp et al. 2011; Boike et al. 2015). In Arctic climates, lakes without bottom-fast ice are generally those with a depth less than 2–2.5 m (e.g. Ling and Zhang 2003). In winter, the presence of cold water with temperatures up to 3–4 °C below lake ice prevents freezing of the bottom sediment. Modelling of heat fluxes through the bottom of lakes in the Lena Delta in Northern Siberia suggests that on an annual basis, the heat flux through the sediment is approximately zero. During the summer, downward heat fluxes up to 5 W m^{-2} warm the bottom sediment, while in winter

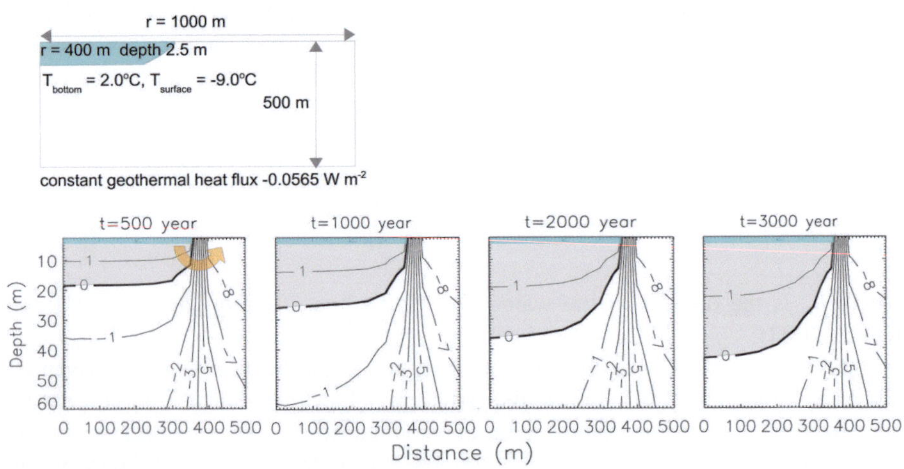

Fig. 2.19 Talik development under a lake with radius r = 400 m, in a finite element heat flow model by Ling and Zhang (2003). Above: model configuration (vertical cross section, not to scale). Below: vertical cross sections with permafrost isotherms below and next to the lake at successive time intervals after lake formation (500, 1000, 2000 and 3000 years). Grey area: talik; orange arrow indicates direction of heat flow near the lake bank. Republished with permission of John Wiley and Sons Inc., from Ling and Zhang (2003): Numerical simulation of permafrost thermal regime and talik development under shallow thaw lakes on the Alaskan Arctic Coastal Plain. Journal of Geophysical Research, doi:https://doi.org/10.1029/2002JD003014; permission conveyed through Copyright Clearance Center, Inc.

slow heat release from the sediment towards the water column occurs (Boike et al. 2015). Even considering a very small magnitude of the bottom heat flux, it is sufficient to warm the permafrost below the lakes (Boike et al. 2015). Stevens et al. (2010) measured and modelled permafrost below near-shore seafloor with water depths less than 2 m at the Mackenzie River delta. The mean winter sediment surface temperature relates exponentially to the number of days that bottom-fast ice is present, and decreases also with decreasing snow cover thickness on the ice. Given average snow and air temperature conditions, the critical water depth for the presence of permafrost was 84 cm, corresponding with presence of bottom-fast ice for 142 days.

Ling and Zhang (2003) modelled permafrost temperatures below, and adjacent to lakes in northern Alaska. When the lake bottom temperature is above 0 °C, a lake talik develops, which increases in thickness over time, at a rate which slows over time. Not only vertical heat flow occurs, but also lateral heat flow, which warms the permafrost next to the lake banks. Figure 2.19 shows modelled permafrost temperatures under a lake, in successive years after its development (Ling and Zhang 2003). The vertical and lateral heat flow is visible in the isotherms of the permafrost. At the lake bank, an approximately circular heat flow outward from the lake occurs (Williams and Smith 1991 and references therein). The development of thaw lakes in ice-rich permafrost is driven by this downward and lateral heat transfer, in combination with geomorphic processes like wave erosion, mass wasting and re-sedimentation (West and Plug 2008). Hydrology of permafrost and the formation of thaw lakes is discussed in Chap. 3.

References

AMAP (2015) AMAP Assessment 2015: Black carbon and ozone as Arctic climate forcers. Arctic Monitoring and Assessment Programme (AMAP), 116 p

AMAP (2017) Snow, water, ice and permafrost in the Arctic (SWIPA) 2017. Arctic Monitoring and Assessment Programme (AMAP), Oslo, p 269

Anderson DM, Tice AR (1972) Predicting unfrozen water contents in frozen soils from surface area measurements. Highw Res Rec 393:12–18

Anisimov O, Nelson F (1997) Influence of climate change on permafrost in the Northern Hemisphere. Russ Meteorol Hydrol 5:47–53

Arp CD, Jones BM, Urban FE, Grosse G (2011) Hydrogeomorphic processes of thermokarst lakes with grounded-ice and floating-ice regimes on the Arctic coastal plain, Alaska. Hydrol Process 25(15):2422–2438

Aubinet M, Vesala T, Papale D (2012) Eddy covariance: a practical guide to measurement and data analysis. Springer, Dordrecht, p 438

Balland V, Arp PA (2005) Modeling soil thermal conductivities over a wide range of conditions. J Environ Eng Sci 4:549–558

Bartelt P, Lehning M (2002) A physical SNOWPACK model for the Swiss avalanche warning: part I: numerical model. Cold Reg Sci Technol 35(3):123–145

Beringer J, Chapin FS, Thompson CC, McGuire AD (2005) Surface energy exchanges along a tundra-forest transition and feedbacks to climate. Agric For Meteorol 131(3):143–161

Biskaborn BK, Lanckman J-P, Lantuit H, Elger K, Dmitry S, William C, Vladimir R (2015) The new database of the global terrestrial network for permafrost (GTN-P). Earth Syst Sci Data 7:245–259

Blok D, Heijmans MM, Schaepman-Strub G, Kononov A, Maximov T, Berendse F (2010) Shrub expansion may reduce summer permafrost thaw in Siberian tundra. Glob Chang Biol 16(4):1296–1305

Blok D, Heijmans MMPD, Schaepman-Strub G, van Ruijven J, Parmentier FJW, Maximov TC, Berendse F (2011) The cooling capacity of mosses: controls on water and energy fluxes in a Siberian tundra site. Ecosystems 14(7):1055–1065. https://doi.org/10.1007/s10021-011-9463-5

Boike J, Wille C, Abnizova A (2008) Climatology and summer energy and water balance of polygonal tundra in the Lena River Delta, Siberia. J Geophys Res 113(G3). https://doi.org/10.1029/2007jg000540

Boike J, Langer M, Lantuit H, Muster S, Roth K, Sachs T, Overduin P, Westermann S, McGuire AD (2012) Permafrost–physical aspects, carbon cycling, databases and uncertainties. In: Recarbonization of the biosphere. Springer, Dordrecht, pp 159–185

Boike J, Georgi C, Kirilin G, Muster S, Abramova K, Fedorova I, Chetverova A, Grigoriev M, Bornemann N, Langer M (2015) Thermal processes of thermokarst lakes in the continuous permafrost zone of northern Siberia – observations and modeling (Lena River Delta, Siberia). Biogeosciences 12(20):5941–5965. https://doi.org/10.5194/bg-12-5941-2015

Bokhorst S, Pedersen SH, Brucker L, Anisimov O, Bjerke JW, Brown RD, Ehrich D, Essery RL, Heilig A, Ingvander S (2016) Changing Arctic snow cover: a review of recent developments and assessment of future needs for observations, modelling, and impacts. Ambio 45(5):516–537

Bonan G (2002) Ecological climatology: concepts and applications. 1st edn. Cambridge University Press

Bonan G (2016) Ecological climatology: concepts and applications. 3d edn. Cambridge University Press, Cambridge. 754 p

Bonan GB, Pollard D, Thompson SL (1992) Effects of boreal forest vegetation on global climate. Nature 359(6397):716–718

Budishchev AVHJ, Parmentier FJW, Maximov TC, Dolman AJ (in prep) Using machine learning algorithms to impute energy and carbon eddy covariance flux measurements. to be submitted to JGR Biogeosciences

Burba G (2013) Eddy covariance method for scientific, industrial, agricultural and regulatory applications: a field book on measuring ecosystem gas exchange and areal emission rates. LI-COR Biosciences, Lincoln

Burba GG, McDermitt DK, Grelle A, Anderson DJ, Xu L (2008) Addressing the influence of instrument surface heat exchange on the measurements of CO_2 flux from open-path gas analyzers. Glob Chang Biol 14(8):1854–1876

Chadburn S, Burke E, Essery R, Boike J, Langer M, Heikenfeld M, Cox P, Friedlingstein P (2015) An improved representation of physical permafrost dynamics in the JULES land-surface model. Geosci Model Dev 8(5):1493–1508

De Vries DA (1987) The theory of heat and moisture transfer in porous media revisited. Int J Heat Mass Transf 30(7):1343–1350

Dutra E, Balsamo G, Viterbo P, Miranda PM, Beljaars A, Schär C, Elder K (2010) An improved snow scheme for the ECMWF land surface model: description and offline validation. J Hydrometeorol 11(4):899–916

Dutra E, Viterbo P, Miranda PM, Balsamo G (2012) Complexity of snow schemes in a climate model and its impact on surface energy and hydrology. J Hydrometeorol 13(2):521–538

Eaton AK, Rouse WR, Lafleur PM, Marsh P, Blanken PD (2001) Surface energy balance of the western and Central Canadian subarctic: variations in the energy balance among five major terrain types. J Clim 14(17):3692–3703

Ekici A, Chadburn S, Chaudhary N, Hajdu L, Marmy A, Peng S, Boike J, Burke E, Friend A, Hauck C (2015) Site-level model intercomparison of high latitude and high altitude soil thermal dynamics in tundra and barren landscapes. Cryosphere 9:1343–1361

Elumeeva TG, Soudzilovskaia NA, During HJ, Cornelissen JH (2011) The importance of colony structure versus shoot morphology for the water balance of 22 subarctic bryophyte species. J Veg Sci 22(1):152–164

Essery R, Pomeroy J (2004) Vegetation and topographic control of wind-blown snow distributions in distributed and aggregated simulations for an Arctic tundra basin. J Hydrometeorol 5(5):735–744

Eugster W, Rouse WR, Pielke RA Sr, Mcfadden JP, Baldocchi DD, Kittel TG, Chapin FS, Liston GE, Vidale PL, Vaganov E (2000) Land–atmosphere energy exchange in Arctic tundra and boreal forest: available data and feedbacks to climate. Glob Chang Biol 6(S1):84–115

Euskirchen E, McGuire A, CHAPIN III F (2007) Energy feedbacks of northern high-latitude ecosystems to the climate system due to reduced snow cover during 20th century warming. Glob Chang Biol 13(11):2425–2438

Foken T (2008) Micrometeorology. Springer, Berlin. 306 p

Froese DG, Westgate JA, Reyes AV, Enkin RJ, Preece SJ (2008) Ancient permafrost and a future, warmer Arctic. Science 321(5896):1648–1648

Gornall J, Jónsdóttir I, Woodin S, Van der Wal R (2007) Arctic mosses govern below-ground environment and ecosystem processes. Oecologia 153(4):931–941

Grace J, Allen S, Wilson C (1989) Climate and the meristem temperatures of plant communities near the tree-line. Oecologia 79(2):198–204

Hinkel KM, Hurd JK Jr (2006) Permafrost destabilization and thermokarst following snow fence installation, Barrow, Alaska, USA. Arct Antarct Alp Res 38(4):530–539

Hoekstra P (1966) Moisture movement in soils under temperature gradients with the cold-side temperature below freezing. Water Resour Res 2(2):241–250

Iwahana G, Machimura T, Kobayashi Y, Fedorov AN, Konstantinov PY, Fukuda M (2005) Influence of forest clear-cutting on the thermal and hydrological regime of the active layer near Yakutsk, eastern Siberia. J Geophys Res Biogeo 110(G2):G02004. https://doi.org/10.1029/2005JG000039

Jammet M, Crill P, Dengel S, Friborg T (2015) Large methane emissions from a subarctic lake during spring thaw: mechanisms and landscape significance. J Geophys Res Biogeo 120(11):2289–2305. https://doi.org/10.1002/2015jg003137

Johansson M, Callaghan TV, Bosiö J, Åkerman HJ, Jackowicz-Korczynski M, Christensen TR (2013) Rapid responses of permafrost and vegetation to experimentally increased snow cover in sub-arctic Sweden. Environ Res Lett 8(3):035025. https://doi.org/10.1088/1748-9326/8/3/035025

Jorgenson MT, Shur YL, Pullman ER (2006) Abrupt increase in permafrost degradation in Arctic Alaska. Geophys Res Lett 33(2):L02503. https://doi.org/10.1029/2005gl024960

Juszak I, Erb AM, Maximov TC, Schaepman-Strub G (2014) Arctic shrub effects on NDVI, summer albedo and soil shading. Remote Sens Environ 153:79–89

Juszak I, Eugster W, Heijmans MM, Schaepman-Strub G (2016) Contrasting radiation and soil heat fluxes in Arctic shrub and wet sedge tundra. Biogeosciences 13:4049–4064

Juszak I, Iturrate-Garcia M, Gastellu-Etchegorry J-P, Schaepman ME, Maximov TC, Schaepman-Strub G (2017) Drivers of shortwave radiation fluxes in Arctic tundra across scales. Remote Sens Environ 193:86–102

Kasurinen V, Alfredsen K, Kolari P, Mammarella I, Alekseychik P, Rinne J, Vesala T, Bernier P, Boike J, Langer M, Belelli Marchesini L, van Huissteden K, Dolman H, Sachs T, Ohta T, Varlagin A, Rocha A, Arain A, Oechel W, Lund M, Grelle A, Lindroth A, Black A, Aurela M, Laurila T, Lohila A, Berninger F (2014) Latent heat exchange in the boreal and arctic biomes. Glob Chang Biol 20(11):3439–3456. https://doi.org/10.1111/gcb.12640

Kholodov A, Gilichinsky D, Ostroumov V, Sorokovikov V, Abramov A, Davydov S, Romanovsky V (2012) Regional and local variability of modern natural changes in permafrost temperature in the Yakutian coastal lowlands, Northeastern Siberia. In: Proceedings of the Tenth International Conference on Permafrost, Salekhard, Yamal-Nenets Autonomous District, Russia, 2012. pp 25–29

Kitover D, Renssen H, Van Balen R, Vandenberghe J (2012) Modeling permafrost response of the last glacial termination: first results of the VAMPER model. In: Tenth International Conference on Permafrost, 2012. pp 209–214
Kitover D, Balen R, Roche D, Vandenberghe J, Renssen H (2013) New estimates of permafrost evolution during the last 21 k years in Eurasia using numerical modelling. Permafr Periglac Process 24(4):286–303
Kormann R, Meixner FX (2001) An analytical footprint model for non-neutral stratification. Bound-Layer Meteorol 99(2):207–224
Kudryavtsev V, Garagulya L, Melamed V (1977) Fundamentals of frost forecasting in geological engineering investigations (Osnovy Merzlotnogo Prognoza pri Inzhenerno-Geologicheskikh Issledovaniyakh). Corps of Engineers, US Army, Cold Regions Research and Engineering Lab, Hanover NH USA, 489 p
Kuipers Munneke P, Van den Broeke M, Lenaerts J, Flanner M, Gardner A, Van de Berg W (2011) A new albedo parameterization for use in climate models over the Antarctic ice sheet. J Geophys Res Atmos 116(D5):D05114. https://doi.org/10.1029/2010JD015113
Kurylyk BL (2015) Discussion of 'a simple thaw-freeze algorithm for a multi-layered soil using the Stefan Equation'by Xie and Gough (2013). Permafr Periglac Process 26(2):200–206
Kurylyk BL, Watanabe K (2013) The mathematical representation of freezing and thawing processes in variably-saturated, non-deformable soils. Adv Water Resour 60:160–177
Lachenbruch AH, Marshall BV (1986) Changing climate: geothermal evidence from permafrost in the Alaskan Arctic. Science 234:689–697
Langer M, Westermann S, Muster S, Piel K, Boike J (2011a) The surface energy balance of a polygonal tundra site in northern Siberia-part 1: spring to fall. Cryosphere 5(1):151–171. https://doi.org/10.5194/tc-5-151-2011
Langer M, Westermann S, Muster S, Piel K, Boike J (2011b) The surface energy balance of a polygonal tundra site in Northern Siberia – part 2: winter. Cryosphere 5(2):509–524. https://doi.org/10.5194/tc-5-509-2011
Lehning M, Bartelt P, Brown B, Fierz C (2002a) A physical SNOWPACK model for the Swiss avalanche warning: part III: meteorological forcing, thin layer formation and evaluation. Cold Reg Sci Technol 35(3):169–184
Lehning M, Bartelt P, Brown B, Fierz C, Satyawali P (2002b) A physical SNOWPACK model for the Swiss avalanche warning: part II. Snow microstructure. Cold Reg Sci Technol 35(3):147–167
Liljedahl A, Hinzman L, Harazono Y, Zona D, Tweedie C, Hollister RD, Engstrom R, Oechel W (2011) Nonlinear controls on evapotranspiration in arctic coastal wetlands. Biogeosciences 8(11):3375
Ling F, Zhang T (2003) Numerical simulation of permafrost thermal regime and talik development under shallow thaw lakes on the Alaskan Arctic Coastal Plain. J Geophys Res 108(D16):4511. https://doi.org/10.1029/2002JD003014
Loranty MM, Goetz SJ, Beck PS (2011) Tundra vegetation effects on pan-Arctic albedo. Environ Res Lett 6(2):024014. https://doi.org/10.1088/1748-9326/6/2/024014
Lunardini V (1978) Theory of n-factors and correlation of data. In: Proceedings of the Third International Conference on Permafrost, 1978. National Research Council of Canada Ottawa, pp 40–46
Lunardini VJ (1981) Heat transfer in cold climates. Van Nostrand Reinhold Company, 731 p
Lund M, Hansen BU, Pedersen SH, Stiegler C, Tamstorf MP (2014) Characteristics of summertime energy exchange in a high Arctic tundra heath 2000–2010. Tellus B: Chem Phys Meteorol 66(1):21631. https://doi.org/10.3402/tellusb.v66.21631
Matsumoto K, Ohta T, Nakai T, Kuwada T, Daikoku K, Iida S, Yabuki H, Kononov AV, van der Molen MK, Kodama Y (2008) Energy consumption and evapotranspiration at several boreal and temperate forests in the Far East. Agric For Meteorol 148(12):1978–1989
Matyshak G, Goncharova OY, Moskalenko N, Walker D, Epstein H, Shur Y (2017) Contrasting soil thermal regimes in the Forest-tundra transition near Nadym, West Siberia, Russia. Permafr Periglac Process 28(1):108–118

McFadden JP, Chapin FS III, Hollinger DY (1998) Subgrid-scale variability in the surface energy balance of arctic tundra. J Geophys Res Atmos 103(D22):28947–28961

McFadden JP, Eugster W, Chapin FS (2003) A regional study of the controls on water vapor and CO_2 exchange in arctic tundra. Ecology 84(10):2762–2776

Ménégoz M, Krinner G, Balkanski Y, Cozic A, Boucher O, Ciais P (2013) Boreal and temperate snow cover variations induced by black carbon emissions in the middle of the 21st century. Cryosphere 7(2):537–554

Mi Y, van Huissteden J, Parmentier FJW, Gallagher A, Budishchev A, Berridge CT, Dolman AJ (2014) Improving a plot-scale methane emission model and its performance at a north-eastern Siberian tundra site. Biogeosciences 11(14):3985–3999. https://doi.org/10.5194/bg-11-3985-2014

Myers-Smith IH, Forbes BC, Wilmking M, Hallinger M, Lantz T, Blok D, Tape KD, Macias-Fauria M, Sass-Klaassen U, Lévesque E (2011) Shrub expansion in tundra ecosystems: dynamics, impacts and research priorities. Environ Res Lett 6(4):045509

Nauta AL, Heijmans MMPD, Blok D, Limpens J, Elberling B, Gallagher A, Li B, Petrov RE, Maximov TC, van Huissteden J, Berendse F (2014) Permafrost collapse after shrub removal shifts tundra ecosystem to a methane source. Nat Clim Chang 5(1):67–70. https://doi.org/10.1038/nclimate2446

Ohta T, Maximov TC, Dolman AJ, Nakai T, van der Molen MK, Kononov AV, Maximov AP, Hiyama T, Iijima Y, Moors EJ (2008) Interannual variation of water balance and summer evapotranspiration in an eastern Siberian larch forest over a 7-year period (1998–2006). Agric For Meteorol 148(12):1941–1953

Osterkamp TE (2005) The recent warming of permafrost in Alaska. Glob Planet Chang 49(3):187–202

Osterkamp T, Gosink JP (1991) Variations in permafrost thickness in response to changes in Paleoclimate. J Geophys Res 96(B3):4423–4434

Osterkamp TE, Payne MW (1981) Estimates of permafrost thickness from well logs in northern Alaska. Cold Reg Sci Technol 5(1):13–27

Oswald CJ, Rouse WR (2004) Thermal characteristics and energy balance of various-size Canadian Shield lakes in the Mackenzie River Basin. J Hydrometeorol 5:129–144

Pinzer B, Schneebeli M, Kaempfer T (2012) Vapor flux and recrystallization during dry snow metamorphism under a steady temperature gradient as observed by time-lapse micro-tomography. Cryosphere 6(5):1141–1155. https://doi.org/10.5194/tc-6-1141-2012

Porada P, Ekici A, Beer C (2016) Effects of bryophyte and lichen cover on permafrost soil temperature at large scale. Cryosphere 10(5):2291–2315. https://doi.org/10.5194/tc-10-2291-2016

Read JS, Hamilton DP, Jones ID, Muraoka K, Winslow LA, Kroiss R, Wu CH, Gaiser E (2011) Derivation of lake mixing and stratification indices from high-resolution lake buoy data. Environ Model Softw 26(11):1325–1336

Romanovsky V, Osterkamp T (1997) Thawing of the active layer on the coastal plain of the Alaskan Arctic. Permafr Periglac Process 8(1):1–22

Romanovsky V, Osterkamp T (2000) Effects of unfrozen water on heat and mass transport processes in the active layer and permafrost. Permafr Periglac Process 11(3):219–239

Romanovsky V, Burgess M, Smith S, Yoshikawa K, Brown J (2002) Permafrost temperature records: indicators of climate change. EOS Trans Am Geophys Union 83(50):589–594

Saito K, Yamaguchi S, Iwata H, Harazono Y, Kosugi K, Lehning M, Shulski M (2012) Climatic physical snowpack properties for large-scale modeling examined by observations and a physical model. Pol Sci 6(1):79–95

Sazonova T, Romanovsky V (2003) A model for regional-scale estimation of temporal and spatial variability of active layer thickness and mean annual ground temperatures. Permafr Periglac Process 14(2):125–139

Serreze MC, Barry RG (2014) The Arctic climate system. Cambridge University Press, Cambridge. 385 p

Smith M, Riseborough D (1996) Permafrost monitoring and detection of climate change. Permafr Periglac Process 7(4):301–309
Smith M, Riseborough D (2002) Climate and the limits of permafrost: a zonal analysis. Permafr Periglac Process 13(1):1–15
Soudzilovskaia NA, Bodegom PM, Cornelissen JH (2013) Dominant bryophyte control over high-latitude soil temperature fluctuations predicted by heat transfer traits, field moisture regime and laws of thermal insulation. Funct Ecol 27(6):1442–1454
Spence C, Rouse WR, Worth D, Oswald CJ (2003) Energy budget processes of a small northern lake. J Hydrometeorol 4:694–701
Stefan J (1891) Über die Theorie der Eisbildung, insbesondere über die Eisbildung im Polarmeere. Ann Phys 278(2):269–286
Stevens CW, Moorman BJ, Solomon SM (2010) Modeling ground thermal conditions and the limit of permafrost within the nearshore zone of the Mackenzie Delta, Canada. J Geophys Res Earth 115:F4
Stiegler C, Johansson M, Christensen TR, Mastepanov M, Lindroth A (2016) Tundra permafrost thaw causes significant shifts in energy partitioning. Tellus B: Chem Phys Meteorol 68(1):30467
Stieglitz M, Déry S, Romanovsky V, Osterkamp T (2003) The role of snow cover in the warming of arctic permafrost. Geophys Res Lett 30(13):1721. https://doi.org/10.1029/2003GL017337
Sturm M, Holmgren J, McFadden JP, Liston GE, Chapin FS III, Racine CH (2001) Snow–shrub interactions in Arctic tundra: a hypothesis with climatic implications. J Clim 14(3):336–344
Sturm M, Douglas T, Racine C, Liston GE (2005) Changing snow and shrub conditions affect albedo with global implications. J Geophys Res Biogeo 110(G1):G01004. https://doi.org/10.1029/2005JG000013
Szewczyk J, Nawrocki J (2011) Deep-seated relict permafrost in northeastern Poland. Boreas 40(3):385–388
Tanaka H, Hiyama T, Kobayashi N, Yabuki H, Ishii Y, Desyatkin RV, Maximov TC, Ohta T (2008) Energy balance and its closure over a young larch forest in eastern Siberia. Agric For Meteorol 148(12):1954–1967
Thompson C, Beringer J, Chapin F III, McGuire A (2004) Structural complexity and land-surface energy exchange along a gradient from arctic tundra to boreal forest. J Veg Sci 15(3):397–406
Van der Molen M, Van Huissteden J, Parmentier F, Petrescu A, Dolman A, Maximov T, Kononov A, Karsanaev S, Suzdalov D (2007) The growing season greenhouse gas balance of a continental tundra site in the Indigirka lowlands, NE Siberia. Biogeosciences 4(6):985–1003. https://doi.org/10.5194/bg-4-985-2007
Van Huissteden J, Vandenberghe J, Pollard D (2003) Palaeotemperature reconstructions of the European permafrost zone during marine oxygen isotope stage 3 compared with climate model results. J Quat Sci 18(5):453–464. https://doi.org/10.1002/jqs.766
Vionnet V, Brun E, Morin S, Boone A, Faroux S, Le Moigne P, Martin E, Willemet J (2012) The detailed snowpack scheme Crocus and its implementation in SURFEX v7. 2. Geosci Model Dev 5:773–791
West J, Plug LJ (2008) Time-dependent morphology of thaw lakes and taliks in deep and shallow ground ice. J Geophys Res Earth 113(F1):F01009. https://doi.org/10.1029/2006JF000696
Westermann S, Lüers J, Langer M, Piel K, Boike J (2009) The annual surface energy budget of a high-arctic permafrost site on Svalbard, Norway. Cryosphere 3(2):245–263
Williams PJ, Smith MW (1991) The frozen earth. Fundamentals of geocryology. Cambridge University Press, Cambridge. 306 p
Wilson J (2015) Computing the flux footprint. Boundary Layer Meteorol 156:1–14. https://doi.org/10.1007/s10546-015-0017-9
WMO (2008) Guide to Meteorological Instruments and Methods of Observation (CIMO Guide). vol WMO-No. 8
Woo M-K (2012) Permafrost hydrology. Springer, Berlin. 563 p
Yosida Z (1955) Physical studies on deposited snow. I.; thermal properties. Contrib Inst Low Temp Sci 7:19–74

Chapter 3
The Role of Ground Ice

Abstract The ice content of permafrost soils has a large impact on processes that occur when permafrost thaws; exposed ground ice is the primary cause of destructive abrupt thaw processes. This chapter presents the origin of common forms of ground ice: segregation ice, ice wedges and larger ground ice bodies. The presence of ground ice is expressed at the surface as geomorphological features such as ice wedge polygon networks, palsas and pingos. A processes that is important for transfer of carbon in permafrost soils is cryoturbation; it is also linked to micro-relief at the soil surface in the shape of various forms of patterned ground. The processes and landforms that result from thawing ground ice are discussed: erosional forms such as cryogenic landslides. Next, permafrost has a profound effect on the hydrology; a frozen subsurface acts as a barrier to groundwater movement. This affects also run-off and river discharge, which are further determined by water storage as snow in winter, and its sudden release in spring during snowmelt. Permafrost thaw results in hydrological changes: the development of thaw lakes and increased groundwater flow. The chapter concludes with a short presentation of geophysical methods to detect and map ground ice and permafrost in the subsurface.

It is the presence of ice in permafrost soils, which makes their carbon cycle differ from that of other soils. Before embarking on the permafrost carbon cycle, an overview of permafrost ice distribution and soil processes that occur on freezing and thawing is necessary.

When permafrost thaws, the distribution of ice in the soil (ground ice) exerts a strong influence on the development of landforms, erosion processes and hydrology. The presence of large amounts of ground ice enhances the susceptibility of soils to erosion and subsidence. These processes disrupt the ecosystem and transfer large amounts of soil organic matter towards other environments, causing changes in decomposition processes.

Ice also profoundly affects the soil environment. Ice affects the hydrological processes, heat transfer and temperature in the soil, in turn influencing biochemical processes rates. Freezing drastically slows decomposition processes, conserving soil carbon, and waterlogging of soils resulting from the presence of ice in soil pores

J. van Huissteden, *Thawing Permafrost*,
https://doi.org/10.1007/978-3-030-31379-1_3

affects the biochemistry of decomposition (Chap. 4). Furthermore, freeze and thaw processes cause displacement of organic material within the soil.

A normal mineral soil has a porosity and water holding capacity in the order of 20–40% by volume. Mineral soils in permafrost regions may contain much higher volume percentages of ice. That is not only caused by the expansion of water on freezing, but also by processes that concentrate ice, or add ice to the soil. The most common processes are ice segregation and crack development, which concentrate ice in lenses and veins. In peat soils the natural porosity is higher than that of mineral soils (80% is not uncommon; e.g. Weiss et al. 2016), but also in peat soils the ice volume may exceed that of the unfrozen porosity. Other processes that contribute to accumulation of ground ice are the creation of ice wedges, intrusive ice formed by injection of water from the subsurface, and burial of lake, river and glacier ice (although buried ice is often not included in ground ice, French 2018; see below). These processes create ice bodies with a low content of mineral or organic matter, below the active layer.

Permafrost with a high ice content is common in sedimentary lowlands or other areas with a thick sedimentary cover or fine-grained rocks or in peatlands (Fig. 3.1). About 8–9% of the permafrost surface area is estimated to have a high ice content (Zhang et al. 2008). The total average volumetric ice content of exposed sediments along the North Alaskan coast amounts 77% (Kanevskiy et al. 2013).

The spatial and depth distribution and type of ground ice is determined by geological and geomorphological processes that shaped the landscape (e.g. French 2018). These determine the type and duration of ice accumulation processes. In turn, the ground ice distribution controls the geomorphological process rates that occur during thaw of permafrost (Van Huissteden and Dolman 2012). These range from abrupt, catastrophic pulse disturbances, often driven by erosion, to more gradual press disturbances (Grosse et al. 2011). These process rates also affect the carbon cycle by destabilization of the environment in which soil carbon is preserved.

Frost heave is a phenomenon that results from the accumulation of ground ice (e.g. segregation ice, injection ice). Unequal spatial distribution of ice in the soil creates various forms of micro-relief that may influence the surface hydrology by impeding drainage. This micro-relief may be amplified by vegetation differences. Larger relief forms are created by injection of ice, for instance ice cored mounds like pingos. Furthermore, upon freezing and thawing of the active layer, pressure differences are created in the soil, resulting in soil movements that deform soil horizons and assist in the burial of carbon in the soil (Kaiser et al. 2007).

3.1 Basic Soil Ice Characteristics

High ice content soils have excess ice: there is more ice in the soil than can be explained by its normal, unfrozen, pore volume plus the expansion of ice when water freezes (Sect. 1.1). The visible presence of ice lenses does not always indicate that the soil has excess ice; in between the visible ice, soil material may be under-

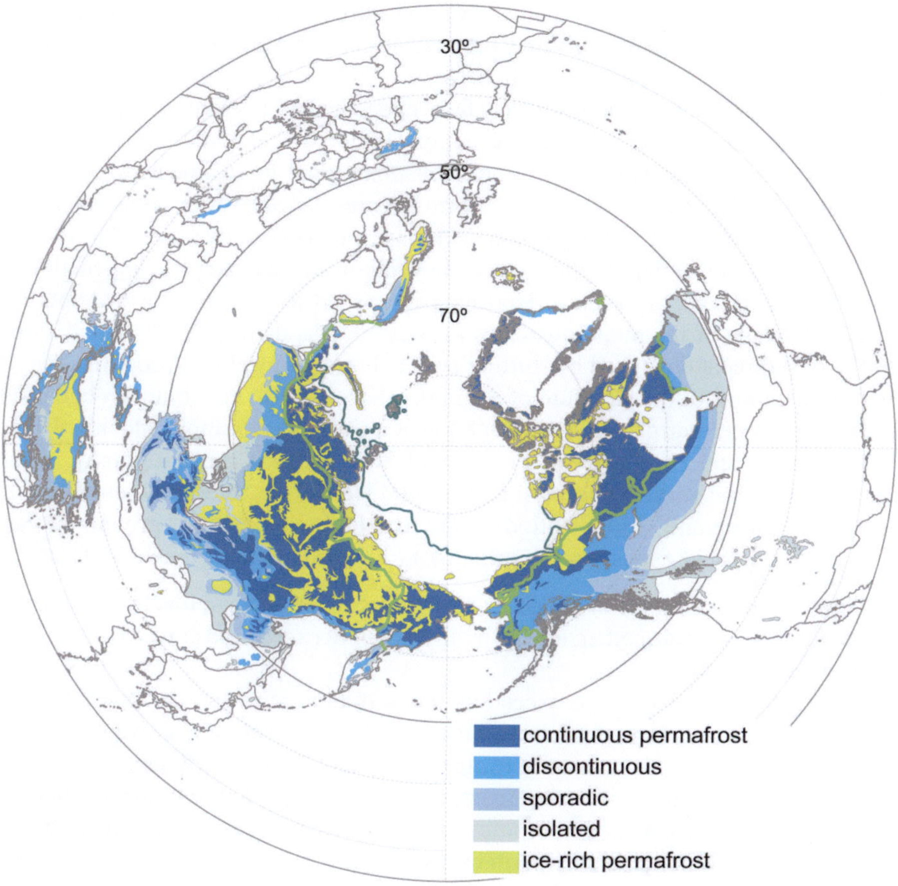

Fig. 3.1 Distribution of ice-rich permafrost. Based on Brown et al. (2002); GIS data from National Snow and Ice Data Center, accessed July 2019, http://nsidc.org/data/ggd318)

saturated with ice, with soil pores not completely filled with ice. The ice content of permafrost may be expressed volumetrically or by weight percentage. The ice content in weight percentage of soils is the weight of ice over dry soil (French 2018) and therefore varies considerably with the nature of the dry matter of the soil - low density peat, or high density mineral soil. A volume percent definition has the advantage that it tells more about the loss of volume of the soil that can be expected when the ice thaws. For instance, mineral and organic floodplain soils in Northeast Siberia have a volumetric ice content between 40% and 96% with an average of 75% (Iwahana et al. 2014); assuming a normal unfrozen porosity of 40% indicates an average volume loss of the soil upon thaw of 35% for mineral soils.

On the circum-arctic permafrost map (Brown et al. 2002; Zhang et al. 2008) a volumetric definition is used, giving the amount of excess ice in the upper 10 to 20 m of the permafrost. High ice content permafrost has more than 20 volume % of excess ice. Expressed as weight percentage, low ice content soils have an ice content up to 50%, high ice content soils may have an ice content well above 100%, depending on the type of soil solids (French 2018).

The ice in sedimentary deposits may be syngenetic ice - freezing occurred *during* sedimentation at times that the sedimentation surface was exposed to subzero temperatures - or epigenetic ice – ice formed *sometimes after* the sediments were deposited (French 2018). Syngenetic ice may form in terrestrial depositional environments with frequent, ongoing sedimentation such as river floodplains and eolian (wind) deposits; the Yedoma deposits introduced in Chap. 1 are an example. Epigenetic ice formation occurs when aqueous deposits are exposed to freezing, for example by natural drainage of the numerous lakes in Arctic lowlands (e.g. Mackay and Burn 2002), or sediment deposited by shifting river courses, or by climate cooling.

Classifications of the structure of ice in permafrost are useful to unravel genesis and history of permafrost and sedimentation at a particular location (Mackay 1971, 1972; Murton and French 1994; French and Shur 2010). Cryostratigraphy is the description of the vertical succession of ground ice, and its interpretation terms of freezing, thawing and sedimentation events. The following types of ground ice can be discerned in field exposures of permafrost (description after French 2018):

- Pore ice: interstitial ice in the pores between sediment grains; it acts as 'cement' holding the grains together, and generally does not exceed the pore volume of the thawed sediment.
- Segregation ice*:* excess ice exceeding the normal pore volume of thawed soils, visible as lenses of ice of a few millimetres thick, up to more massive ice bodies a few metres thick. It is usually formed as a result of thermodynamic and pressure gradients in the soil during freezing, drawing water to the freezing front.
- Vein ice*:* ice formed in open fissures formed by cracking of the soil, from snow, meltwater or rain. The cracks form by thermal contraction of the soil at low temperatures. It varies from single veins a few millimetres thick, to thick ice wedges formed by repeated cracking over many years. Typically it has a vertical laminated appearance.
- Intrusive ice*:* massive ice bodies formed by water intrusion under pressure, for instance by ground water pressure or by pore water expulsion during the freezing of taliks. The ice bodies may form sills analogous to volcanic intrusions, usually with sharp contacts with the surrounding rock or sediment. Another typical examples of intrusive ice bodies are pingos, isolated hills formed by localized groundwater upwelling, creating a lens-shaped body of layered intrusive ice.
- Buried ice*:* pool, lake and river ice, usually linked to to geomorphological processes like mass wasting, erosion and sedimentation, and buried glacier ice.
- Cave ice formed by percolation of water into cave systems or into tunnels formed by severe gully erosion.

Just below the active layer, ice formation processes are most active in the permafrost, in an ice-rich transient layer at the top of the permafrost (French 2018; Fig. 3.2). In summer, at temperatures just below 0 °C, this transient layer may contain considerable amounts of unfrozen water (e.g. Walvoord and Kurylyk 2016).

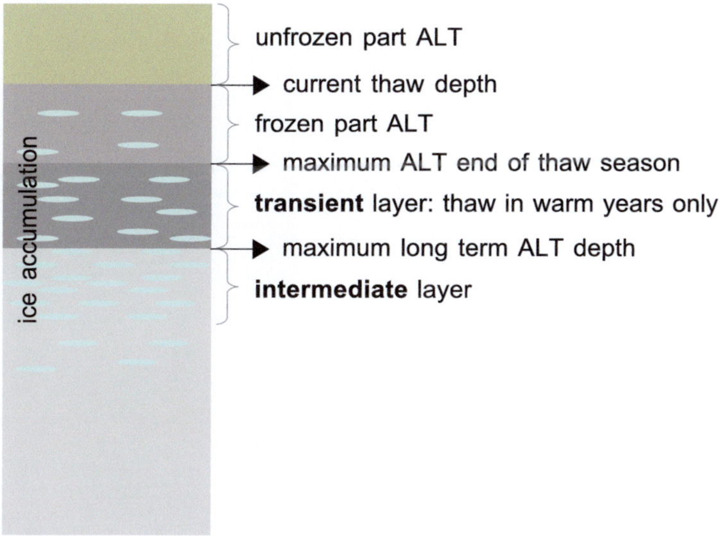

Fig. 3.2 Transition of active layer to permafrost with transient and intermediate layers. (After Kanevskiy et al. 2017)

Under influence of the negative temperature gradient in summer, water moves downward towards the top of the permafrost, causing an increase of the ice content and frost heave in late summer (Mackay 1983; French 2018). It is also the zone where the position of the top of the permafrost fluctuates due to year-to-year variability in thaw depth. In warmer years the active layer thaws deeper than than in colder years, and accumulated ice thaws. If drainage of the extra meltwater is slow, the increasing water pressure in the active layer may trigger mass movements on slopes (French 2018; Sect. 3.6). Below the transient layer, an intermediate layer is sometimes distinguished (Fig. 3.2), which is also generally ice-rich. It results from upward aggradation of the permafrost, by sedimentation or accumulation of organic matter and vegetation on top of the soil, which gradually includes the former transient layer into the top of the permafrost (French and Shur 2010; Kanevskiy et al. 2017).

3.2 Ice Segregation and Frost Heave

Permafrost soils are often subject to frost heave: vertical movements of the soil when freezing or frozen. Frost heave is responsible for the formation of landforms like palsas, and the formation of various patterned ground features, e.g. earth hummocks. In particular fine-grained soils (silts and clay) are sensitive to frost heave (Mackay 1972). Cores taken from such soils show the presence of ice lenses or layers (Fig. 3.3). These ice lenses represent segregation ice and contribute to the excess

Fig. 3.3 Soil core with ice lenses and layer of nearly pure ice. (Indigirka lowlands, Northeast Siberia, photo by Dr. C. Berrittella)

ice in permafrost soils. Segregation ice often makes up the bulk of the soil ice content (e.g. Kanevskiy et al. 2013). Frost heave results in engineering problems to buildings and infrastructure.

3.2.1 Ice Segregation Process

The expansion of water when it freezes is not the main cause of frost heave. Instead, it is the growth of segregation ice, which causes substantially more expansion (Williams and Smith 1991). Ice lenses are formed by water drawn to ice-water interface in the soil by the process of cryosuction. The force drawing at the water derives from a thermodynamic difference between soil water pressure P_w and the pressure of the ice, P_i. This pressure difference is created by forces acting in the soil on a microscopic and molecular scale. In unfrozen soil, the potential energy of water is known as the soil water potential. This potential energy is created by capillary forces drawing water into soil pores, by adhesive forces between water molecules and soil particles, and by osmotic forces resulting from the presence of solutes (Hillel 1998). The force drawing water to the ice-water interface adds to this hydraulic potential of soil water.

To move water to the ice-water interface in soils at temperatures below zero, it should be liquid. The presence of solutes depresses the freezing point of the soil water with respect to pure bulk water. Furthermore, the freezing point is depressed by the Gibbs-Thomson effect, which describes the decrease of freezing temperature of water in conditions of high water surface curvature and surface energy such as occur in small soil pores (Dash et al. 2006; Rempel 2007). Additionally, some liquid water remains in equilibrium with ice at subzero temperatures by a phenomenon known as pre-melting: the presence of thin films of water at the ice - mineral interface at temperatures near the freezing point (Dash et al. 2006; Rempel 2007). The presence of liquid water at subzero temperatures results in water to be drawn down a thermal gradient from higher to lower temperatures.

In a system containing both water and ice at pressures P_w and P_i, the pressure difference P_i - P_w relates to the temperature T according to the Clapeyron equation at thermodynamic equilibrium:

$$P_i - P_w = \frac{\rho_w L_f}{T_m}(T_m - T) \tag{3.1}$$

in which T_m is the freezing temperature of free bulk water, ρ_w the density of water, L_f the latent heat of freezing at T_m and atmospheric pressure, P_w the pressure of water at the ice-water interface (Williams and Smith 1991; Peppin and Style 2013; a simplified version of the equation is given here). If the temperature T is lowered while P_i remains constant, P_w at the ice-water interface will decrease. This increases the water potential and will result in water flow towards the ice-water interface: cryosuction.

Various models have been advanced for the resulting movement of water to form ice lenses (Williams and Smith 1991; Rempel 2007; Peppin and Style 2013). Many questions about the fundamental process driving frost heave still remain, despite much experimental and theoretical work. It is difficult to quantify temperature, water flow and ice or water content at the small scale inside freezing soil material. An overview is given by Peppin and Style (2013).

In most models, ice lenses form somewhat behind the freezing front (where soil temperature passes 0 °C), not exactly at the location where the first ice forms in the soil pores. The forces drawing water to the ice are not restricted to soil freezing only. When soil thaws from the top, the temperature gradient is reversed, and ice lens growth may occur by drawing water from the active layer towards the still frozen soil. Frozen soils are part of a continuous hydrologic system in which flows of water and changes in water content depend on the direction of the temperature gradient. Hydraulic potentials in freezing soils are large; a temperature gradient of only one degree per metre can result in a gradient of the hydraulic potential of 1.2 MPa m^{-1} (Williams and Smith 1991).

For the formation of segregation ice in ice lenses the following conditions are necessary. First, water movement through the freezing or frozen soil should be possible; despite the presence of ice, the soil should still be permeable. Second, water should be available for ice lens formation. Third, the forces exerted by a growing ice lens should be larger than the forces holding the soil together - the overburden pressure and the cohesion of the soil (Mackay 1972). Fourth, the ice lenses should be preserved - which occurs preferentially in the transient and intermediate layers at the top of the permafrost.

Permeability (the distance that water moves per time unit through a medium, measured in m s^{-1}) of unfrozen soil is related to soil texture, and decreases with temperature below 0 °C by the progressive formation of ice in soil pores. However, water movement in frozen soil is still possible, even at quite low temperatures, albeit that the permeability is very low, comparable to that of a dense clay layer. Water films on soil particles have been detected down to temperatures of −10 °C (Romanovsky and Osterkamp 2000). The water transport is facilitated by a high pressure gradient despite the low permeability. At the freezing front, a frozen fringe (Fig. 3.4) is assumed to exist where the soil is partially frozen (Rempel 2007). Within this frozen frozen fringe, permeability decreases rapidly, causing the moving water to 'pile up', with as result the formation of an ice lens at some distance behind the freezing front. With downward freezing, a series of ice lenses will form, provided sufficient water can be drawn from the surrounding soil (Williams and Smith 1991; Rempel 2007). The amount of water drawn to the freezing front and stored in the ice lenses are the cause of the frost heave.

Ice lens development stops when there is insufficient water available. In this case, the soil near the freezing front may dry out. Drying of fine-grained, clayey

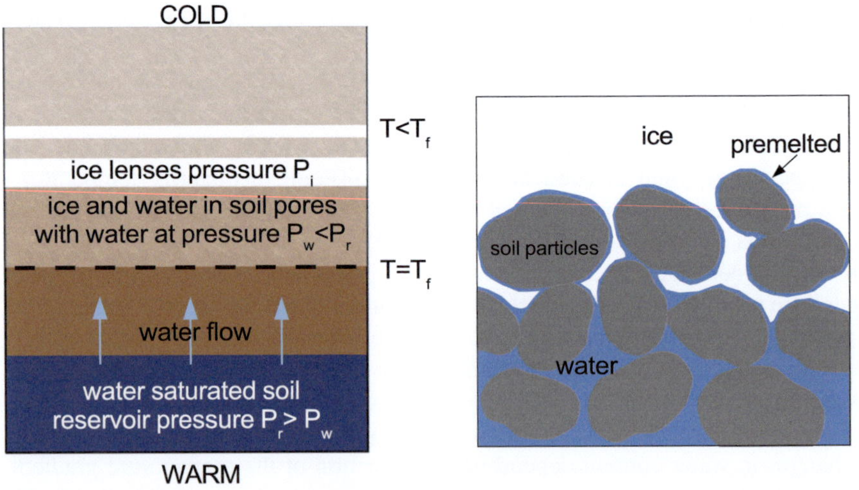

Fig. 3.4 Left: Schematic soil column freezing from the top, with ice lens development. At the base soil pores are saturated with groundwater with reservoir pressure P_r. Below the ice lenses is a zone with ice formation in the soil pores but water flow is still possible. The temperature of the soil T is lower than the bulk freezing temperature of water T_f due to freezing point depression. The water has a lower pressure P_w than that of the groundwater reservoir, inducing water flow to the partly frozen zone and the ice lens. Right: schematic representation of the 'frozen fringe' with premelting at the ice-soil particle contact. (After Peppin and Style 2013)

soils by withdrawal of water by freezing has an effect on the soil structure: it will cause consolidation (compression of the soil by decreasing pore volume and re-arrangement of the clay particles) and lateral shrinkage. In the case of high overburden pressure, ice lenses will not form and all water will freeze in the soil pores (Rempel 2007).

3.2.2 Environmental Conditions for Ice Segregation

The environmental conditions under which extensive ice lens formation and frost heave occur are determined by soil texture (grain size) and water availability; water saturated, fine-grained material is most suitable (Romanovsky et al. 2008). Likely environments are water saturated active layers in fine-grained sedimentary deposits such as loesses and alluvial sediments. However, ice lens formation also occurs in rocks (Walder and Hallet 1985; Hallet et al. 1991; Murton et al. 2006). In particular fissile rocks such as shale, weathered chalk, and other fine-grained porous rocks are susceptible (Murton and Lautridou 2003). Besides the expansion of water upon freezing, ice lens development in rocks drives most of the physical weathering of rocks in the Arctic.

For segregation ice to form in peat, it should contain some mineral matter; pure peat does not show frost heave (Williams and Smith 1991). Ice segregation and frost heave is only partly responsible for the formation of the typical landforms of palsas in peatlands (Sect. 3.4). Nevertheless, the ice volume in frost mounds in peat is generally over 50% (Calmels and Allard 2008); a striking amount of gas is present in the ice, with a volume up to 5%.

The top of the permafrost is commonly rich in ice (Pollard and French 1980; Williams and Smith 1991). This is not only caused by upward movement of water during soil freezing in winter, but by water addition in summer. In summer, the thermal gradient in the soil reverses, inducing a reversed gradient of the P_i - P_w pressure difference, and a resulting movement of water to the top of the permafrost. At the same time, additional water will be supplied by precipitation and snowmelt water (Cheng 1983). The repetition of active layer freeze-thaw cycles results in a gradual accumulation of *aggradational ice* in the top of the permafrost (Williams and Smith 1991).

Ice lens formation is the the most common expression of ice segregation, but other structures occur also (French and Shur 2010, Fig. 3.5). In fine-grained, clayey sediments that shrink on drying, shrinkage cracks may develop when water is drawn to ice lenses. These cracks become filled with ice in turn, leading to a *reticulate* network structure. In heterogenous sediments with gravel or stones, crusts of ice develop on the course fragments. With accumulation of higher amounts of ice such as aggradational ice at the base of the active layer, or when ice segregation is combined with water intrusion, layers of ice may develop. Frequently, an *ataxitic* structure develops (Fig. 3.5a), in which angular fragments of frozen sediment are suspended in ice; the origins of this structure are not well known. Ice segregation in

Fig. 3.5 Cryostructure of segregation ice in cores taken at the Alaskan Beaufort Sea Coast. (**a**) ataxitic (suspended) structure with angular and sub-angular mineral blocks; (**b**) reticulate structure with sub-angular mineral blocks; (**c**) layered, lenticular and braided (interconnected layers); (**d**) crustal structure (combined with suspended) with ice crusts on coarse fragments; (**e**) porous (pore filling) structure; (**f**) suspended organic-matrix with some coarser ice fragments (porphyritic structure). Site locations in Kanevskiy et al. (2013). (Reprinted from Ground ice in the upper permafrost of the Beaufort Sea coast of Alaska Cold Regions Science and Technology, 85:56–70, by Kanevskiy et al. (2013), © 2013, with permission from Elsevier)

combination with additional water supply may lead to the growth of massive ice bodies which are several metres thick (Mackay 1972).

Ice lens formation and repeated freezing and thawing affects soil structure. The structure of pore space and soil aggregates changes (Konishchev and Rogov 1993). Cryogenic processes result in distinctive soil morphology at the macroscopic and microscopic scale (Van Vliet-Lanoë 1985; Van Vliet-Lanoe et al. 2004). Common features are stable platy soil aggregates formed by the pressure exerted by ice lens growth.

Lastly, ice accumulation influences the thermal properties of the soil significantly (Sect. 2.5). In soils with a high ice content, a high amount of the ground heat flux is converted into latent heat of thawing (Boike et al. 2012; Sect. 2.5). An ice-rich transient layer is therefore a buffer against deep thawing of the soil in extremely warm years (Shur et al. 2005; Kanevskiy et al. 2017). It is also an important aspect in modelling of future permafrost thaw (Chap. 8).

3.3 Cracking and Wedging

A polygonal pattern of ice wedges is ubiquitously present in Arctic permafrost areas, in particular in areas underlain by unconsolidated sediments with continuous permafrost (Fig. 1.21). This pattern results from the cracking of soil in winter. After formation of a crack, it is filled in with drifting snow or meltwater, which adds a vein of ice to the soil. Since an existing crack is a weakness zone (the cohesion of ice is less than that of frozen mineral soil), repeated cracking over many years is likely (Lachenbruch 1962). Each time the crack opens, a vein of ice will be added (Fig. 3.6). The resulting vertical wedges of ice contribute to the ice volume of permafrost (e.g. Schirrmeister et al. 2011). Fig. 1.21e shows a fresh crack at the surface.

Infilling of cracks not only includes ice and snow. In dry environments with a thin winter snow cover and abundant eolian surface transport of sand, also sand is included in the infilling of cracks, resulting in a wedge of vertically laminated sand (French 2018 and references therein).

The ice wedge micro-relief is of great significance for the surface hydrology of permafrost terrain, as will be discussed below. Basically two types of ice wedge polygons exist (French 2018): Low-centred polygons, where the polygon consists

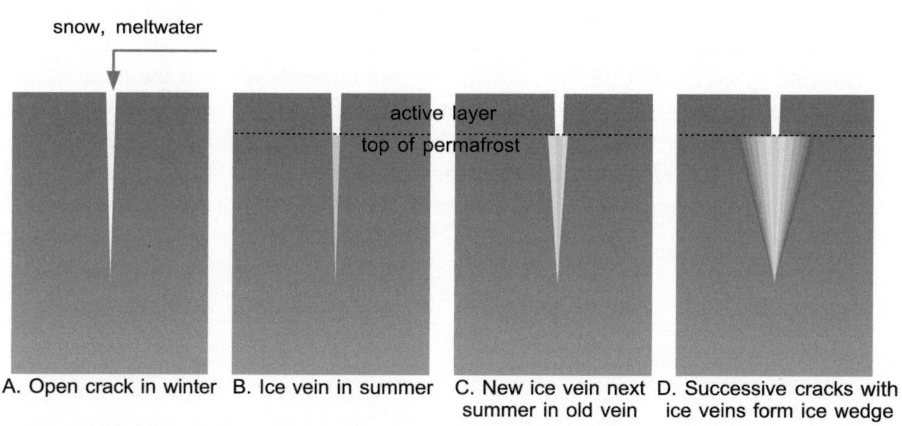

Fig. 3.6 Formation of ice wedges by repeated soil cracking in winter (vertical cross section). (**a**) open crack in winter, which is infilled with drifting snow in winter or snow meltwater in spring. (**b**) summer situation; the infilling of the crack below the permafrost table is frozen. (**c**) next year; a new crack has formed in winter within the old crack; (**d**) successive cracking and infilling at the same location forms an ice wedge, consisting of a large number of ice veins. Grey scale in the wedge indicates the approximate age of the ice (in reality, age relations in the wedge ice may be more complex)

of a central depression, surrounded by rims along the ice wedges, and high-centred polygons, where the polygon centre has a higher elevation than that of the ice wedges. Low-centred polygons tend to have an impeding effect on drainage, while high-centred polygons facilitate drainage (Woo 2012; Liljedahl et al. 2016).

3.3.1 Processes of Ice Wedge Formation

Crack development is the result of the horizontal component of tensile stress (pulling force within the soil) that develops in the soil as a result of thermal contraction of the soil upon cooling. Thermal contraction is a strain (deformation) imposed on the soil (Lachenbruch 1962; Fig. 3.7). The linear coefficient of thermal expansion/contraction of ice at −30 °C is 50×10^{-6} °C^{-1}. However, Mackay and Burn (2002) reported values from field observations in frozen soils that are 6–8 times higher.

The magnitude of the thermal contraction strain, and the rate at which it develops, is important for crack formation. Depending on the strain rate (contraction changing over time), this deformation may be accomplished in three ways. Over short time and with low tensile stress, deformation consists of elastic (reversible) deformation. With higher tensile stress and longer duration, slow creep deformation follows, which deforms the soil irreversibly. If the tensile stress builds up rapidly by a high strain rate, it cannot be accommodated by creep, and the soil will rupture. The resistance to rupture depends on the cohesion forces within the soil; in frozen soil these cohesion forces are higher than in unfrozen soil. As long as the strain can be accommodated by creep, no rupture will occur (Fig. 3.7). As mentioned above, cracking is likely to occur frequently in the same place. Plant roots crossing the crack will be broken and may attest of recent activity (Kokelj et al. 2007a).

The rate of creep deformation before rupture depends also on ice content and temperature. At near-zero temperature, creep deformation proceeds more readily because the presence of larger amounts unfrozen water, making deformation by melting of ice and refreezing of water possible (see above; Williams and Smith 1991). At lower temperatures, creep is slower. Rapid cooling of the soil then results

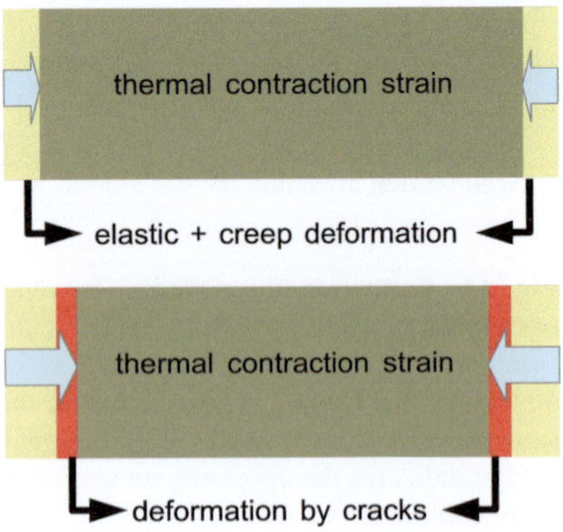

Fig. 3.7 Frozen soil deformation by thermal contraction. Above: the thermal contraction strain can be accommodated by slow elastic and creep deformation. Below: the strain rate is too high to be accommodated by creep only, and cracks develop. Deformation and strain not to scale

in a higher strain rate, while at the same time creep deformation becomes slower. The deformation cannot resolve fully by creep, and rupture will occur. Therefore, the incidence of crack formation is higher at low temperatures and rapid temperature drops in the soil. Cold periods with rapidly dropping air and soil temperatures are favourable for thermal contraction cracking (Lachenbruch 1962). A thick snow cover affects cracking by damping the cooling rate of the soil.

Crack formation occurs at the top of the permafrost and propagates upward and downward; this is concluded from the observation that cracks in ice wedges can propagate through overlying active layer and pond ice overlying the wedge (Mackay 1989). Early reports of ice wedge cracking testify of explosive sounds at crack formation (De Leffingwell 1915).

The incidence of ice wedge cracking is highly variable between years and over short distances; a thicker snow cover in one year may inhibit cracking (Mackay 1992a). At a test area in northern Canada, crack frequency at one site varied between 8 and 42% in 12 years, and for a nearby site from 22 to 75% in 20 years. At a drained lake site in Canada, vegetation growth and resulting snow entrapment caused crack formation to stop (Mackay and Burn 2002).

Cracking generally occurs at air temperatures between −25 °C and − 40 °C and soil temperatures between −15 °C to −25 °C (French 2018). Mackay (1993) reports air temperature drops of 1.8 °C day^{-1} over 4 days as favourable. However, based on extensive field research, Mackay (Mackay 1992a, b) found a poor correlation between cracking frequency and low air temperatures. Nevertheless, ice wedge cracking is broadly related to climate. French (2018) summarizes research on crack frequency in relation to climate in various Arctic environments. This indicate a MAAT of 6 °C and colder, air temperatures ≤ −25 °C, air temperature cooling rates >0.1 °C $hr.^{-1}$, ground surface temperatures ≤ −15 °C and soil thermal gradients ≤ −10 °C m^{-1}. More than a decennium of observations on ice wedge cracking in Svalbard indicates that ice wedge cracking at this location needs ground surface cooling below −20 °C and a thermal gradient steeper than −10 °C m^{-1} (Matsuoka et al. 2018).

These data clearly show that low winter temperatures with rapid temperature drops, and a relatively thin snow cover are the best conditions for ice wedge formation. A warmer climate or thicker snow cover may halt ice wedge activity and growth; ice wedges become relict. A method to test for ice wedge activity is to analyse tritium (^3H) levels in the wedge ice; tritium in the atmosphere has been produced by atomic bomb testing during the 1950's and 1960's. Kokelj et al. (2007a) noted the absence of ice wedge activity, from wedge ice ^3H levels and root growth across ice wedges, under Spruce forest in the MacKenzie Delta. In a nearby tundra peatland with the same winter air temperatures frequent cracking occurred. The difference between the two areas were a deeper snow cover and higher temperatures at the top of the permafrost (−1.8 °C to −2.9 °C vs. −6 °C) in the forest.

Next to climate, also the substrate has a significant influence. Ice-wedge cracking in clay substrates may occur with MAAT as high as −2 °C, while in gravel MAAT needs to be as low as −6 °C (Romanovskii 1985). Burn (1990) reported evidence of recent ice wedge activity in fine-grained sediments from elevated ^3H concentrations in wedge ice in an area in Northern Canada with a MAAT of −4 °C.

The infilling of cracks by ice results in gradual sideways expansion of the ice wedge, inducing strain on the wedge itself and the surrounding soil. This results in an upward creep movement of the soil next to the wedge, and shearing of the wedge ice. The upward soil creep is shown by upturning of soil layers next to the wedge. Mackay and Burn (2002) found slickensides and stepwise upward displacement of frozen mud next to young ice wedges (slickensides are slide planes marked by a smoothly polished surface, with striations in the direction of movement). Moreover, since the relatively pure ice in the wedge is lighter than the surrounding soil, there is an upward buoyancy force with a tendency to lift the wedge up (Mackay 1980b, 1990; Kokelj and Burn 2004). In summer, the soil expands again, resulting in a sideways strain towards the wedge. In combination the resulting forces and soil movements cause the formation of a conspicuous rim over the ice wedge, and on either side of the wedge (Fig. 3.8A).

The development of micro-relief and associated vegetation differences in turn feeds back on the soil heat balance. Winter soil cooling in polygons in Svalbard occurred mainly through the rims developed along the ice wedges, which were mostly snow-free, while the ice wedge troughs were partly filled with snow (Christiansen 2005). The scale of ice wedge polygons is in the order of 10–30 m across. Individual cracks may extend over much longer distances, with lengths of tens to hundreds of meters (Plug and Werner 2002). The intersections between fractures tend to be orthogonal (Lachenbruch 1962). However, soils are heterogeneous and therefore the cooling rate, generation of tensile stress and cohesion vary in space, influencing the direction and rate of crack propagation.

Cracks do not develop all at once, and every new crack is influenced by the presence of other cracks. This causes a self-organizing pattern of polygonal cracks, with varying angles of intersection (Plug and Werner 2001). Initial cracks are long, with irregular bends due to soil heterogeneity. Based on a model of ice wedge polygon development, Plug and Werner (2002) conclude that wedge spacing and width in ice wedge networks reflect infrequent episodes of rapidly falling ground temperatures, rather than mean climatic conditions. The pattern of ice wedge polygons adapts to patterns of subsoil heterogeneity, as is observed for instance in ice wedge polygons on river floodplains, where they align to patterns created by sedimentation caused by channel migration (Fig. 3.9).

The development of ridges along ice wedges results in impeded drainage in the polygon centre, where often a pool can be found, while the ridges are dryer (Fig. 1.21). Generally, the active layer is somewhat thicker in the polygon centre than on the ridges. The vegetation adapts to the soil moisture conditions, with dry plant communities on the ridges, and sedge/moss-dominated wetland communities in the polygon centre (Minke et al. 2009).

Thaw of the wedge ice in summer may result in the development of a trough over the ice wedge accompanied by ridges on either side (Fig. 3.8b, 1.21d). Kanevskiy et al. (2017) explain the vulnerability of wedge ice to thaw by thinning of the soil layer on top of the wedges. This is the result of the upward buoyancy force and

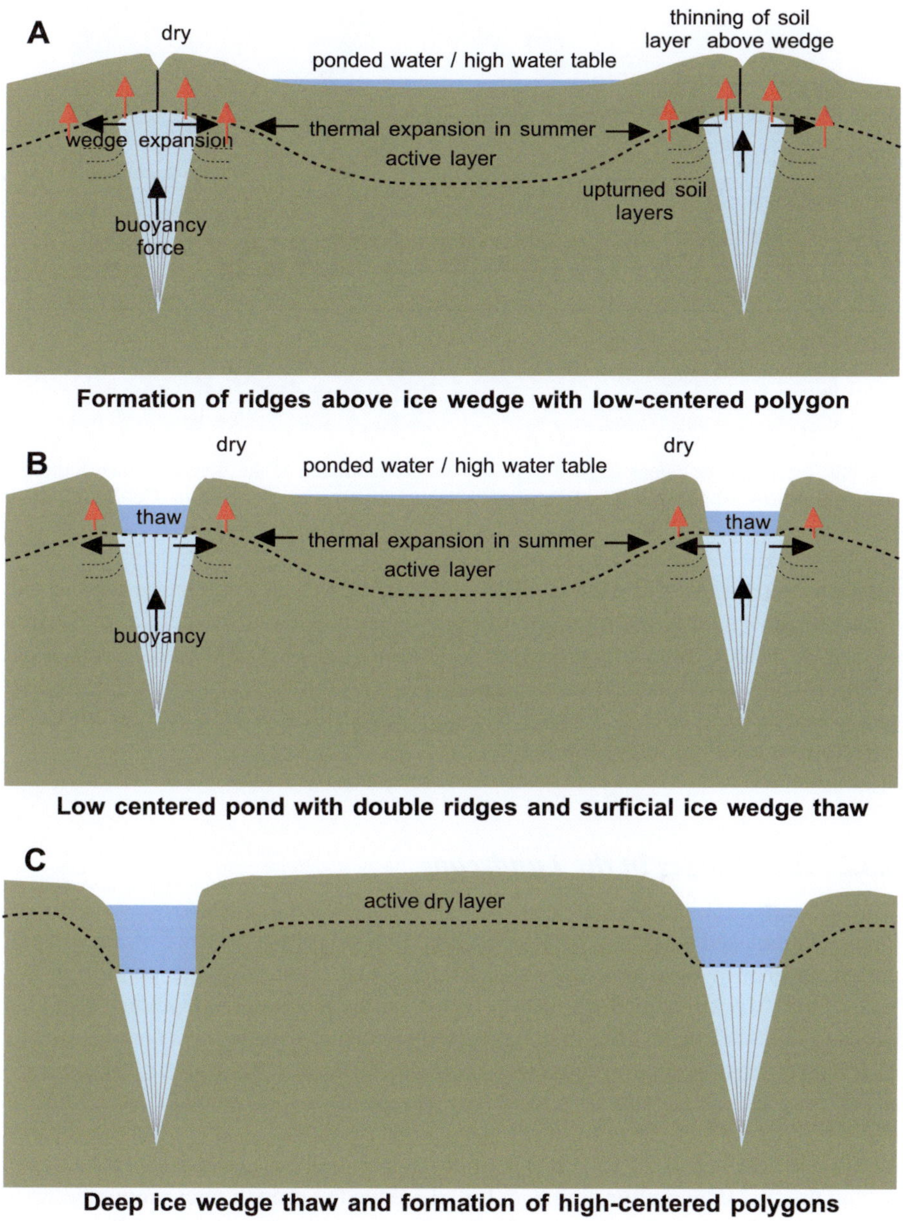

Fig. 3.8 (**a**) Strain and deformation of the soil causing the formation of a ridge along ice wedges, and resulting micro-relief. Black arrows: strain direction, red arrows: soil movement. (**b**) similar, but with simultaneous summer thaw of ice wedge (compare Fig. 1.20). (**c**) progressive thaw of ice wedges creates a high-centred polygon

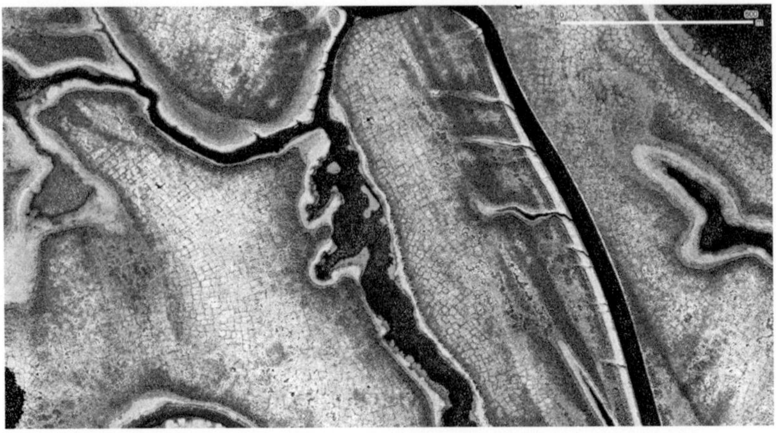

Fig. 3.9 Ice wedge polygons aligned to point bars on floodplain of the Berelegh river, Indigirka lowlands, Northeast Siberia. (GeoEye image, August 2010; scale bar 600 m. GeoEye © 2019 MAXAR)

sideways expansion of the ice wedges, as sketched in Fig. 3.8. On the other hand, also soil movement in the direction of the wedge may occur by thermal expansion of the permafrost in the polygon centre in summer (Mackay 2000, Fig. 3.8b). If net thinning of the soil layer brings the wedge ice within the reach of the ALT, the top of the wedge ice will start thawing. Hence, vulnerability to thaw is inherent to the process of ice wedge formation itself.

3.3.2 *Ice Wedges in the Landscape*

On newly exposed surfaces, the development of a polygonal network of ice wedges, and their growth in width, can be extremely fast when climatic conditions are favourable. Mackay and Burn (2002) report on ice wedge development in freshly exposed lake sediments after drainage of a thaw lake in Northern Canada. Drainage of lakes by erosion occurs frequently in Arctic lowlands and drained lake bottoms are widespread (e.g. Hinkel et al. 2007). As soon as the drained lake bottom is exposed, permafrost starts to develop if climate conditions allow. At the drained lake site of Mackay and Burn (2002), an 80 m long thermal crack system developed during the first winter after drainage. Initial ice wedge growth rates amounted up to 3 cm/year; an order of magnitude higher than growth rates measured in older ice wedges. Cracks opening in winter were up to 15 cm wide, but crack width decreased rapidly over the years. Over the years, rapid vegetation growth of high forbs and willows resulted in a thicker snow depth, and a 10–20 °C higher temperature at the top of the permafrost. Ultimately, cracking ceased.

Just as with ice formation in general, an important distinction can be made in the age relations of ice wedges with their host material (Mackay 1990; Fig. 3.10).

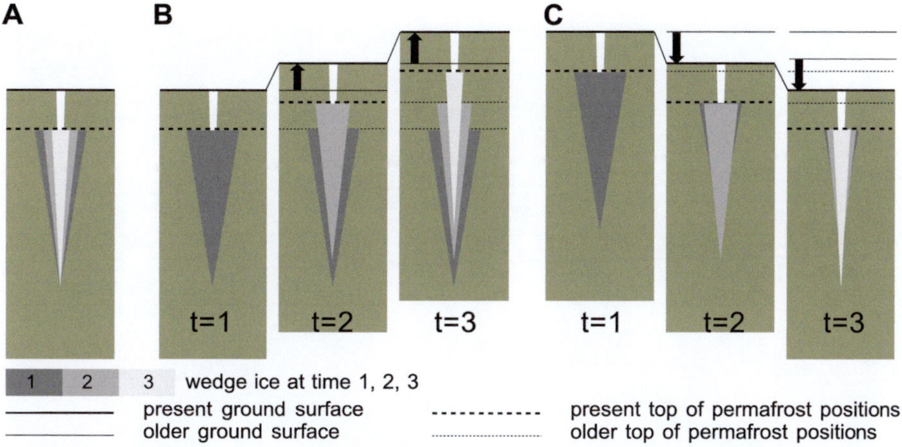

Fig. 3.10 (**a**) epigenetic ice wedge developed in older sediment/rock. (**b**) syngenetic ice wedge developing during active sedimentation of fluvial or eolian sediment (assuming no change in active layer thickness). Continuous lines indicate soil surface positions, dotted lines top of permafrost positions. (**c**) anti-syngenetic ice wedge development, on an eroding surface. (After Mackay 1990).

Epigenetic ice wedges develop in previously unfrozen sediments, after their deposition and as a result of exposure to cold climatic conditions. Syngenetic ice wedges develop alongside with sedimentation processes in a sedimentary environment which is exposed to cold climate, usually in fluvial or eolian sedimentary environments. This requires continuous or intermittent accumulation of sediment at the surface, concurrent with upwards aggradation of permafrost, which follows the upward growth of the sedimentation surface. This results in deeper wedges than those formed by epigenetic wedge formation (Mackay 1990; Ellis and Rochefort 2004). The Yedoma deposits of Pleistocene age (Sect. 1.4.5, Chap. 4) are an example (Murton et al. 2015). In river floodplains also syngenetic ice wedge growth occurs. Mackay (1990) also distinguishes anti-syngenetic wedges, occurring on eroding surfaces. Here, the surface above the wedges is truncated by erosion, followed by deeper wedge growth.

The micro-relief created by ice wedges contributes to abundant ponded water where low-centred polygons dominate. In the polygons described by Minke et al. (2009), the active layer thickness is generally larger in the low polygon centres (Fig. 3.11). That of the ridges is more variable; the largest thaw depth is found in one of the ridges. The vegetation differences between polygon centre and ridges accentuate the micro-relief. The polygon centres in Fig. 3.11 have low stature plant communities that grow in water-saturated or submerged conditions, while ridges allow shrub growth. A palaeobotanical record of ice wedge rim and pond sediments from two sites in Northeastern Siberia testify of the importance of rim development for drainage conditions during the Late Holocene between ca 3000 and 1600 years BP (Schirrmeister et al. 2018). In a site in the Kolyma Delta, pond development occurred simultaneously with rim growth. In a site in the Indigirka Lowlands, rim degradation,

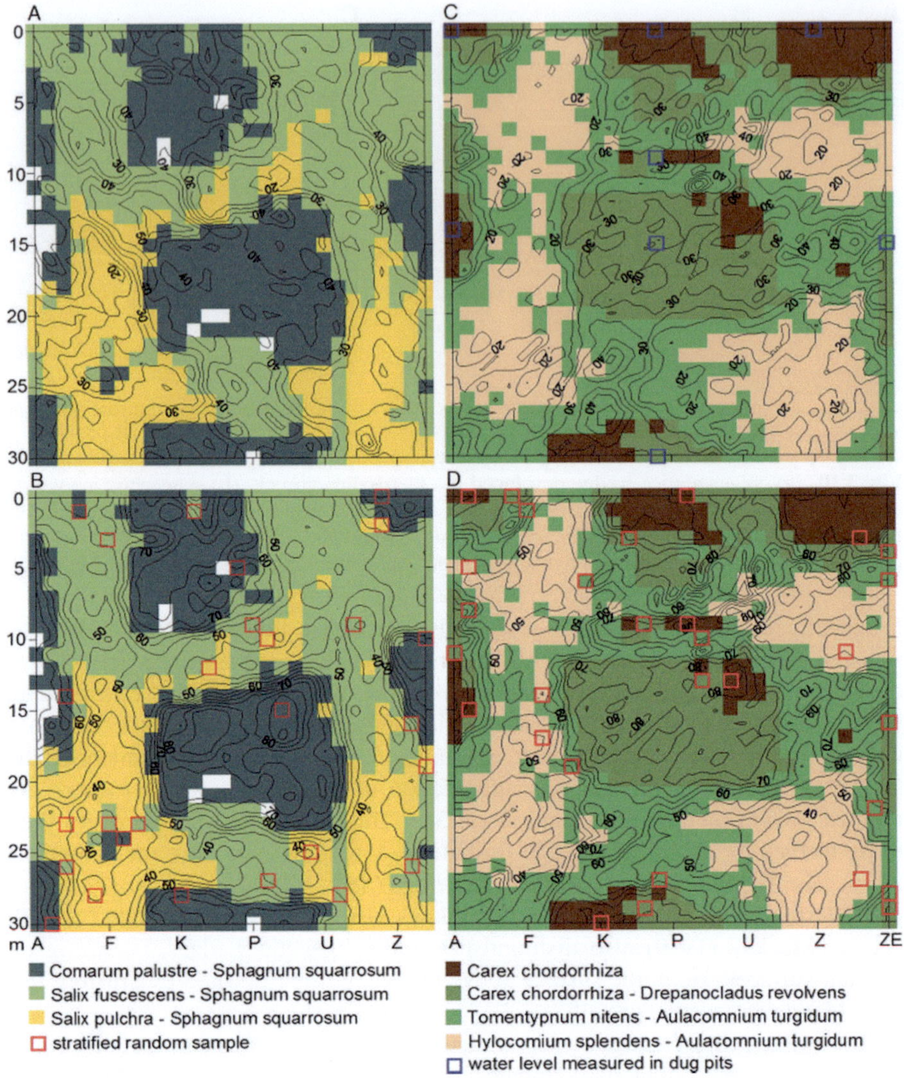

Fig. 3.11 Detailed vegetation maps of two low-centered polygons in Siberian tundra near Chokurdagh, Indigirka Lowlands. (**a**) Vegetation types and thaw depth of polygon Lc03 (number in original publication); (**b**) vegetation types, ground surface height and stratified random samples of Lc03; (**c**) vegetation types, thaw depth and position of water level measurements of polygon Lc04; (**d**) vegetation types, ground surface height and stratified random samples of Lc04. (White denotes missing data). Profiles on the axes have been labelled every 5 metres. Republished with permission of John Wiley and Sons Inc., from Patterns in vegetation composition, surface height and thaw depth in polygon mires in the Yakutian Arctic (NE Siberia): a microtopographical characterisation of the active layer by Minke et al. (2009). Permafrost and Periglacial Processes 20(4) ©2009; permission conveyed through Copyright Clearance Center, Inc.

a change from polygon ponds to wetlands, and development of high centre polygons occurred during the same period, indicating improved drainage of the site.

The polygon rims may retard runoff, but they are not impervious to shallow groundwater flow in the active layer (Woo and Guan 2006). Boike et al. (2008a) found that the change in water storage in polygon ponds was much higher than could result from evaporation only, and therefore drainage by lateral flow should exist. A dye tracer experiment demonstrated subsurface flow between two polygons; the lateral flow approximately matched the polygon pond water balance deficit. Frost cracks also serve as conduits for water at times of high runoff volumes during snowmelt and heavy rain, and promote surface drainage in drier conditions (Abnizova and Young 2008). Also the polygons in Fig. 3.11 contain preferred flow paths through the polygon rims.

The water level in polygons follow a 'fill and spill' process (Woo 2012), in which water fills depressions in the permafrost top surface; with a rising water table resulting from a precipitation surplus, sills between these depressions are successively overtopped by shallow groundwater flow and runoff. At the same time the subsurface flow also feeds back on permafrost by shaping the topography of the top of the permafrost by thawing and eventually surface erosion (Wright et al. 2009).

Ice-wedge polygons are by no means static phenomena. An age classification of drained thaw lake basins suggests that over time, ice wedge polygons evolve from essentially flat-centred polygons to low-centred polygons. Over time scales of centuries to millennia the ridges become higher. In ancient thaw lake basins, low polygon centres may coalesce into larger ponds and small lakes (Hinkel et al. 2003). A paleo-ecological study of polygons on the Beaufort Sea coast by Wolter et al. (2018) shows that terrain with low-centred ice-wedge polygons may persist for millennia. The low centre polygon development in the Kolyma Delta described by Schirrmeister et al. (2018) took ±1500 years. The conditions that contribute to this stability are determined by both geomorphology and climatic stability. Waterlogged ground and cool summers promote stability of low-centred polygons. However, on a detailed level, there is continuous variation. Detailed palaeobotanical work on polygon mires in Eastern Siberia has shown that the height of polygon ridges changes over time, with collapse and regrowth of ridges, influencing surface water ponding (De Klerk et al. 2011; Teltewskoi et al. 2016).

The development from low-centred to high-centred polygons has been linked to self-organizing processes. Lateral movement of material adjacent to ice wedges toward the polygon centre, and vegetation growth and preferential peat formation in the polygon centre may result in a higher polygon centre (Lachenbruch 1962; Ellis and Rochefort 2004; Wolter et al. 2018).

However, thaw of the ice wedges, which lowers the wedge with respect to the centre, is a more generally accepted interpretation (previous section). High-centred polygons are often associated with slightly sloping terrain (Fig. 1.21a, b), or are associated with expanding wedge thaw, creating water bodies above the ice wedges. If troughs above ice wedges start functioning as water discharge routes, soil material may be removed above the wedges, resulting in further thaw and widening or deepening of the trough. This happens in particular when ice wedge troughs have sufficient interconnectivity to form a drainage system (Jorgenson et al. 2015; Liljedahl et al. 2016; Wolter et al. 2018). The paleo-ecological data of Wolter et al. (2018) show that high-centre polygon development and ice-wedge degradation is associated with a warming climate, and Jorgenson et al. (2015) link widespread thaw of ice wedges to exceptionally warm summers.

Water flow along ice wedge troughs can be substantial, even without erosion. In Fig. 3.12, the only outlet flow of an 51 ha large thaw lake in the Siberian tundra occurs through the ice wedge polygon network. In case of larger gradients and strong runoff, erosion and gully formation can occur along the ice wedge network (Liljedahl et al. 2016; Wolter et al. 2018). Water flow and deepening of ice wedge troughs may either stabilize through deepening, or result in deeper and more extensive erosion, depending on terrain slope and water discharge.

Stabilization occurs when material eroded from the sides of the troughs is deposited within the troughs, in combination with vegetation succession and associated accumulation of organic matter, forming a new insulating soil layer above the wedge ice (Jorgenson et al. 2015; Kanevskiy et al. 2017). Formation of rapid downcutting gullies, erosion scarps, waterfalls and tunnels in the ice wedges is the other possible pathway (Fortier et al. 2007). Tunnel formation is a common feature during erosion of ice wedges; refreezing results in a body of clear ice within an ice wedge which points to past thaw and erosion activity (French and Shur 2010; the cave ice mentioned above).

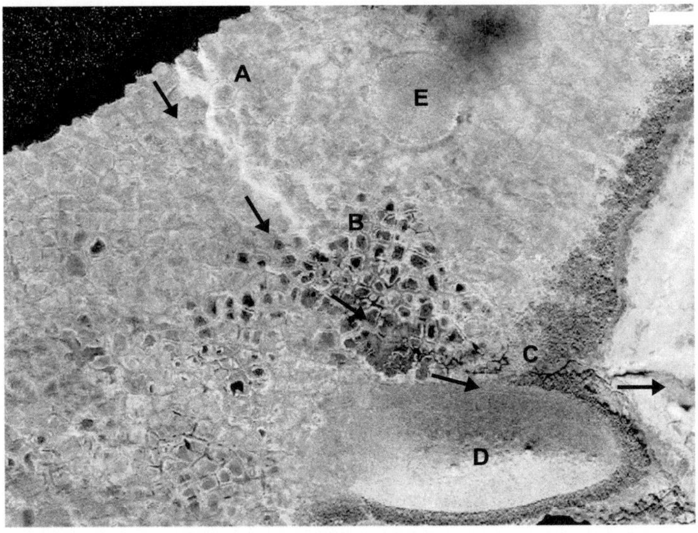

Fig. 3.12 GeoEye satellite image showing lake drainage (see arrows) of a thaw lake in the Siberian tundra through the ice wedge polygon network. The upper track at A follows a few ice wedges eroded to a depth of 0.5–1 m below the surface, lined with lush sedge vegetation (lighter streaks leading from the lake). This is followed at B by diffuse drainage through a wider polygon network, in turn followed by a single channel at the outlet to a next lake, which has a low water table because of a low stand of the river to which it is linked (C). The ice wedges are the only outlet of the 51 ha lake. D and E are pingos. Kytalyk Research Station, Indigirka Lowlands.GeoEye © 2019 MAXAR

3.4 Frost Mounds

In permafrost terrain various forms of temporary and perennial mounds or small hills exist with a core of segregation ice, intrusive ice or both, known under various names: palsas, peat plateaus, pingos, (also known as hydrolaccoliths) and frost blisters (e.g. Fig. 3.12). These frost mounds are produced by ground freezing combined with groundwater movement, or the migration of soil moisture (Harris et al. 1988). The largest of these permafrost landforms are palsas and related forms in bogs, and pingos. These are perennial frost mounds which exist over at least several years. Smaller features exist also; some of these are temporary features which grow in winter and disappear in summer (e.g. frost blisters, earth hummocks, French 2018). Here only the larger, perennial features are discussed.

3.4.1 Palsas and Similar Features

Palsas (single: palsa, Sami for elevated part of a bog) are typical features of peatlands in the discontinuous permafrost zone. They usually occur as elevated, dry islands in otherwise unfrozen bogs, known as palsa mires (Sect. 1.4.3, Fig. 1.18). The size is in the order of 10–150 m, the height above the surroundings 1–10 m. Segregation ice is responsible for much of the elevation, but also buoyancy forces contribute. Seppälä (2011) reviews the conditions for their formation; long-lasting winter air temperature below 0 °C, thin snow cover, and low summer precipitation. MAAT should be below 0 °C but more likely below −2 °C, and a sufficiently thick peat layer (at least 0.5 m) should be present. Human disturbance of the surface hydrology also can cause palsa formation; Kuhry (2008) reports palsa formation related to modified wetland hydrology by railway construction. Larger frost heave features in peat bogs, extending over hundreds of metres, are known as peat plateaus or represent large solitary permafrost peat islands. These are often surrounded by minerotrophic fens, and occur generally in areas where surface water is less abundant and permafrost more extensive (e.g. Vitt 2006).

An equivalent in mineral soil without protective peat cover is called lithalsa (Pissart 2002; Calmels and Allard 2008). Lithalsa development generally requires colder climatic conditions than palsas (Pissart 2002). There has been considerable discussion on the terminology of palsas, lithalsas and similar features in Arctic and Subarctic wetlands (Pissart 2002).

Crucial for the formation of palsas is the contrast in thermal diffusivity between dry and water saturated frozen peat (see Table 2.2; Seppälä 2011). Peat has a low thermal diffusivity, in particular when dry after a dry summer period. Its insulating top layer protects underlying frozen material to summer heat penetration. When the peat becomes saturated by autumn rains and freezes, its thermal diffusivity is an order of magnitude higher, allowing deep penetration of winter frost.

Figure 3.13 shows palsa development schematically. If a part of the bog surface has a thinner winter snow cover as a result of wind action or other local conditions, frost can penetrate deeper than elsewhere. A frozen peat core may be formed which results in frost heave. Because of the elevated surface, the peat above this initial frozen core is drier. This amplifies the insulating effect of the peat layer, preserving the ice beneath it. At the same time, the elevated surface is also favourable for a

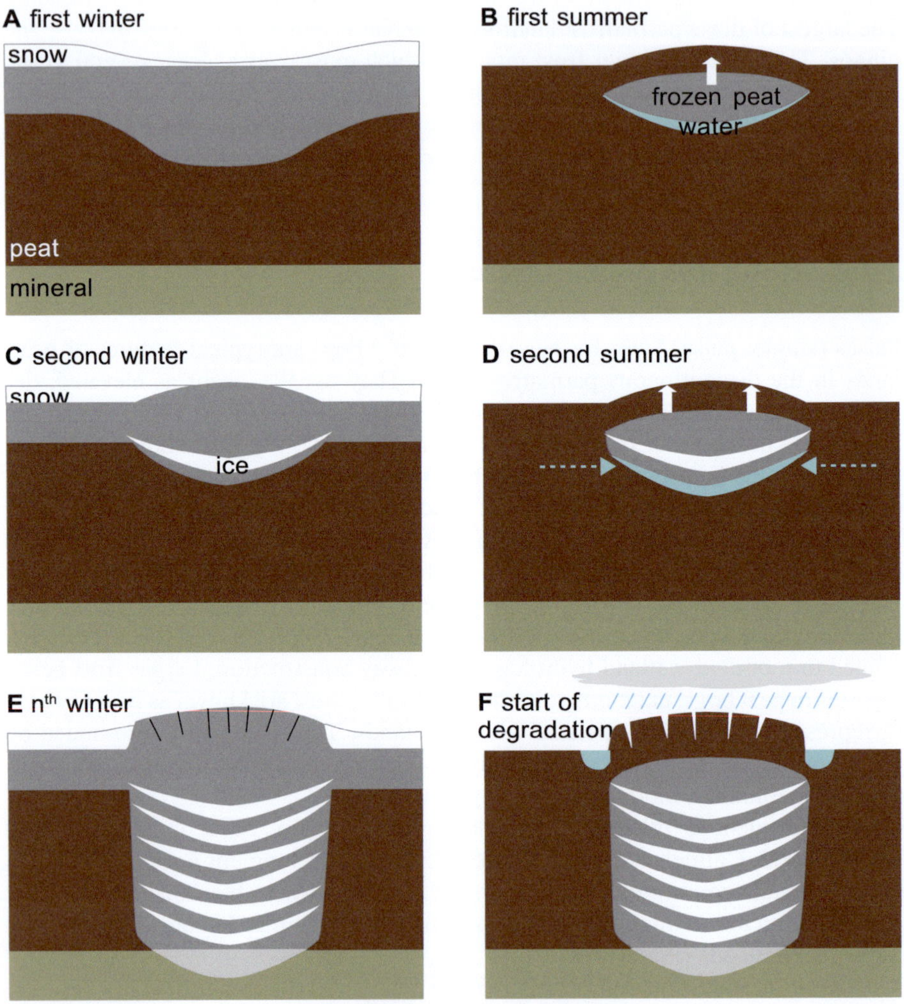

Fig. 3.13 Palsa formation, simplified after Seppälä (2011). Vertical scale not exact. (**a**) initial formation by deeper ground freezing under locally thin snow cover. (**b**) Frozen core remains through summer, water layer forms by buoyant upward movement. (**c**) Next winter: formation of ice lens and again deeper freezing caused by thinner snow cover. (**d**) next summer, as **b**. (**e**) After a number of winters a thick frozen core exists, which may reach the base of the peat deposits. Cracks have developed at the surface. (**f**) Degradation processes by cracking, exposing frozen core to warming, wetting of peat by summer rains, sideways thawing

thinner snow cover in winter, enhancing frost penetration. These self-amplifying effects promote the growth of the frozen core into a perennial palsa. Buoyancy forces of the ice in the water-saturated peat further add to the elevation of the palsa in summer. It also results in a pressurized water layer at the base of the palsa below the frozen core (Seppälä and Kujala 2009). In successive years, more ice will form, largely by freezing of the water layer at the base of the frozen peat core, resulting in a higher elevation of the palsa (Seppälä 2011). Over time, the vegetation also will also adapt to the dry conditions of the palsa surface, with dominance of low shrubs (*e.g. Ledum, Betula, Empetrum*), lichens and dry mosses. In the mature stage of palsa development, the frozen core also reaches the base of the peat and grows into the mineral subsoil (Seppälä and Kujala 2009).

According to Seppälä (2011), ice segregation does not play a large role in the first stages of formation of palsas, because pure peat is not susceptible to ice segregation (see above). It is the layer of water that is drawn to the base of the frozen peat core by the buoyancy of the frozen peat body that forms most of the palsa ice (Seppälä 2011; Fig. 3.13). However, most peats contain a certain amount of mineral matter that may induce the formation of segregation ice. Therefore, other authors (e.g. Pissart 2002) include ice segregation. When the mineral subsoil of the peat is involved, ice segregation becomes a considerable component (Seppälä and Kujala 2009). A larger role for ice segregation is seen in lithalsas (Calmels and Allard 2008).

Calmels and Allard (2008) studied ice lenses in frost-heaved mineral and organic soils in Nunavik, east of the Hudson bay, using CT scans (X-ray computed tomography images) from sediment cores. The cores show abundant ice lenses and layers of various thickness, accompanied by cracks and faults generated by shear stresses related to heaving forces and extension. Inclusion of gas bubbles in the ice shows the direction of aggradation: downward with respect to the surface. In palsas the ice lenses are often inclined subparallel to the surface (including the sloping sides), while in peat plateaus lenses are generally approximately horizontal. The ice volume is generally over 50%. A striking amount of gas bubbles is present in the ice, with a volume up to 5%.

Degradation of palsas, resulting in the formation of water-filled thaw depressions is common. In the mature stage of palsa development (Fig. 3.13e, f), the higher elevation above the surface results in thicker snow accumulation at the sides of a palsa, resulting in thaw of the permafrost and formation of pools surrounding the palsa. This changes the local heat balance and induces further thaw. Also, erosion from sides by flowing water may contribute to further degradation by thermal erosion.

The high elevation induces also mechanical instability and the formation of dilatation cracks, allowing water and heat penetration into the core of the palsa, inducing its collapse (Luoto et al. 2004; Seppälä and Kujala 2009). The process may be initiated or aggravated by wet summers, when a precipitation surplus saturates the dry peat on top of the palsa, decreasing its insulating properties (Seppälä 2011; Fig. 3.13f). The result is a landscape where thaw ponds, palsas and remnants of palsas alternate (Kirpotin et al. 2011). Seppälä (2011) emphasises the cyclical nature of palsa development and decay; the fens and lakes formed by palsa degradation will terrestrialise by peat growth over time, after which new palsa formation may start.

3.4.2 Pingos

Pingos are mounds with a core of ice that develop in permafrost areas by seepage of groundwater through the permafrost (Mackay 1998; Fig. 3.12 and 1.23a). Pingos are a few metres to up to a few tens of metres high. They contain a core of ice covered with a layer of soil. The core is susceptible to thaw which results in the creation of small lakes. Because of their relation with groundwater seepage in permafrost, pingos are indicators of groundwater flow through permafrost and hydrological changes (Woo 2012; Grosse and Jones 2011; Jones et al. 2012).

Pingos differ from palsas by the origin of the water of the ice core, which is derived from deeper groundwater flow. Palsas withdraw their water from the surrounding peat. Pingos are largely restricted to continuous permafrost, while palsas occur in discontinuous permafrost areas. The ice core of pingos consists of intrusive ice. The intrusive ice contains annual layers several centimetres thick (Mackay 1998). The ice has intruded as water in summer and freezes in winter. It often contains gas bubbles. Below the ice core of actively growing pingos, a water lens is present with water under a high hydraulic pressure (Mackay 1998). Springs may be present at the sides and top of the pingo. Variations in water supply results in water pressure variations, which causes a pulsating elevation of active pingos (Mackay 1998).

The groundwater flow that feeds pingos can have two sources: sub-permafrost water under hydraulic pressure (hydraulic pingos or open system pingos), or intra-permafrost water from closed taliks under hydrostatic pressure (hydrostatic pingos or closed system pingos, Fig. 3.14). The water pressure in open system pingos results from a hydraulic gradient in an aquifer below the permafrost, usually driven by a terrain slope. This implies also infiltration of water upslope, e.g. through an open talik below a lake or river, or subglacial water. The water pressure in closed system pingos is created by aggrading permafrost in permeable sediments; the volume expansion of ice results in expulsion of water (e.g. Mackay 1998; Woo 2012).

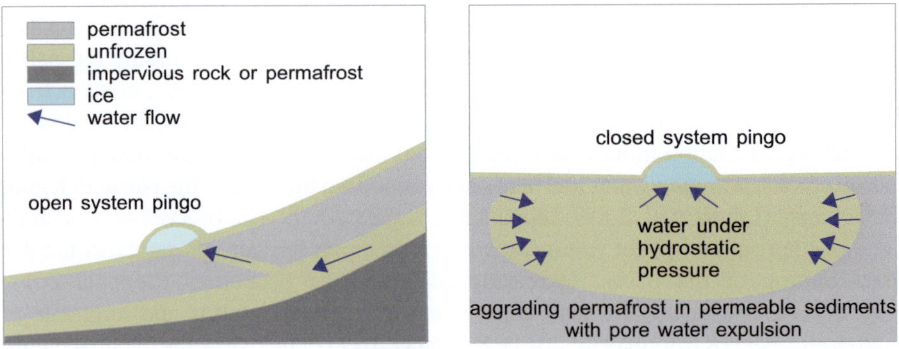

Fig. 3.14 Left: Open system pingo fed by permafrost groundwater under hydraulic head. Right: Closed system pingo fed by water under hydraulic pressure in a refreezing talik

The presence of an aquifer consisting of permeable, coarse-grained deposits is an important condition for pingo formation (Mackay 1998).

An example of open system pingos are pingos in valleys in Svalbard, e.g. Adventdalen (Yoshikawa and Harada 1995). Here, the conditions for pingo growth are the presence of water under considerable hydraulic pressure (artesian water) beneath the permafrost, combined with permafrost that is sufficiently thick to form intrusive ice, but not too thick and continuous to prevent penetration of pressurized water. Yoshikawa et al. (2003) reported a water pressure of 50 kPa under an open system pingo in Alaska. In Svalbard, the hydraulic pressure develops in three types of situations: the presence of bedrock faults which supply water, sloping aquifers (e.g. glacial outwash fans), and aggrading permafrost in recently uplifted seafloor areas in fjords. The water is often derived from subglacial meltwater (Etzelmüller and Hagen 2005).

Closed system pingos are most often situated in the bottom of drained lakes, where permafrost is aggrading into the talik in sands below the former lake (Fig. 3.15). Mackay (1998) suggests that pingos form at locations where the aggrading permafrost in the drained lake bottoms is relatively thin, allowing the formation of water conduits. This happens for instance in and around residual lakes. This is illustrated in Fig. 3.16 from Northeastern Siberia. As shown in Fig. 3.16, pingos also occur on a river floodplain, where meander migration probably has created a similar hydrological situation as under a drained lake bottom. Beneath the river channel, a talik exists in sandy fluvial channel deposits. If the channel migrates, permafrost will aggrade again in the point bar sediments left behind by the migrating channel. If permafrost aggradation disrupts contact between the talik and the present river channel, a closed talik may develop also. In Siberia, 12% of the pingos is located on fluvial deposits (Grosse and Jones 2011).

Pingos are often ruptured at the top by dilatation cracks which also may contain springs. Dilatation cracking is a common process in actively growing pingos. If these cracks expose ice, the ice core may start to thaw, leading to pingo collapse. When the ice core of a pingo is exposed to thaw, a small lake is formed. The soil on the pingo slopes may slide downward from the core to form a rampart around the lake (Fig. 3.15; Mackay 1998). Alternatively, pingos may become inactive when groundwater supply ceases, which is most likely the case for closed system pingos. Many of the pingos in Fig. 3.16 are probably inactive today, since they often lack dilatation cracks or spring activity.

Grosse and Jones (2011) mapped 6049 pingos; 64% of the pingos occur in continuous permafrost with high ice content; another 19% in continuous permafrost with moderate ice content. Only 8% of these pingos did exceed a height of more than 8 m; the highest was 37 m. Most pingos occur in regions with mean annual air temperatures between −7 and − 18 °C, or MAGT between −3 and − 11 °C; at MAGT below −11 °C there is a sharp decline in the number of pingos; a small number of pingos exist at MAGT > −3 °C. To a smaller extent, pingos are found in valleys and plateaus in mountain permafrost regions. Similar results have been obtained by Jones et al. (2012) for Northern Alaska.

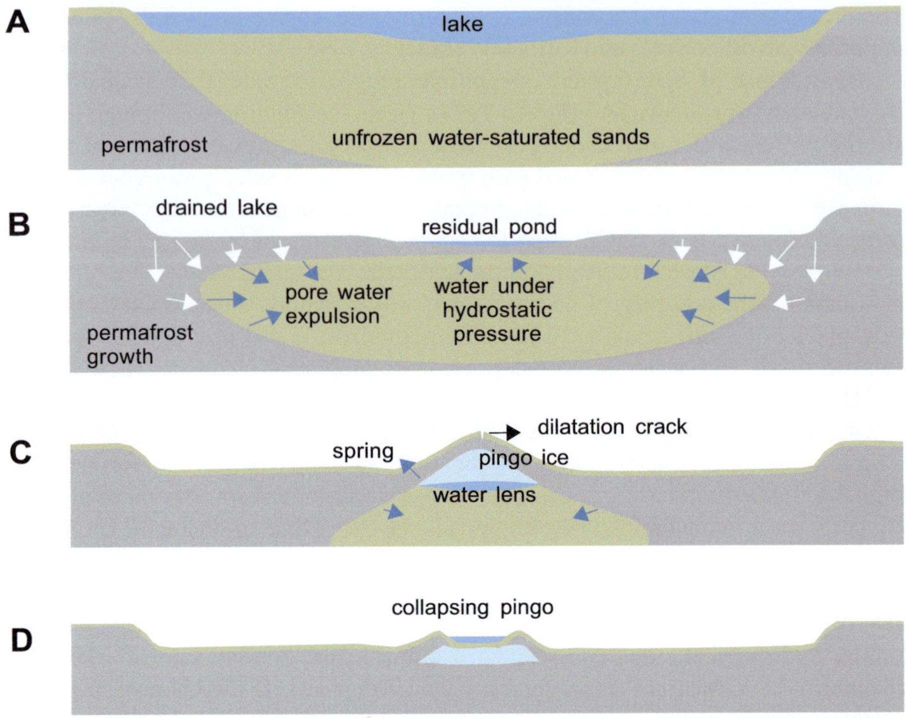

Fig. 3.15 Closed system pingo life cycle after Mackay (1998). (**a**) Large lake underlain by a talik with unfrozen sands. (**b**) The lake is drained by erosion of a drainage threshold, which is usually a rapid process. A shallow residual pond is left. Permafrost starts to aggrade in the unfrozen sands below the former lake bed; if the pond depth is smaller than the maximum winter ice thickness, also permafrost aggrades beneath the pond. In the now confined talik, hydrostatic pressure builds up by pore water expulsion at the freezing front. (**c**) Water expulsion from the talik to the surface in winter (usually where surface permafrost is thinnest) results in a growing intrusive ice core overlain by soil. Ice is added each winter, and the ice core shows growth bands. A pressurized water lens is present below the ice core, which may result in spring flow. Because of freezing point depression, the water has subzero temperatures. At the surface of the pingo, dilatation cracks are common; these are the results of the upward pressure of the growing ice core. (**d**) Over time the pingo may collapse by thawing of the ice core, leaving a small lake surrounded by a rampart of soil, eroded from the pingo slopes

By far the most pingos of Grosse and Jones (2011) occur in continuous permafrost with thick unconsolidated sediments (Fig. 3.17). The largest concentrations of pingos are found in the sedimentary lowlands, occurring in areas which are rich in lakes and drained lake basins. Other areas favourable for pingo growth are fluvioglacial and glacial moraine deposits. A southerly area with many pingos is located in Central Yakutia, where pingos form in drained lake basins (alases) on terraces of the Lena river. Most of the lowland pingos are closed system pingos, while those in mountainous areas are more likely to be open system pingos.

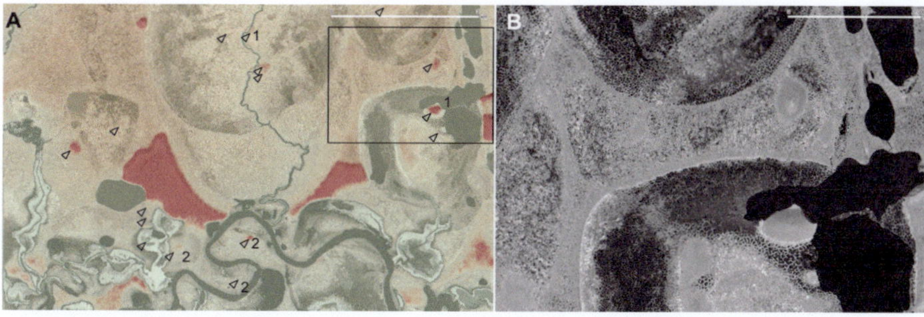

Fig. 3.16 Pingos on drained lake basins in the Indigirka Lowlands, Northeast Siberia. A: Worldview panchromatic satellite image (0.4 m spatial resolution), transparently overlain by ALOS digital elevation model (30 m spatial resolution, July 102,015). Triangles indicate locations of pingos. Red colour indicates elevation over 25 m; scale bare 3 km. The southernmost pingos are located on a river floodplain and near lakes that have been drained by the river (at 2). B: Detail of rectangle in A, showing overlapping drained lake bottoms and a large, partly eroded pingo (1 in A) near a residual lake of a drained thaw lake. The larger red coloured areas in A are yedoma remnants; 1: eroded pingo; 2: pingo located on a river floodplain. Image processing and interpretation by the author. WorldView © 2019 MAXAR

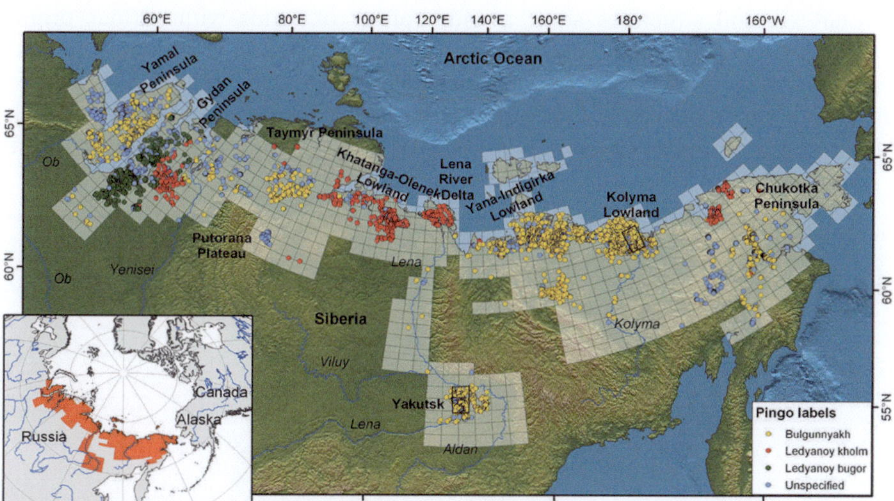

Fig. 3.17 Distribution of 6059 pingos in Siberia, digitized from 675 topographic maps 1:200000 (red area in map inset and light semitransparent frames in large map). The legend refers to the terminology with which the pingos are marked on topographical maps. Pingo labels are names under which the pingos are marked on Russian topographical maps. From Grosse and Jones (2011), Spatial distribution of pingos in northern Asia, The Cryosphere 5(1), doi:https://doi.org/10.5194/tc-5-13-2011, published under Creative Commons License 4.0 https://creativecommons.org/licenses/by/4.0/; image provided courteously by the authors

Phenomena related to pingos are icings, since they are also indicators for groundwater flow, but need not be related to permafrost. Icings are surface ice crusts formed in winter by water seepage in springs, and generally thaw again in summer (Woo 2012). They are found often in river valleys; the presence of coarse-grained alluvial deposits beneath rivers can conduct water to the surface to produce springs in summer, and and in winter icings. Also other situations (faults, surface contacts of aquifers) may give rise to the formation of icings. Springs at pingos also may lead to the formation of icings (Yoshikawa and Harada 1995).

3.5 Cryoturbation and Patterned Ground

Cryoturbation is the collective name for disturbance of soil horizons caused by freeze-thaw processes (e.g. Vandenberghe 1988). This includes movement of topsoil material towards deeper levels in the soil, within the active layer. Therefore, cryoturbation contributes to preservation of soil carbon deeper in the active layer, in an environment where organic matter is less vulnerable to decomposition (Kaiser et al. 2007). Cryoturbation also affects vegetation by bringing minerals and nutrients from deeper in the soil profile to the surface (Broll et al. 1999; Walker et al. 2004).

Various theories have been advanced for the processes behind cryoturbation, and cryoturbatic soil structures show a wide variety of forms with various origins (Bockheim and Tarnocai 1998; Vandenberghe 1988, 2013). In soil profiles, cryoturbation shows as more or less irregular streaks and larger masses of soil material of different colour and texture, that appear to have been displaced by downward sinking or upward movements (Fig. 3.18). Often, the soil material looks as if having

Fig. 3.18 (**a**) Cryoturbated soil in Yamal Peninsula near Bovanenko (histic cryosol), showing irregular, down-warped streaks of topsoil organic matter, resulting from cryostatic heave. Height of profile 0.7 m. (Photo by D. Balykin, Russian Academy of Science Institute IWEP, Barnaul). (**b**) intensely cryoturbated (by periglacial loading) Late Pleistocene soil, Gorsselse Heide, Netherlands. Vertical height of section 0.8 m. (Photo by the author)

been in a semi-fluid state during movement. All processes related to frost action may cause cryoturbation: ice segregation, thermal cracking, processes such as collapse and water-saturated movement of soil mass during active layer or permafrost thaw. The presence of cryoturbation is a diagnostic criterion in Arctic soil classifications (Jones et al. 2010). In Pleistocene sedimentary successions, cryoturbations are used as evidence of former permafrost. Cryoturbation in present-day environments has been studied in relation to micro-relief.

Vandenberghe (1988, 2013) has developed a classification of cryoturbation structures, based on whether or not they show a spatial regularity with a repetitive pattern, the wavelength and amplitude of these repetitions, the type of movement that appears to have occurred (downward or upward) and the processes that may have caused these descriptive properties. These forms are linked to three groups of processes.

- Periglacial loading: resulting from high excess pore water pressure in the thawed layer at the top of ice-rich permafrost. This causes liquefaction (loss of shear strength by high pore water pressure, causing the sediment to behave like fluid). Moreover, a density gradient may develop, since the soil near the top of the permafrost is less dense because of its very high water content, while the near-surface soil may have become denser because of earlier thaw and loss of water. This is in particular the case in fine-grained soils of low permeability (Hallet 2013), although it also may occur in gravelly deposits at high excess pore water pressure (Vandenberghe et al. 2016). The density gradient causes the top of the active layer to sink downward, forming wave-like or drop-like structures, often with a flat bottom indicating the top of the permafrost. All types of soil material may be involved in the down-sinking movement, from gravels to organic horizons. An important condition is flat terrain with poor drainage. These cryoturbations may occur in three dimensional networks over larger areas, or locally as individual forms. Larger forms (0.5–2 m amplitude) are linked by Vandenberghe (1988) to permafrost thaw. Small forms are due to localized excess pore-water pressures at specific points in the subsoil. Liquefaction also occurs as a result of earthquakes and are often accompanied by sand volcanoes; Vandenberghe et al. (2016) mention criteria to distinguish cryoturbation-related from seismicity-related liquefaction.
- Cryohydrostatic flow: this is the result of increasing water pressure in a water-saturated thawed layer, for instance between a downward advancing freezing front and an impervious layer or the top of the permafrost, or otherwise by lateral water supply within a confined thawed layer. This may cause injection of fluidized soil material from below into weakness zones, like desiccation cracks or thermal contraction cracks. This creates a more wavy, smaller amplitude pattern of cryoturbation.
- Cryostatic heave: the result of the volumetric expansion of freezing water and the growth of ice lenses, which may induce cell-like deformation of the soil. These are frequently related to the development of hummocky patterned ground and vegetation patterns (e.g. Mackay 1980a; Bockheim and Tarnocai 1998; Walker et al. 2004; Kokelj et al. 2007b; see below).

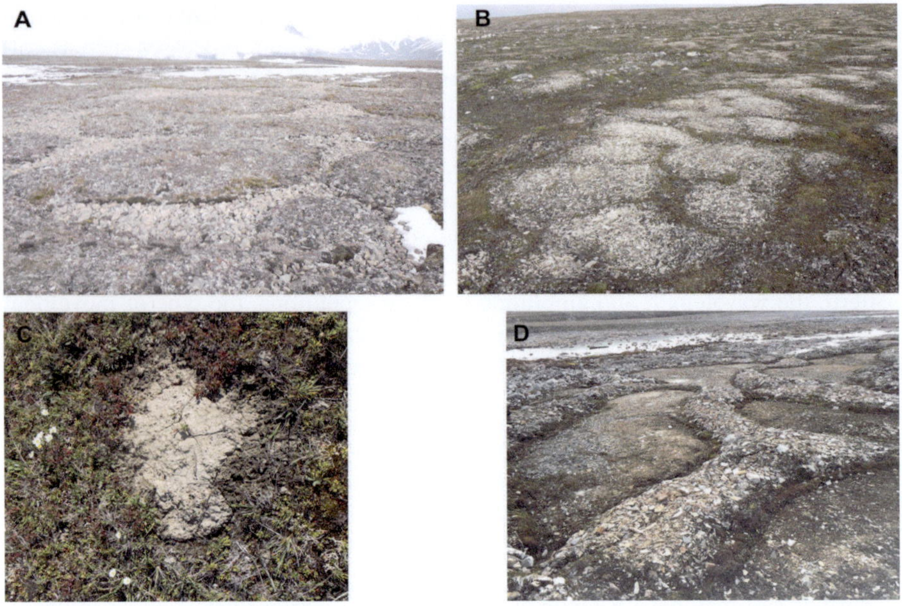

Fig. 3.19 (**a**) sorted circles, with concentrations of coarse rock fragments on the sides of polygon centres with finer material, diameter ± 1.5 m. Spitsbergen, near Longyearbyen. (**b**) non-sorted circles, similar diameter as in **a**, separated by vegetation. Kap Linne, Spitsbergen. (**c**) 'Mud boil', non-sorted circle where water-saturated fine-grained silt from the active layer has been broken through the surface. Yedoma plateau, Indigirka Lowlands, Siberia. (**d**) Irregular stone circles, Kap Linne, Spitsbergen. The sharp, converging border between the central fine-grained patch and the stone rims is clearly visible, and shows accumulation of organic matter and mosses. (Photos by the author).

This classification focusses on the appearance of cryoturbation in a vertical cross section such as in Fig. 3.18. The surface expression of cryoturbation may be a form of patterned ground, although the relation with soil horizon deformation and surface pattern is not always clear and various processes cause cryoturbation and surface pattern formation. Besides the ice wedge polygons discussed in Sect. 3.4, smaller scale patterned ground features are common at the surface of permafrost soils, usually expressed as hummocks, polygons or soil circles in the order of a few metres down to 0.5 metre in size (Fig. 1.20).

Washburn (1956, 1980) developed a classification of surface patterns, which is independent of their presumed origin, and cites 19 possible formation mechanisms. In the classification of patterned ground the main distinction is between sorted patterns, where sorting and separating of rock fragments on the basis of their size has occurred, and non-sorted patterns, which are expressed mostly by vegetation patterns and cracks in the soil (Figs. 1.20 and 3.19a, b). On flat areas, sorted or non-sorted circles and polygon nets are common, while on slopes they develop into sorted or non-sorted stripes in the downslope direction. Other, less regular forms

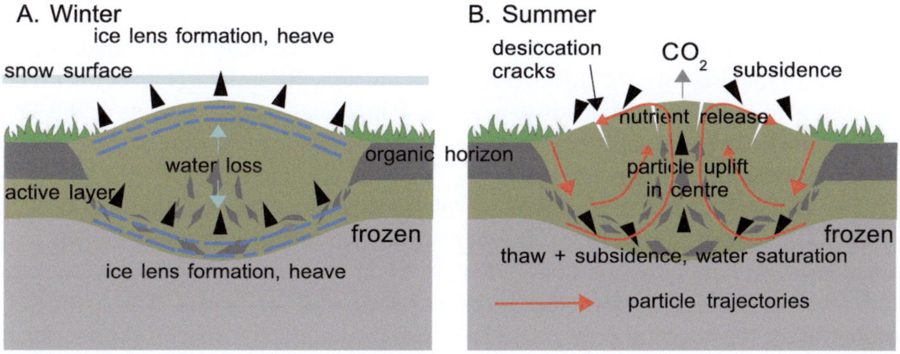

Fig. 3.20 Conceptual diagram of circulation cell-like movement of soil particles and organic matter beneath a frost boil. A: ice lens formation and froste heave in winter. B: Summer. Due to high water content as a result of ice lens thaw in summer, soil material in the centre of the frost boil behaves as a viscous liquid and may be displaced upwards and sideways. The differences in vegetation cover result in slower thaw in the vegetated rim. Derived from Mackay (1980b), Peterson et al. (2003) and Walker et al. (2004). Decomposition of organic material in the centre of the frost boil releases nutrients and CO_2

showing sorting of coarse-grained material and fines are known as labyrinths (Hallet 2013).

The most ubiquitous example of patterned ground is the non-sorted circle, also known as mud boil or frost boil which cover many flat areas with fine to medium textured permafrost soils. They consist of roughly circular vegetation-free spots, are typically 0.5–3 m wide, and tend to grade in areas with increasing vegetation cover into vegetated hummocks (Mackay 1980a). Walker et al. (2004) describe a transition from non-sorted circles to hummocks along a large north-south transect in Alaska. Their description includes the cryoturbation structure in vertical cross section of these patterned ground forms. The northernmost site is covered for 25% by unvegetated non-sorted circles, which declined to 1% in the southernmost sites where vegetated hummocks dominate. Vegetated earth hummocks are also known by the Icelandic word thufur (Schunke and Zoltai 1988).

Both non-sorted circles and hummocks are underlain by a bowl-shaped, thicker active layer than in the areas in between (Fig. 3.20). At the base of this active layer, an intermediate layer with abundant ice lenses occurs. Mackay (1980a) demonstrated that the alternation of frost heave by ice lens formation in winter and thawing in summer results in a cell-like circulation of particles. A mathematical model was developed by Peterson et al. (2003) and Peterson and Krantz (2008). In winter, the surface of the circle is uplifted by the growth of ice lenses at the base of the active layer and near the top. In summer, thaw of the ice lenses results in water saturation and viscous behaviour of the soil mass. When the soil settles, there is a tendency to sideways movement at the top of the material to the sides of the circle. This results in a net upward movement in the centre of the non-sorted circle, which must be compensated by lateral movements at the base and top of the active layer.

The rate of movement is slow, in the order of 1 mm/year, and episodic (Walker et al. 2004). It is facilitated when the liquid limit of the sediment is exceeded, and in those cases material may be injected upwards towards the surface (Fig. 3.19c). The development over time also leads to a decrease in diameter, and increase of the surface relief (Kokelj et al. 2007b). Dilatation or desiccation cracks may develop in the unsorted circle (Fig. 3.20).

The vegetation plays a conspicuous role in this cryoturbation process (Walker et al. 2004; Kokelj et al. 2007b). It ranges from thin layers of mosses, lichens and algae between largely barren non-sorted circles in High Arctic barrens, to a nearly full cover of mosses, graminoids, forbs and shrubs in the tundra. The insulating properties of the vegetation results in less deep penetration of thaw in summer than below the bare surface of the frost boil centre, maintaining the bowl-shaped surface of the active layer. In winter, the slower penetration of frost results in differential frost heave. This difference is enhanced by the thinner snow cover above the higher frost boil or hummock.

The vegetation also adds organic matter to the soil, altering its thermal properties by decreasing its thermal diffusivity. The organic matter in the areas between the circles tends to be transported downwards to the base of the active layer and in the centre of the circle upwards again (Fig. 3.20). The downward movement results in storage of the organic material at the base of the active layer where low temperature, and low oxygen availability reduces decomposition (Kaiser et al. 2007; Chap. 4). Subsequent upward movement in the centre of the circle results in enhanced decomposition and release of nutrients in a more oxygen-rich and warmer soil environment.

Going from barren conditions to densely vegetated areas, the role of vegetation and organic matter becomes more important. Although the surface relief may be masked by a denser vegetation, the bowl-shaped base of the active layer and the cryoturbation activity still is visible (Walker et al. 2004). Usually a distinct microtopography of vegetation hummocks develops; some tussock-forming species, such as *Eriophorum*, contribute strongly to this micro-relief. This hummock formation is not restricted to tundra vegetations. Kokelj et al. (2007b) describe hummocks overgrown with Black Spruce trees, which show the effect of tilting due to hummock activity by the uneven development of tree rings. Fire and climate change leads to active layer deepening and collapse of hummocks.

Detailed analysis of the heat transport, nutrients and organic matter of a non-sorted circle in Svalbard by Boike et al. (2008b) has confirmed the observations of Walker et al. (2004) and their hypothesis on the biogeochemical significance of cryoturbation. Lupachev and Gubin (2012) studied cryoturbation and micro-relief in northern Yakutia, and found that a sufficient thickness of organic material in the troughs between non-sorted circles (nanopolygons in their terminology) is a necessary condition for lateral and downward transport of organic matter by cryoturbation.

Although the bowl-shaped permafrost table explains the cryoturbation process below non-sorted circles and hummocks, it does not explain how it is initiated. Kokelj et al. (2007b) see the aggradation of ice-rich permafrost as an important factor. The model by Peterson et al. (2003) shows that an initial ground surface perturbation of millimetre to centimetre scale can grow into a decimetre high frost boil in a matter of years, through amplifying effects of differential heave. However, due to the growing micro-relief, soil creep (which flattens the micro-relief) will ultimately balance the height growth.

The sorting of stones into polygons or stripes is the result of similar circulating soil movements as in non-sorted circles and hummocks, and are likewise driven by differential frost heave and is therefore diagnostic of cryoturbatic soil movements (Hallet 2013). In winter, fine-grained areas heave more than areas with coarse-material at the surface, which results in a displacement of coarse fragments at the surface of a fine-grained area to its edges, causing an accumulation of stones at these edges. The difference in heave is amplified by a change in thermal properties of fine-grained and coarse-grained soil as the sorting process progresses.

Kessler et al. (2012) developed a model that highlighted the presence of self-organization in the formation of stone polygons and stone stripes. In their model a layer of randomly placed stones organizes itself into a distinct pattern during repeated freeze-thaw cycles. Similar self-organization processes may start the formation of non-sorted circles and hummocks. Small perturbations in the top of the fine-grained material result in rapid development of sorted polygons - and if the surface has a slight slope, in development of stripes. A circulating soil movement as in Fig. 3.20 develops in the fine-grained areas, where heave is strongest. However, where the stony rims are thick as in the stone circles on Svalbard (Fig. 3.19d), also an opposite circulation develops in these rims, converging on the edge between coarse- and fine-grained areas (Hallet 2013). In this type of patterned ground, organic matter also accumulates along the converging borders and will be transported downward (Fig. 3.19d).

Stones in the soil tend to concentrate at the surface. The upward freezing of stones is a common process in stony soils, even in non-permafrost soils in temperate latitudes. In permafrost environments it contributes to patterned ground formation (Hallet 2013). Two processes contribute, 'frost pull' and 'frost push' (Williams and Smith 1991):

- Frost pull results from the larger adhesion of frozen soil to any object compared to unfrozen soil. During freezing of the soil, any solid object that extends for some depth in the soil will be lifted by the forces exerted by the frozen soil attached to its upper part, if the soil at the lower part is still unfrozen. The lifted objects do not reach their original position after thaw due to partial collapse of the hollow that formed below the object during lifting. As a result, net 'up-freezing' of the stone occurs.
- Frost push is caused by preferential ice segregation below the stone. The thermal diffusivity of solid rock is higher than that of the surrounding soil material. Moreover, dense rock will contain less moisture, which could release latent heat during freezing. This results in deeper frost penetration through the stone than through the surrounding soil, and ice lens formation beneath the stone, lifting it up. Some of the water for ice lens formation will be withdrawn from the soil beside the stone, resulting in additional consolidation of the soil on the stone's sides. It will depend on the resistance of the overlying frozen material to uplift, how much movement occurs.

Fig. 3.21 Ice wedge of Middle Weichselian age showing down-sinking organic soil material on as a result of ice wedge thaw. The top of the former wedge has been filled by sediment deposited from running water and lake deposits. Hengelo, Netherlands. Height of section 1.5 m. (From Van Huissteden (1990); photo by the author)

Lastly, soil movements related to ice wedge growth (Sect. 3.3) also contribute to cryoturbation (Bockheim and Tarnocai 1998). The lateral expansion of wedges pushes soil sideways and upward (Fig. 3.8); thawing of wedge ice may cause over-saturation with water with deformation of soil layers, or sliding of soil material into cavities created by thawing ice (Fig. 3.21).

3.6 Slope Process: Solifluction and Cryogenic Landslides

So far, the discussion of cryoturbation and potential burial of soil carbon has been restricted to flat topography. However, also on hill and mountain slopes soil carbon may be buried by solifluction (Elliott 1996; Palmtag et al. 2015). Solifluction is slow mass movement associated with freeze–thaw action in soils on slopes (Matsuoka 2001; French 2018). Rapid mass wasting processes (discussed below) are excluded from the the definition of solifluction. Several processes contribute to solifluction (Matsuoka 2001):

- Needle ice growth moves particles upwards perpendicular to the slope, followed by vertical downward movement upon thaw, together creating a net downslope movement (Fig. 3.22a). Needle ice is the growth of ice needles of a few cm length, carrying some soil material on top, in a water-rich soil.
- Frost heave followed by thaw similarly produces a downslope movement (Fig. 3.22b).
- Gelifluction, the downslope flow of water-saturated soil material. For gelifluction, ice lens thaw and snow meltwater or rain contributes to the water saturated conditions (Fig. 3.22c).
- Plug-like flow by confined movement movement of the thawed part of the active layer between a frozen topsoil and base (Fig. 3.22d).

Needle ice creep and frost creep are typically diurnal movements, confined to the upper centimetres of the soil, while gelifluction and plug-like flow are seasonal movements over larger depths of the soil, in the order of decimetres. The movement of the slope material is generally fastest at the top of the soil profile (Matsuoka 2001; Fig. 3.22). Solifluction is not restricted to permafrost.

Solifluction occurs mostly on slopes steeper than 5°, although Matsuoka (2001) lists slopes with a gradients down to 2°. The depth of movement is in the order of 15 to 100 cm. The depth and amount of ice lenses and the freeze–thaw frequency control the solifluction processes. Diurnal freeze–thaw action in non-permafrost conditions results in shallow but fast movement of a superficial layer. Annual frost creep and gelifluction occurs in climates with deep seasonal frost or permafrost, resulting in a thicker (<60 cm) layer moving at a rate of centimetres per year. In cold permafrost regions, ice-lens formation is more effective both at base and top of the active layer, and can induce plug-like flow of thicker soil masses.

In particular on lower slopes, solifluction produces characteristic landforms with lobes of slope material with a riser in front, roughly transverse to the slope direction. Often, coarser rock fragments are included in the solifluction material (Fig. 3.23). When a lobe moves downslope, it may override soil material downslope. Elliott (1996) shows a complete inversion of a podzolic soil profile in a solifluction lobe on a mountain slope in Norway, as a result of a caterpillar track-like movement of the solifluction lobe. However, because the carbon density in soils in mountain areas is lower than that in lowlands, Fuchs et al. (2015) found no evidence of significant amounts of soil organic matter buried by solifluction in a mountain valley in Northern Sweden.

Compared to solifluction, cryogenic landslides are moving an order of magnitude faster. In cryogenic landslides, the presence of permafrost ice and water released from it by thawing plays a large role (Leibman et al. 2003, 2014). Water liberated by thawing permafrost enhances mass movement by increasing the pore water pressure with the soil, resulting in loss of shear strength (French 2018; Williams and Smith 1991). In permafrost areas, two forms of cryogenic landslides occur frequently: active layer detachment slides and retrogressive thaw slumps (Fig. 3.24a; French 2018).

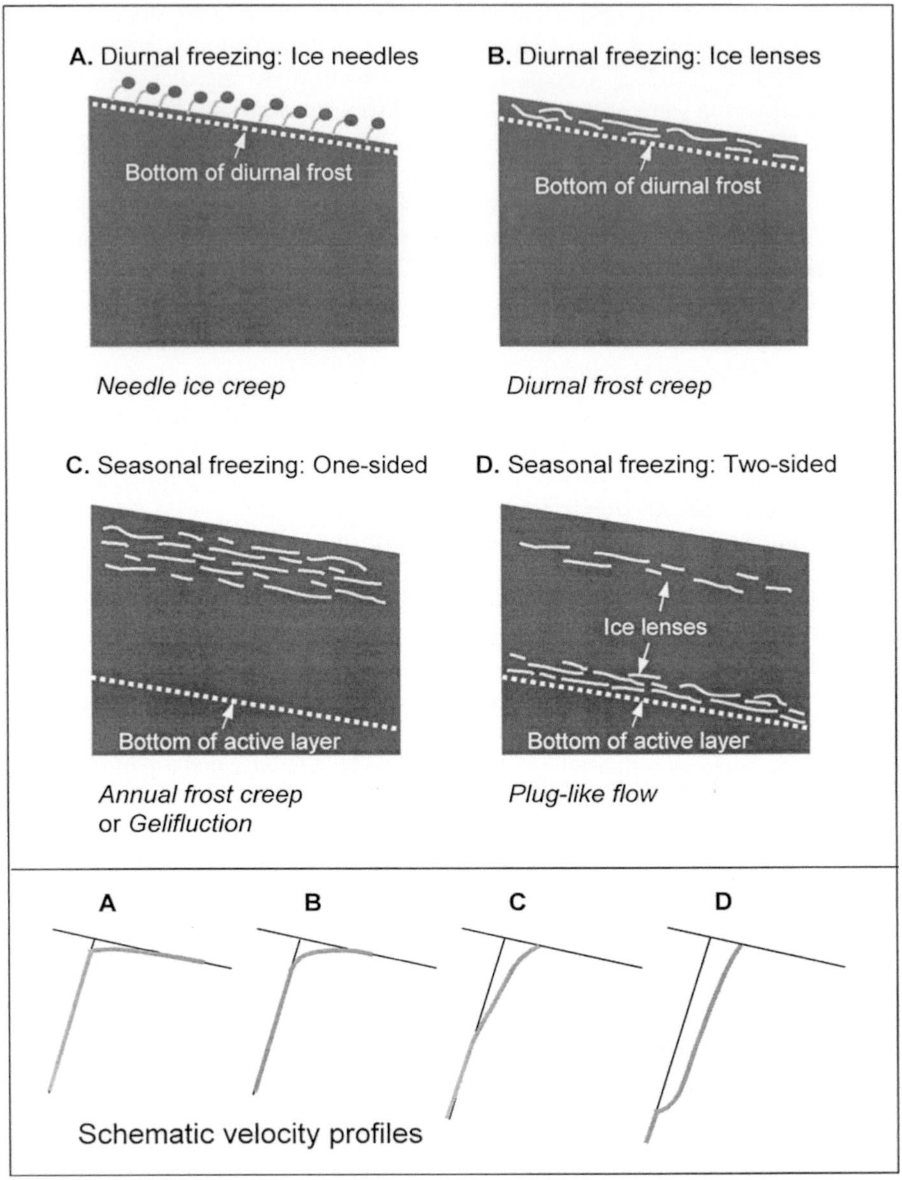

Fig. 3.22 Types of processes contributing to solifluction. Below typical velocity profiles. Reprinted from: Solifluction rates, processes and landforms: a global review, Earth-Science Reviews 55(1–2) by N. Matsuoka, © 2001, with permission from Elsevier; permission conveyed through CopyrightClearance Center, Inc.

Fig. 3.23 (**a**) Series of solifluction lobes on a hill slope in Northern Sweden, Padjelanta National Park. (**b**) Detail of solifluction lobes, Abisko area, Northern Sweden. (Photos by the author)

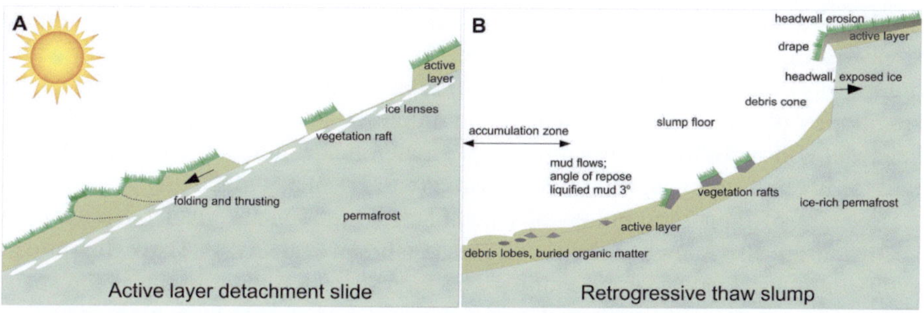

Fig. 3.24 Cross sections of typical rapid mass wasting features on permafrost. A: active layer detachment slide, with sliding of the active layer over the top of the permafrost, where shear strength is lost by high pore water pressure generated by thawing ice lenses. B: Retrogressive thaw slump, generated by thawing of ice-rich permafrost, characterized by a backwards retreating headwall, often with exposed permafrost ice. Based on Lewkowicz (2007), Lacelle et al. (2010) and Abbott and Jones (2015)

Active layer detachment slides are relatively thin, comprising the active layer only. More or less cohesive rafts of active layer material slide over the water-saturated boundary between active layer and permafrost. High pore water pressures created by thawing of ice lenses at the top of the permafrost aid movement by decreasing the shear strength of the soil (e.g. Lewkowicz and Harris 2005; Lewkowicz 2007). Active layer slides can reach lengths of tens to hundreds of meters and need only modest slopes. Gooseff et al. (2009) show a slide with a length of 800 m on a slope of only 4°.

Retrogressive thaw slumps develop on ice-rich permafrost and commonly occur on steeper slopes, for instance on valley sides, near lake and river banks (e.g. Figs. 1.23 and 3.24b; Lacelle et al. 2010; Khomutov and Leibman 2014). They are characterized by a near-vertical headwall of a few to tens of meters high. In this headwall, permafrost ice is exposed. It retreats by thaw of permafrost in the headwall, generating liquified mud and lumps of soil and vegetation. In front of the

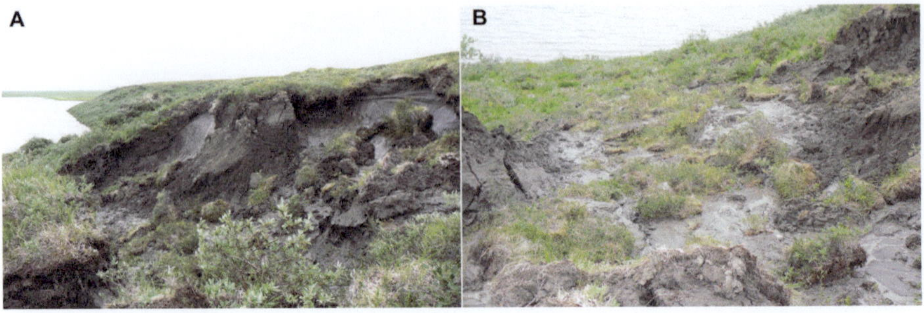

Fig. 3.25 (**a**) Headwall of a small retrogressive thaw slump on a thaw lake bank in the Indigirka Lowlands, Northeast Siberia. (**b**) Slump floor of the same slump, with vegetation rafts broken from the headwall and flowing mud (look direction is downslope). (Photos by the author)

headwall, a slump floor is found with a slope less than that of the original terrain slope. Along this slump floor, thawed material from the headwall is transported by mudflows (Burn 2000; Lacelle et al. 2010; Figs. 3.24b and 3.25). A gradient of the slump floor of 5% is sufficient to maintain mudflows and remove material from the headwall (Burn 2000). Further downslope, material accumulates in debris lobes or fans, or is removed by wave or river erosion (e.g. Lacelle et al. 2010).

Because of their morphological resemblance to mountain glacier cirques, the headwalls are also known as thermocirques (Leibman et al. 2014). Thaw slumps can evolve into spectacular large features. The largest known thaw slump is the spectacular Batagay 'crater' or mega-slump with headwalls of up to 86 m high, in a thick succession of Yedoma sediments (Günther et al. 2015; Murton et al. 2017).

3.7 Contribution of Ice to Rock Weathering

The low soil temperatures combined with repeated ice formation influences rock weathering. This consists of physical weathering of rocks into smaller fragments, and chemical weathering that dissolves and transforms minerals. Both are important for soil geochemical characteristics, and therefore also for conservation of organic material. Common wisdom holds that in cold climates physical weathering dominates over chemical weathering due to low temperatures or ice. Chemical weathering process rates should be low because of low temperatures and solute transport. However, recent research has moderated this picture (Thorn et al. 2001; Hall et al. 2002). Hall et al. (2002) stress that the availability of water, not the temperature, is limiting for chemical weathering. Specific parent material properties, such as the presence of pyrite-rich rocks, also can contribute to high chemical weathering rates in the Arctic, by generating sulphuric acid upon weathering (Thorn et al. 2001).

French (2018) cites three processes in physical weathering in cold environments: the first is expansion of water upon freezing, the second is ice lens formation as described in Sect. 3.2, the third is shattering of rock due to tensions created by thermal stresses. Frost weathering may be enhanced by salt weathering (salt crystal growth) when pore water contains salts. The amount and intensity of diurnal and annual freeze thaw cycles and rock type determine the intensity of cryogenic weathering (Konishchev and Rogov 1993). A fourth physical rock weathering mechanism is the growth of endolithic lichens and algae (growing inside the rocks) which can exert wedging pressures by penetration of hyphae and swelling of thalli (Chen et al. 2000; Guglielmin et al. 2005). This combines physical and chemical weathering, as the organisms also excrete organic acids which dissolve minerals and chelate metallic cations.

Murton et al. (2006) and Matsuoka and Murton (2008) demonstrated the importance of ice segregation and frost heave in fracturing of bedrock under permafrost conditions. Long term observations (19 years) by Mackay (1999) of subaerial weathering of glacial erratics in Canada showed that freezing, ice lens formation and thermal stresses all were instrumental in shattering these rock fragments. However, during the observation period only 5% of the rocks showed visible physical weathering, suggesting that the rate of frost shattering of rocks is not particularly large. Ice segregation was the usual cause of disintegration, in particular for rocks that show high fissility by lamination or orientation of mineral grains, or otherwise high permeability that facilitate the entry of water.

On the scale of mineral grains, cryogenic weathering has a distinctly mineralogy-dependent influence. Quartz, the most abundant mineral in sands, is much less resistant to cryogenic weathering than other minerals. Although quartz is chemically stable in cold and warm climate conditions, quartz grains are physically vulnerable because they shatter easily along imperfections in the crystal lattice, producing silt-sized material in the grain-size range of 4–64 μm in cold climates (Konishchev and Rogov 1993; Dixon et al. 2008; Schwamborn et al. 2012). This contributes to a high silt loam content in fine-grained soils and deposits, e.g. the Yedoma deposits (Schirrmeister et al. 2011). Also clay mineralogy in fine-grained deposits may change; Konishchev and Rogov (1993) report on experiments at which chlorites were destroyed and formation of montmorillonite occurred. Freeze-thaw experiments also have resulted in changes in pH and cation exchange capacity, which increases.

Chemical weathering rates are not necessarily low in Arctic environments. The solubility of $CaCO_3$ in water increases at low temperatures (French 2018). In an Arctic valley in Northern Sweden, the transport of solutes and solids in streams indicates an at least 6 times higher denudation rate by chemical dissolution than by mechanical fluvial denudation (Beylich et al. 2005). In the Arctic alpine Kärkevagge valley in Northern Sweden, the contribution of chemical weathering to landscape lowering is as important as in other environments (Dixon et al. 2008). Weathering rinds on rocks created by surficial chemical alteration are common, and the solutes carried by by streams testify of chemical weathering similar to that in other climates. Moisture supply and water movement, not air temperature, is the dominant control of rock and soil weathering at all spatial scales in this valley (Thorn et al. 2001; Dixon et al. 2008).

Moreover, just as in a warm climate, cold climate vegetation, including the smallest life forms such as lichens, generate sufficient organic acids to maintain chemical weathering (Etienne 2002; Thorn et al. 2006, 2007; Dixon et al. 2008). Thorn et al. (2006) observed variations in weathering with vegetation type. Mass loss of buried dolomite discs under heath vegetations were lower than under the other vegetation types.

3.8 Ice and Hydrology

In climates with seasonal frost and permafrost, soil pores may contain solid ice, side by side with the water and gas phase. This restricts water flow in the soil. In general, soil in frozen condition is considered as a confining layer or acquiclude, a layer with very low permeability for water flow. Soil freezing results in a drastic reduction of the permeability (Woo 2012; Walvoord and Kurylyk 2016, and references therein).

However, this impermeability of permafrost is not absolute. Frozen soil that is unsaturated with ice allows flow of water, at least until this water freezes, and unfrozen water may occur at temperatures well below zero (Boike et al. 1998; Woo 2012). The basic equation for groundwater flow is Darcy's law:

$$D_{water} = -K\,dh / dx \tag{3.2}$$

in which D_{water} is the water discharge through a unit surface area (m s^{-1}), K is the saturated hydraulic conductivity factor (m s^{-1}) depending on the permeability of the soil, and dh/dx the water pressure gradient (m m^{-1}).

K depends on properties of the soil (structure, texture) and on the water content of the soil pores. K is measured under fully water-saturated conditions. When the water content is reduced by freezing or drying of the soil, K is also reduced. Freezing of the soil reduces the hydraulic conductivity factor K of the soil because of accumulation of ice in the pore space, drying reduces K because of air entry into the pores. Kurylyk and Watanabe (2013) discuss several theoretical and empirical formula relating K to temperature, or to the soil freezing curve that relates the unfrozen water content in the soil pores to temperature, discussed in Sect. 2.5.

Permeable zones that allow groundwater flow are the active layer in summer, and supra- or intra-permafrost taliks (Fig. 3.26). Permafrost separates groundwater flow into a deep groundwater system below the permafrost, and a surficial groundwater system in active layer and supra-permafrost taliks. In discontinuous permafrost areas these systems connect with each other through open taliks, that reach through the permafrost (Bense et al. 2009; Woo 2012; Walvoord and Kurylyk 2016). In continuous permafrost areas, open taliks may develop as a result of specific geohydrological conditions, such as warm springs and groundwater with high salt concentrations (Van Everdingen 1990; Woo 2012). Seepage of water through the permafrost results in the formation of pingos and icings as described in Sect. 3.4.

Water movement in the soil is coupled to heat transport, depending on the water temperature and that of the surrounding soil. Whenever water flow is possible, it

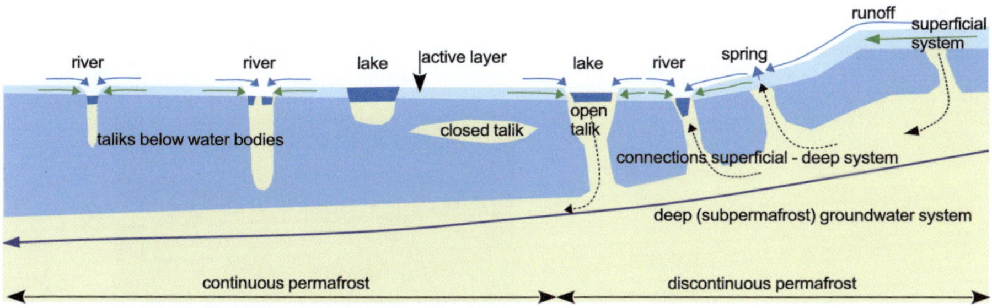

Fig. 3.26 Groundwater flow in continuous an discontinuous permafrost. The groundwater flow splits into a superficial system with groundwater flow through the active layer when thawed, and a deep system below the permafrost and through taliks. (After Walvoord and Kurylyk 2016)

also transports heat, that may enhance thaw, or in the case of supercooled water with dissolved salts, enhance freezing (Woo 2012).

3.8.1 Active Layer Hydrological Processes

The hydrology of the active layer is extremely important for the carbon cycle as it determines moisture supply, transport of carbon and nutrients by groundwater, and the decomposition processes in the soil (Koch et al. 2013). Groundwater flow through the active layer furthermore contributes to lateral heat flow in the soil, enhancing active layer thaw (Wright et al. 2009; Boike et al. 2008a; Jorgenson et al. 2010; Woo 2012). Figure 3.27 shows a general schematic of the water balance of permafrost soils, Fig. 3.28 shows a more detailed schematic of water movements induced by freezing and thawing. Active layer hydrology shows a strong seasonality. In the summer, the active layer thickness is increasing, and water is supplied by snow meltwater, rain, and thaw of excess ice in the ice-rich top of the permafrost. Water is lost by evapotranspiration. In winter, water movement is restricted and evaporation rate is very low, although not absent.

The active layer acts as an aquifer in summer that contributes with groundwater flow over the permafrost table to stream discharge. The hydraulic gradient driving this flow depends on the local surface slope. The discharge is determined by the hydraulic gradient and the permeability of the active layer, its thickness and water saturation. Unsaturated soils have a lower permeability factor for water flow than saturated soils. Climate-driven increasing active layer thickness is expected to enhance surface drainage and summer discharge of rivers (Walvoord and Kurylyk 2016; Chap. 5).

The depth of the permafrost table below the surface varies through differential thaw, with depressions of various depth and size. When the groundwater table is low, active layer groundwater is stored in isolated pockets where the permafrost top is lowest, and groundwater flow is retarded by thresholds in the permafrost table

Fig. 3.27 General water balance of permafrost soils

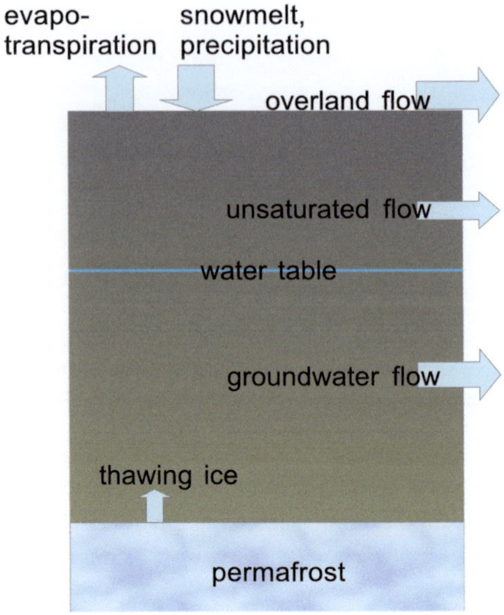

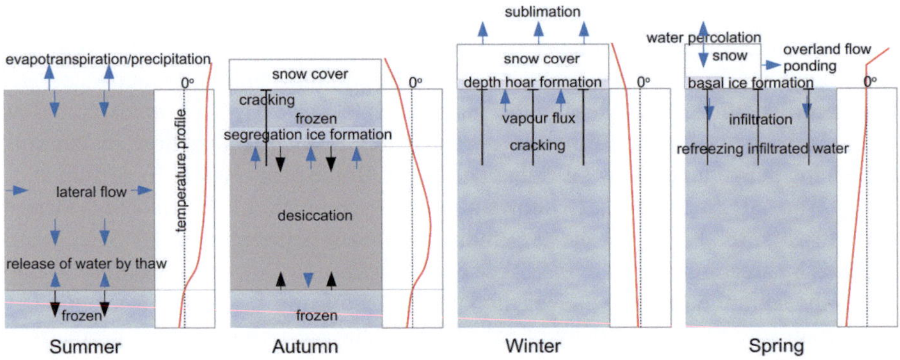

Fig. 3.28 Schematic of active layer processes affecting water transport throughout the year, with approximate soil temperature profiles. After Woo 2012. Summer: water added to soil by thawing permafrost ice; autumn: segregation ice formation and cracking; winter: water vapour flux from soil to base of snow cover; spring: snowmelt and overland flow, infiltration of water into soil. See text for further explanation

(Fig. 3.29). With a rising water table, successively higher frost table thresholds will be overtopped, restoring flow connectivity between the depressions. This fill-and-spill mechanism (Wright et al. 2009; Woo 2012) influences the timing and magnitude of groundwater discharge after input of meltwater or rain.

During soil freezing, the freezing front progresses downward from the top, but with a lower speed also from the top of the frozen soil at the base of the active layer upwards (Mackay 1983). Ice lens formation behind the progressing freezing fronts withdraws water from the still unfrozen soil (Sect. 3.2; Fig. 3.28), causing desicca-

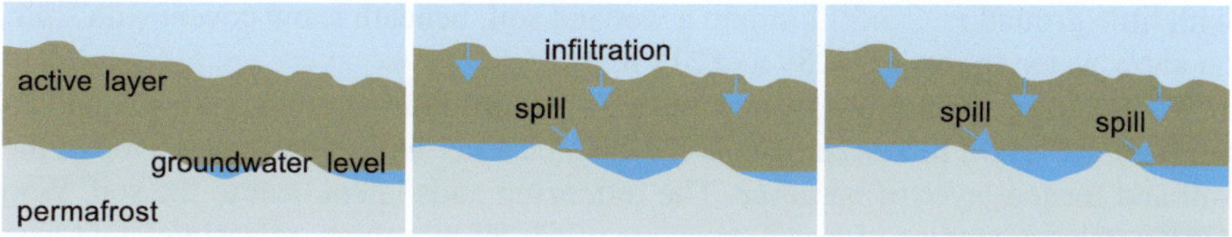

Fig. 3.29 Fill-and-spill concept. Gradual filling of depressions in the top of the permafrost by infiltrating water, followed by spill of groundwater over thresholds in the top of the permafrost. After Woo, 2012

tion and undersaturation of the soil pores with ice (Woo 2012). Cracking of the soil by dilatation upon frost heave, thermal contraction and desiccation also may occur at the same time (Woo 2012; French 2018); cracks may promote exchange of gases with soil and atmosphere in winter (Chap. 4).

However, lateral flow of water in the active layer may delay ground freezing by transporting heat from elsewhere, and adding water that releases latent heat upon freezing. In water-saturated soils, confinement of water between the permafrost and the freezing front can build up hydrostatic pressures (Sect. 3.5), in particular when lateral water flow adds water. This may break to the surface and form layers of ice (icing) on top of the soil in wetlands (Woo 2012).

During the winter, a large part of the annual precipitation is stored in the snow cover. Vapour transport from the soil to the snow cover results in water loss from the soil, which is stored as depth hoar at the base of the snow cover. The rate of vapour transport depends on the gradients of water vapour pressure and soil temperature. Depth hoar formation in the order of 2–4 millimeter of water equivalent have been observed (Woo 2012). Hinkel et al. (1996) found evidence of drying in winter in soil cores, averaging 17%.

In spring, the snow melt period transfers water again into the soil. The melt period may last for several weeks. The energy balance at the surface, in turn depending on weather and surface conditions, determines the melt rate. Cold spells can slow melting, or add additional snow; warm weather and rain increase the melt rate. The total amount of meltwater produced depends on the snow cover water equivalent, and the balance of sublimation/evaporation and precipitation during the melt period (Woo 2012). Thick snow accumulations on the lee side of slopes or in tight valley heads may continue to produce meltwater throughout the summer.

Water percolating through the snowpack refreezes at depth. The resulting higher snow density alters the thermal properties of the snowpack. When the water reaches the soil surface, part of it infiltrates, although the active layer is still frozen with temperatures well below zero. Besides the normal factors that govern the infiltration capacity for unfrozen soils such as soil texture and structure, the infiltration capacity for frozen soils also depends on its ice content. If a soil has been saturated with water before freezing, its infiltration capacity will be low. Crack formation and desiccation of soil horizons during freezing enhances infiltration in spring. Infiltration proceeds rapidly at first, and then slows down to a lower steady state condition (Woo 2012). Marsh and Woo (1993) measured infiltration of 38 mm on gravelly soil with little ground ice, and 12 mm in a wetland soil, beneath snow covers with similar snow water equivalent (289 and 217 mm).

When the meltwater production exceeds the infiltration rate into the soil and its temperature is still below zero, a part of the meltwater will refreeze at the top of the soil and form a layer of basal ice. The refreezing adds latent heat to the soil. Woo (2012) gives examples of basal ice layers of 63–89 mm thick, releasing latent heat between 9.5 and 13 MJ m^{-2}; the thickest was an ice layer produced from a relatively shallow snow pack of 0.4 m in a wetland soil with low infiltration rate. Basal ice formation also decreases the infiltration into the soil strongly, resulting in a larger amount of water available for runoff generation or pond formation at the surface.

3.8.2 Runoff and River Discharge

Surface runoff is strongly determined by the micro-relief, created by vegetation and frost action (Sect. 3.5). Ice wedge polygons can have opposite effects on drainage, either ponding of water in low-centred polygons, or enhanced runoff through ice wedge troughs (Sect. 3.3). Figure 3.11 illustrates the importance of drainage through ice wedges. Much of the drainage in flat, vegetated terrain tends to be diffuse drainage rather than channelized drainage. Smaller drainage channels in flat, densely vegetated permafrost terrain can hardly be recognizable as such, being not much more than a series of depressions with a dense sedge vegetation through which water flows, both below and above the surface, alternating with flow through ice wedge depressions. Another typical feature is beaded drainage, when ponds develop in small streams above ice wedge junctions, like beads on a chain (Fig. 3.30). The combination of labyrinths of flow paths created by micro-relief and vegetation delays surface flow in permafrost wetlands. This retention of water at the surface promotes evaporation (Woo 2012).

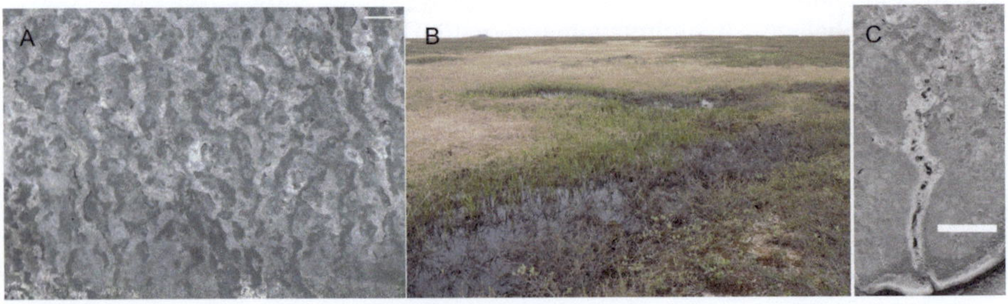

Fig. 3.30 (**a**) pattern of sedge-filled drainage lines (light tones) on a gently sloping surface of a drained thaw lake bottom in Northeastern Siberia (Kytalyk Research Station). Scale bar (top right) is 50 m. (**b**) Ground view of one of the drainage channels in **a**, bordered by Betula nana heath on either side. (Photo by the author). (**c**) Beaded drainage channel, scale bar 100 m. (Satellite images: GeoEye, August 2010, GeoEye © 2019 MAXAR)

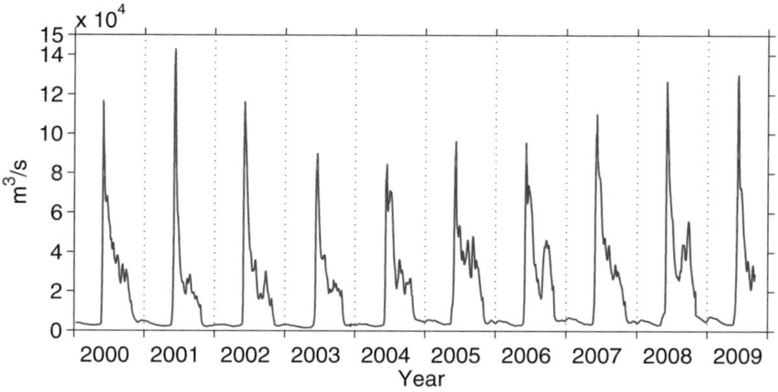

Fig. 3.31 Discharge of the Lena river at Kusur, close to the river delta, showing a pronounced spring discharge peak. (Data from Russia Arctic Net/Roshydromet, http://www.r- arcticnet.sr.unh.edu/ObservedAndNaturalizedDischarge-Website. See Shiklomanov and Lammers 2009. Republished with permission of Elsevier Inc. from: Periglacial Fluvial Sediments and Forms, Van Huissteden et al. (2013), In: Elias S.A. (ed.) Encyclopedia of Quaternary Science Vol. 3, © 2013; permission conveyed through Copyright Clearance Center, Inc)

The variability of river discharge throughout the year is known as the discharge regime or annual hydrograph of the river. The rapid release of snow meltwater in spring, in combination with restricted infiltration in the frozen topsoil and low evapotranspiration, strongly influences the annual distribution of the discharge of high-latitude rivers. This characteristically shows a major discharge peak in late spring (Fig. 3.31). The snowmelt runoff may represent most of the yearly precipitation in dry summer climates (Woo 1990). The importance of snowmelt runoff decreases southward, while the effects of rainfall runoff and groundwater discharge increases. Weather conditions during snowmelt are important for the magnitude of the spring discharge peak (Woo 2012). High temperature and rain on snow may significantly speed up snow melt and cause high discharge peaks, lower temperatures cause delay of snowmelt. Antecedent conditions such as a wet summer or early fall with much rain may fill lake and soil storage before the winter, leaving less storage space for meltwater in the next spring.

River and stream discharge is notoriously difficult to measure, in particular during the major hydrologic event of the year, the spring breakup when river ice breaks up and river water rises rapidly. Then high flow speeds occur, and abundant ice floating in the rivers or very coarse sediment transport may preclude streamflow measurements. Also, for wetland basins with abundant multiple channels, diffuse drainage and poorly defined drainage basins, reliable discharge characterization is difficult (Woo 2012). Nevertheless, river discharge throughout the year is an important component of the permafrost carbon cycle, because of the riverine transport, decomposition and redeposition of carbon from permafrost soils (Vonk and Gustafsson 2013).

Four main types of Arctic and subarctic discharge regimes are distinguished (Church 1988): the Arctic nival regime, the Subarctic nival regime, the proglacial regime and the wetland regime. Additionally, these discharge regimes are modified by the presence of springs, large lakes which act as a water buffer, or by rainfall regimes in mountainous catchments (Woo 2012).

- The Arctic nival discharge regime is characterized by the presence of a pronounced spring flood and decreasing flow in summer. The spring floods may be exacerbated by ice jams (Woo 2012). Winter ice cover in cold continental rivers may be a metre or more thick, and creates large and strongly erosive ice jams (e.g. Costard et al. 2014). In High Arctic watersheds, the spring flow is most pronounced. In basins underlain by continuous permafrost, groundwater discharge towards the river is very low or absent. Only the larger rivers sustain significant winter discharge from groundwater flow.
- In subarctic river basins, which have a larger proportion of discontinuous permafrost, groundwater flow (base flow) contributes more to the river discharge, and summer rainstorms may produce flood peaks of higher magnitude than the spring flood. Nevertheless, the spring snowmelt discharge is still pronounced.
- Proglacial rivers derive their discharge from melting glacier ice; the discharge continues to rise throughout the summer, is highest in August and shows diurnal rise and fall in summer.
- The wetland discharge regime is typical for poorly drained basins and peatlands. The spring snowmelt discharge is pronounced. However, in summer water storage in lakes, peat soils and vegetation contributes to a high evaporation and slow response of the river to rainstorms (Roulet and Woo 1988; Woo 2012).

It is expected that a climate-driven shift from continuous to increasingly discontinuous permafrost will enhance the contribution of groundwater flow towards river system, resulting in a larger baseflow. This is accompanied by changes in the height and timing of flood discharges (Frey and McClelland 2009; Walvoord and Kurylyk 2016). This will affect carbon and solute transport by rivers, and the biogeochemistry of floodplains in permafrost regions. The effects of climate - related river discharge changes are the subject of discussion in Chap. 5.

3.9 Thaw Lakes

Thaw lakes and ponds have been introduced shortly in Chap. 1, as a landform that results from surface subsidence by permafrost thaw in terrain underlain by ice-rich permafrost. As discussed above, ice-rich permafrost has a larger pore volume filled with ice, than would normally exist in a water-saturated soil or sediment of the same material. If excess ice thaws, settlement will occur by downward sinking of particles and upward movement of water, or by removal of water by groundwater flow and evaporation. Thaw settlement results in surface subsidence. The amount of subsidence depends on the thickness of the layer that has been thawed, its porosity/ice content before thaw, and its porosity after settlement (French 2018). The surface subsidence creates depressions, collecting water with formation of ponds and lakes.

The thermal effect of a larger sized water body feeds back on the permafrost temperature and induces further thaw by enhancing heat flow into the permafrost below the lake bed and and the lake banks (Williams and Smith 1991; Ling and Zhang 2003; Sect. 2.7, Fig. 2.19).

Thaw lakes, their significance, genesis and forms have been discussed in an illustrative book chapter by Grosse et al. (2013). Approximately one quarter of all lakes on Earth occur in the High North (Fig. 3.32) as shown by the Global Lake and Wetland Database (GLWD) of Lehner and Döll (2004). A large part of the high-latitude lakes in the areas of Pleistocene glaciations (Canada, Scandinavia) occurs in bedrock terrain and originates from glacial erosion of the rock surface. However, many of the lakes in sedimentary lowlands underlain by permafrost, are thaw lakes. The GLWD does not map lakes that are smaller than 0.1 km^2, and has been validated only down to a lake size of 1 km^2. Many thaw lakes are smaller than that, and the GLWD probably underestimates the high-latitude lake area. Grosse et al. (2013) estimate the total thaw lake area between 250,000 km^2 and 380,000 km^2.

The water balance of a thaw lake can be summarized as.

$$\Delta W = P + R + I + G_{\sup} + G_{deep} - \left(E_{water} + E_{snow}\right) \tag{3.3}$$

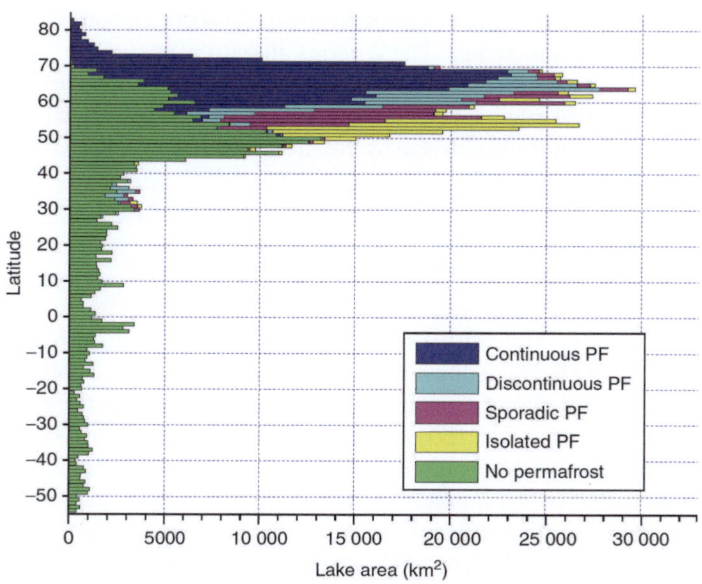

Fig. 3.32 Global lake area distribution as a function of latitude. Most of the Earth's lake area occurs in high northern latitudes, and the majority of these lakes are located in permafrost regions (bins are 1° latitude wide). Data from Global Lakes and Wetland database, Lehner and Döll (2004). (Republished with permission of Elsevier Inc. from: Thermokarst Lakes, Drainage, and Drained Basins, Grosse et al. (2013), In: Shroder et al. (eds.) Treatise on Geomorphology Vol. 8, © 2013; permission conveyed through Copyright Clearance Center, Inc)

Here ΔW is the water volume change, P is the precipitation (snow and rain), R is the net surface runoff input and output (diffuse overland flow and streams), I the water volume from ground ice thaw, G_{sup} the supra-permafrost groundwater flow through the active layer, G_{deep} the intra-permafrost groundwater flow through taliks and sub-permafrost groundwater loss or addition, E_{water} the net evaporation/condensation of the lake water surface and E_{snow} the net evaporation at the snow surface (modified after Fedorov et al. 2014).

The contribution from ground ice thaw I can be considerable, in particular in early stages of permafrost thaw. Fedorov et al. (2014) describe the development of a thaw lake near Yakutsk, which expanded in area by subsidence from a pond with an area of 195 m^2 in 1993 to a lake of 3135 m^2 in 2008. About one third of the water volume in the lake was derived from ground ice thaw. The I term for older lakes is relatively constant over time due to the slower heat transfer into the permafrost through an increasingly thicker sediment cover and talik (see next section). In continuous permafrost areas, the G_{deep} term will be small or negligible. However, if the talik below the lake has contact with sub-permafrost groundwater, its importance increases (Woo 2012; Walvoord and Kurylyk 2016). A change in the runoff term, for instance by a change in the outflow volume through erosive downcutting of the outflow affects lake water volume strongly. Kravtsova and Bystrova (2009) found in northern Siberia also evidence of the reverse process; enlargement of lake size by increased inflow from the drainage system.

Changes in precipitation and evaporation (P and E) occur on both short (weather) and long (climatic) time spans. These terms therefore contribute to short term changes in lake water volume and area (Plug et al. 2008; Labrecque et al. 2009). Labrecque et al. (2009) relate lake area variations in the Old Crow basin (Northwest Canada) to variations in P and E driven by the Arctic and Pacific Decadal Oscillations.

3.9.1 Thaw Lake Formation and Geomorphology

Thaw lakes are not restricted to continuous permafrost. They can result from the thaw of palsas in peatlands in discontinuous permafrost areas (Kirpotin et al. 2009, 2011). In their early stage of formation, lakes grow by coalescence from isolated ponds (Fig. 3.33). Grosse et al. (2013) cite a number of mechanisms: coalescence of irregular subsidence ponds; coalescence of thawing ice wedge ponds, and coalescence of low-centred polygon ponds by thawing of the ridges along ice wedges. A change to a more positive water balance - a larger precipitation surplus - also contributes (Jorgenson and Shur 2007). This has been suggested for the development of thaw lakes during Last Glacial Termination (Walter et al. 2007). A strict definition of thaw lakes holds that subsidence is the main mechanism of formation (Jorgenson and Shur 2007). However, other types of lakes, e.g. lakes on river flood plains, may develop features of thaw lakes by deepening through subsidence and thaw processes along banks (Hopkins 1949; Van Huissteden et al. 2013; Fig. 1.24).

The existence of even shallow ponds retards freezing of the active layer in winter and induces deeper thaw in summer (Langer et al. 2015). Once the water body

Fig. 3.33 Landscape change by thaw lake development at the Yukechi monitoring site in Central Siberia, 50 km SE from Yakutsk. From Fedorov et al. 2014. Upper left: hummocky surface created by initial stage ice wedge degradation. Upper right: coalescing pond formation in ice wedge troughs. Lower left: a small lake has formed and a talik is developing under the lake. Lower right: A large thaw lake with deep talik has formed. (Republished with permission of John Wiley and Sons Inc. from: Estimating the water balance of a thermokarst lake in the middle of the Lena River basin, eastern Siberia, by Fedorov et al. (2014), Ecohydrology 7, © 2014; permission conveyed through Copyright Clearance Center, Inc)

reaches a depth at which it does not freeze to the bottom in winter, the bottom temperature of the lake remains above 0 °C, with continuous permafrost thaw as a result (see also Sect. 2.7). This generally requires depths over 2–2.5 m (Burn 2002, 2005; Arp et al. 2012, 2016; Boike et al. 2015). In shallow parts of the lakes (areas shallower than the winter ice thickness) winter cooling and freezing maintains lower average bottom temperatures (Burn 2005). Kanevskiy et al. (2017) describe stabilisation mechanisms (vegetation growth, sediment accumulation) that may suppress thaw lake expansion/depth growth in this first stage (Chap. 6). They suggest that in areas of epigenetic ice wedges, where large amounts of ground ice are restricted to the top metres of the permafrost, the probability of large-scale thaw lake formation from thawing ice wedges is very small due to these stabilisation mechanisms and the restricted amount of ground ice.

The water surface in summer has a low albedo and absorbs more longwave radiation than the surrounding surface. Mixing of the lake water in summer enhances heat transfer to the lake bottom. Boike et al. (2015) report summer temperatures up to 15 °C in thaw lake bottom water in Northern Siberia; winter temperatures range up to 4 °C. These water temperatures are considerably higher than the mean annual air temperature. The heat capacity of the water is about two times higher than ice, and up to four times higher than frozen soil (Table 2.2), and therefore represents a large heat reservoir in the landscape (Jeffries et al. 1999; Boike et al. 2015). The energy density (MJ m^{-3}) of a lake water body in Siberia was about six times higher than that of the surrounding permafrost. Heat fluxes through lakes with floating ice were more than ten times higher than through a normal tundra surface, and more than three times higher than through a lake with grounded ice (Jeffries et al. 1999). Lake development therefore alters the land surface heat budget.

By continued heat transfer from the lake water into the underlying permafrost, a talik is formed below the lake (Burn 2002, 2005). Thawing of the permafrost requires a net yearly heat flux downward through unfrozen lake sediment downward to the underlying permafrost. This heat flux shows seasonal variations, including reversals of summer-stored heat flowing back to the sediment surface and water column in autumn (Boike et al. 2015). The depth of the talik depends on the lake bottom water temperature, the thermal conductivity of lake sediment and permafrost, and the age of the lake, as demonstrated by heat transfer modelling (Ling and Zhang 2003). For mean lake bottom temperatures of 1°, 2°, and 3 °C, the maximum modelled thickness of the talik reached thicknesses of 28, 43, and 53 m in 3000 years. A model by West and Plug et al. (2008) showed the effects of thermal properties of the permafrost. Talik thickness increases with higher thermal diffusivities. The lake talik development proceeds fast at first, and after time slows down and becomes approximately linear with age (Ling and Zhang 2003, West and Plug 2008; Fig. 3.34). West and Plug et al. (2008) showed that lake depth relates to the square root of time. Lake depth also increases with ice content of the permafrost (Plug and West 2009). Schwamborn et al. (2002) recorded a talik depth of 95 m below a 7000 year old lake in a Lena Delta terrace.

The permafrost temperature near the lake banks rises also (Ling and Zhang 2003; Fig. 2.19). Further heat transfer towards the banks results from wind driven surface currents, transferring warmed surface water towards windward banks. Thermal and

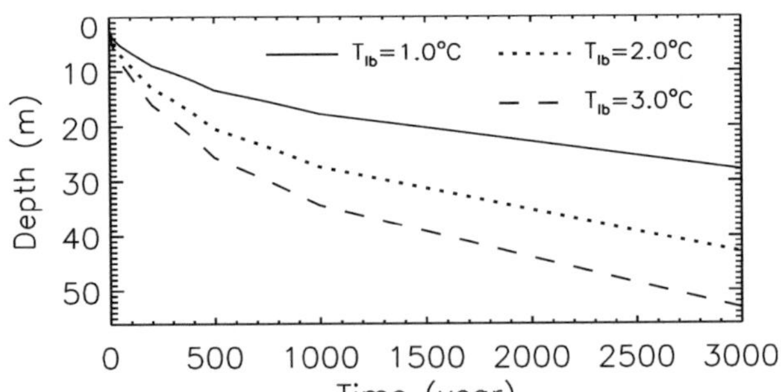

Fig. 3.34 Modelled talik thickness increase over time for thaw lakes with average bottom temperatures of 1.0 °C, 2.0 °C and 3.0 °C. (Republished with permission of John Wiley and Sons Inc., from Ling and Zhang (2003): Numerical simulation of permafrost thermal regime and talik development under shallow thaw lakes on the Alaskan Arctic Coastal Plain. Journal of Geophysical Research, doi:https://doi.org/10.1029/2002JD003014; permission conveyed through Copyright Clearance Center, Inc)

physical erosion of the lake banks can result in rapid expansion of the lakes, in particular in areas where the banks have a high ground ice content. In lakes with sufficient fetch, wave erosion combines with the warm surface water to produce thermal and physical erosion of the banks (Hopkins 1949; Grosse et al. 2013).

Grosse et al. (2013) list several types of lake bank erosion phenomena: wave erosion, over-steepening of banks due to thaw subsidence, thaw slumps on steep banks that expose permafrost ice and produce mudflows, bulldozing action by moving of ice masses over the banks during the spring breakup (ice shove), incorporation of polygon centre ponds or ice wedge troughs on lake banks. A spectacular phenomenon of wave and thermal erosion are thermo-erosional niches, deep cavings at the waterline in banks that consist of ice-bound non-cohesive material, which is thawed and washed away by the waves (Fig. 3.35).

Retrogressive thaw slumps on steep banks produce large amounts of sediment, including vegetation and organic-rich soil horizons that are washed into the lake. This is biogeochemically an important process, since soil organic material is moved from a relatively well aerated environment into an anoxic environment where it may fuel CH_4 production (Walter et al. 2006; Kessler et al. 2012; Chap. 4). These mass wasting processes are also influenced by weather conditions, becoming more active on warm or rainy days (French 2018). Thaw slumps also change the heat balance of

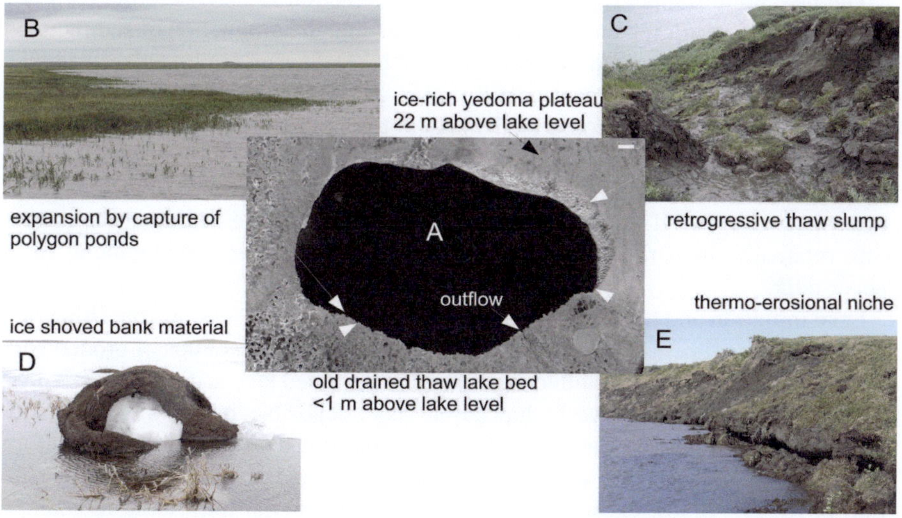

Fig. 3.35 Examples of lake bank erosion along a thaw lake near Kytalyk Research Station (**a**) Worldview panchromatic image 0.4 m resolution, July 10, 2015, Indigirka Lowlands, Eastern Siberia. The lake is actively expanding in a ± 22 m high ice-rich yedoma plateau by wave and thermal erosion, with development thermal erosional niches (**e**), thaw slump with flowing mud and vegetation rafts (**c**). On the southern and eastern side the lake borders on the bed of an older, drained thaw lake, with a surface elevation only slightly higher than the lake level. Here, the lake expands by ice shove of the vegetation mat during spring breakup (**d**) and capturing of polygon ponds (**b**). The ridges protruding into the lake are ice wedge ridges. (Photos by the author. WorldView © 2019 MAXAR)

the soil by removing vegetation and catching a thicker snow cover, further warming the permafrost and enhancing lateral expansion of lakes (Kokelj et al. 2009).

Lake expansion rates cited in literature vary strongly; high values up to 10 m $year^{-1}$ have been reported (references in Plug and West 2009). Expansion rates determined over longer time periods are small, in the range 0.3–0.8 m $year^{-1}$ (Grosse et al. 2013); larger values are likely short-lived and local phenomena. A complicating factor for determining lake size is, that lake size also depends on the water balance which varies from year to year with precipitation and evaporation (Plug et al. 2008).

Plug and West (2009) and Kessler et al. (2012) incorporated mass wasting and sediment redistribution across the lake floor in a model of thaw lake expansion. Plug and West (2009) tested two scenarios for ground ice configuration: permafrost with a thick ice-rich layer and permafrost with a thinner epigenetic ice-rich layer on top. Lake expansion progressed three times faster in the thicker ice-rich permafrost, because more ground ice allows more subsidence and deeper lakes, with more storage space for the sediment which is eroded from the banks. The maximal modelled expansion rate was 0.26 m $year^{-1}$. Thawed material stored on the banks and near-shore lake bottom shields the underlying permafrost by slowing heat flow (Brouchkov et al. 2004); removal of stored sediment to deeper parts of the lake is therefore essential for lake expansion (Kessler et al. 2012).

The lake bottom topography is shaped by interaction between ground ice thaw and subsidence rate of the underlying permafrost, and the production of sediment on lake banks and its redistribution throughout the lakes. During the first phases of lake development, talik development is still minor, and the lake sediment is frequently disturbed by unequal subsidence of the lake bottom due to ongoing thaw of excess ice and ice wedges (Farquharson et al. 2016). Below the water surface, coarser sediment is transported by debris flows to deeper lake parts, while fine-grained material is transported by wind-driven currents before it settles. Lake sediment deposition with relatively undisturbed lamination occurs only after development of a deeper talik.

Lake initiation and expansion results in a sedimentary succession consisting of a basal layer of sediment with a wide range of grain sizes, with lumps of eroded soil, and pockets of slumped or faulted laminated fine-grained lake sediment. These heterogeneous basal deposits overly thawed-out older deposits with ice-wedge pseudomorphs (traces of thawed-out ice wedges). The basal lake sediment layer is in turn overlain by well-laminated lake sediment (Hopkins and Kidd 1988 Farquharson et al. 2016; Fig. 3.36). Close to the banks, coarser material may settle in shallow shelves, often with a delta-like internal structure with a horizontally laminated top and a cross-laminated base (Van Huissteden 1990; Murton 1996, 2001). Shallow water depth may retard permafrost thaw on these shelves (Burn 2002, 2005).

In the centre of thaw lake basins, often a topographical bulge is present (West and Plug 2008). This bulge is created by the interaction between thaw subsidence and sediment accumulation as is demonstrated by a model by Kessler et al. (2012). Sediment accumulation is thickest in the centre of the lake basin, which is generally also the oldest part of the basin. Here, thaw subsidence slows down over time,

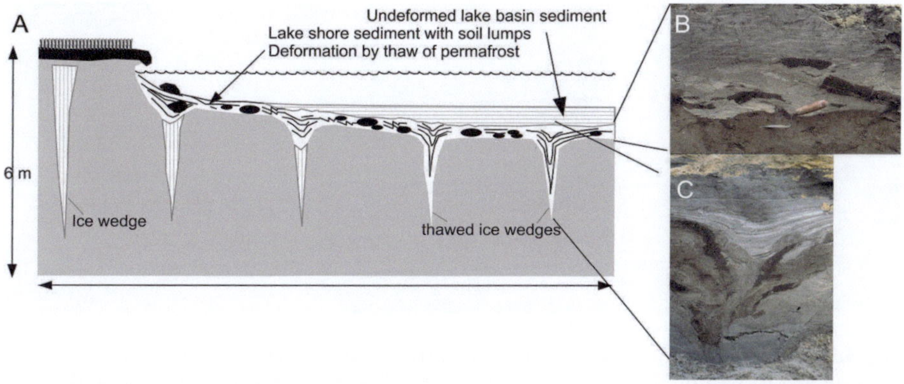

Fig. 3.36 (**a**) Sedimentary succession in a thaw lake, redrawn after Hopkins and Kidd (1988). (**b**, **c**) Mid Weichselian age thaw lake sediments in the Netherlands (Hengelo). (**b**) detail of erosive contact between thawed out subsurface material (loess) and overlying lake sediment; the basal material of the lake sediment consist of transported sand, silt and peat lumps, overlain by laminated silt. (**c**) Thawed out ice wedge (see also Fig. 3.21) below the lake sediment in the same succession. (Photos by the author; Van Huissteden 1990)

because of slow heat transfer through the thick, unfrozen sediment mass and lake talik, which shields the underlying permafrost from further thaw. Closer to the banks, sediment accumulation started later and subsidence progresses fastest. This central bulge is still seen in many drained thaw lakes, where the central area is higher than the areas closer to the former banks (Fig. 1.24, partially drained lakes on the right side of the image).

In several regions, thaw lakes show an intriguing variation of distinct lake shapes and lake orientation (alignment of the long axis of lakes along a particular direction). A more or less smoothed outline shaped by wave action and longshore sediment transport is common, regardless of overall shape. Iconical examples of lake orientation are the highly elliptical, parallel oriented lakes on the Alaskan North Slope. Typical repeating shapes are elliptical, egg-shaped, triangular, rectangular, or D-shaped lakes (Grosse et al. 2013; Fig. 3.37). A digital field trip using Google Earth along the northern coasts of Siberia and North America will show the reader a beautiful sample of lake shapes.

The causes of lake orientation are debated and likely involve a combination of processes. Explanations involve:

- Dominant summer wind directions that result in wind-driven current patterns that cause preferential erosion of particular lake bank sections (e.g. Carson and Hussey 1962; French and Harry 1983);
- Subsurface structure, such as the orientation of ice wedges and other ground ice bodies, or lithological inhomogeneities inherited from past sedimentation processes (e.g. dune formation, beach ridges, river channels, e.g. Sjöberg et al. 2013);
- Sediment redistribution within the lake itself such as the creation of littoral shelves (Grosse et al. 2013).

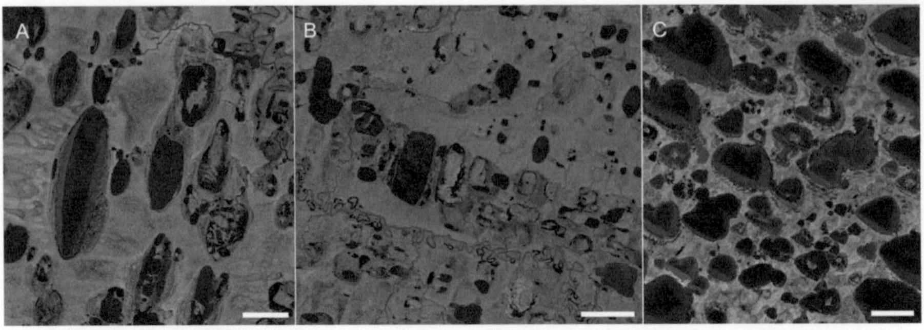

Fig. 3.37 Thaw lake shapes. (**a**) elliptical lakes; (**b**) rectangular lakes; (**c**) triangular/D shaped lakes. (**a** and **b**) East of Yana river delta, (**c**) West of Kolyma river delta, NE Siberia. (Landsat images on Google Maps. ©2019 Landsat/Copernicus, map data ©2019 Google)

The wind direction in the elliptical lakes Northern Alaska is approximately perpendicular to the length axis of the lakes. Wind-driven waves in thaw lakes create circulation cells with the strongest currents along shore sections, that are oriented at angles ±50° from the direction of wave propagation. In other shoreline sections currents are slower, allowing sedimentation. This tends to elongate the lakes in an elliptical fashion, perpendicular to persistent wind directions (Carson and Hussey 1962; Grosse et al. 2013). Lake orientation is enhanced by long exposure to the dominant wind directions (or lake age) and a uniform composition of the lake subsurface material. Morgenstern et al. (2008) analysed lake orientation on three river terraces of the Lena river delta. On the first terrace, lake orientation is poorly developed as most lake shapes and orientation are determined by recent fluvial channels. On the second terrace, which consists of homogeneous fine- to medium-grained Late Pleistocene fluvial sediments, lake orientation was best developed. The third terrace, covered by Yedoma sediments with an inhomogeneous ice content, showed less orientation.

3.9.2 Thaw Lake Disappearance

The end of the lake's lifetime is a biogeochemically important transformation of the lake basin: from an inundated, permanently unfrozen sedimentary environment with anoxic conditions in the sediment, to a terrestrial environment, where new permafrost is established if the climate is sufficiently cold.

The end of a thaw lake's lifetime is marked by drainage, drying or infilling with sediment (mineral or peat) or combinations of these. Drainage is the most frequent. Thaw lake areas generally show many, often overlapping, drained thaw lake basins (DLTB hereafter), suggesting repeated formation and drainage of thaw lakes (e.g. Fig. 1.24). These basins are known under various local names such as 'khasirei' in the peatlands in Western Siberia, and 'alas' in Eastern Siberia. Drainage or infilling need not be complete, a remnant lake may be left behind in an older, wider basin. Drying of lakes is most likely in continental areas (Grosse et al. 2013). Infilling with sediment - peat, fluvial or eolian sediment - is likely for lakes in active sedimentary environments such as river floodplains and dune fields. Infilling with vegetation and peat is common (Sannel and Kuhry 2011). In areas of active erosion by thaw slumps, lakes are filled with sediment derived from these erosion phenomena (Lewkowicz and Way 2019).

Lake drainage is the most common process. In the ~ 35,000 km^2 area of the western Arctic Coastal Plain in Alaska, 50 lakes were entirely or partially drained in approximately 25 years. Hinkel et al. (2007) list several natural causes of lake drainage by surface geomorphological processes: headward erosion of stream channels, lateral erosion by meandering streams, establishment of a connection with another lake at a lower level, and coastal erosion. Fig. 1.24 shows two thaw lakes that have been connected to the river system by erosive action of channel meandering. Both progressive lake bank erosion, and the erosive dynamics of the drainage system contribute to this type of surface drainage. The density of the drainage system, and local relief that influences fluvial erosion rates restricts lake size; lakes grow larger in size where the fluvial channel system is less dense (Morgenstern et al. 2011).

Surface lake drainage by headward erosion of drainage channels is favoured by the presence of ice wedge polygons (Mackay 1988; Marsh et al. 2009). Lake drainage is usually triggered by high water levels in the lake, resulting from high precipitation or snowmelt events, or snow or ice dams in the normal lake outlet (Mackay 1988). Lake water flowing through an ice wedge system is capable of creating deep outlet channels that can drain a lake in a matter of hours, and therefore this type of lake drainage tends to be a catastrophic process (Marsh and Neumann 2001). Formation of tunnels by erosional thaw of ice wedge ice is not uncommon at drainage. Brewer et al. (1993) ascribe the drainage of ten lakes in the Alaskan Arctic Coastal Plain to a wetter than normal summer; Marsh et al. (2009) mention a warm summer, combined with a moderately high lake level as a cause of drainage. The incidence of lake drainage events is related to climate by extreme precipitation and temperature events (Grosse et al. 2013).

In areas with thin continuous permafrost or discontinuous permafrost, the development of open taliks can connect lakes to the subsurface groundwater system (Yoshikawa and Hinzman 2003; Woo 2012; Fig. 3.26). Briggs et al. (2014) and Minsley et al. (2012) show the effect of lake drainage through a talik, including a geophysical survey of the extent of the talik below the lake. Also recharge of lake water may be the result (Kane and Slaughter 1973; Walvoord and Kurylyk 2016). The disappearance of lakes at the southern parts of the permafrost region in Siberia has been attributed to subsurface drainage (Smith et al. 2005). However, an airborne geophysical survey of lake talik configuration and statistics on lake water levels in the Alaskan Yukon Flats did not find a significant links with lake level and the presence of open taliks. Presumably, most of the lake level fluctuations were linked to changes of groundwater flow in the upper part of the permafrost (Jepsen et al. 2013;

Chap. 5). Furthermore, also human-caused lake drainage occur frequently, either intentionally by digging drainage channels or unintentionally by activities that enhance erosion (Mackay 1992b; Grosse et al. 2013).

Once a lake is drained, the talik below the lake bottom refreezes if the climate is sufficiently cold. An experimental drainage of a thaw lake described by Mackay (1997) and Mackay and Burn (2002) shows details of the refreezing process. First, the soft lake-bottom sediments are consolidated by water loss. Frost heave by the growth of aggradational ice follows. At depth, downward and upward freezing occurs within the talik. This results in solute rejection of dissolved salts, pore-water expulsion from the freezing of saturated lake-bottom sands, and convective heat transfer from groundwater flow. Solute rejection results in increasingly saline groundwater in the talik, with below-zero temperatures because of freezing-point depression. Flow of this low temperature saline water accelerates the rate of permafrost growth in the flow direction.

The refreezing of the talik occurs much faster than its formation by thaw. In the thermal model of Ling and Zhang (2003, 2004), taliks formed over 3000 model years, but refreeze again in 40–157 years, depending on their depth. This fast refreezing occurs despite the large amount of latent heat stored in the interstitial water of the talik. Refreezing occurs from the top downwards and bottom upwards, but the rate of downward freezing is 6–8 times higher than the upward freezing. However, in particular in discontinuous permafrost, new permafrost does not necessarily aggrade again. Briggs et al. (2014) describe newly formed permafrost after partial lake drainage, but this permafrost appears to have grown because of development of a protective ecosystem consisting of willow bushes.

Ice wedges develop rapidly at newly exposed lake beds if climate permits (Mackay and Burn 2002; Grosse et al. 2013; Fig. 3.38). In the experimental lake drainage site above, ice-wedge growth on the vegetation-free lake bottom started in the first winter after drainage and progressed rapidly, with wedge thickness growth rates up to 3 cm year^{-1}. However, it slowed down and practically ceased in the course of twelve years. This is attributed to rapid vegetation growth and snow entrapment in the vegetation, an increase in winter ground temperatures, and changes in the thermal contraction properties resulting from progressive consolidation of lake bottom sediments. The growth of ice wedges and ice lenses in the top of the former lake talik can be seen as a recharge of the soil with ice. Once the lake talik is confined by permafrost above and below, increasing water pressure develops within the talik, and pingos may develop (e.g. Mackay 1998; Morgenstern et al. 2011; Sect. 3.4).

Since DLTB's are generally poorly drained and flat areas, carbon accumulating wetlands tend to develop over time (Hinkel et al. 2003; Van der Molen et al. 2007). Hinkel et al. (2003) found a positive correlation between basin age (after drainage) and organic topsoil layer thickness and degree of organic matter decomposition. The amount of soil organic matter in the topsoil tends to increase with basin age, but the relation is obscured for the upper metre of soil due to cryoturbation, decomposition, and ice enrichment. Hinkel et al. (2003) also present a useful classification, relating basin age to micro-relief, soil and vegetation properties. In particular the

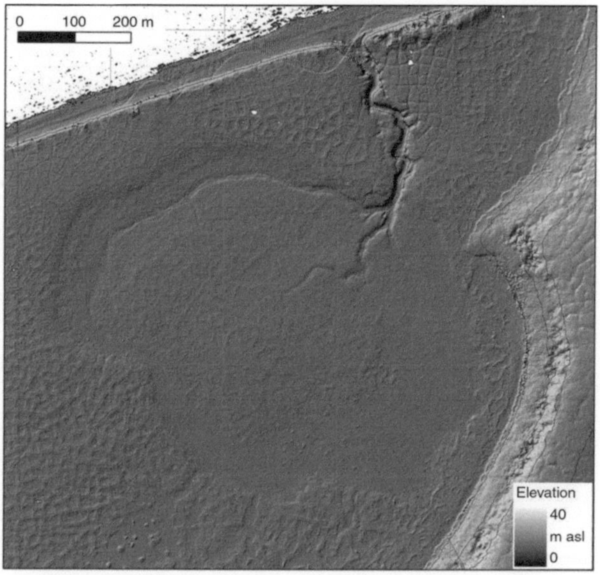

Fig. 3.38 Detailed, LIDAR-based digital elevation map of a relatively young drained lake basin within an older drained basin. Note the deeply incised drainage channel toward the sea coast in the north, the well-developed ice-wedge polygons on the surrounding older basin surface, and the very weakly developed polygons in the younger drained basin. (Republished with permission of Elsevier Inc. from: Thermokarst Lakes, Drainage, and Drained Basins, Grosse et al. (2013), In: Shroder et al. (eds.) Treatise on Geomorphology Vol. 8, © 2013; permission conveyed through Copyright Clearance Center, Inc)

development of ice wedge polygons, soil profile and vegetation diversity increase over time (Table 3.1).

Thaw lake formation is not necessarily related to the present warming climate. Radiocarbon dated thaw lake sediments show that thaw lake formation was most intense during the Last Glacial Termination, starting at about 13,000 year. BP, and Early Holocene (Romanovskii et al. 2000; Katamura et al. 2006; Walter et al. 2007; Shilo et al. 2007). However, many thaw lake areas show series of overlapping DLTB's (e.g. in Fig. 1.24), suggesting continued, cyclical thaw lake formation. It is therefore tempting to assume the existence of a thaw lake cycle, consisting of a phases of lake initiation, growth, drainage, regrowth of excess ice in the lake bed, and initiation of new lakes in the ice-rich sediments (e.g. Hopkins 1949; Billings and Peterson 1980). Central to the thaw lake cycle hypothesis is the development of a sufficient amount of ground ice in a short time, to allow recurring deep thaw subsidence (Grosse et al. 2013). Jorgenson and Shur (2007) doubt the existence of a thaw lake cycle, based on their observations in the northern coastal regions of Alaska and Canada. They point out that the rate of ground ice accumulation and frost heave during the Holocene is too low to allow repeated cycles.

A geomorphological and geological study of thaw lakes in Northeastern Siberia by Morgenstern et al. (2013) is illustrative. It showed a succession of rapid develop-

Table 3.1 Development of surface characters in drained thaw lake basins with age

Criteria	Young	Medium	Old	Ancient
Age	< 50	50–300	300–2000	> 2000
Surface organic horizon thickness	None – very thin (few cm)	Few decimetres	Few - several decimetres	Few - several decimetres
Ice content upper permafrost	Low			High
Organic matter decomposition degree	No higly decomposed material			Highy decomposed material present
Ice wedge polygons form	None, remnant high Centre	Flat Centre	Flat - low Centre	Low Centre, coalescing ponds, some high Centre
Ice wedge polygons rim height	None - incipient formation	Indistinct	Intermediate	High
Vegetation	Low differentiation, wetland graminoids, some mosses			Strong differention, species-rich, dry habitat species on polygon rims, incl. Lichens
Plant vigor/ nutrient status	High			Low
Lakes and ponds	Few remnant lakes/ ponds			Polygon Centre ponds, ice wedge thaw ponds, coalescing
Soil	Poorly developed fluvisol			Horizon differentiation, histosol, cryoturbation

Adapted from Hinkel et al. (2003)

ment of a large thaw lake starting during the Last Glacial Termination. This lake was drained by erosion of a drainage channel about 5700 year BP, leaving a remnant lake. This was followed by repeated permafrost aggradation and degradation, with renewed thaw lake activity that was largely restricted to the area inside the Early Holocene lake basin. These permafrost aggradation and degradation cycles were related to regional climate change, and local hydrological and geomorphological changes. This indicates that in Eastern Siberia, the massive thaw lake development from the Last Glacial Termination into the Early Holocene was not repeated at the same spatial scale. Instead, at smaller scale repeated lake expansion and shrinkage cycles did occur, creating overlapping drained thaw lake basins. The cycles may not need a full return to the high ground ice conditions before a new lake can start to grow (Billings and Peterson 1980; Grosse et al. 2013).

There may be self-cyclicity without a climate driver. A simplified model of regional scale thaw lake development showed self-cyclicity depending on the rate of ice recharge of drained thaw lake sediments (Van Huissteden et al. 2011). The existence of a possible self-cyclicity in thaw lake areas caution to draw conclusions about the relation of thaw lake change with climate. Nevertheless, various processes in the life cycle of thaw lakes are clearly climate-related, driven by average climate change or extreme events. The relation of present thaw lake development with climate and its feedbacks on the carbon cycle will be explored further in Chap. 5.

3.10 Mapping Ice Content

The subsurface ice content is important to understand the effect of future thaw processes and to understand permafrost history. On al large scale, there is very limited information on ground ice content of permafrost. Therefore, mapping of ice content is an increasingly important part of permafrost studies.

The simplest methods of ground ice mapping are based on taking drill core samples or digging observation pits. However, deeper surveys than a few metres in permafrost are expensive. Geophysical prospecting methods have been increasingly successful because they are less expensive. On the other hand, these methods do not asses the ice content directly, but use a geophysical parameter as a proxy value for the permafrost property being investigated, e.g. electrical resistivity for ice content. Often a numerical model needs to be used for interpretation of the measurement data (e.g. Reynolds 1997; Kneisel et al. 2008). A book edited by Hauck and Kneisel (2008) on applied geophysical methods in permafrost environments gives an extensive overview of methods that have been used for subsurface mapping in permafrost. Here a short introduction is given.

Subsurface variation in relevant geophysical parameters depends on water/ice content, ice distribution, porosity, salinity of the pore water, ground temperature and overburden pressure. Parameters that have proven their merits in geophysical prospection of permafrost are electrical resistivity, dielectric constant and seismic wave propagation velocity (Kneisel et al. 2008).

Electrical resistivity is the resistance (Ωm) of a material with a unit length and cross section. Because the phase change of water to ice results in a strong increase of the resistivity, resistivity surveys can be used to locate ice bodies and estimate ice content. However, resistivity of rocks and soils also depends on mineral properties, pore water/ice saturation, solute concentration, and temperature. Resistivities of frozen material can vary over a few orders magnitude (Kneisel et al. 2008; Reynolds 1997). The salinity strongly affects resistivity. In marine sediments, the resistivity of permafrost can be extremely low due to the high conductivity of saline pore water, even at temperatures well below 0 °C, due to freezing point depression. Clays and peats have generally a lower resistivity than sands, given a similar pore water quantity and composition.

Geo-electrical resistivity surveys (Reynolds 1997) measure the subsurface resistivities by applying an electrical potential at two closely spaced potential electrodes inserted into the soil, and measuring the current at two wider spaced current electrodes to measure an apparent (bulk) resistivity of the soil. This apparent resistivity is determined by the resistivities and thickness of the layers below the surface. The wider the spacing of the current electrodes, the deeper the penetration of the electrical current, and the more the apparent resistivity is influenced by the properties of

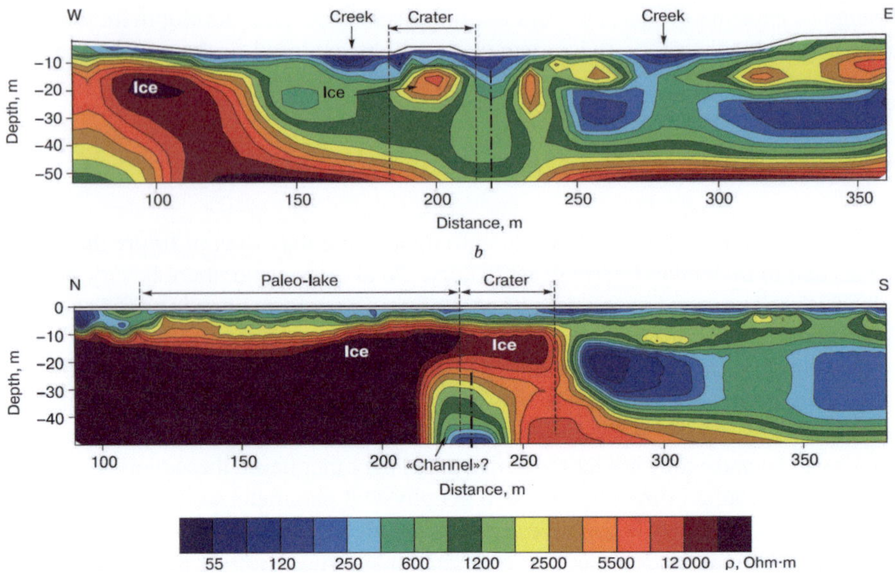

Fig. 3.39 Geo-electric profiles of a site in Yamal for mapping of ground ice configuration around a crater where methane has been emitted from the deeper subsurface. Zones with a resistivity of 100–250 Ωm correspond to frozen clays and silts, zones with resistivity over 2000 Ωm are interpreted as ice-rich sediments and ice. (Republished with permission from Kriosfera Zemli and the author, © 2015, from Olenchenko et al. 2015)

deeper layers. If, for instance, a low resistivity layer is situated at a few metres below the surface under a high resistivity layer, the apparent resistivity will become lower when the current electrodes are spaced further apart.

Two survey modes are distinguished: vertical soundings and resistivity profiles. Vertical soundings are being made by widening the distance between the current electrodes, and leaving the potential electrodes at the same spot. This results in a relation between current electrode distance and apparent resistivity, that is used to fit a model of layer thicknesses and resistivities. Horizontal profiles use a fixed configuration of the potential and current electrodes; both the potential and current electrodes are moved along transects. 2D and 3D modelling then can be used to derive a subsurface model of resistivities (Kneisel et al. 2008). Figure 3.39 shows an example of geo-electrical resistivity profiles. Lewkowicz et al. (2011) present examples of discontinuous permafrost mapping using geo-electrical profiles in various geomorphological settings. Kneisel et al. (2008) give further details on the methods used.

Electromagnetic methods (EM), such as ground penetrating radar (GPR), depend on the dielectric constant (or relative dielectric permittivity, a dimensionless quantity) of soil material. This varies for frozen and unfrozen soil materials, with values of 3–4 for ice, around 6 for frozen sediments, around 25 for unfrozen sediments and

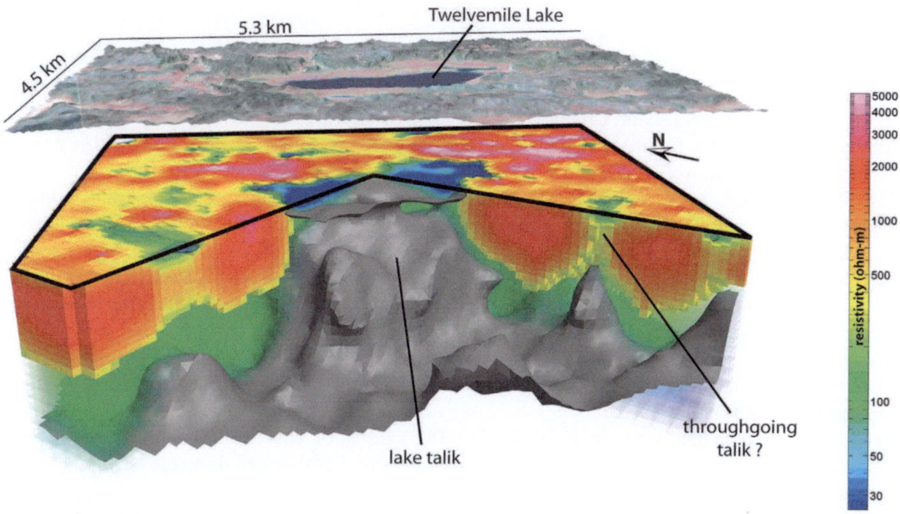

Fig. 3.40 Three-dimensional cutout view of the resistivity model in the vicinity of Twelvemile Lake in Alaska; the talik below the lake penetrates the permafrost. The grey isosurfaces are interpreted to indicate the base of permafrost in the subsurface. The upper image is a Landsat view of the region. Vertical exaggeration is 12:1. (Republished with permission of John Wiley and Sons Inc, from 'Airborne electromagnetic imaging of discontinuous permafrost' by Minsley et al. (2012), Geophysical Research Letters 39(2), © 2012; permission conveyed through Copyright Clearance Center, Inc)

80 for fresh water. The high electrical resistivity of frozen soils is favourable for use of GPR, because high resistivity results in less attenuation of the EM signal, and consequently deeper penetration. The application of GPR is limited by low resistivity layers (Kneisel et al. 2008). EM equipment does not need contact with the soil, and consist of a set of transmitting and receiving antennas for the EM signal. The equipment can be mounted on small vehicles or can be used airborne. Airborne EM has been applied to hydrologic surveys of permafrost areas, creating three-dimensional resistivity tomography of the permafrost (Fig. 3.40). See Yoshikawa et al. (2006) and Minsley et al. (2012) for examples of applications of geophysical methods, and Kneisel et al. (2008) for a more detailed discussion of methods used.

Seismic methods also have proven their merits in subsurface permafrost mapping. These employ changes in the compressional wave (P wave) and shear wave (S wave) velocities of rocks. Seismic wave velocities increase sharply upon freezing of earth materials; the increase is most marked in coarse-grained sediments. The increase in velocity is closely related to the decrease in unfrozen water content during freezing (Kneisel et al. 2008). Seismic waves propagating through a layered subsurface are reflected and refracted at interfaces between different materials in the subsoil. Schwamborn et al. (2012) show an example of a combination of GPR and seismic mapping of a lake talik in the Lena delta.

References

Abbott BW, Jones JB (2015) Permafrost collapse alters soil carbon stocks, respiration, CH_4, and N_2O in upland tundra. Glob Chang Biol 21(12):4570–4587

Abnizova A, Young KL (2008) Hillslope hydrological linkages: importance to ponds within a polar desert high Arctic wetland. Hydrol Res 39(4):309–321

Arp CD, Jones BM, Lu Z, Whitman MS (2012) Shifting balance of thermokarst lake ice regimes across the Arctic coastal plain of northern Alaska. Geophys Res Lett 39(16):L16503. https://doi.org/10.1029/2012GL052518

Arp CD, Jones BM, Grosse G, Bondurant AC, Romanovsky VE, Hinkel KM, Parsekian AD (2016) Threshold sensitivity of shallow Arctic lakes and sublake permafrost to changing winter climate. Geophys Res Lett 43(12):6358–6365

Bense V, Ferguson G, Kooi H (2009) Evolution of shallow groundwater flow systems in areas of degrading permafrost. Geophys Res Lett 36(22):L22041. https://doi.org/10.1029/2009GL039225

Beylich AA, Molau U, Luthbom K, Gintz D (2005) Rates of chemical and mechanical fluvial denudation in an arctic oceanic periglacial environment, Latnjavagge drainage basin, northernmost Swedish Lapland. Arct Antarct Alp Res 37(1):75–87

Billings W, Peterson K (1980) Vegetational change and ice-wedge polygons through the thaw-lake cycle in Arctic Alaska. Arct Alp Res 12(4):413–432

Bockheim JG, Tarnocai C (1998) Recognition of cryoturbation for classifying permafrost-affected soils. Geoderma 81(3–4):281–293

Boike J, Roth K, Overduin PP (1998) Thermal and hydrologic dynamics of the active layer at a continuous permafrost site (Taymyr Peninsula, Siberia). Water Resour Res 34(3):355–363

Boike J, Wille C, Abnizova A (2008a) Climatology and summer energy and water balance of polygonal tundra in the Lena River Delta, Siberia. J Geophys Res 113(G3). https://doi.org/10.1029/2007jg000540

Boike J, Ippisch O, Overduin PP, Hagedorn B, Roth K (2008b) Water, heat and solute dynamics of a mud boil, Spitsbergen. Geomorphology 95(1–2):61–73

Boike J, Langer M, Lantuit H, Muster S, Roth K, Sachs T, Overduin P, Westermann S, McGuire AD (2012) Permafrost–physical aspects, carbon cycling, databases and uncertainties. In: Recarbonization of the biosphere. Springer, pp 159–185

Boike J, Georgi C, Kirilin G, Muster S, Abramova K, Fedorova I, Chetverova A, Grigoriev M, Bornemann N, Langer M (2015) Thermal processes of thermokarst lakes in the continuous permafrost zone of northern Siberia – observations and modeling (Lena River Delta, Siberia). Biogeosciences 12(20):5941–5965. https://doi.org/10.5194/bg-12-5941-2015

Brewer M, Carter, LD, Glenn, R (1993) Sudden drainage of a thaw lake on the Alaskan Arctic Coastal Plain. In: Sixth International Conference on Permafrost, 1993. Proceedings of the Sixth International Conference on Permafrost. South China University of Technology Press, Beijing, pp 48–53

Briggs MA, Walvoord MA, McKenzie JM, Voss CI, Day-Lewis FD, Lane JW (2014) New permafrost is forming around shrinking Arctic lakes, but will it last? Geophys Res Lett 41(5):1585–1592

Broll G, Tarnocai C, Mueller G (1999) Interactions between vegetation, nutrients and moisture in soils in the Pangnirtung pass area, Baffin Island, Canada. Permafr Periglac Process 10(3):265–277

Brouchkov A, Fukuda M, Fedorov A, Konstantinov P, Iwahana G (2004) Thermokarst as a short-term permafrost disturbance, Central Yakutia. Permafr Periglac Process 15(1):81–87. https://doi.org/10.1002/ppp.473

Brown J, Ferrians O, Heginbottom JA, Melnikov E (2002) Circum-Arctic map of permafrost and ground-ice conditions, version 2. NSIDC. National Snow and Ice Data Center, Boulder

Burn C (1990) Implications for palaeoenvironmental reconstruction of recent ice-wedge development at Mayo, Yukon Territory. Permafr Periglac Process 1(1):3–14

Burn C (2000) The thermal regime of a retrogressive thaw slump near Mayo, Yukon Territory. Can J Earth Sci 37(7):967–981
Burn C (2002) Tundra lakes and permafrost, Richards Island, western Arctic coast, Canada. Can J Earth Sci 39(8):1281–1298
Burn CR (2005) Lake-bottom thermal regimes, western Arctic coast, Canada. Permafr Periglac Process 16(4):355–367
Calmels F, Allard M (2008) Segregated ice structures in various heaved permafrost landforms through CT Scan. Earth Surf Process Landf 33(2):209–225
Carson CE, Hussey KM (1962) The oriented lakes of Arctic Alaska. J Geol 70(4):417–439
Chen J, Blume H-P, Beyer L (2000) Weathering of rocks induced by lichen colonization—a review. Catena 39(2):121–146
Cheng G (1983) The mechanism of repeated-segregation for the formation of thick layered ground ice. Cold Reg Sci Technol 8(1):57–66
Christiansen HH (2005) Thermal regime of ice-wedge cracking in Adventdalen, Svalbard. Permafr Periglac Process 16(1):87–98
Church M (1988) Floods in cold climates. Flood geomorphology. Wiley, New York, pp 205–229
Costard F, Gautier E, Fedorov A, Konstantinov P, Dupeyrat L (2014) An assessment of the erosion potential of the fluvial thermal process during ice breakups of the Lena River (Siberia). Permafr Periglac Process 25(3):162–171
Dash J, Rempel A, Wettlaufer J (2006) The physics of premelted ice and its geophysical consequences. Rev Mod Phys 78(3):695
De Klerk P, Donner N, Karpov NS, Minke M, Joosten H (2011) Short-term dynamics of a low-centred ice-wedge polygon near Chokurdakh (NE Yakutia, NE Siberia) and climate change during the last ca 1250 years. Quat Sci Rev 30(21–22):3013–3031
De Leffingwell E (1915) Ground-ice wedges: the dominant form of ground-ice on the north coast of Alaska. J Geol 23(7):635–654
Dixon JC, Thorn CE, Darmody RG (2008) Spatial scale and chemical weathering in Kärkevagge: influences on landscape evolution. Zeitschrift für Geomorphologie 52(1):27–49
Elliott G (1996) Microfabric evidence for podzolic soil inversion by solifluction processes. Earth Surf Process Landf 21(5):467–476
Ellis CJ, Rochefort L (2004) Century-scale development of polygon-patterned tundra wetland, Bylot Island (73° N, 80° W). Ecology 85(4):963–978
Etienne S (2002) The role of biological weathering in periglacial areas: a study of weathering rinds in South Iceland. Geomorphology 47(1):75–86
Etzelmüller B, Hagen JO (2005) Glacier-permafrost interaction in Arctic and alpine mountain environments with examples from southern Norway and Svalbard. Geol Soc Lond, Spec Publ 242(1):11–27
Farquharson L, Anthony KW, Bigelow N, Edwards M, Grosse G (2016) Facies analysis of yedoma thermokarst lakes on the northern Seward Peninsula, Alaska. Sediment Geol 340:25–37
Fedorov A, Gavriliev P, Konstantinov P, Hiyama T, Iijima Y, Iwahana G (2014) Estimating the water balance of a thermokarst lake in the middle of the Lena River basin, eastern Siberia. Ecohydrology 7(2):188–196
Fortier D, Allard M, Shur Y (2007) Observation of rapid drainage system development by thermal erosion of ice wedges on Bylot Island, Canadian Arctic Archipelago. Permafr Periglac Process 18(3):229–243
French HM (2018) The periglacial environment. Wiley, Hoboken, p 515
French H, Harry D (1983) Ground ice conditions and thaw lakes, Sachs River lowlands, Banks Island, Canada. Mesoformen des Reliefs im heutigen Periglazialraum:70–81
French H, Shur Y (2010) The principles of cryostratigraphy. Earth Sci Rev 101(3–4):190–206
Frey KE, McClelland JW (2009) Impacts of permafrost degradation on arctic river biogeochemistry. Hydrol Process 23(1):169–182
Fuchs M, Kuhry P, Hugelius G (2015) Low below-ground organic carbon storage in a subarctic alpine permafrost environment. Cryosphere 9(2):427–438

Gooseff MN, Balser A, Bowden WB, Jones JB (2009) Effects of hillslope thermokarst in northern Alaska. EOS Trans Am Geophys Union 90(4):29–30

Grosse G, Jones B (2011) Spatial distribution of pingos in Northern Asia. Cryosphere 5(1):13

Grosse G, Harden J, Turetsky M, McGuire AD, Camill P, Tarnocai C, Frolking S, Schuur EAG, Jorgenson T, Marchenko S, Romanovsky V, Wickland KP, French N, Waldrop M, Bourgeau-Chavez L, Striegl RG (2011) Vulnerability of high-latitude soil organic carbon in North America to disturbance. J Geophys Res 116. https://doi.org/10.1029/2010jg001507

Grosse G, Jones B, Arp C (2013) Thermokarst Lakes, drainage, and drained basins. In: Shroder J, Giardino, R,Harbor, J (ed) Glacial and periglacial geomorphology, vol 8. Treatise on Geomorphology. pp 325–353. https://doi.org/10.1016/b978-0-12-374739-6.00216-5

Guglielmin M, Cannone N, Strini A, Lewkowicz AG (2005) Biotic and abiotic processes on granite weathering landforms in a cryotic environment, Northern Victoria land, Antarctica. Permafr Periglac Process 16(1):69–85

Günther F, Grosse G, Wetterich S, Jones BM, Kunitsky VV, Kienast F, Schirrmeister L (2015) The Batagay mega thaw slump, Yana Uplands, Yakutia, Russia: permafrost thaw dynamics on decadal time scale. TERRA NOSTRA-Schriften der GeoUnion Alfred-Wegener-Stiftung 2015(1):45–46

Hall K, Thorn CE, Matsuoka N, Prick A (2002) Weathering in cold regions: some thoughts and perspectives. Prog Phys Geogr 26(4):577–603

Hallet B (2013) Stone circles: form and soil kinematics. Phil Trans R Soc A 371(2004):20120357. https://doi.org/10.1098/rsta.2012.0357

Hallet B, Walder J, Stubbs C (1991) Weathering by segregation ice growth in microcracks at sustained subzero temperatures: verification from an experimental study using acoustic emissions. Permafr Periglac Process 2(4):283–300

Harris SA, French HM, Heginbottom JA, Johnston GH, Ladanyi B, Sego DC, van Everdingen RO (1988) Glossary of permafrost and related ground-ice terms. Tech Memorandum Natl Res Counc Canada 142:157

Hauck C, Kneisel C (2008) Applied geophysics in Periglacial environments. Cambridge University Press, Cambridge. https://doi.org/10.1017/CBO9780511535628

Hillel D (1998) Environmental soil physics: fundamentals, applications, and environmental considerations. Elsevier, Burlington. 771 p

Hinkel K, Nelson F, Shur Y, Brown J, Everett K (1996) Temporal changes in moisture content of the active layer and near-surface permafrost at Barrow, Alaska, USA: 1962–1994. Arct Alp Res 28(3):300–310

Hinkel KM, Eisner WR, Bockheim JG, Nelson FE, Peterson KM, Dai X (2003) Spatial extent, age, and carbon stocks in drained thaw lake basins on the Barrow Peninsula, Alaska. Arct Antarct Alp Res 35(3):291–300

Hinkel KM, Jones BM, Eisner WR, Cuomo CJ, Beck RA, Frohn R (2007) Methods to assess natural and anthropogenic thaw lake drainage on the western Arctic coastal plain of northern Alaska. J Geophys Res Earth 112:F2

Hopkins DM (1949) Thaw lakes and thaw sinks in the Imuruk Lake area, Seward Peninsula, Alaska. J Geol 57(2):119–131

Hopkins D, Kidd J (1988) Thaw lake sediments and sedimentary environments. In: Proceedings of the 5th international permafrost conference, 1988. pp 790–795

Iwahana G, Takano S, Petrov RE, Tei S, Shingubara R, Maximov TC, Fedorov AN, Desyatkin AR, Nikolaev AN, Desyatkin RV (2014) Geocryological characteristics of the upper permafrost in a tundra-forest transition of the Indigirka River Valley, Russia. Pol Sci 8(2):96–113

Jeffries MO, Zhang T, Frey K, Kozlenko N (1999) Estimating late-winter heat flow to the atmosphere from the lake-dominated Alaskan North Slope. J Glaciol 45(150):315–324

Jepsen SM, Voss CI, Walvoord MA, Minsley BJ, Rover J (2013) Linkages between lake shrinkage/expansion and sublacustrine permafrost distribution determined from remote sensing of interior Alaska, USA. Geophys Res Lett 40(5):882–887

Jones A, Stolbovay V, Tarnocai C, Broll G, Spaargaren O, Montanarella L (2010) Soil atlas of the northern circumpolar region. European Commission, Publications office of the European Union, Luxembourg, p 144

Jones BM, Grosse G, Hinkel KM, Arp CD, Walker S, Beck RA, Galloway JP (2012) Assessment of pingo distribution and morphometry using an IfSAR derived digital surface model, western Arctic coastal plain, Northern Alaska. Geomorphology 138(1):1–14

Jorgenson MT, Shur Y (2007) Evolution of lakes and basins in northern Alaska and discussion of the thaw lake cycle. J Geophys Res Earth Surface 112(F2):F02S17. https://doi.org/10.1029/2006JF000531

Jorgenson MT, Romanovsky V, Harden J, Shur Y, O'Donnell J, Schuur EA, Kanevskiy M, Marchenko S (2010) Resilience and vulnerability of permafrost to climate change. Can J For Res 40(7):1219–1236

Jorgenson MT, Kanevskiy M, Shur Y, Moskalenko N, Brown D, Wickland K, Striegl R, Koch J (2015) Role of ground ice dynamics and ecological feedbacks in recent ice wedge degradation and stabilization. J Geophys Res Earth 120(11):2280–2297

Kaiser C, Meyer H, Biasi C, Rusalimova O, Barsukov P, Richter A (2007) Conservation of soil organic matter through cryoturbation in arctic soils in Siberia. J Geophys Res Biogeo 112(G2):G02017. https://doi.org/10.1029/2006JG000258

Kane K, Slaughter C (1973) Recharge of a central Alaska lake by subpermafrost groundwater. In: International Conference on Permafrost, 2nd, Yakutsk, Siberia. Papers, 1973

Kanevskiy M, Shur Y, Jorgenson M, Ping C-L, Michaelson G, Fortier D, Stephani E, Dillon M, Tumskoy V (2013) Ground ice in the upper permafrost of the Beaufort Sea coast of Alaska. Cold Reg Sci Technol 85:56–70

Kanevskiy M, Shur Y, Jorgenson T, Brown DNR, Moskalenko N, Brown J, Walker DA, Raynolds MK, Buchhorn M (2017) Degradation and stabilization of ice wedges: implications for assessing risk of thermokarst in Northern Alaska. Geomorphology 297:20–42

Katamura F, Fukuda M, Bosikov NP, Desyatkin RV, Nakamura T, Moriizumi J (2006) Thermokarst formation and vegetation dynamics inferred from a palynological study in Central Yakutia, Eastern Siberia, Russia. Arct Antarct Alp Res 38(4):561–570

Kessler M, Plug LJ, Walter Anthony K (2012) Simulating the decadal-to millennial-scale dynamics of morphology and sequestered carbon mobilization of two thermokarst lakes in NW Alaska. J Geophys Res Biogeo 117(G2):G00M06. https://doi.org/10.1029/2011JG001796

Khomutov A, Leibman M (2014) Assessment of landslide hazards in a typical tundra of Central Yamal, Russia. In: Landslides in cold regions in the context of climate change. Springer, Cham, pp 271–290

Kirpotin SN, Polishchuk Y, Bryksina N (2009) Abrupt changes of thermokarst lakes in Western Siberia: impacts of climatic warming on permafrost melting. Int J Environ Stud 66(4):423–431. https://doi.org/10.1080/00207230902758287

Kirpotin S, Polishchuk Y, Bryksina N, Sugaipova A, Kouraev A, Zakharova E, Pokrovsky OS, Shirokova L, Kolmakova M, Manassypov R, Dupre B (2011) West Siberian palsa peatlands: distribution, typology, cyclic development, present day climate-driven changes, seasonal hydrology and impact on CO_2 cycle. Int J Environ Stud 68(5):603–623. https://doi.org/10.1080/00207233.2011.593901

Kneisel C, Hauck C, Fortier R, Moorman B (2008) Advances in geophysical methods for permafrost investigations. Permafr Periglac Process 19(2):157–178

Koch JC, Runkel RL, Striegl R, McKnight DM (2013) Hydrologic controls on the transport and cycling of carbon and nitrogen in a boreal catchment underlain by continuous permafrost. J Geophys Res Biogeo 118(2):698–712

Kokelj S, Burn C (2004) Tilt of spruce trees near ice wedges, Mackenzie Delta, Northwest Territories, Canada. Arct Antarct Alp Res 36(4):615–623

Kokelj S, Pisaric MF, Burn C (2007a) Cessation of ice-wedge development during the 20th century in spruce forests of eastern Mackenzie Delta, Northwest Territories, Canada. Can J Earth Sci 44(11):1503–1515

Kokelj S, Burn C, Tarnocai C (2007b) The structure and dynamics of earth hummocks in the subarctic forest near Inuvik, Northwest Territories, Canada. Arct Antarct Alp Res 39(1):99–109
Kokelj SV, Lantz TC, Kanigan J, Smith S, Coutts R (2009) Origin and polycyclic behaviour of tundra thaw slumps, Mackenzie Delta region, Northwest Territories, Canada. Permafr Periglac Process 20(2):173–184
Konishchev V, Rogov V (1993) Investigations of cryogenic weathering in Europe and Northern Asia. Permafr Periglac Process 4(1):49–64
Kravtsova V, Bystrova A (2009) Changes in thermokarst lake sizes in different regions of Russia for the last 30 years (Изменение размеров термокарстовых озер в различных районах россии за последние 30 лет). Kriosfera Zemli (Earth Cryosphere) 13:16–26
Kuhry P (2008) Palsa and peat plateau development in the Hudson Bay Lowlands, Canada: timing, pathways and causes. Boreas 37(2):316–327
Kurylyk BL, Watanabe K (2013) The mathematical representation of freezing and thawing processes in variably-saturated, non-deformable soils. Adv Water Resour 60:160–177
Labrecque S, Lacelle D, Duguay CR, Lauriol B, Hawkings J (2009) Contemporary (1951-2001) evolution of lakes in the old Crow Basin, Northern Yukon, Canada: remote sensing, numerical modeling, and stable isotope analysis. Arctic:225–238
Lacelle D, Bjornson J, Lauriol B (2010) Climatic and geomorphic factors affecting contemporary (1950–2004) activity of retrogressive thaw slumps on the Aklavik Plateau, Richardson Mountains, NWT, Canada. Permafr Periglac Process 21(1):1–15
Lachenbruch AH (1962) Mechanics of thermal contraction cracks and ice-wedge polygons in permafrost, vol 70. Special GSA Papers. Geological Society of America, 65 p
Langer M, Westermann S, Walter Anthony K, Wischnewski K, Boike J (2015) Frozen ponds: production and storage of methane during the Arctic winter in a lowland tundra landscape in northern Siberia, Lena River delta. Biogeosciences 12(4):977–990. https://doi.org/10.5194/bg-12-977-2015
Lehner B, Döll P (2004) Development and validation of a global database of lakes, reservoirs and wetlands. J Hydrol 296(1–4):1–22
Leibman M, Kizakov A, Sulerzhitsky L, Zaretskaia N (2003) Dynamics of landslide slopes and their development on Yamal peninsula. In: Permafrost. Proceedings of the 8th international conference on permafrost. Swets and Zeitlinger, Lisse, pp 651–656
Leibman M, Khomutov A, Kizyakov A (2014) Cryogenic landslides in the West-Siberian plain of Russia: classification, mechanisms, and landforms. In: Landslides in cold regions in the context of climate change. Springer, Cham, pp 143–162
Lewkowicz AG (2007) Dynamics of active-layer detachment failures, Fosheim peninsula, Ellesmere Island, Nunavut, Canada. Permafr Periglac Process 18(1):89–103
Lewkowicz AG, Harris C (2005) Frequency and magnitude of active-layer detachment failures in discontinuous and continuous permafrost, northern Canada. Permafr Periglac Process 16(1):115–130
Lewkowicz AG, Way RG (2019) Extremes of summer climate trigger thousands of thermokarst landslides in a high Arctic environment. Nat Commun 10(1):1329. https://doi.org/10.1038/s41467-019-09314-7
Lewkowicz AG, Etzelmüller B, Smith SL (2011) Characteristics of discontinuous permafrost based on ground temperature measurements and electrical resistivity tomography, southern Yukon, Canada. Permafr Periglac Process 22(4):320–342
Liljedahl AK, Boike J, Daanen RP, Fedorov AN, Frost GV, Grosse G, Hinzman LD, Iijma Y, Jorgenson JC, Matveyeva N, Necsoiu M, Raynolds MK, Romanovsky VE, Schulla J, Tape KD, Walker DA, Wilson CJ, Yabuki H, Zona D (2016) Pan-Arctic ice-wedge degradation in warming permafrost and its influence on tundra hydrology. Nat Geosci 9(4):312–318. https://doi.org/10.1038/ngeo2674
Ling F, Zhang T (2003) Numerical simulation of permafrost thermal regime and talik development under shallow thaw lakes on the Alaskan Arctic Coastal Plain. J Geophys Res 108(D16):4511. https://doi.org/10.1029/2002JD003014

Ling F, Zhang T (2004) Modeling study of talik freeze-up and permafrost response under drained thaw lakes on the Alaskan Arctic Coastal Plain. J Geophys Res Atmos 109. (D1:D01111. https://doi.org/10.1029/2003JD003886

Luoto M, Heikkinen RK, Carter TR (2004) Loss of palsa mires in Europe and biological consequences. Environ Conserv 31(1):30–37

Lupachev A, Gubin S (2012) Suprapermafrost organic-accumulative horizons in the tundra cryozems of northern Yakutia. Eurasian Soil Sci 45(1):45–55

Mackay JR (1971) The origin of massive icy beds in permafrost, western Arctic coast, Canada. Can J Earth Sci 8(4):397–422

Mackay JR (1972) The world of underground ice. Ann Assoc Am Geogr 62(1):1–22

Mackay JR (1980a) The origin of hummocks, western Arctic coast, Canada. Can J Earth Sci 17(8):996–1006

Mackay JR (1980b) Deformation of ice-wedge polygons, Garry Island, Northwest Territories. Curr Res, Part A Geol Surv Canada, Paper:287–291

Mackay JR (1983) Downward water movement into frozen ground, western arctic coast, Canada. Can J Earth Sci 20(1):120–134

Mackay J (1988) Catastrophic lake drainage, Tuktoyaktuk peninsula area, district of Mackenzie. Curr Res Part D Geol Surv Canada, Paper:83–90

Mackay JR (1989) Ice-wedge cracks, Western Arctic coast. The Canadian Geographer/Le Géographe canadien 33(4):365–368

Mackay JR (1990) Some observations on the growth and deformation of epigenetic, syngenetic and anti-syngenetic ice wedges. Permafr Periglac Process 1(1):15–29

Mackay JR (1992a) The frequency of ice-wedge cracking (1967–1987) at Garry Island, western Arctic coast, Canada. Can J Earth Sci 29(2):236–248

Mackay JR (1992b) Lake stability in an ice-rich permafrost environment: examples from the western Arctic coast. In: Aquatic ecosystems in semi-arid regions: implications for resource management. NHRI Symposium Series. pp 1–26

Mackay JR (1993) Air temperature, snow cover, creep of frozen ground, and the time of ice-wedge cracking, western Arctic coast. Can J Earth Sci 30(8):1720–1729

Mackay JR (1997) A full-scale field experiment (1978–1995) on the growth of permafrost by means of lake drainage, western Arctic coast: a discussion of the method and some results. Can J Earth Sci 34(1):17–33

Mackay JR (1998) Pingo growth and collapse, Tuktoyaktuk Peninsula area, western Arctic coast, Canada: a long-term field study. Géog Phys Quatern 52(3):271–323

Mackay JR (1999) Cold-climate shattering (1974 to 1993) of 200 glacial erratics on the exposed bottom of a recently drained arctic lake, Western Arctic coast, Canada. Permafr Periglac Process 10(2):125–136

Mackay JR (2000) Thermally induced movements in ice-wedge polygons, western Arctic coast: a long-term study. Géog Phys Quatern 54(1):41–68

Mackay JR, Burn C (2002) The first 20 years (1978-1979 to 1998–1999) of ice-wedge growth at the Illisarvik experimental drained lake site, western Arctic coast, Canada. Can J Earth Sci 39(1):95–111

Marsh P, Neumann NN (2001) Processes controlling the rapid drainage of two ice-rich permafrost-dammed lakes in NW Canada. Hydrol Process 15(18):3433–3446

Marsh P, Woo M (1993) Infiltration of meltwater into frozen soils in a continuous permafrost environment. In: Proceedings of the sixth international conference on permafrost. South China University of Technology Press, Beijing, pp 443–448

Marsh P, Russell M, Pohl S, Haywood H, Onclin C (2009) Changes in thaw lake drainage in the Western Canadian Arctic from 1950 to 2000. Hydrol Process Int J 23(1):145–158

Matsuoka N (2001) Solifluction rates, processes and landforms: a global review. Earth Sci Rev 55(1–2):107–134

Matsuoka N, Murton J (2008) Frost weathering: recent advances and future directions. Permafr Periglac Process 19(2):195–210

Matsuoka N, Christiansen HH, Watanabe T (2018) Ice-wedge polygon dynamics in Svalbard: lessons from a decade of automated multi-sensor monitoring. Permafr Periglac Process 29(3):210–227
Minke M, Donner N, Karpov N, de Klerk P, Joosten H (2009) Patterns in vegetation composition, surface height and thaw depth in polygon mires in the Yakutian Arctic (NE Siberia): a microtopographical characterisation of the active layer. Permafr Periglac Process 20(4):357–368
Minsley BJ, Abraham JD, Smith BD, Cannia JC, Voss CI, Jorgenson MT, Walvoord MA, Wylie BK, Anderson L, Ball LB (2012) Airborne electromagnetic imaging of discontinuous permafrost. Geophys Res Lett 39(2):L02503. https://doi.org/10.1029/2011GL050079
Morgenstern A, Grosse G, Schirrmeister L (2008) Genetic, morphological, and statistical characterization of lakes in the permafrost-dominated Lena Delta. In: Proceedings, ninth international conference on permafrost, 2008. pp 1239–1244
Morgenstern A, Grosse G, Günther F, Fedorova I, Schirrmeister L (2011) Spatial analyses of thermokarst lakes and basins in Yedoma landscapes of the Lena Delta. Cryosphere 5(4):849–867. https://doi.org/10.5194/tc-5-849-2011
Morgenstern A, Ulrich M, Günther F, Roessler S, Fedorova IV, Rudaya NA, Wetterich S, Boike J, Schirrmeister L (2013) Evolution of thermokarst in East Siberian ice-rich permafrost: a case study. Geomorphology 201:363–379. https://doi.org/10.1016/j.geomorph.2013.07.011
Murton JB (1996) Thermokarst-lake-basin sediments, Tuktoyaktuk Coastlands, western arctic Canada. Sedimentology 43(4):737–760
Murton JB (2001) Thermokarst sediments and sedimentary structures, Tuktoyaktuk Coastlands, western Arctic Canada. Glob Planet Chang 28(1–4):175–192
Murton JB, French HM (1994) Cryostructures in permafrost, Tuktoyaktuk coastlands, western arctic Canada. Can J Earth Sci 31(4):737–747
Murton JB, Lautridou JP (2003) Recent advances in the understanding of quaternary periglacial features of the English Channel coastlands. J Quat Sci 18(3–4):301–307
Murton JB, Peterson R, Ozouf J-C (2006) Bedrock fracture by ice segregation in cold regions. Science 314(5802):1127–1129
Murton JB, Goslar T, Edwards ME, Bateman MD, Danilov PP, Savvinov GN, Gubin SV, Ghaleb B, Haile J, Kanevskiy M (2015) Palaeoenvironmental interpretation of Yedoma silt (ice complex) deposition as cold-climate loess, Duvanny Yar, Northeast Siberia. Permafr Periglac Process 26(3):208–288
Murton JB, Edwards ME, Lozhkin AV, Anderson PM, Savvinov GN, Bakulina N, Bondarenko OV, Cherepanova MV, Danilov PP, Boeskorov V (2017) Preliminary paleoenvironmental analysis of permafrost deposits at Batagaika megaslump, Yana Uplands, northeast Siberia. Quat Res 87(2):314–330
Olenchenko V, Sinitsky A, Antonov E, Eltsov I, Kushnarenko O, Plotnikov A, Potapov V, Epov M (2015) Results of geophysical surveys of the area of "Yamal Crater", the new geological structure. Methods 19(4):84–95
Palmtag J, Hugelius G, Lashchinskiy N, Tamstorf MP, Richter A, Elberling B, Kuhry P (2015) Storage, landscape distribution, and burial history of soil organic matter in contrasting areas of continuous permafrost. Arct Antarct Alp Res 47(1):71–88
Peppin SS, Style RW (2013) The physics of frost heave and ice-lens growth. Vadose Zone J 12(1)
Peterson R, Krantz W (2008) Differential frost heave model for patterned ground formation: corroboration with observations along a North American arctic transect. J Geophys Res Biogeo 113(G3):G03S04. https://doi.org/10.1029/2007JG000559
Peterson R, Walker D, Romanovsky V, Knudson J, Raynolds M, Krantz W (2003) A differential frost heave model: cryoturbation-vegetation interactions. In: Proceedings of the eighth international conference on permafrost, 2003. pp 885–890
Pissart A (2002) Palsas, lithalsas and remnants of these periglacial mounds. A progress report. Prog Phys Geogr 26(4):605–621
Plug LJ, Werner B (2001) Fracture networks in frozen ground. J Geophys Res Solid Earth 106(B5):8599–8613

Plug LJ, Werner B (2002) Nonlinear dynamics of ice-wedge networks and resulting sensitivity to severe cooling events. Nature 417(6892):929

Plug LW, West JJ (2009) Thaw lake expansion in a two-dimensional coupled model of heat transfer, thaw subsidence, and mass movement. J Geophys Res Earth 114(F1):F01002. https://doi.org/10.1029/2006JF000740

Plug LJ, Walls C, Scott B (2008) Tundra lake changes from 1978 to 2001 on the Tuktoyaktuk Peninsula, western Canadian Arctic. Geophys Res Lett 35(3):L03502. https://doi.org/10.1029/2007GL032303

Pollard W, French H (1980) A first approximation of the volume of ground ice, Richards Island, Pleistocene Mackenzie Delta, Northwest Territories, Canada. Can Geotech J 17(4):509–516

Rempel A (2007) Formation of ice lenses and frost heave. J Geophys Res Earth 112(F2):F02S21. https://doi.org/10.1029/2006JF000525

Reynolds JM (1997) An introduction to applied environmental geophysics. Wiley, Hoboken, p 796

Romanovskii N (1985) Distribution of recently active ice and soil wedges in the USSR. Field and Theory: Lectures in Geocryology, pp 154–165

Romanovskii N, Hubberten HW, Gavrilov A, Tumskoy V, Tipenko G, Grigoriev M, Siegert C (2000) Thermokarst and land–ocean interactions, Laptev Sea region, Russia. Permafr Periglac Process 11(2):137–152

Romanovsky V, Osterkamp T (2000) Effects of unfrozen water on heat and mass transport processes in the active layer and permafrost. Permafr Periglac Process 11(3):219–239

Romanovsky V, Marchenko S, Daanen R, Sergeev D, Walker D (2008) Soil climate and frost heave along the permafrost/ecological North American Arctic transect. In: Proceedings of the ninth international conference on permafrost. Institute of Northern Engineering, Fairbanks, pp 1519–1524

Roulet NT, Woo M-K (1988) Runoff generation in a low arctic drainage basin. J Hydrol 101(1–4):213–226

Sannel A, Kuhry P (2011) Warming-induced destabilization of peat plateau/thermokarst lake complexes. J Geophys Res Biogeo 116(G3):G03035. https://doi.org/10.1029/2010JG001635

Schirrmeister L, Kunitsky V, Grosse G, Wetterich S, Meyer H, Schwamborn G, Babiy O, Derevyagin A, Siegert C (2011) Sedimentary characteristics and origin of the late Pleistocene ice complex on north-east Siberian Arctic coastal lowlands and islands – a review. Quat Int 241(1–2):3–25. https://doi.org/10.1016/j.quaint.2010.04.004

Schirrmeister L, Bobrov A, Raschke E, Herzschuh U, Strauss J, Pestryakova LA, Wetterich S (2018) Late Holocene ice-wedge polygon dynamics in northeastern Siberian coastal lowlands. Arct Antarct Alp Res 50(1):e1462595. https://doi.org/10.1080/15230430.2018.1462595

Schunke E, Zoltai SC (1988) Earth hummocks (thufur). In: Clark M (ed) Advances in Periglacial geomorphology. Wiley, Chichester, pp 231–245

Schwamborn G, Dix J, Bull J, Rachold V (2002) High-resolution seismic and ground penetrating radar–geophysical profiling of a thermokarst lake in the western Lena Delta, Northern Siberia. Permafr Periglac Process 13(4):259–269

Schwamborn G, Schirrmeister L, Frütsch F, Diekmann B (2012) Quartz weathering in freeze–thaw cycles: experiment and application to the El'Gygytgyn crater lake record for tracing Siberian permafrost history. Geogr Ann Ser B 94(4):481–499

Seppälä M (2011) Synthesis of studies of palsa formation underlining the importance of local environmental and physical characteristics. Quat Res 75(2):366–370

Seppälä M, Kujala K (2009) The role of buoyancy in palsa formation. Geol Soc Lond, Spec Publ 320(1):51–56

Shiklomanov A, Lammers R (2009) Record Russian river discharge in 2007 and the limits of analysis. Environ Res Lett 4(4):045015

Shilo N, Lozhkin A, Anderson P (2007) Radiocarbon dates of evolution cycles of thermokarst lakes on the Kolyma lowland. In: Doklady earth sciences, vol 1. Springer, New York, pp 259–261

Shur Y, Hinkel KM, Nelson FE (2005) The transient layer: implications for geocryology and climate-change science. Permafr Periglac Process 16(1):5–17

Sjöberg Y, Hugelius G, Kuhry P (2013) Thermokarst lake morphometry and erosion features in two peat plateau areas of northeast European Russia. Permafr Periglac Process 24(1):75–81

Smith LC, Sheng Y, MacDonald G, Hinzman L (2005) Disappearing arctic lakes. Science 308(5727):1429–1429

Teltewskoi A, Beermann F, Beil I, Bobrov A, De Klerk P, Lorenz S, Lüder A, Michaelis D, Joosten H (2016) 4000 years of changing wetness in a permafrost polygon Peatland (Kytalyk, NE Siberia): a comparative high-resolution multi-proxy study. Permafr Periglac Process 27(1):76–95

Thorn CE, Darmody RG, Dixon JC, Schlyter P (2001) The chemical weathering regime of Kärkevagge, arctic–alpine Sweden. Geomorphology 41(1):37–52

Thorn CE, Dixon JC, Darmody RG, Allen CE (2006) Ten years (1994–2004) of 'potential' weathering in Kärkevagge, Swedish Lapland. Earth Surf Process Landf 31(8):992–1002

Thorn CE, Darmody RG, Campbell SW, Allen CE, Dixon JC (2007) Microvariability in the early stages of cobble weathering by microenvironment on a glacier foreland, Storbreen, Jotunheimen, Norway. Earth Surf Process Landf 32(14):2199–2211

Van der Molen M, Van Huissteden J, Parmentier F, Petrescu A, Dolman A, Maximov T, Kononov A, Karsanaev S, Suzdalov D (2007) The growing season greenhouse gas balance of a continental tundra site in the Indigirka lowlands, NE Siberia. Biogeosciences 4(6):985–1003. https://doi.org/10.5194/bg-4-985-2007

Van Everdingen R (1990) Ground-water hydrology. In: Prowse TD OC (ed) Northern hydrology: Canadian perspectives, vol 1. NHRI Science Report. National Hydrology Research Institute, Environment Canada, pp 77–101

Van Huissteden J (1990) Tundra rivers of the last glacial: sedimentation and geomorphological processes during the Middle Pleniglacial in Twente, Eastern Netherlands. In: Van Amerom H (ed) Mededelingen Rijks Geologische Dienst, vol 44–3. Rijks Geologische Dienst, 1–136

Van Huissteden J, Dolman AJ (2012) Soil carbon in the Arctic and the permafrost carbon feedback. Curr Opin Environ Sustain 4(5):545–551. https://doi.org/10.1016/j.cosust.2012.09.008

Van Huissteden J, Berrittella C, Parmentier F, Mi Y, Maximov T, Dolman A (2011) Methane emissions from permafrost thaw lakes limited by lake drainage. Nat Clim Chang 1(2):119–123

Van Huissteden J, Vandenberghe J, Gibbard PL, Lewin J (2013) Permafrost and Periglacial features | Periglacial fluvial sediments and forms. In: Elias SA (ed) Encyclopedia of the quaternary, vol 3. Elsevier, Amsterdam, pp 490–499. https://doi.org/10.1016/b978-0-444-53643-3.00108-4

Van Vliet-Lanoë B (1985) Frost effects in soils. Soils and Quaternary Landscape Evolution:117–158

Van Vliet-Lanoe B, Fox CA, Gubin SV (2004) Micromorphology of cryosols. In: Cryosols. Springer, New York, pp 365–390

Vandenberghe J (1988) Cryoturbations. In: Clark M (ed) Advances in Periglacial geomorphology. Wiley, New York, pp 179–198

Vandenberghe J (2013) Cryoturbation structures. Encyclopedia of quaternary science, vol 3, 2nd edn. Elsevier, Amsterdam, pp 430–435

Vandenberghe J, Wang X, Vandenberghe D (2016) Very large cryoturbation structures of last permafrost maximum age at the foot of the Qilian Mountains (NE Tibet Plateau, China). Permafr Periglac Process 27(1):138–143

Vitt DH (2006) Functional characteristics and indicators of boreal peatlands. In: Wieder R, Vitt DH (eds) Boreal peatland ecosystems. Springer, pp 9–24

Vonk JE, Gustafsson Ö (2013) Permafrost-carbon complexities. Nat Geosci 6(9):675

Walder J, Hallet B (1985) A theoretical model of the fracture of rock during freezing. Geol Soc Am Bull 96(3):336–346

Walker DA, Epstein HE, Gould WA, Kelley AM, Kade AN, Knudson JA, Krantz WB, Michaelson G, Peterson RA, Ping CL (2004) Frost-boil ecosystems: complex interactions between landforms, soils, vegetation and climate. Permafr Periglac Process 15(2):171–188

Walter KM, Zimov SA, Chanton JP, Verbyla D, Chapin FS 3rd (2006) Methane bubbling from Siberian thaw lakes as a positive feedback to climate warming. Nature 443(7107):71–75. https://doi.org/10.1038/nature05040

Walter K, Edwards M, Grosse G, Zimov S, Chapin F (2007) Thermokarst lakes as a source of atmospheric CH_4 during the last deglaciation. Science 318(5850):633–636
Walvoord MA, Kurylyk BL (2016) Hydrologic impacts of thawing permafrost—a review. Vadose Zone J 15(6):0. https://doi.org/10.2136/vzj2016.01.0010
Washburn AL (1956) Classification of patterned ground and review of suggested origins. Geol Soc Am Bull 67(7):823–866
Washburn AL (1980) Geocryology: a survey of periglacial processes and environments. Wiley, New York
Weiss N, Blok D, Elberling B, Hugelius G, Jørgensen CJ, Siewert MB, Kuhry P (2016) Thermokarst dynamics and soil organic matter characteristics controlling initial carbon release from permafrost soils in the Siberian Yedoma region. Sediment Geol 340:38–48. https://doi.org/10.1016/j.sedgeo.2015.12.004
West J, Plug LJ (2008) Time-dependent morphology of thaw lakes and taliks in deep and shallow ground ice. J Geophys Res Earth 113(F1):F01009. https://doi.org/10.1029/2006JF000696
Williams PJ, Smith MW (1991) The frozen earth. Fundamentals of geocryology. Cambridge University Press, Cambridge. 306 p
Wolter J, Lantuit H, Wetterich S, Rethemeyer J, Fritz M (2018) Climatic, geomorphologic and hydrologic perturbations as drivers for mid-to late Holocene development of ice-wedge polygons in the western Canadian Arctic. Permafr Periglac Process:1–18. https://doi.org/10.1002/ppp.1977
Woo M-k (1990) Consequences of climatic change for hydrology in permafrost zones. J Cold Reg Eng 4(1):15–20
Woo M-k (2012) Permafrost hydrology. Springer, Berlin, p 563
Woo MK, Guan XJ (2006) Hydrological connectivity and seasonal storage change of tundra ponds in a polar oasis environment, Canadian High Arctic. Permafr Periglac Process 17(4):309–323
Wright N, Hayashi M, Quinton WL (2009) Spatial and temporal variations in active layer thawing and their implication on runoff generation in peat-covered permafrost terrain. Water Resour Res 45(5)
Yoshikawa K, Harada K (1995) Observations on nearshore pingo growth, Adventdalen, Spitsbergen. Permafr Periglac Process 6(4):361–372
Yoshikawa K, Hinzman LD (2003) Shrinking thermokarst ponds and groundwater dynamics in discontinuous permafrost near council, Alaska. Permafr Periglac Process 14(2):151–160
Yoshikawa K, White D, Hinzman L, Goering D, Petrone K, Bolton W, Ishikawa N, Plummer N (2003) Water in permafrost; case study of aufeis and pingo hydrology in discontinuous permafrost. In: Proceedings of the international conference on permafrost. Zurich
Yoshikawa K, Leuschen C, Ikeda A, Harada K, Gogineni P, Hoekstra P, Hinzman L, Sawada Y, Matsuoka N (2006) Comparison of geophysical investigations for detection of massive ground ice (pingo ice). J Geophys Res Planets 111(E6):E06S19. https://doi.org/10.1029/2005JE002573
Zhang T, Barry R, Knowles K, Heginbottom J, Brown J (2008) Statistics and characteristics of permafrost and ground-ice distribution in the Northern Hemisphere. Polar Geogr 31(1–2):47–68

Chapter 4
Permafrost Carbon Quantities and Fluxes

Abstract The permafrost soil and ecosystem carbon storage is the net result of carbon uptake from the atmosphere or water by photosynthesis, and organic matter decomposition by respiration of ecosystem organisms. In this chapter, the ecosystem and soil carbon balance are treated in more detail. First, the basic components of the ecosystem carbon balance are discussed. Recent quantifications of the permafrost carbon store, and how the ecosystem carbon balance contributed in the past to these carbon stores, is reviewed. The processes of carbon decomposition in soils are discussed, with emphasis on the effects of changes in the soil environment that affect decomposition rates. Anaerobic conditions created by high water tables are an important component of carbon preservation in permafrost soils next to low temperatures. Redox processes in soils, and ecosystem methane emission therefore are emphasized. Organic carbon decomposition in soils also releases its nutrients. Slow decomposition of organic matter is an important factor contributing to nutrient limitation in permafrost ecosystems; nevertheless, leaks in the nutrient cycle may occur which result in emission of the strong greenhouse gas N_2O. The chapter concludes with an overview of measurement methods of carbon fluxes, and the magnitude of these fluxes in various permafrost ecosystems.

Keywords Carbon balance · Carbon cycle · Permafrost carbon · Paleo-permafrost · Greenhouse gases · Soil carbon decomposition · Nutrients · Methane · Flux measurements

It is not the carbon balance of permafrost soils that is decisive for the strength of the permafrost carbon feedback, but the emission of carbon-containing greenhouse gases CO_2 and CH_4, and the strong non-carbon greenhouse gas N_2O. The carbon balance and the greenhouse gas balance are therefore not synonymous. The carbon balance of an ecosystem is determined by the net fluxes of carbon compounds to atmosphere, water and adjacent ecosystems. The greenhouse gas balance is determined by the net fluxes of greenhouse gases to the atmosphere.

For the permafrost carbon feedback we therefore need to consider two balances: the carbon balance of the ecosystem, which determines the size of the ecosystem and soil carbon store, and the greenhouse gas balance, which is the balance between

J. van Huissteden, *Thawing Permafrost*,
https://doi.org/10.1007/978-3-030-31379-1_4

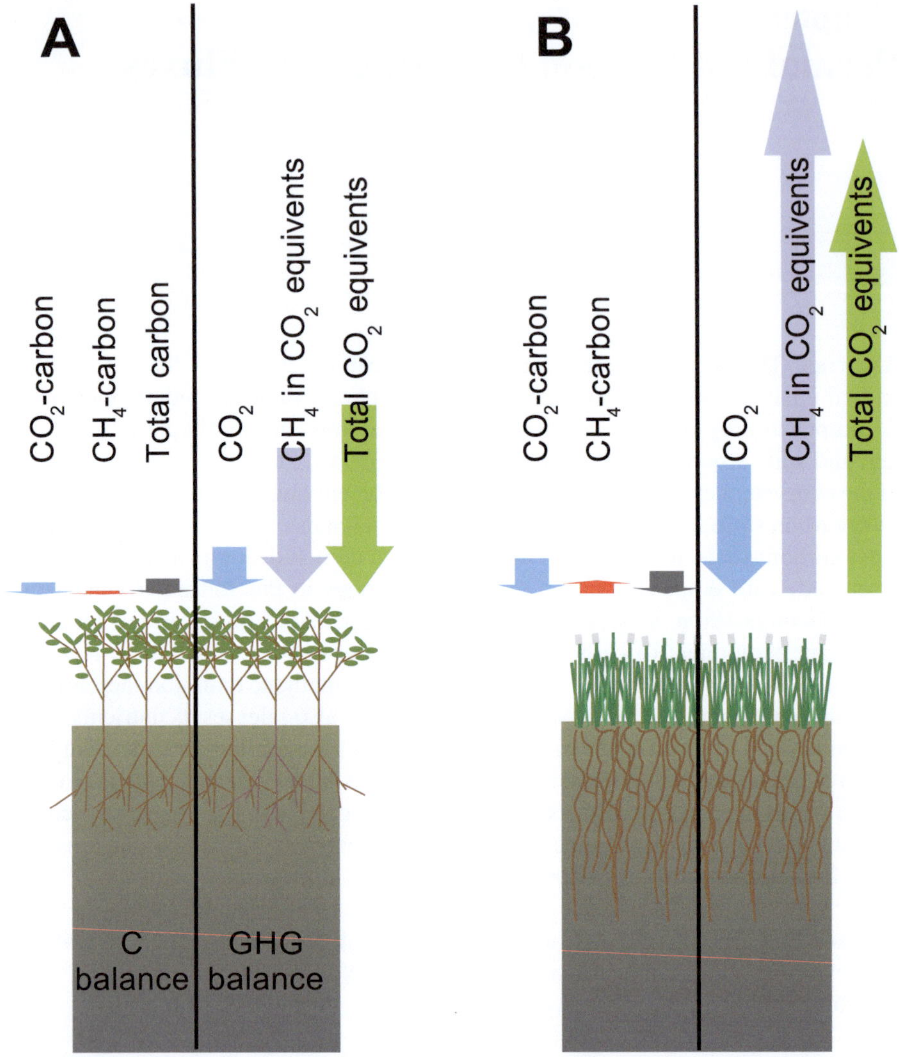

Fig. 4.1 Difference between carbon and greenhouse gas or radiative forcing balance. (**a**) an ecosystem with a small net CO_2-carbon uptake from the atmosphere and a small uptake of CH_4-carbon; the left three arrows represent the carbon balance (CO_2 carbon sink, CH_4 carbon sink and total carbon sink in carbon mass); the right three arrows represent the greenhouse gas balance in CO_2-equivalents, based on a global warming potential for CH_4 of 34 cf. IPCC (2013). (**b**) an ecosystem with a 3 times higher net CO_2-carbon uptake from the atmosphere and emission of CH_4-carbon

greenhouse gases withdrawn from the atmosphere by the ecosystem, and the gases emitted to the atmosphere. CO_2, CH_4, and N_2O have varying radiative properties, as is usually expressed by their global warming potential, which needs to be accounted for in the greenhouse gas balance. An ecosystem can have net storage of carbon, but still have a warming effect on the climate by emission of a radiatively unfavourable mix of greenhouse gases (Fig. 4.1).

The present state of the carbon and greenhouse gas balances of permafrost soils and ecosystems is the main scope of this chapter. In the next chapters, the effects of climate change on the permafrost carbon cycle components are discussed. Here, the basic processes of the carbon and greenhouse gas balance are considered. A discussion of the present inventories of carbon in permafrost soils and frozen sediments is included. These stores consist of recent organic horizons in soil profiles, and organics in sediments deposited during the Holocene and older geologic periods. To understand the carbon accumulation process and the nature of the older soil carbon, we need to consider also the paleo-environments in which it accumulated. Second, it is important to understand why and how it is preserved. A third aspect is the lability of this carbon, or its vulnerability to decomposition. Fourth, we need to understand responses to climate in decomposer communities, both micro-organisms and soil fauna (Aerts 2006).

An important and confusing sign convention needs to be mentioned here: atmospheric and climate scientists assign a positive sign to a net flux of greenhouse gas (e.g. CO_2) *to* the atmosphere, and a negative sign to its withdrawal *from* the atmosphere (for instance, when plants withdraw CO_2 from the atmosphere through photosynthesis). However, from the soil or ecosystem perspective, a positive sign for withdrawal of carbon from the atmosphere and storage in the ecosystem is more logical (Chapin et al. 2006). To avoid confusion, the atmospheric sign convention will be followed here: A flow of carbon in CO_2 or CH_4 from atmosphere to ecosystem is marked with a −, the other way a**+**.

4.1 The Ecosystem Carbon Balance

Chapin et al. (2006) developed a conceptual framework for carbon cycle terminology, which is adopted here. The gross conversion of CO_2 into organic carbon by photosynthesis in an ecosystem is the gross primary production (*GPP*) or gross photosynthesis. This is partly balanced by the amount of C that is respired again to the atmosphere as CO_2 (ecosystem respiration, R_{eco}). The net result, $GPP - R_{eco}$, is the Net Ecosystem Production (*NEP*). The ecosystem respiration is the sum of two components: the autotrophic respiration, which is the CO_2 respiration of living plants including roots, stems and leaves, and the heterotrophic respiration, which is the respiration of all decomposers of organic matter (bacteria, fungi, fauna) in the ecosystem. The distinction between autotrophic and heterotrophic respiration is not sharp – in the soil environment around the plant roots (the rhizosphere) lives a host of bacteria and fungi in symbiosis with the roots. These decompose organic compounds excreted by the root system, and provide nutrients to the roots in turn. Because of their intense interaction with the root system, the respiration of these symbionts is sometimes included with the autotrophic respiration, although it is conceptually more logical to include them in the heterotrophic respiration. The distinction in measurements is very difficult, if not impossible.

A concept that is superficially similar to *NEP* is the net ecosystem exchange (*NEE*). *NEE* is what atmospheric measurements of ecosystem carbon exchange see – the exchange of CO_2 between the ecosystem air volume and the atmosphere around it. With daylight, *NEE* is generally negative, because CO_2 is consumed by photosynthesis and flows into the ecosystem air volume from outside. During the night, when photosynthesis stops because of too low light intensity, *NEE* is positive and equals R_{eco}. *NEP* and *NEE* have opposite signs. When *NEE* is negative, indicating removal of carbon from the atmosphere into the ecosystem, *NEP* accumulates the same carbon into the ecosystem.

The absolute magnitude of *NEE* and *NEP* need not be the same. Consider the case of negative *NEE* during daytime, when CO_2 uptake by photosynthesis is larger than R_{eco}. *NEE* is measured by some measurement system in the atmosphere, covering a certain area of the ecosystem (for instance, the eddy covariance method described in Sect. 2.2.2; see also Sect. 4.6). Ideally, if the entire CO_2 flux is captured by the measurement system, the absolute values of *NEE* and *NEP* should be equal.

However, not all respired carbon is returned to the atmosphere as CO_2, but some CO_2 dissolves in groundwater and is transported by water flow out of the ecosystem as dissolved inorganic carbon (DIC, consisting of dissolved CO_2, carbonate and bicarbonate ions) and may be returned to the atmosphere outside the ecosystem. In that case the absolute value of *NEE* is larger than that of *NEP* for the ecosystem that lost this carbon; *NEE* measurements (in the atmosphere) do not 'see' all of the R_{eco}-CO_2 since it is not detected within the spatial range of the measurement system. CO_2 also may be missed or added by advection, which is sideways transport of CO_2 in the air within the vegetation canopy, that is not measured by the measurement system (e.g. Aubinet et al. 2005).

Carbon transport by water is common in terrestrial wetland ecosystems, where DIC transport may be in the order of 20% of *NEP* (Chapin et al. 2006). In aquatic ecosystems which receive this transported DIC, it may return to the atmosphere as CO_2, resulting there in a higher *NEE*. Advection occurs for instance during quiet conditions at night in forests on sloping terrain, where CO_2 is transported by light winds down valleys. In practice, *NEE* is most often the ecosystem property that is quantified, but a full accounting of the carbon or greenhouse gas balance should also include carbon fluxes by groundwater, surface water, or by advection.

The ecosystem property that really matters in this chapter, is the change in the amount of carbon in the ecosystem over time, the *net ecosystem carbon balance* (*NECB*), *dC/dt* (Fig. 4.2). *NECB* includes carbon that is stored in ecosystem biomass and carbon in soils. This is related to *NEE* and other ecosystem carbon fluxes as:

$$dC/dt = -NEE + F_{\mathrm{DIC}} + F_{\mathrm{DOC}} + F_{\mathrm{POC}} + F_{\mathrm{VOC}} + F_{\mathrm{CO}} + F_{\mathrm{CH4}} \tag{4.1}$$

F_{DIC} is the dissolved inorganic carbon flux discussed above. F_{DOC} is water dissolved organic carbon (abbreviated DOC), which gives the brownish colour to peatland streams, F_{POC} is particulate organic matter (POC), ranging from floating tree trunks to finely divided plant fragments and organic material associated with mineral particles. These three fluxes are mostly water-borne, although wind erosion of peat soils and soot from fires may sometimes contribute to F_{POC} (e.g. Seppälä 2003).

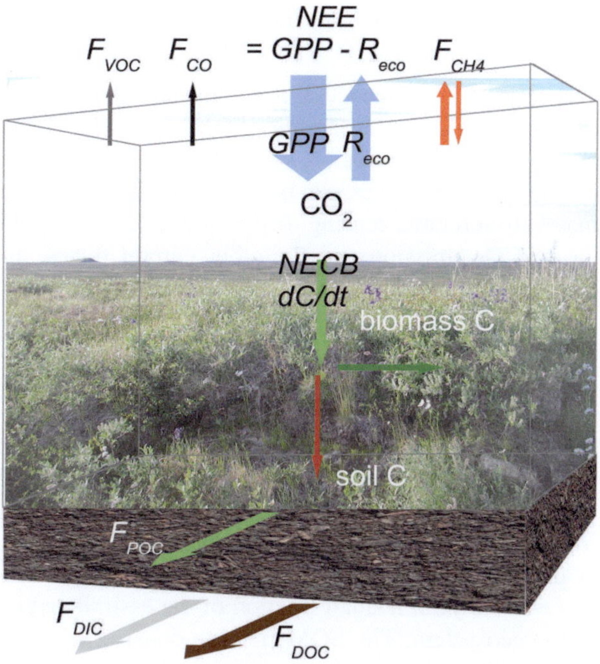

Fig. 4.2 Net ecosystem carbon balance with its components (see text). *NEE*: net ecosystem exchange of CO_2; *NECB* Net Ecosystem Carbon Balance, F_{DIC} net flux of dissolved inorganic carbon, F_{DOC} net flux of dissolved organic carbon, F_{POC} net flux of particulate organic carbon, F_{VOC} net flux of volatile organic carbon, F_{CO} carbon monoxide flux, F_{CH4} net flux of methane

The next three fluxes are air-borne. F_{VOC} is a flux of biogenic volatile organic compounds (VOC) emitted to the atmosphere by plants. VOC consist of thousands of volatile organic compounds, the most well known are isoprenes, terpenoids, alkanes, alkenes, alcohols and esters. They are responsible for the typical 'forest scent' of forests; also tundra plants emit VOC. Other producers of VOC are soil organisms (Guenther et al. 2000). An important factor in the emission of VOC from plants is plant stress (e.g. climate stress, herbivores). VOC emission has a large impact on atmospheric chemistry and is climatically significant since it interacts with greenhouse gases (Arneth et al. 2016; Thonat et al. 2017). F_{CO} (carbon monoxide) is also largely derived from plants, and is emitted from biomass burning. F_{CH4} is the methane flux, which derives from water-saturated soils with anaerobically (oxygen-free) decomposing organic material; below it is discussed in more detail as the ecosystem methane flux. However, F_{CH4} is a two-way flux; dry soils also may consume CH_4 by bacterial oxidation (e.g. Van Huissteden et al. 2008; Juncher Jørgensen et al. 2014).

Large carbon pools in high-latitude environments are generally the result of low losses by ecosystem respiration (R_{eco}) and other carbon fluxes, rather than high carbon inputs from *GPP* (Grosse et al. 2011). *GPP* is limited by the short growing season; R_{eco} is physically limited by low temperatures and a generally high water table. The various carbon pools, accumulation mechanisms and carbon accumulation rates in permafrost environments will be discussed in more detail in the next sections.

4.1.1 Terrestrial Environments

The change in the ecosystem carbon balance (*dC/dt*) includes both biomass and carbon in soil organic matter (SOM), see Fig. 4.2. If positive, it represents an increase of biomass carbon, soil carbon or both. The amount of carbon that is stored in vegetation, e.g. trees, can be assessed with repeated biomass inventories and estimates from remote sensing. If the biomass dies, it will be decomposed and contribute to CO_2 emission and the other carbon fluxes, and contribute to soil carbon storage.

The yearly change of the carbon balance can be quantified by measuring the fluxes in Eq. 4.1. However, these measurements are subject to measurement error, and may be incomplete. Often, one or more fluxes (e.g. CH_4, DIC, DOC) have not been measured. In Arctic sites, winter fluxes may be missed due to the extreme operating environment for instrumentation. These winter fluxes are not negligible (Zona et al. 2016; Sect. 4.6). Since the carbon balance is a small difference between large fluxes with considerable uncertainty, its uncertainty is also large relative to its magnitude (e.g. Van der Molen et al. 2007). Moreover, the year-to-year variability is also large (e.g. Roulet et al. 2007; Parmentier et al. 2011a, c).

Over the year, a small fraction of the assimilated carbon will be stored in SOM in the ecosystem. The SOM fraction consists of a large range of organic components, with various decomposition rates and residence time in the soil. The most labile carbon components decompose fastest, within one season, and therefore participate in the respiration and other carbon fluxes of the ecosystem belonging to the yearly balance. More stable carbon components decompose over much longer time scales of years to decades and more. Trumbore (2009) lists four stabilization mechanisms:

- climatic stabilization by freezing (in permafrost);
- storage in water-saturated soils, which decreases access by oxygen and results in slower, anaerobic decomposition;
- carbon which is stabilized by a specific chemical structure (e.g. compounds pyrolysed by fire, such as charcoal);
- physical stabilization by sorption to mineral surfaces, which inhibits accessibility by microbes.

Burial in sedimentary successions such as peat beds and floodplain sediments is similar to storage in water-saturated soils. The stored carbon enters an environment where it may remain stored under oxygen-free conditions over long time periods, unless erosion of the deposited sediment occurs. Storage in permafrost occurs when the permafrost table aggrades upward with the sedimentation surface. In this way, large amounts of organic carbon have been preserved in the Pleistocene Yedoma sediments (Schirrmeister et al. 2011a, b, 2013; Sect. 4.4). However, the stabilization in sediments and permafrost is never absolute; under waterlogged conditions still some anaerobic decomposition occurs (Clymo 1984) and microbial activity continues at subzero temperatures, albeit at a very low rate (e.g. Rivkina et al. 2000).

Figure 4.3 shows the large yearly variation of *NEE* at a High Arctic site in Northeast Greenland (Lund et al. 2012), and the environmental variables that drive *NEE*. The daily standard deviation between the years is nearly as high as the daily average *NEE*, reflecting this variability. This figure also shows that negative values of *NEE*, reflecting a net uptake of CO_2 from the atmosphere, are restricted to only a short time of the year in this High Arctic site. Despite the short growing season, this site has a negative *NEE*, representing a sink for CO_2 (which may differ from ecosystem *NEP*, see above).

Because of the year-to-year variability of the carbon balance and the relatively small fraction that ultimately enters the soil, decadal to millennial estimates of the carbon balance are useful, next to flux measurements. For sedimentary deposits like peat and lake sediments this is determined from a well-dated sediment succession. Dating of a sedimentary succession using radiocarbon datings or other types of age datings at various levels within the deposit, combined with measurements of the carbon mass per volume of sediment, results in an estimate of net carbon sedimentation rates, also known as LORCA (Long term ORganic Carbon Accumulation rate, e.g. Roulet et al. 2007). Besides radiocarbon dating, a large number of other dating methods can be applied, e.g. radiometric methods like 210Pb dating, or tephrachronology, using volcanic ash layers (Walker and Walker 2005).

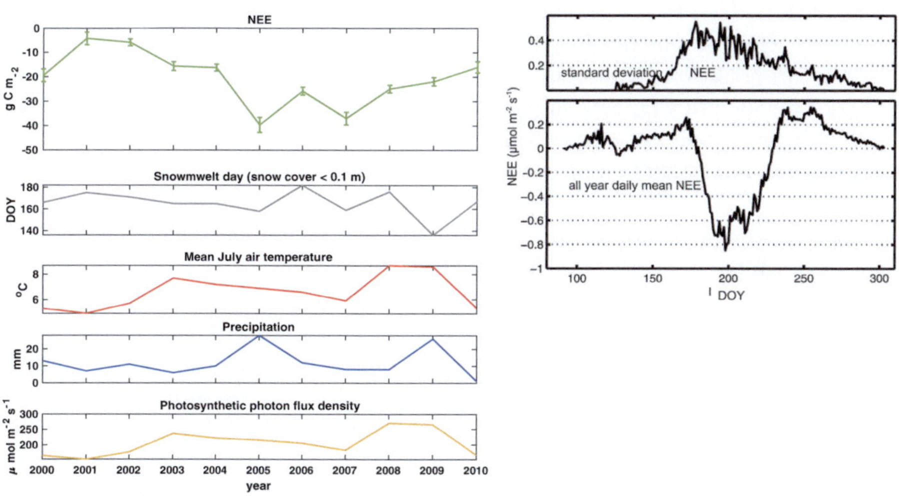

Fig. 4.3 Left: total yearly net ecosystem exchange (NEE) of Arctic Cassiope heath at Zackenberg, Northeast Greenland, and variation of average yearly weather variables from 2000 to 2010. Snowmelt day is the first day of the year that the snow cover is thinner than 10 cm, *DOY* day number (day of year). The photosynthetic photon flux density is the amount of light available for photosynthesis. Redrawn from Lund et al. (2012). Right: daily mean NEE over all years (below) with standard deviation between years. Republished with permission of John Wiley and Sons Inc, 'Trends in CO_2 exchange in a High Arctic tundra heath, 2000–2010' by Lund et al. (2012), Journal of Geophysical Research: Biogeosciences 117(G2) © 2012; permission conveyed through Copyright Clearance Center, Inc.

In terrestrial environments without sedimentation, carbon is stored by accumulation in soil surface and subsurface horizons. Because these consist of a mix of materials of various ages, estimation of the carbon accumulation rate is less straightforward. Radiocarbon dating of the soil carbon provides a mean age of the organic material. Over sub-recent time spans, carbon accumulation in soils can be assessed by making use of the atmospheric ^{14}C spike produced by atomic bomb testing (Trumbore 2009).

LORCA may differ considerably from yearly measurements of the ecosystem carbon balance. LORCA integrates over long time spans, which may include undetected hiatuses in carbon accumulation, e.g. erosion or non-deposition in the case of sedimentary deposits. Second, considerable year to year variability of carbon accumulation measurements is commonplace; measurements over a time span of a statistically reliable number of years are necessary to arrive at an average carbon accumulation rate that can be compared reasonably with LORCA data.

4.1.2 Lakes

The carbon balance of aquatic environments contains the same fluxes and stores as terrestrial environments (Eq. 4.1), although the magnitude of the fluxes may differ (Cole et al. 2007). In lakes, anaerobic conditions are more ubiquitous in bottom sediments, causing a higher F_{CH4} flux. This flux may amount 20–60% of the overall organic carbon decomposition in lakes (Bastviken et al. 2008) Approximately two thirds of the natural CH_4 emissions north of 50°N originates from lakes (Wik et al. 2016). The F_{DOC} and F_{POC} fluxes into the lake ecosystem are generally larger than the outgoing fluxes; terrestrial ecosystems may lose carbon to aquatic systems by these fluxes.

The incoming F_{DIC} flux includes contributions from groundwater, which is usually supersaturated with CO_2. The net influx of inorganic and organic carbon is compensated by a higher outgoing CH_4 flux and CO_2 flux from ecosystem respiration. This respiration of carbon imported from terrestrial ecosystems adds to the respiration of the lake ecosystem itself (e.g. Kling et al. 1991; Tranvik et al. 2009). Hence, lakes tend to have a positive *NEE* (source of CO_2 to the atmosphere) on the short term, although during summer periods with high photosynthesis rates of water vegetation and phytoplankton, *NEE* may be negative. Storage of carbon in lake sediments is usually large; organic carbon in lake sediments may be conserved over thousands of years, by contrast to terrestrial ecosystems (Cole et al. 2007). The long-term storage of carbon from lake *GPP* can create a net sink of carbon over longer time frames (Walter Anthony et al. Walter Anthony et al. 2014). Lake sediments are also a sink for carbon that has been eroded from terrestrial ecosystems (Vonk and Gustafsson 2013).

An important mechanism that determines the presence of anoxic environments in lake sediments and eventually the lake water column, is thermal stratification (Sect. 2.7). Thermal stratification prevents fast turbulent transport of gases (e.g. O_2); the transport rates by diffusion across the thermocline is much slower. In shallow lakes, turbulent mixing by waves prevents development of stratification.

4.1.3 The Greenhouse Gas Balance

For assessing the effects of greenhouse gas balance of an ecosystem on climate, we need to compare the radiative effect of greenhouse gases based on their radiative properties and lifetime in the atmosphere. A simple metric that is often used for anthropogenic greenhouse gas emissions is the Global Warming Potential (GWP). GWP is the time-integrated radiative forcing of a pulse emission (short duration) of a mass of non-CO_2 greenhouse gas with respect to that of the same mass of CO_2 (IPCC 2013). The GWP's of the most important non-CO_2 greenhouse gases from permafrost, CH_4 and N_2O, have been mentioned in Sect. 1.3, and are considerably larger than that of CO_2. An ecosystem with a negative carbon balance (removing net carbon from the atmosphere) may have a net warming effect on the climate, if its emission of CH_4 or N_2O is high (Fig. 4.1).

However, the GWP is an unsatisfactory metric for ecosystems, because ecosystems generally do not produce *pulse* emissions, but *sustained* emissions over many years, with gradual or step changes in the emission level. Moreover, the usual choice of the time period (20 or 100 years) is rather arbitrary for an ecosystem. Neubauer and Megonigal (2015) developed a metric for sustained ecosystem emissions which also differentiates between sources and sinks of greenhouse gases, respectively the Sustained Global Warming Potential (SGWP) and the Sustained Global Cooling Potential (SGCP). For instance, for CH_4, the 100 year GWP is 34, while the 100 year SGWP (for a CH_4 source) is 45, and the SGCP (for a CH_4 sink) is 203.

The main problem of the approach of Neubauer and Megonigal (2015) is that it does not circumvent the shortcomings of selecting a rather arbitrary time frame. Because of the varying lifetime of greenhouse gases, an ecosystem can switch over time from a net source to a sink of greenhouse gases (Neubauer and Megonigal 2015). For instance, CH_4 has a much shorter lifetime in the atmosphere than CO_2. Consider again the ecosystem example in the first paragraph of this section, which removes net CO_2 from the atmosphere, but has a net warming effect because of its high CH_4 emission. However, over a time period of many years, it may have removed so much long-lived CO_2 from the atmosphere that the warming effect of the short-lived CH_4 is compensated. This switch can take decades or centuries depending on the sink and source strengths of the greenhouses gases emitted or taken up by the system. The time it takes is known as the switchover time.

Another approach is that of Frolking and Roulet (2007) who compute the radiative forcing of greenhouse gases over the entire lifetime of an ecosystem. This avoids the selection of an arbitrary time frame, but replaces the discussion towards the often unknown history and lifetime of the ecosystem. The approach of Neubauer and Megonigal (2015) will be applied here when discussing the greenhouse gas balance (e.g. Sect. 4.6), unless otherwise stated; the GWP is still used quite often in publications on ecosystem emissions.

4.2 Vegetation Primary Production

The source of carbon in all ecosystems is vegetation primary production. The produced organic matter is stored in vegetation biomass (stems and leaves), litter on top of the soil, in roots below the soil surface and finally in the soil organic matter reservoir. There it is decomposed by soil fauna and micro-organisms; the decomposition rate determines how much of the original carbon remains. Primary production in cold environments needs adaptation to this environment, and is strongly influenced by these environmental restrictions. In this section, primary production will be considered in more detail.

4.2.1 Photosynthesis and Carbon Allocation

Assimilation of carbon by the vegetation cover in permafrost ecosystems is determined by:

- The biological potential of the vegetation canopy to intercept light and fix carbon,
- An environmental potential determined by water and nutrient availability, optimal temperatures for enzyme activity, the quantity of photosynthetically active radiation (PAR) and CO_2 concentration.

Arctic plant communities are strongly restricted by this environmental potential because of low temperatures and light for much of the year. Additionally, nutrient availability is often limiting; in areas with higher nutrient availability, denser and more productive vegetation occurs (e.g. Chapin et al. 1995; Williams and Rastetter 1999). Nitrogen is usually limiting in Arctic ecosystems. For example, at the taiga-tundra transition in northern Siberia, NH_4^+ in soils is correlated positively with leaf nitrogen content and carbon assimilation in Larch trees (Liang et al. 2014).

The overall reaction equation of photosynthesis is:

$$nCO_2 + 2nH_2O \xrightarrow{PAR} (C_2HO)_n + nO_2 + nH_2O \tag{4.2}$$

The wavelength of PAR lies between 0.4 and 0.7 μm. Sunlight at higher wavelengths, e.g. the near infrared (NIR) band, is reflected, which prevents overheating of the plants. This strong reflection in the NIR band is a hallmark of living vegetation, and it has been used extensively in mapping vegetation using remote sensing techniques. In remote sensing, various vegetation indices have been developed by combining information in different wavelength bands of the electromagnetic radiation spectrum (e.g. Schultz et al. 2016). The most well known is Normalized Difference Vegetation Index (NDVI); it is calculated from the difference between the reflectance in the red and near infrared parts of the spectrum:

$$\mathrm{NDVI} = (\text{near infrared} - \text{red}) / (\text{near infrared} + \text{red}) \tag{4.3}$$

NDVI has been related to various indices for living green biomass and chlorophyll abundance (Myneni et al. 1995), for instance the leaf are index (LAI), the leaf area per unit ground area ($m^2\ m^{-2}$). Figure 4.4 shows a Circum-Arctic vegetation map of treeless vegetation, derived from remote sensing data (Walker et al. 2005); an agreement between the major vegetation transitions between tundra and barren, and the Circum-Arctic NDVI map in Fig. 4.5 is easily seen.

The first step in photosynthesis is the absorption of light photons in pigment molecules contained in chloroplasts, specialized organelles in cells of plants and algae. The most well-known photosynthesis pigment is green chlorophyll, others are reddish carotenoids. It is the amount of photons received by the plant that determines the rate of photosynthesis, not their energy. The amount of photons is expressed in moles, like other chemical species involved in the chain of biochemical reactions behind Eq. 4.2. Here the details of these reactions will be discussed only insofar they are important to understand their contribution to the permafrost carbon cycle.

In the photosynthesis reaction chain two groups of reactions occur: first the light reactions, which use eight photons to split two water molecules and generate energy for the next step. The next step comprises the dark reactions in which is CO_2 is reduced. These dark reactions are known as the Calvin cycle. An important enzyme in the Calvin cycle is the enzyme rubisco, which is short for ribulose bi-phosphate carboxylase-oxygenase. See e.g. Bonan (Bonan 2016) for a more extensive description.

Two plant groups can be discerned on the basis of the dark reactions. In C3 plants, first a 3-carbon compound is formed, by reaction of a 5-carbon sugar (ribulose- 1,5-bi-phosphate, or short *RuBP*) with CO_2 to form a 3-carbon compound, phosphoglyceric acid. In C4 plants 4-carbon acids are formed, which occurs in different cells and requires transport between cells, taking more energy. However, the C_4 process is more efficient in the use of light, CO_2 and water; C_4 plants are better adapted to high light levels and high temperatures. In the C_3 process, some oxidation of RuBP by O_2 may occur, making the CO_2-reduction step in photosynthesis less efficient – the net CO_2 reduction may be lower by 30–50%. This is known as photorespiration – CO_2 production by the photosynthesis process itself. In C_4 plants the photorespiration is practically absent (Bonan 2016).

The carbon isotope ratio of photosynthesised material is affected by C_3 or C_4 photosynthesis; C_3 photosynthesis results in a more negative 13C isotope ratio (−33 to −24‰) than C_4 photosynthesis (−16 to −10‰). C_4 plants are also more tolerant to the low atmospheric CO_2 concentration that existed during glacials in the past,

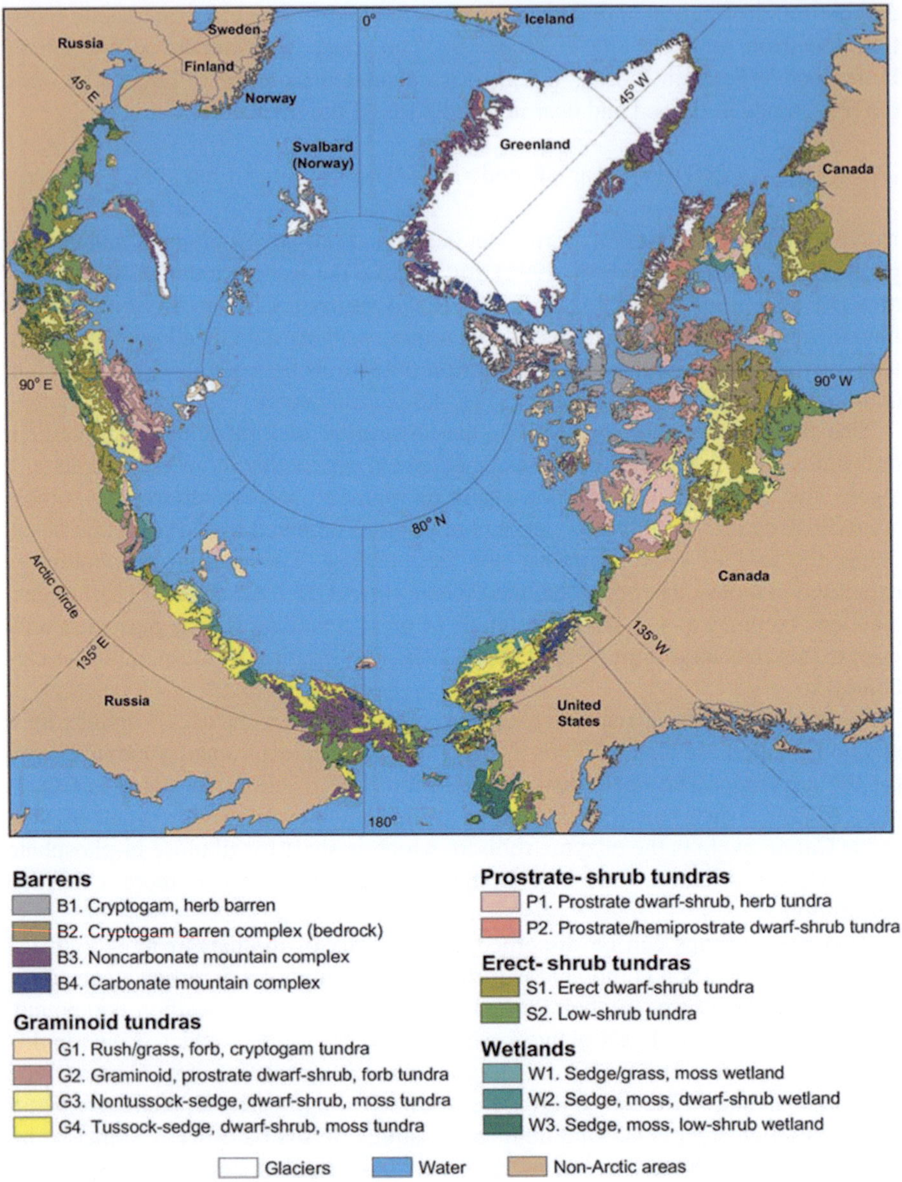

Fig. 4.4 Circum-Arctic vegetation map of treeless vegetations north of the tree-line. Prostrate shrubs are shrubs of which the branches lie upon the ground, as opposite to erect shrubs which have standing branches. Walker et al. (2005), CAVM mapping team; reprinted with permission of the authors; https://www.geobotany.uaf.edu/cavm/

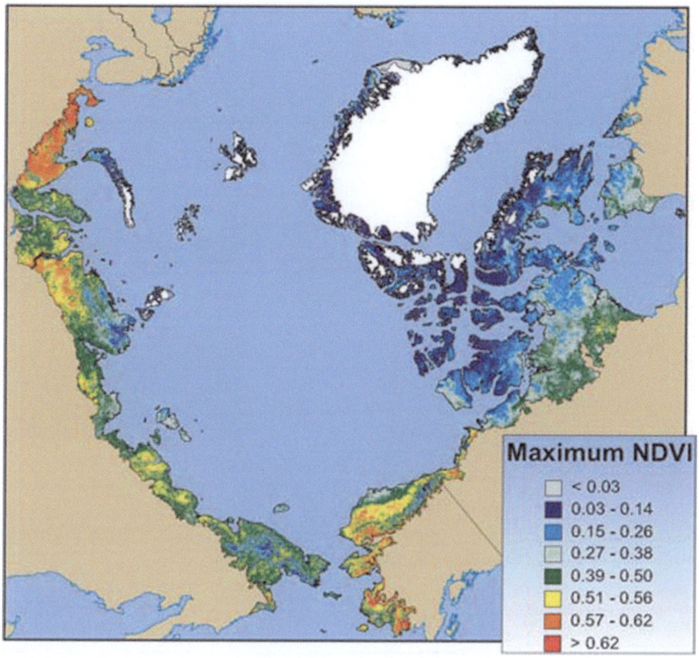

Fig. 4.5 Circum-Arctic NDVI map. Walker et al. (2005), CAVM mapping team; reprinted with permission of the authors; https://www.geobotany.uaf.edu/cavm/

while C_3 plants likely profit more from present high CO_2 levels. Therefore, during glacials C_4 plants may have had a wider geographic distribution. C_4 photosynthesis is common in grasses and sedges. However, in present cold climate ecosystems C_4 plants are practically absent (Still et al. 2003; Ehleringer et al. 1997).

The net carbon balance of a plant is the amount of carbon fixed by photosynthesis, minus the respiration of the plant, which can be subdivided in maintenance respiration and growth respiration. The maintenance respiration is the CO_2 released by cell metabolic processes. The growth respiration is carbon which is lost during biosynthesis of plant tissue. Maintenance and growth respiration add to the carbon lost by photorespiration. Taken together, maintenance, growth and photorespiration are the autotrophic respiration.

Photosynthesis rates are influenced by light level, temperature, leaf nutrient availability, CO_2 concentration and water stress. The opening and closing of the leaf stomata in response to these variables governs the stomatal conductance (measured in mmol CO_2 m^{-2} s^{-1}; Bonan 2016). The stomatal conductance increases with increasing light levels (photosynthetic photon flux density, PPFD, in μmol photons m^{-2} s^{-1}) until a saturation level is reached (Fig. 4.6). At low light levels, some photosynthesis may occur, but it does not exceed the respiration of the plant. The PPFD where photosynthesis equals respiration is known as the light compensation point (Fig. 4.7), typically ranging between 10–40 μmol photons m^{-2} s^{-1}.

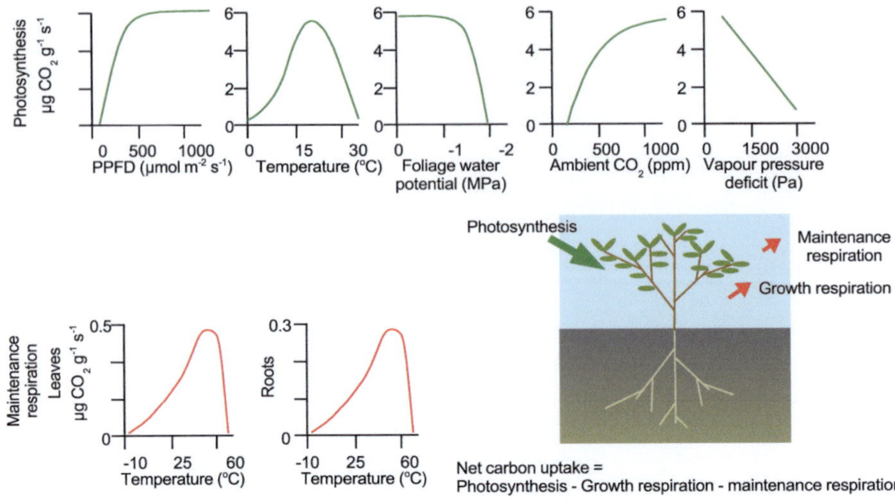

Fig. 4.6 The carbon balance of trees and approximate relations with the environment that influence the carbon uptake by photosynthesis. (After Bonan 2002)

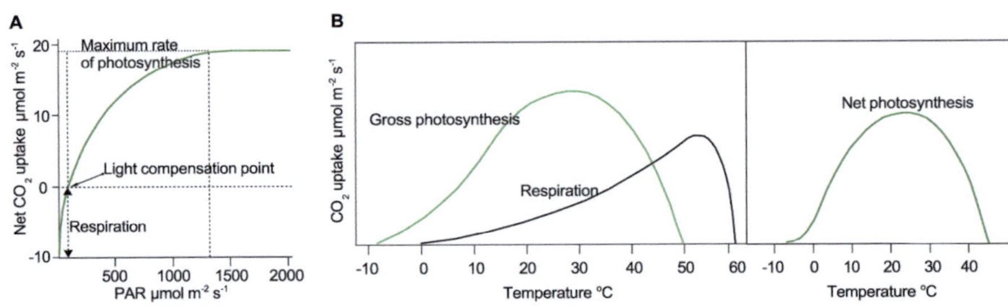

Fig. 4.7 (**a**) Net photosynthesis related to photosynthetic active radiation (PAR); the PAR quantity where the gross photosynthesis equals respiration (= zero net photosynthesis) is the light compensation point. (**b**) Temperature sensitivity of gross photosynthesis and respiration (left), and the resulting net photosynthesis (right). After Bonan (2002)

Second, photosynthesis depends on temperature, with a photosynthesis optimum temperature. The optimum is in the order of 25 °C but differs among plant groups; in particular mosses have lower optimum temperatures and can photosynthesise at temperatures near 0 °C, and even at temperatures down to −10 °C (He et al. 2016). Plant respiration equally depends on temperature. It increases exponentially up to a maximum, which is higher than that of photosynthesis. The net result of the temperature sensitivity of photosynthesis and respiration is shown in Fig. 4.7.

Photosynthesis also depends on water availability, expressed in Fig. 4.6 in the relations with foliage water potential and air vapour pressure deficit. A low leaf water potential occurs when transpiration of water cannot be replenished by uptake of water from the soil, e.g. if soil water content drops to very low levels (the wilting point). A low leaf water potential occurs under conditions of a high vapour pressure deficit, which is the difference between the actual amount of water vapour in the air and the amount of water vapour in saturated air. At low moisture availability, photosynthesis rapidly decreases because the stomata are closed to prevent water loss. The trade-off between water loss by transpiration and uptake of CO_2 through stomata is an important compromise for plants. A high rate of photosynthesis requires opening of the stomata, but also increases water loss.

The CO_2 concentration of the air influences photosynthesis rates, up to a saturation level, which is higher for C_3 plants than for C_4 plants (Bonan 2016). Similar to light availability, there is also a CO_2 compensation point at low CO_2 concentrations (30–50 ppm), where photosynthesis equals respiration. Light and CO_2 concentration interact strongly, in particular in C_3 plants. With higher CO_2 concentration, an increase of irradiance increases the photosynthesis rate more effectively than at low CO_2 concentration. For C_4 plants, photosynthesis saturates at a CO_2 concentration of 400 ppm irrespective of light levels. Likewise, there is also a relation of atmospheric CO_2 concentration with the amount of water used by the plants during photosynthesis (Water Use Efficiency, WUE). Higher CO_2 concentrations require a lower stomatal conductance, hence also lower transpiration. The relation of CO_2 concentration with photosynthesis and other environmental variables is important for understanding primary production in future high CO_2 atmospheres, and during glacial climates when CO_2 concentration decreased to less than half of those at present.

The assimilated carbohydrates are allocated to the various parts of the plants: leaves, stems including tree trunks and branches, and roots (non-vascular mosses have no root system). In all plant parts, respiration losses of CO_2 occur. Plant carbon is transferred to the soil via litter fall and the root system (Fig. 4.8). A part of the produced organic compounds is excreted: first, the volatile organic compounds mentioned in Sect. 4.1 (F_{VOC}) which are excreted by leaves and stems, and second, root exudates excreted by plant roots. In cold climates, plants invest more carbon in root growth (Shaver and Billings 1975; Shaver and Cutler 1979).

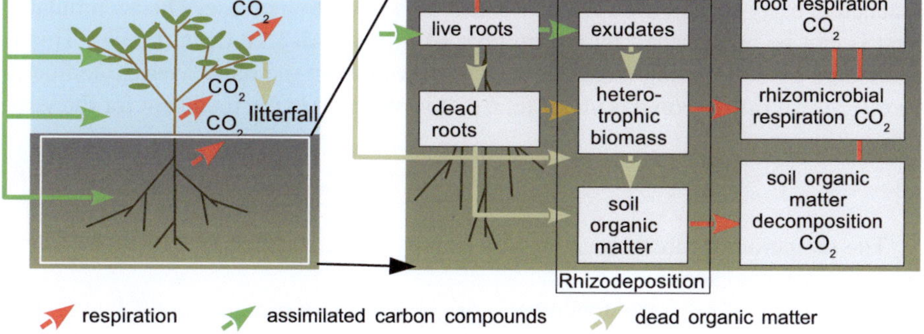

Fig. 4.8 Transfer of photosynthesised carbon from plant to soil into soil carbon pools and losses to CO_2 by respiration

Although these root exudates are a fraction of the photosynthesised carbon compounds, they are highly important for microbial processes and the below-ground carbon fluxes, since these exudates generally consist of readily decomposable compounds such as mucilage, ectoenzymes, organic acids, phenolic compounds, sugars and amino acids (Ström et al. 2005). Other compounds are toxins to protect the plants to pathogens or pests. Root exudates fuel the microbial population associated with the plant roots. In turn, this microbial population contributes to nutrient availability of plants, via a symbiotic relation known as mycorrhiza. With endomycorrhiza, fungal hyphae penetrate the plant roots, ectomycorrhiza is a more loose relationship where fungi do not penetrate into the roots (Hawkes et al. 2007). Common Arctic species which are known to have such symbiotic relationships are *Betula*, *Larix*, *Pinus*, *Salix*, *Dryas*, *Cassiope*, *Polygonum* and *Kobresia* (Callaghan et al. 2004). The combination of exudates and other organic matter originating from the plant roots such as dead root cells, is known as rhizodeposition (Fig. 4.8).

Rhizodeposition becomes available for decomposition by the soil fauna and microbial population. The resulting rhizomicrobial respiration of freshly supplied organic matter is difficult to separate from the respiration of the roots and the decomposition of other soil organic matter, because of the intimate association between roots, microbial population and soil carbon. ^{14}C labelling experiments have shed some light on the below-ground carbon transfers. Typically, 20–50% of the net assimilated carbon in agricultural species is transferred below ground to contribute to root biomass (~50%), of which ~30% is rhizosphere-derived CO_2, and ~20% organic residues. These figures also depend on plant species, age, micro-organism population, soil texture, and nutrient availability (Cheng and Gershenson 2007). However, in cold climates and nutrient-poor environments, plants invest more in root growth, nutrient uptake and respiration. In Arctic tundra, up to 70% of plant biomass is situated below ground surface (Iversen et al. 2015). A review by Wang et al. (2016) has shown that the proportion of below-ground biomass relative to above-ground biomass increases with decreasing MAAT.

The yearly pattern of photosynthesis depends on the phenology or life cycle of the plants, in particular for deciduous species which shed their leaves in winter. Bud break in spring does not occur before a certain amount of accumulated heat has been reached in early spring, after which the leaf area starts to increase. The accumulated heat is can be approached by the number of growing-degree-days (GDD). This is calculated by subtracting a temperature threshold from the daily average temperature $T_{average}$; if the outcome is larger than 0, it is included in the total sum for the year:

$$GDD = \Sigma max\left(T_{average} - T_{threshold}, 0\right) \tag{4.4}$$

The temperature threshold $T_{threshold}$ is usually set at 10 °C at temperate latitudes; however, see next section.

The relation of gross photosynthesis or gross primary production *GPP* to PAR is often modelled by a function which is known as the rectangular hyperbola (e.g. Laurila et al. 2001; Kutzbach et al. 2007a):

$$GPP = \frac{P_{\max} \alpha PAR}{P_{\max} + \alpha PAR} \tag{4.5}$$

Here α (μmol m^{-2} s^{-1}) is the initial slope of the *GPP* − *PAR* curve at *PAR* = 0; P_{max} is the theoretical maximal photosynthesis rate at infinite *PAR* values. The slope of the curve is the quantum efficiency, the amount of CO_2 that is assimilated for each micromole of *PAR* photons; the quantum efficiency decreases with higher *PAR*. Figure 4.9 gives an example of applications of Eq. 4.5 applied to analysis of CO_2 fluxes in Arctic ecosystems. P_{max} is usually derived from data, by fitting Eq. 4.5 to the data. By adding R_{eco} to the right hand side of Eq. 4.4, an estimate of *NEE* is acquired (Laurila et al. 2001). Another light response equation that has been used for high-latitude ecosystems, is the Misterlich function (Falge et al. 2001; Mbufong et al. 2014):

$$NEE = -\left(P_{\max} + R_{eco}\right)\left(1 - e^{\frac{\alpha PAR}{P_{\max} + R_{eco}}}\right) + R_{eco} \tag{4.6}$$

The parameters of Eq. 4.6 have been shown to correlate with peak summer season LAI across many Arctic sites (Mbufong et al. 2014). This is the case for dark respiration (R_{eco}) in particular. Maximum potential photosynthesis at light saturation P_{max} decreases with increasing LAI, α increases.

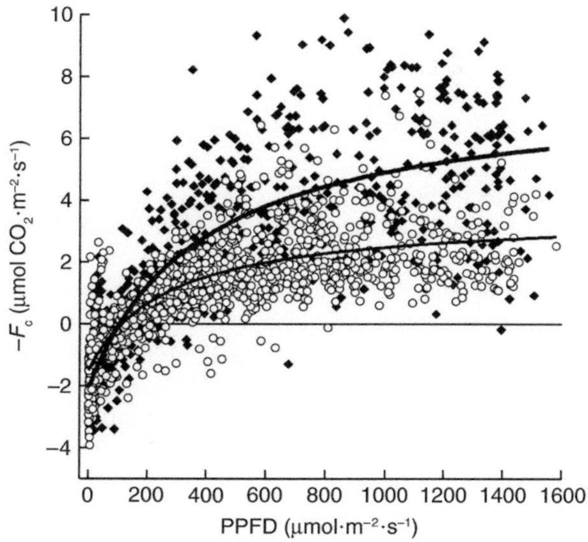

Fig. 4.9 Rectangular hyperbola function fitted to 30-minute average values of CO_2 flux measured in shrub tundra sites (solid diamonds) and moist tundra sites (open circles) in Alaska. The heavy, upper curve is for shrub tundra and the light, lower curve is for moist tundra. (Republished with permission of the Ecological Society of America, from McFadden et al. (2003), © 2003; permission conveyed through Copyright Clearance Center, Inc.)

4.2.2 Primary Production in a Cold Climate

In permafrost environments, the growing season is strongly restricted in time by temperature and light availability. On the other hand, the Arctic the day length during the growing season is long. The growing season also starts much later in the year than at temperate latitudes; low temperatures and snowpack may linger until close to the summer solstice. In particular in areas with a maritime climate, low clouds influence the light distribution for part of the summer.

At high summer above the Arctic circle the sun stays above the horizon; for example, the maximum midnight radiation at midnight in midsummer at Zackenberg in northeast Greenland is 130 W m^2 (Laurila et al. 2001; Mbufong et al. 2014). Aurela et al. (2001) estimated that the midsummer nighttime assimilation rates in a subarctic mountain birch ecosystem amounted 0.03 mg CO_2 m^{-2} s^{-1}. Nevertheless, the amount of PAR at night is low, because of the low sun elevation angle. Even if nighttime photosynthesis occurs, it does not overcome the ecosystem respiration, resulting in positive *NEE* fluxes at night (e.g. Laurila et al. 2001; Kutzbach et al. 2007a; Van der Molen et al. 2007).

In lower latitudes, the ecosystem respiration component of the carbon balance (R_{eco}) is often estimated from nighttime CO_2 flux measurements, when there is no photosynthesis. Because of the nighttime Arctic photosynthesis, this method is problematic for high summer (Eugster et al. 2005). Therefore, it is usually estimated from *NEE* at very low values of PAR (PPFD <20–50 μmol m^{-2} s^{-1}, Kutzbach et al. 2007a, Euskirchen et al. 2012). An alternative method is using daytime measured values of *NEE*, estimating gross photosynthesis from a light response curve like Eq. 4.5 and subtracting it from measured *NEE* to arrive at an estimate of R_{eco} (Lasslop et al. 2010; Parmentier et al. 2011c).

The presence of a snow cover does not exclude photosynthesis. Starr and Oberbauer (2003) found significant photosynthesis of Arctic evergreen species (the graminoid *Eriophorum vaginatum*, and dwarf shrubs *Vaccinium vitis-idaea, Ledum palustre, Cassiope tetragona*) under snow on sunny days in spring, when PPFD fluxes were sufficient to drive photosynthesis in the top 30 cm of the snow cover. Additional favourable conditions are the relatively high temperatures and CO_2 concentrations within the snow cover. Nevertheless, this is of minor importance for the *NEE*; in general, the start of negative *NEE* correlates with the timing of snowmelt (Lund et al. 2012).

A continuous low cloud cover reduces the solar radiation by more than a half, with its largest effect around noon. Above tundra vegetation in Siberia, reduction in the daily incoming shortwave radiation in the peak growing season with full cloud cover was about 58% (Juszak et al. 2016). On the other hand, the diffuse light under clouds penetrates deeper into the vegetation into lower parts of the canopy, and could potentially increase photosynthesis there (Shaver et al. 2007; Kutzbach et al. 2007a).

The effect of temperature on photosynthesis of Arctic ecosystems is evident; for the seasonal course it is closely tied to phenology. Both temperature and phenology determine the onset of CO_2 uptake of the ecosystem in spring. In the Siberian tundra, photosynthesis from mosses and grasses starts during snowmelt, but substantial CO_2 uptake by the ecosystem starts only after bud break of the shrubs, which is driven by air temperature buildup (Parmentier et al. 2011a, c). In particular heat penetration into the soil takes time in spring; low temperatures strongly restrict root growth (Alvarez-Uria and Körner 2007). Warm periods in early summer, even if short, are important for starting plant growth (Laurila et al. 2001). Blok et al. (2011a) found in northeast Siberia a positive correlation between tree ring width of the shrub *Betula nana* and *Salix* and early summer temperature. The use of growing-degree-days has been introduced in the previous section. However, because of genetic adaptions of plants to low temperatures the often-used threshold at temperate latitudes of 10 °C is too high for permafrost environments. In subarctic climates like northern Scandinavia, 5 °C is a more appropriate temperature (Aurela et al. 2001), which may be even too high for High Arctic ecosystems (Laurila et al. 2001). As mentioned above, mosses may start photosynthesis even at below-zero temperatures.

The onset of dormancy in autumn appears to be controlled more by light availability than by temperature (Van der Molen et al. 2007; Euskirchen et al. 2012). Temperature also acts independently of phenology during cold snaps in summer. Euskirchen et al. (2012) observed that the ecosystem switches temporarily to a source of CO_2 at steep temperature drops in summer. The low air temperature immediately decreases photosynthesis, while the large soil component of the ecosystem respiration reacts much slower to temperature decrease. Moreover, increased cloud cover associated with the passage of a cold front, reduces photosynthesis.

Adaptations of plants to low temperatures cover a range of other inconveniences of a cold climate. These consist of protection to freezing and ice crystal growth and dehydration, and adaptations to a short growing season. Many common Arctic plants are not restricted to the Arctic but extend their range to temperate latitudes, e.g. the cotton grasses *Eriophorum.* Others are restricted to the Arctic only, e.g. the Arctic bell-heather *Cassiope tetragona.* In particular for the first group, phenotypical variation contributes to adaptation. Callaghan et al. (2004) list a large number of traits and mechanisms by which plants in permafrost environments adapt to their cold environment:

- Tolerance to extremely low temperatures (−40 °C and lower). Antifreeze proteins in cells have an important role in this type of cold adaptation (Griffith and Yaish 2004); low stature allows plants to profit from the protective snow cover in winter.
- Low summer temperatures induce increased investment in root growth, nutrient uptake and respiration (e.g. Shaver and Billings 1975; Shaver and Cutler 1979; Iversen et al. 2015; Wang et al. 2016).
- Adaptations in growth form (e.g. cushion-like growth form, formation of tussocks, small compact leaves to minimize heat loss).
- The short and late growing season requires stored resources for the long dormant period and restricts flowering and germination, clonal reproduction is therefore common.

- Encasing of plant parts with ice accretion in some weather conditions creates anoxia; non-wetland High Arctic species have a higher tolerance for this type of anoxia (Crawford et al. 1994)

Drought stress may seem to be less important in permafrost environments, but because low precipitation amounts many permafrost areas are classify as semi-deserts or deserts (e.g. Callaghan et al. 2004; Walker et al. 2005). Freezing also results in a lack of available water. Besides tolerance to low temperatures, many Arctic plants are adapted to extremely dry environments. In particular lichens are able to survive under such conditions. In a study on CO_2 exchange of Low Arctic tundra in Canada, *GPP* increased significantly and R_{eco} decreased after a rainfall event (Nobrega and Grogan 2008).

However, Gold and Bliss (1995) did not find large differences in soil water availability and drought stress between sites with continuous vegetation cover, fed by runoff from thawing snow fields, and barren sites with patchy vegetation. At the meltwater-fed sites, a cover of mosses and lichens created also favourable conditions for vascular plants, because mosses and lichens enhanced the nutrient and organic matter content of the soil. Keuper et al. (2012) found no response of total annual productivity to increased rainfall at a Siberian tundra site and a northern Swedish bog tundra. However, individual species (the shrubs *Betula nana*, *Salix pulchra,* the grass *Arctagrostis latifolia*) did show positive response by length growth and leaf growth, indicating that these species were water-limited. Blok et al. (2011a) found positive correlation of tree ring width of the same shrub species with summer precipitation. Jespersen et al. (2018) found a 40–50% contribution from snowmelt water to growth of deciduous shrubs from isotopic evidence. In boreal forests, midsummer depletion of water sources is a common feature (Kotani et al. 2014).

Thawing permafrost ice in summer is also a source of water for plants; for Siberian Larch forest this was shown to be a water source in summer (Sugimoto et al. 2002; Saito et al. 2007; Ohta et al. 2008). On the other hand, permafrost thaw, creating water-saturated soils, decreases productivity of species that do not tolerate waterlogged soils, e.g. Larch (Iijima et al. 2014) and Dwarf Birch (Nauta et al. 2014; Li et al. 2017).

4.3 Vegetation Composition: Effects on the Carbon Cycle

Mosses and lichens (cryptogams) are an important component of the vegetation biomass in Arctic tundra, where they dominate in wetland habitats (e.g. Walker et al. 2005). In forest floors, mosses and lichens are likewise of common occurrence. In suitable environments, mosses facilitate long term carbon accumulation by peat formation (Turetsky et al. 2012). Mosses differ from vascular plants in several important aspects. They have no root system and lack specialized water transport cells; they depend on capillary transport of external water on their outer surfaces.

Mosses are capable of losing their cell water without dying, and resume photosynthesis when re-wetted, although their tolerance to drying differs between species that live in shaded locations such as forest floors and species that are commonly exposed to full sunlight (He et al. 2016). Kutzbach et al. (2007a) observed in polygonal tundra that mosses in dry periods rapidly desiccated, depressing their photosynthetic potential P_{max} and with that, also the P_{max} of the entire vegetation canopy. After rainfall, moss photosynthesis was quickly restored.

The water-holding capacity of mosses depends on species group and colony morphology. The peat forming *Sphagnum* mosses are capable of storing large amounts of water in their cells. Colony morphology is a trade-off between light interception and water holding capacity; dense colonies can store much water at the cost of light interception, and are found mostly in exposed sites (Elumeeva et al. 2011). Mosses also have lower temperature optima for photosynthesis and assimilate CO_2 under lower irradiation conditions (Turetsky et al. 2012; He et al. 2016).

Tundra bryophytes have temperature optima for photosynthesis between 5 and 15 °C, and can photosynthesise at temperatures well below 0 °C. These traits create an advantage over vascular plants in early spring, which cannot start photosynthesis before considerable temperature rise and bud break. Moreover, mosses continue to photosynthesise until the last weeks of Arctic summer when light levels become low and many vascular plants shed leaves (Arndal et al. 2009).

Another aspect of mosses is their relation with vascular plant species. Competition for light and other resources are the most obvious relations. However, also relations such as suppression of growth or facilitation of seedlings of other species exist (He et al. 2016). In particular *Sphagnum* species are ecosystem engineers (Heijmans et al. 2002). *Sphagnum* mosses suppress growth of sedges, but on the other hand facilitate seedlings of *Betula nana* (Li et al. 2017). Furthermore, mosses play an important role in the nitrogen budget of northern ecosystems (Turetsky et al. 2012; Sect. 4.5.5).

A third effect of a moss layer on the carbon cycle is its effect on soil temperature (Chap. 2), which also influences organic matter decomposition rates. Under a thick bryophyte layer, soil temperatures are lower in summer, and the diurnal variation of temperature is smaller. Furthermore, the soil tends to be somewhat drier. Gornall et al. (2007) showed that the effect of a thick (12 cm) bryophyte mat on soil temperature reduced the growing season for vascular plants, and decreased microbial biomass by 40%, with respect to a thinner (3–6 cm) moss layer. Removal experiments with mosses in the Siberian tundra showed that understory evaporation increases, and more heat penetrates into the ground (Blok et al. 2011b).

Lichens consist of a symbiosis between photosynthetic algae or cyanobacteria and fungi. They have various shapes and colours, often as crusts on rocks, or with a leafy plant-like appearance in between other vegetation. As with mosses, they rely on external water for growth (He et al. 2016). In some Arctic vegetations, mostly on drier sites, they make up a significant part of the biomass, providing staple food for fauna such as reindeer. Furthermore, they host N-fixing bacteria which contribute to the nutrient supply of the ecosystem, and contribute to rock weathering (Cornelissen et al. 2007a).

The complexity of individual species has been classified into broader groups of plants with similar ecology, using the concept of plant functional types (PFT's). This has been used extensively in climate, vegetation and carbon cycle modelling (Chapin et al. 1996; Bonan et al. 2002; Iturrate-Garcia et al. 2016; see Van der Kolk et al. 2016 for an example). PFT's are classifications of species based on their phylogenetics, physical characteristics and life cycle, which interact in a similar way with environmental conditions. Figure 4.10 is an example of a PFT classification. The detail in such a classification is usually adapted to the modelling problem at hand. Criticism on PFT classifications is, that they can be rather arbitrary, and that the ecological and climatological parameters of PFT's are often not well known. This holds in particular for mosses which are often lumped into a single PFT in models (Turetsky et al. 2012; Iturrate-Garcia et al. 2016). A more recent trend is the direct use of plant traits rather than PFT classes in modelling of vegetation (Kattge et al. 2011; Scheiter et al. 2013; Van Bodegom et al. 2014).

Diversity in vegetation distribution and species composition in permafrost environments is large (Virtanen et al. 2013). This variation in plant cover in turn drives variation in the carbon and greenhouse gas balance. The carbon pools of the vegetation, and gross ecosystem production vary strongly among vegetation units. For instance, the carbon pool above the soil surface in shrub-dominated vegetation consist for a large part of woody stems (e.g. Arndal et al. 2009). In a seemingly featureless tundra in Northeast Siberia, large differences in species composition and carbon fluxes were found at small scales (e.g. Fig. 1.16 mid); these vegetation differences proved good predictors for CO_2 and CH_4 fluxes (Van der Molen et al. 2007; Budishchev et al. 2014; Mi et al. 2014). Plant species also control soil organic matter decomposition via the composition of their litter (Sect. 4.5.2).

In natural ecosystems, spatial variability is caused by interaction with soils, microclimate and competition between species. Important soil characteristics that drive spatial variability of vegetation are differences in soil acidity, nutrient avail-

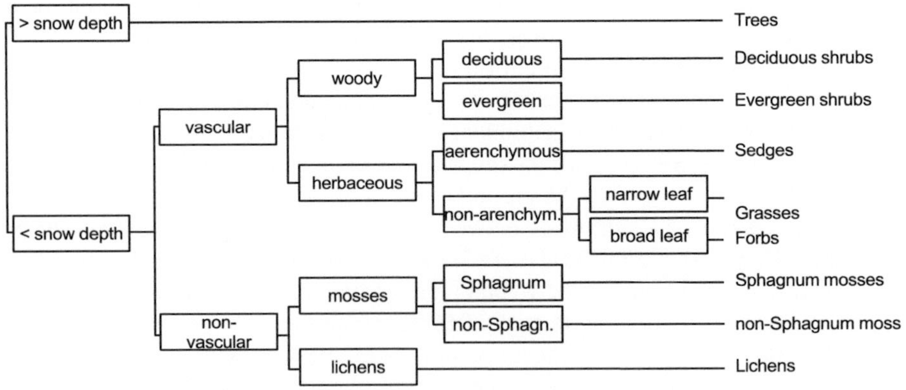

Fig. 4.10 Example of a classification of plant functional types designed for modelling high-latitude responses to climate change. (After Chapin et al. 1996)

ability, and soil moisture. Specifically in permafrost environments, the interaction with the micro-relief created by freeze-thaw processes (Chap. 3) is important.

However, not only environmental factors are important causes of spatial variation of vegetation; also biotic relations, such as competition and the suppression or facilitation of vascular plants mentioned above. Iturrate-Garcia et al. (2016) conducted a multivariate statistical analysis of vegetation cover and species diversity in two different landscape units in Northeast Siberia: a Yedoma plateau and an ancient drained thaw lake bed. The relations with soil variables differ among plant functional types. The mosses were more sensitive to soil variation than vascular plants. However, less than half of the variability in the plant community composition was explained by soil- and permafrost-related factors in multivariate statistical analysis. This suggests that most of the variability is explained by other factors, notably biotic factors such as competition and suppression/facilitation of seed germination. A study on vascular plant and bryophyte biodiversity by Virtanen et al. (2013) also indicates the importance of competitive interactions on productivity; species diversity is low in highly productive tundra, while tundra with low productivity has a high species diversity.

4.4 Carbon Quantity in Permafrost Soils and Frozen Deposits

World soils down to 2 m depth contain about three times as much carbon (~2400 Gt C, Gt = Gigatonne, 10^{12} kg) as CO_2 carbon (~830 Gt C) in the atmosphere (Batjes 1996; Paustian et al. 2016). Most of this carbon is stored in Boreal and Arctic soils, including permafrost soils. ‘Soil’ is used here in a wide sense. To soil scientists, soil is the upper few metres of the Earth surface, which is relevant for plant growth, has a distinct profile of soil horizons created by soil forming processes such as weathering and leaching of minerals and organics (Sect. 1.4.4). However, since permafrost thaw may progress deeper than that, also organic carbon-containing sediments below the surface soil need to be included (Tarnocai et al. 2009).

Carbon accumulation in permafrost soils vary with climate, landscape position and age. Organic horizons are typically thin in dry environments or where vegetation cover is thin, and thickest in wetland environments with water-saturated soils (e.g. Palmtag et al. 2015; see next section). Over time, soils accumulate more carbon given otherwise environmentally favourable conditions. Hinkel et al. (2003) describe soil development over time in drained thaw lake basins in Northern Alaska. The average organic horizon thickness at the soil surface increases with age, from less than 10 cm on average in the young lake basins to over 30 cm in the ancient basins; cryoturbation contributes to organic matter storage in older soils.

A method to estimate soil carbon amounts in permafrost environments is to estimate primary production and soil carbon turnover in ecosystems. However, this ignores carbon storage by sedimentation processes in the past and present, and the process of cryoturbation that enhances carbon storage in permafrost soils (Mishra et al. 2013). Several empirical inventories of permafrost soil organic carbon (SOC) have been made based on soil profile data, with improving accuracy over time due to increasing data availability. A northern circumpolar soil carbon database (NCSCD) has been developed to assist future inventories and permafrost carbon research (Tarnocai et al. 2009; Hugelius et al. 2013a, 2013b, 2014, Fig. 4.11). The first estimate of SOC in the tundra zone arrived at 191.8 Gt of SOC down to a depth of 1 m, based on 30 samples, an estimate which represents only a subset of permafrost soils (Post et al. 1982; Tarnocai et al. 2009).

Tarnocai et al. (2003) estimated an amount of 268 Gt C in the upper metre of Cryosols (permafrost-affected soils). This estimate was updated by Tarnocai et al. (2009) based on 1322 soil profiles in the NCSCD, including the soil depth down to 3 m, deeper deltaic sediments and Yedoma sediments. This includes the entire permafrost region. This estimate is considerably higher than previous estimates: 1672 Gt in total, covering approximately 18,782 × 10^3 km^2, nearly twice the amount of CO_2 carbon in the atmosphere. This includes 496 Gt organic carbon in the first metre, 1024 Gt up to 3 m depth, and 648 Gt in Yedoma deposits and deltaic sediments below 3 m. About 27% of the carbon in the upper 3 m is stored in peat soils, which cover 19% of the area of all soils included. Not all of this organic material is frozen; 1466 Gt (88%) of the total carbon stock is stored in frozen condition in permafrost (also non-frozen soils in the discontinuous and sporadic permafrost zones are included).

Further improved estimates revised the results of Tarnocai et al. (2009) downward (Hugelius et al. 2014). This revision included more soil profile data, improved data from the Yedoma deposits (Strauss et al. 2013) and fluvial and deltaic deposits, using better statistical upscaling methods. This updated estimate amounts 1300–1370 Gt C (depending on the method of upscaling from point observations, and with an uncertainty range of 930–1690 Gt). The largest part of this total, 974–991 Gt, is stored in permafrost terrain. Assuming a larger than 0.3 m thick active layer, less than 819–836 Gt or less than 61–63% of the total permafrost carbon stock is in frozen condition. A remaining source of uncertainty is the lack of data on deeper peat deposits below 3 m depth; the depth of Boreal and Arctic peatlands is generally poorly known (Grosse et al. 2011). Furthermore, only the deltas of the large Arctic rivers have been included, not those of the many smaller rivers. Any carbon stored in other types of sedimentary successions remains unaccounted for. Figure 4.12 shows the permafrost carbon reservoirs in relation to other carbon reservoirs.

4.4.1 Yedoma Deposits

The Yedoma carbon pool in East Siberia and northern Alaska has attracted much public and scientific attention because it is a large carbon reservoir which is vulnerable for thaw because of its high ground ice content (Zimov et al. 1997; Walter et al. 2006; Fig. 4.13). The total amount of carbon in Yedoma deposits is estimated at

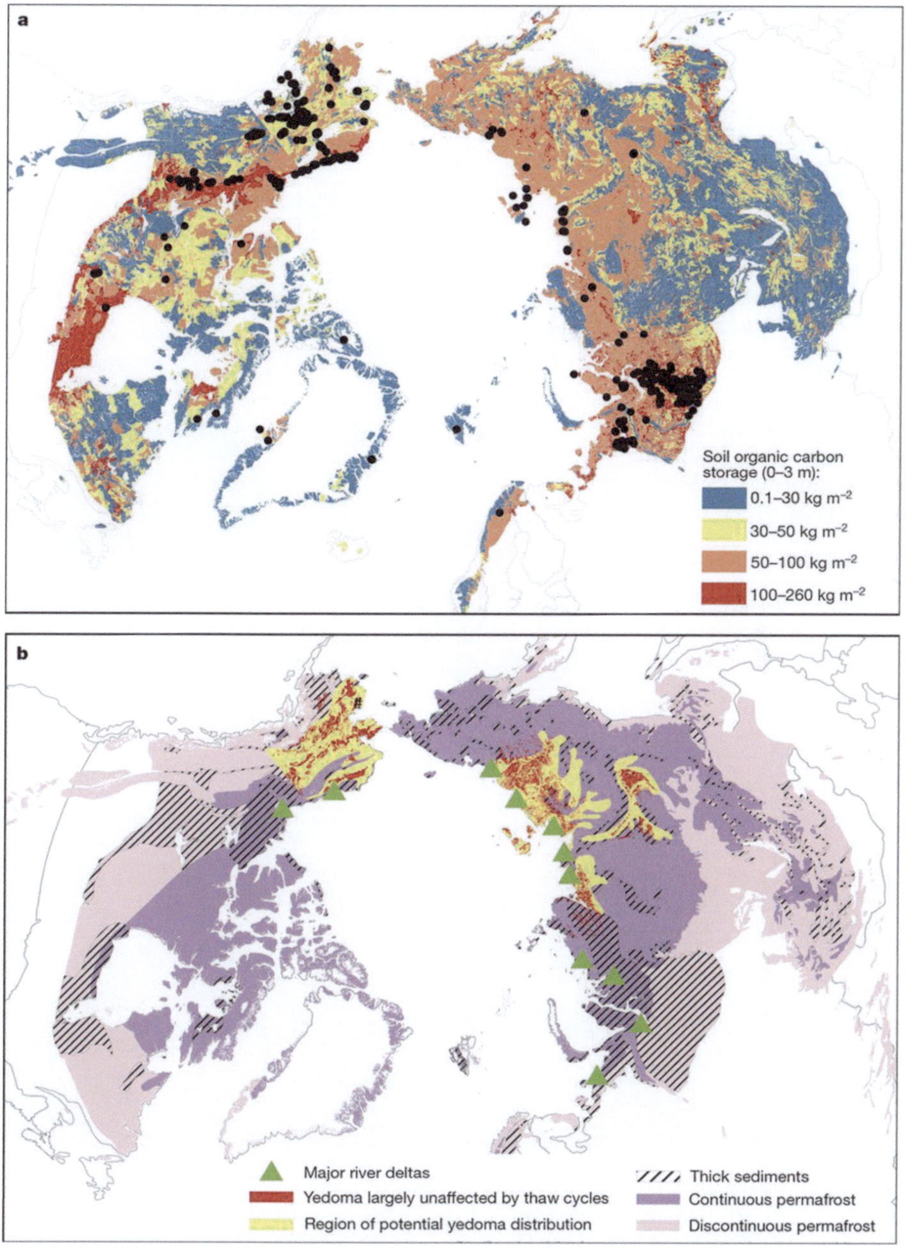

Fig. 4.11 A. Soil organic carbon pool (kg C m^{-2}) contained in the 0–3 m depth interval of the northern circumpolar permafrost zone (Hugelius et al. 2014). Points show field site locations with 0–3 m depth carbon inventory measurements. B. Deep permafrost carbon pools (>3 m), including the location of major permafrost-affected river deltas (green triangles), the extent of the yedoma region previously used to estimate the carbon content of these deposits (Zimov et al. 2006; yellow), the current extent of yedoma region soils largely unaffected by thaw-lake cycles that alter the original carbon content (Grosse et al. 2013, red), and the extent of thick sediments overlying bedrock (black hashed). The base map layer shows continuous and discontinuous permafrost. Reprinted with permission from Springer Nature, from Schuur et al. (2015): Climate change and the permafrost carbon feedback, Nature Vol. 520 © 2015

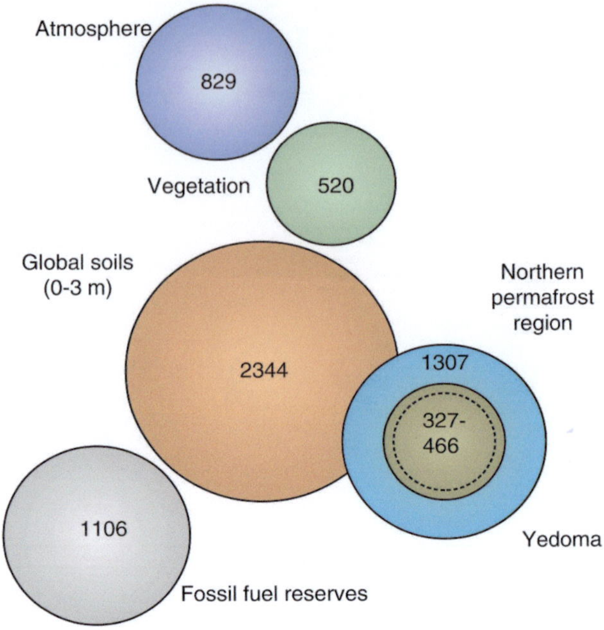

Fig. 4.12 Terrestrial carbon reservoirs and atmospheric carbon in relation to the carbon stored in permafrost and the Yedoma domain (in Gt). The surface area of the circles represents the size of the carbon stock. (Reprinted with permission from Elsevier from Earth Science Reviews 172:75–86, Strauss et al. (2017): Deep Yedoma permafrost: A synthesis of depositional characteristics and carbon vulnerability, ©2017)

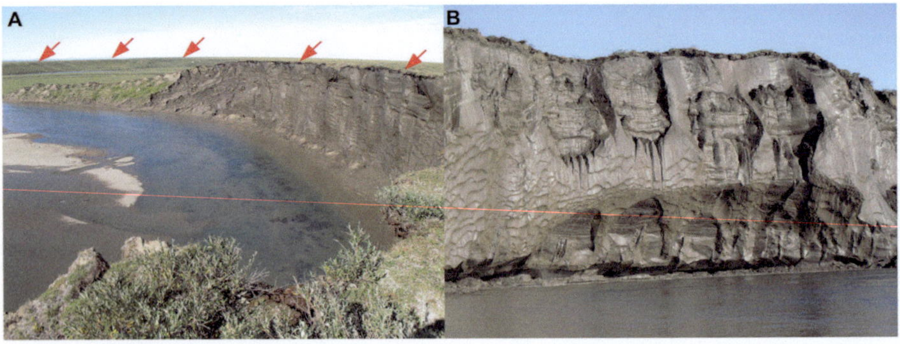

Fig. 4.13 Yedoma silt with large syngenetic ice wedges exposed in the 34 m high bank of the Itkillik River, northern Alaska (August 2011). Note the relatively flat yedoma plain in (A) indicated by arrows. (B) shows large syngenetic ice wedges (with scalloped surfaces) penetrating yedoma silt. Note the horizontal bands within the yedoma silt between the ice wedges, which represent former vegetation surfaces between the ice wedges. (Republished with permission of John Wiley and Sons Inc., from 'Palaeoenvironmental Interpretation of Yedoma Silt (Ice Complex) Deposition as Cold-Climate Loess, Duvanny Yar, Northeast Siberia' by Murton et al. (2015), Permafrost and Periglacial Processes 26(3),© 2015; permission conveyed through Copyright Clearance Center, Inc. High resolution images courteously provided by Mihail Kanevsky)

367–466 Gt (Strauss et al. 2013, 2017) including thaw lake deposits and thawed Yedoma beneath lakes. This is more than 25% of the frozen carbon content in the permafrost region, while covering only 7% of soils in the permafrost region (Strauss et al. 2017).

Schirrmeister et al. (2011a, b, 2013) and Murton et al. (2015, 2017) have described the stratigraphy and sedimentary environment of the Yedoma deposits, which occur over large areas in Eastern Siberia and Alaska (Fig. 4.11). As already mentioned in Sect. 1.4.5, interpretations of the sedimentary environment of the deposits varies from loesses, wind-deposited silt (Murton et al. 2015, 2017) to polygenetic loess/alluvial/colluvial deposition (Schirrmeister et al. 2011a, b). The distinction between both is not sharp, if local reworking of loess by water and slope processes is included in the loess depositional environment (Strauss et al. 2017). Murton et al. (2015) characterize the sediment as a cryopedolith, a sediment that has experienced incipient soil formation while aggrading, along with syngenetic freezing. Soil formation is shown by the characteristics of the organic detritus: signs of humification, and mineral particles with iron-humus coatings. The main source of carbon has been a vegetation of grasses, sedges and willow (Schirrmeister et al. 2011b; Murton et al. 2015). The vegetation composition is dominated by *Poaceae* and forbs (Murton et al. 2015).

Whatever the exact mode of deposition, the Yedoma deposits are a major example of carbon stabilization by upwards aggradation of permafrost. The buried carbon consists of paleosol horizons, grasses, sedges, mosses, in situ vegetation roots, thin peat beds, transported and re-deposited plant detritus, infilled animal burrows, and faunal remains (Fig. 4.14). During the cold climate stages of the Pleistocene, the vegetation has been similar to the present-day meadow steppe occurring in the Yana Highlands; in the northern lowlands a tundra-steppe vegetation dominated. Meadow steppes are dominated by herbaceous species (forbs) instead of grasses and indicate warm summers; tundra steppe contains more sedge-like species and indicates lower summer temperatures (Ashastina et al. 2018). Preservation of

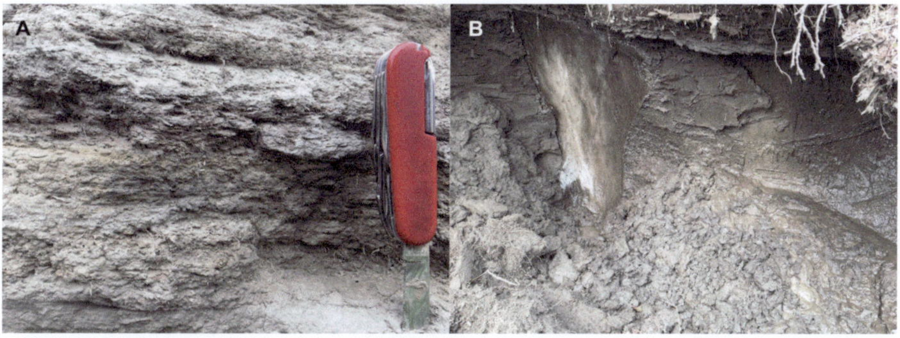

Fig. 4.14 (**a**) Paleo-vegetation horizons consisting of layers of roots and other plant material, in Yedoma exposure. Size of pocket knife 9 cm. (**b**) Ice wedge and warped layers with roots in Yedoma exposure. Kytalyk research station, Indigirka lowlands, Northeast Siberia. (Photos by the author)

organic material in the Yedoma sediments even has allowed the revival of 30,000 year old seeds (Yashina et al. 2012).

The deposits date mainly from the last two glacial cycles (Schirrmeister et al. 2011a): based on the oxygen isotope stratigraphy of the Quaternary, the deposits belong to Marine Isotope Stage (MIS) 6 (glacial conditions), MIS 5 (Last Interglacial), late MIS 5 to MIS 2 (glacial conditions). The Holocene is characterized by widespread reworking of Yedoma deposits by thaw, and deposition of thaw lake sediments. During the Last Interglacial, climatic conditions were similar to the Holocene, with predominantly lake and fluvial deposits, while during the colder stages the more typical ice-rich and loessic deposits (Ice-Complex deposits) were formed. These attest of upward aggradation of permafrost, following the sedimentation surface. Ice wedges were growing syngenetically upwards, including polygon centre depressions in between the wedges. Within the polygon centres vegetation horizons were formed; the waterlogged conditions have supported carbon sequestration (Figs. 4.13 and 4.14). In the succession also taberites occur; these are sediments that have been deeply thawed below thaw lakes, and have subsequently been refrozen. During the last glacial, climatic conditions also varied, from the extreme cold MIS 4 and MIS 2 to the milder and less dry MIS 3.

The carbon content also varies with depositional environment and climate (Table 4.1); the carbon content of the sediments deposited during the cold and dry MIS 2 are low compared to those of MIS 3, while Holocene and interglacial lake deposits have a higher organic carbon content. The highest carbon content is found in Holocene peat deposits. The Holocene lake and alluvial deposits also have a relatively high carbon content, since they accumulate carbon from eroded organic material, and from primary productivity of vegetation in the lakes and floodplains. At the same time lakes also contribute to anaerobic decomposition of organic matter from thawed permafrost, producing CH_4 (Schirrmeister et al. 2011a, b; Walter Anthony et al. 2014; Chap. 5).

Table 4.1 Ice and carbon content of stratigraphic units and sedimentary facies in the Yedoma succession in Northeast Siberia

Stratigraphic unit / sedimentary facies	Ice content weight %	Total organic carbon weight %	Carbon content (kg C m^{-3})
Holocene valley deposits	44.2 ± 9.0	5.3 ± 4.9	41.42 ± 40.87
Holocene thaw lake deposits	44.4 ± 16.0	6.9 ± 9.0	53.51 ± 77.22
Holocene peat cover	47.4 ± 14.5	10.9 ± 12.9	74.73 ± 96.26
Taberites	28.8 ± 4.8	2.7 ± 1.4	33.55 ± 17.82
MIS2 ice complex	38.3 ± 12.5	2.2 ± 0.9	21.08 ± 11.92
MIS3 ice complex	40.5 ± 12.8	3.7 ± 4.1	33.23 ± 40.07
MIS 5–3 fluvial deposits	22.4 ± 11.3	0.5 ± 1.4	7.17 ± 18.72
MIS 5 lake deposits	29 ± 8.3	3.2 ± 4.2	39.57 ± 50.10
MIS 6 floodplain deposits	32.6 ± 8.3	1.0 ± 0.8	11.29 ± 8.28
MIS 6 ice-rich deposits	58.7 ± 20.1	5.3 ± 4.3	18.41 ± 34.98

Adapted from Schirrmeister et al. (2011a)

4.4.2 Peat

Extensive Holocene peat deposition started after deglaciation in North America, Europe and Siberia. The peat deposits are on average of 2–3 m thick but occasionally up to 10 m thick (Grosse et al. 2011). Analysis of the ^{14}C age of over 1500 basal layer samples from peat deposits throughout the Arctic and Subarctic has shown that peat growth started ~16.5 kyr ago outside the glacial limits, and expanded rapidly with deglaciation between 12 and 8 kyr (MacDonald et al. 2006). This rapid and widespread peat growth contributed to an Early Holocene rise in atmospheric CH_4 concentration, and a small decline of CO_2 concentrations. Peat deposits are estimated to contain about 302 Gt C, peat deposits in North America are thicker than those in Eurasia (Hugelius et al. 2014). An overview of major peatland types in permafrost areas has been given in Sect. 1.4.3. The largest northern peatland areas are those of the West Siberian Lowland and the Hudson Bay Lowland.

There is still considerable uncertainty in these estimates. For many peat soil profiles extrapolation procedures have to be applied to extend shorter soil profiles over the full 3 m depth of the inventory. For an inventory of peat deposits overlying irregular bedrock topography, many borehole profiles are needed. Existing data are often selective in their locations, selecting deeper profiles or profiles that can be accessed easily (Kuhry and Turunen 2006).

Although peat growth inception expanded rapidly in the Early Holocene, the largest rate of peat accumulation occurred during the Late Holocene (Grosse et al. 2011). Warm and dry conditions in the Early Holocene have restricted peat growth in continental Canada and Siberia. Furthermore, peat growth inception has been determined by deglaciation history and emergence of coastal areas by glacio-isostatic rebound (land uplift after disappearing of the ice caps; examples are the Hudson Bay Lowland and Western Finland). In the southern parts of the permafrost zone, peat development started without presence of permafrost during warmer than present conditions; permafrost was established later during cold intervals of the late Holocene, e.g. the Little Ice age (Zoltai 1995; Kuhry and Turunen 2006).

Peat formation depends on climatic and other environmental factors (temperature affecting R_{eco} and *GPP*, water balance and groundwater flow, topography and geochemistry). A common succession of peat formation starts with open water conditions in shallow lake basins where finely divided, partly decomposed aquatic vegetation debris accumulates as gyttja. This is followed by accumulation of plant debris under more terrestrial conditions, such as swamp forests resulting in deposition of woody peat, or peat formation by sedges. If the peat layer becomes thick enough to isolate the surface from groundwater, peat formation dominated by *Sphagnum* occurs. This succession is known as terrestrialization (Charman 2002).

Alternatively, paludification (transition of dry soil to wetland conditions) gradually changes dry soils into wetland soils with peat accumulation. Paludification may be the result of a changed water balance by climate change or drainage changes. Drainage water from nearby mires aids paludification by increasing soil wetness and lowering the pH. Peat formation feeds back on itself by changing hydrological conditions by creating a surface layer of peat with low permeability, impeding drainage by groundwater flow. Conditions at the inception of peat growth may therefore differ from those during peat expansion (Charman 2002; Kuhry and Turunen 2006).

The water sources for the peatlands (Sect. 1.4.3) change during their development. After an open water or paludification phase, a transition towards a fen vegetation occurs, where peat formation is governed by a graminoid vegetation, largely consisting of sedges. In the fen stage, input of nutrients and minerals from groundwater and surface water is still present; nutrient status at this stage is characterized as mesotrophic. This in turn changes into a bog dominated by *Sphagnum* mosses, when rain water becomes the main water source. By gradual hydrologic isolation of a growing peat deposit, nutrient supply is strongly limited, and nutrient-poor, ombrotrophic conditions result. At this stage, nutrient supply supporting vegetation growth is limited to deposition from the atmosphere and bacterial N fixation, converting atmospheric N_2 to N species that can be taken up by plants (Charman 2002; Kuhry and Turunen 2006; Vile et al. 2014). In this stage, also an important enhancement of carbon conservation occurs; *Sphagnum* remains are more recalcitrant to decomposition than most vascular plant remains, and contribute to acidification of surface water which also restricts decomposability (Verhoeven and Toth 1995; Grosse et al. 2011).

Peat bog soil profiles are characterized by a surface layer (catotelm) in which the water table fluctuates, the decomposition rate of freshly deposited plant material is high, and also permeability for water and air flow is relatively high. Below the catotelm, the acrotelm is permanently water-saturated, anaerobic, denser and has a low permeability. This zone is characterized by slow anaerobic decomposition (Clymo 1984).

The amount of decomposition of peats can be estimated based on visible characteristics. Fibric material is poorly decomposed and consists of at least two thirds of recognizable plant fibres; the remainder is mostly amorphous organic material. Sapric material is on the other end of the scale, consisting of less than one third of recognizable plant material. Hemic is intermediate, between one third and two thirds of plant fibres. A simple field trick is to squeeze a handful of peat in the hand, and to see how much clear water or mucky amorphous material can be squeezed out between the fingers; on this method the Von Post scale with 10 humification classes is based (Charman 2002).

Although permafrost establishment results in better conservation of peat organic matter, it also may disrupt peat formation. Ice segregation causes frost heave, resulting in the formation of palsas and peat plateaus, where dryer surface conditions prevails, and vegetation is invaded is by lichens and moss species growing under dry conditions. At these sites, the conditions for peat formation become unfavourable. Subsequent thaw of ground ice, initiated by auto-cyclic processes of vegetation disruption (Sect. 3.4) or climate change may result in the formation of thaw depressions and lakes with fen-like vegetations (Kuhry and Turunen 2006; Turetsky et al. 2007). Other processes that may disrupt peat accumulation rates are forest fires and erosion by fluvial processes or wind (Kuhry and Turunen 2006).

4.4.3 Alluvial and Lake Sediments

In the estimates of the permafrost carbon pool, a considerable part of the alluvial carbon pool is probably not included. Hugelius et al. (2014) include only the deltas of the large arctic rivers, which store 91 ± 39 Gt C below 3 m depth. Depending on river type, river sediments upstream of these deltas may contain large amounts of carbon, comparable to deltas. In river valleys which are wide enough to harbour wetlands and lakes next to the channel belt, the organic matter derived from the local ecosystem carbon balance is stored in sediments and preserved in waterlogged conditions. Depending on local sedimentation rate, this burial may be rapid, preventing decomposition (Shur and Jorgenson 2007; Van Huissteden et al. 2013; Fig. 4.15). Primary productivity in floodplain ecosystems is enhanced because of nutrients added by flooding (Squires et al. 2009).

In-channel deposition of organic matter occurs also, in particular on sandbars in braided rivers and point bars of meandering rivers, depending on the dynamism of the channel system. This is largely plant debris eroded upstream. On higher vegetated surfaces of bars in the river system also local primary production adds carbon. Sediment re-entrainment within the floodplains and deltas through bank and channel erosion and subsequent deposition on point bars can be substantial (Carson et al. 1999). Therefore, these channel-related deposits store carbon at a shorter time scale than fine-grained floodplain deposits which are less vulnerable to reworking of sediments by erosion.

On river floodplains, the activity of erosion and (re-)deposition processes depend on the river floodplain morphology which is in turn related to discharge regime and sediment supply (e.g. Van Huissteden et al. 2013). In particular the mobility of channels and mode of floodplain deposition is important for carbon storage (Fig. 4.16). A classical textbook subdivision of river channel and floodplain configuration (Schumm 1985) distinguishes:

Fig. 4.15 (**a**) Floodplain backswamp deposits overlying pointbar deposits, exposed in a creek at low river stage along the Berelegh river (Indigirka lowlands, Northeast Siberia). White arrows: paleo-vegetation horizons. Vegetation on floodplain surface: Arctophila fulva grass. (**b**) Flood mark left at the edge of the Berelegh river floodplain by high spring flood. (Photos by the author)

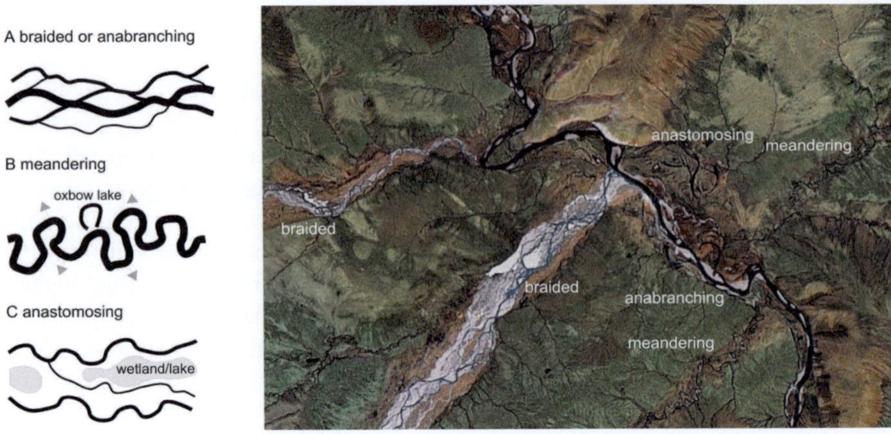

Fig. 4.16 Left: Three broad classes of river pattern. Arrows at meandering pattern indicate the direction of lateral migration of the river channel; frequent meander bend cutoffs occur, leaving oxbow lakes behind. Right: Braided (*A*), meandering (*B*), and anastomosing (*C*) river channels in Eastern Siberia, approx. 68° 54′ N, 140° 4′ E. (Image ©2010 Google™, Right side republished with permission from Elsevier, Encyclopedia of Quaternary Science, Permafrost and Periglacial Features | Periglacial Fluvial Sediments and Forms by Van Huissteden et al. (2013))

- braided rivers (multi-channel rivers with, highly mobile channels and islands or bars, with a high sediment load relative to water discharge) or anabranching (multi-channel rivers with more stable islands);
- meandering rivers (single channel, high sinuosity rivers of medium mobility, with a characteristic pattern of expansion of channel loops by erosion on the outer bends)
- anastomosing rivers (generally marshy, low gradient floodplains or deltas with low mobility channels, usually more than one, which migrate periodically by sudden displacement of entire channel sections – called avulsions – triggered by extreme floods).

A wide variety of rivers is found throughout the permafrost realm (Van Huissteden et al. 2013). The river type is mainly determined by local relief and sediment and water discharge. In all cases, the spring flood dynamics strongly determine channel mobility and floodplain sedimentation. Since water and sediment discharge is also related to drainage basin climate (Sects. 3.6 and 3.8) river morphology and sedimentation / erosion patterns may change by climate change, as occurred during the Pleistocene (e.g. Van Huissteden 1990; Vandenberghe 2002).

Braided rivers are often associated with glacial streams and rivers in polar desert environments, and because of their highly mobile and extreme environment do not have a large amount of primary production; frequent reworking of deposited sediment and vegetation is common. By contrast, deltas, and anastomosing rivers have wide floodplains with a large percentage of productive wetlands and lakes, which may escape reworking by erosion for a longer time because of comparatively stable channels. Meandering and anabranching rivers have an intermediate position; the probability of reworking of earlier deposited sediments depends on floodplain width and channel mobility.

During spring break-up in cold climate rivers, massive transport of fragmented river ice occurs, which strongly enhances the erosive power of floods by mechanical pressure exerted by the moving ice on river banks and the heads of islands. Erosion of river banks by floating ice adds large amounts of organic debris to the river (e.g. Costard et al. 2014). The river ice – which may be 2 metres thick in the anabranching Lena river section described by Costard et al. (2014) – starts to fracture and float as soon as a flood wave from upstream arrives. The water level rise may be exacerbated by the formation of ice jams, mixed with uprooted trees. One of the break-up events described by Costard et al. (2014) occurred at very rapid water level rise of several metres in one night, pushing up river ice in large jams on banks and the heads of islands (Fig. 4.17). The bulldozing action of the ice jams removed large swaths of forest and created ice-shove ridges of sediment. Although the water level is high, ice accumulations tend to prevent further erosion until their removal by thaw.

Fig. 4.17 Ice breakup of the Lena River in 2010 as seen from 7 km distance. One day separates the two pictures. Upper: Just before the breakup, the main channel has a 1.7 m thick ice cover. Lower: One day later, the breakup causes a large accumulation of ice up to 10 m high at the head of the island. The high stage of the water level causes the ice to glide up the banks. (Republished with permission of John Wiley and Sons Inc, from Permafrost and Periglacial Processes 25(3), An Assessment of the Erosion Potential of the Fluvial Thermal Process during Ice Breakups of the Lena River (Siberia) by Costard et al. (2014) © 2014, permission conveyed through Copyright Clearance Center, Inc.)

The river ice also carries coarse material derived from the river bed, frozen into the ice in winter.

The break-up phase is followed by normal flooding when the ice is melting and snowmelt proceeds. In this phase, thermal and mechanical erosion of frozen, ice cemented bank material becomes important, creating thermal erosional niches and erosion along ice wedges. The rate of erosion (bank retreat) is increased by the duration of submersion and the water temperature, rather than flood height or discharge. The water temperature rises strongly when shallow flood water on the river plain is warmed by the sun (Emmerton et al. 2007). Thermal erosion may continue later in summer (Forbes and Lamoureux 2005; Costard et al. 2014). Thermal erosion erodes large amounts of sediment on the river channel banks. Thermo-erosional activity concentrates within a water level interval that corresponds to water levels observed normally during spring meltwater peak and summer floods (Tananaev 2016).

Over-bank deposition of sediments on the floodplain wetlands occurs mostly during the spring flood. Deposition rates are highest near the channel bank where coarser sediment is deposited. In anastomosed systems (e.g. Fig. 3.34), floodplain wetlands occupy large areas and have a high *GPP* in summer by nutrient import from floodwater. During floods, also large amounts of particulate organic matter is carried by the rivers, from tree trunks to large masses of dead grasses, which are usually deposited in flood marks (Fig. 4.15b). Burial by deposition of river sediments on the floodplain further conserves this organic matter.

Unfortunately, not much quantitative data are known about net organic carbon burial on floodplains in permafrost regions. The sedimentation rate determined from ^{14}C dates on a levee location in the Northeast Siberian tundra floodplain of Fig. 3.33 amounted 0.5–2 cm $year^{-1}$; the sediment consisted of silt with ice lenses and layers of grasses and sedges. Holocene valley deposits in Northern Siberia contain on average 41 kg C m^{-3}, which is considerably more than the Weichselian age Yedoma deposits from which they are derived (21–33 kg C m^{-3}), highlighting again the role of GPP of floodplain ecosystems and redeposition of eroded organic matter from soils (Schirrmeister et al. 2011a).

Lakes receive organic carbon from bank erosion, runoff and stream or river transport as dissolved and particulate organic matter, and organic carbon produced by primary production of the ecosystem. Next to organic matter, the lakes receive considerable amounts of mineral matter (Fig. 4.18). Losses of carbon occur via ecosystem respiration and transport of water-borne organic matter through the lake outlet. The remaining organic matter stays within the lake ecosystem and sinks together with mineral matter towards the bottom, where most of it is conserved because of anoxic conditions. In Table 4.1, which lists the carbon content of sedimentary units within the Yedoma successions, the lake sediments stand out because of their higher carbon content. The variability of the carbon content of lake sediments depends on the relative amount of mineral influx, compared to the net ecosystem carbon balance (Sect. 4.1).

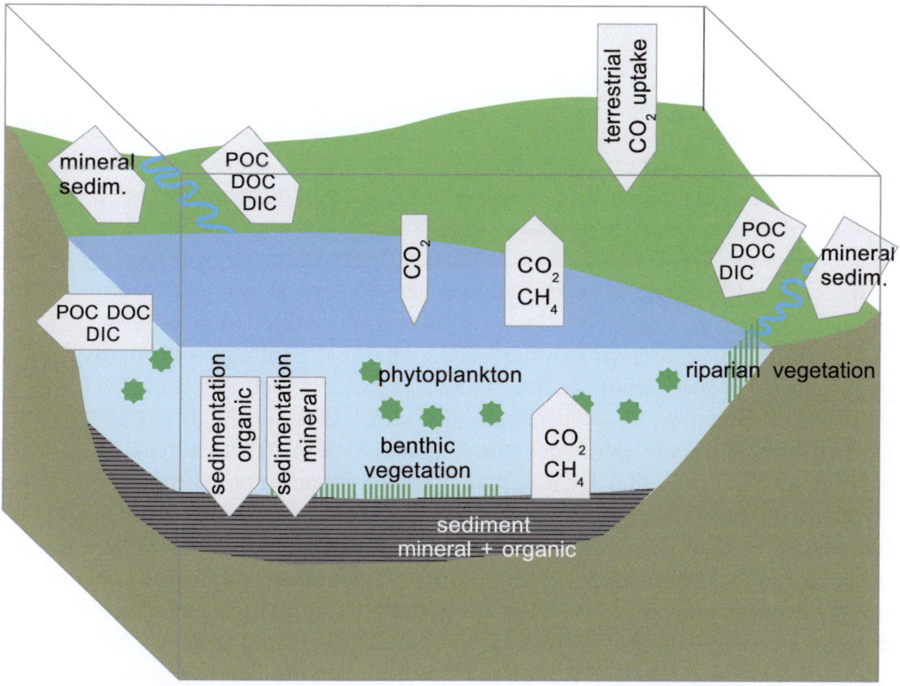

Fig. 4.18 Sedimentation of organic-rich sediment in lakes. The carbon content of the sediment is derived from the organic matter imported into the lake by bank erosion and river transport (dissolved organic carbon, DOC, and particulate organic carbon, POC) and from the primary production of the vegetation in the lake (bottom or benthic vegetation, riparian vegetation and phytoplankton). DIC is dissolved inorganic carbon (dissolved CO_2 and carbonates). Carbon losses occur via respiration of the lake ecosystem in sediment and water column, and via river transport at the lake outlet

Carbon accumulation rates in lake sediments may be high. Walter Anthony et al. (2014) derived carbon accumulation rates in Holocene thaw lake sediments. For Siberian thaw lakes, 47 ± 10 g C m^{2} year^{1} were found. These rates are some 2.5-fold higher than the average for northern peatlands. The mineral + organic sedimentation rate in lakes in the Mackenzie delta varies between 1.2 and 5.3 kg m^{-2} year^{-1}, which is derived from ^{137}Cs dated lake sediment cores by Marsh et al. (1999). Lakes which are in permanent contact with river channels have the highest sedimentation rates, lakes which are only infrequently reached by river floods, the lowest. Given that these lake sediments contain less than 7% organic matter (no exact data given), this would amount to similar or up to five times as high carbon sedimentation rates as reported by Walter Anthony et al. (2014).

4.4.4 Landscape-Scale Variation of the Soil Organic Carbon Stock

Carbon stocks vary strongly at the scales of landscapes and major vegetation units therein. Table 4.2 shows estimates of the SOC stock in the upper metre the soil from a range of regional studies; all are based on soil carbon quantification of a number of soil profiles, followed by upscaling using remote sensing mapping of major vegetation units or NDVI. All studies in Table 4.2 are estimates over a larger region with considerable variation in vegetation cover, but several also give information on differences between vegetation classes. Peatlands are excluded in Table 4.2.

The High Arctic sites in Northern Greenland and Svalbard with partly barren ground have low soil carbon stocks, below 20 kg C m^{-2} (Palmtag et al. 2015; Weiss et al. 2017). However, at a local scale this varies strongly; High Arctic wetlands have higher SOC stocks (Horwath Burnham and Sletten 2010; Palmtag et al. 2015; Weiss et al. 2017). Similar local variation is also found in other vegetation zones. Throughout Arctic vegetation zones, the areal percentage of barren ground with low SOC, and that of wetlands with a high SOC store, strongly influences the regional estimates of SOC (Palmtag et al. 2015). Horwath Burnham and Sletten (2010) point to significant discrepancies for their study area in Northern Greenland with other global-scale studies on permafrost carbon stores (e.g. Tarnocai et al. 2009). Their data show that the amount of carbon in High Arctic cryoturbated soils may be overestimated, while the amount of carbon in the most barren polar desert soils is severely underestimated.

Tundra SOC ranges between 14 and 62 kg C m^{-2}; most of the variation is again attributed to local vegetation differences and site history. Fen vegetations in drained thaw lake basins have in general high SOC stocks because of peat deposition; dryer sites have lower values. Recent floodplain deposits also have lower SOC because of active sedimentation of mineral sediments (Zubrzycki et al. 2013; Siewert et al. 2015).Vegetation succession history is an important variable, resulting in thicker organic horizons. However, in Table 4.2, the differences in SOC in the upper metre between age classes of drained thaw lake basins (Hinkel et al. 2003) appears insignificant, while there is a clear increase in *thickness* of the organic top horizon of the soils (previous section). This poor relation between upper metre SOC and lake basin age is attributed to varying amounts of reworked organic material in the lake sediments below the top organic horizon, progressive ice enrichment in the soil which decreases the carbon mass per unit area, cryoturbation which moves soil carbon to deeper levels, and organic matter decomposition. This illustrates the necessity to distinguish between various organic matter reservoirs and depths.

Taiga / boreal forest SOC tends to be slightly smaller than that of tundra. However, only a few taiga sites are included in Table 4.2. At the Spasskaya Pad site, the dry and relatively open Scots Pine forest has the lowest SOC stock. Also here, the wetlands have the highest C stocks (Siewert et al. 2015).

Most publications on the permafrost soil carbon stock distinguish between SOC in the top organic horizon of the soil, and SOC in the mineral layer below the top

Table 4.2 Landscape scale soil organic carbon stocks in the upper metre of the soil profile in various continuous permafrost ecosystems

Site	Vegetation	MAAT (°C)	Annual precip. (mm)	SOC kg C m^{-2}	References
Thule region, NW Greenland	Polar desert	−11.9	108	2.2 (0.5–5.3)	Horwath Burnham and Sletten (2010)
	Polar semidesert			7.9 (1.5–19.5)	
	Mire			15.4 (2.7–26.4)	
Spitsbergen, Svalbard	Tundra-barren	−6 to −15	200–600	6.26 ± 1.47	Weiss et al. (2017)
Zackenberg, NE Greenland	Tundra-barren	−9.2	200	8.3 ± 1.8	Palmtag et al. (2015)
Tulemalu lake, Canadian Arctic	Tundra	−9.4 to −14.3	<300	33.8 (82–12)	Hugelius et al. (2010)
Lena delta floodplain	Tundra	−12.5	125	14 ± 7	Zubrzycki et al. (2013)
Lena delta terrace	Tundra	−12.5	125	29 ± 10	
Shalaurovo NE Siberia	Tundra			30.0 ± 2.0	Palmtag et al. (2015)
N. Alaska, thaw lake basins, young	Tundra	−12.2	113	43.6	Hinkel et al. (2003)
Thaw lake basins, medium age	Tundra			36.2	
Thaw lake basins, old	Tundra			51.6	
Thaw lake basins, ancient	Tundra			41.8	
Non-basin tundra	Tundra			57.5	
Barrow, N. Alaska, coastal plain	Tundra	−12.2	113	62 ± 17	Michaelson et al. (1996)
N. Alaska, foothills	Tundra			44 ± 23	Michaelson et al. (1996)
Kytalyk, NE Siberia	Tundra	−13.4	196	29.2 ± 2.9	Siewert et al. (2015)
Kytalyk yedoma hill	Frost boil, tussock			27.6 ± 11.3	
Kytalyk DLTB	Sph + sedge fen			33.9 ± 14.9	
Kytalyk floodplain	Willow, grass, sedge			22.2 ± 2.2	
Cherskiy, NE Siberia	Taiga	−11.3	290	20.3 ± 2.2	Palmtag et al. (2015)
Spasskaya pad, central Yakutia	Taiga	−8.8	236	23.8 ± 11.3	Siewert et al. (2015)

(continued)

Table 4.2 (continued)

Site	Vegetation	MAAT (°C)	Annual precip. (mm)	SOC kg C m^{-2}	References
Spasskaya pad forest	Larch			24.4 ± 15.8	
	Pine			12.1 ± 1.9	
	Birch			16.5 ± 3.8	
	Wetland			27.6 ± 1.8	
Spasskaya pad alas/meadow	Grassland, wetland			27.5 ± 9	

Sph *Sphagnum, DLTB* drained thaw lake basin

organic horizon. In general, most SOC is stored in the top organic layer in wetland soils in both the tundra and taiga region. For instance, at the Kytalyk site in Northeast Siberia, the amount of SOC in the topsoil of relatively dry tundra (tussock and non-tussock tundra) is 13–14% of total SOC in the top metre, while for the fens (dry dwarf-shrub, *Sphagnum* and sedge fens) this is 30 and 46%. At the Spasskaya Pad forest site, the organic layer SOC in the non-wetland vegetation types (Larch, Pine, Birch forest, dry alas / drained thaw lake basin grassland) ranges between 7% and 19% of the total, while the wetland types (forested wetlands, Alas wetland) range between 19% and 69% (Siewert et al. 2015). Similar differences are seen in the other publications cited in Table 4.2.

Cryoturbation is known to enhance SOC storage (Sect. 3.5). Siewert et al. (2015) made a distinction between SOC in the mineral layer and SOC enriched by cryoturbation at their study sites. At the Kytalyk site, between 18% and 46% of total upper metre SOC is cryoturbated, the highest amounts are found in the tussock tundra and Sphagnum fen. In this location also old carbon should be involved in cryoturbation, since the tussock tundra sites of Siewert et al. (2015) are situated on a Yedoma plateau In the nearby drained thaw lake basin, also older peat beds are found (Teltewskoi et al. Teltewskoi et al. 2016). In the Spasskaya Pad forest site up to 30% of total SOC is stored in permafrost.

A further carbon stock quantified by Siewert et al. (2015) is the carbon stock in vegetation (phytomass carbon, PC) as shown in Fig. 4.19; together with the SOC, the phytomass carbon is the total ecosystem carbon stock (Ecosystem C in Fig. 4.19). At the tundra site (Kytalyk), by far the most carbon is stored below-ground, less than 2.5% is stored in vegetation, except in the densely vegetated high willow zone along river banks where 16% of the ecosystem carbon stock is phytomass carbon (however, this excludes fine roots). In the taiga forest at Spasskaya Pad, the phytomass carbon ranges between 11% and 21% for the forest.

4.5 Soil Organic Matter Quality and Decomposition

In the following discussion, the term soil organic matter (SOM) is preferred rather than soil organic cabon, because not only the carbon content is important for the substances discussed here. Next to soil organic matter (SOM) quantity, quality of SOM is important for potential carbon release to the atmosphere by decomposition. Loosely defined, quality is the fitness of SOM as substrate for microbial decomposition. SOM that is readily consumed by microbial decomposition is known as labile SOM, the opposite – very slow or no decomposition – is recalcitrant SOM. Furthermore, the temperature sensitivity of microbial reaction rates is important to understand the change in respiration rate of permafrost soils over the season, or to consider the effects of climate change. This temperature sensitivity increases with molecular complexity of the substrate (Davidson and Janssens 2006; Schmidt et al. 2011). Microbes can only assimilate soluble, low molecular weight compounds and therefore more complex, high molecular weight and insoluble compounds need to be broken up, or depolymerized, to access the nutrients and energy. This is done by extracellular enzymes that degrade these substances by hydrolysis or oxidation, producing dissolved organic matter (Conant et al. 2011).

Often, the lability of soil organic matter components is seen as an intrinsic property determined by their molecular structure. However, the current understanding of the notions of 'labile' and 'recalcitrant' organic matter is shifting, from intrinsic properties of organic matter, to properties that include the environment of the organic matter, which may or may not facilitate decomposition. See below for a more extensive discussion.

4.5.1 Organic Matter Quality in Permafrost

Several studies have stressed high intrinsic lability of permafrost SOM, making it readily available for decomposition by bacterial processes upon thaw of permafrost (e.g. Zimov et al. 2006; Walter et al. 2006; Walter et al. 2007). SOM lability has been related to climatic conditions and sedimentary environment under which permafrost carbon has been accumulated (e.g. Grosse et a., 2011; Schirrmeister et al. 2011a, b) or vegetation (e.g. Sjögersten et al. 2016). Pleistocene SOM is supposed to be more labile than younger SOM. Lability varies with plant species from which the organic matter has been derived; litter derived from graminoid species (grasses, sedges) is more labile than that of woody species such as shrubs, due to the presence of lignin (e.g. Hobbie 1996; Aerts 2006).

However, the concept of lability, or its opposite recalcitrance to decomposition, is not straightforward. Lability is partly an intrinsic property of SOM determined by its chemical composition – an intrinsic lability. Measured C losses may represent an apparent lability which is the result of the actual microbial population, enzyme

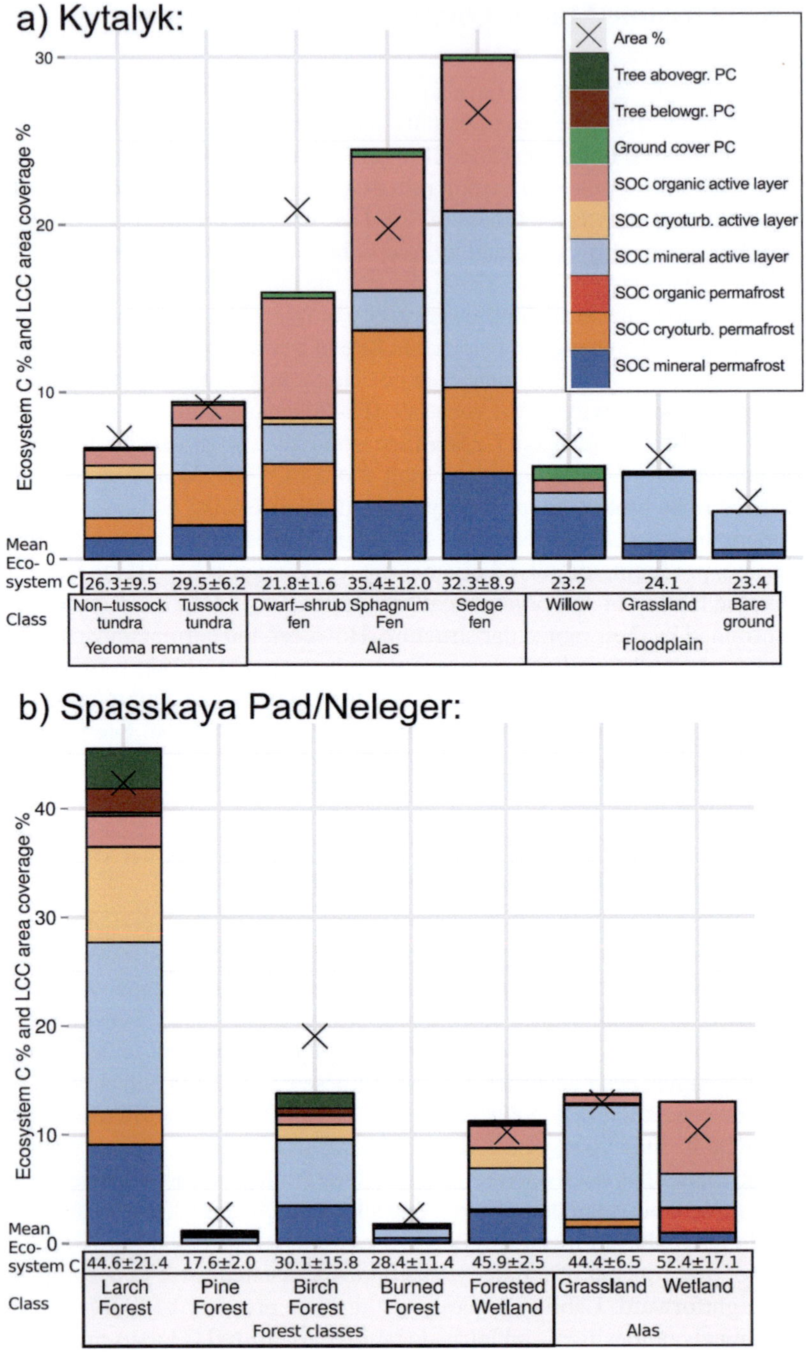

Fig. 4.19 Detailed ecosystem C partitioning for Kytalyk and Spasskaya Pad/Neleger sites. Note that bar lengths indicate organic matter reservoirs weighted by area of landcover classes in the

abundance, oxygen availability, nutrient availability and substrate accessibility (e.g. Waldrop et al. 2010).

The lability of SOM is inferred from chemical characteristics such as stable isotope ratio of carbon ($\delta^{13}C$), C/N ratio and molecular structure (e.g. Schirrmeister et al. 2011b). Several of these chemical parameters are only relative indicators of decomposition. In general, $\delta^{13}C$ increases, and C/N ratio decreases by loss of C during composition, while N is retained, but these parameters are also known to depend on the original plant material with which the decomposition started (Kuhry and Vitt 1996). Another way to asses lability is the use of incubation experiments in which the carbon loss or the emission of CO_2- or CH_4-carbon from the samples is measured under controlled conditions (e.g. Dutta et al. 2006; Vonk et al. 2013; Knoblauch et al. 2013).

An example is given by an extensive analysis of SOM lability using various chemical parameters and incubation experiments at three permafrost sites in Alaskan boreal forest with various C storage of largely Holocene-aged SOM by Waldrop et al. (2010). Lability was measured using a range of parameters: C/N ratio, $\delta^{13}C$, $\delta^{15}N$, DOC quantity and aromaticity (using specific ultraviolet absorbance), and ^{13}C NMR spectra, which gives information on functional groups of carbon compounds (alkyl, aromatic, phenolic, amide and carboxyl groups). This is combined with assessment of bacterial, fungal, and methanogenic population size and enzymes, and aerobic and anaerobic incubation at −5 °C and 5 °C.

The chemical and NMR data in this study indicate that the SOM from frozen soil below the permafrost table is intrinsically more labile and less humified, than surface soil SOM. However, enzyme concentrations and the size of microbial populations were also lower in permafrost compared with surface soil samples, in combination with a low temperature sensitivity of decomposition. Despite the higher intrinsic lability of the permafrost organic matter, the decomposition of SOM was restricted by low bacterial and enzymatic activity. CH_4 production from the permafrost samples was also low. Notwithstanding the microbial limitations, respiration per unit soil C was higher in permafrost SOM, compared with the active layer soil material. The low enzyme activity could point to substrate limitation, or other factors such as pH and available electron acceptors.

A meta-analysis of anaerobic soil incubations of permafrost organic material by Treat et al. (2015) also shows that substrate type plays a secondary role, and did not confirm a specifically high lability of permafrost SOM. CH_4 production per gram SOM was two times larger in soil collected from sites without permafrost than from

Fig. 4.19 (continued) respective study areas. The height of the bars represents the percentage of the total ecosystem C storage. The subdivision shows the proportion of organic, cryoturbated, and mineral soil, which is further distinguished into active layer and permafrost. The crosses show the percentage of land cover of the respective class in the entire landscape in the study area. PC: phytomass carbon; SOC: soil organic carbon. (Republished with permission of John Wiley and Sons Inc., from 'Comparing carbon storage of Siberian tundra and taiga permafrost ecosystems at very high spatial resolution' by Siewert et al. (2015), Journal of geophysical research. Biogeosciences 120(10) © 2015; permission conveyed through Copyright Clearance Center, Inc.)

permafrost sites. The maximum CH_4 production of SOM from the active layer was four times higher than that of permafrost below. Anaerobic CO_2 production was also higher in sites without permafrost. In statistical analysis, landscape level controls (biome, permafrost, climate) explained most of the variability, followed by substrate controls (vegetation and soil type), and incubation temperature. The variability of the anaerobic CO_2 production was predicted best from these variables (75% variability explained), while predictability of CH_4 production was much poorer (39% of the variance explained).

The higher lability of Pleistocene SOM (e.g. Zimov et al. 2006) is not always clear and shows conflicting results in incubation experiments. Knoblauch et al. (2013) found no significant differences in C flux in long term aerobic and anaerobic incubations from Pleistocene and Holocene permafrost SOM in Northern Siberia. Weiss et al. (2016) compared carbon release (CO_2 and CH_4) upon short term (days) incubation, to indices of the degree of decomposition of SOM (C/N ratio, $\delta^{13}C$, $\delta^{15}N$) of the top of the permafrost of Pleistocene Yedoma and Holocene thaw lake sediments in Northeastern Siberia. They found that older, more decomposed soil material shows higher C respiration rates per unit incubated C, than younger and less decomposed samples with higher C content. For Yedoma deposits, the soil thin section data of Murton et al. (2015) show that during deposition, Yedoma organic matter has been decomposed significantly prior to preservation, as testified by organic coatings on mineral grains. However, Vonk et al. (2013) incubated water thawed out from the Yedoma deposits at the same river bank section. Radiocarbon datings of the DOC in the sediment confirmed a Pleistocene age of the organic matter. It proved to be highly labile; on a 14 day incubation under oxygenated conditions at ambient river water temperatures, 34% was lost. Similar rapid DOC losses were found in collapsing permafrost (thaw slumps) on the North Slope of Alaska (Abbott et al. 2014).

A factor that may contribute to faster decomposition of soil carbon is photobleaching by UV light. This occurs in particular when soil carbon is exposed to sunlight as happens at erosion sites such as river bank and lake bank cliffs, landslides and other erosion features (Sect. 5.3). In particular DOC is sensitive to UV radiation. This stimulates oxidation, resulting in the release of CO_2 and CO, and susceptibility to microbial degradation is enhanced (Cory et al. 2013). Exposure of DOC from thaw slumps to natural sunlight stimulated bacterial activity by 46.8%.

The studies cited above clearly indicate that lability of SOM and carbon loss by decomposition after thaw of permafrost is not determined by a high intrinsic lability of the permafrost SOM, but also by activity of the microbial population (e.g. Waldrop et al. 2010) and other environmental conditions. This agrees well with recent insights in SOM decomposition in which lability or recalcitrance is seen more as an ecosystem property, rather than as an intrinsic property of the organic matter itself. Older views on soil organic matter decomposition hold that decomposition removes labile organic compounds, while the remaining pools of soil carbon consist of stable recalcitrant material and a pool of humic substances, formed by condensation reactions as decomposition products. The recent dynamic view that

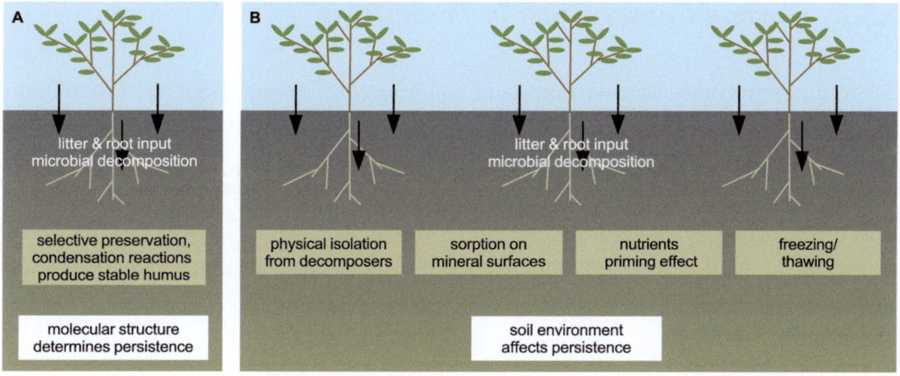

Fig. 4.20 A synopsis of recent insights, contrasting 'historical' (**a**) and emerging views of soil carbon cycling (**b**). Modified after Schmidt et al. (2011). The historical view (**a**) stable organic matter is seen to comprise mainly selectively preserved plant inputs and de novo synthesis products like humic substances, whose chemical complexity and composition render them nearly inert relative to microbial degradation. The emerging view is that the molecular structure of organic material does not necessarily determine its stability in soil. Rather, organic matter cycling is governed by multiple processes, shaped by environmental conditions that may increase or decrease the lifetime of soil organic matter

has emerged from studies on organic geochemistry and microbiology puts more emphasis on environmental conditions and biological processes in the soil environment, rather than on molecular structure of the SOM (Fig. 4.20). In the soil environment, various 'recalcitrant' plant-derived compounds, such as lignin, and 'labile' compounds may have quite similar residence times, suggesting that there is no sharp distinction between 'labile' and 'stable' SOM (Schmidt et al. 2011).

Furthermore, a meta-analysis of long-term soil incubation experiments of organic and mineral soils puts the importance of lability in another perspective. Schädel et al. (2014) separated the soil organic matter pool in these incubation experiments in a labile, fast decomposing pool, a slow decomposing pool and passive soil carbon pool, by fitting incubation results to a three-pool SOM decomposition model. For the most labile pool, very short turnover times well below 1 year indicated high decompositions rates. The slow pool had turnover times between 5 and 15 years and a higher C/N ratio than the labile pool. However, for the longer term decomposition of permafrost SOM (order of years) the most labile material does not determine the carbon fluxes. The slow pool is larger than the fast pool; the fast decomposing pool represents less than 5% of the SOM in most cases. Although there may be labile SOM in permafrost, it is the larger pool of more slowly decomposing material that is responsible for sustained carbon loss. SOM decomposition on time scales much longer than those of typical incubation experiments of a few years at most, determines carbon release from permafrost SOM. The results of incubation experiments by Knoblauch et al. (2018) also show the importance of longer duration of these experiments for CH_4 production.

4.5.2 Carbon Conservation in Permafrost

Given the importance of environmental conditions versus soil organic carbon composition for preserving soil carbon, the environmental conditions for soil carbon in permafrost need to be considered more closely. Factors in the soil environment that influence decomposition rates are (Davidson and Janssens 2006; Schmidt et al. 2011):

- Physical isolation of organic matter from bacterial activity, such as in dense soil aggregates, sorption of organic matter on mineral surfaces;
- Freezing/thawing;
- Nutrient limitation;
- Litter quality;
- Adaptation of the soil food web to the type of litter produced by the vegetation (home-field effect);
- Energy availability effects such as priming;
- Absence of electron acceptors to perform oxidation of organic matter;
- Moisture availability;
- The activity of soil fauna (Aerts 2006),

Physical isolation is an important factor, but not unique to permafrost soils. SOM may be physically or chemically protected from decomposition, for instance by being physically locked in soil aggregates (clumps of soil particles that are bound together by organic or mineral precipitates), or by adsorption to mineral surfaces, mostly clay minerals (Davidson and Janssens 2006). In incubation experiments by Treat et al. (2014), homogenisation of the samples, instead of using intact cores, increased CO_2 production by about 30% which indicates mechanisms restricting access by microbes to organic matter in the non-homogenized material.

The litter quality is largely determined by its nutrient content (e.g. Hobbie 1996; Aerts 2006; Cornelissen et al. 2007b) and by the type of organic molecules. All decomposers need basic nutrients as N and P; if these nutrients are in limited supply in the substrate, decomposition will be slower. Shrub and tree litter is generally thought to be more recalcitrant due to the presence of lignin (Hobbie 1996; Cornelissen et al. 2007b). A comparison of organic chemistry of peat from the Hudson Bay Lowlands in Canada and Northern Sweden shows differences in functional groups of organic matter related to vegetation (Sjögersten et al. 2016). The peat from Canada, dominated by lichens at the surface, contained more aromatic compounds, which are considered recalcitrant. The aromaticity of the organic matter influences the microbial community; higher aromaticity was related to more fungal biomass (Sjögersten et al. 2016). The Swedish peat, with a *Sphagnum*-dominated vegetation, had higher concentrations of carbohydrates and low amounts of lignin. Approximately 35% of the peat at the Swedish site and 26% at the Canadian site is easily degradable. These differences affected CO_2 emission from the peat in incubation experiments, and the microbial community in the soil. Lichens may also contain antimicrobial compounds (Yoshikawa et al. 2008).

The home-field effect demonstrates adaptation of the soil microbial population to the type of litter produced by the local vegetation. For vegetation on permafrost soils, this was demonstrated by Wang et al. (2017). Litter from *Eriophorum vaginatum* and *Betula nana* decomposed significantly faster in its 'home' soil below the species from which the litter originated, than when transplanted below another vegetation type.

The priming effect means that a high carbon input rate, e.g. fresh organic matter supplied by plant roots, enhances overall decomposition rate by a higher energy availability and enhanced microbial population growth and enzyme activity (Kuzyakov 2010; Bernal et al. 2016). Absence of electron acceptors refers in particular to absence of O_2; decomposition rates are considerably slower in anaerobic conditions when alternative electron acceptors are used (see next section). A review of a range of soil warming experiments has shown the importance of soil moisture availability. Low soil moisture may restrict decomposition rates (Robinson et al. 1995; Aerts 2006). Furthermore, soil fauna has a significant role in organic matter decomposition in lower latitudes by fragmenting plant litter. It is poorly known to what extent climate-driven changes in the faunal part of the soil food web may increase decomposition rates (Aerts 2006).

With permafrost thaw, the soil environment changes dramatically, not only with respect to temperature, but also in other aspects such as moisture, nutrient, oxygen availability and vegetation composition. Conversely, changes in the soil environment towards conditions that are favourable for SOM conservation are the main reasons for high carbon storage in permafrost soils. Processes common in non-permafrost environments are burial of SOM in anoxic conditions by changes in the water table or by sedimentation in anoxic environments, and formation of organic-mineral complexes. Several mechanisms are specific to permafrost, some of which already have been discussed above:

1. Cryoturbation (Sect. 3.5) results in soil movements that bring organic matter deeper into the active layer. This conserves organic matter by lower oxygen availability or anoxic conditions and low temperatures, including freezing. Cryoturbated organic material in the permafrost in Northeast Siberia showed ^{14}C ages between 3303 and 10,003 cal. Years BP; cryoturbated organic material in the present active layer showed considerably younger ages than the surrounding SOM (388 vs. 2230 cal. Year BP) showing both old and recent cryoturbatic activity (Palmtag et al. 2015), continuing throughout the Holocene.
2. Permafrost aggradation by upwards growth of a sedimentation surface. Carbon burial by sedimentation is a common process in permafrost and non-permafrost environments alike (Sect. 4.4). However, in permafrost landscapes the upwards growth of the permafrost table, following the upwards growth of the sedimentation surface adds a strong conserving factor.
3. Permafrost aggradation by climatic or other environmental change, e.g. vegetation cover change, which adds the lower part of the active layer to the permafrost. The cryoturbated organic material in permafrost in Northeast Siberia described by Palmtag et al. (2015) is an example. Stratigraphic analysis of

permafrost sections often reveals periods of permafrost thaw with thickening of the active layer, followed by re-freezing (French and Shur 2010). The organic material that accumulated in a once thicker active layer becomes incorporated in permafrost.

4. A change in hydrological conditions leading towards wetter soils, resulting in anoxic conditions. Such a change may result from permafrost thaw, causing thawing of ground ice and soil subsidence. However, although this may favour net carbon conservation, some of the fluxes from the ecosystem may increase such as F_{DOC} and F_{CH4} (e.g. Schädel et al. 2016). Moreover, in dry soil environments more moisture supply may enhance decomposition (Robinson et al. 1995; Aerts 2006).

4.5.3 Decomposer Communities in Cold and Waterlogged Soils

As in all soils, the biochemical pathways of organic matter degradation that are carried out by microorganisms in permafrost soils depend temperature, the soil being frozen or not, and on the redox conditions. In this section, key aspects of microbial life related to permafrost and redox conditions in waterlogged soils will be discussed. For a more in-depth treatment of soil microbial ecology and biogeochemistry, specialized literature is recommended, e.g. Paul (2014).

With availability of oxygen, decomposition of organic matter is driven by a complex food web of soil fauna and microbes including bacteria and fungi. Although recent research focusses much on the microbial life, the larger soil fauna should not be underestimated. Even high-arctic soil environments have an invertebrate and protozoan fauna that plays an important role in decomposition of organic matter. Sørensen et al. (2006) found a diverse soil fauna of mites, collembolans, enchytraeids, diptera larvae, nematodes and protozoa in Zackenberg (Northeast Greenland). The driest soils had the richest fauna. Many of these species are predating on the microbial community, but other species enhance bacterial and fungal access by comminution of plant remains.

In waterlogged soils where oxygen diffusion is slow, specialized microbial organisms take over. Because of the presence of a nearly impervious permafrost table and micro-relief that may impede surface drainage, many permafrost soils are waterlogged for much of the growing season. The slow diffusion of oxygen in the water-saturated soil pores results in rapid depletion of oxygen by microbial organic matter decomposition, creating anaerobic conditions. Therefore, anaerobic microbial reactions which use other electron acceptors instead of O_2 are very important in the carbon cycle of these soils.

Subzero temperatures also challenge adaptation mechanisms of microbial life. Many organisms in permafrost soils are cold-adapted and are known as psychrophiles – or synonymous – cryophiles: species that can grow and reproduce at low temperatures, including below-zero temperatures. Specific problems that microorganisms face at low temperature extremes, besides low thermal energy, are high viscosity. Proteins are less flexible, cell membranes may lose their fluidity, and nucleic acids are more stable, affecting reproduction, and ice crystal growth may occur within the cell (Jansson and Taş 2014). Mechanisms to cope with these conditions are development of special proteins or a transfer to a dormant state (e.g. formation of spores).

Fluid water is commonly present in the shape of water films around soil particles down to temperatures of at least −10 °C, allowing bacterial activity (Sect. 3.2; Romanovsky and Osterkamp 2000). Bacterial activity at subzero temperatures permafrost is common (Panikov and Dedysh 2000; Rivkina et al. 2000) and has been demonstrated at temperatures down to −17 °C (Feller and Gerday 2003; Rivkina et al. 2004, 2007) or even down to −40 °C (Price and Sowers 2004), although biochemical processes are very slow under these circumstances.

Three types of metabolic rates can be discerned: a rate for growth, a rate sufficient for maintenance of functions but at environmental conditions not allowing growth; and a rate for survival of communities, in for instance glacial or permafrost ice in which cells can repair macromolecular damage but are probably largely dormant. The three rates have metabolic rates that scale as 10^6:10^3:1 (Price and Sowers 2004). Processes detected by metagenomics/metaproteomics analysis were similar in both permafrost and active layer soil, but rates of these processes measured by incubation were low in the frozen soil (Hultman et al. 2015). Metagenomics is the direct sequencing of DNA from the soil environment; metaproteomics analyses which proteins are being produced. Both allow the analysis of genes from the environment that characterize biochemical pathways and their associated enzymatic processes (Jansson and Taş 2014; Hultman et al. 2015). The cold adaptation and viability of microorganisms in permafrost has become an important subject, not only for climate change effects on permafrost, but also for the search for extraterrestrial life, since microorganisms may live under cold conditions on other planets than Earth. Jansson and Taş (2014) give a review of recent results on the microbiology, showing the diversity of microbial populations found in permafrost soils from various high-latitude sites. Taxonomic distribution of bacteria in different permafrost environments based on their 16S rRNA gene classifications vary strogly among circumarctic sample locations, with a large variety of dominant phyla.

The inventory of species of soil bacteria is biased to species that can be isolated and cultured. However, metagenomics/metaproteomics analysis have revolutionized the insight in the presence and activity of soil microorganisms. New organisms have been isolated from permafrost soils (Jansson and Taş 2014), and evidence for activity of other unknown organisms is common (Hultman et al. 2015). However, a problem in the research on permafrost microbial life is that permafrost is also an excellent preserver of old inactive DNA (Jansson and Taş 2014).

Despite the cold environment, the microbial diversity in permafrost soils is high, with various aerobic and anaerobic decomposers (see below). The active layer shows the highest diversity with more species. However, a metagenomics analysis by Hultman et al. (2015) from permafrost including the active layer under Black Spruce vegetation found only a small fraction of Fungi; Bacteria and Archaea dominated. All elements of the nitrogen cycle are present: nitrogen fixation, nitrification (oxidation of NH_4^+ derived from organic matter decomposition into NO_3 and other N oxides, including N_2O) and denitrification (reduction of NO_3 to N_2 and N_2O).

In the permanently frozen soil microbes are active also. Viable bacteria (*Exiguobacterium sibiricum*) have been recovered from two to three million-year-old permafrost at a depth of 43.6 m (Vishnivetskaya et al. 2000). However, the abundance of microorganisms in permafrost is lower – Yergeau et al. (2010) reported 20 times less cells in permafrost compared to the active layer. Metagenomic analysis has shown that after thaw, multiple genes involved in cycling of C and nitrogen shift rapidly during thaw (Mackelprang et al. 2011).

The diversity and quantity of active cells in a polygonal tundra soil in Northeast Siberia was strongly related to the organic matter content of the soil (Kobabe et al. 2004; Wagner and Liebner 2009). The bacterial species diversity in permafrost is generally higher than the species diversity of archaea or fungi. In ice wedges, also microorganisms are found, derived from the surrounding soil communities (Wilhelm et al. 2012). Many functional groups represent microorganisms that tolerate, or thrive under, anaerobic conditions. Also the spatial variability in microbial community structure is high, showing variation in community structure over short distances (Jansson and Taş 2014).

Increasing anaerobic conditions in the soil are expressed by a decreasing redox potential. With progressively lower redox potential, a sequence of anaerobic microbial reactions occurs, known as the redox ladder. These reactions are characterized by the alternative electron acceptors that are being used in the oxidation of organic matter (Table 4.3) and have a progressively lower Gibbs free energy (Schlesinger 1997, and references therein). The organic matter decomposition reaction rates in anoxic conditions are considerably lower than those in the presence of oxygen.

After oxygen has been depleted, the first alternative electron acceptor that is being used is nitrate (NO_3^-) in the denitrification reaction, that produces N_2 and CO_2, with by-products NO (nitric oxide) and N_2O (nitrous oxide). This is performed by several facultative anaerobic bacteria, which use O_2 when available, but switch to nitrate reduction otherwise. The nitrification reaction (oxidation of NH_4^+ to NO_3^-) also produces NO and N_2O as byproducts from incomplete reaction. Metagenomics has shown that most of the genes that encode for proteins in various steps of the nitrogen cycle are present in permafrost (Yergeau et al. 2010; Jansson and Taş 2014). But gene abundance is low for the last steps in the denitrification pathway, suggesting that nitrate may not be completely reduced to N_2. This could result either in accumulation of N_2O, or the assimilation of nitrate in biomass. Because of its strong radiative effect, N_2O production is a relevant component of the permafrost carbon cycle (Sect. 1.3).

Denitrification reactions occur abundantly in fertilized agricultural soils, but are less common in the relatively pristine permafrost landscapes where nitrogen deposition and agricultural activities are restricted. Nevertheless, in specific conditions high N_2O emissions have been measured, over unvegetated frost boils on a peat plateau in Northern Russia and palsa mires in Northern Finland (Repo et al. 2009; Marushchak et al. 2011) and erosion of permafrost (Abbott and Jones 2015). These emissions are discussed in Sect. 4.5.5 in relation to the nutrient cycle.

At lower redox potential, reduction of Mn^{4+} and Fe^{3+} takes over (Table 4.3). Fe(III) reduction belongs to the realm of obligate anaerobes, while Mn reduction still can be performed by facultative anaerobes. A number of microbes can oxidise organic matter directly with Mn and Fe reduction. However, usually this is performed by a consortium of microbes, one group splits organic matter in organic acids, CO_2 and H_2, while a second group oxidises H_2 and reduces the metal oxides (Schlesinger 1997). There is little overlap between the zones of Mn and Fe reduction, because the reduction reaction of Mn is thermodynamically more favourable. Fe(III) reduction has been shown to be an important agent of organic matter decomposition in Arctic peat soils. High availability of Fe(III) will outcompete methanogens (Lipson et al. 2010).

Below an E_h of −221 mV, sulfate reduction starts (Table 4.3). Most of the H_2S that is formed by sulfate reduction, will react with available Fe^{2+} and Fe oxides to form the iron sulphides FeS and FeS_2. H_2S also reacts with organic matter, forming carbon-bonded sulphur; often most of the sulphur in wetland soils is bound to organic matter, rather than in inorganic mineral compounds (Wieder and Lang 1988; Gauci et al. 2002). Sulfate reduction depends on the availability of sulfate, which is generally in low supply in freshwater environments. Sources of sulfate may be flooding by sea water, marine deposits, and rocks containing sulphur minerals such as pyrite, which is oxidized to sulfate and Fe oxides upon subaerial weathering. At lower water table in wetlands, bacterial oxidation of iron sulphides produces sulfate, which is used again in sulfate reduction on return of anaerobic conditions (Schlesinger 1997).

Methanogenesis is the next step at an E_h of −244 mV. Methanogens belong to the Archeae. Sulfate reducing bacteria outcompete methanogens for substrate, therefore presence of sulfate will depress methanogenesis (Gauci et al. 2002). Two reaction pathways lead to the formation of CH_4 (Table 4.3): the acetoclastic pathway (acetate fermentation) using methyl groups of acetate, and CO_2 reduction with H_2 (hydrogenotrophy), producing equal amounts of CO_2 and CH_4 (e.g. Sugimoto and Wada 1993; Corbett et al. 2015). Recently, it has been shown that various other

Table 4.3 Sequence of microbial organic matter decomposition reactions at progressively lower redox potentials (E_h, in millivolts, at pH 7.0), with gaseous products and byproducts

Reaction name	Redox potential (mV)	Overall reaction equation	Gaseous (by) products
denitrification	< 747	$5CH_2O + 4\,H^+ + 4NO_3^- \rightarrow 2N_2 + 5CO_2 + 7H_2O$	N_2, NO, **N_2O, CO_2**
fermentation + Mn reduction	< 526	$C_6H_{12}O_6 \rightarrow CH_3COOH + CH_3CH_2COOH + CO_2 + H_2$ $2Mn^{4+} + H_2 \rightarrow 2Mn^{3+} + 2H^+$	H_2, **CO_2**
fermentation + Fe reduction	< −47	$C_6H_{12}O_6 \rightarrow CH_3COOH + CH_3CH_2COOH + CO_2 + H_2$ $2Fe^{3+} + H_2 \rightarrow 2Fe^{2+} + 2H^+$	H_2, **CO_2**
sulfate reduction	<−221	$2(CH_2O) + 2H^+ + SO_4^{2-} \rightarrow H_2S + 2CO_2 + 2H_2O$	H_2S, **CO_2**
methanogenesis	<−244	acetate splitting: $CH_3COOH \rightarrow CO_2 + CH_4$ CO_2 reduction: $CO_2 + 4H_2 \rightarrow H_2O + CH_4$	**CH_4, CO_2**

Boldface: greenhouse gases. Shaded: reactions by obligate anaerobic microbes methylated substrates are being used for methanogenesis in wetland soils (Zalman et al. 2018).

It has been assumed that CO_2 reduction is the main process in high-latitude wetlands since acetate accumulates over over the season (Hines et al. 2001). However, Ström et al. (2003, 2015) and Metje and Frenzel (2007) have shown that acetate is a major substrate for methanogenesis in high-latitude wetland soils, although CO_2 reduction is locally important (Vaughn et al. 2016). Other studies provide isotopic evidence for shifts in dominant reaction pathway driven by seasonal changes, from CO_2 reduction in the cool season to acetate fermentation during the growing season (Avery et al. 1999).

Despite the low temperature in permafrost soils, the abundance and composition of the methanogenic microbial population is similar to that of temperate wetland soils (Wagner and Liebner 2009). A distinct vertical zonation is found in permafrost, with more psychrophilic archaea at depth in the soil, close to the permafrost table. Activity of psychrophilic methanogenic archaea has been recorded at subzero temperatures (Rivkina et al. 2000).

Several groups of bacteria oxidise CH_4 for carbon assimilation when O_2 is available, a reaction pathway known as methanotrophy (e.g. Frenzel and Karofeld 2000; Wagner and Liebner 2009). Many species of these methanotrophic bacteria have been found in permafrost (Rivkina et al. 1998; Trotsenko and Khmelenina 2005), that metabolize methane derived from anaerobic parts of the soil (see below). There are two types of methanotrophs (I, belonging to the *Gammaproteobacteria*, and II, belonging to the *Alphaproteobacteria*). The groups are distinguished by the enzymatic pathway of CH_4 oxidation in soils; both have been found in permafrost soils. The amount of CH_4 formed in the soil which ultimately reaches the atmosphere, is strongly determined by the effectiveness of methanotrophy in the soil and vegetation (Parmentier et al. 2011b; Knoblauch et al. 2015); Sect. 4.5.6.

Various forms of anaerobic methane oxidation may occur in freshwater environments, based on nitrate reduction (e.g. Van Breukelen and Griffioen 2004; Zhu et al. 2012), Fe reduction (Sivan et al. 2011) and sulfate reduction (Schubert et al. 2011). The evidence is mostly based on lake bottom sediments and groundwater. In several cases specific environmental conditions apply that provide the electron acceptors (groundwater pollution, sulfate supply via sulphide-rich rocks or sulfate-rich rock). The role of anaerobic methane oxidation is uncertain; previously it has been thought that it is virtually absent in soils. The process is more difficult to isolate in soils because of the complex array of biogeochemical processes occurring at short distances from each other (Smemo and Yavitt 2011). Zhu et al. (2012) provide evidence for anaerobic methane oxidation depending on nitrite (NO_2^-) in a temperate peatland that has been infiltrated by nitrate-rich water from agricultural origin. In Siberian thaw lake sediments, De Jong et al. (2018) found no evidence for anaerobic methane oxidation with nitrite or nitrate. It is likely that anaerobic CH_4 oxidation is not a common or important process in permafrost wetlands, because they are generally far from N sources, but in situations where a supply of nitrate or sulfate is possible, it cannot be excluded. See also Sect. 4.5.5 for the potential effects of increasing anthropogenic sources of reactive N compounds.

The redox ladder as discussed above appears deceivingly simple and should be used as a very general model only. Recent microbiological research has painted a more complicated picture. First, methanogens prefer small and simple molecules such as hydrogen, carbon dioxide and acetate which are end products or intermediate products of the metabolism of other microbes, e.g. iron reducers. Therefore, limitations elsewhere in this metabolic chain, for instance limitations in oxidized Fe availability, can influence the rate of methanogenesis (Metje and Frenzel 2007; Ernakovich et al. 2017). Second, soils are complex 3-dimensional bodies with large variations in redox potential, oxygen availability and carbon availability at very short distances. Methanogenesis has been observed to occur above the water table (Wagner et al. 1999; Angle et al. 2017). For permafrost soils it has been shown that zones of methanotrophy and methanogenesis can overlap (Knoblauch et al. 2015). Soil redox potential was not an important explaining variable in multivariate analysis of environmental factors for methanogenesis (Ernakovich et al. 2017).

Methane emission from permafrost soils has become one of the most widely known effects of climate change and permafrost thaw. However, it is a matter of debate whether CO_2 produced by both aerobic and anaerobic metabolism, or CH_4 emitted from water-saturated soils will produce the strongest climate feedback of warming permafrost (Schädel et al. 2016). More on this subject in the next chapter. Also the emission process of CH_4 from wetlands and lakes needs further consideration, because that determines how much CH_4 escapes consumption by methanotrophs and escapes to the atmosphere. This is treated in more detail in Sect. 4.5.6.

4.5.4 Organic Matter Decomposition Reaction Rates and Their Dependence on Temperature

Bacterial reaction rates in soil organic matter decomposition depend on enzyme kinetics and a number of environmental factors: temperature, accessibility of substrate, nutrient availability, and soil chemical conditions such as redox potential and pH. For permafrost soils, the effect of temperature on bacterial reaction rates is arguably a very important factor (e.g. Hobbie 1996). However, the effects of reaction kinetics is often difficult to separate from other changes induced by temperature change, such as microbial community changes (Knoblauch et al. 2018). Other factors that influence decomposition rates, for instance soil moisture and nutrient availability, may mask the effect of temperature on decomposition rates in experiments (Aerts 2006). Furthermore, also the methods by which temperature sensitivity of decomposition rates are determined have an effect on the results, because they integrate various sets of organic matter – temperature – biota interactions besides temperature, depending on the method used. Based on a model study, Koven et al. (2017) introduce a climatological temperature sensitivity, which combines several temperature effects besides the direct effect of temperature on reaction kinetics. This includes the effect of climate on soil temperature (thawed season length and thaw depth) and microbial population interactions. They show, that SOM decomposition rate in cold climates is much more strongly depending on temperature than in warm climates.

The basic reaction rate of microbial decomposition, given a bacterial biomass of a certain size, depends on the available substrate, the soil organic matter. In the simplest case the reaction rate is constant, known as zero-order kinetics (e.g. Parton et al. 2014):

$$\frac{dS}{dt} = -k \tag{4.7}$$

where S is the substrate concentration (e.g. mol per kg soil or comparable metrics), dS/dt is the change of soil organic matter quantity over time, and k is a constant. After integration, the substrate concentration as a function of time becomes $S(t) = S_0 - kt$. S_0 is the initial concentration of the substrate. Zero-order kinetics can be expected when there is an abundant supply of substrate with respect to the size of the microbial decomposer biomass.

However, if the substrate is in more limited supply, the reaction rate depends also on the amount of substrate; a low substrate concentration limits decomposition. In that case, first order kinetics hold:

$$\frac{dS}{dt} = -kS \tag{4.8}$$

which on integration becomes

$$S(t) = S_0 e^{-kt} \tag{4.9a}$$

The rate constant has units of inverse time, e.g. per hour, h^{-1}. A quantity that is often used in this context is the half life $t_{1/2}$, the time required to transform one half of S_0, which can be calculated from Eq. 4.9 by $t_{1/2} = ln\ 2 / k$. Another often used metric is the mean residence time, which is $1/k$ for first-order reactions and S_0/k for zero-order reactions.

The chemical reactions of microbial decomposition are mediated by enzymes, specialized proteins which catalyse the reactions by lowering the amount of energy required to initiate the reaction, the activation energy. These consist of two steps; in the first step, enzymes bind to the substrate, in the second step, the enzyme releases a reaction product, e.g. CO_2. Zero-order and first order kinetics are end members of a more general model of enzyme reaction kinetics, the Michaelis-Menten equation:

$$\frac{dS}{dt} = V_m \frac{S}{K_m + S} \tag{4.9b}$$

Here V_m is the maximum possible reaction rate which is proportional to the total mass of enzyme in the soil. K_m is the half-saturation concentration, at which the reaction proceeds at half the maximum rate. K_m relates inversely to the capability of the enzymes to bind to the substrate, or its affinity. When the affinity is high and enzymes bind easily to the substrate, K_m is low. At high substrate concentrations S is much higher than K_m, and the term $S/(K_m + S)$ will approach to 1. This results in zero-order kinetics, since the reaction rate practically equals the constant V_m. At low substrate concentrations, the term $S/(K_m + S)$ approaches S/K_m, and Eq. 4.5 reduces to $dS/dt = (V_m/K_m)S$, which is first order kinetics with rate constant $k = V_m/K_m$.

This is a simplified approach so far, since the equations do not account for the growth of the microbial population itself, which follows its own dynamics such as exponential or logistic growth, (Parton et al. 2014). This will not be discussed here, but the environmental factors that modify the reaction rates will be considered.

The first environmental factor is temperature. The relation between temperature and chemical reaction proposed by Arrhenius is (Voroney and Heck 2014):

$$k_T = Ae^{\frac{-E_a}{RT}} \tag{4.10}$$

Here, k_T is the reaction rate constant due to temperature (dimension $time^{-1}$), A is a factor which relates to the frequency of molecular collisions which can start a reaction of the chemical species involved in the reaction, E_a the activation energy required to initiate the reaction, R the gas constant (8.314 J mol^{-1} T^{-1}) and T the absolute temperature in °K. E_a for organic matter decomposition is in the order of 50–150 kJ mol^{-1}, which results in an exponential increase of k with temperature. For organic substances in the soil, E_a depends on molecular structure; simple molecules tend to have a lower activation energy than large and complex molecules (Davidson and Janssens 2006).

For temperature effects on enzyme activity a similar equation is used:

$$k_{cat} = \kappa\left(k_B T / h\right)e^{-\frac{\Delta G\#}{RT}} \tag{4.11}$$

where T and R are the same as in Eq. 4.10, k_{cat} is the reaction rate, κ a transmission coefficient, k_B the Boltzmann constant, h the Planck constant and $\Delta G^{\#}$ the activation energy. The transmission coefficient is related to the viscosity of the soil solution, which itself is temperature dependent – in the temperature range between 10 °C and 0 °C it increases by almost 40%.

However, the complexity of microbial populations in the soil are difficult to grasp by these equations. As temperatures change, microbial communities also change, and adaptations in the physiology of species may occur that influence reaction rates. The relative sensitivity to temperature is therefore often expressed by a simpler relation, the Q10 relation, which is often based on soil incubation tests or field observations of soil respiration:

$$Q_{10} = (k_2 / k_1)^{[10/(T_2 - T_1)]} \tag{4.12}$$

where k_1 is the reaction rate at temperature T_1 and k_2 is the reaction rate at temperature T_2. This formula represents the increase in reaction rate with 10 °C warming of the soil. For aerobic heterotrophic CO_2 respiration from soils at temperatures around 20 °C, the value of Q_{10} lies between of 2.0–3.0. However, at lower temperatures, psychrophiles will contribute increasingly to decomposition of soil organics. Because of their adaptation to low temperatures, this changes the temperature sensitivity of the decomposition reactions. A common adaptation of psychrophiles is, that the activation energies of their enzymes is lower and specific activity is much higher (Voroney and Heck 2014). As a result, the actual reaction rates at low temperatures may be low, but the *change* of reaction rate at change of temperature as expressed by Q_{10} is much higher than that at higher temperatures (Kirschbaum 1995, 2006; Fig. 4.21). Q_{10} may be a factor 2–4 higher in the 0–5 °C range than in the 20–25 °C range, which amplifies the effect of soil warming on soil carbon in permafrost areas. This agrees well with the higher climatological temperature sensitivity of SOM decomposition in cold climates, found by Koven et al. (2017), who found a $Q_{10} > 5$ in cold climates, contrasting with a Q_{10} of 1 (no temperature sensitivity) in hot climates.

Figure 4.21 reveals another problem with temperature sensitivity: it varies also with the methods used to determine it. Laboratory incubations result in a higher temperature sensitivity at low temperatures than methods based on in situ measurements. These results are seemingly conflicting, but under in situ soil conditions,

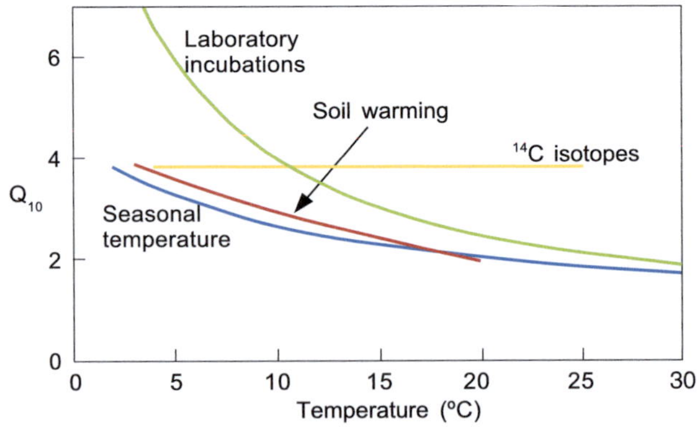

Fig. 4.21 Temperature dependence of soil-carbon efflux rates estimated with different methodologies. Laboratory incubations represent experiments under highly controlled laboratory conditions, soil warming refers to Q_{10} values from situ soil warming experiments, seasonal temperature refers to Q_{10} values derived from seasonal CO_2 emission data, ^{14}C isotopes refers to Q_{10} values based on ^{14}C isotope analysis. For the source data of this figure, see original publication.Reprinted from 'The temperature dependence of organic-matter decomposition—still a topic of debate' by Kirschbaum 2006; in Soil Biology and Biochemistry 38(9):2510–2518 © 2006 with permission from Elsevier

many factors other than substrate properties influence temperature sensitivity, for instance plant root respiration with its own temperature sensitivity, the production of organic matter which is also temperature sensitive, depletion of easily accessible substrate, microbial community effects, and soil environmental factors that restrict microbial access to substrate (previous section).

In field studies with soil warming experiments, increased rates of soil respiration in the first year do not persist, suggesting adaptation of the microbial community to the new temperature conditions or rapid loss of easily decomposable substrate. Two groups of field warming experiments in a review by Aerts (2006) – open top chambers, which are enclosures of transparent material open at the top, and experiments with warming lamps – gave opposite results: open top chambers reduced decomposition rates, and warming lamps stimulated decomposition. Moreover, both groups of experiments had relatively modest responses to warming. This is attributed to lower soil moisture in the open top chamber experiments; the effect was strongest on dry sites.

These effects are supposed to be absent in laboratory incubations (Kirschbaum 2006; Conant et al. 2011), but these have their own biases. They measure mostly short-term decomposition of materials with a short residence time, and studies of decomposition of bulk soil organic matter integrate the effects of multiple processes (Conant et al. 2011). Long term incubation experiments are rare, but reveal that adjustments of the microbial community over time significantly influence CO_2 and CH_4 production (Knoblauch et al. 2018). Many experiments do not use realistic temperature ranges for permafrost, and do not include subzero temperatures (Treat et al. 2014). Nevertheless, the majority of incubation studies show that the decomposition rate of slowly decomposing soil organic matter is more sensitive to temperature than that of rapid decomposing material, and show an increase of temperature sensitivity at low temperatures (Conant et al. 2011; Kirschbaum 2006).

Conant et al. (2011) and Zheng et al. (2018) conclude that several groups of processes determine temperature sensitivity and decomposition rates: de-polymerization of biochemically complex compounds; microbial enzyme production; and processes that limit the availability of soil OM by interactions with minerals (adsorption/desorption and aggregate binding). The first two groups of processes show high temperature sensitivity for complex organic compounds, the temperature sensitivity of the last group is poorly known, and sometimes opposite (slowing down at high temperatures). It shows that temperature sensitivity of microbial decomposition of SOM is a complex phenomenon. The higher temperature sensitivity of slowly decomposing SOM puts the importance of labile SOM in permafrost soils again in another perspective (Sect. 4.4). It will be the long-term decomposition rates that determine soil responses to climate change, besides short-lived transient responses that occur under conditions of rapid changes.

The discussion above was restricted to oxidation of SOM in an oxic environment. The problems of determining temperature sensitivity for anoxic environments are the same, as is demonstrated by the Q10 value of methanogenesis (Zheng et al. 2018). A review by Segers (1998) shows a very wide range of Q_{10} values. For northern wetlands the range in Q_{10} values found by incubation experiments, field measurements and model inversions is also wide. In general, there are no significant differences between Q_{10} values from permafrost wetlands than from non-permafrost northern wetlands (Lupascu et al. 2012). However, there are clear Q_{10} differences between various subarctic wetland vegetation types at the site in Northern Sweden studied by these authors. Samples below sedge vegetation have a higher CH_4 production rate, but lower Q_{10} (on average 4.3), while sites with *Sphagnum* have a lower production rate, but higher Q_{10} (on average 8). The mean CH_4 production rates R_0 per day per gram dry matter (extrapolated to the production rate at 0 °C) are respectively 5.3 μg CH_4 d^{-1} g^{-1} for sedge vegetation and 0.5 μg CH_4 d^{-1} g^{-1} for *Sphagnum*. A literature survey by Lupascu et al. (2012) shows a broad range of Q_{10} values for CH_4 production in Sphagnum-dominated sites, with outliers around 30.

These differences are attributed to differences in substrate; the temperature sensitivity of the more recalcitrant material (produced by *Sphagnum*) is assumed to be higher than that of the organic material produced by sedge roots (Verhoeven and Toth 1995). Another important effect is the higher CH_4 production from samples taken close to the soil surface (Lupascu et al. 2012). Close to the soil surface the input of fresh organic matter from roots and litter is largest (Joabsson and Christensen 2001). Similar observations apply to thaw lake sediments. For thaw lake sediments from Alaska, De Jong et al. (2018) found high Q_{10} values, between 3.3 and 9.8 in incubation experiments with application of different substrates. These Q_{10} values were depending on the substrate that was added to the incubated sediment; trimethylamine amendment showed the highest Q_{10} and CH_4 production. Furthermore, these Q_{10} values decreased over time (279 days) towards values around 1. This shows that methanogenic activity and its temperature sensitivity is highly substrate-dependent.

Although the incubation experiments give insight in the basic temperature sensitivity of the CH_4 production chain in the soil, this is not the same as the intrinsic temperature sensitivity of the methanogenic archeal metabolism itself. As mentioned in the previous section, methanogenesis uses substrates that result from other aerobic and anaerobic reactions (e.g. denitrification, Fe and Mn reduction); the rate of these reactions may limit the rate of production and temperature sensitivity of the final step. E.g. Fe(III) is an important intermediate in conversion of SOM to CH_4 and at the same time, Fe(III) reduction competes with methanogenesis for organic acid substrate (Zheng et al. 2018; Fig. 4.22). Not much is known of the effect of these limitations on overall methanogenesis Q_{10} (Ernakovich et al. 2017). Second, the effect of methanotrophy is often not included; methanotrophy and methanogenesis may occur within the same sample or soil horizon (Treat et al. 2015; Zheng et al. 2018), indicating that methanotrophy may occur close to methane production.

An important difference in temperature sensitivity is the difference in Q_{10} between between methanogenesis and methanotrophy. Q_{10} of methanogenesis is higher than Q_{10} of methanotrophy (Segers 1998; De Jong et al. 2018); hence, on warming of anaerobic soils, methanogenesis will react stronger on temperature change. Zheng et al. (2018) explored this in more detail for tundra polygon soils,

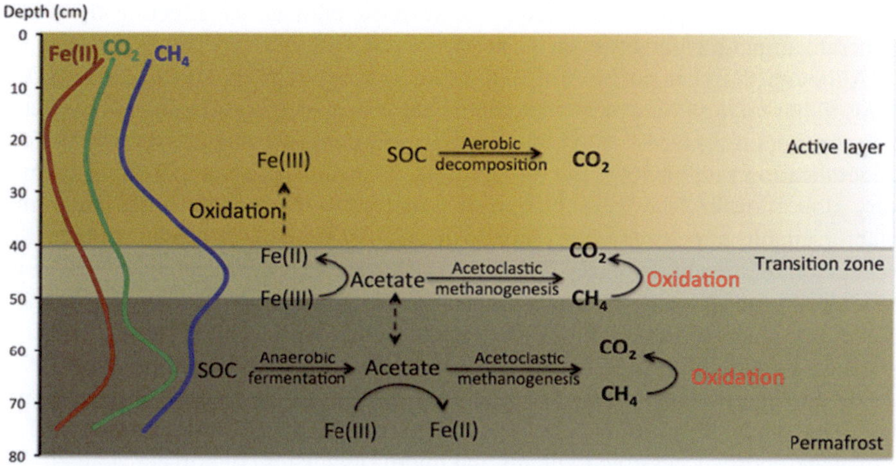

Fig. 4.22 Conceptual model of aerobic and anaerobic soil organic carbon decomposition pathways and the release of CO_2 and CH_4 in flat-centred polygon (FCP) soils, based on anaerobic incubations of active layer, transient layer and permafrost horizons at temperatures from −2 to +8 °C by Zheng et al. (2018). Lines on the left marked as Fe(II), CO_2, and CH_4 represent qualitative gradients through the soil column. The median height of the water table in FCP centres is at the ground surface. The transition zone is episodically thawed. From Zheng et al. (2018): Impacts of temperature and soil characteristics on methane production and oxidation in Arctic tundra, Biogeosciences 15(21), Fig. 9, https://doi.org/10.5194/bg-15-6621-2018 published under Creative Commons License 4.0, https://creativecommons.org/licenses/by/4.0/

with incubations in the temperature range of −2 °C to 8 °C. They found an average Q_{10} for methanogenesis of 4.2, and a Q_{10} for methanotrophy of 2.0, which is consistent with other values reported from peat soils (Segers 1998). However, Juncher Jørgenson et al. (2014) reported for dry, high-arctic mineral soils a methanotrophic Q_{10} which is similar to the Q_{10} for methanogenesis.

4.5.5 *Nutrient Cycles and Nitrous Oxide*

Most permafrost ecosystems are more or less pristine and do not receive a large input of nutrients. Nutrient addition experiments result in increased nutrient uptake by the vegetation, showing that it is nutrient-limited (Jonasson et al. 2001). Atmospheric reactive nitrogen deposition from fossil fuel combustion and agricultural activities in higher latitudes is generally less than 1 kg N ha^{-1} $year^{-1}$ (Phoenix et al. 2006; Dentener 2006). However, N deposition is rising globally and the signal of rising anthropogenic N deposition has been detected from the ^{15}N:^{14}N stable isotope ratio in Arctic lake sediments as far north as Svalbard and Baffin land (Holtgrieve et al. 2011). Besides an increased anthropogenic source of reactive N, nutrient cycles are also influenced by a warming climate, by affecting the SOM mineralization rate (Jonasson et al. 2001).

Although the N deposition in high-latitude ecosystems is small, Arctic SOM tends to have a high C/N ratio, typically about 20 for soils and about 60 for vegetation. An important factor is the low mineralization rate of organic matter. The low mineralization rate of SOM in cold climates (Nadelhoffer et al. 1991) results in a very slow transfer of organic N (which is normally unavailable for plants), into plant-available mineral N. E.g. Zona et al. (2010) found that productivity of the vegetation in drained thaw lake basins in Alaska decreases with age of the basin, as available nutrients are increasingly locked up in vegetation and dead organic matter. Moreover, plants have to compete for nutrients with the microbial population. This accounts for the low N content of vegetation relative to soil (Jonasson et al. 1999, 2001). In Siberian tundra, only 5% of the total N was available in biologically active fractions, while the soil showed a high N:P ratio, indicative for large amounts of N which are poorly accessible to plants and micro-organisms (Beermann et al. 2015). However, given the high N store in Arctic SOM, a source from N fixation from atmospheric N_2 in the ecosystem is likely (see below).

The form of N (NO_3^-, NH_4^+, organic N) which is taken up preferably by plants is species-specific (Iversen et al. 2015). There is evidence for N uptake in organic compounds (amino acids) rather than mineral N, as an adaptation to low mineral N availability (Jonasson et al. 2001; Weintraub and Schimel 2005). Organic N is a favourable N source if it can be exploited; plants can bypass N mineralization in this way. The C/N ratio of amino acids is very low (Hobbie et al. 2002). For shrubs and trees and many forbs and graminoids, mycorrhiza assist in rapid transfer of N from decomposing organic material to the plant. Plants like sedges mostly lack mycorrhiza and show evidence of N uptake in organic compounds (Chapin et al. 1993). Nevertheless, clear evidence of N limitation by low decomposition rates abounds.

The distribution of N in vegetation and soil in high-latitude vegetation appears to be quite uniform in various areas. Along a transect in Alaska, the relation between foliar N (g N per m^2 ground surface) and the leaf area index of vascular plants proved to be simple linear and uniform throughout varying vegetation (Williams and Rastetter 1999). However, mosses were not included in this study; Arndal et al. (2009) showed that mosses dominate the N pool and the gross ecosystem production of most Arctic vegetation types in Greenland. The gross primary production correlated with leaf N content.

N fixation (fixing N_2 into plant-available N) in tundra ecosystems is estimated at 25–80% of total N input and 1–20% of annual plant N uptake (Chapin and Bledsoe 1992; Hobbie et al. 2002). Free-living cyanobacteria or cyanobacteria in lichens fix N, also in High Arctic environments. This phototrophic N fixation is highly sensitive to temperature, light, and moisture (e.g. Chapin et al. 1991; Hobara et al. 2006). In an Alaskan tundra watershed, N input by this type of N fixation was estimated to amount 80 to 131 mg N m^2 yr^1, indicating that N-fixation contributed 85 to 90% of total watershed N inputs.

In the moss layer, several species groups such as feather mosses and *Sphagnum* facilitate N fixation (Turetsky et al. 2012; Beermann et al. 2015). In boreal forests feather mosses contribute to the N budget (DeLuca et al. 2002). An additional likely source is methanotrophic bacteria. *Sphagnum* and other mosses have an association with methanotrophic bacteria (Basiliko et al. 2004; Raghoebarsing et al. 2005; Kip et al. 2010; Knoblauch et al. 2015; see next section). Many methanotrophic bacteria contain genes that code for N_2 fixing pathways. In an age sequence of peat development in Finland, Larmola et al. (2014) found indications of significant N fixation associated with methanotrophs in *Sphagnum* mosses, in particular in the younger peat development stages of mesotrophic fens. N fixation by methanotrophs contributed about 40% to the total N_2 fixing in these environments. The time scale over which fixed N becomes available to the host plants of the methanotrophs ranges from days, to years in case of N derived from decomposition. There is also a link to the other macronutrient, phosphorus: N_2 fixation is highest when N/P ratios are lowest, i.e. where N is in short supply relative to P.

It is not certain to what extent methanotrophs directly contribute to N_2 fixation because the relations between N-fixing heterotrophic decomposers, phototrophic and methanotrophic N fixers are complex; their mutual relations may provide more than one N_2 fixing pathway (Fig. 4.23), including phototrophic N fixers and a potential favourable effect of methanotrophy on N fixing by reduction of O_2 tension in the micro-environment within the moss layer. However, the link of N_2 fixation with

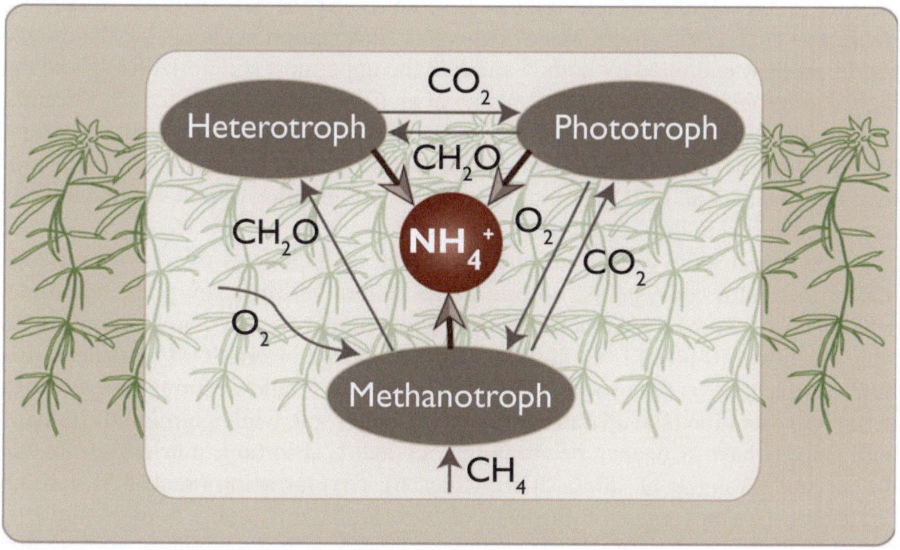

Fig. 4.23 The interplay of N_2-fixing bacteria in the upper moss vegetation. Phototrophic, heterotrophic, and methanotrophic bacteria all take part in the fixation of atmospheric N_2 to plant available N (NH_4^+). The supply of CH_4 may support the direct methanotrophic N_2 fixation, but it may also provide additional CO_2 for phototrophic microbes, as well as for the moss itself. Both phototrophs and methanotrophs provide organic molecules (CH_2O) to heterotrophic N_2 fixers. Methanotrophic activity may also decrease the oxygen tension, providing a more suitable micro-environment for the oxygen-sensitive nitrogenase enzyme complex. (From Larmola et al. (2014); republished with permission from PNAS)

methanotrophy has clearly been demonstrated (Larmola et al. 2014). Given that large swaths of permafrost soils are covered with mesotrophic tundra wetlands where methanotrophy is prominent (Kip et al. 2010; Parmentier et al. 2011b; Knoblauch et al. 2015) this is a likely large source of N in these environments.

The N cycle has further connections to methane. Both methanogenesis and methanotrophy are affected by N fertilization. N fertilization in upland forest soils has been shown to reduce CH_4 oxidation and N enrichment of ecosystems. Fertilization may inhibit groups of methane oxidisers, while others are stimulated; some methane oxidisers can switch from CH_4 oxidation to NH_4 oxidation (Bodelier 2011). Fertilization may seem less relevant for most Arctic ecosystems that are far from important N sources (see above), but increasing atmospheric transport of reactive N species and increasing human activity in the Arctic may enhance reactive N deposition in the future, with the effect of stimulating CH_4 emission. Furthermore, these fertilization effects could affect CH_4 emission in floodplain environments that receive nutrient input from flood water (Corradi et al. 2005; Van Huissteden et al. 2005, 2009).

In contrast to N, the macronutrients P and K have no significant atmospheric component except dust deposition. Next to recycling by SOM mineralization, rock weathering contributes to availability of these nutrients. Nutrient co-limitation of N and P is common (Chapin et al. 1995; Beermann et al. 2015). Nutrient limitation may differ for vegetation and microbial biomass. At the Siberian site studied by Beermann et al. (2015) there was evidence for N limitation in plants, but the microbial biomass was limited by both N and P in the uppermost soil horizons, where the highest microbial activity is found. Wang et al. (2017) found evidence for P limitation of decomposition of dwarf birch and cotton grass roots at the same location; mass loss in litter decomposition experiments was positively correlated with P content. Beermann et al. (2015) found no evidence of limitation of potassium, which is largely adsorbed to organic matter.

Plants invest more carbon in root biomass in cold climates than in warmer climates. In tundra ecosystems, rooting depth differs between key plant functional types (PFTs): shrubs have much shallower roots (upper 15 cm) than graminoids. To what extent plants profit from nutrients that are released from SOM by decomposition also depends on their root system characteristics. Shallow rooting dwarf shrubs in tundra vegetation take up nutrients from the upper soil, while graminoids (grasses and sedges) have a deeper penetrating root mass, absorbing nutrients from the deeper soil (Wang et al. 2016, 2018; Chap. 6). Cryoturbation (Sect. 3.5) with its circulation-like movements contributes also to vertical differences in nutrient distribution (e.g. Boike et al. 2008). Palmtag et al. (2015) found a common vertical distribution of C/N ratios, which were higher in the topsoil, lower in cryoturbated pockets of SOM, and lowest in mineral soils, presumably indicating that stronger mineralization at depth exerts a strong influence on decomposition (Aerts 2006; Chap. 6). Shrub litter is generally more recalcitrant than graminoid litter because of its higher lignin content. Dead plant materials shows lower nutrient content then live material, and root litter, in particular dead roots, are more recalcitrant than leaf litter (Cornelissen et al. 2007a, b; Wang et al. 2017). However, a higher nutrient content offsets this effect. The data from Wang et al. (2017) show fivefold differences in mass loss by decomposition between between different types of litter, and a 17-fold difference in P content across all analysed litter types in Siberian tundra.

The N cycle is also important as a source of the strong greenhouse gas nitrous oxide, N_2O. High N2O emissions have been reported recently from peat circles on peat plateaus and palsas (Repo et al. 2009; Marushchak et al. 2011) and other erosion features (Abbott and Jones 2015). These emissions represent a leak in the N cycle which is bound to specific soil surface conditions: soils rich in organic matter but without vegetation cover, which is lost by erosion (Voigt et al. 2017). Examples are palsa and peat plateau surfaces which have lost their plant cover by a combination of intense cryoturbation (responsible for the circular shape) and wind erosion of the peat surface (Fig. 4.24), or permafrost erosion features (Abbott and Jones 2015). Since this Arctic peat has a low C/N ratio, abundant N is available for mineralization. The absence of vegetation results in faster warming of the soil in summer, resulting in higher peat decomposition and N mineralization rates, transferring organic N to NO_3^- and NH_4^+. However, this N is not taken up by plants, and is available for microbial nitrification and denitrification; the process that releases the most N_2O is in the peat circles is denitrification (Voigt et al. 2017). Sparsely vegetated

Fig. 4.24 Peat plateau with peat circles in Northern Russia (Seida site). (**a**) Close-up of a peat circle and vegetation on the surrounding peat plateau. (**b**) View over the landscape with peat circles in the bottom left corner. (From Repo et al. 2009, Nature Geoscience 2, DOI: https://doi.org/10.1038/NGEO434)

peat mounds had considerably lower N_2O emission compared to bare, unvegetated spots (Marushchak et al. 2011).

The high N_2O emissions are a new and rather unexpected feature of the generally nutrient-limited permafrost ecosystems, but their role could increase in the future. Voigt et al. (2017) infer that similar processes also could occur when vegetation is stressed by drought and heat in a future warmer climate, increase of erosion features (Chap. 5) likewise could increase N_2O emissions (Abbott and Jones 2015).

4.5.6 Ecosystem Methane Emission

In Sect. 4.5.3 the conditions for methane production in the soil have been considered, and in Sect. 4.5.5 the effect of temperature on both methanogenic and methanotrophic bacterial activity. However, methane emission from wetland soils and lakes is more complex. This complexity is caused by the presence of CH_4 production and consumption in the same soil or sediment column, and the various transport pathways of CH_4 from the soil to the atmosphere.

There are four pathways by which gases from the soil are transported to the atmosphere (Fig. 4.25):

- Gas diffusion: the slow movement of gas molecules through the pores in the soil, either as dissolved gas in the soil solution or through air-filled pores.
- Convective flow generated by pressure differences between soil air and atmosphere.
- Ebullition: the movement of gas bubbles through water-filled soil pores or the water column until they reach the soil surface or the water table.
- Plant transport: movement of gas through hollow roots and stems of wetland plants.

For most gases, diffusion is the normal transport process in the soil. Diffusive gas transport in the soil depends on the soil gas concentration gradient in the soil (e.g. Hillel 1998; Jury and Horton 2004):

$$a\frac{\left(\partial C_g\right)}{\left(\partial t\right)} = -\xi_g D_g \frac{\left(\partial C_g\right)}{\left(\partial z\right)} + r_g \tag{4.13}$$

a is the soil air volume, C_g the concentration (mass of gas g per unit volume of soil air), t time (in sec), D_g is the gas diffusion coefficient for gas g in free air ($m^2\ s^{-1}$), z the depth (in m), r_g a sink term, which represents chemical processes by which the gas is converted by a chemical reaction (negative sign), or is produced (positive sign). The diffusion coefficient is specific for each gas and furthermore depends on temperature and pressure. ξ_g (<1) is a tortuosity factor, to downscale the diffusion coefficient in free air, to account for slower diffusion in the tortuous pores in the soil. In soil physics text books, several equations are given to derive this tortuosity factor from the soil air volume or the pore water saturation, θ. The value of ξ_g is in the order of 0.1 to 0.7. If soil pores are saturated, the diffusion coefficient D_g in air for the gas has to be replaced for that of water. For most gases, the diffusion coefficient in water is four orders of magnitude smaller (for CH_4 in air with standard temperature and pressure 2.1×10^{-5} m² s⁻¹ and 1.5×10^{-9} m² s⁻¹ in water).

Equation 4.13 is used to calculate the evolution of gas concentrations in the soil over depth and time. The diffusion in a water-saturated soil is extremely slow. This slow diffusion also holds for oxygen, and is the main reason why water-saturated soils become anaerobic; the oxygen demand by decomposition cannot be matched by a supply from the atmosphere via diffusion. The sink term r_g for CH_4, is consumption of CH_4 by methanotrophic bacteria in an aerated topsoil, giving a concentration profile as sketched in Fig. 4.25; for CO_2 there is no sink within the soil.

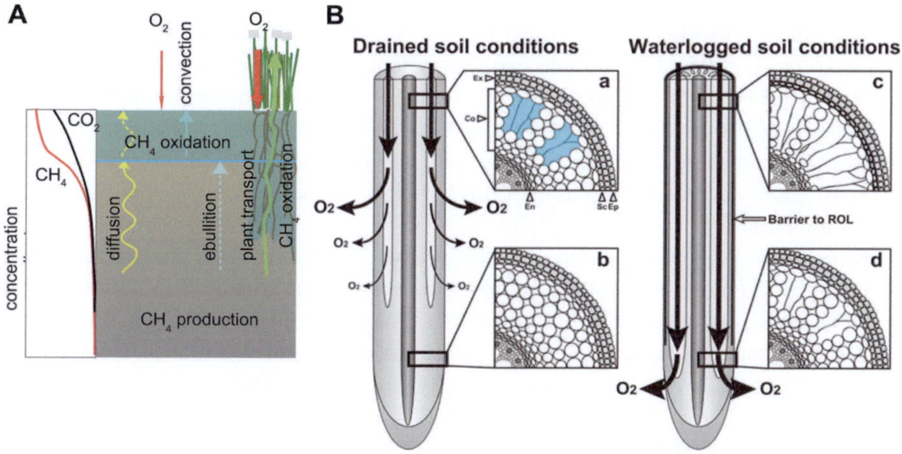

Fig. 4.25 (**a**) Four transport pathways for CH_4 and CO_2 to the soil surface, and their relation with the aerated zone above the water table where CH_4 oxidation occurs. In the blue shaded areas, methane consumption by methanotrophy is to be expected. Left side: typical vertical CH_4 and CO_2 concentration profiles in the soil, resulting from diffusion and production/consumption. (**b**) Aerenchyma and root O_2 transport in rice roots, from Nishiuchi et al. (2012). (**b**) *a*, *b*, *c* and *d* are quarter cross sections of roots; aerenchyma are marked blue in a. ROL stands for Radial Oxygen Loss. Under drained soil conditions, lysigenous aerenchyma (aerenchyma formed by die-off and lysis of cells) is formed, but O_2 can diffuse from the roots into the soil, a barrier to ROL is not formed. Under waterlogged soil conditions aerenchyma formation is enhanced and a barrier to ROL is induced. This results in the promotion of O_2 diffusion to the root apex, where root growth occurs. Under drained soil conditions, aerenchyma is constitutively formed at the basal part of the roots (*a*), but it is not usually formed at the apical part of the roots (*b*). Under waterlogged soil conditions, aerenchyma is induced at the basal part (*c*) and the apical part (*d*) of the roots. Lysigenous aerenchyma is more highly developed at the basal part of the roots (*a*, *c*) than at the apical part (*b*, *d*). Arrow thickness reflects the amount of O_2 available. *Ep* epidermis, *Ex* exodermis, *Sc* sclerenchyma, *Co* cortex, *En* endodermis. Figure 4.25b: © 2012 Nishiuchi et al; reprinted under Creative Commons Attribution License (http://creativecommons.org/licenses/by/2.0)

The second transport pathway is convective flow by pressure gradients in the soil. Similar to the flow of water, it can be described by an equation in which the flow is proportional to the pressure gradient:

$$F_c = -K_a \frac{dP}{dz} \tag{4.14}$$

Here F_c is the convective gas flux density (m s^{-1}), dP/dz is a pressure gradient, and K_a the air conductivity, and z the depth. Pressure gradients are typically caused by wind eddies moving over the soil surface, temperature gradients in the soil, barometric pressure changes and changes in pore water content by evapotranspiration, infiltrating rain or soil freezing. In general, the convective transport creates only a minor flux.

Ebullition is an important CH_4 transport pathway (Walter and Heimann 2000; Stamp et al. 2013). The rate of emission is similar to that of the diffusive flux, but may result in episodically high rates in soft, water-logged soils (Baird et al. 2004; Turetsky et al. 2008). CH_4 bubble formation depends on the poor solubility of CH_4 in water; the Henry's law solubility constant for CO_2 at standard temperature and pressure is 3.3 x 10^{-4} mol m^{-3} Pa^{-1}, for CH_4 1.4 x 10^{-5} mol m^{-3} Pa^{-1} at standard atmospheric pressure, an order of magnitude lower (Sander 2015). Moreover, solubility of gases decreases with temperature, making it a climate-related parameter in warming soils.

Strack et al. (2005) review the processes that govern ebullition and summarize it in a conceptual model. Bubbles form in the methanogenic zone in the soil where soil water tends to be over-saturated with CH_4. They have to rise through soil pores; where pore space is large enough they may rise to the surface, but otherwise may become trapped in layers of lower porosity, forming layers with overpressurised gas. The presence of trapped gas decreases groundwater flow and contributes to buoyancy of peat soils. As soon as overpressure exceeds the threshold for interparticle stresses in the soil, the confining layer gives way to further rise of the bubbles. This may also create preferred pathways for easier upward bubble movement. Release of trapped bubble gas is further provoked by variations in atmospheric and hydrostatic pressure (Waddington and Roulet 1997). Lowering of air pressure, or dry spells during which the water table is lowered, are examples. Surface air turbulence and atmospheric instability also induce ebullition and convective flow, as has been shown by turbulence-related eddy covariance fluxes of CH_4 (Wille et al. 2008; Parmentier et al. 2011c); higher turbulence and unstable conditions result in higher fluxes.

Ebullition is a serious problem for measuring CH_4 fluxes (e.g. Turetsky et al. 2008; Stamp et al. 2013). The release of bubbles is spatially highly variable and erratic, and in particular chamber measurements (see below) need a sufficient number of replicates to capture this variability. Furthermore, ebullition is very easily promoted by artificial disturbance of soils. As anyone who has walked through a wetland will have seen, the mere presence of a researcher may induce bubble fluxes, which results in erroneous overestimation of CH_4 fluxes (see Sect. 4.6).

Plant roots need oxygen for their cell metabolism and growth, which is in short supply in wetland soils. Wetland plants have developed adaptation mechanisms to supply their roots with oxygen; an important adaptation to anaerobic soil conditions is the development of aerenchyma: gas transport channels in roots and stems (Fig. 4.25). Many wetland plants have thick, well visible air channels in stems and roots; roots of reeds like *Phragmites australis* or *Typha sp.* may consist of more than a centimetre thick hollow tubes. Aerenchyma serve to transport reduced gases such as CH_4 from the soil, and deliver O_2 to the roots; in particular growing root tips need oxygen supply. The transport of gases in aerenchyma is driven by diffusion or convection through the aerenchyma. Plants can adapt their aerenchymous tissue during root growth to the anaerobic conditions in the soil (Nishiuchi et al. 2012; Fig. 4.25).

Aerenchyma in wetland plants provide by far the largest transport pathway of CH_4 to the atmosphere; between 30 and 100% of the CH_4 flux. This flux is species-dependent. Graminoids transport more methane (±80%) than forbs (±60%) (Bhullar et al. 2013). In sedges, the lower 15 cm of the above-ground plant parts is responsible for most of the emission (Kelker and Chanton 1997). Also wetland trees transport CH_4 via their root system, which is emitted by their stems (Gauci et al. 2010). In ponds in northern Siberia, plant-mediated transport of CH_4 by vegetation protruding above the water table amounts between 70% and 90% of the total CH_4 flux (Kutzbach et al. 2004; Knoblauch et al. 2015).

However, the oxygen supply to roots also supplies oxygen for methanotrophic bacteria within the rhizosphere; a varying percentage of the plant transported CH_4 is oxidized in this way (Walter and Heimann 2000). On the other hand, the O_2 transported into the soil via aerenchyma regenerates alternative electron acceptors, such as Fe oxides, that contribute to the anaerobic organic matter decomposition processes that produce the substrates for methanogenesis. The Interaction between exudation of methanogenic substrate by roots, oxidation of CH_4 in the rhizosphere and transport of CH_4 is species-specific: *Carex rostrata* has higher CH_4 emissions due to lower CH_4 oxidation than *Eriophorum vaginatum* and *Juncus effusus* (Ström et al. 2005).

The pathways for CH_4 transport determine to a large extent the quantity of CH_4 which is emitted by the ecosystem, since they determine to what extent CH_4 can escape oxidation by methanotrophic bacteria. The zones where aerobic methanotrophs thrive are the aerated top of the soil profile (if present), the root zone of aerenchymous plants, and within moss vegetation. Anaerobic methanotrophy is in most permafrost wetlands probably of minor significance (Sect. 4.5.3). The diffusion through an aerated topsoil is slow and is responsible for most of the CH_4 consumption by methanotrophs. If the aerated topsoil is sufficiently thick, most CH_4 may be consumed and the soil even may become a small sink for CH_4 by diffusion of CH_4 into the soil from the atmosphere. In peat soils, an unsaturated layer of some 20–30 cm is sufficient to reduce CH_4 emission to very low levels. This is shown by low levels of CH_4 emission or negative fluxes on drier soils in many studies (e.g. Frenzel and Karofeld 2000; Kutzbach et al. 2004; Van Huissteden et al. 2005; Natali et al. 2015). However, the relation of CH_4 flux with water table is often not linear and depends on soil and ecosystem type (Van Huissteden et al. 2008; Turetsky et al.

2008; Olefeldt et al. 2013). Incubation experiments by Zheng et al. (2018) have shown that the highest methanotrophic acivity occurs close to the soil horizons where CH_4 is produced, in lower, mostly anaerobic horizons instead of well aerated topsoil. This corroborates well with CH_4 consumption in boreal forest soils with water tables less than 10 cm below the surface observed by the author (Van Huissteden et al. 2008).

Ebullition is subject to negligible methanotrophy below the water table, but once the bubbles arrive in the aerated zone, the CH_4 released by bubbles is subject to bacterial oxidation. Plant transport is the most effective way to escape methanotrophy, despite CH_4 oxidation in the root zone (Ström et al. 2005). At the surface, further CH_4 consumption may occur within the vegetation cover. As noted above, *Sphagnum* and other mosses have a symbiotic association with methanotrophic bacteria living inside the leaves of the mosses. In particular when the plants are submerged, the CO_2 produced by CH_4 oxidation becomes available for photosynthesis by the plants (Basiliko et al. 2004; Raghoebarsing et al. 2005; Kip et al. 2010; Larmola et al. 2010; Knoblauch et al. 2015). Species that have been associated with these endophytic methanotrophs are various *Sphagnum* species and brown mosses as *Scorpidium scorpioides*, which are abundant in ice wedge polygon centre ponds (Knoblauch et al. 2015). Some of the highest CH_4 consumption rates have been measured in Sphagnum from Arctic sites (Kip et al. 2010) .This may explain the often low CH_4 fluxes from polygon ponds with a stable submerged moss vegetation (Parmentier et al. 2011b). Next, mosses have no root system, and therefore do not engage in plant-mediated transport of CH_4. Figure 4.26 summarizes the relations between CH_4 fluxes, biochemical cycling and vegetation.

The presence of a water layer above the soil surface may suppress CH_4 fluxes because of the slow diffusion through the water column. Zona et al. (2009) observed lower CH_4 fluxes above flooded soil, with a sudden rise of fluxes as soon as the soil surface emerged above the water table. The presence of emergent vascular plants, which transport CH_4 via aerenchymous tissue, is dominant if the soil surface is submerged. However, this only holds when the plant parts that emit most of the CH_4 are not covered by water. In the flooding experiment described by Zona et al. (2009), the low CH_4 fluxes were explained by submergence of *Carex* plant parts where most of emission occurs near the above-ground base of the plant.

At first sight, one would expect that favourable conditions for methanogenesis (anoxic conditions, sufficient labile substrate) should control the spatial variability of emission of CH_4 from ecosystems. This is true in the sense that wetlands are the most likely places where CH_4 emission is to be expected (Table 4.4). However, the considerations above show that vegetation, and the small-scale relief determining water table and flooding, should be responsible for most of the spatial variation of CH_4 emission *within* wetlands. This holds in particular in permafrost areas, where conspicuous micro-relief is created by ice wedge polygons, palsas and other ground ice features. The effect of micro-relief and vegetation has been the subject of many studies in permafrost ecosystems (e.g. Wagner et al. 2005; Kutzbach et al. 2004; Van Huissteden et al. 2005, 2009; Knoblauch et al. 2015). Additionally, gradients in nutrient availability have an indirect effect on methanogenesis, by stimulating veg-

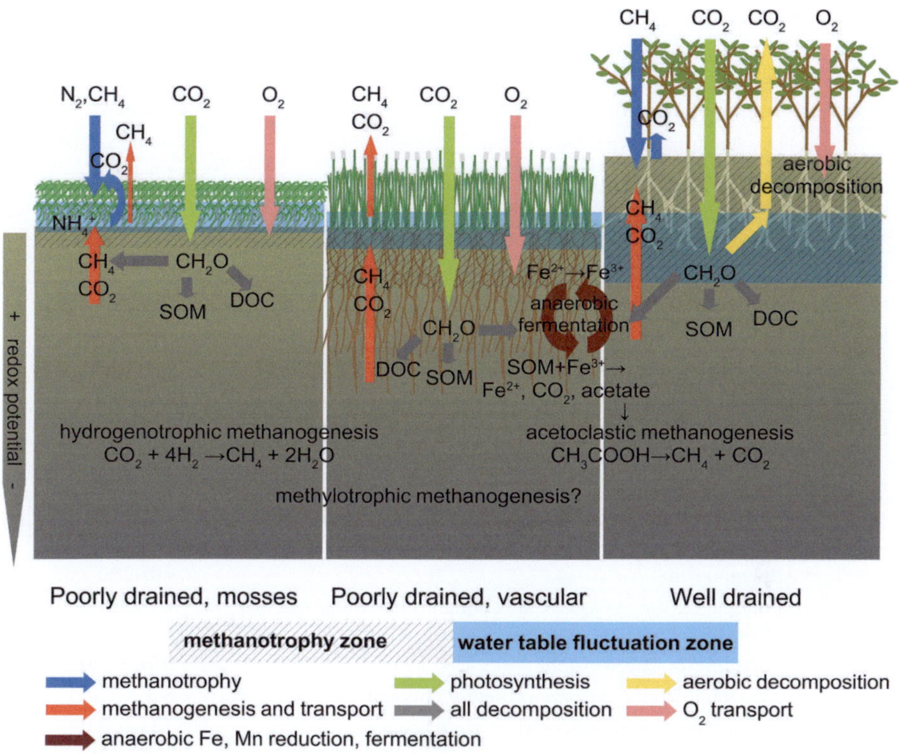

Fig. 4.26 Biochemical processes involved in CH formation and oxidation in soil environments with varying water table and vegetation (non-vascular mosses without root system vs. vascular plants). Indicated processes are: photosynthesis and transfer of organic compounds (CH_2O) to the soil, methanogenesis pathways, CH_4 oxidation by methanotrophy including N fixation (methanotrophy in mosses), transfer of O_2 to the soil, and anaerobic fermentation by Fe reduction producing substrate for methanogens, including recycling of Fe^{2+} to Fe^{3+} by oxidation (similar anaerobic reduction involves Mn, not shown here). SOM: soil organic matter; DOC: dissolved organic carbon. From Van Huissteden (2019)

etation productivity, and thus the availability of labile organic compounds as substrate for methanogens.

If growing season fluxes in various ecosystems are compared (Table 4.4), the following pattern with respect to vegetation emerges. Graminoid wetland vegetations with high water table, usually dominated by *Carex* and *Eriophorum spp*. and grasses, e.g. *Arctophila fulva*, show the highest fluxes. Mesotrophic to eutrophic wetlands in floodplains are the strongest emitters of CH_4. Oligotrophic sites show lower fluxes with otherwise similar water table conditions. Thaw ponds formed by ground ice thaw show a similar magnitude of fluxes as graminoid vegetations. Stable ice wedge polygon centre ponds generally show lower fluxes than fresh thaw ponds. *Sphagnum*-rich and other mosses-rich ecosystems show the lowest fluxes. Dry environments, in particular on peat plateaus, palsas, and polygon rims tend to

Table 4.4 Representative CH_4 emission and uptake in various terrestrial wetland environments in Siberia

Reference	Vegetation / micro-relief	CH_4 flux mg m^{-2} h^{-1}
1	Dry floodplain levee, *Salix* brush	−0.04 ± 0.04
1	Dry floodplain, grasses	1.3 ± 1.8
1	Floodplain wetland, *Arctophila fulva* grass and mosses	37.0 ± 5.3
1	Floodplain wetland, *Carex, Eriophorum*, grasses	12.6 ± 10.1
1	Dry tundra, *Betula nana*, mosses, lichens	−0.22 ± 0.34
1	Dry tundra, *Eriophorum* tussocks	−0.09 ± 0.05
1	Dry tundra, grasses	0.71 ± 1.1
1	Tundra wetland, usually inundated, *Carex, Eriophorum*	8.0 ± 3.0
1	Tundra wetland, *Sphagnum* hummocks	2.0 ± 1.7
1	Tundra wetland, inundated polygon pond with *Sphagnum* or mosses	4.0 ± 1.9
1	Ice wedge thaw pond	4.4 ± 2.8
2	Polygon centres with mosses (Scorpidium, Drepanocladus, Carex),	±4.2*
2	Polygon rim	<0.42*
3	Polygon Centre, mosses, Carex aquatilis and other vascular plants	1.2 ± 2*
3	Polygon rim, species-rich	0.18 ± 0.03*
4	Floodplain with *Carex appendiculata* tussocks	8.2 ± 3.4*
5	Floodplain with *Carex appendiculata* tussocks	13.8 ± 7.0*
6	Mesotrophic fen in larch forest, *Molinea* tussocks	1.3 ± 1.7
6	Eutrophic lake bank, grasses	11.9 ± 15.5
6	Moist forest floor, water table 10–27 cm below surface, *Betula*	−0.26 ± 0.45
6	Dry forest floor, *Larix* and *Pinus* with shrub and lichen understory	−0.37 ± 0.47
7	Mineral forest soil, no permafrost, *Betula/Picea/Larix,* shrub understory	−0.06 – 0*
7	Mineral forest soil with permafrost	−0.04 – 0*
7	Peat plateau, permafrost, *Betula nana, Ledum palustre, Sphagnum*, mosses	−0.025 – 0*
7	Permafrost thaw pond	12 and larger

References: *1*: Van Huissteden et al. (2009); *2*: Sachs et al. 2008; *3*: Kutzbach et al. 2004; *4*: Corradi et al. 2005; *5*: Kwon et al. 2017; *6*: Van Huissteden et al. 2008; Corradi et al. 2005; *7*: Flessa et al. 2008. Stars (*) mark values that have been recalculated to mg m^{-2} h^{-1}; negative values represent uptake from the atmosphere into the soils. From Van Huissteden (2019)

be a sink of CH_4. A more extensive evaluation of specific plant traits that influence CH_4 emission is given in Chap. 6.

A flux data synthesis by Olefeldt et al. (2013) sheds light on the effect of environmental and vegetation parameters on CH_4 fluxes, using data from 303 sites described in 65 studies. Figure 4.27 shows the magnitude of the fluxes measured under various conditions. The following relations were found in the data:

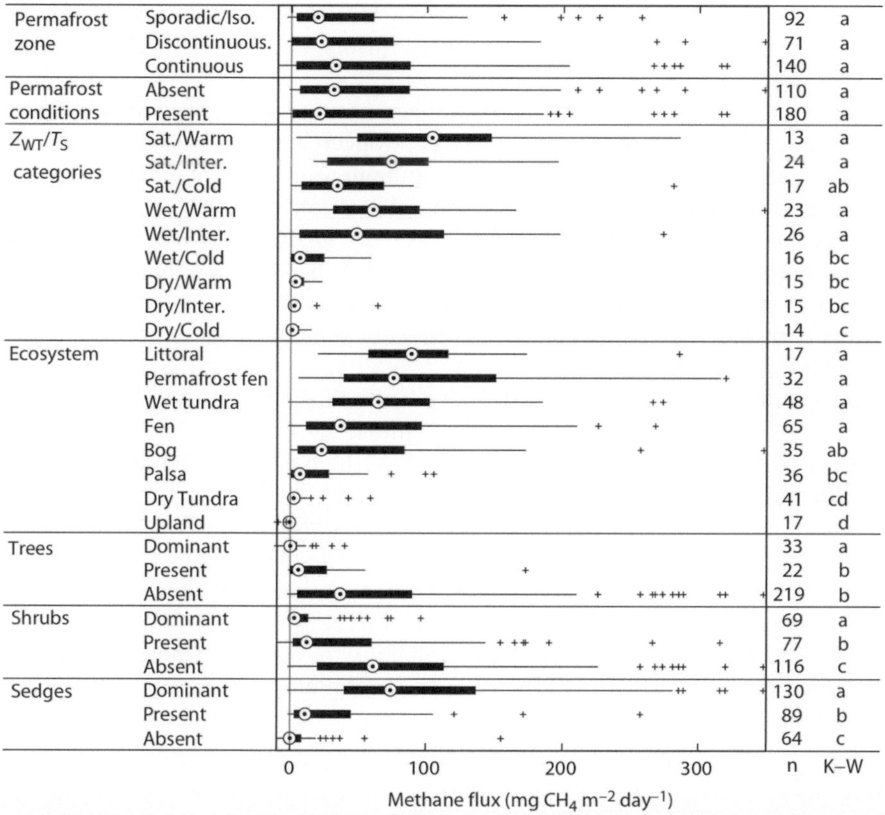

Fig. 4.27 Site CH_4 fluxes classified by several categorical variables. Circles indicate median flux, boxes show the range between the 25th and 75th percentiles and plus signs represent outliers. The number of sites within each category is indicated on the right (n), and differences among categories within each variable are denoted on the far right (Kruskal-Wallis, K-W, test followed by Tukey HSD, $P < 0.05$). Three permafrost zones are included: continuous, discontinuous, and a combined sporadic/isolated zone. Permafrost conditions refer to the presence or absence of permafrost in the upper 2 m of soils. Classification of water table/soil temperature (Z_{WT}/T_S) categories is based on separations for Z_{WT} at −15 and −2.5 cm (yielding Dry, Wet, and Saturated categories) and for T_S at 5 and 10 °C (yielding Cold, Intermediate, and Warm categories). (Republished with permission of John Wiley and Sons Inc., from Environmental and physical controls on northern terrestrial methane emissions across permafrost zones, by Olefeldt et al. (2013), Global Change Biology 19(2) © 2013; permission conveyed through Copyright Clearance Center, Inc.)

- The three dominating parameters are vegetation cover (*GPP* and amount of sedge cover), water table and soil temperature.
- Sites with a dense sedge cover (mainly tall sedges) show higher fluxes than other sites with a comparable water table. This highlights the importance of sedges, with their aerenchymous CH_4 transport and generally deep root system.
- Sphagnum coverage had no statistically discernable influence (data not shown in Fig. 4.27).

- The CH_4 flux as percentage of total ecosystem respiration varies strongly among vegetation types. The highest percentages (average ± 5%) occur in wet sedge ecosystems, percentages >15% occur.
- There is clear interaction between site parameters: sedge cover with soil temperature (the effect of a sedge cover is stronger with low soil temperatures), water table with soil temperature (fluxes of high water table sites are more sensitive to soil temperature than lower water table sites).
- Fluxes at lower water table sites (e.g. palsa, dry tundra, dryer bog and fen sites) are more sensitive to water table table position than fluxes at high water table sites (wet tundra, fen underlain by permafrost, littoral sites). The relation of CH_4 flux with water table is not linear; the effect of water table increases with higher water table.
- Sites with near-surface permafrost have lower fluxes than sites without permafrost but otherwise similar water table position.
- Sites with permafrost thaw depressions have higher fluxes than adjacent sites with intact permafrost.
- There is no significant difference between sites with permafrost thaw depressions and comparable sites without permafrost. This clearly indicates that permafrost thaw in itself is not a reason for high CH_4 emissions; only the local vegetation and soil environment controls the emissions.
- No evidence emerged that increased availability of organic matter from thawing permafrost leads to higher CH_4 fluxes; the determining parameters were the vegetation, soil wetness and temperature.

The synthesis of Olefeldt et al. (2013) and the ecosystem effects above clearly show the dominating effect of ecosystem type, hence ecosystem methane emission is a most appropriate term. It is not permafrost thaw by itself that results in high CH_4 emissions. Permafrost thaw only creates conditions for the establishment of ecosystems that emit high amounts of CH_4. These ecosystems also exist independently of permafrost thaw.

In lakes, the ecosystem CH_4 emission processes differ in some respects from those of wetlands, although the transition from lake to land in the riparian zone is similar to wetlands (Fig. 4.28). In deeper lakes, bottom waters may be anoxic due to thermal stratification (Bastviken et al. 2008). This impedes the exchange of gases, other than by slow diffusion or ebullition. These deeper, anoxic bottom layers of the lakes (the hypolimnion) are the main CH_4 production zone, and also store dissolved CH_4 as long as the stratification remains intact. Seasonal overturning of the stratification may release large amounts of CH_4 from the hypolimnion.

In many lakes the main transport pathway of CH_4 to the surface is ebullition (Bastviken et al. 2008; Wik et al. 2018). The ebullition is driven by methanogenesis in the bottom sediment and is related to energy input into the lakes. Wik et al. (2014) has found relations between ebullition rates and ice-free period, incoming shortwave radiation and water/sediment temperature. The presence of coarse detritus from redeposited peat and water vegetation in the lake sediment facilitates ebullition (Wik et al. 2018). In winter, the ebullition from the lake sediment continues if

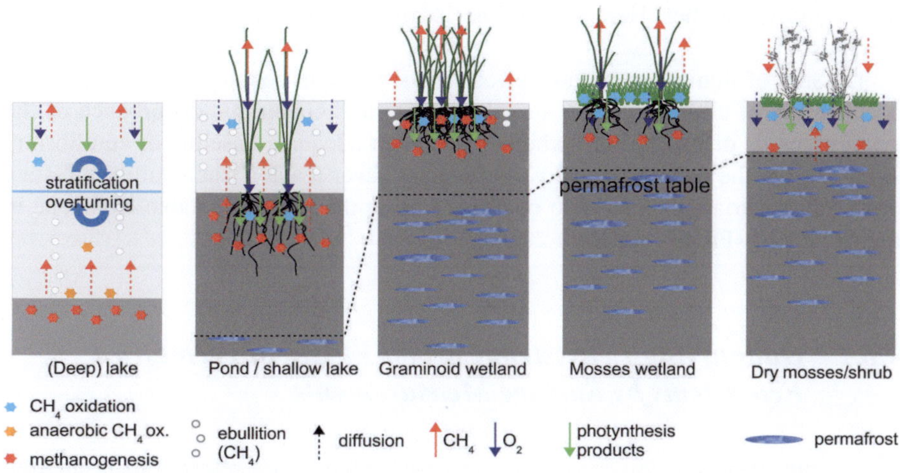

Fig. 4.28 Conceptual diagram of interactions between plants, substrate supply for methanogenesis, methanotrophy and methane transport pathways in various lake and wetland environments in permafrost areas, from lake to dry soil. The main locations of methane formation and methanotrophy are indicated with stars. Modified after Van Huissteden (2019)

the lake is not frozen to the bottom; the ice cover may store considerable amounts of CH_4. Above hotspots, where localized, strong ebullition occurs, lake freezing may be prevented in early winter (Walter et al. 2007).

Shallower lake zones may emit more CH_4 to the atmosphere than deep lake parts. In these shallower zones, turbulence and waves enhance CH_4 transport to the atmosphere, bypassing methanotrophy. In the littoral zone with emergent plants, plant transport also contributes depending on the density of the vegetation (Bastviken et al. 2008). The proportion of produced CH_4 that is oxidized in sediments and water column may amount 30–99%, and includes anaerobic CH4 oxidation coupled to denitrification or sulfate reduction (e.g. Sect. 4.5.3). Anaerobic CH_4 oxidation can effectively block CH_4 emission from lake sediments (e.g. Deutzmann et al. 2014). The oxidation of CH_4 contributes carbon to the food web in lakes, influencing its carbon isotopic signature (Bastviken et al. 2003, Van Hardenbroek et al. 2012; Sect. 5.3.2).

4.6 Ecosystem Carbon Flux Data

The quantification of ecosystem carbon flux data in permafrost regions has been started in the last decennia of the previous century and has been accelerated by the development of better instrumentation and methods in this century. Nevertheless, there are still considerable knowledge gaps resulting from the inaccessibility of many permafrost areas and the environment, which puts high demands on instruments and researcher. Here, an outline is given on the most frequently applied methods.

The start of significant research efforts in the previous century occurred when the acceleration of climate change already had started. Given this disturbance of the ecosystems, is difficult to tell whether this is an adequate baseline or equilibrium situation. On the other hand, ecosystems may never reach a true equilibrium state and climate variations have also occurred during the relatively stable Holocene in the Arctic (e.g. French and Shur 2010).

4.6.1 Quantifying Ecosystem Carbon Fluxes of Permafrost Ecosystems by Surface Measurements

Here, only measured fluxes will be discussed, models are referred to in another chapter. Gas fluxes to the atmosphere are measured usually by two methods, eddy covariance (see Sect. 2.2.2 for the principles an application for energy balance studies) and flux chamber methods. The focus is here on practical considerations; each method has its own merits and uncertainties.

With eddy covariance, the concentration of a substance in the air and vertical wind speed variations are measured at a high frequency (10 Hz or more); the correlation of both quantities is a measure of the flux (Sect. 2.2.2; Fig. 2.5). Atmospheric CO_2 concentration measurements at this high frequency have been possible for several decades. However, high frequency measurements of other atmospheric trace gases which occur in lower concentrations than CO_2, such as CH_4 and N_2O, have become available in the last two decades. At first, these instruments were very demanding in terms of operating conditions and power supply, which makes application in Arctic environments difficult. For CH_4 the situation has improved by availability of instruments with low power demand (Fig. 2.5).

An important distinction between instruments for measuring gas concentration is, whether they are closed path instruments or open path instruments. All instruments apply some form of measurement of absorption of infrared radiation by the greenhouse gas that is measured. This absorption is dependent on the length of the path that the infrared radiation travels through the gas mixture. Open path instruments measure directly in the open air between a radiation source and detector. This restricts the path length; eventually the path length is extended using a mirror system (for example the CH_4 instrument in Fig. 2.5). Closed path instruments take samples to a measurement cell using an air pump, and require therefore correction for travel time of the air through the pumping system. However, they are less vulnerable to pollution than open path systems. The capacity of the air pump depends on the measurement cell volume and measurement frequency. It puts extra demands on the power supply of the measurement system.

The instantaneous location and size of the source area varies with height of the instruments above the surface (generally a few metres to a few tens of metres), wind speed, wind direction and turbulence properties of the atmosphere. To derive a statistical definition of the source area for comparison with other measurements, e.g. chamber measurements, footprint models are used (Sect. 2.2.2; Budishchev et al. 2014). Usually the fluxes are summarized into half-hourly values. Eddy covariance cannot be applied in all terrain conditions; the area should be flat and have approximately homogenous surface conditions in terms of aerodynamic roughness.

Another restriction of eddy covariance is, that it measures only the net ecosystem exchange *NEE* to the atmosphere, and that additional measurements (e.g. chamber flux measurements) or modelling is necessary to estimate component fluxes such as R_{eco} and *GPP*. Equation 4.4 (Sect. 4.2) is often used for estimating *GPP*; equations that relate R_{eco} to soil temperature (Sect. 4.5) are often used to model R_{eco}.

R_{eco} also can be estimated from nighttime fluxes; however, high-latitude nighttime light levels and lower temperatures at night influence such an estimate (Sect. 4.2). For modelling R_{eco}, the model parameters and structure vary throughout the growing season, in particular the increasing active layer thickness throughout the growing season adds important variability (e.g. Van der Molen et al. 2007; Euskirchen et al. 2012; Runkle et al. 2013; Trucco et al. 2012). Runkle et al. (2013) developed a multi-step procedure to estimate R_{eco} and *GPP* which takes other factors into account than temperature alone, such as plant physiology and light conditions. See also the references in Table 4.5. for various approaches to modelling. Models used for the growing season are not directly applicable for the cold seasons (Oechel et al. 2014).

Differences between instrument configurations and micrometeorological quality considerations of eddy covariance measurements further cause variation between measurement locations. Various corrections need to be applied to the raw data (e.g. Aubinet et al. 2012). Spectral corrections are needed for loss of high- and low frequency components of the signal, corrections for air density fluctuations due to temperature and humidity variations, and sensor-specific corrections. Under calm conditions, in particular at nighttime, storage of CO_2 may occur close to the surface, which is released again when the wind picks up. To quantify these storage effects, it may be necessary to measure vertical concentration profiles (Parmentier et al. 2011c). A correction that is specifically important in cold environments is a correction for heating of the air by instruments which was an issue with older CO_2/H_2O sensors (Sect. 2.2.2; Burba et al. 2008).

Chamber flux measurements (Fig. 4.29) consist of the placement of an enclosure over the soil surface. For CO_2 measurements, transparent chambers measure *NEE*, if the chamber walls are opaque or measurements are taken at night, they measure R_{eco}. The increase or decrease of the target gas is measured by taking gas samples (for static chambers) or by measuring concentration differences between inflowing and outflowing air (dynamic chambers). The advantage of chambers is that the footprint of the system is exactly known. Disadvantages are the small surface area (generally less than 1 m^2), and measuring over taller vegetation than the chamber height is impossible without cutting vegetation – which is a serious disturbance of CO_2 and and CH_4 fluxes.

Table 4.5 Recently published carbon flux data from various permafrost ecosystems

Region	Site + reference	biome / vegetation / landform	nr years	nr sites	met-hod	RM	NEE	R_{eco}	GPP	CH_4	DOC	VOC	GHG 100	NECB total
N Canada	Ellsmere 1	polar desert wetland	3		EC, C					0.46				
N Canada	Ellsmere 1	polar desert dry	5		C					-0.5				
NE Siberia	Indigirka lowl Kytalyk 2	tundra	4	1	EC, C (CH_4)		−92	141	−232	3.1			48	−89
NE Siberia	Kytalyk 3	tundra	8 (3 CH_4)	1	EC	2,4,5	−95 - −69	74 - 129	−211 - −158	2.7	3.1		67 - 93	−89 - −63
E Siberia	Yakutsk, Spasskaya Pad 4	taiga, Larch, Scots Pine	3	1	EC	3	−140	19	159	−0.3			−432	−140
W Siberia	Tura 5	taiga forest	1	1	EC		-76							-76
Alaska	Imnavait Watershed 6	tussock tundra	3	1	EC	2	−62 - −84	46 - 112	−191 - −108					−62 - −84
NE Greenland	Zackenberg valley 7	tundra - dry Cassiope heath	10	1	EC	7	−21±11	79±14	53±11					−21±11
NE Greenland	Zackenberg valley 8	High Arctic wetland	2	1	EC		−50 - −88	159 - 211	−216 - −305	6.5 - 7.6			−23.5 - +242	−43 - 83
N Siberia	Lena delta, Samoylov 9	tundra, floodplain	2	1	EC, C	1	*−20*	*99*	*−118*	*2.4*			*69.8*	*−17.2*
Alaska	Imnavait Watershed 6	tundra heath	3	1	EC	2	21-61 −58 - −51	173 - 269	−220 - −138					21 - 61
Alaska	Imnavait Watershed 6	wet sedge tundra	3	1	EC	2	2 - 82 −95 - −59	194 - 264	−164 - −206					2 - 82
Alaska N Slope	Atqasuk 10	moist acidic tussock tundra	1	1	EC		13.6, −24.3							23.6
NE Siberia	Kolyma Floodplain 11	floodplain with sedge tussocks	4	2	EC	6	−42			7			266	−35
N. Sweden	Abisko, Stordalen* 12	palsa mire	2	1	EC, C	2	−38 - −62			2	2.8	0.65	−64	−33 - −56
N. Russia	Vorkuta, Seida* 13	tundra, palsa mire	1	1	EC, C		−79 - −41	197 - 212	−294 - −238	5			10 - 149	−74 - 36
N. Russia	West Sibera lakes* 14	Lake	1	3	C		22 - 61			0.7 - 5.3			172 - 539	23 - 66

RM remarks, *DOC* dissolved organic carbon water, *VOC* volatile organic carbon, *NEE*, R_{eco}, *GPP*, CH_4, DOC, VOC fluxes and NECB in g C m^{-2} $year^{-1}$; GHG balance in g CO_2 eq m^{-2} $year^{-1}$. The 100 year greenhouse gas (GHG) balances deviate from the often used IPCC Global Warming Potentials, but have been calculated cf. Neubauer and Megonigal (2015). GHG balance calculations have been performed only when CH_4 flux data are available. For readability of the table, data uncertainty has not been added to the table. Red figures: data from the growing season only, black: year-round or nearly year-round data. For sites with a few years of data, the range of observations is indicated, for long time series averages and standard deviations are given. The signs of the fluxes follow the 'atmospheric' convention: negative fluxes are fluxes from atmosphere to ecosystem, positive fluxes are fluxes from ecosystem to atmosphere or water. * after site name: discontinuous permafrost, all others are underlain by continuous permafrost. Method: EC is eddy covariance, C chambers. Remarks column: 1 – estimated full yearly balance using models and autumn data; winter data incomplete; 2 – data range of multiple years; 3 – CH_4 data from other year than CO_2 data; 4 CH_4 average 3 years assuming 2 months growing season; 5 – DOC from unpublished data Belelli et al. in prep; 6 – only data from control experiment used here; 7 – multi-year mean and standard deviation.; 8 – red numbers are growing season fluxes. Colors in the column biome/vegetation/landform indicate broad biome groupings: white – polar desert, light green – moist to wet tundra, dark green – forest, yellow – dry arctic heath vegetation, purple – palsa mire, blue – lake. Literature references: 1 – Emmerton et al. (2014); 2 – Van der Molen et al. (2007); 3 – Parmentier et al. (2011a, b); 4 – Dolman et al. (2004) and Van Huissteden et al. (2008); 5 – Nakai et al. (2008); 6 – Euskirchen et al. (2012); 7 – Lund et al. (2012); 8 – Tagesson et al. (2012); 9 – Kutzbach et al. (2007a, b) and Wille et al. (2008); 10 – Oechel et al. (2014); 11 – Kittler et al. (2017); 12 – Olefeldt et al. (2012); 13 – Marushchak et al. (2013, 2016); 14 – Repo et al. (2007)

Fig. 4.29 Light, transportable automated flux chamber system. The front chamber is closed. The connecting hoses and cables are for transport of air from the chamber to a gas analyser, for operating the pneumatic opening and closing of the chamber lids and for data transport to a data logger. The chambers open and close in turn. Made by VU University Technical Department; photo by the author

The emplacement and presence of the chamber also disturbs the soil to some extent. In particular during CH_4 flux measurements in soft sediments, induced ebullition (accidental release of gas by bubbles) may occur (Silva et al. 2015; previous section). To overcome the spatial and temporal representativeness problems, multiple, spatially distributed measurements have to be taken and repeated over time, preferably with automatically operated systems (e.g. Trucco et al. 2012). However, automatic chambers that are left in place over more than a few days have the disadvantage that the microclimate within the chambers soon starts influencing vegetation and soil temperatures. Kutzbach et al. (2007b) and Silva et al. (2015) give an account of uncertainties in flux calculations of chamber systems.

Fluxes over lakes can be measured with eddy covariance (e.g. Jammet et al. 2015), with floating chambers, or by measuring the gas concentration gradient between surface water and air (e.g. Repo et al. 2007; Bastviken et al. 2008). The same problems of spatial representativeness apply here for eddy covariance and chamber measurements as for terrestrial applications. In particular ebullition is problematic as it is highly localized in space, and episodic in time. Attempts to quantify ebullition have been made using funnel-shaped bubble traps and bubble counting in lake ice, but these are likewise subject to large uncertainty (Walter et al. 2007; Wik et al. 2011).

Lateral fluxes of dissolved or particulate carbon in water are measured by measuring water discharge in streams, combined with water sampling or continuous measurements of dissolved and particular substances. Here, the uncertainties arise from the discharge measurement itself; in particular in wetlands, which have often a diffuse drainage instead of well delineated channels, discharge is difficult to quantify and drainage basin boundaries are difficult to quantify. Additionally, the transition from soil pore water to stream release is under-studied; associated dissolved carbon loss along this pathway may be significant.

Upscaling of fluxes from measurement stations to larger regions is usually done with help of land cover or vegetation classifications developed from remote sensing material. Present-day commercial satellite imagery can provide high resolution images down a spatial resolution of less than 1 m, which can be used for detailed land cover classification. By relating measured fluxes to land cover, upscaled fluxes over larger areas can be developed (e.g. Tagesson et al. 2012; Marushchak et al. 2013). Budishchev et al. (2014) explore the relations between land cover types in the footprint of an eddy covariance tower tower and upscaling of chamber fluxes.

4.6.2 *Temporal and Spatial Variability of Permafrost Ecosystem Carbon Fluxes*

A significant problem with flux measurements in permafrost environments is a lack of winter data. Flux measurement equipment is difficult to operate under harsh winter conditions, and it has often been assumed that the summer is the most active period, while in winter emissions were deemed negligible. However, the summer months cover only 30–20% of the year in most permafrost environments (Zona et al. 2016). Measurements continuing into the shoulder seasons (late autumn and early spring) or winter have shown continued activity of soil respiration, and also storage effects such as the release of CO_2 and CH_4 stored in the soil, snow cover or ice have been recorded (e.g. Whalen and Reeburgh 1988; Zimov et al. 1993; Mastepanov et al. 2008, 2013; Jammet et al. 2015; Zona et al. 2016). At the start of snowmelt significant emission from gases stored in the snow cover may occur (Wille et al. 2008; Tagesson et al. 2012). Similarly, lake ice thaw results in elevated emissions from CH_4 stored as bubbles in the ice or in the water column (Jammet et al. 2015).

At a tussock tundra site in northeastern Siberia, the winter emissions amounted to 60% of the CO_2 uptake in summer; 30% of the annual emission of CH_4 was emitted outside the growing season (Kittler et al. 2017). Zona et al. (2016) measured summer and winter fluxes of CH_4 at five tundra sites on the Alaskan North Slope using eddy covariance. Generally, the warm season flux peaks were highest, but the cold season fluxes averaged over these site amounted about 50% (37–64%) of the yearly total, with an average flux of 2.3 g CH_4 m^{-2}. The winter fluxes were highest at the driest sites. The high cold season emissions occur predominantly during the zero curtain period (Sect. 2.5) in the soil in the fall and early winter. In particular in wetter sites, the soil temperature remains near zero for a considerably period, more than 90 days. The active layer thickness in summer proved a good predictor of both summer and winter emissions.

Three groups of processes have been invoked to explain these ongoing emissions of CH_4 and CO_2 in fall-winter from permafrost soils. First, there unfrozen water available which may support bacterial metabolism, even at temperatures below 0 °C. Metabolism has been shown to continue at temperatures well below the freezing point (Sect. 4.5; Zona et al. 2016). Second, the release from gases stored in soil is supposed to occur when the soils start to freeze. The advancing freezing fronts from the permafrost table upwards, and from the soil surface downwards, increases pressure on the stored soil gases (Mastepanov et al. 2008, 2013). The stored gas is released through cracks in the freezing soil or via plant root systems and stems. The resulting bursts of both CH_4 and CO_2 from soil storage during soil freezing are more infrequent than the autumn-winter emission related to ongoing metabolism (Tagesson et al. 2012; Mastepanov et al. 2013; Zona et al. 2016). Third, emissions from soil and vegetation cavities may be induced by atmospheric turbulence (Tagesson et al. 2012).

Table 4.5 lists a number of recent carbon and greenhouse gas flux studies from ecosystems on permafrost soils. Note that differences in instrumentation and data processing among the sites may cause differences in measured fluxes as discussed in the previous section; there is no correction for these differences in the table. The fluxes that represent growing season fluxes only, are marked in red.

As seen from the *NEE* data, the importance of year-round winter data collection is immediately clear. While growing season *NEE* is always negative, indicating a CO_2 sink, at several sites with year-round data this summer sink is more than compensated by positive fluxes in the cold season, resulting in a CO_2 source (see e.g. the data from Euskirchen et al. 2012). Oechel et al. (2014) recorded in 1 year a release of CO_2 outside the growing season, that was 1.6 times as high as the summer uptake.

Inter-annual variability is also prominent in Table 4.5, shown by the *NEE* data range for sites with multiple observation years. Several tens of percents of difference are not uncommon. This implies that one or a few years of data are merely snapshots of the carbon fluxes. The only data set with a time series of more than a few years is that of Lund et al. (2012), and has a standard deviation on growing season *NEE* that exceeds half of the average. The severe lack of winter observations and long observation time series throughout the permafrost realm, results in a large uncertainty on ecosystem carbon fluxes. It frustrates attempts to derive ongoing changes of the permafrost carbon fluxes from direct observations, and hampers validation of predictive models.

Variability in growing season *NEE* and *GPP* is attributed for most sites to temperature variation and temperature-related variables such as growing-degree days (e.g Parmentier et al. 2011a; Lund et al. 2012; Kittler et al. 2017). The timing of the start of the growing season, when ecosystems turn from a carbon source to a sink, is also determined by the number of growing-degree-days (Euskirchen et al. 2012).

In the tundra bud break occurs some 4 weeks after snowmelt, and requires build-up of heat energy up to a critical limit (Pop et al. 2000; Parmentier et al. 2011a, c). For the taiga forest in Table 4.5, humidity (high vapour pressure deficit) in summer is potentially a restrictive factor on *GPP* (Sugimoto et al. 2002; Dolman et al. 2004).

GPP is potentially facilitated by longer growing seasons and higher temperature. However, Parmentier et al. (2011a) did not find a significant correlation with growing season length and *GPP*, and Lund et al. (2012) see a levelling off of *GPP* at higher temperatures. The relation of *NEE* with growing season temperature is even more uncertain, since R_{eco} also increases with higher growing season temperature (Parmentier et al. 2011a; Lund et al. 2012). Besides the effect of higher temperatures on soil microbial activity, the increasing active layer thickness contributes also (e.g. Van der Molen et al. 2007). For the Siberian tundra of Parmentier et al. (2011a), the net result is absence of a significant relation of temperature with *NEE*.

For CO_2 release outside the growing season, Q_{10} relations of R_{eco} with temperature are insufficient to predict fluxes (Oechel et al. 2014). The winter *NEE,* fed by continued soil respiration, is negatively correlated with snow cover depth, and positively with wind speed and drops in atmospheric pressure (Euskirchen et al. 2012). This suggests storage of CO_2 in the snow cover or soil; release is promoted by specific weather conditions. Other factors that have been reported are time, and the size of the vegetation and soil carbon pool (Elberling 2007). In particular the spring and autumn seasons are complicated, when snow is accumulating or melting, and repeated freezing and thawing affects the snow density (Kwon et al. 2006; Trucco et al. 2012; Oechel et al. 2014).

The variation among sites in Table 4.5 is also large, although this is to be expected given the variation in latitude and biomes – from High Arctic heath and tundra to boreal forest. Nevertheless, within the same biome also the variation is large; e.g. for the two boreal forest sites with Larch, one site has almost a double growing season *NEE* compared to the other. This is attributed to differences in tree density, expressed by leaf area index (LAI) differences (Nakai et al. 2008). Within the East Greenland site Zackenberg, the wetland has a 2.5–4 times lower *NEE* (more carbon uptake) than the dry Cassiope heath. A similar contrast (although smaller) between dry tundra heath and wetland is seen in the tundra *NEE* data in Alaska of Euskirchen et al. (2012). For the sites with year-round flux data, there is a contrast between the tundra heath, sedge and tussock sites which are small carbon sources (Euskirchen et al. 2012; Oechel et al. 2014), and floodplain sites and subarctic mire sites, which are carbon sinks.

The variation in *NEE* throughout high-latitude permafrost areas is the result of variation in climatic parameters such as temperature, precipitation, growing season length, the amount of PAR, and local factors that affect productivity: moisture and nutrient availability (Williams and Rastetter 1999; Williams et al. 2000; McGuire et al. 2002). LAI integrates the effects of local soil conditions. High Arctic sites tend to have a lower *NEE* than low Arctic sites (Lafleur et al. 2012) and *NEE* is largest in the wet sedge ecosystem and smallest in dry heath (Nobrega and Grogan 2008). Shaver et al. (2013) found that a regression model including air temperature, photosynthetic active radiation and vascular plant LAI explained 75% in the variation of chamber-measured *NEE* across the Arctic. The role of the moss biomass appears to be less important than that of vascular plant for model-based prediction of *NEE* (Williams et al. 2000).

In particular at the scale of measurement plots as recorded by chamber flux measurements, a strong spatial variation related to vegetation units is seen. For instance, Marushchak et al. (2016) found that vascular LAI explained ±90% of the variability in the *GPP* and *NEE* in their chamber flux data, and for R_{eco} 67% of the variability was explained by LAI. However, other variables such as moisture and nutrient availability likely drive these differences in LAI.

Interestingly, successional changes in vegetation also determine the carbon exchange. Zona et al. (2010) characterized differences in *GPP* and R_{eco} in drained thaw lake basins of various ages. Recently drained basins (±50 years ago) showed a higher *GPP* (11.4 g CO_2 m^{-2} d^{-1}) and R_{eco} (7.3 g CO_2 m^{-2} d^{-1}) than old basins (>300 year; *GPP* 7.2 g CO_2 m^{-2} d^{-1}, R_{eco} 4 g CO_2 m^{-2} d^{-1}). The difference is attributed to increasing storage of nutrients in organic material, where it is only slowly released because of low decomposition rates.

For most sites the carbon balance has not been measured exhaustively. In most cases data on CH_4 fluxes are available. Although the losses by the CH_4 flux amount only 1–2% of the *GPP*, it has a large effect on the greenhouse gas balance; most sites are greenhouse gas sources on a 100 year time scale if the GWP calculation cf. Neubauer and Megonigal (2015) is applied. The polar desert environment in Ellsmere Island, Northern Canada shows significant negative CH_4 fluxes for dry soils, and very low positive fluxes for a stream wetland (Emmerton et al. 2014).

Only in two sites in Table 4.5. the dissolved organic matter flux in runoff water is known (note that this is DOC only – considerable amounts of particulate material may be transported during spring snowmelt flooding or soil erosion). For these two sites, this carbon loss is only a few percent of the *GPP*. In Table 4.5 (and in most publications) the DOC nor POC flux has not been included in the GHG balance. Whether water-transported carbon contributes to greenhouse gas fluxes, strongly depends on the fate of the transported organic matter: decomposition during transport followed by outgassing of CO_2 and CH_4 to the atmosphere, or sequestering of the carbon in river, lake or marine sediments (Vonk and Gustafsson 2013; Chap. 5).

The lakes in Table 4.5 are net sources of carbon to the atmosphere (Repo et al. 2007). This is generally the case for lakes. Although vegetation and algae may contribute to a high *GPP* of the lake, they receive an additional net carbon flux from other ecosystems by water discharge to the lake. This includes CH_4-saturated groundwater flow from the active layer of soils surrounding the lakes (Paytan et al. 2015). Part of this carbon is stored in the sediment, but part is processed to CO_2 in the sediment and water column (e.g. Kling et al. 1991; Tranvik et al. 2009). Furthermore, anoxia in lake sediments contributes to high CH_4 production (e.g. Bastviken et al. 2008). For instance in Toolik lake, Alaska, 20% of the incoming organic matter and inorganic carbon is respired as CO_2, and 2% is stored in sediment, and 78% leaves the lake again as water-borne carbon fluxes (DIC, dissolved inorganic carbon, DOC and POC) (Tranvik et al. 2009 and references therein).

Temperature stratification and ice cover influence lake fluxes of CO_2 and CH_4 to the atmosphere (previous section, Sect. 2.7). Jammet et al. (2015) describe a detailed sequence of CH_4 emission events during winter and spring on a subarctic lake in Sweden. In winter, CH_4 emissions from the lake are very low as they are impeded by the ice cover. When the snowmelt starts, CH_4 emission rises as CH_4 stored in the snow layer is released, following a daily pattern induced by temperature. When the ice cover breaks up, CH_4 emission increases, both from open water and thawing ice in which bubbles are stored. This varies with wind speed, resulting from turbulent exchange at the water surface. However, the largest emissions occur at overturning of thermal stratification of the water column.

References

Abbott BW, Jones JB (2015) Permafrost collapse alters soil carbon stocks, respiration, CH_4, and N_2O in upland tundra. Glob Chang Biol 21(12):4570–4587

Abbott BW, Larouche JR, Jones JB, Bowden WB, Balser AW (2014) Elevated dissolved organic carbon biodegradability from thawing and collapsing permafrost. J Geophys Res Biogeo 119(10):2049–2063

Aerts R (2006) The freezer defrosting: global warming and litter decomposition rates in cold biomes. J Ecol 94(4):713–724

Alvarez-Uria P, Körner C (2007) Low temperature limits of root growth in deciduous and evergreen temperate tree species. Funct Ecol 21(2):211–218

Angle JC, Morin TH, Solden LM, Narrowe AB, Smith GJ, Borton MA, Rey-Sanchez C, Daly RA, Mirfenderesgi G, Hoyt DW (2017) Methanogenesis in oxygenated soils is a substantial fraction of wetland methane emissions. Nat Commun 8(1):1567

Arndal MF, Illeris L, Michelsen A, Albert K, Tamstorf M, Hansen B (2009) Seasonal variation in gross ecosystem production, plant biomass, and carbon and nitrogen pools in five high arctic vegetation types. Arct Antarct Alp Res 41(2):164–173

Arneth A, Makkonen R, Olin S, Paasonen P, Holst T, Kajos MK, Kulmala M, Maximov T, Miller PA, Schurgers G (2016) Future vegetation–climate interactions in Eastern Siberia: an assessment of the competing effects of CO_2 and secondary organic aerosols. Atmos Chem Phys 16(8):5243–5262. https://doi.org/10.5194/acp-16-5243-2016

Ashastina K, Kuzmina S, Rudaya N, Troeva E, Schoch WH, Römermann C, Reinecke J, Otte V, Savvinov G, Wesche K (2018) Woodlands and steppes: Pleistocene vegetation in Yakutia's most continental part recorded in the Batagay permafrost sequence. Quat Sci Rev 196:38–61

Aubinet M, Berbigier P, Bernhofer C, Cescatti A, Feigenwinter C, Granier A, Grünwald T, Havrankova K, Heinesch B, Longdoz B (2005) Comparing CO_2 storage and advection conditions at night at different carboeuroflux sites. Bound-Layer Meteorol 116(1):63–93

Aubinet M, Vesala T, Papale D (2012) Eddy covariance: a practical guide to measurement and data analysis. Springer, Dordrecht, 438 p

Aurela M, Tuovinen J-P, Laurila T (2001) Net CO_2 exchange of a subarctic mountain birch ecosystem. Theor Appl Climatol 70(1–4):135–148

Avery GB, Shannon RD, White JR, Martens CS, Alperin MJ (1999) Effect of seasonal changes in the pathways of methanogenesis on the $\delta^{13}C$ values of pore water methane in a Michigan peatland. Glob Biogeochem Cycles 13(2):475–484

Baird AJ, Beckwith CW, Waldron S, Waddington J (2004) Ebullition of methane-containing gas bubbles from near-surface Sphagnum peat. Geophys Res Lett 31(21):L21505. https://doi.org/10.1029/2004GL021157

Basiliko N, Knowles R, Moore TR (2004) Roles of moss species and habitat in methane consumption potential in a northern peatland. Wetlands 24(1):178

Bastviken D, Ejlertsson J, Sundh I, Tranvik L (2003) Methane as a source of carbon and energy for lake pelagic food webs. Ecology 84(4):969–981

Bastviken D, Cole JJ, Pace ML, Van de Bogert MC (2008) Fates of methane from different lake habitats: connecting whole-lake budgets and CH_4 emissions. J Geophys Res Biogeo 113(G2):G02024. https://doi.org/10.1029/2007JG000608

Batjes NH (1996) Total carbon and nitrogen in the soils of the world. Eur J Soil Sci 47(2):151–163

Beermann F, Teltewskoi A, Fiencke C, Pfeiffer E-M, Kutzbach L (2015) Stoichiometric analysis of nutrient availability (N, P, K) within soils of polygonal tundra. Biogeochemistry 122(2–3):211–227

Bernal B, McKinley DC, Hungate BA, White PM, Mozdzer TJ, Megonigal JP (2016) Limits to soil carbon stability; deep, ancient soil carbon decomposition stimulated by new labile organic inputs. Soil Biol Biochem 98:85–94

Bhullar GS, Edwards PJ, Olde Venterink H (2013) Variation in the plant-mediated methane transport and its importance for methane emission from intact wetland peat mesocosms. J Plant Ecol 6(4):298–304. https://doi.org/10.1093/jpe/rts045

Blok D, Sass-Klaassen U, Schaepman-Strub G, Heijmans M, Sauren P, Berendse F (2011a) What are the main climate drivers for shrub growth in Northeastern Siberian tundra? Biogeosciences 8(5):1169–1179

Blok D, Heijmans MMPD, Schaepman-Strub G, van Ruijven J, Parmentier FJW, Maximov TC, Berendse F (2011b) The cooling capacity of mosses: controls on water and energy fluxes in a Siberian tundra site. Ecosystems 14(7):1055–1065. https://doi.org/10.1007/s10021-011-9463-5

Bodelier PL (2011) Interactions between nitrogenous fertilizers and methane cycling in wetland and upland soils. Curr Opin Environ Sustain 3(5):379–388

Boike J, Ippisch O, Overduin PP, Hagedorn B, Roth K (2008) Water, heat and solute dynamics of a mud boil, Spitsbergen. Geomorphology 95(1):61–73

Bonan G (2002) Ecological climatology: concepts and applications, 1st edn. Cambridge University Press, Cambridge

Bonan G (2016) Ecological climatology: concepts and applications, 3rd edn. Cambridge University Press, Cambridge

Bonan GB, Levis S, Kergoat L, Oleson KW (2002) Landscapes as patches of plant functional types: an integrating concept for climate and ecosystem models. Glob Biogeochem Cycles 16(2). https://doi.org/10.1029/2000GB001360

Budishchev A, Mi Y, van Huissteden J, Belelli-Marchesini L, Schaepman-Strub G, Parmentier FJW, Fratini G, Gallagher A, Maximov TC, Dolman AJ (2014) Evaluation of a plot-scale methane emission model using eddy covariance observations and footprint modelling. Biogeosciences 11(17):4651–4664. https://doi.org/10.5194/bg-11-4651-2014

Burba GG, McDermitt DK, Grelle A, Anderson DJ, Xu L (2008) Addressing the influence of instrument surface heat exchange on the measurements of CO_2 flux from open-path gas analyzers. Glob Chang Biol 14(8):1854–1876

Callaghan TV, Björn LO, Chernov Y, Chapin T, Christensen TR, Huntley B, Ims RA, Johansson M, Jolly D, Jonasson S (2004) Biodiversity, distributions and adaptations of Arctic species in the context of environmental change. AMBIO J Hum Environ 33(7):404–417

Carson M, Conly FM, Jasper J (1999) Riverine sediment balance of the Mackenzie delta, Northwest Territories, Canada. Hydrol Process 13(16):2499–2518

Chapin DM, Bledsoe CS (1992) Nitrogen fixation in arctic plant communities. In: Chapin F, Jefferies RL, Reynolds JF, Shaver GR (eds) Arctic ecosystems in a changing climate: an ecophysiological perspective. Academic, San Diego, pp 301–319

Chapin FS III, Moilanen L, Kielland K (1993) Preferential use of organic nitrogen for growth by a non-mycorrhizal arctic sedge. Nature 361(6408):150

Chapin FS III, Bret-Harte MS, Hobbie SE, Zhong H (1996) Plant functional types as predictors of transient responses of arctic vegetation to global change. J Veg Sci 7(3):347–358

Chapin DM, Bliss L, Bledsoe L (1991) Environmental regulation of nitrogen fixation in a high arctic lowland ecosystem. Can J Bot 69(12):2744–2755

Chapin FS, Shaver GR, Giblin AE, Nadelhoffer KJ, Laundre JA (1995) Responses of arctic tundra to experimental and observed changes in climate. Ecology 76(3):694–711

Chapin FS, Woodwell GM, Randerson JT, Rastetter EB, Lovett GM, Baldocchi DD, Clark DA, Harmon ME, Schimel DS, Valentini R (2006) Reconciling carbon-cycle concepts, terminology, and methods. Ecosystems 9(7):1041–1050

Charman D (2002) Peatlands and environmental change. Wiley, Chichester, 301 p

Cheng W, Gershenson, A. (2007) Carbon fluxes in the rhizosphere. In: Cardon ZG, Whitbeck JL (eds) The Rhizosphere, an ecological perspective. Elsevier Academic Press, Burlington, pp 1–30

Clymo R (1984) The limits to peat bog growth. Philosophical Transactions of the Royal Society of London B: Biological Sciences 303(1117):605–654

Cole JJ, Prairie YT, Caraco NF, McDowell WH, Tranvik LJ, Striegl RG, Duarte CM, Kortelainen P, Downing JA, Middelburg JJ (2007) Plumbing the global carbon cycle: integrating inland waters into the terrestrial carbon budget. Ecosystems 10(1):172–185

Conant RT, Ryan MG, Ågren GI, Birge HE, Davidson EA, Eliasson PE, Evans SE, Frey SD, Giardina CP, Hopkins FM (2011) Temperature and soil organic matter decomposition rates–synthesis of current knowledge and a way forward. Glob Chang Biol 17(11):3392–3404

Corbett JE, Tfaily MM, Burdige DJ, Glaser PH, Chanton JP (2015) The relative importance of methanogenesis in the decomposition of organic matter in northern peatlands. J Geophys Res Biogeo 120(2):280–293

Cornelissen JH, Lang SI, Soudzilovskaia NA, During HJ (2007a) Comparative cryptogam ecology: a review of bryophyte and lichen traits that drive biogeochemistry. Ann Bot 99(5):987–1001

Cornelissen JH, Van Bodegom PM, Aerts R, Callaghan TV, Van Logtestijn RS, Alatalo J, Stuart Chapin F, Gerdol R, Gudmundsson J, Gwynn-Jones D (2007b) Global negative vegetation feedback to climate warming responses of leaf litter decomposition rates in cold biomes. Ecol Lett 10(7):619–627

Corradi C, Kolle O, Walter K, Zimov SA, Schulze ED (2005) Carbon dioxide and methane exchange of a north-east Siberian tussock tundra. Global Change Biol 0(0):051115033519001-??? https://doi.org/10.1111/j.1365-2486.2005.01023.x

Cory RM, Crump BC, Dobkowski JA, Kling GW (2013) Surface exposure to sunlight stimulates CO_2 release from permafrost soil carbon in the Arctic. Proc Natl Acad Sci 110(9):3429–3434

Costard F, Gautier E, Fedorov A, Konstantinov P, Dupeyrat L (2014) An assessment of the erosion potential of the fluvial thermal process during ice breakups of the Lena River (Siberia). Permafr Periglac Process 25(3):162–171

Crawford R, Chapman H, Hodge H (1994) Anoxia tolerance in high Arctic vegetation. Arct Alp Res 26(3):308–312

Davidson EA, Janssens IA (2006) Temperature sensitivity of soil carbon decomposition and feedbacks to climate change. Nature 440:165. https://doi.org/10.1038/nature04514

de Jong AE, in't Zandt MH, Meisel OH, Jetten MS, Dean JF, Rasigraf O, Welte CU (2018) Increases in temperature and nutrient availability positively affect methane-cycling microorganisms in Arctic thermokarst lake sediments. Environ Microbiol 20(12):4314–4327

DeLuca TH, Zackrisson O, Nilsson M-C, Sellstedt A (2002) Quantifying nitrogen-fixation in feather moss carpets of boreal forests. Nature 419(6910):917

Dentener F (2006) Global maps of atmospheric nitrogen deposition, 1860, 1993, and 2050. ORNL DAAC, Oak Ridge. https://doi.org/10.3334/ORNLDAAC/830

Deutzmann JS, Stief P, Brandes J, Schink B (2014) Anaerobic methane oxidation coupled to denitrification is the dominant methane sink in a deep lake. Proc Natl Acad Sci 111(51):18273–18278

Dolman A, Maximov T, Moors E, Maximov A, Elbers J, Kononov A, Waterloo M, Van der Molen M (2004) Net ecosystem exchange of carbon dioxide and water of far eastern Siberian Larch (Larix cajanderii) on permafrost. Biogeosciences 1(2):133–146

Dutta K, Schuur E, Neff J, Zimov S (2006) Potential carbon release from permafrost soils of northeastern Siberia. Glob Chang Biol 12(12):2336–2351

Ehleringer JR, Cerling TE, Helliker BR (1997) C4 photosynthesis, atmospheric CO_2, and climate. Oecologia 112(3):285–299

Elberling B (2007) Annual soil CO_2 effluxes in the high Arctic: the role of snow thickness and vegetation type. Soil Biol Biochem 39(2):646–654

Elumeeva TG, Soudzilovskaia NA, During HJ, Cornelissen JH (2011) The importance of colony structure versus shoot morphology for the water balance of 22 subarctic bryophyte species. J Veg Sci 22(1):152–164

Emmerton CA, Lesack LF, Marsh P (2007) Lake abundance, potential water storage, and habitat distribution in the Mackenzie River Delta, western Canadian Arctic. Water Resour Res 43(5):W05419. https://doi.org/10.1029/2006WR005139

Emmerton CA, St Louis V, Lehnherr I, Humphreys E, Rydz E, Kosolofski HR (2014) The net exchange of methane with high Arctic landscapes during the summer growing season. Biogeosciences 11(12):3095–3106

Ernakovich JG, Lynch LM, Brewer PE, Calderon FJ, Wallenstein MD (2017) Redox and temperature-sensitive changes in microbial communities and soil chemistry dictate greenhouse gas loss from thawed permafrost. Biogeochemistry 134(1–2):183–200

Eugster W, McFadden JP, Stuart Chapin IIIF (2005) Differences in surface roughness, energy, and CO_2 fluxes in two moist tundra vegetation types, Kuparuk watershed, Alaska, USA. Arct Antarct Alp Res 37(1):61–67

Euskirchen E, Bret-Harte MS, Scott G, Edgar C, Shaver GR (2012) Seasonal patterns of carbon dioxide and water fluxes in three representative tundra ecosystems in northern Alaska. Ecosphere 3(1):1–19

Falge E, Baldocchi D, Olson R, Anthoni P, Aubinet M, Bernhofer C, Burba G, Ceulemans R, Clement R, Dolman H (2001) Gap filling strategies for defensible annual sums of net ecosystem exchange. Agric For Meteorol 107(1):43–69

Feller G, Gerday C (2003) Psychrophilic enzymes: hot topics in cold adaptation. Nat Rev Microbiol 1(3):200

Flessa H, Rodionov A, Guggenberger G, Fuchs H, Magdon P, Shibistova O, Zrazhevskaya G, Mikheyeva N, Kasansky OA, Blodau C (2008) Landscape controls of CH_4 fluxes in a catchment of the forest tundra ecotone in northern Siberia. Glob Chang Biol 14(9):2040–2056. https://doi.org/10.1111/j.1365-2486.2008.01633.x

Forbes AC, Lamoureux SF (2005) Climatic controls on streamflow and suspended sediment transport in three large middle arctic catchments, Boothia peninsula, Nunavut, Canada. Arct Antarct Alp Res 37(3):304–315

French H, Shur Y (2010) The principles of cryostratigraphy. Earth Sci Rev 101(3–4):190–206

Frenzel P, Karofeld E (2000) CH_4 emission from a hollow-ridge complex in a raised bog: the role of CH_4 production and oxidation. Biogeochemistry 51(1):91–112

Frolking S, Roulet NT (2007) Holocene radiative forcing impact of northern peatland carbon accumulation and methane emissions. Glob Chang Biol 13(5):1079–1088

Gauci V, Dise N, Fowler D (2002) Controls on suppression of methane flux from a peat bog subjected to simulated acid rain sulfate deposition. Glob Biogeochem Cycles 16(1). https://doi.org/10.1029/2000GB001370

Gauci V, Gowing DJ, Hornibrook ER, Davis JM, Dise NB (2010) Woody stem methane emission in mature wetland alder trees. Atmos Environ 44(17):2157–2160

Gold WG, Bliss LC (1995) Water limitations and plant community development in a polar desert. Ecology 76(5):1558–1568

Gornall J, Jónsdóttir I, Woodin S, Van der Wal R (2007) Arctic mosses govern below-ground environment and ecosystem processes. Oecologia 153(4):931–941

Griffith M, Yaish MW (2004) Antifreeze proteins in overwintering plants: a tale of two activities. Trends Plant Sci 9(8):399–405

Grosse G, Harden J, Turetsky M, McGuire AD, Camill P, Tarnocai C, Frolking S, Schuur EAG, Jorgenson T, Marchenko S, Romanovsky V, Wickland KP, French N, Waldrop M, Bourgeau-Chavez L, Striegl RG (2011) Vulnerability of high-latitude soil organic carbon in North America to disturbance. J Geophys Res 116. https://doi.org/10.1029/2010jg001507

Grosse G, Jones B, Arp C (2013) Thermokarst lakes, drainage, and drained basins. In: Shroder J, Giardino, R, Harbor, J (eds) Glacial and periglacial geomorphology, Treatise on geomorphology, vol 8, pp 325–353. https://doi.org/10.1016/b978-0-12-374739-6.00216-5

Guenther A, Geron C, Pierce T, Lamb B, Harley P, Fall R (2000) Natural emissions of non-methane volatile organic compounds, carbon monoxide, and oxides of nitrogen from North America. Atmos Environ 34(12):2205–2230

Hawkes CV, De Angelis, K.M., Firestone, M.K. (2007) Root interactions with soil microbial communities. In: Cardon ZG, Whitbeck JL. (eds) The rhizosphere, an ecological perspective. Elsevier Academic Press, New York, pp 1–30

He X, He KS, Hyvönen J (2016) Will bryophytes survive in a warming world? Perspect Plant Ecol Evol System 19:49–60

Heijmans MM, Klees H, Berendse F (2002) Competition between Sphagnum magellanicum and *Eriophorum angustifolium* as affected by raised CO_2 and increased N deposition. Oikos 97(3):415–425

Hillel D (1998) Environmental soil physics: fundamentals, applications, and environmental considerations. Elsevier, Burlington, 771 p

Hines ME, Duddleston KN, Kiene RP (2001) Carbon flow to acetate and C1 compounds in northern wetlands. Geophys Res Lett 28(22):4251–4254

Hinkel KM, Eisner WR, Bockheim JG, Nelson FE, Peterson KM, Dai X (2003) Spatial extent, age, and carbon stocks in drained thaw lake basins on the Barrow peninsula, Alaska. Arct Antarct Alp Res 35(3):291–300

Hobara S, McCalley C, Koba K, Giblin AE, Weiss MS, Gettel GM, Shaver GR (2006) Nitrogen fixation in surface soils and vegetation in an Arctic tundra watershed: a key source of atmospheric nitrogen. Arct Antarct Alp Res 38(3):363–372

Hobbie SE (1996) Temperature and plant species control over litter decomposition in Alaskan tundra. Ecol Monogr 66(4):503–522

Hobbie SE, Nadelhoffer KJ, Högberg P (2002) A synthesis: the role of nutrients as constraints on carbon balances in boreal and arctic regions. Plant Soil 242(1):163–170

Holtgrieve GW, Schindler DE, Hobbs WO, Leavitt PR, Ward EJ, Bunting L, Chen G, Finney BP, Gregory-Eaves I, Holmgren S (2011) A coherent signature of anthropogenic nitrogen deposition to remote watersheds of the northern hemisphere. Science 334(6062):1545–1548

Horwath Burnham J, Sletten RS (2010) Spatial distribution of soil organic carbon in Northwest Greenland and underestimates of high Arctic carbon stores. Glob Biogeochem Cycles 24(3):GB3012. https://doi.org/10.1029/2009GB003660

Hugelius G, Kuhry P, Tarnocai C, Virtanen T (2010) Soil organic carbon pools in a periglacial landscape: a case study from the Central Canadian Arctic. Permafr Periglac Process 21(1):16–29

Hugelius G, Bockheim JG, Camill P, Eberling B, Grosse G, Harden JW, Johnson K, Jorgenson T, Koven C, Kuhry P (2013a) A new data set for estimating organic carbon storage to 3 m depth in soils of the northern circumpolar permafrost region. Earth Syst Sci Data 5:393–402

Hugelius G, Tarnocai C, Broll G, Canadell J, Kuhry P, Swanson D (2013b) The northern circumpolar soil carbon database: spatially distributed datasets of soil coverage and soil carbon storage in the northern permafrost regions. Earth Syst Sci Data 5(1):3–13

Hugelius G, Strauss J, Zubrzycki S, Harden JW, Schuur EAG, Ping CL, Schirrmeister L, Grosse G, Michaelson GJ, Koven CD, O'Donnell JA, Elberling B, Mishra U, Camill P, Yu Z, Palmtag J, Kuhry P (2014) Estimated stocks of circumpolar permafrost carbon with quantified uncertainty ranges and identified data gaps. Biogeosciences 11:6573–6593. https://doi.org/10.5194/bg-11-6573-2014

Hultman J, Waldrop MP, Mackelprang R, David MM, McFarland J, Blazewicz SJ, Harden J, Turetsky MR, McGuire AD, Shah MB (2015) Multi-omics of permafrost, active layer and thermokarst bog soil microbiomes. Nature 521(7551):208

Iijima Y, Ohta T, Kotani A, Fedorov AN, Kodama Y, Maximov TC (2014) Sap flow changes in relation to permafrost degradation under increasing precipitation in an eastern Siberian larch forest. Ecohydrology 7(2):177–187

IPCC (2013) Climate change 2013: the physical science basis. Contribution of working group I to the fifth assessment report of the Intergovernmental Panel on Climate Change. Cambridge University Press. https://doi.org/10.1017/CBO9781107415324

Iturrate-Garcia M, O'Brien MJ, Khitun O, Abiven S, Niklaus PA, Schaepman-Strub G (2016) Interactive effects between plant functional types and soil factors on tundra species diversity and community composition. Ecol Evol 6(22):8126–8137

Iversen CM, Sloan VL, Sullivan PF, Euskirchen ES, McGuire AD, Norby RJ, Walker AP, Warren JM, Wullschleger SD (2015) The unseen iceberg: plant roots in arctic tundra. New Phytol 205(1):34–58

Jammet M, Crill P, Dengel S, Friborg T (2015) Large methane emissions from a subarctic lake during spring thaw: mechanisms and landscape significance. J Geophys Res Biogeo 120(11):2289–2305. https://doi.org/10.1002/2015jg003137

Jansson JK, Taş N (2014) The microbial ecology of permafrost. Nat Rev Microbiol 12(6):414

Jespersen RG, Leffler AJ, Oberbauer SF, Welker JM (2018) Arctic plant ecophysiology and water source utilization in response to altered snow: isotopic (δ 18 O and δ 2 H) evidence for meltwater subsidies to deciduous shrubs. Oecologia:1–15

Joabsson A, Christensen TR (2001) Methane emissions from wetlands and their relationship with vascular plants: an Arctic example. Glob Chang Biol 7(8):919–932

Jonasson S, Michelsen A, Schmidt IK (1999) Coupling of nutrient cycling and carbon dynamics in the Arctic, integration of soil microbial and plant processes. Appl Soil Ecol 11(2–3):135–146

Jonasson S, Chapin F, Shaver G (2001) Biogeochemistry in the Arctic: patterns, processes, and controls. In: Global biogeochemical cycles in the climate system. Elsevier, Amsterdam, pp 139–150

Juncher Jørgensen C, Lund Johansen KM, Westergaard-Nielsen A, Elberling B (2014) Net regional methane sink in high Arctic soils of Northeast Greenland. Nat Geosci 8(1):20–23. https://doi.org/10.1038/ngeo2305

Jury WA, Horton R (2004) Soil physics, 6th edn. Wiley, Hoboken

Juszak I, Eugster W, Heijmans MM, Schaepman-Strub G (2016) Contrasting radiation and soil heat fluxes in Arctic shrub and wet sedge tundra. Biogeosciences 13:4049–4064

Kattge J, Diaz S, Lavorel S, Prentice IC, Leadley P, Bönisch G, Garnier E, Westoby M, Reich PB, Wright IJ (2011) TRY–a global database of plant traits. Glob Chang Biol 17(9):2905–2935

Kelker D, Chanton J (1997) The effect of clipping on methane emissions from Carex. Biogeochemistry 39(1):37–44

Keuper F, Parmentier FJ, Blok D, van Bodegom PM, Dorrepaal E, van Hal JR, van Logtestijn RS, Aerts R (2012) Tundra in the rain: differential vegetation responses to three years of experimentally doubled summer precipitation in Siberian shrub and Swedish bog tundra. Ambio 41(Suppl 3):269–280. https://doi.org/10.1007/s13280-012-0305-2

Kip N, van Winden JF, Pan Y, Bodrossy L, Reichart G-J, Smolders AJP, Jetten MSM, Damsté JSS, Op den Camp HJM (2010) Global prevalence of methane oxidation by symbiotic bacteria in peat-moss ecosystems. Nat Geosci 3(9):617–621. https://doi.org/10.1038/ngeo939

Kirschbaum MU (1995) The temperature dependence of soil organic matter decomposition, and the effect of global warming on soil organic C storage. Soil Biol Biochem 27(6):753–760

Kirschbaum MUF (2006) The temperature dependence of organic-matter decomposition—still a topic of debate. Soil Biol Biochem 38(9):2510–2518

Kittler F, Heimann M, Kolle O, Zimov N, Zimov S, Göckede M (2017) Long-term drainage reduces CO2 uptake and CH4 emissions in a Siberian permafrost ecosystem. Glob Biogeochem Cycles 31:1704–1717. https://doi.org/10.1002/2017GB005774

Kling GW, Kipphut GW, Miller MC (1991) Arctic lakes and streams as gas conduits to the atmosphere: implications for tundra carbon budgets. Science 251(4991):298–301

Knoblauch C, Beer C, Sosnin A, Wagner D, Pfeiffer EM (2013) Predicting long-term carbon mineralization and trace gas production from thawing permafrost of N ortheast S iberia. Glob Chang Biol 19(4):1160–1172

Knoblauch C, Spott O, Evgrafova S, Kutzbach L, Pfeiffer EM (2015) Regulation of methane production, oxidation, and emission by vascular plants and bryophytes in ponds of the northeast Siberian polygonal tundra. J Geophys Res Biogeo 120(12):2525–2541

Knoblauch C, Beer C, Liebner S, Grigoriev MN, Pfeiffer E-M (2018) Methane production as key to the greenhouse gas budget of thawing permafrost. Nat Clim Chang 8(4):309

Kobabe S, Wagner D, Pfeiffer EM (2004) Characterisation of microbial community composition of a Siberian tundra soil by fluorescence in situ hybridisation. FEMS Microbiol Ecol 50(1):13–23. https://doi.org/10.1016/j.femsec.2004.05.003

Kotani A, Kononov AV, Ohta T, Maximov TC (2014) Temporal variations in the linkage between the net ecosystem exchange of water vapour and CO_2 over boreal forests in eastern Siberia. Ecohydrology 7(2):209–225

Koven C, Hugelius G, Lawrence DM, Wieder WR (2017) Higher climatological temperature sensitivity of soil carbon in cold than warm climates. Nat Clim Chang 7:817–824. https://doi.org/10.1038/NCLIMATE3421

Kuhry P, Turunen J (2006) The postglacial development of boreal and subarctic peatlands. In: Wieder R, Vitt DH (eds) Boreal peatland ecosystems. Springer, Heidelberg, pp 25–46

Kuhry P, Vitt DH (1996) Fossil carbon/nitrogen ratios as a measure of peat decomposition. Ecology 77(1):271–275

Kutzbach L, Wagner D, Pfeiffer E-M (2004) Effect of microrelief and vegetation on methane emission from wet polygonal tundra, Lena Delta, northern Siberia. Biogeochemistry 69(3):341–362

Kutzbach L, Wille C, Pfeiffer E-M (2007a) The exchange of carbon dioxide between wet arctic tundra and the atmosphere at the Lena River Delta, Northern Siberia. Biogeosciences 4(5):869–890

Kutzbach L, Schneider J, Sachs T, Giebels M, Nykänen H, Shurpali N, Martikainen P, Alm J, Wilmking M (2007b) CO_2 flux determination by closed-chamber methods can be seriously biased by inappropriate application of linear regression. Biogeosciences 4(6):1005–1025

Kuzyakov Y (2010) Priming effects: interactions between living and dead organic matter. Soil Biol Biochem 42(9):1363–1371

Kwon HJ, Oechel WC, Zulueta RC, Hastings SJ (2006) Effects of climate variability on carbon sequestration among adjacent wet sedge tundra and moist tussock tundra ecosystems. J Geophys Res Biogeo 111(G3):G03014. https://doi.org/10.1029/2005JG000036

Kwon MJ, Beulig F, Ilie I, Wildner M, Kusel K, Merbold L, Mahecha MD, Zimov N, Zimov SA, Heimann M, Schuur EAG, Kostka JE, Kolle O, Hilke I, Gockede M (2017) Plants, microorganisms, and soil temperatures contribute to a decrease in methane fluxes on a drained Arctic floodplain. Glob Chang Biol 23(6):2396–2412. https://doi.org/10.1111/gcb.13558

Lafleur PM, Humphreys ER, St. Louis VL, Myklebust MC, Papakyriakou T, Poissant L, Barker JD, Pilote M, Swystun KA (2012) Variation in peak growing season net ecosystem production across the Canadian Arctic. Environ Sci Technol 46(15):7971–7977

Larmola T, Tuittila E-S, Tiirola M, Nykänen H, Martikainen PJ, Yrjälä K, Tuomivirta T, Fritze H (2010) The role of Sphagnum mosses in the methane cycling of a boreal mire. Ecology 91(8):2356–2365

Larmola T, Leppänen SM, Tuittila E-S, Aarva M, Merilä P, Fritze H, Tiirola M (2014) Methanotrophy induces nitrogen fixation during peatland development. Proc Natl Acad Sci 111(2):734–739

Lasslop G, Reichstein M, Papale D, Richardson AD, Arneth A, Barr A, Stoy P, Wohlfahrt G (2010) Separation of net ecosystem exchange into assimilation and respiration using a light response curve approach: critical issues and global evaluation. Glob Chang Biol 16(1):187–208

Laurila T, Soegaard H, Lloyd C, Aurela M, Tuovinen J-P, Nordstroem C (2001) Seasonal variations of net CO 2 exchange in European Arctic ecosystems. Theor Appl Climatol 70(1–4):183–201

Li B, Heijmans MM, Blok D, Wang P, Karsanaev SV, Maximov TC, van Huissteden J, Berendse F (2017) Thaw pond development and initial vegetation succession in experimental plots at a Siberian lowland tundra site. Plant Soil 420(1–2):147–162

Liang M, Sugimoto A, Tei S, Bragin IV, Takano S, Morozumi T, Shingubara R, Maximov TC, Kiyashko SI, Velivetskaya TA (2014) Importance of soil moisture and N availability to larch growth and distribution in the Arctic taiga-tundra boundary ecosystem, northeastern Siberia. Pol Sci 8(4):327–341

Lipson DA, Jha M, Raab TK, Oechel WC (2010) Reduction of iron (III) and humic substances plays a major role in anaerobic respiration in an Arctic peat soil. J Geophys Res Biogeosci 115(G4):G00I06. https://doi.org/10.1029/2009JG001147

Lund M, Falk JM, Friborg T, Mbufong HN, Sigsgaard C, Soegaard H, Tamstorf MP (2012) Trends in CO_2 exchange in a high Arctic tundra heath, 2000–2010. J Geophys Res Biogeo 117(G2):G02001. https://doi.org/10.1029/2011JG001901

Lupascu M, Wadham J, Hornibrook E, Pancost R (2012) Temperature sensitivity of methane production in the permafrost active layer at Stordalen, Sweden: a comparison with non-permafrost northern wetlands. Arct Antarct Alp Res 44(4):469–482

MacDonald GM, Beilman DW, Kremenetski KV, Sheng Y, Smith LC, Velichko AA (2006) Rapid early development of circumarctic peatlands and atmospheric CH_4 and CO_2 variations. Science 314(5797):285–288

Mackelprang R, Waldrop MP, DeAngelis KM, David MM, Chavarria KL, Blazewicz SJ, Rubin EM, Jansson JK (2011) Metagenomic analysis of a permafrost microbial community reveals a rapid response to thaw. Nature 480(7377):368

Marsh P, Lesack L, Roberts A (1999) Lake sedimentation in the Mackenzie Delta, NWT. Hydrol Process 13(16):2519–2536

Marushchak M, Pitkämäki A, Koponen H, Biasi C, Seppälä M, Martikainen P (2011) Hot spots for nitrous oxide emissions found in different types of permafrost peatlands. Glob Chang Biol 17(8):2601–2614

Marushchak M, Kiepe I, Biasi C, Elsakov V, Friborg T, Johansson T, Soegaard H, Virtanen T, Martikainen P (2013) Carbon dioxide balance of subarctic tundra from plot to regional scales. Biogeosciences 10:437–452. https://doi.org/10.5194/bg-10-437-2013

Marushchak M, Friborg T, Biasi C, Herbst M, Johansson T, Kiepe I, Liimatainen M, Lind S, Martikainen P, Virtanen T (2016) Methane dynamics in the subarctic tundra: combining stable isotope analyses, plot-and ecosystem-scale flux measurements. Biogeosciences 13(2):597–608

Mastepanov M, Sigsgaard C, Dlugokencky EJ, Houweling S, Strom L, Tamstorf MP, Christensen TR (2008) Large tundra methane burst during onset of freezing. Nature 456(7222):628–630. https://doi.org/10.1038/nature07464

Mastepanov M, Sigsgaard C, Tagesson T, Ström L, Tamstorf MP, Lund M, Christensen T (2013) Revisiting factors controlling methane emissions from high-Arctic tundra. Biogeosciences 10(7):5139–5158. https://doi.org/10.5194/bg-10-5139-2013

Mbufong HN, Lund M, Aurela M, Christensen TR, Eugster W, Friborg T, Hansen BU, Humphreys ER, Jackowicz-Korczynski M, Kutzbach L, Lafleur PM, Oechel WC, Parmentier FJW, Rasse DP, Rocha AV, Sachs T, van der Molen MK, Tamstorf MP (2014) Assessing the spatial variability in peak season CO_2 exchange characteristics across the Arctic tundra using a light response curve parameterization. Biogeosciences 11(17):4897–4912. https://doi.org/10.5194/bg-11-4897-2014

McFadden JP, Eugster W, Chapin FS (2003) A regional study of the controls on water vapor and CO_2 exchange in arctic tundra. Ecology 84(10):2762–2776

McGuire A, Wirth C, Apps M, Beringer J, Clein J, Epstein H, Kicklighter D, Bhatti J, Chapin Iii F, De Groot B (2002) Environmental variation, vegetation distribution, carbon dynamics and water/energy exchange at high latitudes. J Veg Sci 13(3):301–314

Metje M, Frenzel P (2007) Methanogenesis and methanogenic pathways in a peat from subarctic permafrost. Environ Microbiol 9(4):954–964. https://doi.org/10.1111/j.1462-2920.2006.01217.x

Mi Y, van Huissteden J, Parmentier FJW, Gallagher A, Budishchev A, Berridge CT, Dolman AJ (2014) Improving a plot-scale methane emission model and its performance at a northeastern Siberian tundra site. Biogeosciences 11(14):3985–3999. https://doi.org/10.5194/bg-11-3985-2014

Michaelson GJ, Ping C, Kimble J (1996) Carbon storage and distribution in tundra soils of Arctic Alaska, USA. Arct Alp Res 28:414–424

Mishra U, Jastrow J, Matamala R, Hugelius G, Koven C, Harden JW, Ping C, Michaelson G, Fan Z, Miller R (2013) Empirical estimates to reduce modeling uncertainties of soil organic carbon in permafrost regions: a review of recent progress and remaining challenges. Environ Res Lett 8(3):035020

Murton JB, Goslar T, Edwards ME, Bateman MD, Danilov PP, Savvinov GN, Gubin SV, Ghaleb B, Haile J, Kanevskiy M (2015) Palaeoenvironmental interpretation of Yedoma silt (ice complex) deposition as cold-climate loess, Duvanny Yar, Northeast Siberia. Permafr Periglac Process 26(3):208–288

Murton JB, Edwards ME, Lozhkin AV, Anderson PM, Savvinov GN, Bakulina N, Bondarenko OV, Cherepanova MV, Danilov PP, Boeskorov V (2017) Preliminary paleoenvironmental analysis of permafrost deposits at Batagaika megaslump, Yana uplands, Northeast Siberia. Quat Res 87(2):314–330

Myneni RB, Hall FG, Sellers PJ, Marshak AL (1995) The interpretation of spectral vegetation indexes. IEEE Trans Geosci Remote Sens 33(2):481–486

Nadelhoffer KJ, Giblin A, Shaver GR, Linkins A (1991) Microbial processes and plant nutrient availability in arctic soils. In: Arctic ecosystems in a changing climate. Elsevier, St. Louis, pp 281–300

Nakai Y, Matsuura Y, Kajimoto T, Abaimov A, Yamamoto S, Zyryanova O (2008) Eddy covariance CO_2 flux above a Gmelin larch forest on continuous permafrost in Central Siberia during a growing season. Theor Appl Climatol 93(3–4):133–147

Natali SM, Schuur EAG, Mauritz M, Schade JD, Celis G, Crummer KG, Johnston C, Krapek J, Pegoraro E, Salmon VG, Webb EE (2015) Permafrost thaw and soil moisture driving CO_2 and CH_4 release from upland tundra. J Geophys Res Biogeo 120(3):525–537. https://doi.org/10.1002/2014jg002872

Nauta AL, Heijmans MMPD, Blok D, Limpens J, Elberling B, Gallagher A, Li B, Petrov RE, Maximov TC, van Huissteden J, Berendse F (2014) Permafrost collapse after shrub removal shifts tundra ecosystem to a methane source. Nat Clim Chang 5(1):67–70. https://doi.org/10.1038/nclimate2446

Neubauer SC, Megonigal JP (2015) Moving beyond global warming potentials to quantify the climatic role of ecosystems. Ecosystems 18(6):1000–1013

Nishiuchi S, Yamauchi T, Takahashi H, Kotula L, Nakazono M (2012) Mechanisms for coping with submergence and waterlogging in rice. Rice 5(1):2

Nobrega S, Grogan P (2008) Landscape and ecosystem-level controls on net carbon dioxide exchange along a natural moisture gradient in Canadian low arctic tundra. Ecosystems 11(3):377–396

Oechel WC, Laskowski CA, Burba G, Gioli B, Kalhori AA (2014) Annual patterns and budget of CO_2 flux in an Arctic tussock tundra ecosystem. J Geophys Res Biogeo 119(3):323–339

Ohta T, Maximov TC, Dolman AJ, Nakai T, van der Molen MK, Kononov AV, Maximov AP, Hiyama T, Iijima Y, Moors EJ (2008) Interannual variation of water balance and summer evapotranspiration in an eastern Siberian larch forest over a 7-year period (1998–2006). Agric For Meteorol 148(12):1941–1953

Olefeldt D, Roulet NT, Bergeron O, Crill P, Bäckstrand K, Christensen TR (2012) Net carbon accumulation of a high-latitude permafrost palsa mire similar to permafrost-free peatlands. Geophys Res Lett 39(3):L03501. https://doi.org/10.1029/2011GL050355

Olefeldt D, Turetsky MR, Crill PM, McGuire AD (2013) Environmental and physical controls on northern terrestrial methane emissions across permafrost zones. Glob Chang Biol 19(2):589–603

Palmtag J, Hugelius G, Lashchinskiy N, Tamstorf MP, Richter A, Elberling B, Kuhry P (2015) Storage, landscape distribution, and burial history of soil organic matter in contrasting areas of continuous permafrost. Arct Antarct Alp Res 47(1):71–88

Panikov NS, Dedysh S (2000) Cold season CH_4 and CO_2 emission from boreal peat bogs (West Siberia): winter fluxes and thaw activation dynamics. Glob Biogeochem Cycles 14(4):1071–1080

Parmentier F, Van Der Molen M, Van Huissteden J, Karsanaev S, Kononov A, Suzdalov D, Maximov T, Dolman A (2011a) Longer growing seasons do not increase net carbon uptake in the northeastern Siberian tundra. J Geophys Res Biogeo 116(G4):G04013. https://doi.org/10.1029/2011JG001653

Parmentier FJW, van Huissteden J, Kip N, Op den Camp HJM, Jetten MSM, Maximov TC, Dolman AJ (2011b) The role of endophytic methane-oxidizing bacteria in submerged *Sphagnum* in determining methane emissions of Northeastern Siberian tundra. Biogeosciences 8(5):1267–1278. https://doi.org/10.5194/bg-8-1267-2011

Parmentier FJW, van Huissteden J, van der Molen MK, Schaepman-Strub G, Karsanaev SA, Maximov TC, Dolman AJ (2011c) Spatial and temporal dynamics in eddy covariance observations of methane fluxes at a tundra site in northeastern Siberia. J Geophys Res 116(G3). https://doi.org/10.1029/2010jg001637

Parton WJ, Del Grosso SJ, Plante AF, Adair EC, Lutz SM (2014) Modeling the dynamics of soil organic matter and nutrient cycling. In: Paul EA (ed) Soil microbiology, ecology and biochemistry. Academic, London

Paul EA (2014) Soil microbiology, ecology and biochemistry. Academic

Paustian K, Lehmann J, Ogle S, Reay D, Robertson GP, Smith P (2016) Climate-smart soils. Nature 532(7597):49

Paytan A, Lecher AL, Dimova N, Sparrow KJ, Kodovska FG-T, Murray J, Tulaczyk S, Kessler JD (2015) Methane transport from the active layer to lakes in the Arctic using Toolik Lake, Alaska, as a case study. Proc Natl Acad Sci 112(12):3636–3640

Phoenix GK, Hicks WK, Cinderby S, Kuylenstierna JC, Stock WD, Dentener FJ, Giller KE, Austin AT, Lefroy RD, Gimeno BS (2006) Atmospheric nitrogen deposition in world biodiversity hotspots: the need for a greater global perspective in assessing N deposition impacts. Glob Chang Biol 12(3):470–476

Pop EW, Oberbauer SF, Starr G (2000) Predicting vegetative bud break in two arctic deciduous shrub species, Salix pulchra and Betula nana. Oecologia 124(2):176–184

Post WM, Emanuel WR, Zinke PJ, Stangenberger AG (1982) Soil carbon pools and world life zones. Nature 298(5870):156

Price PB, Sowers T (2004) Temperature dependence of metabolic rates for microbial growth, maintenance, and survival. Proc Natl Acad Sci 101(13):4631–4636

Raghoebarsing AA, Smolders AJ, Schmid MC, Rijpstra WIC, Wolters-Arts M, Derksen J, Jetten MS, Schouten S, Damsté JSS, Lamers LP (2005) Methanotrophic symbionts provide carbon for photosynthesis in peat bogs. Nature 436(7054):1153

Repo E, Huttunen J, Naumov A, Chichulin A, Lapshina E, Bleuten W, Martikainen P (2007) Release of CO_2 and CH_4 from small wetland lakes in western Siberia. Tellus B Chem Phys Meteorol 59(5):788–796

Repo ME, Susiluoto S, Lind SE, Jokinen S, Elsakov V, Biasi C, Virtanen T, Martikainen PJ (2009) Large N_2O emissions from cryoturbated peat soil in tundra. Nat Geosci 2(3):189

Rivkina E, Gilichinsky D, Wagener S, Tiedje J, McGrath J (1998) Biogeochemical activity of anaerobic microorganisms from buried permafrost sediments. Geomicrobiol J 15(3):187–193

Rivkina E, Friedmann E, McKay C, Gilichinsky D (2000) Metabolic activity of permafrost bacteria below the freezing point. Appl Environ Microbiol 66(8):3230–3233

Rivkina E, Laurinavichius K, McGrath J, Tiedje J, Shcherbakova V, Gilichinsky D (2004) Microbial life in permafrost. Adv Space Res 33(8):1215–1221

Rivkina E, Shcherbakova V, Laurinavichius K, Petrovskaya L, Krivushin K, Kraev G, Pecheritsina S, Gilichinsky D (2007) Biogeochemistry of methane and methanogenic archaea in permafrost. FEMS Microbiol Ecol 61(1):1–15. https://doi.org/10.1111/j.1574-6941.2007.00315.x

Robinson C, Wookey P, Parsons A, Potter J, Callaghan T, Lee J, Press M, Welker J (1995) Responses of plant litter decomposition and nitrogen mineralisation to simulated environmental change in a high arctic polar semi-desert and a subarctic dwarf shrub heath. Oikos 74:503–512

Romanovsky V, Osterkamp T (2000) Effects of unfrozen water on heat and mass transport processes in the active layer and permafrost. Permafr Periglac Process 11(3):219–239

Roulet NT, Lafleur PM, Richard PJ, Moore TR, Humphreys ER, Bubier J (2007) Contemporary carbon balance and late Holocene carbon accumulation in a northern peatland. Glob Chang Biol 13(2):397–411

Runkle B, Sachs T, Wille C, Pfeiffer E-M, Kutzbach L (2013) Bulk partitioning the growing season net ecosystem exchange of CO_2 in Siberian tundra reveals the seasonality of its carbon sequestration strength. Biogeosciences 10(3):1337–1349

Sachs T, Wille C, Boike J, Kutzbach L (2008) Environmental controls on ecosystem-scale CH_4 emission from polygonal tundra in the Lena River Delta. Siberia Journal of Geophysical Research 113. https://doi.org/10.1029/2007jg000505

Saito H, Kobayashi Y, Shirota T, Iwahana G, Maximov T, Fukuda M (2007) Interannual environmental-soil thawing rate variation and its control on transpiration from Larix cajanderi, Central Yakutia, Eastern Siberia. J Hydrol 338(3–4):251–260

Sander R (2015) Compilation of Henry's law constants (version 4.0) for water as solvent. Atmos Chem Phys 15(8):4399–4981

Schädel C, Schuur EA, Bracho R, Elberling B, Knoblauch C, Lee H, Luo Y, Shaver GR, Turetsky MR (2014) Circumpolar assessment of permafrost C quality and its vulnerability over time using long-term incubation data. Glob Chang Biol 20(2):641–652

Schädel C, Bader MK-F, Schuur EA, Biasi C, Bracho R, Čapek P, De Baets S, Diáková K, Ernakovich J, Estop-Aragones C (2016) Potential carbon emissions dominated by carbon dioxide from thawed permafrost soils. Nat Clim Chang 6(10):950

Scheiter S, Langan L, Higgins SI (2013) Next-generation dynamic global vegetation models: learning from community ecology. New Phytol 198(3):957–969

Schirrmeister L, Kunitsky V, Grosse G, Wetterich S, Meyer H, Schwamborn G, Babiy O, Derevyagin A, Siegert C (2011a) Sedimentary characteristics and origin of the late Pleistocene ice complex on north-east Siberian Arctic coastal lowlands and islands – a review. Quat Int 241(1–2):3–25. https://doi.org/10.1016/j.quaint.2010.04.004

Schirrmeister L, Grosse G, Wetterich S, Overduin PP, Strauss J, Schuur EAG, Hubberten H-W (2011b) Fossil organic matter characteristics in permafrost deposits of the northeast Siberian Arctic. J Geophys Res 116. https://doi.org/10.1029/2011jg001647

Schirrmeister L, Froese D, Tumskoy V, Grosse G, Wetterich S (2013) Permafrost and Periglacial features | Yedoma: late Pleistocene ice-rich Syngenetic permafrost of Beringia. In: Elias S (ed) Encyclopedia of the quaternary, vol 3. Elsevier, Amsterdam, pp 542–552. https://doi.org/10.1016/b978-0-444-53643-3.00106-0

Schlesinger WH (1997) Biogeochemistry. Academic, San Diego

Schmidt MW, Torn MS, Abiven S, Dittmar T, Guggenberger G, Janssens IA, Kleber M, Kögel-Knabner I, Lehmann J, Manning DA (2011) Persistence of soil organic matter as an ecosystem property. Nature 478(7367):49

Schubert CJ, Vazquez F, Lösekann-Behrens T, Knittel K, Tonolla M, Boetius A (2011) Evidence for anaerobic oxidation of methane in sediments of a freshwater system (Lago di Cadagno). FEMS Microbiol Ecol 76(1):26–38

Schultz M, Clevers JG, Carter S, Verbesselt J, Avitabile V, Quang HV, Herold M (2016) Performance of vegetation indices from Landsat time series in deforestation monitoring. Int J Appl Earth Obs Geoinf 52:318–327

Schumm SA (1985) Patterns of alluvial rivers. Annu Rev Earth Planet Sci 13(1):5–27

Schuur EA, McGuire AD, Schadel C, Grosse G, Harden JW, Hayes DJ, Hugelius G, Koven CD, Kuhry P, Lawrence DM, Natali SM, Olefeldt D, Romanovsky VE, Schaefer K, Turetsky MR, Treat CC, Vonk JE (2015) Climate change and the permafrost carbon feedback. Nature 520(7546):171–179. https://doi.org/10.1038/nature14338

Segers R (1998) Methane production and methane consumption: a review of processes underlying wetland methane fluxes. Biogeochemistry 41(1):23–51

Seppälä M (2003) Surface abrasion of palsas by wind action in Finnish Lapland. Geomorphology 52(1–2):141–148

Shaver G, Billings W (1975) Root production and root turnover in a wet tundra ecosystem, Barrow, Alaska. Ecology 56(2):401–409

Shaver G, Cutler J (1979) The vertical distribution of live vascular phytomass in cottongrass tussock tundra. Arct Alp Res 11(3):335–342

Shaver G, Street L, Rastetter E, Van Wijk M, Williams M (2007) Functional convergence in regulation of net CO_2 flux in heterogeneous tundra landscapes in Alaska and Sweden. J Ecol 95(4):802–817

Shaver GR, Rastetter EB, Salmon V, Street LE, van de Weg MJ, Rocha A, van Wijk MT, Williams M (2013) Pan-Arctic modelling of net ecosystem exchange of CO_2. Philos Trans R Soc Lond B Biol Sci 368(1624):20120485

Shur Y, Jorgenson M (2007) Patterns of permafrost formation and degradation in relation to climate and ecosystems. Permafr Periglac Process 18(1):7–19

Siewert MB, Hanisch J, Weiss N, Kuhry P, Maximov TC, Hugelius G (2015) Comparing carbon storage of Siberian tundra and taiga permafrost ecosystems at very high spatial resolution. J Geophys Res Biogeo 120(10):1973–1994

Silva JP, Lasso A, Lubberding HJ, Peña MR, Gijzen HJ (2015) Biases in greenhouse gases static chambers measurements in stabilization ponds: comparison of flux estimation using linear and non-linear models. Atmos Environ 109:130–138

Sivan O, Adler M, Pearson A, Gelman F, Bar-Or I, John SG, Eckert W (2011) Geochemical evidence for iron-mediated anaerobic oxidation of methane. Limnol Oceanogr 56(4):1536–1544

Sjögersten S, Caul S, Daniell T, Jurd A, O'Sullivan O, Stapleton C, Titman JJ (2016) Organic matter chemistry controls greenhouse gas emissions from permafrost peatlands. Soil Biol Biochem 98:42–53

Smemo K, Yavitt J (2011) Anaerobic oxidation of methane: an underappreciated aspect of methane cycling in peatland ecosystems? Biogeosciences 8(3):779–793

Sørensen LI, Holmstrup M, Maraldo K, Christensen S, Christensen B (2006) Soil fauna communities and microbial respiration in high Arctic tundra soils at Zackenberg, Northeast Greenland. Polar Biol 29(3):189–195

Squires MM, Lesack LF, Hecky RE, Guildford SJ, Ramlal P, Higgins SN (2009) Primary production and carbon dioxide metabolic balance of a lake-rich arctic river floodplain: partitioning of phytoplankton, epipelon, macrophyte, and epiphyton production among lakes on the Mackenzie Delta. Ecosystems 12(5):853–872

Stamp I, Baird AJ, Heppell CM (2013) The importance of ebullition as a mechanism of methane (CH_4) loss to the atmosphere in a northern peatland. Geophys Res Lett 40(10):2087–2090

Starr G, Oberbauer SF (2003) Photosynthesis of arctic evergreens under snow: implications for tundra ecosystem carbon balance. Ecology 84(6):1415–1420

Still CJ, Berry JA, Collatz GJ, DeFries RS (2003) Global distribution of C3 and C4 vegetation: carbon cycle implications. Glob Biogeochem Cycles 17(1)

Strack M, Kellner E, Waddington J (2005) Dynamics of biogenic gas bubbles in peat and their effects on peatland biogeochemistry. Glob Biogeochem Cycles 19(1):GB1003. https://doi.org/10.1029/2004GB002330

Strauss J, Schirrmeister L, Grosse G, Wetterich S, Ulrich M, Herzschuh U, Hubberten HW (2013) The deep permafrost carbon pool of the Yedoma region in Siberia and Alaska. Geophys Res Lett 40(23):6165–6170. https://doi.org/10.1002/2013GL058088

Strauss J, Schirrmeister L, Grosse G, Fortier D, Hugelius G, Knoblauch C, Romanovsky V, Schädel C, von Deimling TS, Schuur EA (2017) Deep Yedoma permafrost: a synthesis of depositional characteristics and carbon vulnerability. Earth Sci Rev 172:75–86

Ström L, Ekberg A, Mastepanov M, Røjle Christensen T (2003) The effect of vascular plants on carbon turnover and methane emissions from a tundra wetland. Glob Chang Biol 9(8):1185–1192

Ström L, Mastepanov M, Christensen TR (2005) Species-specific effects of vascular plants on carbon turnover and methane emissions from wetlands. Biogeochemistry 75(1):65–82

Ström L, Falk JM, Skov K, Jackowicz-Korczynski M, Mastepanov M, Christensen TR, Lund M, Schmidt NM (2015) Controls of spatial and temporal variability in CH_4 flux in a high arctic fen over three years. Biogeochemistry 125(1):21–35

Sugimoto A, Wada E (1993) Carbon isotopic composition of bacterial methane in a soil incubation experiment: contributions of acetate and CO_2H_2. Geochim Cosmochim Acta 57(16):4015–4027

Sugimoto A, Yanagisawa N, Naito D, Fujita N, Maximov TC (2002) Importance of permafrost as a source of water for plants in east Siberian taiga. Ecol Res 17(4):493–503

Tagesson T, Mölder M, Mastepanov M, Sigsgaard C, Tamstorf MP, Lund M, Falk JM, Lindroth A, Christensen TR, Ström L (2012) Land-atmosphere exchange of methane from soil thawing to soil freezing in a high-a rctic wet tundra ecosystem. Glob Chang Biol 18(6):1928–1940

Tananaev NI (2016) Hydrological and sedimentary controls over fluvial thermal erosion, the Lena River, central Yakutia. Geomorphology 253:524–533

Tarnocai C, Kimble J, Broll G (2003) Determining carbon stocks in cryosols using the northern and mid latitudes soil database. Swets and Zeitlinger, Lisse, pp 1129–1134

Tarnocai C, Canadell JG, Schuur EAG, Kuhry P, Mazhitova G, Zimov S (2009) Soil organic carbon pools in the northern circumpolar permafrost region. Glob Biogeochem Cycles 23(2):GB2023. https://doi.org/10.1029/2008gb003327

Teltewskoi A, Beermann F, Beil I, Bobrov A, De Klerk P, Lorenz S, Lüder A, Michaelis D, Joosten H (2016) 4000 years of changing wetness in a permafrost polygon Peatland (Kytalyk, NE Siberia): a comparative high-resolution multi-proxy study. Permafr Periglac Process 27(1):76–95

Thonat T, Saunois M, Bousquet P, Pison I, Tan Z, Zhuang Q, Crill PM, Thornton BF, Bastviken D, Dlugokencky EJ (2017) Detectability of Arctic methane sources at six sites performing continuous atmospheric measurements. Atmos Chem Phys 17(13):8371–8394

Tranvik LJ, Downing JA, Cotner JB, Loiselle SA, Striegl RG, Ballatore TJ, Dillon P, Finlay K, Fortino K, Knoll LB (2009) Lakes and reservoirs as regulators of carbon cycling and climate. Limnol Oceanogr 54:2298–2314. https://doi.org/10.4319/lo.2009.54.6_part_2.2298

Treat C, Wollheim WM, Varner R, Grandy AS, Talbot J, Frolking S (2014) Temperature and peat type control CO_2 and CH_4 production in Alaskan permafrost peats. Glob Chang Biol 20(8):2674–2686

Treat CC, Natali SM, Ernakovich J, Iversen CM, Lupascu M, McGuire AD, Norby RJ, Roy Chowdhury T, Richter A, Šantrůčková H (2015) A pan-Arctic synthesis of CH_4 and CO_2 production from anoxic soil incubations. Glob Chang Biol 21(7):2787–2803

Trotsenko YA, Khmelenina VN (2005) Aerobic methanotrophic bacteria of cold ecosystems. FEMS Microbiol Ecol 53(1):15–26

Trucco C, Schuur EAG, Natali SM, Belshe EF, Bracho R, Vogel J (2012) Seven-year trends of CO_2 exchange in a tundra ecosystem affected by long-term permafrost thaw. J Geophys Res Biogeo 117(G2):G02031. https://doi.org/10.1029/2011jg001907

Trumbore S (2009) Radiocarbon and soil carbon dynamics. Annu Rev Earth Planet Sci 37:47–66

Turetsky M, Wieder R, Vitt D, Evans R, Scott K (2007) The disappearance of relict permafrost in boreal North America: effects on peatland carbon storage and fluxes. Glob Chang Biol 13(9):1922–1934

Turetsky M, Treat C, Waldrop M, Waddington J, Harden J, McGuire A (2008) Short-term response of methane fluxes and methanogen activity to water table and soil warming manipulations in an Alaskan peatland: J Geophys Res Biogeosci 113(G3):G00A10. https://doi.org/10.1029/2007JG000496

Turetsky MR, Bond-Lamberty B, Euskirchen E, Talbot J, Frolking S, McGuire AD, Tuittila ES (2012) The resilience and functional role of moss in boreal and arctic ecosystems. New Phytol 196(1):49–67

Van Bodegom PM, Douma JC, Verheijen LM (2014) A fully traits-based approach to modeling global vegetation distribution. Proc Natl Acad Sci 111(38):13733–13738

Van Breukelen BM, Griffioen J (2004) Biogeochemical processes at the fringe of a landfill leachate pollution plume: potential for dissolved organic carbon, Fe (II), Mn (II), NH_4, and CH_4 oxidation. J Contam Hydrol 73(1–4):181–205

Van der Kolk H-J, Heijmans MMPD, van Huissteden J, Pullens JWM, Berendse F (2016) Potential Arctic tundra vegetation shifts in response to changing temperature, precipitation and permafrost thaw. Biogeosciences 13(22):6229–6245. https://doi.org/10.5194/bg-13-6229-2016

Van der Molen M, Van Huissteden J, Parmentier F, Petrescu A, Dolman A, Maximov T, Kononov A, Karsanaev S, Suzdalov D (2007) The growing season greenhouse gas balance of a continental tundra site in the Indigirka lowlands, NE Siberia. Biogeosciences 4(6):985–1003

Van Hardenbroek M, Lotter AF, Bastviken D, Duc N, Heiri O (2012) Relationship between δ13C of chironomid remains and methane flux in Swedish lakes. Freshw Biol 57(1):166–177

Van Huissteden J (1990) Tundra rivers of the last glacial: sedimentation and geomorphological processes during the Middle Pleniglacial in Twente, Eastern Netherlands. In: Van Amerom H (ed) Mededelingen Rijks Geologische Dienst, vol 44-3. Rijks Geologische Dienst, pp 1–138

Van Huissteden J (2019) Methane and BVOC. In: Ohta T (ed) Water-carbon dynamics in Eastern Siberia. Springer, Singapore, pp 101–134

Van Huissteden J, Maximov TC, Dolman AJ (2005) High methane flux from an arctic floodplain (Indigirka lowlands, eastern Siberia). J Geophys Res Biogeo 110(G2). https://doi.org/10.1029/2005jg000010

Van Huissteden J, Maximov TC, Kononov AV, Dolman AJ (2008) Summer soil CH_4 emission and uptake in taiga forest near Yakutsk, Eastern Siberia. Agric For Meteorol 148(12):2006–2012. https://doi.org/10.1016/j.agrformet.2008.08.008

Van Huissteden J, Maximov TC, Dolman AJ (2009) Correction to "high methane flux from an arctic floodplain (Indigirka lowlands, eastern Siberia)". J Geophys Res Biogeo 114(G2):G02002. https://doi.org/10.1029/2009jg001040

Van Huissteden J, Vandenberghe J, Gibbard PL, Lewin J (2013) Permafrost and Periglacial features | Periglacial fluvial sediments and forms. In: Elias SA (ed) Encyclopedia of the quaternary, vol 3. Elsevier, Amsterdam, pp 490–499. https://doi.org/10.1016/b978-0-444-53643-3.00108-4

Vandenberghe J (2002) The relation between climate and river processes, landforms and deposits during the quaternary. Quat Int 91(1):17–23

Vaughn LJ, Conrad ME, Bill M, Torn MS (2016) Isotopic insights into methane production, oxidation, and emissions in Arctic polygon tundra. Glob Chang Biol 22(10):3487–3502. https://doi.org/10.1111/gcb.13281

Verhoeven J, Toth E (1995) Decomposition of Carex and Sphagnum litter in fens: effect of litter quality and inhibition by living tissue homogenates. Soil Biol Biochem 27(3):271–275

Vile MA, Wieder RK, Živković T, Scott KD, Vitt DH, Hartsock JA, Iosue CL, Quinn JC, Petix M, Fillingim HM (2014) N2-fixation by methanotrophs sustains carbon and nitrogen accumulation in pristine peatlands. Biogeochemistry 121(2):317–328

Virtanen R, Grytnes JA, Lenoir J, Luoto M, Oksanen J, Oksanen L, Svenning JC (2013) Productivity–diversity patterns in arctic tundra vegetation. Ecography 36(3):331–341

Vishnivetskaya T, Kathariou S, McGrath J, Gilichinsky D, Tiedje JM (2000) Low-temperature recovery strategies for the isolation of bacteria from ancient permafrost sediments. Extremophiles 4(3):165–173

Voigt C, Lamprecht RE, Marushchak ME, Lind SE, Novakovskiy A, Aurela M, Martikainen PJ, Biasi C (2017) Warming of subarctic tundra increases emissions of all three important greenhouse gases–carbon dioxide, methane, and nitrous oxide. Glob Chang Biol 23(8):3121–3138

Vonk JE, Gustafsson Ö (2013) Permafrost-carbon complexities. Nat Geosci 6(9):675–676. https://doi.org/10.1038/ngeo1937

Vonk JE, Mann PJ, Davydov S, Davydova A, Spencer RGM, Schade J, Sobczak WV, Zimov N, Zimov S, Bulygina E, Eglinton TI, Holmes RM (2013) High biolability of ancient permafrost carbon upon thaw. Geophys Res Lett 40(11):2689–2693. https://doi.org/10.1002/grl.50348
Voroney RP, Heck RJ (2014) The soil habitat. In: Paul EA (ed) Soil microbiology, ecology and biochemistry. Academic, London, pp 15–40
Waddington J, Roulet N (1997) Groundwater flow and dissolved carbon movement in a boreal peatland. J Hydrol 191(1–4):122–138
Wagner D, Liebner S (2009) Global warming and carbon dynamics in permafrost soils: methane production and oxidation. In: Permafrost soils. Springer, Berlin, pp 219–236
Wagner D, Pfeiffer E-M, Bock E (1999) Methane production in aerated marshland and model soils: effects of microflora and soil texture. Soil Biol Biochem 31(7):999–1006
Wagner D, Lipski A, Embacher A, Gattinger A (2005) Methane fluxes in permafrost habitats of the Lena Delta: effects of microbial community structure and organic matter quality. Environ Microbiol 7(10):1582–1592
Waldrop MP, Wickland KP, White Iii R, Berhe AA, Harden JW, Romanovsky VE (2010) Molecular investigations into a globally important carbon pool: permafrost-protected carbon in Alaskan soils. Glob Chang Biol 16(9):2543–2554
Walker M, Walker MJC (2005) Quaternary dating methods. Wiley, New York, 286 p
Walker DA, Raynolds MK, Daniëls FJ, Einarsson E, Elvebakk A, Gould WA, Katenin AE, Kholod SS, Markon CJ, Melnikov ES (2005) The circumpolar Arctic vegetation map. J Veg Sci 16(3):267–282
Walter Anthony KM, Zimov SA, Grosse G, Jones MC, Anthony PM, Chapin FS 3rd, Finlay JC, Mack MC, Davydov S, Frenzel P, Frolking S (2014) A shift of thermokarst lakes from carbon sources to sinks during the Holocene epoch. Nature 511(7510):452–456. https://doi.org/10.1038/nature13560
Walter BP, Heimann M (2000) A process-based, climate-sensitive model to derive methane emissions from natural wetlands: application to five wetland sites, sensitivity to model parameters, and climate. Glob Biogeochem Cycles 14(3):745–765
Walter KM, Zimov SA, Chanton JP, Verbyla D, Chapin FS 3rd (2006) Methane bubbling from Siberian thaw lakes as a positive feedback to climate warming. Nature 443(7107):71–75. https://doi.org/10.1038/nature05040
Walter KM, Smith LC, Chapin FS 3rd (2007) Methane bubbling from northern lakes: present and future contributions to the global methane budget. Philos Trans A Math Phys Eng Sci 365(1856):1657–1676. https://doi.org/10.1098/rsta.2007.2036
Wang P, Heijmans MM, Mommer L, van Ruijven J, Maximov TC, Berendse F (2016) Belowground plant biomass allocation in tundra ecosystems and its relationship with temperature. Environ Res Lett 11(5):055003
Wang P, van Ruijven J, Heijmans MM, Berendse F, Maksimov A, Maximov T, Mommer L (2017) Short-term root and leaf decomposition of two dominant plant species in a Siberian tundra. Pedobiologia 65:68–76
Wang P, Limpens J, Nauta A, van Huissteden C, Quirina van Rijssel S, Mommer L, de Kroon H, Maximov TC, Heijmans MM (2018) Depth-based differentiation in nitrogen uptake between graminoids and shrubs in an Arctic tundra plant community. J Veg Sci 29(1):34–41
Weintraub MN, Schimel JP (2005) Nitrogen cycling and the spread of shrubs control changes in the carbon balance of arctic tundra ecosystems. Bioscience 55(5):408–415
Weiss N, Blok D, Elberling B, Hugelius G, Jørgensen CJ, Siewert MB, Kuhry P (2016) Thermokarst dynamics and soil organic matter characteristics controlling initial carbon release from permafrost soils in the Siberian Yedoma region. Sediment Geol 340:38–48. https://doi.org/10.1016/j.sedgeo.2015.12.004
Weiss N, Faucherre S, Lampiris N, Wojcik R (2017) Elevation-based upscaling of organic carbon stocks in high-Arctic permafrost terrain: a storage and distribution assessment for Spitsbergen, Svalbard. Polar Res 36(1):1400363

Whalen SC, Reeburgh WS (1988) A methane flux time series for tundra environments. Glob Biogeochem Cycles 2(4):399–409
Wieder RK, Lang GE (1988) Cycling of inorganic and organic sulfur in peat from Big Run Bog, West Virginia. Biogeochemistry 5(2):221–242
Wik M, Crill PM, Bastviken D, Danielsson Å, Norbäck E (2011) Bubbles trapped in arctic lake ice: potential implications for methane emissions. J Geophys Res Biogeo 116(G3):G03044. https://doi.org/10.1029/2011JG001761
Wik M, Thornton BF, Bastviken D, MacIntyre S, Varner RK, Crill PM (2014) Energy input is primary controller of methane bubbling in subarctic lakes. Geophys Res Lett 41(2):555–560
Wik M, Varner RK, Anthony KW, MacIntyre S, Bastviken D (2016) Climate-sensitive northern lakes and ponds are critical components of methane release. Nat Geosci 9(2):99
Wik M, Johnson JE, Crill PM, DeStasio JP, Erickson L, Halloran MJ, Fahnestock MF, Crawford MK, Phillips SC, Varner RK (2018) Sediment characteristics and methane ebullition in three subarctic lakes. J Geophys Res Biogeo 123(8):2399–2411
Wilhelm RC, Radtke KJ, Mykytczuk NC, Greer CW, Whyte LG (2012) Life at the wedge: the activity and diversity of Arctic ice wedge microbial communities. Astrobiology 12(4):347–360
Wille C, Kutzbach L, Sachs T, Wagner D, Pfeiffer E-M (2008) Methane emission from Siberian arctic polygonal tundra: eddy covariance measurements and modeling. Glob Chang Biol 14(6):1395–1408. https://doi.org/10.1111/j.1365-2486.2008.01586.x
Williams M, Rastetter EB (1999) Vegetation characteristics and primary productivity along an arctic transect: implications for scaling-up. J Ecol 87(5):885–898
Williams M, Eugster W, Rastetter EB, Mcfadden JP, Chapin Iii FS (2000) The controls on net ecosystem productivity along an Arctic transect: a model comparison with flux measurements. Glob Chang Biol 6(S1):116–126
Yashina S, Gubin S, Maksimovich S, Yashina A, Gakhova E, Gilichinsky D (2012) Regeneration of whole fertile plants from 30,000-y-old fruit tissue buried in Siberian permafrost. Proc Natl Acad Sci 109(10):4008–4013
Yergeau E, Hogues H, Whyte LG, Greer CW (2010) The functional potential of high Arctic permafrost revealed by metagenomic sequencing, qPCR and microarray analyses. ISME J 4(9):1206
Yoshikawa K, Kokudo N, Tanaka M, Nakano T, Shibata H, Aragaki N, Higuchi T, Hashimoto T (2008) Novel abietane diterpenoids and aromatic compounds from Cladonia rangiferina and their antimicrobial activity against antibiotics resistant bacteria. Chem Pharm Bull 56(1):89–92
Zalman C, Meade N, Chanton J, Kostka J, Bridgham S, Keller J (2018) Methylotrophic methanogenesis in Sphagnum-dominated peatland soils. Soil Biol Biochem 118:156–160
Zheng J, RoyChowdhury T, Yang Z, Gu B, Wullschleger SD, Graham DE (2018) Impacts of temperature and soil characteristics on methane production and oxidation in Arctic tundra. Biogeosciences 15(21):6621–6635. https://doi.org/10.5194/bg-15-6621-2018
Zhu B, van Dijk G, Fritz C, Smolders AJ, Pol A, Jetten MS, Ettwig KF (2012) Anaerobic oxidization of methane in a minerotrophic peatland: enrichment of nitrite-dependent methane-oxidizing bacteria. Appl Environ Microbiol AEM:02102-02112
Zimov S, Semiletov I, Daviodov S, Voropaev YV, Prosyannikov S, Wong C, Chan Y-H (1993) Wintertime CO_2 emission from soils of northeastern Siberia. Arctic 46(3):197–204
Zimov S, Voropaev YV, Semiletov I, Davidov S, Prosiannikov S, Chapin FS, Chapin M, Trumbore S, Tyler S (1997) North Siberian lakes: a methane source fueled by Pleistocene carbon. Science 277(5327):800–802
Zimov SA, Davydov SP, Zimova GM, Davydova AI, Schuur EAG, Dutta K, Chapin FS (2006) Permafrost carbon: Stock and decomposability of a globally significant carbon pool. Geophys Res Lett 33(20). https://doi.org/10.1029/2006gl027484
Zoltai S (1995) Permafrost distribution in peatlands of west-Central Canada during the Holocene warm period 6000 years BP. Géog Phys Quatern 49(1):45–54
Zona D, Oechel W, Kochendorfer J, Paw U, Salyuk A, Olivas P, Oberbauer S, Lipson D (2009) Methane fluxes during the initiation of a large-scale water table manipulation experiment in the Alaskan Arctic tundra. Glob Biogeochem Cycles 23(2):GB2013. https://doi.org/10.1029/2009GB003487

Zona D, Oechel W, Peterson K, Clements R, PAW UK, Ustin S (2010) Characterization of the carbon fluxes of a vegetated drained lake basin chronosequence on the Alaskan Arctic Coastal Plain. Glob Chang Biol 16(6):1870–1882

Zona D, Gioli B, Commane R, Lindaas J, Wofsy SC, Miller CE, Dinardo SJ, Dengel S, Sweeney C, Karion A, Chang RY, Henderson JM, Murphy PC, Goodrich JP, Moreaux V, Liljedahl A, Watts JD, Kimball JS, Lipson DA, Oechel WC (2016) Cold season emissions dominate the Arctic tundra methane budget. Proc Natl Acad Sci U S A 113(1):40–45. https://doi.org/10.1073/pnas.1516017113

Zubrzycki S, Kutzbach L, Grosse G, Desyatkin A, Pfeiffer E-M (2013) Organic carbon and total nitrogen stocks in soils of the Lena River Delta. Biogeosciences 10(6):3507

Chapter 5
Permafrost in Transition

Abstract Climate change results in physical changes in permafrost soils: active layer thickness, temperature, soil hydrology and abrupt thaw features in ice-rich soils. Abrupt thaw features create new landforms such as ponds, lakes and erosion phenomena. In this chapter, current observations of physical changes in permafrost soils are discussed, including their effect on the soil carbon cycle. For the carbon cycle changes, the results of observations and experimental studies are emphasized. First, the effects of soil warming without further geomorphological change is considered. The potential effect of self-amplifying soil warming by heat production from bacterial production is discussed. Next, the changes in geomorphological processes expressed by formation of thaw ponds, lakes and erosion features are considered. These contribute to an increase of CO_2 and non-CO_2 greenhouse gas emissions. Hydrological changes include the effects of permafrost thaw on the water cycle via groundwater flow and directly climate-driven changes in precipitation and evapotranspiration. These result in river discharge changes with effects on floodplains, and influence transport of carbon from permafrost regions to the Arctic Ocean. Soil hydrology changes – wetting or drying – induce changes in the pattern of greenhouse gas emissions of permafrost soils.

Keywords Permafrost thaw · Active layer · Subsidence · Ponds · Thaw lakes · Erosion phenoma · Landslides · Gully erosion · Fluvial carbon transport

The last section of Chap. 1 addressed climate change, and the dependence of its rate and magnitude on the ability of our society to curb greenhouse gas emissions. Permafrost extent in the future, and the way in which changes of permafrost manifest themselves are strongly dependent on this rate and magnitude of climate change. The key aspects are temperature and precipitation; for the land areas the MAAT is expected to increase strongly, but the temperature increase in winter is strongest. The same holds for precipitation; winter precipitation increases more strongly in many permafrost regions, while summer precipitation increases much less (Figs. 1.26 and 1.27).

A recent survey of permafrost temperature changes in deep boreholes (Biskaborn et al. 2019) shows that permafrost temperature increased recently (2007–2016) by 0.39 ± 0.15 °C in continuous permafrost areas and 0.20 ± 0.1 °C in discontinuous

J. van Huissteden, *Thawing Permafrost*,
https://doi.org/10.1007/978-3-030-31379-1_5

permafrost areas (measured at the depth of zero annual temperature amplitude). The trend follows that of air warming, and for discontinuous permafrost, snow cover thickness. The highest rates of temperature change were observed in Western and Northern Siberia.

Climate change is by definition a change in the average weather over 30 years. However, the concurrent increasing incidence of extreme weather conditions is also important, and may have long-lasting effects on land surface ecosystems (e.g. Frank et al. 2015). In particular the effect of one or a few hot summers are triggers of abrupt thaw processes such as widespread cryogenic landslide development, extensive ice wedge thaw and pond development (Jorgenson et al. 2006, 2015; Lewkowicz and Way 2019). For eroding ice wedges, extreme winter snowfall and subsequent high meltwater discharge, or summer precipitation extremes are also potential drivers (Liljedahl et al. 2016).

5.1 Which Changes?

Nearly every aspect of high-latitude ecosystems is subject to change, not only permafrost and soil temperature, but also vegetation, hydrological and biogeochemical cycles (Hinzman et al. 2005). Higher winter temperatures and insulation by snow result in a strong reduction of heat loss in the soil in winter, combined with a longer and warmer thaw season. Changes in snow cover differ regionally (Park et al. 2012). In most of the permafrost region, the trend is a decrease of snow depth and duration, but in areas bordering the Arctic ocean snow cover is increasing (Sect. 5.4.2). Recent model simulations indicate losses of permafrost between 3.2 and 4.7 million km^2 for the moderate RCP4.5 climate projections and between 5.7 and 16.1 million km^2 for the IPCC RCP8.5 climate projections by the year 2300, in less than three centuries (McGuire et al. 2018).

It is difficult to imagine which effects this pace of permafrost warming will have on permafrost landscapes, ecosystems and the carbon cycle. Walter et al. (2007b) have shown what the Last Glacial Termination has meant for permafrost landscapes; they demonstrated a widespread and rapid lake development between 15 and 9 kyr calendar years before present, based on radiocarbon dates of thaw lake sediments. This was not restricted to the margins of the permafrost realm, but included areas in Siberia and Alaska that are located presently in the continuous permafrost zone.

Furthermore, our imagination is blurred by the complicated interaction of processes – vegetation change, soil carbon processes, thaw and soil temperature change – that produce multiple feedbacks. Climate change will induce change in the permafrost temperature, but at the same time changes in vegetation, which influence in turn the surface heat balance and evaporation, feeding back on climate change. Zhang et al. (2018) showed with modelling of land surface energy fluxes that climate warming is amplified by expansion of shrub and trees, leading to higher air temperatures and precipitation. However, shrub expansion or establishment of a moss layer can potentially change the heat balance of the soil in such a way that the net result is lower, rather than higher permafrost temperatures (Blok et al. 2010, 2011). The relations between climate, vegetation, heat balance and soil temperature are not simple and unidirectional, and trigger processes on various time scales (Chap. 6).

Single processes should be seen in the context of other processes, some of which may increase loss of carbon, others foster carbon storage, or may result in a change of the ratio of carbon emission pathways, e.g. the ratio between CO_2 and CH_4 emission. For example, longer growing seasons may enhance CO_2 uptake by ecosystems and potentially enhance carbon storage (e.g. suggested by the modelling work of McGuire et al. 2018), but at the same time enhanced soil respiration offsets extra CO_2 uptake by enhanced respiration of soil carbon, at least on the short term (Parmentier et al. 2011; Trucco et al. 2012).

Next, processes may be either gradual, or catastrophic, often showing threshold behaviour, also known as tipping points – once a system is pushed over a threshold level of one of its state variables, rapid, irreversible and potentially destructive changes or regime shifts may occur (Scheffer et al. 2009; Prowse et al. 2011; Carstensen and Weydmann 2012). Grosse et al. (2011) make for permafrost a distinction between slow but gradual and widespread press disturbances, and rapid, generally local and destructive pulse disturbances. The potential for drastic changes should always be considered.

Climate change is an indirect source of anthropogenic disturbance of permafrost ecosystems. However, also direct effects of increasing human presence in the Arctic may further aggravate its effects. Industrial activities, mining exploration and development, forestry and transport are increasing (French 2018). Disturbance of the vegetation by human activities, such as deforestation, ploughing, digging, vehicle traffic, fires, or even simply trampling of vegetation by repeated walking, can start thaw (Figs. 1.17 and 5.1). On the other hand, permafrost thaw can be a hazard to

Fig. 5.1 Active permafrost thaw and lake expansion after forest clearance in the 1960s, south of Yakutsk. (Photo by the author)

infrastructure. These additional effects of human disturbance are discussed in Chap. 6.

In this chapter, the effects of climate change on permafrost temperature and active layer thickness and its geomorphological and hydrological effects will be discussed, in the next chapter changes of vegetation and the soil ecosystem. For every aspect, its effect on the carbon cycle and greenhouse gas emission of permafrost will be considered. The focus in this chapter will be on process studies and observations of effects on the carbon cycle, rather than on models. Modelling is an art which deserves a chapter on its own (Chap. 8).

To consider changes based on observations, one would preferably see a long record of continuous observations of good quality. However, only a few consistent records of long-term changes in permafrost ecosystems are available. Science has several instruments to circumvent this, each with it own merits and disadvantages:

- Field experiments may study the effects of artificial warming or precipitation change; an example is the use of Open Top Chambers (OTC) which induce local air warming by small greenhouses open at the top (e.g. Oberbauer et al. 2007).
- Time-for-space studies make use of the climatic zoning of permafrost to study differences in processes in climate-related permafrost and vegetation zones.
- The year-to-year variability in weather can be used to infer the effects of climate change, if a sufficiently long data record (in the order of 10 years) is available (e.g. Parmentier et al. 2011).
- Remote sensing studies can help to study spatial changes of permafrost and ecosystems over time, provided enough material is available with a sufficient quality.
- Organic carbon transported from river basins integrate processes of carbon release throughout the drainage basin; analysis of biomarkers and radiocarbon age of the transported organic matter can provide information on the release of thawed old carbon from soils (e.g. Vonk et al. 2010; Gustaffson et al. 2011).
- Paleo-records of change of vegetation, sediments and cryoturbation features can shed light on history of climate, ecosystem and permafrost change over time (e.g. French and Shur 2010; Teltewskoi et al. 2016).

All disturbances of the soil heat balance can induce permafrost thaw. Human activities have been listed above; other natural disturbances are erosion, flooding and fires. Even if climate change is the main cause, the actual drivers of thaw may be multiple. The direct effects are the higher temperatures in summer or winter, changing snow cover thickness and duration in winter (e.g. Stieglitz et al. 2003; Park et al. 2013). However, it triggers also indirect pathways by which the surface energy balance changes: thermal conductivity changes of the soil, increased incidence of high snowmelt runoff and flooding.

The local impact of thaw on permafrost varies strongly, depending vegetation, soil type, ground ice fraction, discontinuous or continuous permafrost, and topographic position (French 2018). The simplest scenario is a gradual increase of the active layer thickness. Ground ice is unevenly distributed in the soil, and ground ice thaw therefore is likely to cause local subsidence of the soil surface, often resulting in relief inversions (e.g. Turetsky et al. 2007 and Seppälä 2011 for bogs with discontinuous permafrost; Jorgenson et al. 2006 and Liljedahl et al. 2016 for continuous permafrost). This may have contrasting effects, depending on the configuration of ground ice bodies and the extent of surface subsidence: ponding of water, or increased surface drainage. Furthermore, it affects vegetation (e.g. Turetsky et al. 2007; Nauta et al. 2014) and mobilizes nutrients and organic matter (e.g. Vonk et al. 2015; Koch et al. 2018).

Shur and Jorgenson (2007) developed a permafrost classification system that accounts for ecosystem effects on permafrost preservation. They distinguish five situations in which permafrost may exist:

Climate-driven permafrost is typical for recently developed permafrost in the continuous permafrost zone, where permafrost forms readily, after a land surface is exposed to the atmosphere (e.g. by lake drainage).

Climate-driven, ecosystem-modified permafrost occurs within the continuous permafrost zone; it is characterized by considerable ecosystem and soil development and accumulation of ground ice, including ice wedges. Vegetation and organic-rich topsoil contribute to the protection of this ice-rich permafrost to thaw, but disruption of this ecosystem protection leads to thaw and loss of ground ice. Once thaw has occurred, permafrost re-establishes under the present climatic conditions.

Climate-driven, ecosystem protected permafrost has grown under favourable climatic conditions as ecosystem-modified permafrost and survives under warmer conditions, protected by the ecosystem and soil properties, in the discontinuous permafrost zone. Once degraded, it cannot re-establish under the present climatic conditions. During warming climates, climate-driven permafrost can persist in the discontinuous permafrost zone for a long time as ecosystem-protected permafrost; Froese et al. (2008) describe an example of ancient (740,000 years old) ice-rich permafrost that survives today in the discontinuous permafrost zone in Alaska.

Ecosystem-driven permafrost forms in the discontinuous permafrost zone in favourable situations for permafrost establishment: peatlands where vegetation and peat layers influence the soil thermal regime, and topographical factors such as north-facing slopes.

Ecosystem-protected permafrost exists solely because of protection by the ecosystem. It occurs as sporadic permafrost, but after thaw it cannot regrow.

This classification helps to realize, to what extent permafrost and ecosystems depend on each other, and in which situations permafrost is likely to be preserved or disappear.

Permafrost thaw justifies some remarks on how it can be recognized. Superficially, an increase of the active layer thickness (ALT) could be taken as evidence for permafrost thaw. ALTs are measured as averages, over a large number of individual points on a regular grid, averaging the effect of small-scale local variations in ALT caused by vegetation and soil composition differences. The maximum ALT occurs near the end of the thawing season and is usually measured on a grid of measurement points with a probe to record the top of the permafrost. Unless stated other otherwise, the ALT in this section is this maximum ALT. The Circumpolar Active Layer Monitoring program systematically records ALT of a large number of observation stations in the Arctic (https://www2.gwu.edu/~calm/; Brown et al. 2000, Fig. 1.29).

However, matters are more complicated than an increase of the ALT. It starts with an increasing ALT in summer, while the soil still freezes completely in winter. Ongoing thickening of the active layer will result in incomplete freeze back of the active layer in winter, resulting in the formation of a layer on top of the permafrost which remains unfrozen in winter (a supra-permafrost talik; French 2018, Connon et al. 2018; Fig. 5.2). This phase of permafrost thaw cannot be detected with conventional ALT measurements at the end of the summer, but can only be detected by winter measurements with soil temperature sensors or probes. Development of supra-permafrost taliks is at first intermittent (not occurring every year) but with progress of thaw it becomes permanent. This also affects the ALT measured at the end the summer: it may include a supra-permafrost talik. In this case it is better to use the term supra-permafrost layer (SPL) instead of ALT (Connon et al. 2018). For practical reasons, 'ALT' is generally used here if there is no a clear reason to use 'SPL'. If the supra-permafrost talik thickens further, the permafrost becomes essentially relict or disappears entirely, and ALT changes into the seasonally frozen layer thickness. The top-of-permafrost depth at which the permafrost becomes relict in permafrost models is often set at 3 m below the soil surface for modelling purposes (Chap. 8).

The presence of large amounts of excess ice may delay the development a supra-permafrost talik. In this case, an increase of ALT is compensated by thaw of ground ice at the top of the active layer, in combination with subsidence. If water from

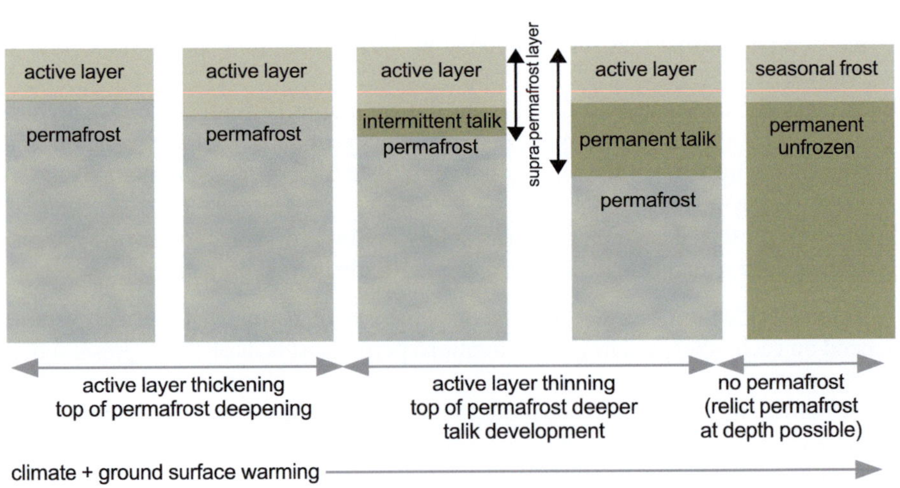

Fig. 5.2 Development of a supra-permafrost talik by thaw of permafrost. Once the permafrost has disappeared the active layer becomes a seasonally frozen layer. After Connon et al. (2018). Disappeared permafrost does not exclude the presence of relict permafrost at larger depth

thawed ground ice is drained or evaporated, the loss of ground ice and associated subsidence may not result in a significant ALT increase. Moreover, ground ice acts as a buffer against deep thaw because of the latent heat of thawing of the ice. This at least postpones the development of a supra-permafrost talik. However, uneven distribution of ground ice is likely to produce visible relief changes.

5.2 Diffuse Permafrost Thaw

The active layer thickness (ALT) and soil temperature are important variables of the permafrost soil environment. Vegetation, hydrology and the carbon cycle are strongly influenced by ALT (Nelson et al. 1998; Brutsaert and Hiyama 2012; Jorgenson et al. 2013). Increase or decrease of the ALT results from a changing energy balance at the soil surface, which may be climate-related, but also may relate to natural or human-induced vegetation change. In this section, the relation of ALT and soil temperature with climate change is discussed, without consideration of the effects of uneven ground ice thaw and its geomorphological and hydrological effects, which are discussed in the next sections. This is followed by a discussion of the carbon cycle effects of ALT and soil temperature change.

5.2.1 Observations of Active Layer Thickness and Surface Subsidence

The spatial heterogeneity of ALT differs in various permafrost landscapes due to variation in vegetation, soils and topography. For instance, on the Alaskan coastal plain the maximum spatial variation occurs over distances of hundreds of meters, while in the northern foothills there is more shorter distance variation. Consideration of this spatial variability using spatial statistics, and adaptation of sampling schemes to landscape characteristics is necessary to quantify ALT changes. In terrain where the spatial variability of ALT is high, other sampling schemes such as hierarchical sampling, or nested sampling based on landcover classes should be considered (Nelson et al. 1998). Figure 5.3 shows typical variation of ALT in tundra in north-eastern Siberia.

Large scale mapping of ALT is based on land surface models or soil permafrost models, using soil and climate information and remote sensing information on vegetation (e.g. Sazonova and Romanovsky 2003; Lawrence et al. 2008; Park et al. 2013; Chap. 8). The use of remote sensing information can be extended with remotely sensed information on snow cover and soil moisture (Yi et al. 2018). A method to estimate average active layer thickness in river drainage basins is based on hydrological modelling of the base flow of the river (Brutsaert and Hiyama 2012). Base flow is the river discharge during periods without precipitation and

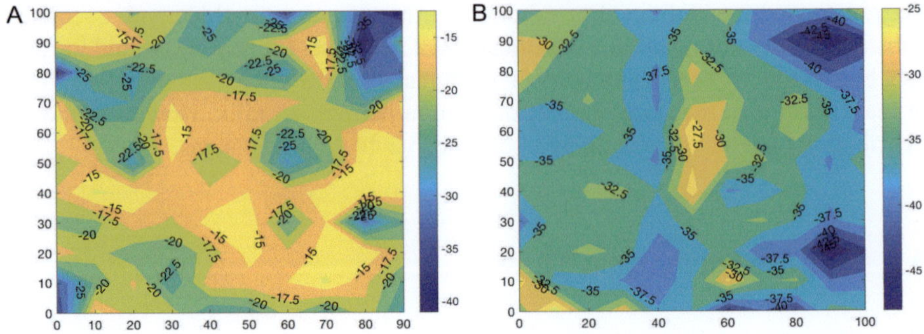

Fig. 5.3 Active layer thickness variation on different spatial scales on a former thaw lake bottom in Northeastern Siberia (Kytalyk research station, Indigirka lowlands). (**a**) Midsummer (July) ALT on a 100 × 100 m grid over moist tussock tundra; upper left corner contains thaw ponds over ice wedges. (**b**) Midsummer (July) ALT on a 10 × 10 m grid below a Sphagnum -sedge hummock vegetation

Table 5.1 Observed active layer thickness trends.

Region	ALT trend cm yr^{-1}	References
Northern Alaska	0.32 ± 1.18	Yi et al. (2018)
Interior Alaska	>3	Yi et al. (2018)
Northeast Greenland	1	Hollesen et al. (2011)
Northern Sweden	0.7–1.3; accelerating	Åkerman and Johansson (2008)
Siberia, Lena basin	0.9	Zhang et al. (2005)
Siberia, Lena basin discont. Permafrost	0.3–1	Brutsaert and Hiyama (2012)
Arctic Russia	1.7–5.5	Mazhitova et al. (2008)

depends on the outflow of aquifers along the river banks. In continuous permafrost areas, the active layer is practically the only aquifer for base flow during the summer. Without going in the details of the model, it should be intuitively clear that thicker active layers in a river basin could discharge more water into the river.

Trends in the ALT vary throughout the northern permafrost region (Hinzman et al. 2013; Table 5.1). Year-to-year variability in ALT is large due to differences in weather conditions between years. Romanovsky and Osterkamp (1997) report ALT variation in ALT time series in the North Alaskan coastal plain of a factor two. Walvoord and Kurylyk et al. (2016) summarized several studies that report ALT changes: increasing trends were found Northern Sweden, Arctic Russia, Eastern Siberia; variable ALT without clear trends in Alaska and the entire Pan-Arctic permafrost region. Osterkamp (2005) did not find a consistent trend of ALT in Northern Alaska, despite increasing air temperature and deeper permafrost temperature. Long term records of ALT from the Alaskan coastal plain at Barrow show that active layer thicknesses from the 1960s are generally larger than those from the 1990s, and are very similar to those of the 2000s; a long term trend is absent (Shiklomanov et al. 2010).

However, records of ALT are complicated by subsidence caused by thawing of ice-rich permafrost in the transient layer at the top of the permafrost (Liu et al. 2010). Integration of field observations with radar remote sensing by Yi et al. (2018) show a widespread ALT increase since 2001, with smaller increases in northern Alaska (mean trend 0.32 ± 1.18 cm year^{-1}) and a tenfold larger increase (>3 cm year^{-1}) across interior and southern Alaska. The uncertainty in this trend for discontinuous permafrost areas is larger than in continuous permafrost areas. For the Lena basin in Eastern Siberia in the period 1950–2008, Brutsaert and Hiyama (2012) derive an increasing ALT thickness trend of 0.3 to 1 cm year^{-1} in areas with discontinuous permafrost and about half as large in areas with continuous permafrost, based on hydrologic modelling.

In general, ALT time series on the Eurasian continent show an increasing trend, while in North America trends are lacking or less prominent, which is attributed to variation in interaction between air temperature trends and snow cover thickness trends (Park et al. 2013). The lack of ALT trends in North America is attributed to thinner snow depth in winter and drier soils (with lower thermal conductivity) in summer, whereas in particular in West Siberia snow cover thickness increases (Bulygina et al. 2011). In general, ALT trends found in discontinuous permafrost areas are larger (Brutsaert and Hiyama 2012; Yi et al. 2018).

Surface subsidence due to ALT increase may be very gradual and not readily visible. Thaw of excess ice at the top of the permafrost just below the active layer will result in seasonal subsidence (thaw settlement). During the summer, a part of the water from the thawed ice will be lost by drainage and evapotranspiration. The next winter, water that has not been removed from the soil will refreeze, causing frost heave (Fig. 5.4). This results in a cycle of subsidence in summer and heave in winter. If the energy budget at the soil surface allows downward progressing thaw at the top of the permafrost over the years, a net loss of excess ice from the soil will occur by the combination of thawing and drainage/evaporation. This causes a trend of net subsidence superposed on the yearly cycle, the secular subsidence (Liu et al. 2010).

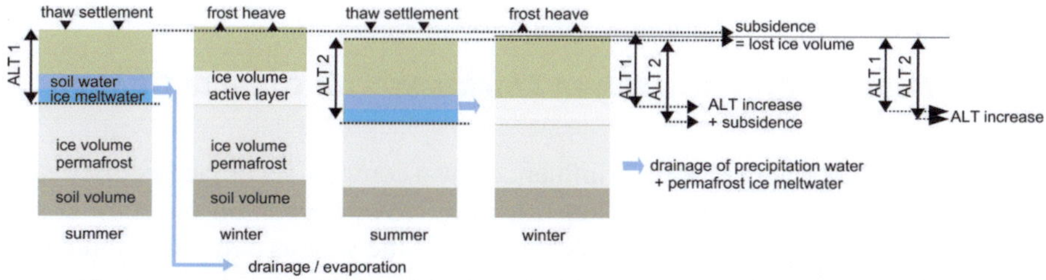

Fig. 5.4 Thawing of ground ice at the base of the active layer, combined with drainage of water in summer results in net (secular) surface subsidence, and reduces the deepening of the active layer in a warming climate. Thaw in summer results in settlement and surface subsidence of the active layer, and drainage of ground ice meltwater. In the next winter, refreezing of the active layer results in frost heave. The next summer, further thaw of ground ice and drainage of meltwater occurs, and renewed settlement to a lower level than the previous summer (secular subsidence). Right: difference between net ALT growth and ALT growth including subsidence. (Based on Liu et al. 2010)

This secular subsidence may mask an ALT increase, depending on the fraction of meltwater from excess ice that is lost from the soil (Fig. 5.4). Subsidence therefore does not depend on the ice content of the permafrost only, but also on hydrological conditions that determine the amount of meltwater loss by drainage and evapotranspiration.

The secular subsidence is difficult to quantify by precise land surveying. However, radar interferometry using synthetic aperture radar (SAR) satellites can map small surface deformations. InSAR (Interferometric SAR) compares two or more radar images. By using the phase difference of the radar signal, mapping of millimetre-scale surface deformations with respect to stable points is possible (e.g. Hanssen 2001; Liu et al. 2010). Typical problems causing uncertainties in InSAR in permafrost areas are changes in surface properties, which influence the phase of the radar signal, such as vegetation change, variations in soil moisture influencing the dielectric properties of the soil and snow cover (Liu et al. 2010).

Liu et al. (2010) detected in northern Alaska seasonally varying vertical displacements of 1–4 cm, and a secular subsidence of 0.1–0.4 cm $year^{-1}$, measured over 8 years (1992–2000). This is attributed to thawing of ice-rich permafrost, and could explain the lack of a clear trend in ALT during the 1990s (Shiklomanov et al. 2010), despite observed increase in air temperatures during that time. Liu et al. (2012) used InSAR to estimate the active layer thickness of a part of the Alaskan North Slope, based on the magnitude of the seasonal thaw settlement and information on soil properties. The results were within the uncertain range of observations, and showed an ALT of 30–50 cm over moist tundra areas, and an ALT of 50–80 cm over wet tundra areas.

A component of surface subsidence in peat soils is the decomposition of peat; it is a well-known feature of agriculturally used peat soils. This subsidence component is rarelely assessed separately from subsidence by ground ice thaw. In peat deposits, ALT increase results in a thicker peat column that is subject to decomposition; the associated volume loss of the peat should contribute to subsidence (e.g. Audry et al. 2011; O'Donnel et al. 2012a, b).

5.2.2 *Relation of Active Layer Thickness with Climate Change*

The relation of ALT with climate is influenced by a complex set of factors that determine the surface energy balance (Chap. 2), modified by vegetation and snow cover, and soil thermal properties which vary throughout the year as a result of soil moisture variations and ice content. The effect of climate change on ALT and soil temperature is strongly modified by vegetation change, that is also driven by climate change (Loranty et al. 2018). Several vegetation characteristics counteract the effect of climate warming (Hinzman et al. 2013). Nonlinear responses of ALT to climate change, which moreover vary in time and space, should be considered as normal in a situation of climate warming in permafrost ecosystems.

The summer temperature has the strongest influence on ALT (e.g. Zhang et al. 2005; Park et al. 2013). However, models calculating thaw depths based on the Stefan equation (Eq. 2.33), which only contains the summer thaw index, do not reproduce observed ALT's very well (Romanovsky and Osterkamp 1997; Zhang et al. 2005). The penetration of winter cooling into the soil, influenced by winter air temperatures and snow cover thickness also affects soil temperatures in the thaw season. This memory effect of previous winter conditions (Park et al. 2013) cannot be neglected, in particular for the early summer season. Chadburn et al. (2015) found that model simulations of winter and spring/autumn season soil temperatures were very sensitive to snow cover thickness and snow density. A sufficient model soil depth (15 m) is also necessary to improve modelled temperatures, indicating the influence of deeper soil heat storage on ALT.

Several experimental and observational studies have confirmed the effect of snow cover on ALT (Hinkel and Hurd 2006; Nowinski et al. 2010; Johansson et al. 2013), although these were sometimes quite drastic experiments with snow fences used to trap snow. These changed the winter snow depth up to twofold (Johansson et al. 2013) or up to eightfold (Hinkel and Hurd 2006, Nowinski et al. 2010). Besides the insulating effect in winter, snow depth affects ALT also in summer. Snow melt water contributes to higher soil moisture content, which increases the thermal conductivity of the soil in spring summer summer, and enhances penetration of summer warmth (Zhang et al. 2005; Park et al. 2013). On the other hand, thick snow accumulations may persist during a part of the summer, delaying soil warming (Hinkel and Hurd 2006).

In general, precipitation is increasing over most of the permafrost area (Vihma et al. 2016). However, snow depth is approximately stable or decreasing over large areas, together with a decrease of the length of the snow cover period. In areas bordering the Arctic Ocean snow depth is increasing as a result of more winter precipitation (Park et al. 2012). A further discussion of the soil water balance is included Sect. 5.4.

Next, the spatially varying thermal properties of the soil affect ALT. In particular the thickness of the organic top horizon of the soil influences thaw depth (Shiklomanov et al. 2010; Fisher et al. 2016). Shiklomanov et al. (2010) suggest the existence of a landscape-specific thawing regime determined by differences in organic-layer thickness, topography, snow depth, soil moisture regime, and vegetation and soil substrate conditions. Significant differences can exist in the response to change in climate variables due to these differences in thawing regime.

Vegetation structure interferes strongly with the heat balance at the surface (Chap. 2). Change of the vegetation cover by climatic warming will induce changes in ALT. However, this occurs in a less straightforward way than may be expected, often resulting in a trend that counteracts the direct effect of climate warming. Walker et al. (2003) shows that vegetation zones in the tundra along a summer climate gradient in Alaska are strongly linked with ALT. Warmer summer air temperatures result in a thicker ALT, but a denser vegetation canopy, a thicker moss cover and thicker organic soil horizons that develop with warmer climate oppose this trend. Some of the deepest ALTs were recorded in the northernmost sites which are largely barren. Indirectly, via the vegetation cover, even soil chemical characteristics proved to exert an influence on ALT. Acidic soils have a denser vegetation cover than non-acidic soils, which in turn affects the heat balance and results in a thinner ALT in acidic soils in the same vegetation zone.

The presence of a layer of mosses and shrubs has a strong effect on the penetration of summer thaw (Blok et al. 2010, 2011; Soudzilovskaia et al. 2013; Chadburn et al. 2015). Shrub and tree expansion into tundra ecosystems is an important effect of climate warming on the vegetation (e.g. Sturm et al. 2001; Harsch et al. 2009; Myers-Smith et al. 2011). An increasing shrub cover has the effect of shading of the ground surface; more of the incoming radiation is caught by the vegetation and transferred to the atmosphere as sensible heat, while less energy is available for warming the soil. On the other hand, thicker winter snow accumulation between the vegetation results in higher winter soil temperatures (Loranty et al. 2018).

In the boreal forest zone in Canada, Fisher et al. (2016) also demonstrated a decrease of ALT with tree canopy leaf area index (LAI). This relation with canopy LAI with ALT in boreal forests acts potentially in three ways (Fisher et al. 2016). First, the tree canopy of evergreen trees can intercept a part of the snowfall and reduce the snow depth that covers the ground surface, which allows deeper frost penetration in winter. Second, a denser tree cover also transpires more water, reducing soil moisture and thermal conductivity in winter. Third, a denser tree cover results in more shading of the ground surface, reducing incoming solar radiation. Fisher et al. (2016) conclude that the combination of shading and transpiration effects dominates.

Soil moisture change in combination with soil material is also a factor which affects ALT. Compare for instance the thermal conductivities of a moss layer, peat and mineral soil (sand) at fully water saturated conditions and at 50% water saturation (Table 2.2):

- moss layer: 10 times higher thermal conductivity at saturated conditions than at 50% saturation;
- peat: 5 times higher thermal conductivity, thermal diffusivity 1.2 times higher;
- sand: 1.5 times higher thermal conductivity, thermal diffusivity 1.1 times higher.

This influences ALT by stronger heat penetration in summer, and stronger cooling in winter if a snow cover is thin, in particular for organic soils. Conversely, warm summer conditions may not cause deep thaw, if coinciding with dry conditions (Woo 2012). The effect of soil moisture change in a surface organic horizon is larger than that of deeper mineral horizons.The strong effect of soil moisture is illustrated by a combined water table manipulation and warming experiment on peat soils in a fen in Alaska (Turetsky et al. 2008). A raised water water table warmed the soil by about 4 °C; this was much larger than the soil warming treatment (±1 °C).

Furthermore, the latent heat effects on soil temperature in autumn and spring are larger in wet soils than in drier soils, resulting in later freezing in autumn and later thaw in spring (Iwahana et al. 2005; Langer et al. 2011). This latent heat effect also introduces a memory effect of precipitation in the soil, which is demonstrated by measurements of soil moisture in two consecutive years by Sugimoto et al. (2003)

for soil profiles in a Siberian Larch forest. In the first year (1999), the summer precipitation total was 177 mm, which is a wet summer for the region. Heavy rainfall in August stored much water in the soil, followed by a high ice content in the next spring. This resulted in a warmer soil, slower freezing and a strong zero-curtain effect in autumn (Sect. 2.5.1), and slow thaw and slow percolation of snow meltwater the next spring.

This indicates a two-sided effect precipitation change; on the one hand, increased precipitation (snow cover and summer precipitation) raises soil temperatures, but at least in early summer, it may result in a shallower ALT, while later in summer, wetter soils allow deeper penetration of summer warmth. The ice content of the transient layer at the top of the permafrost (Sect. 3.1), is also important; accumulation of ice in the transient layer over a number of years with normal thaw depth introduces a memory effect by controlling the thaw depth in warmer years (Burn and Nelson 2006).

Adding to the complexity, vegetation also interacts with soil moisture through evapotranspiration, while soil moisture changes also may induce vegetation change. An increasing vascular plant cover (shrubification in tundra, or tree growth) will tend to increase evapotranspiration, opposing the effect of larger precipitation (e.g. Iwahana et al. 2005; Fisher et al. 2016). On the other hand, vegetation also affects soil moisture via its effect on snow cover (Fisher et al. 2016). Enhanced trapping of snow by a dense shrub cover results in higher soil moisture.

Figure 5.5 summarizes the relation of ALT change with climate change derived from the literature cited above. The effect of climate on ALT is influenced most directly by the components of the surface heat balance (e.g. Woo 2012), which determine the heat flux available for soil warming. Next, the heat balance is linked to the soil water balance, which influences the soil thermal properties by soil moisture. Climate indirectly influences ALT by changing vegetation structure which in turn changes the surface heat balance. Vegetation also affects soil thermal properties by changes of organic matter content. This may produce counter-intuitive effects – including potential decrease of ALT with a denser vegetation. Clearly, this should not be taken as evidence of absence of effects of climate warming on permafrost, but as evidence of climate change affecting the vegetation.

5.2.3 Carbon Cycle Effects of Active Layer and Soil Temperature Change

An increase of ALT results in general in more organic matter being available for aerobic or anaerobic decomposition of soil organic matter. This can be seen in the increase of ecosystem respiration R_{eco} in carbon flux measurements in permafrost ecosystems throughout the summer (e.g. Van der Molen et al. 2007). However, changes of soil hydrology and vegetation tend to occur simultaneously with soil warming, and feed back on each other. For instance, subsidence may result in wetter

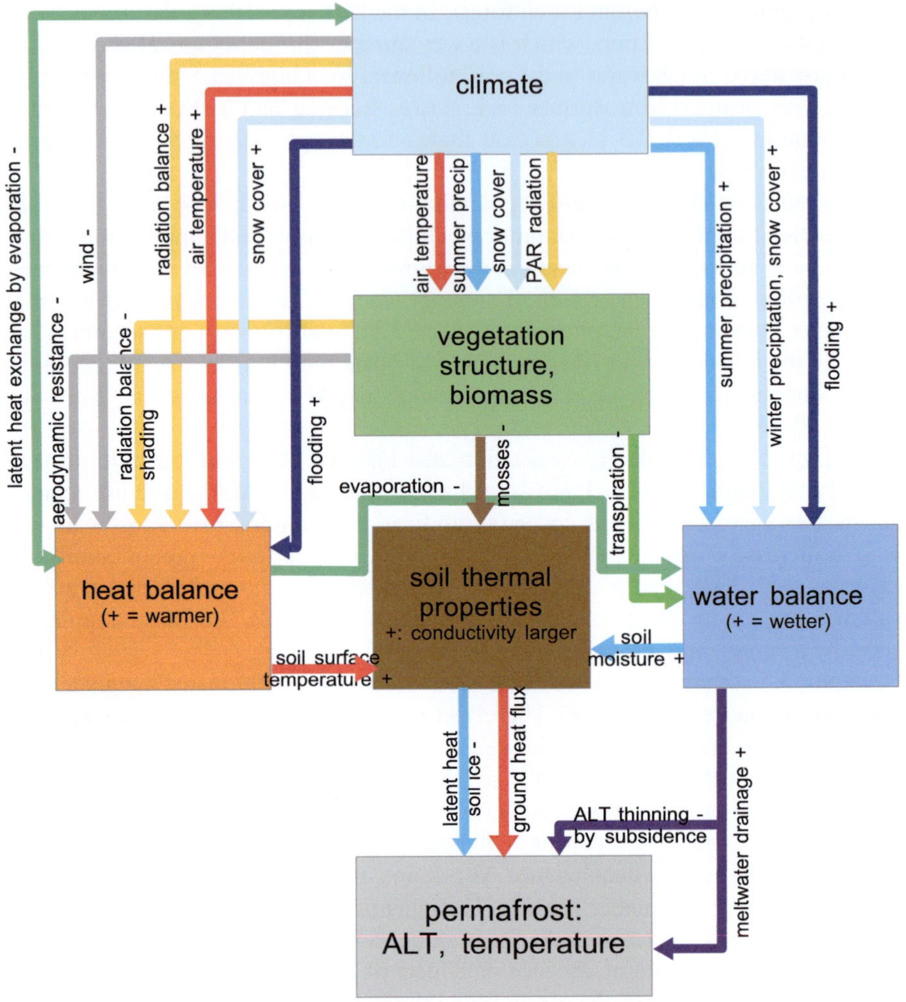

Fig. 5.5 Relations of active layer thickness and permafrost temperature with climate change. The arrows indicate relations influencing heat transfer to the soil, + indicates a positive influence towards warming soil. Climate change influences the soil heat balance both directly by changing surface heat fluxes to the atmosphere, and indirectly through vegetation which influences the surface heat and water balance and soil thermal properties. Thaw of ground ice by warming also feeds back on ALT by subsidence. In this diagram, change in soil carbon content by decomposition is not included

soils with more anoxic conditions, increasing the anaerobic decomposition at the expense of aerobic decomposition and decreasing the overall decomposition rate (e.g. Natali et al. 2015), or the reverse may happen, soil drying resulting in more aerobic decomposition. Increased decomposition of soil organic matter (SOM) liberates nutrients that enhance vegetation productivity. Warming also changes ecosystem species composition and the ratio of below-ground to above-ground biomass (e.g. Trucco et al. 2012; Sistla et al. 2013; Hicks Pries et al. 2015; Wang et al. 2016). To complicate matters further, the increase in labile organic compounds by enhanced vegetation productivity may drive a priming effect, which enhances decomposition of more recalcitrant soil organics (Sect. 4.5.2).

The basic question is here, whether a thicker ALT and warmer soil results in larger R_{eco} or CH_4 production, and more importantly – which component of R_{eco} changes (Fig. 5.6). R_{eco} change may be driven by enhanced decomposition of older soil organic matter. However, an increase of primary production also increases autotrophic vegetation respiration and heterotrophic respiration of recently produced organic material (litter, roots, root exudates), without increasing the decomposition of older soil carbon (Dorrepaal et al. 2009). It is the release of old carbon stored in permafrost soils that basically represents the permafrost carbon feedback, which should be separated from R_{eco} change by other organic (recent) matter sources. Furthermore, changes in all carbon reservoirs, including ecosystem biomass, should

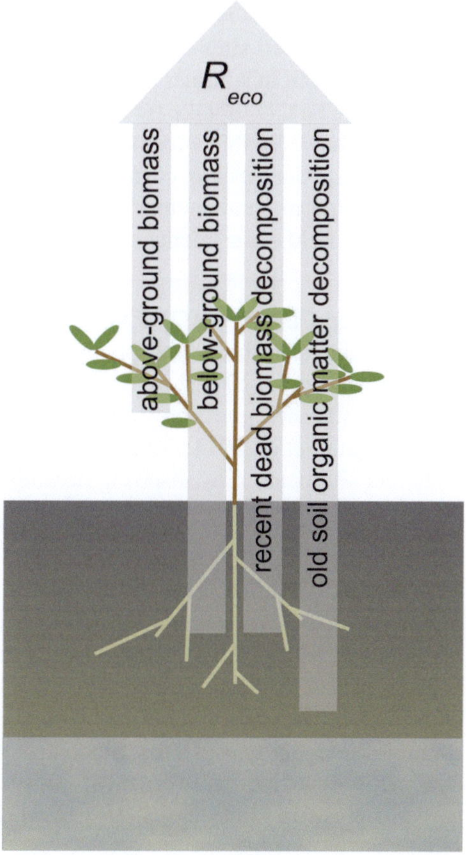

Fig. 5.6 Sources of CO_2 contributing to ecosystem respiration (R_{eco})

be accounted for (Schuur et al. 2008, 2015). Separation of vegetation and soil temperature effects is difficult.

Studies of spatial variability of soil carbon fluxes have found systematic variations of CO_2 and CH_4 fluxes with ALT or soil temperature (e.g. Christensen et al. 1995, 2003; Van Huissteden et al. 2005, 2009), but have the disadvantage that this spatial variability co-varies with soil wetness and vegetation. Some field experiments suggest that climate-driven changes in CH_4 and CO_2 flux are influenced more by changes in vegetation than by direct temperature effects on the soil (Verville et al. 1998). Besides vegetation effects, also the effect of peat composition is large. In a warming experiment on peat columns of two peatland sites in Canada and Sweden, Sjögersten et al. (2016) demonstrated that the peat organic chemistry strongly controls CO_2 emission from the peat.

Methods that are able to discriminate R_{eco} sources are necessary for separating vegetation and soil warming effects. Schuur et al. (2009) compared winter and summer R_{eco} from three sites with different gradations of permafrost thaw, expressed by ALT increase and surface subsidence; the contribution of old carbon to R_{eco} was estimated by ^{14}C of the respired CO_2. This allows to distinguish CO_2 derived from old soil carbon from respiration of recently assimilated carbon. The moderate and extensively thawed sites had higher R_{eco} from old soil carbon than the minimally thawed site, both in summer and winter. The losses of old SOM were 40% higher at these sites. The CO_2 uptake by vegetation had increased at the thawed sites, but for the oldest thaw site this was more than compensated by the loss of old soil carbon over the years that thaw occurred.

Other experimental methods employ artificial warming. Artificial soil and air warming in field experiments can be achieved in various ways:

- Greenhouses (e.g. Verville et al. 1998; Sistla et al. 2013);
- Open Top Chambers (OTC's), similar to greenhouses but open to the air at the top, with less influence on precipitation and temperature than greenhouses (e.g. Oberbauer et al. 2007);
- Artificially increased snow cover in winter by adding snow or using snow fences (e.g. Natali et al. 2011; Johansson et al. 2013);
- Direct soil heating by inserting heating cables in the soil (Wang et al. 2016).

In the experiments by Verville et al. (1998) the effects of air warming (±4 °C) and soil warming (± 2 °C at 5 cm depth) did not result in a significant change of the R_{eco} CO_2 flux and CH_4 flux. The effect of species removal, conducted simultaneously in this experiment, were larger. However, the soil warming in this experiment was restricted to the topsoil, ALT did not change. Oberbauer et al. (2007) compared *NEE* and R_{eco} of several Arctic and High Arctic ecosystems warmed with OTC's in Alaska and Canada. They found a general increase of R_{eco} in all ecosystems, with the highest increase in dry ecosystems. Since also gross ecosystem photosynthesis and vegetation biomass increased in most ecosystems, this resulted in a net increase in carbon uptake in some cases. However, this experiment did not discriminate the effect of ALT change.

Experiments using other indicators of soil organic matter (SOM) decomposition next to measured R_{eco} have been more successful to delineate the effect of warming on old SOM, and on SOM produced by enhanced vegetation growth (Dorrepaal et al. 2009; Natali et al. 2011, 2014, 2015; Hicks Pries et al. 2013a, 2015). The experiments by Natali et al. (2011, 2014, 2015) included warming of the soil in winter by creating a thicker snow cover using snow fences, combined with summer air warming with open top chambers. The additional snow was removed in spring to avoid effects on the soil water balance. At the same time, R_{eco}, *NEE* and CH_4 fluxes (including winter fluxes), soil CO_2 $\delta^{14}C$ and $\delta^{13}C$, and in situ substrate decomposition rates of a standardized organic substrate were measured. CO_2 derived from plant metabolism, decomposition of recently assimilated plant material (recent detritus), and decomposition of older SOM each have a characteristic $\delta^{14}C$ value, resulting from radioactive decay of ^{14}C for old SOM, and an increase in atmospheric ^{14}C caused by nuclear weapons testing for recently produced SOM (Schuur et al. 2009).

Winter warming by the thicker snow cover increased the soil temperature of the upper 40 cm by 1.5 °C and increased the ALT by 10%. The primary productivity of the vegetation increased by 20% due to warming. During the growing season, R_{eco} did not increase, despite the additional soil carbon that became available by the ALT increase. However, the winter CO2 flux increased strongly, almost doubling the annual R_{eco} (from 69 g CO_2-C m^{-2} $year^{-1}$ to 123 g CO_2-C m^{-2} $year^{-1}$). Warming also increased substrate and older SOM decomposition as shown by higher ^{14}C age of the soil air (Natali et al. 2015).

In a similar open top chamber warming experiment in peatlands in Northern Sweden, isotopic analysis of soil CO_2 also showed an increase of the decomposition of old soil carbon at the base of the active layer, and an increase of summer and spring R_{eco} by 52–60%. The contribution of the old carbon to higher R_{eco} was at least 69% (Dorrepaal et al. 2009). Hicks Pries et al. (2013a) found a contribution of old soil carbon to R_{eco} between 6 to 18% during the growing season in tussock tundra on mineral soils in Alaska. This contribution increased linearly with ALT. In Greenland, CO_2 under an experimentally thicker snow cover was older than under normal snow cover (Lupascu et al. 2014). In general increased winter snow cover enhanced decomposition of older soil carbon. However, a thicker snowpack also may reduce the growing season length and summertime R_{eco} (Lupascu et al. 2014).

Altogether, the experiments show enhanced decomposition of older soil carbon by soil warming and ALT increase. In the experiments by Natali et al. (2011, 2014, 2015), significant winter warming was caused by a thicker snow cover, and the highest emissions occurred in winter. These winter emissions completely compensated the increase in vegetation CO_2 uptake at higher temperatures. In the growing season the effects of increasing gross primary production on R_{eco} dominates, by increasing the contribution of autotrophic respiration from plants and soil respiration of recent photosynthesis products (Hicks Pries et al. 2015). Understanding of the processes that contribute to high winter emissions caused by soil warming is therefore important (Sect. 4.6.2.). The effect of carbon release from the soil by warming is more pronounced for peatlands than for mineral soils (Hicks Pries et al. 2015).

These experiments also illustrate how difficult it is to avoid interaction between soil temperature, vegetation and hydrological effects. The insignificant increase of summer R_{eco} in the experiments of Natali et al. (2015) was attributed to waterlogging of the warmed soil, despite the removal of the extra snow used for winter warming to avoid adding extra meltwater. The warmed soil had a higher water table, probably caused by changes in evaporation or soil subsidence. Warming in combination with lower water table by drainage did show a strong increase of growing season R_{eco}, while in experiments without drying also the CH_4 flux increased (Natali et al. 2015). Increased N levels in the leaves of some of the shrub species in the first year of the soil warming experiment shows that a rise of N availability, caused by enhanced soil organic matter decomposition, occurs very rapidly (Natali et al. 2011). In the Swedish peatland experiment by Dorrepaal et al. (2009) and Hicks Pries et al. (2015), enhanced *Sphagnum* growth by warming may have reduced heat flow into the soil.

Long term experiments of decades or more are highly valuable, since experiments that last only a few years might record only transient effects. The sudden increase of warming by an experimental setup may induce adaptation effects that would not occur under normal, more gradual warming conditions. Establishment of a new equilibrium situation may take more than one year (Elmendorf et al. 2012). Such a transient effect was shown in incubation experiments by Dutta et al. (2006); in the first year of the experiment the CO_2 release from the incubated soils was temperature sensitive with a Q_{10} of 1.9, in the following years no temperature sensitivity was found. In the peatland warming experiment of Dorrepaal et al. (2009), the respiration of old carbon increased during the first years of the experiment, but did not increase significantly after 11 years (Hicks Pries et al. 2015).

One long term experiment, a 20 year summer greenhouse warming experiment in Alaskan tussock tundra (Sistla et al. 2013) did not show a change in soil organic matter content, although a redistribution of SOM occurred towards deeper soil horizons. However, in this experiment strong abiotic and biotic interactions occurred between vegetation change (increase of shrub cover), surface heat balance, snow cover and soil ecosystem. Input of litter to the soil increased and litter quality changed, soil shading increased and more snow was trapped in winter by the denser shrub growth. Soil warming was modest in this experiment. In the early years of the experiment, the topsoil warmed in summer by 2 °C, in later years soil warming in summer was negligible. Winter warming of about 2 °C occurred as a result of increased snow cover. While an increase of SOM could be expected from the increased organic matter input to the soil, this likely has been offset by carbon losses resulting from soil warming. However, data on soil microbial and faunal activity show that also the structure and activity of the soil food web did change profoundly during the experiment, with a decrease of microbial respiration in the topsoil and an increase at depth. This may have contributed at least partly to the carbon losses. The summertime soil respiration at onset of the experiment was assumed to be highest but declined during the experiment. This suggests a significant transient effect of experimental warming.

Long term incubation experiments (for 12 years at 5 °C) in Northeast Greenland Arctic heath and grassland by Elberling et al. (2013) indicate significant C loss (9 to 75%) from the incubations, depending on the water saturation of the samples (samples from the dry heather site showed the largest losses). However, a constant 5 °C incubation temperature is high compared to field conditions, where only for a few months of the year the temperature would be above zero. Field soil samples taken over the same time span did not show detectable carbon losses, despite evidence of local warming expressed by an increasing ALT. Similar 1200 day incubations (at 4 °C) of Holocene and Pleistocene permafrost soil in Northern Siberia showed comparable decreases of organic matter as those for wettest samples from Greenland (Knoblauch et al. 2013). In this experiment, anaerobic incubation was included; on average 3.1% of initial C was converted to CO_2 under aerobic conditions, 0.55% was emitted as CO_2 and 0.28% as CH_4 under anaerobic conditions. This demonstrates a strong decrease of overall soil organic matter respiration under anaerobic conditions.

Long term data on R_{eco} change under ambient conditions that avoid the artefacts of experiments or incubations are scanty. Some indications of the effect of warming on R_{eco} can be found in longer eddy covariance records, although these do not allow to discriminate the recent and old carbon components of R_{eco} and generally miss winter data. Some eddy covariance records are long enough to detect trends, or use yearly weather variability as proxy for climate change. In Northeast Siberian tundra, years with a high growing season temperature summation (representing warmer or longer growing seasons) had a higher R_{eco}, but this was offset by higher *GPP* (Parmentier et al. 2011; Fig. 5.7). In the Zackenberg valley in Northeast Greenland, which experienced significant warming and ALT increase, an eddy covariance data set from 2000 to 2011 shows significant correlations of R_{eco} and *GPP* with ALT and summer temperature variables (Lund et al. 2012). *GPP* increased in the first years, but this levelled off later, while R_{eco} was continuously increasing.

Taken together, experiments which separate the components of R_{eco} into an old carbon source and a recent source related to current vegetation growth, show a trend of increasing old SOM decomposition with warmer soil or thicker active layer. In particular winter emissions of CO_2 are large; higher soil temperatures in winter strongly increase R_{eco}. In summer, the old carbon contribution to R_{eco} is lowest. However, the uncertainties are large, resulting from methodological approaches and length of experiments. There is a lack of long-term studies with a consistent methodology. The long term studies that exist suggest significant transient effects of experimental soil warming. Over time the soil food web and vegetation adapt to new conditions, therefore it is difficult to separate long term warming effects from biotic changes of soil and vegetation. The flux measurement data of Lund et al. (2012) under conditions of increasing ALT confirm an increasing R_{eco}, although this does not necessarily indicate net soil carbon losses.

Winter emissions have been included in very few experimental studies and need better quantification, since they may be a crucial part of the total soil carbon loss. If warming is combined with wetter soils (e.g. by subsidence), R_{eco} may be suppressed by anoxic conditions, in favour of higher CH_4 emission. The effect of a thicker winter snow cover likely enhances decomposition of old soil organic matter in winter

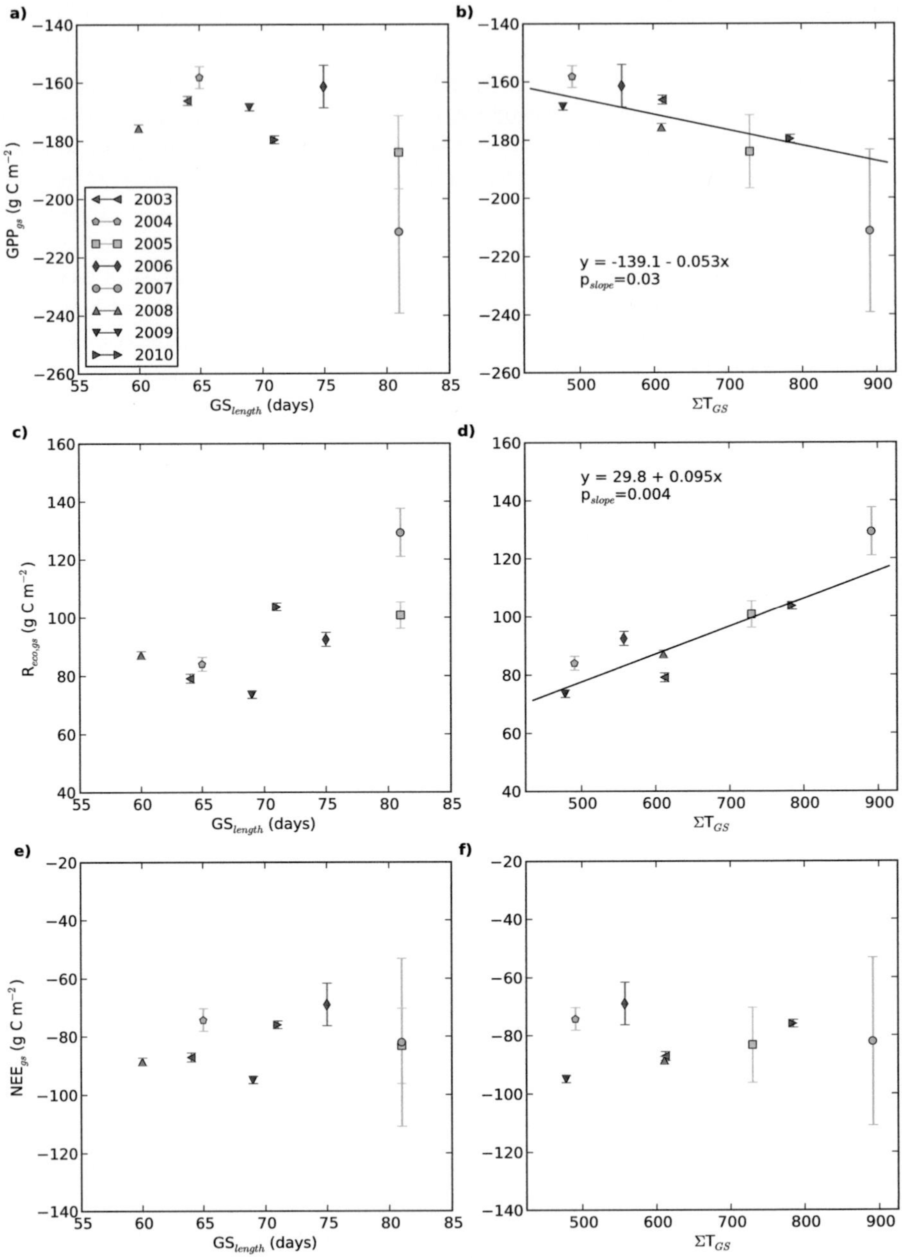

Fig. 5.7 Total eddy covariance fluxes of (a, b) GPP, (c, d) Reco and (e, f) NEE plotted against the growing season length and the temperature sum during the snow-free period for the Kytalyk site in Northeast Siberia. Data from the years 2004, 2008–2010 are complete. The error bars denote measurement and gap-filling uncertainty. For the incomplete years, this also includes uncertainty due to model simulation of the early growing season periods. Regression lines only shown for plots where the slope of the weighted least squares regression is significant ($p < 0.05$). (Republished with permission of John Wiley and Sons Inc, from Parmentier et al. (2011), see reference list; permission conveyed through Copyright Clearance Center, Inc.)

considerably, but much is still unknown of magnitude, processes and sources of winter emissions. Besides soil temperature in winter, a thicker snow cover also interacts with soil hydrology in spring and summer.

5.2.4 Self-Heating Effect

A side effect of bacterial decomposition of organic matter and temperature change that deserves attention is the self-heating effect. Any gardener knows this effect from the compost heap: micro-organisms that decompose organic matter, generate waste heat. If this cannot be dissipated to the atmosphere, self-heating will occur, as long is there is sufficient organic matter to fuel rapid decomposition. Temperatures may rise well above normal soil temperatures; at higher temperature, mesophile and thermophile microbes have taken over the decomposition process, and temperatures up to 70 °C are possible. Even self-combustion is possible (Ryckeboer et al. 2003; Luke and Cox 2011). This self-heating effect also occurs with anaerobic decomposition, e.g. in sewage sludge treatment (Zinder 1990). Zimov et al. (1993) were the first to draw attention to the self-heating effect as a possible explanation of continued winter CO_2 emissions and slow freezing of the active layer in Northern Siberia. Since then, it has been explored with models, which demonstrated it as a viable mechanism to accelerate permafrost thaw (Khvorostyanov et al. 2008a, b).

The model by Khvorostyanov et al. (2008a, b) is applied to dissolved organic carbon in Yedoma deposits, which is labile upon thawing (Zimov et al. 2006; Vonk et al. 2013). It includes self-heating by aerobic decomposition, methanogenesis and methanotrophy, and full gas and heat transfer in the soil. The specific heat released by these processes in the model amounts to 40, 5.5 and 50 MJ kg C^{-1} respectively, which are realistic values (Ryckeboer et al. 2003; Zinder 1990). An illustrative first test of the model (Khvorostyanov et al. 2008a) applied a pulse warming (±3.8 °C) and precipitation increase (0.4 mm d^{-1}) lasting 50 years relative to the present climate for a location in Central Siberia.

In the first 30 years after the warming, most of the deep soil remains frozen and thaws gradually by slow diffusion of heat from the surface. The heat from aerobic decomposition in the topsoil contributes to increasing ALT. After 30 years, a supra-permafrost talik starts to develop, and thaw and decomposition of organic matter accelerate strongly. Intense aerobic respiration of more than 5 g C m^{-3} d^{-1} occurs above the water Table (4 m depth). Below the water table, methanogenesis picks up. CH_4 that migrates upwards is consumed by methanotrophs, which further contribute to heating of the soil; eventually temperatures *up to 40°* are reached. The process declines when the carbon density of the soil falls below a critical value. The soil carbon loss amounts to some 85% in 160 model years.

By contrast, the model shows only gradual ALT thickening if bacterial heating is not included in the model. Further testing of the model used a step change in climate (Khvorostyanov et al. 2008b). Only 7% of the original soil carbon remained after 1000 years of warming by 3.8 °C. The model shows sensitivity for a number of parameters, of which the soil carbon density, methanogenesis rate, microbial heat production and soil porosity are the most prominent. The self heating occurs above a modest soil carbon content of 10 kg C m^{-3}, and microbial heat production of 15 MJ kg C^{-1}.

This model presents an alarming runaway feedback, at temperature input that is not unrealistic, except for their pulse and stepwise function. Luke and Cox (2011) analysed the conditions under which such a runaway effect can occur for peat soils (not specifically permafrost peat), with a simpler soil carbon model and a gradual linear temperature change. Their conclusion is that soil carbon most likely becomes unstable when the heat from microbial respiration is generated more rapidly than it can escape from the soil to the atmosphere. Soils that are most prone to instability of the carbon reservoir are drying peat soils with a high porosity and a cover of an insulating lichen or moss layer – as might occur in boreal peatlands. Furthermore, it is strongly dependent on the rate of global warming. A rate of long-term warming equivalent to 10 °C per century is sufficient to trigger this 'compost bomb' from peat soils. This is within the range of projected average warming for the Arctic under the high emission RCP 8.5 scenario, albeit at the high end of this range (Fig. 1.26).

There is no clear field evidence for the self-heating effect in organic-rich permafrost soils (Luke and Cox 2011). However, the effect needs to be tested in field or laboratory-based experiments that could confirm the potentially unstable behaviour of the soil carbon. Furthermore, the effect of short-lived weather events such as extreme warm and dry summers and the effect of fires also need to be included as triggers. The self-heating effect appears a more important feedback for peat deposits in the northern hemisphere, than specifically for permafrost soils.

5.3 Permafrost Thaw and Geomorphological Change

Permafrost thaw is quite often not a gradual, evenly distributed process, but instead results in localized, rapid and sometimes catastrophic changes of surface geomorphology (Chap. 3). Geomorphological processes that result from thaw of permafrost are formation of thaw depressions or lakes, expansion or drainage of thaw lakes, mass wasting and erosion. Thaw lake formation and the life cycle of thaw lakes has been discussed in Sect. 3.8.3. Together, thaw pond and lake formation and thaw-related erosion are known as thermokarst, which is a somewhat unfortunate term (Sect. 1.4.6) and will not be used here; abrupt thaw is a better alternative.

These processes may occur in a permafrost environment without significant climate change, as a result of other disturbances of the soil heat balance, such as forest and tundra fires, slope erosion processes and disturbance of vegetation and soil by human activities (French 2018; Chap. 6). The subject of this section is the question to what extent, and what rate climate warming is affecting these processes, and how it affects the carbon cycle. If occurring on a sufficiently large scale, abrupt thaw processes also feed back in other ways on the climate and permafrost thaw, by changing land surface characteristics, vegetation and the heat balance.

Geomorphological processes have important effects on the carbon cycle. By moving soil material into other environmental conditions, organic matter decomposition of the

SOM may be enhanced or decreased, and the ratio between aerobic and anaerobic decomposition is changed. Further transport by mass wasting and water flow results in redeposition of the material in an environment where it can refreeze (e.g. in river floodplain sediments), or is exposed to oxygenated or anaerobic conditions (e.g. Walter et al. 2008; Cory et al. 2013; Vonk et al. 2015). If thaw of ice-rich permafrost results in soil subsidence and formation of ponds and lakes, CH_4 emission increases; under better drained conditions, CO_2 emission will dominate (e.g. Schädel et al. 2016). In particular lakes, expanding by permafrost thaw have attracted attention as greenhouse gas sources (e.g. Walter et al. 2006, 2007a). Besides changes in extent and volume of lakes, also changes in ice cover, nutrient and sediment supply and water balance will affect the ecosystem and the carbon balance of these water bodies (Prowse et al. 2011).

5.3.1 Thaw Pond and Fen Development

Much attention in literature has focussed on lakes, but present-day permafrost thaw is more widely shown by the formation of smaller thaw ponds (also known as thaw pits) above shallow ground ice bodies such as ice wedges and palsas (Jorgenson et al. 2006). Pond formation is usually the first and most readily observable sign of active thaw of permafrost. Grosse et al. (2008) showed that lake size follows approximately a power law distribution: Lake number increases exponentially with exponentially smaller size. In remote sensing imagery from space, which is often used to map water bodies in permafrost areas, the smaller ponds are difficult to map, unless high resolution, meter-scale imagery is used (Fig. 5.8). In lake and wetland maps, the area of the large class of small-sized water bodies is likely grossly underestimated (Sect. 3.9). In the Lena Delta, ponds make up roughly half of the water surface area and are responsible for 28–43% of the total landscape scale CO_2 emission, and also contribute significantly to CH_4 emissions (Abnizova et al. 2012). At the same time, ponds are also highly dynamic features, showing changes by thaw and vegetation succession (Nauta et al. 2014; Li et al. 2017a, b) and have a large seasonal dynamics of water level (Boike et al. 2008; Koch et al. 2014; Fig. 5.8).

In this section, a pond is defined as water body that has a surface area smaller than ±100 m^2. These ponds may be of various origin in permafrost areas. Polygon centre ponds and ice wedge thaw ponds are related to the presence of ice wedge polygons (Sect. 3.3). The polygon centre ponds result from impeded drainage and are therefore *not* caused primarily by permafrost thaw, although thaw subsidence can deepen and connect polygon ponds (French 2018). Ice wedge thaw creates deeper, ditch-shaped ponds (Jorgenson et al. 2006; Liljedahl et al. 2016; Fig. 1.21g). Thaw of various other ground ice bodies (ice-cored mounds, palsas) results in more irregular shaped ponds (Fig. 5.8). Lateral erosion of ground ice bodies related to surface water flow is common (Fig. 5.8).

Pond formation alters the heat balance of the surface (Boike et al. 2015). In particular when created by active thaw of permafrost, ponds are biogeochemically highly active due to their effect on soil temperatures, emitting relatively large amounts of greenhouse gases (Sachs et al. 2010; Langer et al. 2015). A positive

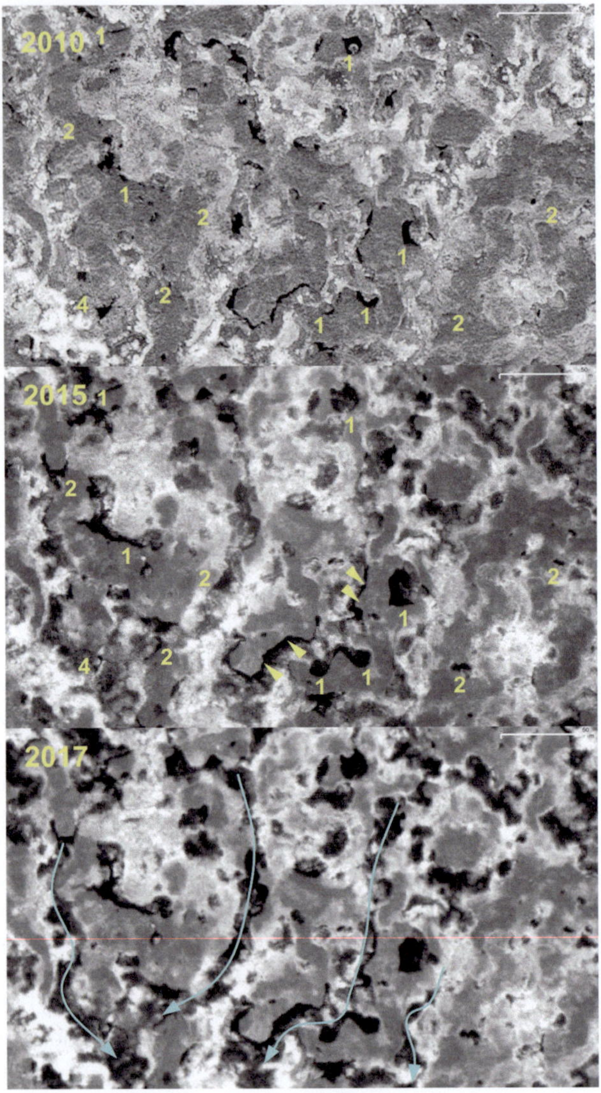

Fig. 5.8 High resolution satellite images from three different years of tundra in an ancient drained thaw lake basin in Northeastern Siberia (Indigirka lowlands). The darkest patches are ponds, dark grey patches are Betula nana heath on dry, slightly elevated surfaces underlain by ground ice bodies (including ice wedges). The lighter areas represent sedge and Sphagnum vegetation. Sedge vegetation usually fills former ponds, or shallow drainage channels. The 2015 image (centre) shows an expansion of pond area (1) or new formation of ponds (2). Less frequently, ponds are overgrown with vegetation (3). The arrows indicate lateral erosion along drainage lines. The lower image dates from 2017, after rapid melt of a thick snow cover and rain, with higher water table in the ponds and activated drainage lines (blue arrows). Scale bar (top right in images) is 50 m. August 2010 image: Geoeye satellite; July 10 2015 and July 2017: Worldview, panchcromatic, 0.4–0.5 m resolution. WorldView © 2019 MAXAR and GeoEye © 2019 MAXAR

feedback on permafrost warming below ponds is also created by the accumulation of a thicker snow cover in the pond depressions in winter (Blok et al. 2010; Nauta et al. 2014). Pond formation and evolution is closely linked to vegetation succession. Thaw of permafrost and creation of a waterlogged depression results in dieback of species that do not tolerate waterlogged soil conditions, and appearance of wetland species such as sedges (Nauta et al. 2014). Over time, vegetation succession in the ponds results in infilling of the ponds (see below).

Jorgenson et al. (2001, 2006) first attracted the attention towards pond formation by ice wedge thaw in permafrost in Northern Alaska. In their study area in Northern Alaska, analysis of air-photos dating between 1945 and 2001 showed an increase in the area (4.4%) and a more than 15-fold increase in density (from 88 to 1336 ponds/km^2) of thaw ponds since 1945 (Jorgenson et al. 2006). This was associated with record warm temperatures during 1989–1998 that increased the mean annual ground temperature by up to 5 °C and the active layer depth up to 25 cm. The increase in pond area was mainly attributed to the thawing of the uppermost parts of ice wedges. Liljedahl et al. (2016) observed at ten sites in Arctic North America and Northern Siberia active, landscape-wide thaw of ice wedges, accompanied by soil subsidence and ponding of water above ice wedge troughs; at an eleventh site, localized ice wedge degradation occurred. This occurred under climatic conditions where active ice wedges should be stable, with MAAT down to −14 °C. This shows that thaw of ice wedges is occurring at a large scale throughout the Arctic. Liljedahl et al. (2016) assume that the degradation is triggered by climate extremes such as an unusual warm summer or thick winter snow cover.

Liljedahl et al. (2016) show that ice wedge thaw may lead do a dryer, rather than a wetter landscape (Fig. 5.9). In the not degraded initial stage, the ice wedges are intact, and marked by ridges and low centre polygons. The first sign of change is vegetation change resulting from subsidence and wetter soil conditions. This is fol-

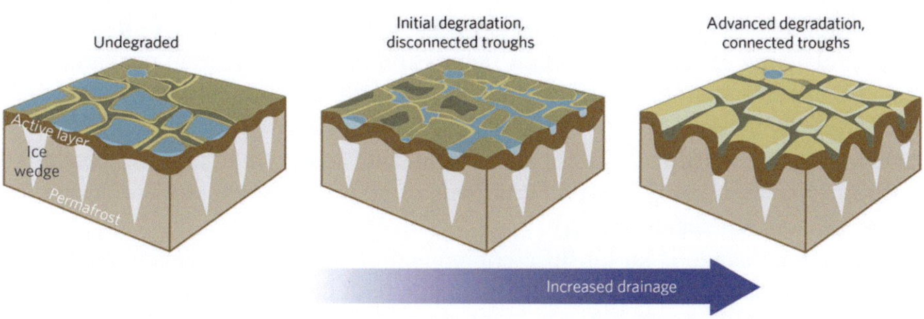

Fig. 5.9 Schematic representation of the ice wedge polygon thaw and drainage succession. Left: the undegraded stage with inundated low-centred polygon centres or non-patterned ground. Centre: the initial degradation stage with disconnected troughs, where the tops of the ice wedges have melted forming trough-ponds. Right: the advanced degradation stage with connected troughs, growing into a laterally connected drainage network, with mounds mounds and high-centred polygons. (Reprinted with permission from Springer Nature from Liljedahl et al. (2016), © 2016, Springer Nature)

lowed by increased trough formation and ponding of water above ice wedges. The next stage is a continued widening and growing connection between troughs. Finally, the troughs start to function as drainage lines, by which the throughs as well as polygon centre ponds are drained. Although the thaw of ice wedges results in an increase of ponded water and wetter soils at first, the improved drainage ultimately leads to a drier landscape, in particular in areas with a sloping topography.

In peatlands, rapid degradation of ice-rich palsas and peat plateaus is commonplace, creating ponds, lakes and fen wetlands by subsidence (e.g. Jorgenson et al. 2001; Payette et al. 2004; Agafonov et al. 2004; Kirpotin et al. 2009, 2011; Turetsky et al. 2007; Borge et al. 2017). In particular isolated permafrost features in boreal peatlands represent relict surface permafrost that is in disequilibrium with the current climate, and therefore extremely sensitive to present climate change (Turetsky et al. 2007). Development of thaw ponds in peatlands represent a significant loss of older soil carbon. O'Donnell et al. (2012a) compared the soil carbon stock in a peat plateau and in thaw depressions with that of recently formed bog peat in the depressions; the formation of thaw depressions caused a net soil carbon reduction of some 40%. Over time, a part of the carbon loss is recovered by new peat growth, but this takes centuries (Sects. 6.2.3 and 6.2.4). In West Siberian peat lands about 70% of the peat is lost by thaw (Audry et al. 2011). The decomposition of peat contributes to subsidence in such cases, besides the subsidence by permafrost ice thaw.

In Northern Norway, the area covered by palsas and peat plateaus has decreased steadily since the 1950s through lateral erosion and formation of new ponds, with a total decrease in area between 33 and 71%; newly formed palsa were not found. Air temperatures increased by 1.0 to 1.5 °C at the same time (Borge et al. 2017). Expansion of thaw ponds in a peatland on the eastern coast of the Hudson Bay in Canada progressed from only 18% of the area in 1957, to 87% in 2003 (Payette et al. 2004). This rapid, large scale down-wasting of a peat plateau was attributed to increased snowfall. It is combined with a self-amplifying thermal effect of increased fragmentation of the peat plateau into isolated palsa islands, which enhances lateral thaw of remaining plateau fragments (Jorgenson et al. 2001, 2013). Agafonov et al. (2004) relate the expansion of a thaw depression in a West Siberian peatland to an increase in precipitation. In the Tanana Flats, Central Alaska, 35% of the area consist of degraded permafrost (Jorgenson et al. 2001). However, in this area, the permafrost degradation already began in the eighteenth century, and most of it occurred before 1950, presumably caused by periods of relatively warm climate during the eighteenth and nineteenth century.

Vegetation change is an intrinsic component of shallow subsidence ponds. The first effect of local subsidence is paludification, with the formation of local wetland vegetation, while existing dryland species decline (e.g. Jorgenson et al. 2001, 2013; Nauta et al. 2014). In general, the vegetation changes from rather dry, nutrient-poor forest or heath vegetation towards a vegetation that is characteristic for water-saturated or submerged soils and higher nutrient availability (e.g. Turetsky et al. 2007; Jorgenson et al. 2001, 2013). Shallow pools are easily recolonized by vegetation (Li et al. 2017a, b; Fig. 5.10). Vegetation succession and deposition of peat and sediment may stabilize further thaw (Jorgenson et al. 2010, 2015; Sect. 6.2.3).

Fig. 5.10 (**a**) Thaw pond on continuous permafrost in Northeastern Siberia, Indigirka Lowlands. In the background right side, a dry mound with Betula nana-dominated vegetation is visible, with on the edges and in the centre of the pool branches of dead Betula nana protruding above the water. On the right side, the pool is colonized by Eriophorum. (**b**) Colonizing Sphagnum patch in Eriophorum-Carex vegetation in a thaw pond, same location. (Photos by the author)

Rapid vegetation succession is driven by release of nutrients from thawed soil, faster decomposition of organic matter (Keuper et al. 2012a), and nutrients provided by surface water and ground water (Koch et al. 2014, 2018; Jorgenson et al. 2001). Nutrient release starts very rapidly after thaw (Reyes and Lougheed 2015). Section 6.2 discusses nutrient and vegetation effects in more detail.

Pond formation liberates soil organic matter for decomposition under higher temperatures and anoxic conditions. Moreover, paludification contributes to the death of previous vegetation that does not tolerate the wet conditions. This additional carbon reservoir contributes to the decomposing carbon pool (Fig. 5.10). Thaw pond water therefore has generally high concentrations of both CH_4 and CO_2 and contributes significantly to landscape greenhouse gas fluxes (Breton et al. 2009; Abnizova et al. 2012). The late summer CO_2 emissions from lakes and ponds in the Lena Delta were an order of magnitude higher than the net CO_2 flux from vegetated tundra surfaces (Abnizova et al. 2012). CO_2 and CH_4 production from bottom sediments continues in early winter as shown by CH_4 bubble distribution in the pond ice (Langer et al. 2015).

The fluxes also depend on pond size, state of thaw activity and water temperature. Pokrovsky et al. (2013) found that CO_2 and CH_4 concentrations were higher in smaller ponds in West Siberia. Langer et al. (2015) found higher CH_4 fluxes in ponds that showed fresh signs of erosion. Pokrovsky et al. (2013) found that the average concentrations of dissolved CO_2 and CH_4 in ponds and thaw lakes increased by factors of 1.4 and 4.9 respectively, during the extremely hot summer of 2012 in Siberia. The concentrations of CO_2 and CH_4 rose by factors of 5 and 150 in the smallest ponds, compared to thaw lakes.

The vegetation strongly modifies the greenhouse gas fluxes (Sects. 6.2.3 and 6.2.4). CO_2 fluxes may be negative when photosynthesis by algae or submerged mosses and vascular plants occurs, but CH_4 fluxes remain generally high. High concentration of particulate or dissolved organic matter may decrease photosynthetic use of dissolved CO_2 (Breton et al. 2009; Laurion et al. 2010). Presence of submerged mosses (Sphagnum, *Scorpidium scorpiodes*) which have methanotrophs as symbionts can reduce CH_4 fluxes strongly (Liebner et al. 2011; Langer et al. 2015). A very pronounced effect of vegetation successional stages on greenhouse gas fluxes in thaw ponds in Siberia was found by Nauta et al. (2014) and Li et al. (2017b). CO_2 and CH_4 fluxes are high in recently formed ponds. CO_2 fluxes decrease to negative (CO_2 uptake) when Eriophorum and other sedges invade the ponds. When *Sphagnum* is established, also the CH_4 flux decreases strongly, probably by CH_4 oxidation in the moss layer (Sects. 6.2.3 and 6.2.4).

Although thaw ponds are significant sources of CH_4 and CO_2, vegetation succession potentially reduces this emission over time. Moreover, the nutrient-rich conditions result in a high *GPP* when vegetation succession fully develops. This stimulates net carbon accumulation in new peat, in contrast to the dry surfaces of uplifted palsa and peat plateaus in boreal and arctic peatlands (Turetsky et al. 2007). Nevertheless, despite the establishment of an actively carbon storing ecosystem in ponds, the thaw pond and lake development represents a significant loss of older peat carbon (Audry et al. 2011; O'Donnell et al. 2012a; Sect. 6.2.3).

Thaw pond development may be the most climatically sensitive reaction to both increasing precipitation and warming in permafrost areas. It produces a significant positive feedback to climate change by storing more summer heat in the soil and enhancing greenhouse fluxes to the atmosphere. These emissions from large areas of high-latitude permafrost are probably strongly underestimated, since ponds are commonly not included in the global land surface classifications that are used for estimating the Arctic carbon budget. Moreover, small variations in the water balance (precipitation, evaporation, or drainage) potentially produce changes in these greenhouse gas emissions, since the water table level strongly affects the volumes and surface areas of ponds (Abnizova et al. 2012).

On the other hand, ponds are relatively short-lived phenomena. As shown by observations cited above (e.g. Turetsky et al. 2007; Li et al. 2017a, b) vegetation succession in thaw ponds may restore the carbon sink function in relatively short time. Resilience of the carbon sink is discussed extensively in Sect. 6.2.4.

As shown in many areas, the rate of pond formation has increased, indicating a net increase of greenhouse gas fluxes (Jorgenson et al. 2006; Liljedahl et al. 2016). Lara et al. (2015) made an estimate of future emissions from thaw ponds for the Barrow area on the north Alaskan Coastal Plain, based on geomorphological terrain mapping, observed present-day *NEE* and CH_4 fluxes in various geomorphic units and current trends in thaw pond development. They concluded that increased thaw pond area led to an increase in CH_4 emission from ponds during the peak summer season, but this was offset by an increase in *NEE*, also when the larger radiative effect of CH_4 was taken into account. However, this is only the peak summer emission; the high emissions of CH_4 and CO_2 outside the peak season (Sect. 4.6.2) were not included in this estimate, and may be substantial.

Permafrost thaw has not been uncommon in the past as shown by thaw unconformities, indicating a previous thicker active layer, dating from the Holocene Thermal Maximum, or from local environmental change such as shift in coastlines or vegetation zones (Burn 1997; Jorgenson et al. 2001, 2006). However, the widespread formation of ice wedge ponds in areas with continuous permafrost, where the current climate is still sufficiently cold to maintain permafrost, indicates that climate change is provoking widespread abrupt thaw (Jorgenson et al. 2006; Liljedahl et al. 2016).

5.3.2 *Thaw Lake Expansion*

At a larger scale, subsidence by permafrost thaw creates lakes. Like smaller ponds, thaw lakes are a source of greenhouse gases (Sect. 4.1). They receive import of organic matter from runoff and thaw of the underlying permafrost, which is respired as CO_2 and CH_4. The anaerobic conditions in the lake bottom sediment induce high rates of CH_4 emission, which continue in winter, when CH_4 is trapped in bubbles in the ice (Walter et al. 2006). Thaw lakes therefore acts as a conduit of permafrost carbon to the atmosphere. However, on very long time scales the net primary production of the lake ecosystem and sequestering of carbon in lake bottom sediments also may result in a CO_2 sink. Moreover, the lifetime of lakes is finite and drained lakes may become sites of carbon accumulation (Sect. 3.9).

Besides their contribution to the permafrost carbon feed back, the lakes also feed back on the local climate by way of the surface energy and water balance (Eugster et al. 2000; Jorgenson et al. 2010; Boike et al. 2015). The lake water below the ice in lakes that do not freeze to the bottom in winter represent a considerable heat reservoir. Jeffries et al. (1999) found that the winter ground heat flux in Alaskan North Slope tundra was 1.5 W m^{-2}, while the heat flux from lakes with floating ice was an order of magnitude higher, 18.6 W m^{-2}. In the lake-rich areas of the North Slope, this heat flux contributes significantly to the regional heat budget. Lakes also strongly influence river discharge peaks by storage of spring snow melt water (Bowling and Lettenmaier 2010; Woo 2012).

Do thaw lakes and ponds expand in area, and can this be attributed to climate change? As seen in Sect. 3.9, there are several potential relations of lake formation and drainage processes with climate that influence lake formation, expansion and disappearance (Table 5.2). The extent to which thaw lakes expand or contract is important for both the permafrost carbon balance and the land surface heat balance in the future. The ratio of CO_2 and CH_4 emission from carbon which is liberated by thawing permafrost, and the potential for preservation of carbon in lake sediments, depend at least partially on thaw lake expansion. However, irrespective of lake expansion, also heat balance changes of lakes affect their carbon and greenhouse balance. An increase in the length of the ice-free season will increase energy input into the lake water that increases lake water temperature, and is expected to increase CH_4 emissions (Wik et al. 2016; next section).

The winter ice thickness is decisive for the presence of a lake talik (Sect. 3.9). If the winter ice is bedfast (frozen to the lake bed) for a sufficiently long time in winter, frost penetrates into the lake bed. A change in the winter ice regime towards a floating ice regime results in a permanent water body above the lake bed, and development of a talik below the lake. The lake bed temperature is strongly related to the

Table 5.2 Climate-related components of the thaw lake system, summarized from Sect. 3.8. The processes that reduce lake area are marked grey. Although bank erosion processes expand lake area, but erosion also enhances the probability of lake drainage

Climate variable	Lake system component change
Increasing summer T	Lake bottom temperature increase → lake deepening → bank erosion
	Lake bottom temperature increase → open talik development → drainage
	Thermal erosion of banks
	Pond coalescence / lake initiation
Increasing winter T	Thinner ice cover → lake deepening → bank erosion
	Lake bottom temperature increase → lake deepening → bank erosion
Increasing summer P	Lake level higher → erosion drainage threshold → drainage
	Increased bank erosion by runoff and mass wasting
	Lake level higher → thermal erosion of banks
	Pond coalescence / lake initiation
Increasing winter P	Thicker snow cover → thinner ice cover → lake deepening → bank erosion
	Summer lake level higher → thermal erosion of banks
	Pond coalescence / lake initiation in spring
Change P – E	Change in lake water volume → area expansion or shrinkage
Wind strength	Increased thermal and physical erosion of banks
All	Erosion of banks → erosion drainage threshold → drainage

P precipitation, *E* Evapotranspiration, *T* temperature

difference between maximum ice thickness and lake depth (the effective lake depth). When this effective lake depth is larger than approximately −0.5 m (floating ice or infrequent bedfast ice), above-zero bed temperatures in winter are common (Arp et al. 2012; Fig. 5.11).

Climate warming and increased winter snow cover thickness result in a thinner lake ice cover. An increasing number of shallow lakes, and shallow parts of lakes, will be exposed to continued permafrost thaw, talik development and lake bottom subsidence (Arp et al. 2016). A study on lake ice in the Alaskan coastal plain has shown that about 16% of lakes that had bedfast ice in 1980, had a floating ice regime after 2003. Temperature measurements have shown increasing lake bottom temperatures in shallower lakes as a result of winter ice thinning (Arp et al. 2012, 2016). Deeper lakes also allow for more rapid lake expansion (Sect. 3.9). Lake initiation is the result of coalescing and deepening of smaller ponds (Grosse et al. 2013). This need not be the result of subsidence by thawing permafrost only, also a changing water balance by increased effective net precipitation (precipitation – evapotranspiration) contributes (Fedorov et al. 2014). Water balance change may have been crucial for the large-scale thaw lake expansion during the termination of the Last Glacial (Jorgenson and Shur 2007).

The future development of the effective net precipitation input to the water balance is uncertain, and is likely to vary throughout permafrost regions. Rawlins et al. (2010) see generally positive trends in precipitation and evapotranspiration from

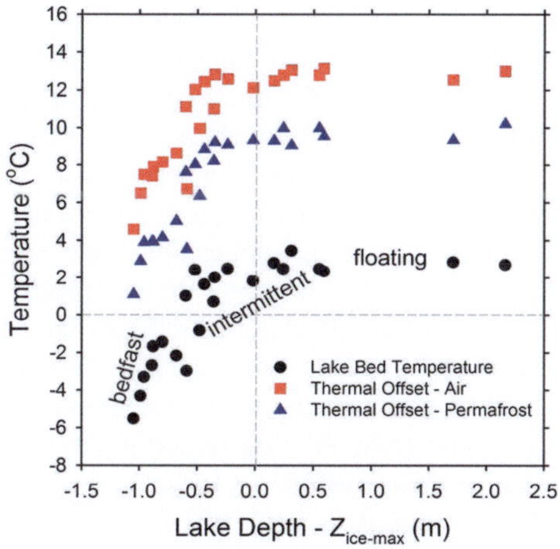

Fig. 5.11 The relation between effective lake depth (depth – maximum ice thickness) and mean annual lake bed temperature, the thermal offset of mean annual lake bed temperature with air temperature (3 m height) and the thermal offset of mean annual lake bed temperature with permafrost at 1.2 m depth for thaw lakes in Northern Alaska. Republished with permission of John Wiley and Sons Inc, from 'Shifting balance of thermokarst lake ice regimes across the Arctic Coastal Plain of northern Alaska' by Arp et al. (2012) in Geophysical Research Letters 39 (L16503) © 2012; permission conveyed through Copyright Clearance Center

climate models and observations, but the uncertainty in the observations limits their robustness. Iijima et al. (2010, 2016) report increased enhancement of cyclones over the Arctic Ocean and increased eastward penetration of storms in late summer and early winter in the central Lena Basin between 2004 and 2008, which enhanced soil temperatures and soil wetness. Fedorov et al. (2014) and Zakharova et al. (2018) report enhanced thaw lake growth in this region for these years, but since 2011, a trend of decreasing lake levels has set in (Zakharova et al. 2018). Pokrovsky et al. (2013) report significant lake shrinkage in Western Siberia during the extremely hot summer of 2012 in that region. The role of natural climate variability in thaw lake initiation by changes in precipitation/evapotranspiration should not be underestimated.

Lake expansion by bank erosion is a complex phenomenon (Sect. 3.9). The lateral heat transport from lake water into permafrost induces permafrost thaw, but erosion and sediment removal are necessary agents for lake expansion. The direct thermal and physical erosion by waves, currents and floating ice remove material from the banks, which is subsequently transported by currents into deeper parts of the lake basin. Therefore, increasing summer water temperatures, high lake water levels, and a high incidence of wave-forming and current-inducing winds should foster lake expansion by erosion. Also, weather conditions (summer warmth, rainfall) that enhance mass wasting on higher, ice-rich lake banks contribute to lake expansion (Sect. 5.3.2).

A necessary condition for continuous lake expansion is the continuous creation of sediment storage space by active thaw and subsidence beneath the lake bottom and a modest sediment supply from lake bank erosion (Plug and West 2009; Kessler et al. 2012). An increasingly thicker sediment layer from bank-eroded material results in slower heat penetration into the underlying permafrost (Brouchkov et al. 2004) and sedimentation can create shallow areas where permafrost eventually can re-establish (Sect. 3.9). A high ice volume of the permafrost and mostly low, ice-rich lake banks that do not supply large amounts of sediment are therefore necessary conditions for continued subsidence; this makes thaw lake expansion dependent on terrain and geological conditions besides any climate sensitivity. Numerous overlapping drained thaw lake basins are present in Arctic lowlands, of which the largest lake initiation and expansion phase dates from the Last Glacial Termination to Mid Holocene (Walter et al. 2007b). However, many of the present-day lakes are secondary lakes, developing in older thaw lake basin sediments, with generally less ice volume than the original ice-rich substrate. A lower ice volume tends to reduce the rate of lake growth (Morgenstern et al. 2011, 2013).

In areas of peat plateaus and palsas with discontinuous permafrost, groundwater movement may enhance permafrost collapse (Jorgenson et al. 2001, 2013). Seepage of groundwater with above-zero temperatures enhances thaw at the edges of palsas, and groundwater movement in turn may be enhanced by enlargement of thawed areas. In thaw lakes in peatlands, decomposition of peat also contributes considerably to subsidence; in West Siberian peatlands up to 70% (Audry et al. 2011).

Changes in lake area and the number of lakes have been reported from several remote sensing studies. Given the many climatic and non-climatic factors that affect lake expansion, it is not surprising that observations of lake area change show both decrease and increase, and vary regionally (Table 5.3).

- First, regional variations that are not directly related to climate, such as local relief, state of the permafrost (discontinuous/continuous, ice content and temperature) and subsurface geology affects these observation (Marsh et al. 2009; Kravtsova and Bystrova 2009).
- Second, areal change of lakes and ponds is difficult to estimate; the area of ponds and lakes is influenced by year-to-year variability of the water balance (e.g. Plug et al. 2008; Fedorov et al. 2014), which masks general trends.
- Third, lake expansion may result in reduction of lake area when drainage sills with other lakes or the river system are eroded away (Van Huissteden et al. 2011; Jones et al. 2011).

The general pattern that emerges from the various studies can be summarized as follows (Table 5.3):

- *Decrease of lake area in discontinuous permafrost areas.* Reported by Nitze et al. (2018) from four transects in Alaska, West Canada, West Siberia and east Siberia; with in particular a decrease at the boundary of continuous and discontinuous permafrost.
- *Decrease of lake area in discontinuous permafrost areas, increase in continuous permafrost.* Smith et al. (2005) reports an overall decrease of lake area by ±6% in Siberia, with in continuous permafrost areal increase of 12%, and a 45%

Table 5.3 Results of thaw lake change studies based on large scale remote sensing, showing changes in the number of lakes, total lake area, lake expansion rate, drainage rate and new lake formation

Area	Lake num-ber	Lake area	Exp rate %	Drain / infill rate	New lakes	Reference	Nr yrs
Seward Peninsula Alaska	+	–	stable	stable	remnant lakes	Jones et al., 2011	
Southern Seward Peninsula Alaska		–				Yoshikawa and Hinzman (2003)	
Alaska Continuous permafrost		−0.24	3.21	3.45		Nitze et al., 2018	15
Alaska Disontinuous pf		−5.96	5.43	11.39		Nitze et al., 2018	15
Tuktoyaktuk Peninsula, NW Canada				–		Marsh et al., 2009	
Tuktoyaktuk Peninsula, NW Canada		+/−; recent −				Plug et al., 2008	
Northern Yukon, NW Canada		–			new ponds	Labrecque et al., 2009	
Boreal forest zone Alaska	–	–				Riordan et al., 2006	
Arctic coastal plain Alaska		stable				Riordan et al., 2006	
Hudson Bay Lowlands	−/stable	+	0.11-0.27	≤0.07	new ponds	Sannel and Kuhry, 2011	
West Canada Continuous pf		−2.02	0.77	2.79		Nitze et al., 2018	15
West Canada Discontinuous pf		−1.50	0.48	1.99		Nitze et al., 2018	15
Sweden	+/−	–	≤0.2	≤0.76		Sannel and Kuhry, 2011	10
Northern Russia	+/−	+/−	<0.1	<0.08		Sannel and Kuhry, 2011	10
Northwest Siberia, Nadym, Pur basins		+/− stable				Karlsson et al., 2014	
Northwest Siberia		–		+		Kravtsova and Bystrova (2009)	
West Siberia Continuous pf		−4.29	1.70	5.99		Nitze et al., 2018	15
West Siberia Discontinuous pf		−7.89	1.18	9.07		Nitze et al., 2018	15
Central Yakutia		+	+		thaw	Kravtsova and Bystrova (2009)	
Central Yakutia	+	+	+	+,−		Boike et al. (2016)	
East Siberia Continuous pf		3.81	7.83	4.02		Nitze et al., 2018	15
East Siberia Discontinuous pf		−3.12	3.16	6.28		Nitze et al., 2018	15
Southeast Siberia		+	+		thaw	Kravtsova & Bystrova (2009)	
West Siberia peatlands		–		+		Kirpotin et al. (2009)	
Northwest Siberia		+				Kirpotin et al. (2009)	
Northern Siberia		+				Smith et al. (2005)	
Southern Siberia		–				Smith et al. (2005)	

All numbers are percentages; negative is decrease. Grey bars: discontinuous permafrost areas, otherwise continuous permafrost. Expansion and drainage/infill rate in area percentage per year

increase in the number of lakes. Kirpotin et al. (2009) observed a similar pattern in West Siberia with a decrease of lake area between 12 and 29% in discontinuous permafrost. In continuous permafrost of Central Yakutia, lake area increased by 11–42% (Boike et al. 2016) coinciding with increasing autumn and spring temperatures, and an increase in soil temperature and precipitation. Fedorov and Konstantinov (2009) and Fedorov et al. (2014) report large expansion of a single thaw lake in Central Yakutia.

- *Lake shrinkage by drainage and infilling.* Lake shrinkage by both drainage and infilling with fen vegetation is common in thaw lakes in peat plateaus (Sannel and Kuhry 2011).
- *No consistent trend.* Karlsson et al. (2014) and Kravtsova and Bystrova (2009) did not find a consistent trend in lake area in Siberia. At most of the study areas of Kravtsova and Bystrova (2009), the lake area was stable, but in Central Siberia lake area was expanding.

- *Decrease of larger lakes by drainage, but increase of smaller lakes.* Jones et al. (2011) observed on the Seward Peninsula in Alaska between 1950 and 2007 an increase in the number of lakes >0.1 ha by 10.7%, while the total lake surface area decreased by 14.9%. In particular large lakes (>40 ha) decreased both in number (−24%) and area (−26%) by drainage. Lake expansion rates (0.35–0.39 m $year^{-1}$) did not vary substantially.
- *Decreasing lake drainage rate.* Marsh et al. (2009) report for the Tuktoyaktuk Peninsula in Northern Canada a decreasing lake drainage rate from 1950 till 2000.
- *Lake area variations related to variations in net effective precipitation.* Labrecque et al. (2009) report for Northwest Canada a recent decrease in lake area, following an increase in lake area before 1972 attributed to wet/dry climate variation related to the Arctic and Pacific Decadal oscillations. Similar results are reported by Plug et al. (2008) for the Tuktoyaktuk Peninsula (increase in lake area before 1992, followed by decrease). Riordan et al. (2006) report a decrease in both lake area and number for the boreal forest zone in Alaska, attributed to both permafrost thaw and a warmer and drier climate with higher evapotranspiration. Fedorov et al. (2014) and Zakharova et al. (2018) report precipitation/evapotranspiration-related lake variations in Central Siberia.
- *Variations based on ice cover as driver for evapotranspiration.* Arp et al. (2011) found that on the Alaskan Arctic coastal plain shallow, bedfast-ice lakes were more sensitive to changes in precipitation/evapotranspiration than deeper, floating-ice lakes; deeper lakes retain their ice cover longer in summer, reducing evaporation.

Net lake area expansion is evidently not commonplace throughout the permafrost zone. In the discontinuous permafrost zone lake area shrinkage (by drainage or evapotranspiration) dominates. In the continuous permafrost zone large regional differences exist. In continuous permafrost areas in Northern America the general trend is a decrease of lake area, related to both climate-driven change in permafrost and change in net precipitation. In continuous permafrost areas in Central Siberia, there is a well documented change towards increasing lake area, which is driven by both permafrost thaw and an increase of net precipitation. Regionally, lake area expansion by creation of new lakes and bank erosion may be more than compensated by the drainage of a few older, large lakes. Changes in net effective precipitation tend to be dominated by pseudo-cyclic changes in atmospheric circulation, such as the Arctic Oscillation. Furthermore, climate related vegetation change (longer growing season, changes in vegetation structure and biomass) which enhance summer evapotranspiration may be an important component (Riordan et al. 2006).

5.3.3 *Thaw Lake Carbon Cycle Change*

Changes in the carbon cycle of thaw lakes are caused by increased permafrost thaw that result in expansion of lakes and lake taliks, exposing more permafrost carbon to thaw. Furthermore, changes in other environmental factors of the lake ecosystem – affecting ecosystem respiration R_{eco}, or primary production *GPP* affect the carbon cycle of lakes. These are caused for instance by changes in lake ice cover, water temperature or water and nutrient supply.

In lake environments, the proportion of anaerobic decomposition is high, which produces both CO_2 and CH_4, next to CO_2 produced by aerobic decomposition. Emission of carbon from wetland lakes in the discontinuous permafrost zone consists for more than 90% of CO_2-C; however, in terms of effect on the climate, the CH_4 emissions dominate (Repo et al. 2007). Much of the CO_2 in lake emissions is also produced anaerobically. Aerobic CO_2 production occurs mainly in the water column, while anaerobic CO_2 production occurs in the sediment layer. Audry et al. (2011) found evidence of nitrate, sulfate and Fe and Mn reduction in West Siberian thaw lakes. Although the focus in many publications on thaw lake greenhouse gas emission is put on CH_4, CO_2 production still is an important component.

Walter et al. (2007a) estimated CH_4 emissions from lakes (ebullition + diffusion) in permafrost areas at 18.2 ± 6.5 Tg CH_4 year^{-1}. In a synthesis of northern lake CH_4 emissions (north of 50°), Wik et al. (2016) estimate emissions from thaw lakes at 4.1 ± 2.2 Tg CH_4 year$^{-1,}$ based on 149 lakes. This is about one fifth of all northern lake emissions (21.3 ± 11.9 Tg CH_4 year^{-1}; the largest sources are glacial lakes). Most of the emission occurs by ebullition (about 79%) during the ice-free season. About one fifth of the emission is stored in winter ice and released during spring (Fig. 5.12).

Compared to ponds, CO_2 and CH_4 emissions from lakes are smaller on an area basis; an expansion of thaw ponds should have a stronger effect on CH_4 emission than lake expansion (Langer et al. 2015; Lara et al. 2015). The CH_4 emission is strongest in the nearshore zones of thaw lakes where the contribution of soil carbon eroded from the banks is highest. Lake CH_4 emissions in the Kolyma lowlands amounted to 25 g CH_4 m^{-2} year^{-1}, but fluxes from the 15 m wide band adjacent to actively eroded banks are more than five times higher: 128 g CH_4 m^{-2} year^{-1}, representing 79% of the total lake flux. This high CH_4 flux is driven by the contribution of active talik deepening and eroded bank material (Walter et al. 2006). Isotopic analysis and radiocarbon dating of the emitted CH_4 from thaw lakes in Siberia and Alaska has shown that most of the CH_4 carbon is derived from old Pleistocene organic matter. Walter et al. (2008) report CH_4 from high-emission hot spots and point sources in Siberian thaw lakes with an age range between 11,355 to 42,900 years, while the low-emission background ebullition emissions have an age

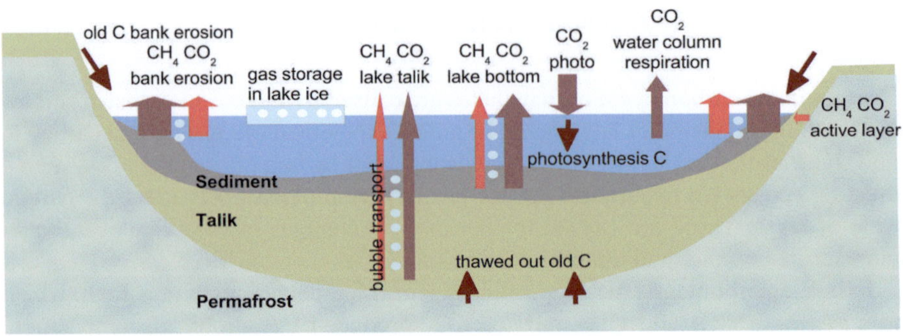

Fig. 5.12 The carbon/greenhouse gas balance of a thaw lake. The thickness of the arrows indicates an approximate relative magnitude of the greenhouse gas flux. CH_4: reddish arrows, CO_2: brownish arrows

between 1345 to 8845 years. The high emission sources were related to active thaw of old yedoma deposits and show an isotopic composition that indicates hydrogenotrophic methanogenesis. Methanogenesis shifts from the acetoclastic pathway towards CO_2 reduction if more old carbon is involved (Zimov et al. 1997; Nakagawa et al. 2002; Walter et al. 2007a).

The CH_4 release from actively expanding thaw lakes is proportional to the amount of thawed permafrost carbon by lake expansion. This is shown by numerical modelling of permafrost thaw and data on nearshore erosion. Radiocarbon dating of emitted CH_4 is nearly identical to the age of the organic matter stored in permafrost near lake banks. This demonstrates the importance of old organic matter for thaw lake emissions (Walter Anthony et al. 2016). An additional increasing source of lake water CH_4 may be increasing CH_4-saturated groundwater flow from thicker active layers of soils surrounding the lakes (Paytan et al. 2015).

According to Walter Anthony et al. (2016), 0.2 to 2.5 Pg of carbon was released as CO_2 and CH_4 directly related to expansion of thaw lakes alone during the last 60 years. If 50% is emitted as CH_4, this would amount to 2.2 to 27.8 Tg CH_4 year^{-1}. The upper limit is difficult to reconcile with the estimate of thaw lake emissions by Walter et al. (2007a) and Wik et al. (2016) and testifies of the large uncertainties of these emission estimates. Walter Anthony et al. (2018) arrive at a smaller model-based estimate of present-day direct emissions from thaw lake expansion of 0.7–4.0 Tg CH_4 year^{-1}, which is more in line with the other estimates of total thaw lake emission.

How this would change with future warming, depends on the future greenhouse gas emission scenarios, and on the interaction with climate of the processes affecting thaw lake expansion, drainage and biogeochemical cycling. Walter et al. (2007a) assumed an acceleration of thaw lake growth in Northeast Siberia resulting from the present warming climate, and extensive reworking of ice-rich Yedoma deposits containing labile organic carbon. However, given that many thaw lakes are secondary lakes in older drained thaw lake basins with old permafrost carbon that has undergone already a decomposition cycle (Morgenstern et al. 2011, 2013), it is uncertain how much labile carbon is really available. Furthermore, the decrease of thaw lake area by drier climatic conditions in North America or by drainage of lakes (Table 5.3) suggests that thaw lake area also may contract in the future. There is large uncertainty on future trends, although all climate models predict wetter conditions in the future (Walter Anthony et al. 2018).

Next to a higher contribution of permafrost carbon from lake expansion, the lake ecosystems will change as well. Lake water and bottom temperatures and the length of the ice-free season are likely to rise (Prowse et al. 2011), which increases the primary production. This may be further enhanced by an increasing supply of nutrients from decomposing permafrost organic matter (Kokelj et al. 2005; Schuur et al. 2007), and from increasing anthropogenic nitrogen deposition in Arctic environments (Dentener 2006; Lougheed et al. 2011). In tundra ponds near Barrow, Alaska, Lougheed et al. (2011) recorded in 40 years a significant increase in water temperatures and nutrients, and a higher phytoplankton biomass. However, also the emission of CH_4 is likely to increase from a larger supply of labile organic matter and increased energy input driving ebullition of CH_4 (Wik et al. 2014, 2016, 2018). Ebullition is not strongly related to the carbon content of lake sediments. Rather, the effect of warmer water and increased heat input to the bottom sediment resulting from longer ice-free seasons are likely to alter future ebullition emissions (Wik et al. 2018).

To what extent an increase of primary productivity of lake ecosystems and burial of organic carbon in sediments may compensate for decomposition of thawed organic matter in lake taliks and from eroded banks, is uncertain. For permafrost thaw in discontinuous permafrost peatlands, Turetsky et al. (2007) show that dry peatlands with a small net ecosystem carbon balance are replaced by wetlands with a high rate of carbon uptake by the creation of thaw depressions. Localized regrowth of permafrost may occur by establishment of new palsa mounds (Payette et al. 2004).

However, as noted above, any renewed carbon uptake is preceded by high carbon losses (Audry et al. 2011; O'Donnel et al. 2012a, b). Moreover, again the distinction between the carbon balance and the greenhouse gas balance is important; high CH_4 emission offsets the increased CO_2 uptake. The new wetlands and lakes created by thaw collapse of relatively dry peat plateau and palsa surfaces have a generally high production of CH_4, while the former dry peat surface may have been a sink for CH_4. Turetsky et al. (2007) found that increased CH_4 emissions to the atmosphere offsets the enhanced peatland carbon sink by permafrost degradation for at least 70 years after degradation occurred. And not only CH_4 is important here. Thawing of frozen peatlands may induce emission of the more potent greenhouse gas N_2O, in particular if thawing causes drainage and rewetting cycles (Elberling et al. 2010; Marushchak et al. 2011).

Drainage of thaw lakes results in the re-establishment of a carbon and CO_2 sink in soil and vegetation (Hinkel et al. 2003; Van der Molen et al. 2007; Jones et al. 2012), although the resulting wetlands still may remain a net greenhouse gas source by CH_4 emission (Zona et al. 2016). Nevertheless, the CH_4 source in drained lake beds is likely to be some 80% lower than that of lakes (based on comparison of data by Walter et al. 2006 and Zona et al. 2016). In drained lake basins in the Seward

Peninsula in Alaska, high peat accumulation rates of 35.2 g C m^{-2} $year^{-1}$ were recorded in recently (<500 years) drained lake basins, which declines to 9 g C m^{-2} $year^{-1}$ in old basins (Jones et al. 2012).

If thaw lakes are not drained, the recovery of a carbon gas sink takes longer; carbon sedimentation rates may be high (Sect. 4.4.3) but usually do not compensate for high CH_4 and CO_2 emission. Walter Anthony et al. (2014) derived carbon accumulation rates in Holocene thaw lake sediments for Siberian thaw lakes, averaging 47 ± 10 g C m^{-2} yr^{-1}. Over time carbon accumulation in lake sediments compensates for the large initial CH_4 source of the lakes (Sect. 6.2). CH_4 emission from thaw lakes may not only originate from biogenic decomposition of permafrost organic matter in the lake talik, but also from deeper geological sources, such as methane seeps along faults. Thaw of permafrost potentially may enhance seeps (Walter Anthony et al. 2012). This will be discussed in Chap. 7.

Several attempts have been made to reconstruct past thaw lake changes and emissions, and link these to known climate change. Lake sediments lend themselves well for reconstruction of both climate change and erosional activity, which are recorded in the sediment by macro- and microfossils and sediment properties. In forested permafrost areas, tree ring dating of tree trunks which have fallen into the lakes by bank erosion can be used to date lake expansion.

Depositional cycles in thaw lake sediments are common (Hopkins and Kidd 1988; Van Huissteden 1990; Murton 1996, 2001). Present lake basins may have a long and complicated history of lake initiation, partial drainage and renewed expansion (Lenz et al. 2016; Schleusner et al. 2015). Depositional cycles need not be related to climate if recorded in a single lake only (Biskaborn et al. 2013). Biskaborn et al. (2013) used geochemical properties to record pulses of sediment flowing into a northern Siberian thaw lake. They found evidence of a cyclicity of bank erosion, expressed by deposition of coarser and finer sand throughout the Holocene; the duration of the cycles was in the order of 500–1000 years. These cycles could be related to climate change: periods of increased summer warmth, precipitation, or increased summer storminess. However, Biskaborn et al. (2013) also propose non-climatic explanations involving slope processes on the lake bank, produced by inhomogeneous distribution of ice in the lake banks, which could produce sediment production pulses.

Evidence for climatic causes of lake changes is stronger when recorded regionally. Radiometric dating of lake sediment cores and geochemical analysis of the lake sediment of two lakes in Northern Siberia showed three episodes of increased permafrost thaw during the last 170 years: 1870 to 1880, 1900 to 1930 and from 1960 till present (Fedotov et al. 2012). Katamura et al. (2006) recorded periods with enhanced lake initiation in the Lena Basin in Central Siberia, coinciding with warm and moist climatic episodes during the Holocene. Jones et al. (2012) and Lenz et al. (2016) recorded a high frequency of lake drainage events in the Seward Peninsula in Alaska during the Medieval warm period.

A proxy for past lake CH_4 production has been developed by Van Hardenbroek et al. (2012). Some invertebrate species in lakes feed on methanotrophic bacteria, which have a light carbon isotopic signature by their consumption of isotopically light biogenic CH_4. In particular the ^{13}C content of Chironomid larvae (non-biting midges) groups, and Daphnia (water flea) were significantly correlated with diffusive methane fluxes. By detecting the ^{13}C content of Chironomid larvae in Siberian thaw lake sediments lake sediments, variations in CH_4 release from the bottom sediment could be detected. The invertebrate $\delta^{13}C$ record over the last 1000 years showed a good agreement with a reconstruction of local July temperatures from tree rings, indicating a decrease of temperature and CH_4 emission over the last thousand years, followed by a recent increase (Fig. 5.13).

5.3.4 *Erosion*

Slow mass wasting processes on slopes, like solifluction, are climate-sensitive since they are affected by freeze-thaw cycles (both diurnal and seasonal cycles) and soil moisture. In permafrost environments with MAAT below −6 °C, surface movement rates are low, but speed increases at higher temperatures (Matsuoka 2001), and therefore mass movement rates can be expected to increase with climate warming, including the rate of downslope carbon transport. On the other hand, increased veg-

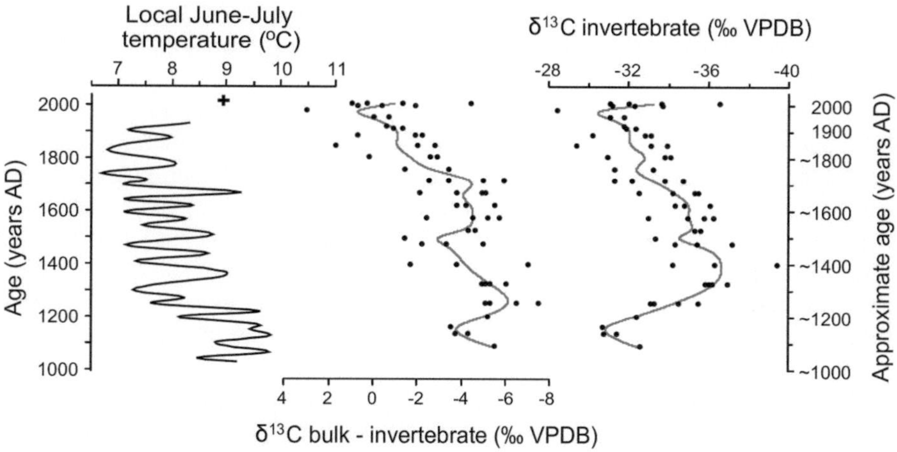

Fig. 5.13 Temperature time series reconstructed from tree rings and $\delta^{13}C$ of invertebrates in thaw lake sediments in Northeastern Siberia, Indigirka lowlands,. Left: June–July temperature reconstruction (smoothed using a 50-year low-pass filter) based on a local tree-ring series from the region around the town of Chokurdagh, northern Yakutia (references in Van Hardenbroek et al. 2012). The '+' indicates the 1997–2006 average June–July temperature, measured at the nearest meteorological station in Chokurdagh. The temperature record is compared with offsets between bulk organic matter $\delta^{13}C$ and $\delta^{13}C$ of Chironomini, Tanytarsini, and Daphnia (centre), and the absolute $\delta^{13}C$ of the same taxa (on right). A smoother filter (bandwidth 0.15) is fitted through the invertebrate data. Note inverted x-axis for $\delta^{13}C$. (Reprinted from Quaternary Science Reviews 66:74–84, Van Hardenbroek et al. (2012): Evidence for past variations in methane availability in a Siberian thermokarst lake based on $\delta^{13}C$ of chitinous invertebrate remains. © 2012 with permission from Elsevier)

etation growth might stabilize slopes. The transfer of carbon generated by solifluction is poorly quantified.

Thaw of ice-rich permafrost in sloping terrain results in rapid mass wasting and erosion phenomena, that are also designated as forms of abrupt thaw and are sources of permafrost carbon to atmosphere and water (Fig. 1.23d). In permafrost areas, two forms of cryogenic landslides occur frequently: active layer detachment slides and retrogressive thaw slumps (Sect. 3.6; Fig. 3.24). These phenomena are typical examples of the 'pulse disturbances' of Grosse et al. (2011). In cryogenic landslides, water liberated by thawing permafrost enhances mass movement by increasing the pore water pressure within the soil, resulting in loss of shear strength (Williams and Smith 1991; Leibman et al. 2003, 2014). System components of both active layer slides and thaw slumps are linked to weather conditions and therefore climate (Sect. 3.6). Ice-rich permafrost, unconsolidated sediments and slopes are the necessary ingredients for the production of cryogenic landslides. Kokelj et al. (2017) have shown that in Northern Canada, glacial sediments at the limits of the Last Glaciation which contain large amounts of ground ice and have a suitable surface topography, are particularly prone to landslide development.

A second rapid transfer of soil carbon is erosion by running water. Running water with temperatures above zero effectively transfers convective heat to the permafrost ice that binds sediment together and therefore is a strong erosive force. This creates thermo-erosional gullies, and is particularly prominent along ice wedges, which result in rapid expansion of the stream drainage system (Sect. 3.3.2; Fortier et al. 2007).

The contribution of water from thawing permafrost to the initiation and activity of cryogenic landslides is important and marks their sensitivity to climate change. Active layer detachment slides are initiated preferentially in warm weather and sunny conditions, or after forest fires. A high rate of thaw at the base of the active layer and ice-rich permafrost are critical factors (Lewkowicz and Harris 2005). The effect of rainfall appears to be a less important factor than warm spells in summer, since a record rainfall in the area studied by these authors generated only a few slides. However, high precipitation events may contribute to pre-conditioning the soil for slides by creating water-saturated conditions.

Retrogressive thaw slumps are triggered when ice-rich permafrost is exposed to summer warmth. Other erosion phenomena like active layer slides, gully erosion, coastal, river or lake bank erosion contribute to initiation of thaw slumps (Lantuit and Pollard 2008). The slumps create a strong thermal disturbance of the permafrost, since protective vegetation is removed in the slump floor. The slope angle of the headwall affects insolation, and the terrain depression created by the slump catches a thick snow cover by snow drift in winter. As a result, the mean annual ground temperatures are 3–4 °C warmer in the slump floor than in the surrounding terrain, and also the active layer is thicker (Burn 2000; Kokelj et al. 2009; Abbott and Jones 2015).

The activity of thaw slumps is strongly determined by weather conditions and the rate of removal of debris at the base of the slump floor. Debris removal from the slump floor at the base of the headwall needs sufficient water from headwall thaw or rain. The thaw rate at the headwall generally produces insufficient water to sustain debris removal, therefore summer rainstorms are a necessary water source (Lacelle et al. 2010). Retreat rates of the headwall are in the order of meters per year, and depend on its orientation; north facing headwalls retreat slower than south facing (Lewkowicz 1987). Lantz and Kokelj (2008) recorded a recent average retreat rate of about 1 m $year^{-1}$ of 50 slumps in the Mackenzie Delta region in Canada. Burn and Friele (1989) report a retreat rate of up to 16 m $year^{-1}$ in the Yukon Territory.

Cryogenic landslides may be active over many years, although their activity is intermittent between years, depending on weather conditions. Time lapse camera videos of the slump of Fig. 3.25 showed increased activity during a warm spell after rainfall; in a following relatively dry and cool summer activity was minor. Stabilization occurs when activity decreases, either by weather conditions or by accumulation of sediment on the slump floor or in debris cones in front of the headwall (Burn and Friele 1989; Lacelle et al. 2010).

Vegetation re-establishment and aggradation of permafrost further contributes to landslide stabilization. However, the ecosystem is strongly altered and re-establishment of vegetation may take several decades. Vegetation characteristics can be taken as an indicator for the time that has elapsed since stabilization (Burn and Friele 1989; Ukraintseva et al. 2014). In tundra environments, old thaw slump sites remain visible for many decennia by the presence of high willow bushes (Ukraintseva et al. 2014). Re-activation of regressive thaw slumps is common, resulting in polycyclic thaw slumps (Lantuit and Pollard 2008; Kokelj et al. 2009; Lacelle et al. 2010; Leibman et al. 2014).

Given the close causal links of cryogenic landslides to permafrost thaw (Fig. 5.14), it is not surprising that recent climate change exerts its influence on number and activity rates of these features. Lantz and Kokelj (2008) record a recent (1973–2004) increase in the MacKenzie Delta region by 40% of retrogressive thaw slumps and an approximate doubling of their headwall retreat rate with respect to the 1950–1973 period; at the same time summer temperatures in the area have increased by 2.2 °C, and the number of days with temperatures over 20 °C have increased by 14 days. Gooseff et al. (2009) found a more than 200% increase in the number of active layer slides and erosion gullies in the Brooks Range in Alaska since 1980. Lewkowicz and Way (2019) report a huge 60-fold increase of retrogressive thaw slumps on Banks Island, Northern Canada, from 63 in 1984 to 4077 in 2013, amounting to an average areal expansion rate of 5.1 ± 0.2 km^2 $year^{-1}$. The increase in number of slumps were mainly driven by three exceptionally warm summers, and showed a log-linear relation with summer temperature. Lacelle et al. (2010) found a less clear relation with temperature in their study area west of the Mackenzie Delta; the net number of slumps has decreased since 1954, although their data show also a recent increase in the initiation of thaw slumps. Rather, the number of slumps follows the precipitation record of the area, showing the importance of higher precipitation next to summer temperature. The strong increase in activity of these erosion features in the reports above clearly indicate recent desta-

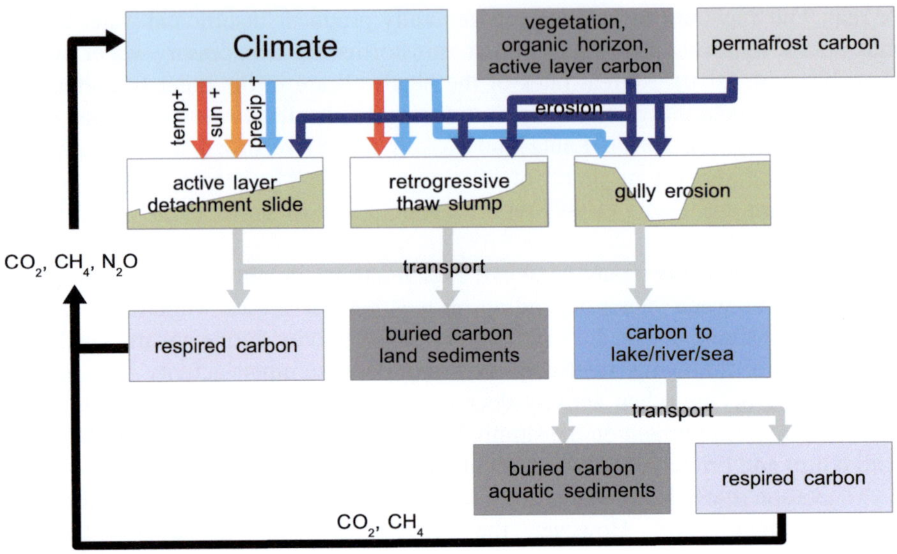

Fig. 5.14 Relation of soil and permafrost carbon displacement by erosion with climate, and downstream fate of removed carbon

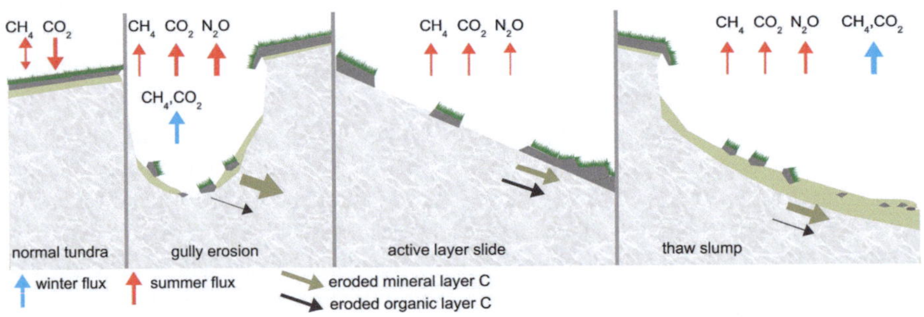

Fig. 5.15 Greenhouse gas emission and organic carbon erosion of the most common upland ice-rich permafrost erosion features. (Adapted to include negative CH_4 fluxes in undisturbed soils, from Abbott and Jones (2015))

bilisation of sloping, ice-rich permafrost terrain, related to increasing summer temperature and precipitation extremes.

Erosion of permafrost results in episodical high rate of movement of large masses of permafrost soil, including its carbon content. Soil carbon is displaced towards a different environment in a relatively short time, where environmental conditions may enhance or retard decomposition processes (Sect. 4.5). It depends on the pathway of the soil carbon after erosion, what kind and rate of decomposition processes will follow (Fig. 5.15). Deposition in lakes results in burial in lake sediments and largely anaerobic, slow decomposition with CH_4 emission (Sect. 5.3.2). Transfer to rivers results in further decomposition in the river system and partly in re-sedimentation; coastal erosion provides material for decomposition and burial in the marine coastal environment (Vonk and Gustafsson 2013; Vonk et al. 2013, 2015; Sect. 5.4.5). In thaw slumps, a part of the removed material may be buried in mudflow deposits on the slump floor and in debris lobes and fans beyond the slump (Lacelle et al. 2010; Leibman et al. 2014).

How the eroded permafrost carbon is partitioned among pathways and depositional reservoirs is still poorly quantified (Pizano et al. 2014). Emission of carbon as CO_2 and CH_4 to the atmosphere in erosion phenomena is difficult to measure because of the inhomogeneity and inaccessibility of the terrain, which precludes application of eddy covariance and makes the application of flux chambers a huge – and muddy – effort. Outflow of organic carbon via water is also difficult to quantify in the often highly diffuse drainage system of erosion phenomena.

In any case, erosion represents a considerable net loss of soil organic matter compared to undisturbed terrain (Lantuit and Pollard 2008; Abbott and Jones 2015). In Northwestern Alaska, upland erosion in cryogenic landslides and erosion gullies removed on average 0.48–2.7 kg C m^{-2} of organic carbon from the organic layer (Abbott and Jones 2015); for slumps, Pizano et al. (2014) report similar values (3.0 kg C m^{-2}). The lowest amount (about 6%) was removed by thermo-erosional gullies; active layer slides and thaw slumps removed respectively 32% and 51% of the original organic horizon soil carbon. For mineral horizons the differences are smaller. However, the the carbon export from the mineral soil horizons could be an order of magnitude higher than that of the organic horizons, since parts of the organic topsoil and and vegetation remain at the erosion sites as vegetation rafts, drapes and buried material (Figs. 3.24, 3.25, 5.15).

Abbott and Jones (2015) measured greenhouse gas fluxes from slides, slumps and gully systems using chambers. Ecosystem respiration R_{eco} in exposed mineral soil and sediment patches inside the erosion phenomena was considerably lower (57–70%) than on undisturbed soil next to the eroded sites. However, the soils inside the erosion phenomena soil lack vegetation and intact root systems, and have a relatively low amount of organic matter. A better comparison results from measuring R_{eco} on remnants of the original vegetation. For vegetation rafts and drapes inside erosion phenomena, R_{eco} was higher than on undisturbed soil; 26% for slump rafts, and 71% higher for gully vegetation drapes. Soil temperature at 15 cm depth was up to 4 °C higher within the erosion phenomena compared to undisturbed soil, and soil moisture was locally higher.

In general, the change in R_{eco} is related to the percentage of original vegetation, remaining as rafts or drapes in the erosion phenomena. A decrease is found when low percentages remained, and an increase when high percentages remain. A second, more remarkable relation found by Abbott and Jones (2015) is an increase of R_{eco} after erosion at sites with an originally low R_{eco}, and a decrease for sites with an originally high R_{eco}. For sites with low R_{eco} the effect of soil warming and change in soil moisture probably dominates, while for high R_{eco} sites impacts of erosion on vegetation such as disruption of plant roots and removal of organic material dominates.

Gas concentrations of CO_2, CH_4 and N_2O within the soil pores are also higher inside the erosion phenomena than in the control sites. CH_4 concentration is higher in gully sediments than in slides and slumps. The high concentrations of N_2O are remarkable. High emissions of N_2O have been found also in bare peat soils in permafrost (Repo et al. 2009; Marushchak et al. 2011). This indicates active N cycling in an environment where plant uptake of N is strongly diminished or absent. In the erosion features of Abbott and Jones (2015), soil moisture conditions and N availability from decomposition likely trigger denitrification, while disruption of the vegetation cover decreases N uptake by vegetation.

The decomposition of organic matter upon erosion is enhanced by exposure to sunlight (photobleaching; Sect. 4.5.1). Cory et al. (2013) showed that exposure to natural sunlight enhances the bacterial decomposition rate of DOC in water from eroding permafrost by more than 40%. This holds in particular for DOC that is recently derived from permafrost. Also under dark conditions, DOC from old permafrost organic matter is highly labile for heterotrophic respiration (Vonk et al. 2013; see Sect. 5.4.4).

Cryogenic landslides are rapidly recolonized by vegetation once activity ceases; recolonization contributes to stabilization (Burn and Friele 1989; Ukraintseva et al. 2014). On young surfaces, first a pioneer vegetation establishes with high grasses and forbs, which is gradually replaced by forest and tundra vegetation. Thaw slumps in forest in the Alaskan Yukon Territory were re-forested with similar forest as before slumping in 35–50 years (Burn and Friele 1989), although soils still remained immature, lacking soil profile development.

In tundra areas, the vegetation is altered for a long time to tall willow shrub tundra (Pizano et al. 2014; Ukraintseva et al. 2014). This results in a higher biomass than that of the original tundra vegetation. Colonization of landslide sites starts with high forbs and grasses. The bulk vegetation biomass on ancient, well vegetated thaw slumps on the Yamal Peninsula is dominated by willows and is much higher (about 1.6 kg m^{-2}) than on not eroded tundra surfaces (about 0.85–1.2 kg m^{-2}). In tundra thaw slumps studied by Pizano et al. (2014) the C and N pools of disturbed sites were similar to those of undisturbed soils after 30–64 years. The N pools recovered more rapidly than expected, which suggests that N fixation, lateral transport of reactive N or a deeper soil N reservoir provides additional N.

Complete restoration of a stable vegetation on thaw slumps in tundra restores the soil stability and adds a higher vegetation carbon pool. However, this ecosystem resilience should not be taken as evidence for a small impact of permafrost erosion on the carbon cycle. Carbon pool restoration is preceded by considerable losses, and the current increase of the incidence and activity of erosion phenomena (including rejuvenation of polycyclic slides) will strongly enhance these carbon losses, irrespective of ecosystem resilience (Van Huissteden and Dolman 2012; Lewkowicz and Way 2019).

5.4 Hydrological Changes

Permafrost thaw alters both surface and subsurface hydrology. Increase of ALT, subsidence and formation of ponds, thaw lakes and erosion are already discussed examples that change the surface hydrology. However, also large-scale changes of the soil water balance are to be expected. Basic categories of changes in the soil water balance related to cause and extent are:

- Wide-spread changes of precipitation, evaporation and snow cover (Figs. 5.16 and 5.17);
- Wide-spread changes driven by change in permafrost, e.g. change in groundwater flow (Fig. 5.18);
- Wide-spread changes driven by vegetation change that alter evapotranspiration;
- Localized change driven by changes in permafrost, e.g. erosion, pond and lake formation, occurring on a large scale and over wide areas.

These changes will affect all aspects of the hydrology of permafrost regions – soil hydrology, groundwater flow and river discharge.

The main reason for high soil wetness in many permafrost areas is impeded infiltration and groundwater flow due to permafrost presence and the low evaporation rates. For the carbon cycle of permafrost soils, it is crucial how the impact of climate change will work out in the future. These changes will differ for the relatively dry High Arctic and the and the less arid Subarctic (e.g. Lupascu et al. 2014).

The effects of river discharge change should not be underestimated. In Arctic lowlands river floodplains make up a large part of the landscape and even modest changes in flooding frequency may affect carbon sequestration or release in floodplain sediments. Moreover, floodplain soils are strong sources of CH_4 compared to soils outside the reach of frequent river flooding (Van Huissteden et al. 2005, 2009). Furthermore, rivers are an important component of the carbon cycle of permafrost because of

- The integrated information that river water biogeochemistry can provide on permafrost changes throughout their drainage basin,
- Their function as conduits for DOC, POC and nutrients derived from thawing permafrost soils towards marine or lacustrine (lake) environments,
- Their effect on flooding, and erosion or sedimentation of organic carbon on river floodplains.

In this section the large-scale hydrological changes and their effects on permafrost soil hydrology and carbon cycle are assessed. Additionally, the role of rivers as transporters of carbon are considered, including the transfer of carbon within the river system.

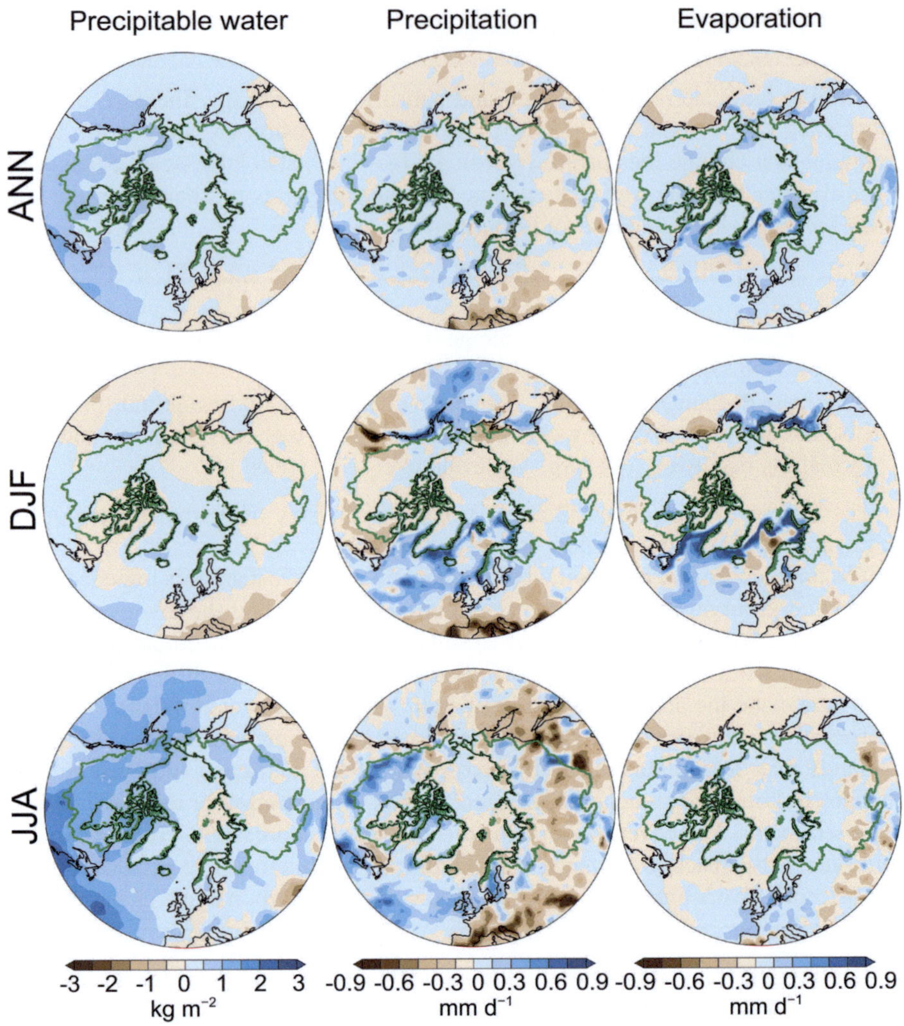

Fig. 5.16 Differences between 1986–2013 and 1958–1985 for precipitable water, precipitation, and evaporation on the basis of JRA-55 re-analysis for annual means, winter (DJF), and summer (JJA). The green lines indicate the boundaries of the Arctic river catchment. (Republished with permission of John Wiley and Sons Inc, from 'The atmospheric role in the Arctic water cycle: A review on processes, past and future changes, and their impacts' in Journal of geophysical research Biogeosciences 121(3) by Vihma et al. (2016) © 2016; permission conveyed through CopyrightClearance Center, Inc.)

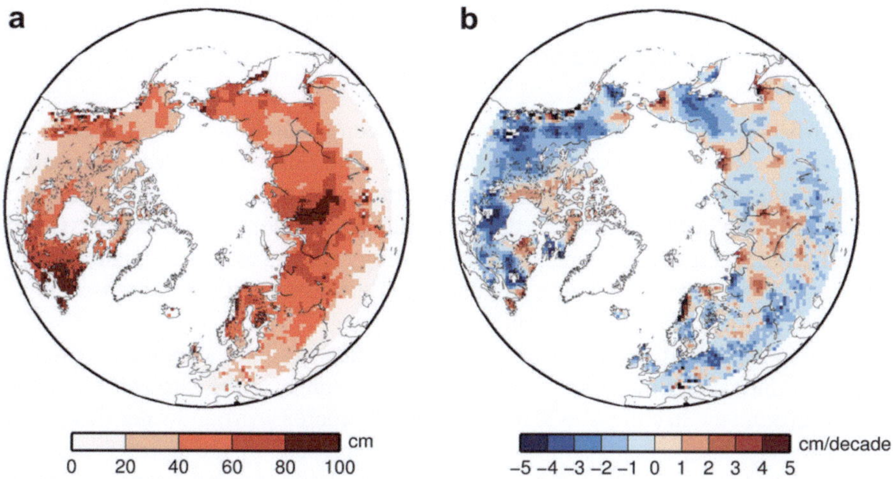

Fig. 5.17 (**a**) Distribution of average snow depth for JFM over the period 1948–2006. (**b**) Trend derived by a linear analysis. (Reprinted from Polar Science 6(1):23–37, Park et al. (2012) Analysis of satellite and model datasets for variability and trends in Arctic snow extent and depth, 1948–2006 © 2012 with permission from Elsevier)

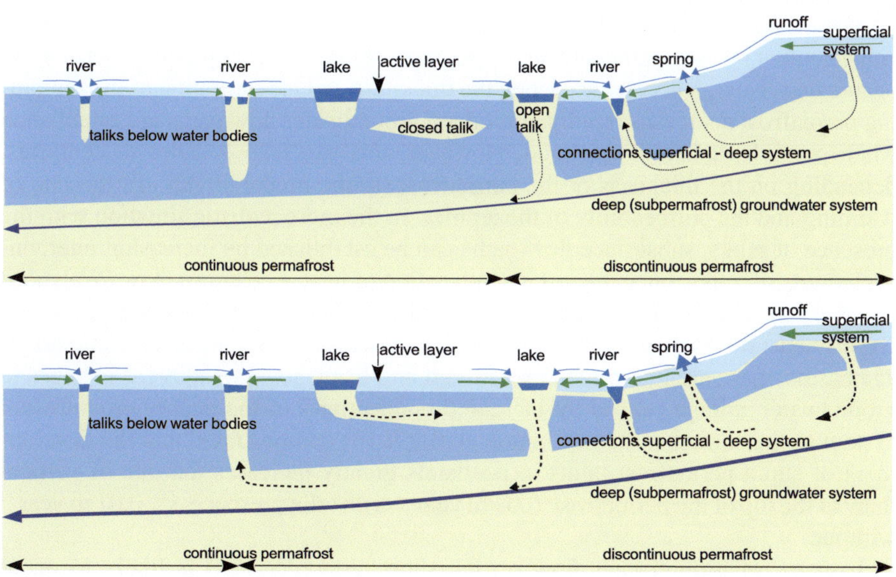

Fig. 5.18 Changes to the permafrost groundwater flow system by increased permafrost thaw

5.4.1 Water Balance: Groundwater Hydrology and Permafrost Thaw

Permafrost separates groundwater flow into a deep groundwater system below the permafrost, and a shallow groundwater system in the active layer and supra-permafrost taliks (Fig. 3.26). Thaw increases the depth of the shallow groundwater system, and the connections between the shallow and deep groundwater system. The effects depend strongly on subsurface geology (Bense et al. 2009) and the subsurface configuration of existing permafrost (Bense et al. 2009; Jorgenson et al. 2013; Walvoord and Kurylyk 2016).

The interaction with heat transport by groundwater results in a nonlinear response of the groundwater flow system to temperature change at the surface. Complete thaw of ground ice in soil pores is not necessary to increase groundwater flow, since at below-zero temperatures, fluid water is still present in the pores (Sect. 3.8; Walvoord and Kurylyk 2016). Advective heat transport by groundwater flow adds to thawing and resulting increase of hydraulic conductivity (Rowland et al. 2011). In general, a change from a hydrological system which is dominated by surface water runoff and shallow groundwater flow, towards a system dominated by deeper groundwater flow is expected to occur with ongoing warming (Fig. 5.18); Walvoord and Kurylyk et al. (2016) list a large number of studies with observations of changes in the hydrology that may indicate a transition.

Modelling work by Bense et al. (2009) shows that shallow groundwater flow enhancement by enlargement of supra-permafrost taliks over continuous permafrost occurs first. Next, it accelerates when the thickness and ice saturation of the remaining permafrost decreases rapidly, until deep groundwater flow paths are established when permafrost has disappeared. However, this can take decades to centuries, depending on the thickness of the permafrost (in the model 30–85 m), the rate of warming and the permeability of the aquifer. In the more realistic situation with the presence of taliks, subsurface flow paths can be established by increasing interconnectedness of taliks; once this occurs, advection of heat accelerates thaw (Walvoord and Kurylyk 2016). Gravity data from the GRACE (Gravity Recovery and Climate Experiment) satellite mission show water mass changes in a number of Arctic drainage basins, these are interpreted by Muskett and Romanovsky (2009) as changes in groundwater storage caused by increased connectivity of the deeper groundwater system to the surface system. Connon et al. (2018) demonstrated that the development of supra-permafrost taliks in peatlands greatly increases the rate of vertical thaw of the top of the permafrost: 0.07 m $year^{-1}$ with talik presence, vs. 0.01 m $year^{-1}$ without.

Increase of groundwater flow on the short term (decades) is likely to occur mainly above the top of the permafrost, in particular in continuous permafrost areas. However, ALT increase does not necessarily generate a large contribution to groundwater flow. As seen in Sect. 5.2.1, a part of the ALT increase may be compensated by surface subsidence. Moreover, many permafrost soils have a relatively permeable organic top layer, overlying a less permeable mineral soil or denser peat. If the soil water table is lowered by drainage or increasing evapotranspiration, subsurface flow is also reduced by these less permeable soil horizons.

The creation of preferred flow paths by advective heat transport by flowing soil water, e.g. the flow paths connecting ponds described by Boike et al. (2008) and Minke et al. (2009), or deeper depressions of the permafrost surface below preferred surface drainage lines (Woo 2012) may be more important than general active layer thickening. Furthermore, as discussed in the previous sections, erosion and thaw of ice wedges creates new surface drainage pathways. Connon et al. (2014) describe also enhancement of surface drainage through a peatland with peat plateaus by expansion of fens, caused by permafrost thaw.

In discontinuous permafrost areas, where contact between the surface and deeper groundwater systems already exist, this contact will be enhanced by ongoing thaw of permafrost (Jorgenson et al. 2013). The exact configuration of subsurface permafrost is generally poorly known, but mapping capabilities are improving (Sect. 3.10). Discontinuous permafrost is generally warmer than continuous permafrost, and at higher temperatures the pores have a higher unfrozen water content. The deeper thaw of discontinuous permafrost is strongly enhanced by advective heat transport by groundwater flow as shown by modelling work of Rowland et al. (2011). In their model, the permafrost thickness depends strongly on the presence or absence of sub-permafrost groundwater flow – without groundwater flow, permafrost is 2 to 5 times thicker than with groundwater flow. The model shows that a lake talik can develop into an going through the permafrost on the time scale of decades. If advective heat transport by groundwater is included besides conductive transport, the time of development of an open talik was reduced by a half. In isolated permafrost areas with peat plateaus, lateral heat transport by groundwater flow results in an increased thaw rate of features as peat plateaus (Connon et al. 2014; Kurylyk et al. 2016).

In permafrost drainage basins, base flow is a relatively small proportion of the total river discharge, that increases with discontinuity of the permafrost. The autumn and winter flow in permafrost river basins consists entirely of base flow; in smaller rivers on continuous permafrost, discharge terminates entirely in winter. Increased groundwater flow in permafrost areas increases the base flow of rivers, which is the proportion of river discharge supported by groundwater flow (Woo 2012). The effect of increasing groundwater contribution to river discharge is likely most important in the discontinuous permafrost zone, where existing links between the deep and surface groundwater systems will be enlarged, and at the boundary between the continuous and discontinuous permafrost zone, where new connections may be established.

Lake water levels are also strongly influenced by groundwater flow. Drainage of lakes has also been attributed to subsurface drainage by increased connectivity of surface and deep groundwater system (e.g. Yoshikawa and Hinzman 2003; Smith et al. 2005; Sects. 3.8 and 5.3.2). The increased subsurface drainage depends on the hydraulic conductivity of the underlying thawed material, which will be generally low in fine-grained deposits (Jorgenson et al. 2013). However, the opposite also occurs – lake water level rise by increased groundwater recharge. Increased connectivity with deeper groundwater flow also will be expressed by differences in water chemistry; the shallow groundwater system contributes organic-rich water, while deeper water systems contribute mineral-rich water and a different organic matter composition (Walvoord and Kurylyk 2016; O'Donnell et al. O'Donnell et al. 2012b).

Jepsen et al. (2013) combined an airborne electromagnetic survey of permafrost and statistics on lake expansion / contraction on the Yukon flats. Using the geophysical data, the configuration of the lake taliks and the underlying permafrost and geology was determined. They found no significant linkage of shrinking lakes with taliks going through the permafrost. However, *expanding* lakes lakes were significantly associated with the presence of impermeable frozen silt underlying the lakes, that prevents contact with the deep groundwater system below the permafrost. Their conclusion is, that additional water supply in the expanding lakes was not derived from below the permafrost, but from increased water supply through the upper tens of meters of the permafrost, via thicker active layers and growth of taliks.

Increased groundwater flow in permafrost also affects land surface drainage and wetland areal extent. Enhanced subsurface interconnectivity by permafrost table lowering will enhance connection between water bodies and wetlands, and increase water discharge via groundwater rather than via surface runoff (Walvoord and Kurylyk 2016). This may result in lower water tables and decreased surface runoff from wetland areas, although locally soil wetness and surface drainage may be enhanced by increased groundwater seepage or water flow from springs. However, also erosion features exert a strong influence on drainage and may become more effective in drainage at the land surface (Sect. 5.3).

5.4.2 Water Balance: Precipitation and Evapotranspiration

The atmospheric part of the water balance, precipitation minus evaporation ($P - ET$) or net precipitation (NP), is subject to change from both climate change and change in vegetation. Climate change includes both anthropogenic climate change and multi-year pseudo-cyclic variations of the Arctic climate such as the Arctic Oscillation (Sect. 1.4.1). With positive AO index, storm tracks are shifted northward and winters are mild in northern Eurasia and eastern United States, but cold in Greenland, Canada, Alaska, and eastern Siberia. When the AO index is negative, storm tracks shift southward and the temperature field reverses (Cohen et al. 2014; Vihma et al. 2016).

The uncertainty in net precipitation (NP) is large. As touched upon in Chap. 1, the observation networks in the Arctic are very sparse. Precipitation measurement, snow depth and snow water equivalent measurements are subject to instrument uncertainty (e.g. precipitation measurements by rain gauges). Evapotranspiration observations of ET are even more sparse and often derived from relatively short measurement campaigns (e.g. from a few eddy covariance stations). The alternatives are model-derived or remote-sensing derived ET estimates, which also have their biases (Kattsov et al. 2007). Furthermore, $NP = P - ET$ is a small difference between two large variables and therefore highly sensitive to uncertainty in these variables (Bring et al. 2016).

A general increase of precipitation has been observed in re-analysis data over permafrost areas, although this increase is regionally variable (Vihma et al. 2016; Fig. 5.16). The increase of precipitation over the Arctic drainage basin has two potential moisture sources – moisture transported northward from lower latitudes, and from the Arctic itself. Climate model analysis has shown that the Arctic moisture source is most important; it is related to a decline of Arctic sea ice cover in late summer, which enhances the evaporation over the ocean surface in autumn and early winter (Bintanja and Selten 2014; Vihma et al. 2016). The observed and climate model-derived precipitation has been increasing also throughout the Arctic, followed by the evapotranspiration which also increased. The upward trend of evapotranspiration may be a result of increasing precipitation, combined with longer growing seasons (Rawlins et al. 2010). However, the observation trends are small and obscured by natural variability. Nevertheless, the model results show consistency. Extreme precipitation events generated by convective clouds (thunderstorms) in spring and summer are also increasing (Groisman et al. 2007; Rawlins et al. 2010). Storm tracks may change also as a result of feedbacks to the atmosphere by disappearing sea ice (Bring et al. 2016).

Despite the increase in winter precipitation, this does not necessarily increase snow depth (with its large effects on soil temperature) throughout most of the permafrost regions. The amount of precipitation that falls as rain is increasing relative to snowfall, resulting in a thinner snow cover in the more southerly permafrost regions. A detailed hydrological – snow cover model driven by weather data from 1948–2006 shows a trend of decreasing snow depth over much of the permafrost area, albeit with high spatial variability (Park et al. 2012). However, in areas bordering the Atlantic and Arctic Ocean snow depth increases (Fig. 5.17). Data from Russia indicate a decreasing snow cover duration since 1966, with a thicker snow cover over West Siberia, approximately stable snow duration and thickness in Central and East Siberia, and a decreasing snow duration and thickness in Far East Siberia (Bulygina et al. 2011).

For most areas in the permafrost zone, *P* is larger than *ET*, except in semi-arid areas in the south (Bring et al. 2016). In general, the trend in *P* is larger than that in *ET*, resulting in a net (but not always significant) increase of *NP*. Data collected from literature and re-analysis data by Vihma et al. (2016) indicate an increase of *NP* over the Circum-Arctic continental landmasses. However, there are regional differences, the trends depend on the period addressed, and on the data sources. Based on the differences in Fig. 5.16, the annual mean *P* has decreased and *ET* has increased in southern parts of the Eurasian permafrost zone, and in some areas in Canada and Alaska, with lower *NP* as result. An increase in *P* with a simultaneous decrease in *ET*, resulting in net increase of *NP*, has taken place further north, in parts of the Canadian Archipelago and Northern Siberia (Vihma et al. 2016).

The CMIP5 climate models project a robust increase of precipitation over the Arctic for the future (Vihma et al. 2016). The modelled Arctic mean winter precipitation (October–March) increases by 35–60% relative to the period 1986–2005, depending on the future greenhouse gas concentration pathways (RCP4.5 and 8.5). The precipitation changes in spring and summer are weaker: 15% and 30%; the largest relative changes are found over the Arctic Ocean (IPCC 2013). The models project an increase in precipitation extremes; the winter precipitation extremes are likely to increase more than the summer extremes (Kharin et al. 2013; Vihma et al. 2016). Table 5.4 gives a regional assessment of future changes in *P*, *ET* en runoff (or *NP*).

Changes in the permafrost land cover also feed back on the precipitation surplus. Evaporation is larger over lakes and ponded water. A reduction of the ice-cover period of lakes further increases their evaporation (Bring et al. 2016). Drier soils reduce evapotranspiration by increasing the bulk surface resistance to the latent heat flux (Liljedahl et al. 2011). A reduction of lake and wetland area and deeper soil drainage by increasing subsurface drainage is likely to reduce evapotranspiration, while wetter conditions created by expansion of ponds, lakes and wetlands will increase evapotranspiration. Vegetation change is occurring throughout the permafrost area. In tundra areas increasing shrub growth is observed widely (Sturm et al. 2001, 2005; Wrona et al. 2016). Taller vegetation increases evapotranspiration (Sect. 2.3). A similar effect may be expected from an advancing tree line. Mosses, lacking roots, have a limited capability to transfer water from soil to leaf, resulting in a reduced evapotranspiration in dry spells (Liljedahl et al. 2011), so a change from moss dominated vegetation to vascular plants also enhances evaporative water loss.

Examples of the complicated interaction between precipitation, vegetation and ALT are emerging. Iijima et al. (2010) noted an increase in soil temperatures and active layer thicknesses in the Central Lena Basin in Eastern Siberia, after 4 years with increased winter and summer precipitation. In terrain depressions, this led to local damage of trees with loss of foliage, and die-back of trees because of waterlogging. The sap flow and stomatal conductance of damaged Larch trees was strongly reduced, resulting in a reduction of transpiration – a positive feedback on soil wetness (Iijima et al. 2014).

5.4.3 River Discharge Changes and Flooding

Floodplains and deltas make up a large part of lowlands bordering the Arctic Ocean. In the middle and lower parts of river basins, floodplains are temporary stores of sediment, including organic carbon. These floodplains are both sources and sinks of carbon by erosion and sedimentation, have generally higher nutrient supply and more productive vegetation than non-flooded areas, and can be hotspots of CH_4 emissions, which are several times higher than those of other wetland vegetations (Corradi et al. 2005; Van Huissteden et al. 2005, 2009). Changes in river discharge

Table 5.4 Projected precipitation, evapotranspiration and runoff changes in the next century, for major hydrophysiographical regions in the Arctic; + indicates increase, – indicates decrease; boldface of the + or – indicates a relatively large increase or decrease. Based on Table 2 in Bring et al. (2016)

Hydrophysiographical Region	P annual	P winter	P summer	E annual	E winter	E summer	Runoff mean	Runoff high flow	Runoff base flow
Tundra, polar desert	**+**	**+**	+	**+**	+	**+**	+	+	+
Boreal forest, shield rock areas	**+**	**+**	+ (–)	**+**	+	**+**	+	+(–)	+(–)
Mountains	**+**	**+**	+ (–)	**+**	+	**+**	+	+	+
Ice sheets, glaciers	**+**	**+**	+	**+**	+		+ → –		

and flooding are likely to influence the carbon cycles on floodplains. For the seasonal cycle of discharge in high-latitude rivers, storage of precipitation as snow and snowmelt is important (Sect. 3.8.2). Other basin characteristics that have a large influence on the annual discharge distribution are the percentage of wetlands and lakes in the drainage basin, the presence of dams, and other human activities such as agriculture. The permafrost coverage of a drainage basin is particularly important for the winter base flow of high-latitude rivers; Table 5.5 lists the permafrost coverage of major Arctic drainage basins (Sect. 3.8.2).

The oldest discharge and water quality monitoring programs have started in Russia in the '30s of the previous century, while those of North American rivers started during the 1970s (Shiklomanov et al. 2007). There is uncertainty on the quality of older runoff and precipitation datasets of Arctic river basins. Comparing discharges over longer time spans needs correction for the effects of human activities such as dam construction and forest clearance. Monitoring discharge in Arctic rivers is logistically difficult, and really problematic during spring break-up when rivers are rising rapidly and filled with floating ice. The estimation of winter discharge below a thick ice cover is prone to errors up to 30%; smaller rivers may freeze to the bottom in winter and have no discharge at all. Until the year 2000, the trends of precipitation in three precipitation datasets and runoff of Siberian rivers was not consistent (Berezovskaya et al. 2004; Shiklomanov et al. 2007). The annual variation in discharge of Arctic rivers is very high (up to a factor of 2) that may obscure trends (Bring et al. 2016). Recent research investments have improved the situation for the six largest Arctic rivers (https://arcticgreatrivers.org/). However, rapid changes in the Arctic drainage basins present a challenging task to provide a baseline for research on future changes (McClelland et al. 2016).

In earlier publications on river discharge increase in the Arctic, it was debated whether river discharge was influenced by climatic or other effects. McClelland et al. (2004) could exclude major effects from forest fires (Chap. 6) and dam construction on increasing discharge of Eurasian rivers, and demonstrated an increase of discharge by climate change. Climate modelling by Wu et al. (2005) confirms that this can be attributed to an intensification of the hydrological cycle caused by anthropogenic climatic warming. Besides the effect of anthropogenic climate change, also the multi-year cyclic variations of the Arctic climate such as the Arctic Oscillation exert a discernible influence (Ye et al. 2004). More recent data show that river discharges from Arctic rivers have consistently increased throughout the last decennia (Bring et al. 2016; Holmes et al. 2018). Trends have continued (Fig. 5.19), with a better statistical significance with respect to natural variation (Rawlins et al. 2010). Since 1976, the increase in river discharge for Eurasian rivers has been 3.3 ± 1.6% per decade, and 2.0 ± 1.8% per decade for North American rivers (Holmes et al. 2018). For the pan-Arctic drainage basin, discharge has increased by 5.3 km^3 $year^{-1}$ (Rawlins et al. 2010).

The most direct climate influence is represented by the water input or net precipitation (*NP*) into the system, $P - ET$. In general, the trend in *P* over permafrost areas is larger than that in ET, resulting in an increase of *NP* (previous section). Remarkably, the trend in annual discharge is nearly double the *NP* trend. This can

Table 5.5 Characteristics of the Arctic's largest watersheds

Watershed	Area 10^6 km^2	Runoff mm year^{-1}	Sediment yield (t km^{-2}year^{-1})	POC (10^9 g year^{-1})	Tun-dra %	Perma-frost %	Conti-nuous perma-frost (%)	Discont. permafr. (%)	Sporadic, Isolated permafr. (%)	Population density people km^{-2}
Ob	2.95	145	6	572 ± 20	1	26	2	4	20	9.51
Yenesei	2.56	263	2	249 ± 3	1	88	33	11	44	3.14
Lena	2.4	245	8	814 ± 52	1	99	79	11	9	0.45
Mackenzie	1.75	181	74	758 ± 66	1	82	16	29	37	0.27
Yukon	0.83	251	72	539 ± 26	6	99	23	66	10	0.18
Kolyma	0.65	209	19	123 ± 19	4	100	100	0	0	0.12
Indigirka	0.34	159	36		7	100	100	0	0	0.09
Pechora	0.31	529	38		23	39	13	4	22	1.78
Sevarnaya Dvina	0.29	369	12		0	0	0	0	0	4.46
Khatanga	0.29	372	–		8	100	100	0	0	0.05
Yana	0.23	170	18		3	100	100	0	0	0.1
Olenek	0.22	218	6		12	100	100	0	0	0.02
Taz	0.15	287	7		0	100	6	64	30	0.1
Pur	0.1	320	7		0	100	4	76	20	2.73
Pan-Arctic WS	16.8	±220	21		15	76	45	13	18	2.52

Compiled from Holmes et al. (2012) and McClelland et al. (2016). Pan-arctic WS: Pan-arctic watershed, total of al watersheds draining towards the Arctic Ocean

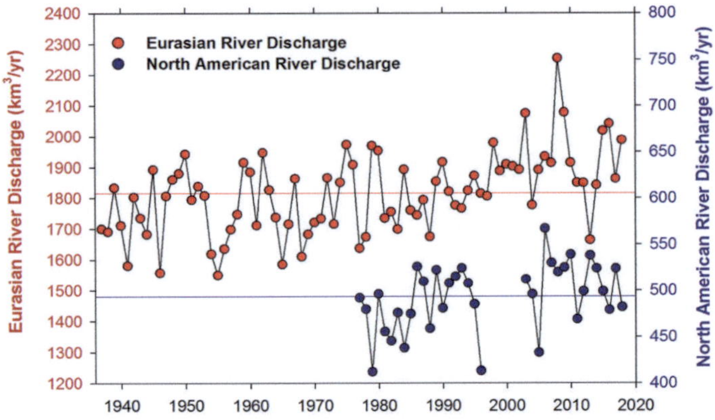

Fig. 5.19 Trends in annual river discharge for Eurasian Arctic rivers (red) and American Arctic rivers (blue), draining towards the Arctic Ocean and Bering Sea. Note the difference in scale for both groups of rivers; the discharge for the Eurasian rivers is 3–4 times larger than that for the American rivers. The reference lines show the long-term averages for the Eurasian rivers (1816 km^3 year^{-1}) and North American rivers (493 km^3 year^{-1}). (Source: NOAA Arctic Report Card 2018; reference Holmes et al. 2018; https://arctic.noaa.gov/Report-Card/Report-Card-2018)

be attributed to increased winter precipitation; the spring snow melt of a thicker snow cover would convert a larger proportion of the precipitation to runoff (Sect. 3.8.2). Another explanation could be drainage from an increasing area of lakes and ponds and permafrost thaw (Rawlins et al. 2010). However, as mentioned in the previous section, increasing snow cover thickness is regionally restricted (Fig. 5.17).

Permafrost thaw by increasing active layer thickness by itself does not contribute significant amounts of water to river discharge. The amount of additional water that can be drained from soils with increasing ALT is too small to explain the observed discharge increase (McClelland et al. 2004). However, the groundwater input and seasonal distribution of discharge is likely to change by permafrost thaw. Increased connectivity of supra- and sub-permafrost groundwater systems increase river base flow in winter, and summer discharge may be affected by a larger groundwater storage in soils (Walvoord and Kurylyk 2016). An increase of base flow discharge has been reported for most Eurasian and North American rivers (Ye et al. 2004; Walvoord and Striegl 2007; Shiklomanov et al. 2007; St. Jacques and Sauchy, St. Jacques and Sauchyn 2009; Lesack et al. 2013). Another effect of permafrost thaw, in particular in wetland basins, is increased connectivity of surface flows, by erosion (e.g. Liljedahl et al. 2016) or thaw of peat plateaus in discontinuous permafrost peatlands (Connon et al. 2014).

Since the spring snowmelt flooding is by far the largest flood peak which causes flooding for Arctic rivers, it is important to understand its future trends. The spring snowmelt flood is determined by the amount of snow in the drainage basin, weather conditions during snowmelt (temperature, precipitation), antecedent weather conditions of the previous year, and physiographic characteristics of the river basin. Ice jams during break up may increase flood height (Sect. 3.8.2). As much as 80% of the snow meltwater may be retained in lakes and wetlands (Bowling and Lettenmaier 2010). Shiklomanov et al. (2007) found no systematic change in the maximum daily discharge or the spring discharge in a large number of Russian Arctic basins. However, the timing of the spring discharge peak also has shifted to earlier dates by several days for many rivers (Shiklomanov et al. 2007; Prowse et al. 2011; Lesack et al. 2014; Holmes et al. 2018). Ice freeze-up and break-up is also shifting, towards later, respectively earlier dates (Park et al. 2016). On the other hand, model simulation of river ice suggests a trend towards a thicker ice cover resulting from a regionally thinner winter snow cover (Park et al. 2016).

Statistical analysis of Siberian river discharges by Ye et al. (2004) showed that spring discharge is strongly influenced by spring warmth. Winter snowfall contributes to spring discharge, but the impact varies from basin to basin. Its influence may range from governing largely the spring discharge (Lena basin) to having an effect on the discharge in other seasons as well, depending on differences in permafrost coverage of the basins. An extreme spring flood year of 2007 in Siberian rivers was related to high precipitation in the middle and northern parts of the the drainage basins. It fitted in the trend of increasing discharge (Shiklomanov and Lammers 2009).

Overeem and Syvitsky (2010) found in a 30 year record for 19 large Arctic rivers a strong 66% increase in the discharge of the 'melt month', the month at which the discharge starts increasing during snowmelt (usually May), while the peak month discharge (the discharge of the month during which peak discharge occurs) decreased by 7%. This is attributed to earlier snowmelt. For the Mackenzie Delta, Lesack et al. (2014) found that the advance of spring break-up dates was related to local spring warming combined with a decline of snow cover. In Northern Quebec, Boucher et al. (2011) found evidence for increasing intensity and variability since 1965, from tree ring data and tree damage on floodplains.

Current trends in peak flood height in spring are uncertain from the literature cited above, although total annual discharge is increasing with an intensified hydrological cycle and precipitation. Nevertheless, there are several signals that point in an upwards direction for the spring flood in some regions: potentially higher temperatures and more precipitation in spring, pre-conditioning of drainage basins for high spring floods by higher summer and autumn precipitation, and a thicker snow cover for regions near the Arctic Ocean. A trend which opposes increasing spring flood peak height is a decrease of the ice cover thickness, that may reduce flood peaks by reducing the impact of ice jams (Costard et al. 2007).

Increases in summer discharge are modest, except for basins with a significant glacier cover. A small majority of 6% of the discharge records in Russia described by Shiklomanov et al. (2007) show an increasing discharge trend for July and August; in only a few cases this is a significant trend. For the autumn months September–November, more of the analysed discharge records show a significantly increasing discharge trend. It is the winter base flow that increases most. From climate model simulations, summer and early autumn precipitation is expected to increase by remote poleward moisture transport from the south, while the autumn and winter precipitation increases by enhanced evaporation over the Arctic Ocean (Bintanja and Selten 2014). An increase in summer runoff may be expected in the

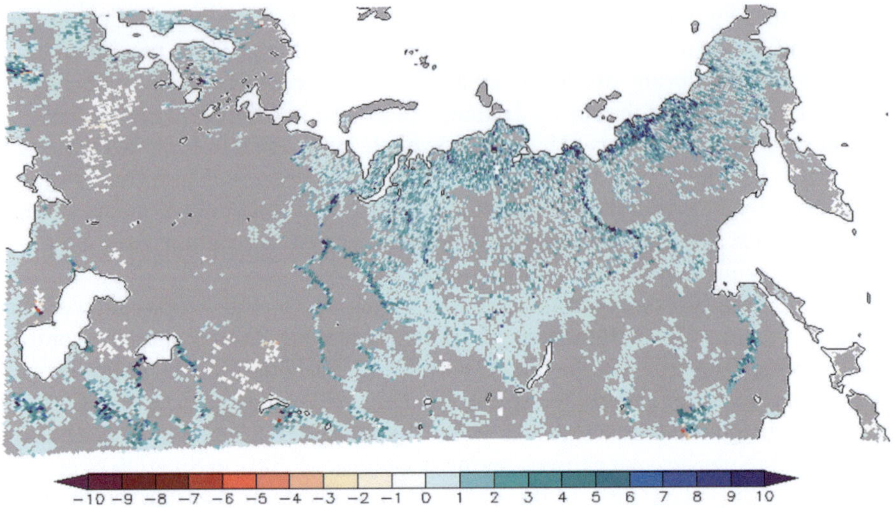

Fig. 5.20 Projected changes in the years 2050–2059 in the fraction (%) of model grid box covered by the mean annual maximum flood with respect to the the baseline period (1990–1999) as simulated by RCM30–CaMa-Flood model under the IPCC RCP8.5 scenario. Values are masked in grey where either the signal-to-noise ratio is lesser than 1 or floods do not occur in the baseline and future periods. (From Shkolnik et al. 2017, in Climate Dynamics 50(1), Copyright © 2017, Springer-Verlag Berlin Heidelberg)

future. For basins with glaciers, thaw of glacier ice determines the summer discharge. Increasing mass loss of Arctic glaciers currently contributes to discharge, but is expected to decrease when glaciers disappear (Bring et al. 2016).

The annual maximum extent of river flooding is expected to expand by 2% to 5% over Siberia by the mid of this century, under the IPCC RCP 8.5 scenario, as shown by hydrological modelling work of Shkolnik et al. (2017). The increased spring flooding is smallest in the western Siberian rivers, and largest along the East Siberian rivers, in the lower part of their drainage basins (Fig. 5.20). The model used by Shkolnik et al. (2017) is somewhat biased towards lower discharge peaks, and does not include the effect of ice jams; potentially higher extremes cannot be excluded.

The main reasons why flooding frequency and flood heights of rivers are important for the permafrost and its carbon cycle, are:

- the effects of advective heat added to flooded permafrost soils by the flood water;
- transfer of nutrient-rich flood water into lacustrine environments and other floodplain ecosystems;
- erosion or deposition processes and associated floodplain ecosystem changes which affect both the heat balance of the soil and carbon storage in floodplain sediments. Radical changes in floodplain geomorphology by extreme floods – reorganization of channel patterns and floodplain sedimentary environments – are possible (Woo 2012; Van Huissteden et al. 2013).

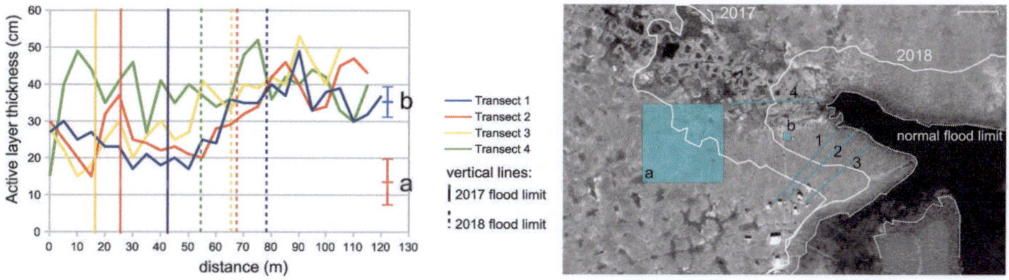

Fig. 5.21 Effect of river flooding of ALT outside the normal floodplain, showing the effects of two consecutive high floods in 2017 and 2018. Left: ALT along four transects approximately perpendicular to the flood limits. Transect 1 to 3 cross both flood limits and show an increase of ALT between the 2017 and 2018 flood limits (indicated by vertical lines). Transect 4 is entirely within the 2017 limit and follows a diffuse drainage line with many pools and Sphagnum cover. The average and standard deviation of ALT of two gridded plots, one largely outside the flood limits (*a*) and one within the 2018 and 2017 limits (*b*), is indicated to the right of the transects (see also Fig. 5.3 for the ALT grid data). Right: satellite image July 2015 indicating the flood limits, location of the grids and the transects. Satellite image copyright WorldView © 2019 MAXAR, right panel created with QGIS

The heat stored in floodwater is advected into the underlying soil. A layer of flood water with its low albedo is rapidly warmed by the sun (Emmerton et al. 2007; Woo 2012). The effect on the active layer thickness is demonstrated in Fig. 5.21, on a floodplain in the Indigirka lowlands in Northern Siberia. Two successive floods in 2017 and 2018 extended well beyond the normal floodplain limits, resulting in a nearly doubling of the ALT below normally non-flooded *Eriophorum* tussock tundra. A fourth transect, in wet sedge-dominated tundra with a usually high water table, was less affected. Additionally, the vegetation of the tussock tundra was affected by flooding; species that were severely affected are *Vaccinium vitis-idea, Betula nana* and *Sphagnum sp*. Repeated flooding my induce further permafrost thaw and subsidence, and hence expansion of the flood-prone area.

Besides carbon, river water also contains nutrients (reactive N and P) derived from organic matter decomposition and soil leaching. Since these amounts are higher than precipitation, a change of flooding frequency or flood height can change the input of nutrients into floodplain ecosystems. In floodplain lakes and thaw lakes, nutrient-enriched river water may influence ecosystem structure and primary productivity (Squires et al. 2009; Mesquita et al. 2010). More frequent river water input increases the primary productivity of floodplain wetland and lake ecosystems. This may increase carbon storage, but also enhance CH_4 emission, given the high CH_4 emission rates of floodplain ecosystems (Corradi et al. 2005; Van Huissteden et al. 2005, 2009). In the Mackenzie Delta, river-connected lakes occur on various floodplain levels; the lakes of which the elevation of the river connection is higher, receive less frequent river water input and nutrients (Prowse et al. (2011); Emmerton et al. (2007). Most floodplains show various levels (river terraces) where lakes occur.

Changing flood regimes change deposition and erosion rates. Costard et al. (2007) report for the Lena river a small increase in spring break-up discharge, a decrease in winter ice thickness, and a pronounced increase of the water temperature in spring. The increasing water temperature strongly enhances thermal erosion, resulting in a (modelled) bank retreat rate increase of 21–29%. Warming therefore is likely to increase channel mobility and erosion in rivers on permafrost, which may be further enhanced by increase in summer discharge and summer flooding. Using isotopic information of ground ice (δD and $\delta^{18}O$) it is potentially possible to reconstruct past flooding events in permafrost soils (Iwahana et al. 2014).

Climate-driven vegetation change is also likely to change sediment supply to rivers. Tape et al. (2011) have found evidence for a decrease of landscape erosion rates in Arctic Alaska since about 1980 in lake sediment cores. This is attributed to increased shrub growth in erosion-prone parts of the tundra landscape, probably combined with a decrease in spring peak discharge. On the other hand, an increase of erosion at the landscape scale by permafrost thaw (Sect. 5.3.4; Gooseff et al. 2009) increases sediment supply to rivers. The large expansion of retrogressive thaw slumps on Banks Island, Canada found by Lewkowicz and Way (2019) causes extensive valley floor sedimentation. The net effect will depend on the basin physiography: ground ice conditions, relief and presence of large amounts of unconsolidated sediments. If the effect is a decrease of sediment supply to the rivers by stabilization of erosion by vegetation, it may enhance the erosive capacity of river channels. Potentially, increasing fluvial erosion will re-mobilize previously deposited carbon in floodplain sediments. If landscape erosion dominates, the increasing sediment supply to rivers may cause higher sedimentation rates (including river transported carbon) instead of erosion rates on floodplains, and changes of the floodplain geomorphology (Vandenberghe 2003, 2008).

5.4.4 Water Transport of Carbon and Nutrients

River water is a conduit of soil-derived carbon; a part of this carbon is transferred to the atmosphere by decomposition in lake and river water, another part is sequestered again in lacustrine and fluvial sediments by sedimentation. The remainder enters deltaic and marine environments where again a partitioning between decomposition / respiration and sedimentation follows (Fig. 5.22). The transport of fresh water and organic carbon by rivers is very important for marine biogeochemistry of the Arctic Ocean, which contains 1% of global ocean volume, but receives more than 10% of global river discharge, from drainage basins with carbon-rich and thaw-vulnerable soils (Frey and McClelland 2009; Semiletov et al. 2012). For the six largest Arctic rivers, the average annual fluxes of particulate organic matter are highest in the Lena River and lowest in the Kolyma River (Table 5.5, Fig. 5.23, McClelland et al. 2016). Organic carbon transport by water (lateral transport) is seen as increasingly important, relative to local decomposition and direct release of carbon to the atmosphere (Battin et al. 2009; Vonk and Gustafsson 2013). Without knowledge of transport of

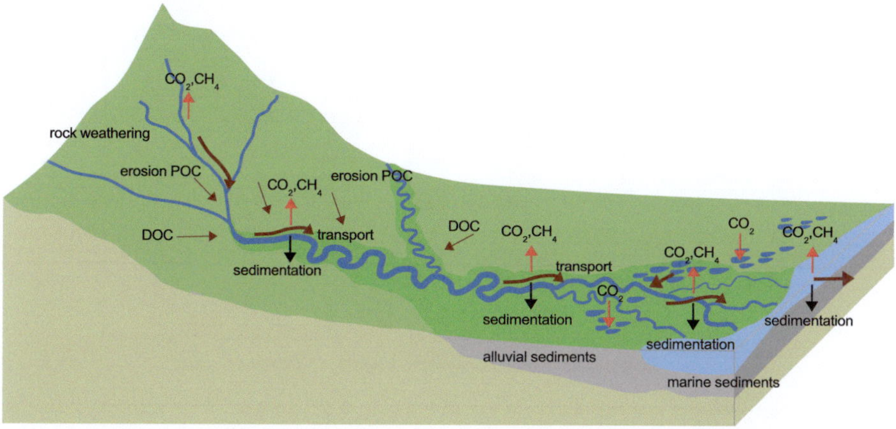

Fig. 5.22 Erosion and transport of permafrost derived organic matter in drainage basins. Eroded particulate organic matter (POC) and dissolved carbon (DOC) is transported by the rivers system; partly transformed to carbon greenhouse gases by decomposition during transport, partly stored in floodplain sediments, or finally in marine sediments. Upward arrows indicate carbon loss to the atmosphere, downward arrows carbon storage

carbon from ecosystems by water, the carbon balance of land ecosystems is incomplete.

Rivers also integrate the effects of soil carbon decomposition in their drainage basin. The water-transported carbon contains biochemical clues about the dominant processes of soil carbon decomposition and changes therein in the drainage basin (Benner et al. 2004; Feng et al. 2013). However, as with discharge data, the measurement time series of the amount of organic matter transport by rivers are short for Arctic rivers, and transport strongly varies from year to year, which complicates the detection of trends (Holmes et al. 2013). CO_2 and CH_4 evasion from river water to the atmosphere is also poorly quantified (Striegl et al. 2012). Recently, new data collection efforts (https://arcticgreatrivers.org/) have provided a baseline with data on particulate organic carbon and nitrogen transport for the six largest Arctic rivers between 2003 and 2013 (McClelland et al. (2016), and innovations in analysing dissolved organic matter using optical methods increase the temporal and spatial coverage of these measurements (Mann et al. 2016).

Erosion of terrestrial organic matter from Arctic drainage basins, followed by deposition in marine sediments, has been seen as a large geological carbon sink, exceeding that of rock weathering (Hilton et al. 2015). However, for each transport step, from erosion at the source, to repeated erosion and deposition in the fluvial environment, to transport into the marine environment, a part of the organic matter is degraded and lost to the atmosphere as CO_2 and CH_4. For instance, the Lena river water is supersaturated with CO_2 and CH_4, by up to 1.5–two fold in summer, and 4–five fold in winter (Semiletov et al. 2011), indicating organic matter decomposition within the river. The amount of erosion, and partitioning between decomposition and sedimentation is poorly known; for each amount of particulate organic

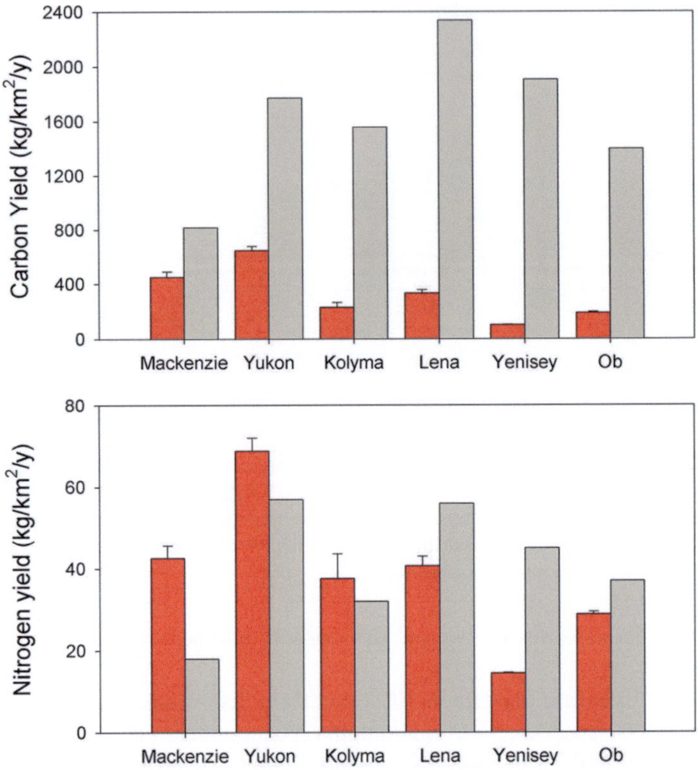

Fig. 5.23 Top: Carbon yields for the major Arctic rivers. Values are annual averages ±1 standard error for suspended POC (particulate organic carbon) from McClelland et al. 2016 and averages for DOC (dissolved organic carbon) from Holmes et al. (2012). Bottom: Nitrogen yields for the major Arctic rivers. PN (particulate organic nitrogen, red bars) from McClelland et al. (2016) and DON (dissolved organic nitrogen, grey bars) from Holmes et al. (2012). Republished with permission of John Wiley and Sons Inc, from 'Particulate organic carbon and nitrogen export from major Arctic rivers', by McClelland et al. (2016), Global biogeochemical cycles 30, © 2016; permission conveyed through Copyright Clearance Center, Inc.

sediment entering the marine bottom sediment, an unknown amount is lost to the atmosphere or converted into dissolved organic matter (Vonk and Gustafsson 2013).

Two classes of organic matter from permafrost soils are transported by the rivers:

- dissolved organic carbon (DOC) leached from soils or solubilized in the river water, which can be divided in low molecular weight DOC (LMW-DOC) and colloidal organic carbon (COC);
- particulate organic matter (POC), derived mostly from erosion of soils (Guo et al. 2007).

LMW-DOC is the most readily available organic matter for microbial degradation (Battin et al. 2009). The organic matter, in particular POC, may be associated with or sorbed to mineral particles. Next to carbon, also nutrients are included in water-transported organic matter. In general, the large Eurasian rivers, have a smaller proportion of particulate organic carbon and nitrogen transport than the large American rivers (McClelland et al. 2016; Fig. 5.23). Usually Total Organic Nitrogen (TON) is reported, eventually split in Particle-bound Nitrogen (PN) and Dissolved Nitrogen (DN), the latter consisting of Dissolved Organic nitrogen (DON) in dissolved organic matter and Dissolved Inorganic Nitrogen (DIN), as NH_4^+ and NO_3^- ions. Besides organic carbon, also inorganic carbon is transported by river water as dissolved CO_2, bicarbonate and carbonate.

The relative amount of these organic matter classes varies with the season (Neff et al. 2006; Raymond et al. 2007; Holmes et al. 2013). The transport of organic matter in Arctic rivers is highest during the spring flood, both in terms of total mass, and concentration in the river water (Neff et al. 2006; Holmes et al. 2008; Frey and McClelland 2009). In the Yukon river, the concentration of Total Organic Carbon (TOC) is highest in spring (mid-May), and decreases about five-fold towards the autumn (September). Here, most of the transported organic carbon is colloidal (±63%), followed by LMW-DOC (±23%) and POC (14%) (Guo and Macdonald 2006). Also POC varies similarly with discharge for the six largest Arctic rivers. However, the relation between river discharge and POC concentration differ markedly among these rivers (McClelland et al. 2016).

Organic carbon enters the river system by erosion, surface runoff and leaching from the active layer by groundwater flow. Up to 78% of the POC delivered to two river basins in the North Canadian Archipelago originated from permafrost erosion features such as cryogenic landslides and gully erosion (Lamoureux and Lafrenière 2014). Leaching experiments (Guo et al. 2007) indicate that soils loose less than 1% of their carbon as DOC by leaching by water percolating through the soil, so erosion should be an important process for liberating both organic matter fractions. Leaching experiments have shown that the top of the permafrost loses more DOC and total N upon thawing than active layer soils (Wickland et al. 2018). Surface runoff contributes considerable amounts of DOC and POC, in particular during snow melt. POC also consists of recent plant material, next to eroded soil material (Guo and Macdonald 2006). A source of DOC is also the erosion of river banks, including erosion of yedoma (Vonk et al. 2013).

The flux of CO_2 and CH_4 has been quantified for a few Arctic river systems. Striegl et al. (2012) quantified riverine fluxes in the Yukon based on measured water CO_2 and CH_4 partial pressures and fluxes. The CO_2 fluxes average 750 g CO_2-C m^{-2} $year^{-1}$ based on water surface area, and 9.0 g C m^{-2} $year^{-1}$ normalized to the river basin area. The CH_4-C emissions are a small fraction of this (0.7%, 5.25 g CH_4-C m^{-2} $year^{-1}$, 6.4% of the combined radiative forcing of CO_2 and CH_4. This comprised about half of the total C export in all water-borne C-fractions (inorganic and organic carbon) of the river system. However, CO_2 and CH_4 emission from rivers are not derived from degradation of transported organic matter alone. Probably only a small fraction (6%) of the emitted C gases are derived from decomposition of transported DOC (Wickland et al. 2012). Other sources of CO_2 and CH_4 are dissolved gases in groundwater flow towards the river bed, in runoff, and carbonate equilibrium reactions with carbonate-containing sediments or rocks in the riverbed.

In particular in the lower reaches there may be considerable primary production, including uptake of CO_2 in lake waters (Striegl et al. 2012; Squires et al. 2009).

CO_2 outgassing also varies throughout the river systems from small to large river tributaries. The summer CO_2 fluxes at low flows from small streams in the Kolyma basin in Siberia are (on area basis) much higher than that of larger tributaries and the main river (Denfeld et al. 2013). The fluxes in the small streams are highly variable (8.6 ± 8.0 g C $m^{-2}d^{-1}$) and an order of magnitude higher than that of the main river (0.5 ± 0.7 g C $m^{-2}d^{-1}$). The CO_2 partial pressure of larger rivers and the main river are closer to equilibrium with the atmosphere, and several of the tributaries are small CO_2 sinks. The differences are attributed to a larger amount of eroded soil carbon and its rapid degradation in the small streams, and primary production with CO_2 uptake by photosynthesis of phytoplankton, combined with slower decomposition downstream for the larger rivers.

In general, the radiocarbon age of the organic matter differs strongly. DOC is of very recent age, reflecting decomposition of recently photosynthesised organic matter, while POC is considerable older, with ages up to thousands of years, reflecting erosion of old organic material from soils and peat deposits (Benner et al. 2004; Raymond et al. 2007; Guo et al. 2007; McClelland et al. 2016). Throughout the thaw season, the age of transported DOC varies; the early season DOC tends to be young, reflecting material transported by surficial runoff, while with progressing thaw the age of DOC increases (Frey and McClelland 2009).

In the North American Yukon and Mackenzie river basins, the age of POC transported by the rivers and in recent sediments ranges from ±4400 to ±8000 year BP, while DOC ranges from 390–1440 year BP. In a third basin (Sagavanirktok) the ages of DOC and POC were more similar, consisting of several thousand years old C (Guo et al. 2007). Old DOC enters the river system by erosion of Pleistocene and Early Holocene deposits along river banks. However, old DOC from Yedoma sediments (age >21,000 year BP) along the Kolyma river proved to be highly labile upon incubation in river water, resulting in rapid decomposition; 34% was lost in a two-week incubation (Vonk et al. 2013). Preferential decomposition of old labile DOC from Pleistocene deposits may explain the relatively young ages of DOC at the lower reaches of the river, despite the large drainage basin reservoir of dissolved old carbon; the material is decomposed and outgassed before arriving at the river mouth (see also Drake et al. 2015).

In the Kolyma river in Siberia, the DOC flux is very high during the spring discharge, containing 86% of the yearly flux. In upstream areas, the DOC age increases throughout the summer season (up to 675 years), in the lower reaches, this trend is absent (Neff et al. 2006). These changes are also expressed in the structure of the organic material, that indicate the presence of relatively undecomposed lignin in the younger DOC and more lignin-poor, aromatic compounds in the older DOC. This reflects increasingly deeper thaw of the permafrost with release of older DOC throughout the growing season. In the Yukon river, the colloidal organic matter is mainly of modern origin, while POC has an average age of 3698 ± 902 years BP, and low molecular weight DOC has intermediate ages of 976 ± 212 years BP (Guo and Macdonald 2006).

As stated by Guo et al. (2007), the age and source difference of DOC and POC imply that massive thawing of permafrost will be signalled by the age and amounts of POC in arctic rivers, whereas DOC will be affected most by changing vegetation and primary production of ecosystems. However, as yet it is difficult to detect trends in the relatively short and highly variable records. Moreover, in these records the signal is complicated by concurrent increase of river discharge, and changing Arctic vegetation (in particular in tundras) that influences also total carbon export and permafrost thaw. Additionally, regional differences in e.g. soil composition will determine whether permafrost thaw will lead to increases in fluvial DOC (when organic rich soils are thawed) or decreases in fluvial DOC (when mineral soils with increasing sorption capacity are thawed) (Vonk et al. 2015).

Gustafsson et al. (2011) and Feng et al. (2013) separated fluvial carbon signals by compound-specific ^{14}C age of specific plant biochemical tracers (*n*-alkanes in wax lipids, lignin phenols, hydroxy phenols) in estuarine sediments of the large Eurasian Arctic rivers. There are differences going from west to east; the alkanes decrease in age from 11,400 BP on average in the Ob basin to 6400 BP in the Kolyma, consistent with a trend of colder climate and a larger basin coverage of continuous permafrost, that limits flow paths through soils. On the other hand, the age of lignin is controlled strongly by mean annual runoff, being youngest in the westernmost basin with the highest runoff rate. On the basis of this analysis, Feng et al. (2013) conclude that recent organic matter input to the rivers has increased, and that the proportion of old carbon has increased by 3–6%. Schreiner et al. (2014) also found very old POC (ranging in ^{14}C age between 7590 and 21,480 BP) in the North Alaskan Colville river, indicating liberation of old carbon from Yedoma deposits in the basin.

How much DOC and POC survives the transport is depending on its lability. There are large differences in the amount and lability of river DOC transported during the spring discharge, and during later lower flows in summer, as shown by data from three Alaskan rivers (Holmes et al. 2008). The DOC concentration is highest during the spring discharge peaks, and spring discharge DOC loses 20–42% of its weight upon a 3 month incubation. Incubation losses were minor ($\leq$ 12%) for DOC sampled in summer. Light penetrating the water column stimulates bacterial decomposition of DOC (Cory et al. 2013). Vonk et al. (2013) and Drake et al. (2015) have shown that old Pleistocene DOC derived from erosion of Yedoma deposits along the Kolyma river is highly labile in river water. The lability of the freshly eroded material was probably related to its sudden liberation from the permafrost by erosion without much in situ degradation, its composition (low molecular weight compounds), nutrient availability, and a boost of metabolic activity of bacteria by rapid temperature rise during thaw.

The biolability of Pleistocene POC is uncertain. It is subject to degradation during transport (Sanchez-Garcia et al. 2011; Schreiner et al. 2014). Physical processes also may remove it preferentially form river water by sedimentation due to ballasting by minerals associated with the organic particles (Vonk et al. 2010, 2014). The re-sedimentation rate and lability of Pleistocene POC and DOC implies that a signal of increasing permafrost thaw and release of old carbon may be difficult to detect in rivers, because of a short residence time of this organic matter in the river water. This is confirmed by Mann et al. (2015), who found a stronger old DOC signal in smaller streams of the Kolyma, close to the source of the organic material. With longer transport, this signal subsequently becomes dominated in larger tributaries and main channel by recent DOC. Consequently, a signal of old permafrost carbon liberated by permafrost thaw, must be sought upstream. The clearest signal of release of old permafrost organic matter is ^{14}C fingerprinting of POC; for the large Eurasian rivers 63 ± 10% is derived from permafrost and peat organic matter (Wild et al. 2019).

It is uncertain how the total organic carbon (TOC) transport will react on climate change; both future increase and decrease is predicted by various studies (Holmes et al. 2013; Vonk et al. 2015), depending mostly on regional factors such as soil type, ice content, topography, and permafrost zonation. In West Siberia, the DOC export is higher in watersheds with a higher mean annual air temperature and in permafrost-free watersheds, suggesting an increase of DOC transport (Frey and Smith 2005; Frey and McClelland 2009; Fig. 5.24). However, Walvoord and Striegl (2007) argue for a future decrease of DOC transport because of an increasing contribution of groundwater flow, which carries lower amounts of organic carbon relative to surface flow. Guo et al. (2007) suggest a future increase of DOC from increasing primary production in Arctic ecosystems, and to a lesser extent from thawing permafrost, while POC could increase by permafrost thaw with increasing erosion.

Besides carbon, also nutrients are exported by the rivers. Fluvial export of particulate organic carbon and nitrogen throughout the pan-Arctic watershed is estimated at 5767×10^9 g of POC and 695×10^9 g of PN per year, respectively (McClelland et al. 2016). In the Upper Yukon, the C/N mass ratios of colloidal DOC (39 ± 2.5) and POC (13 ± 1.2) differ markedly (Guo and Macdonald 2006). The

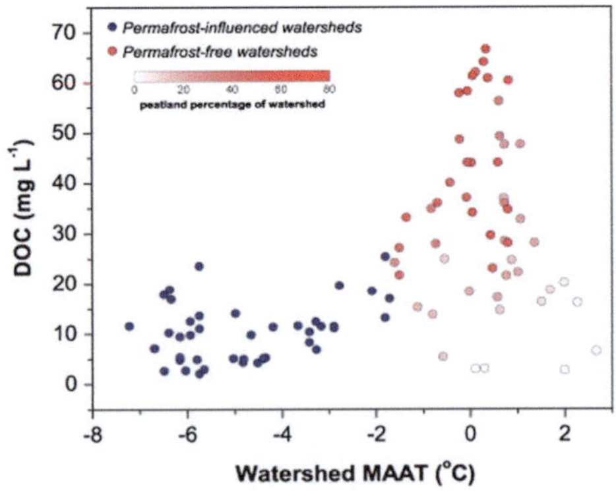

Fig. 5.24 Dependence of DOC concentration on watershed mean annual air temperature (MAAT) in West Siberia. A sharp increase in concentrations occurs in watersheds not impacted by permafrost. Low concentrations in these permafrost-free watersheds are due to sparse peatland coverage. (Republished with permission of John Wiley and Sons Inc, from 'Impacts of permafrost degradation on arctic river biogeochemistry', by Frey and McLelland (2009), Hydrological processes 23, © 2009; permission conveyed through Copyright Clearance Center, Inc.)

amount of N bound to organic material (Total Organic N, TON) consists for about 28% of N in POC, the remainder of N is included in DOC. Next to the organic N and P, the river system also carries dissolved inorganic N (DIN, NO_3^-, NO_2^- and NH_4^+). However, the organic N load for large Arctic rivers is 15–300% higher than the inorganic N (Holmes et al. 2013).

With an increasing organic C input to the rivers, likely also the nutrient load will increase. However, the number of datasets that might demonstrate such an increase is small. Data from the Kuparuk river in Alaska show a five-fold increase of NO_3^- export between 1991 and 2001 (McClelland et al. 2007). In Alaskan watersheds, the NO_3^- concentration and organic N load is an order of magnitude higher in discontinuous permafrost watersheds than in continuous permafrost watersheds (Jones et al. Jones Jr et al. 2005), which may be an effect of deeper groundwater flow paths in discontinuous permafrost areas that facilitate export of larger amounts of nitrate and other mineral nutrients (phosphate). However, in West Siberian watersheds with a similar range of permafrost coverage this difference between discontinuous and continuous permafrost watershed proved absent (Frey et al. 2007a, b). For N, also other watershed characteristics are important, such as the amount of wetlands, that limit remineralization of organic matter and facilitate denitrification (Frey and McClelland 2009). For other mineral nutrients, such as phosphate, the contribution of mineral-rich groundwater to river discharge is important and will increase by permafrost thaw (Frey and McClelland 2009). An increased input of nutrients will influence also the primary productivity of floodplain ecosystems, enhancing both carbon storage and the CH_4 flux from these systems (see above).

The research on river water carbon transport has demonstrated that it provides insights in basin-wide changes in permafrost carbon release. In particular POC ages and quantities have a signal function, next to the DOC fractions and their chemical composition. However, quantification of Arctic river carbon and nutrient fluxes are equally important for closing of the terrestrial carbon balance, and for future ecosystem changes in the wide Arctic lowland floodplains.

5.4.5 *Soil Hydrology Changes – Wetting or Drying?*

The previous sections have made clear that soil hydrological changes are to be expected from warming and thawing permafrost, resulting from subsidence. Superposed on that, are changes in the soil water balance that result from climatic changes in evapotranspiration and precipitation. These can progress in the direction of both wetting or drying. Here, the effects of wetting or drying are discussed in a general sense.

Based on an earth system climate model (UVIC), Avis et al. (2011) suggest a decrease of wetlands by permafrost thaw. However, although this modelling result is not unlikely, their results are more based on a general linkage of permafrost to wetlands, climate indices and terrain slope rather than assessment of changes in water table and drainage conditions. The actual changes will vary according to the

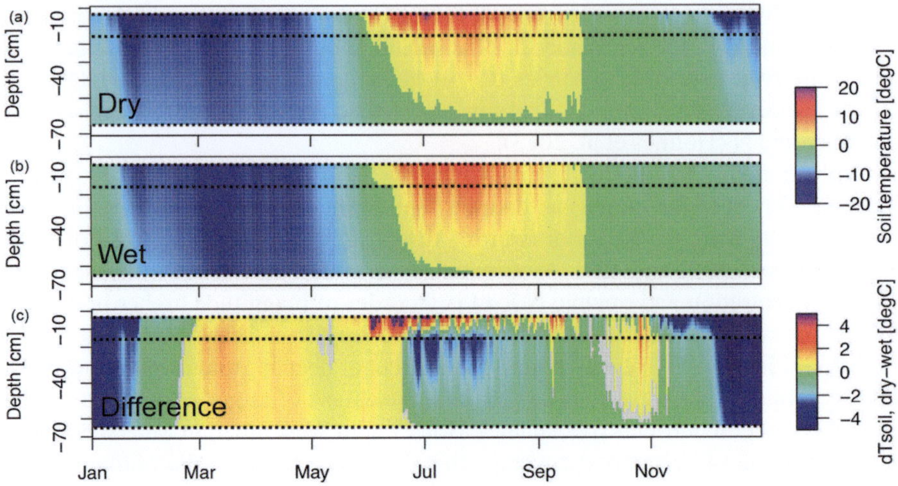

Fig. 5.25 Daily mean soil temperature profiles at a dry (a) and wet (b) micro-site, and the difference (dry–wet, c) for the calendar year 2015. Data for three measurement depths (dotted lines) are linearly interpolated along the vertical profile. Values in the difference plot were truncated to the range [−5; 5] °C to enhance visibility of fine-scale patterns, while values fell within the range [−11.3; 10.5] °C. From Göckede et al. (2017), Shifted energy fluxes, increased Bowen ratios, and reduced thaw depths linked with drainage-induced changes in permafrost ecosystem structure, The Cryosphere 11, https://doi.org/10.5194/tc-11-2975-2017, published under Creative Commons License 3.0 https://creativecommons.org/licenses/by/3.0/; image provided courteously by the authors

present hydrologic conditions, permafrost ice content, geomorphological conditions and vegetation cover.

The effect of drying, wetting or flooding of soils on the soil thermal regime is large. The thermal conductivity and heat capacity of soils increase with water content; both induce a net increase of the thermal diffusivity (Table 2.2). A dry and a wet site on a partially drained floodplain in Northeastern Siberia illustrates the effect of soil wetness (Göckede et al. 2017; Fig. 5.25). In the first half of winter, the wet soil tends to be warmer because of the latent heat of freezing of the larger amount of water in the topsoil. Later in winter, differences diminish. In summer, the topsoil of the dry soil warms faster than the wet soil, but heat transport to deeper layers is slower because of the lower thermal diffusivity of the dry topsoil. Consequently, deeper layers stay cooler. Therefore, the active layer of the dry soil tends to be thinner. Dry soils tend to retard permafrost thawing, while wet soils result in a thicker active layer. Moreover, the biochemical process rates are strongly influenced by these temperature and moisture differences.

Several approximate scenarios are likely if we include effects of precipitation and evapotranspiration changes and changes in permafrost discussed above:

- Presently well drained soils in relatively flat terrain with predominantly continuous permafrost may receive more moisture from snowmelt or rain, combined with a thicker ALT. An increasing vegetation cover causing higher evapotranspiration may counteract this moisture increase partly, and may reduce the growth rate of the ALT. Also, saturated and unsaturated groundwater flow will increase. In the High Arctic barrens, the changes in soil moisture, ALT and vegetation are likely to be most pronounced. If the subsurface is ice-rich, soils will be susceptible to subsidence, creating waterlogged soils or pools and fostering more heat transport into the permafrost; in the vicinity of eroding polygon networks, soil drainage will increase, possibly slowing permafrost thaw.

- Well drained soils on sloping terrain on continuous permafrost will show similar changes. However, soils on slopes are more susceptible to erosion, in particular when ice-rich. Erosion results in loss of vegetation, decreasing evapotranspiration, but on the other hand increases topographical and hydraulic gradients, increasing runoff and soil drainage.
- Soils in poorly drained terrain will tend to become wetter, by a combination of subsidence by permafrost thaw and increased net precipitation. Development of standing water bodies can enhance further ALT increase. Soils on floodplains will be flooded more frequently. However, erosion of drainage thresholds along ice wedges (if present) or other increased connectivity of surface runoff created by permafrost thaw could result in local or more widespread drainage of wetlands (Connon et al. 2014; Liljedahl et al. 2016).
- Soils in boreal peatlands with discontinuous permafrost tend to become wetter with increasing thaw of palsas and peat plateaus, creating terrain depressions with a high water table or standing water.

5.4.6 *Soil Hydrology Changes – Carbon Cycle Effects*

To consider the effects of soil hydrology change on the carbon cycle, a distinction should be made between water table depth, and soil moisture. Below the water table anoxic conditions prevail at which the biogeochemical pathways for decomposition of organic matter change drastically; above the water table, optimal moisture conditions for bacterial activity become potentially important – bacterial activity also decreases at low soil moisture (Sect. 4.5).

Soil moisture and water table influence the ratio of CO_2 over CH_4 emission; wet soils emit more CH_4 and less CO_2. However, the overall carbon turnover by heterotrophic respiration in water saturated soils is lower than in soils with low water table. Schädel et al. (2016) analysed 25 aerobic and anaerobic incubation studies; 10 °C warming results in a factor 2 higher release of CO_2-C and CH_4-C, but under aerobic conditions soils released on average 3.4 times more C than under anaerobic conditions. Even if the stronger radiative effect of CH_4 is taken into account, the effect of CO_2 emission dominates. On the other hand, longer term incubation experiments (7 years) show a comparatively larger CH_4 production after a stable methanogenic bacteria community is established in the samples; equal amounts of CO_2 and CH_4 were produced (Knoblauch et al. 2018). These studies use incubation experiments and not field conditions. Nevertheless, it indicates that hydrological changes upon thawing of permafrost landscapes, which determine whether soil carbon is released under anaerobic or aerobic conditions, control the strength of the permafrost carbon feedback by controlling the relative proportions of CO_2 and CH_4 being released.

Several field experiments have evaluated the effects of drying or wetting of permafrost soils. Just as with soil temperature change, the effects of soil moisture changes are difficult to disentangle from other variables in field experiments, such as vegetation change. In particular soil moisture changes have a large influence on vegetation – either by short term effects on the vegetation photosynthesis rate and evapotranspiration (e.g. Keuper et al. 2012b; Iijima et al. 2014), or by change of vegetation composition, function and structure (e.g. Nauta et al. 2014); see next chapter. These in turn affect the soil carbon cycle and feed back on soil moisture, by changing evapotranspiration.

A three-year water table draw-down experiment by Natali et al. (2011, 2015) showed that soil drying doubled soil organic matter decomposition in the top 20 cm of the soil (measured as mass loss of a standard substrate). Drying also increased decomposition of older soil carbon as shown by ^{14}C-CO_2. During the experiment there was no decline in plant carbon uptake, that could have occurred by drought stress. The CH_4 flux decreased significantly at lower water table depth, switching from positive (emission) to negative (uptake). R_{eco} (including CO_2 from soil respiration) increased with water table depth (Fig. 5.26). Both a decrease of CH_4 production, or increase of CH_4 oxidation should have reduced the CH_4 flux.

This field experiment confirms the conclusions of Schädel et al. (2016) drawn from laboratory incubation experiments. However, the effects of soil moisture at low water table soils are more complicated, because of potential decrease of bacterial activity under dry conditions. In a natural permafrost thaw gradient in the same tundra area as the experiments by Natali et al. (2015), Hicks Pries et al. (2013b) found that *higher* soil moisture *increases* the decomposition rate significantly compared to dry soils. An increase of growing season precipitation from 200 to 300 mm doubled the decomposition rate of a standard substrate. However, in the study by Hicks Pries et al. (2013b) thaw depth, soil temperature, nutrient availability and microbial community effects may have affected the decomposition rate as well as precipitation.

An experiment on High Arctic semi-desert soils in Northwest Greenland also demonstrates an important combined effect of warming and wetting of the soil and vegetation, and further demonstrates that the effect of soil moisture will differ among vegetation/climate zones within the Arctic (Lupascu et al. 2014). Experimental summer warming of 4 °C decreased the summer CO_2 sink of the ecosystem by 55%, while the combination of warming with wetting increased the CO_2 sink by nearly an order of magnitude, in particular for vegetated soils. Warming increased the ecosystem *GPP* only slightly, while R_{eco} increased comparatively more, including the respiration of old soil organic matter. By contrast, warming combined with wetting increased *GPP* threefold and R_{eco} by about 40%, increasing the CO_2 sink. The abundant production of new organic carbon by the higher *GPP*

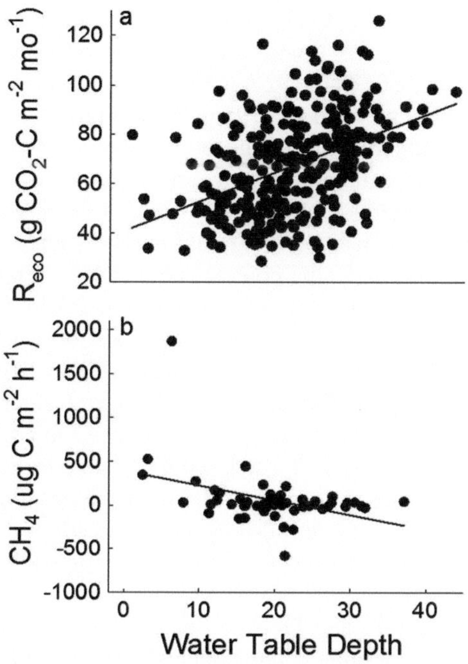

Fig. 5.26 Relationship between water table depth (WTD, distance below ground surface) and (a) monthly ecosystem respiration (Reco) in the field experiments described by Natali et al. (2015). Average over 2011 to 2013 ($R_{eco} = 1.2 \times WTD + 41.2$; $P < 0.001$, $R^2 = 0.19$) and (b) CH_4 emission rates ($CH_4 = -16.7 \times WTD + 393.4$; $P = 0.003$, $R^2 = 0.16$). Note that higher water table values (WTD) represent dryer soil conditions. (Republished with permission of John Wiley and Sons Inc, from 'Permafrost thaw and soil moisture driving CO_2 and CH_4 release from upland tundra', by Natali et al. (2015), Journal of geophysical research Biogeosciences 120, © 2015; permission conveyed through Copyright Clearance Center, Inc.)

also restricted the decomposition of older soil carbon. In this High Arctic semi-desert environment, low soil moisture and low temperature are both an important restriction on vegetation growth and soil carbon decomposition. It is expected that warmer and wetter conditions do not increase emission of old carbon from these high-arctic dry soils, but increase the CO_2 sink in vegetation, at least temporary.

These experiments are relatively short. On the long term, increased drainage proves to increase CO_2 emissions and reduce CH_4 emission as shown by a drainage experiment on the Kolyma river floodplain in Northeastern Siberia (Merbold et al. 2009; Kittler et al. 2017). The site, a periodically flooded floodplain area dominated by *Carex appendiculata* tussocks, was drained in 2005 by constructing a drainage ditch around the measurement site. Drainage lowered the water table by ±30 cm. Before drainage, 3 years of eddy covariance CO_2 flux measurements and chamber CH_4 flux measurements were taken during the growing season, that recorded pre-drainage conditions (Corradi et al. 2005; Merbold et al. 2009). The floodplain site was – before drainage – a relatively large source of greenhouse gases to the atmosphere, dominated by high CH_4 fluxes. It lost carbon to the atmosphere during the growing season (+55 g CO_2 m^{-2}), with a total greenhouse gas emission including CH_4 (in CO_2 equivalents) of 1741 ± 927 gCO_2 eq m^{-2}. After drainage, this reduced to respectively +29 g CO_2 m^{-2} and 84 ± 95 gC CO_2 eq m^{-2}.

About 10 years after the initial drainage, eddy covariance CO_2 and CH_4 measurements have been repeated at the drained site, and at a nearby undrained control site, with year-round measurements including the winter season (Kittler et al. 2017). In 2014–2015, the drained site was on an annual basis a source of CO_2 to the atmosphere (47.3 ± 56.5 g CO_2 m^{-2}), the control site a sink (−136.0 ± 6.2 g CO_2 m^{-2}). Both sites were a source of CH_4 (a small source of 4.8 ± 0.1 g CH_4 m^{-2} for drained, a large source of 8.9 ± 0.7 g CH_4 m^{-2} for control). This followed the trend visible in the first year, albeit that the control site is a sink of CO_2 in 2014–2015 instead of a source. Combining the CO_2 and CH_4 fluxes into CO_2 equivalents based on the IPCC (2013) global warming potentials, the drainage resulted in an extra source in CO_2 equivalents of 44 g CO_2 m^{-2} annually, so the net radiative forcing was increased by drainage. Most of the difference can be attributed to higher R_{eco} in the drained site (Kittler et al. 2017). The experiment did not include carbon export by water, only the fluxes to the atmosphere. However, additional carbon may have left the drained site via the drainage ditch.

Kwon et al. (2016, 2017) and Göckede et al. (2017) analysed the ecosystem effects of the drainage experiment; their analysis shows an intriguing cascade of secondary effects of drainage on the ecosystem. The changes in the carbon fluxes towards the atmosphere were just the tip of the iceberg. Drainage triggered shifts in vegetation structure and other ecosystem properties, snow cover, radiation budget and soil thermal conductivity by moisture changes.

On the drained site, wetland cotton grasses (*Eriophorum angustifolium*) were strongly reduced in abundance, in favour of shrubs (*Betula nana, Salix pulchra*), of which the cover increased from almost absent to 11%, and increased the aerodynamic roughness. The increased shrub cover also decreased the albedo during snowmelt. The latent heat flux was on average lower over the drained area by 2.2 W m^{-2}. A higher Bowen ratio (Sect. 2.2) indicates a shift to sensible heat flux over the drained area at the expense of evapotranspiration. The reduced thermal conductivity of the drier soil resulted in a warmer topsoil, and cooling of the soil below 10 cm depth and a thinner ALT (Göckede et al. 2017; Fig. 5.25). The changes in the soil temperature in turn caused a stabilization of organic matter at the base of the active layer by freezing, and concentration of heterotrophic respiration in the topsoil. Kwon et al. (2016) report an up to 95% increase of the heterotrophic respiration in the topsoil. At the same time the *GPP* was reduced by the dry soil conditions. This also depressed the autotrophic respiration. The net difference in R_{eco} between drained and undrained site was 25%, while the CO_2 uptake was reduced, resulting in the net increase in the ecosystem CO_2 source.

Drainage also reduced CH_4 emission during the growing season by a factor of 20, while in the non-growing season a slightly higher emission was found. The reduction of the growing season CH_4 flux was accompanied by a decrease of both the methanogenic and methanotrophic bacterial community, indicating a strongly decreased CH_4 production. Cooling of the deeper soil reduced methanogenesis in the anoxic soil zone below the water table. At the same time, also the CH_4 transporting aerenchymous plant cover (*E. angustifolium*) was reduced, inducing a shift from plant-mediated transport to diffusive transport through soil pores. Diffusion allows more methanotrophic methane consumption in the oxygenated topsoil (Kwon et al. 2017).

Similar effects on CH_4 fluxes by water table manipulation have been found by Turetsky et al. (2008) in a boreal fen in Alaska, where the water table was raised and lowered in experimental plots by about 2–4 cm, together with experimental warming. A raised water table increased the cumulative summer CH_4 flux by 17–177%, a lowered water table decreased it by 11–44%.

However, an increase of the water table does not always result in higher CH_4 fluxes, attesting of complicated, nonlinear effects of water table on CH_4 fluxes. An experiment that raises the water up to the level of flooding has been published by Zona et al. (2009, 2012). This flooding was achieved by installing water retention dikes in a drained thaw lake basin with polygonal tundra near Barrow, Northern Alaska. The water table between the flooded plot and control plot differed initially about 5 cm, and was left to drop below the surface during the summer season; the flooded plot had the longest and deepest submersion.

Unexpectedly, the CH_4 flux at the flooded site was *lower* during the flood period than that of the control site, while the CO_2 fluxes for the flooded site were slightly positive (emission) and those of the control site negative. In the first year of the experiment, the water table depth had a highly nonlinear influence on the CH_4 flux. The CH_4 flux was lowest at high water above the surface, highest when the water table was at the surface and dropped again to an approximately constant level for water table levels in the range 0 to 15 cm below the surface (Fig. 5.27). A water table above the surface probably diminished the plant-mediated transport of CH_4, because plant parts that emit most CH_4 remained below the water table. Thaw depth throughout the season influenced the CH_4 fluxes in a similar way. At thaw depths shallower than 15 cm, CH_4 fluxes remained low, but increased stepwise at a thaw depth of 15 cm. Soil temperature, not water table, was the best predictor of CH_4 flux in this experiment.

At the continuation of the experiment, the flooded site proved to be a considerable CO_2 source (next to CH_4) despite the high water table and anoxic conditions in the soil. The CO_2/CH_4 ratio in the soil varied between 13 to 20. The site had a larger thaw depth and higher soil temperature than the control site. Incubation experiments, redox measurements and data on Fe(III) concentration showed that anaerobic fermentation at moderately reducing conditions are the drivers of the CO_2 emission (Zona et al. 2012; Lipson et al. 2012). Fe minerals were abundantly present in the soil solids, the high presence of siderophores (biologically produced organic molecules capable of binding Fe(III)) indicated active turnover of Fe(III) by the bacterial population. The high amount of alternative electron acceptors may have competed with methanogenesis. Although the soil was definitively anoxic, the redox potentials were positive in the flooded soil, outside the range for methanogen-

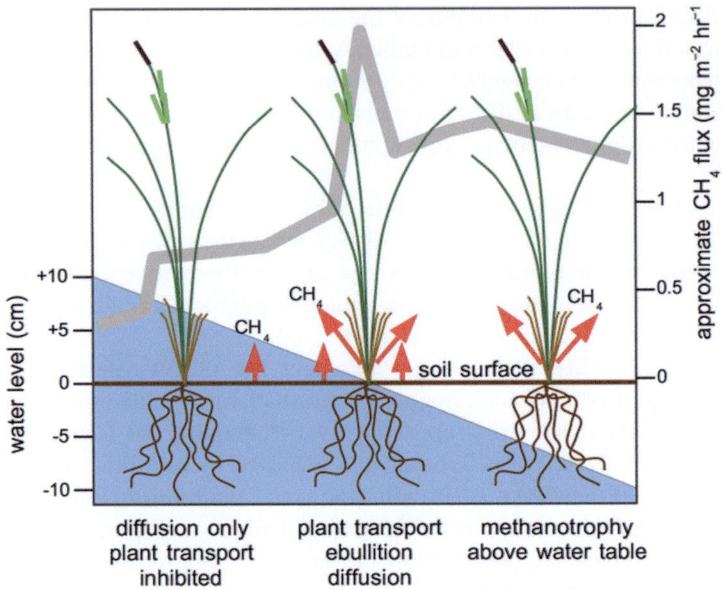

Fig. 5.27 Effect of water table in a flooding experiment in a drained thaw lake basin in Northern Alaska, after Zona et al. (2009). A water table above the surface suppresses CH_4 fluxes from plants by covering the lower parts of the plants (Carex aquatilis) that emit most CH_4. A flux peak occurs when the water table is at the soil surface. The CH_4 flux graph is derived from the data of Zona et al. (2009)

esis (Table 4.3). Accompanying changes in soil chemistry were a higher soil pH, and higher nutrient concentrations in the flooded soil (Lipson et al. 2012).

A laboratory experiment on flooding of peat columns from boreal peatlands confirm an increase in CH_4 emission with 170% over months; longer term inundation did not show a further increase (Sjögersten et al. 2016). However, the CH_4 emission remained relatively low in this experiment, compared to the large rise of CH_4 emission that are found in permafrost collapse sites in boreal peatlands. The pore water chemistry suggested that the amount of substrate for acetoclastic methanogenesis (low molecular weight organic acids) remained at very low levels during the experiment. This implies that the increase in CH_4 production was driven by hydrogenotrophic methanogenesis, or that acetoclastic methanogenesis was limited by acetate supply. It is likely that the strong increase of CH_4 emission, seen in peat plateau and palsa collapse sites, is mostly driven by the establishment of a more productive plant community and associated release of fresh root exudates and litter, and by plant species that rapidly transport CH_4 into the atmosphere (Sjögersten et al. 2016; Ström et al. 2003).

These experiments show that, as with soil drainage, soils do not react in a simple way to flooding as occurs in thaw depressions, because of the cascade of soil chemical and vegetation changes that occur. Although CH_4 emission generally will increase in wetter or flooded soils, CO_2 emission may remain important as well,

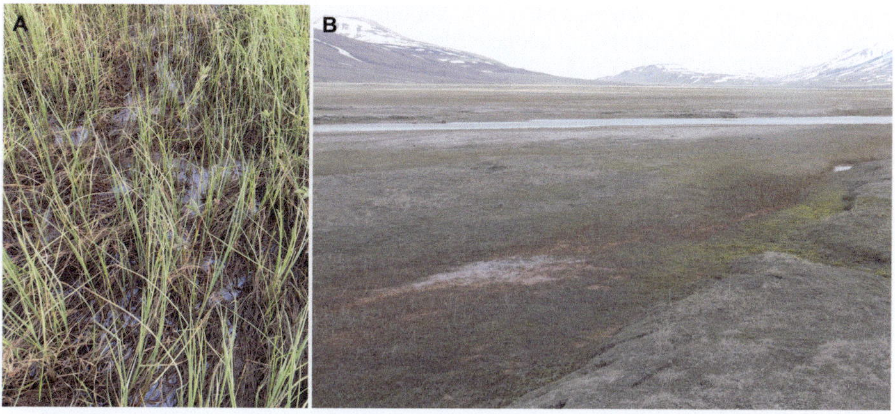

Fig. 5.28 (**a**) Oily sheen of Fe-oxidizing bacteria in sedge vegetation at the edge of a river floodplain, indicating seepage of Fe-rich groundwater. Kytalyk Research station, Northeast Siberia. (**b**) Fe oxides at the surface soil, floodplain, Adventalen, Svalbard. (Photos by the author)

depending on the availability of other electron acceptors (e.g Fe). Areas favourable for accumulation of Fe by groundwater flow are floodplains and depressions created by permafrost thaw (Fig. 5.28). The experiments described by Zona et al. (2009), Sjögersten et al. (2016) and Kwon et al. (2016, 2017) illustrate the important role of vegetation in CH_4 emission above that of water table, by providing substrate for methanogenesis and transport pathways through the soil. The combination of nutrient supply, supporting a high primary production of vegetation with ample release of labile carbon into the soil, and an ample supply of electron acceptors for anoxic organic matter decomposition likely support the high biochemical activity in depressions creating by ground ice thaw, rather than the creation of anaerobic conditions by a higher water table alone.

References

Abbott BW, Jones JB (2015) Permafrost collapse alters soil carbon stocks, respiration, CH_4, and N_2O in upland tundra. Glob Chang Biol 21(12):4570–4587

Abnizova A, Siemens J, Langer M, Boike J (2012) Small ponds with major impact: the relevance of ponds and lakes in permafrost landscapes to carbon dioxide emissions. Glob Biogeochem Cycles 26(2):GB2041. https://doi.org/10.1029/2011GB004237

Agafonov L, Strunk H, Nuber T (2004) Thermokarst dynamics in Western Siberia: insights from dendrochronological research. Palaeogeogr Palaeoclimatol Palaeoecol 209(1–4):183–196

Åkerman HJ, Johansson M (2008) Thawing permafrost and thicker active layers in sub-arctic Sweden. Permafr Periglac Process 19(3):279–292

Arp CD, Jones BM, Urban FE, Grosse G (2011) Hydrogeomorphic processes of thermokarst lakes with grounded-ice and floating-ice regimes on the Arctic coastal plain, Alaska. Hydrol Process 25(15):2422–2438

Arp CD, Jones BM, Lu Z, Whitman MS (2012) Shifting balance of thermokarst lake ice regimes across the Arctic coastal plain of northern Alaska. Geophys Res Lett 39(16):L16503. https://doi.org/10.1029/2012GL052518

Arp CD, Jones BM, Grosse G, Bondurant AC, Romanovsky VE, Hinkel KM, Parsekian AD (2016) Threshold sensitivity of shallow Arctic lakes and sublake permafrost to changing winter climate. Geophys Res Lett 43(12):6358–6365

Audry S, Pokrovsky O, Shirokova L, Kirpotin S, Dupré B (2011) Organic matter mineralization and trace element post-depositional redistribution in Western Siberia thermokarst lake sediments. Biogeosciences 8(11):3341–3358

Avis CA, Weaver AJ, Meissner KJ (2011) Reduction in areal extent of high-latitude wetlands in response to permafrost thaw. Nat Geosci 4(7):444

Battin TJ, Luyssaert S, Kaplan LA, Aufdenkampe AK, Richter A, Tranvik LJ (2009) The boundless carbon cycle. Nat Geosci 2(9):598

Benner R, Benitez-Nelson B, Kaiser K, Amon RM (2004) Export of young terrigenous dissolved organic carbon from rivers to the Arctic Ocean. Geophys Res Lett 31(5):L05305. https://doi.org/10.1029/2003GL019251

Bense V, Ferguson G, Kooi H (2009) Evolution of shallow groundwater flow systems in areas of degrading permafrost. Geophys Res Lett 36(22):L22041. https://doi.org/10.1029/2009GL039225

Berezovskaya S, Yang D, Kane DL (2004) Compatibility analysis of precipitation and runoff trends over the large Siberian watersheds. Geophys Res Lett 31(21):L21502. https://doi.org/10.1029/2004GL021277

Bintanja R, Selten F (2014) Future increases in Arctic precipitation linked to local evaporation and sea-ice retreat. Nature 509(7501):479–483

Biskaborn BK, Herzschuh U, Bolshiyanov DY, Schwamborn G, Diekmann B (2013) Thermokarst processes and depositional events in a tundra lake, northeastern Siberia. Permafr Periglac Process 24(3):160–174

Biskaborn BK, Smith SL, Noetzli J, Matthes H, Vieira G, Streletskiy DA, Schoeneich P, Romanovsky VE, Lewkowicz AG, Abramov A (2019) Permafrost is warming at a global scale. Nat Commun 10(1):264

Blok D, Heijmans MM, Schaepman-Strub G, Kononov A, Maximov T, Berendse F (2010) Shrub expansion may reduce summer permafrost thaw in Siberian tundra. Glob Chang Biol 16(4):1296–1305

Blok D, Heijmans MMPD, Schaepman-Strub G, van Ruijven J, Parmentier FJW, Maximov TC, Berendse F (2011) The cooling capacity of mosses: controls on water and energy fluxes in a Siberian tundra site. Ecosystems 14(7):1055–1065. https://doi.org/10.1007/s10021-011-9463-5

Boike J, Wille C, Abnizova A (2008) Climatology and summer energy and water balance of polygonal tundra in the Lena River Delta, Siberia. J Geophys Res 113(G3). https://doi.org/10.1029/2007jg000540

Boike J, Georgi C, Kirilin G, Muster S, Abramova K, Fedorova I, Chetverova A, Grigoriev M, Bornemann N, Langer M (2015) Thermal processes of thermokarst lakes in the continuous permafrost zone of northern Siberia – observations and modeling (Lena River Delta, Siberia). Biogeosciences 12(20):5941–5965. https://doi.org/10.5194/bg-12-5941-2015

Boike J, Grau T, Heim B, Günther F, Langer M, Muster S, Gouttevin I, Lange S (2016) Satellite-derived changes in the permafrost landscape of central Yakutia, 2000–2011: wetting, drying, and fires. Glob Planet Chang 139:116–127

Borge AF, Westermann S, Solheim I, Etzelmüller B (2017) Strong degradation of palsas and peat plateaus in northern Norway during the last 60 years. Cryosphere 11(1):1–16. https://doi.org/10.5194/tc-11-1-2017

Boucher É, Ouarda TB, Bégin Y, Nicault A (2011) Spring flood reconstruction from continuous and discrete tree ring series. Water Resour Res 47(7):W07516. https://doi.org/10.1029/2010WR010131

Bowling LC, Lettenmaier DP (2010) Modeling the effects of lakes and wetlands on the water balance of Arctic environments. J Hydrometeorol 11(2):276–295

Breton J, Vallieres C, Laurion I (2009) Limnological properties of permafrost thaw ponds in northeastern Canada. Can J Fish Aquat Sci 66(10):1635–1648

Bring A, Fedorova I, Dibike Y, Hinzman L, Mård J, Mernild S, Prowse T, Semenova O, Stuefer SL, Woo MK (2016) Arctic terrestrial hydrology: a synthesis of processes, regional effects, and research challenges. J Geophys Res Biogeo 121(3):621–649

Brouchkov A, Fukuda M, Fedorov A, Konstantinov P, Iwahana G (2004) Thermokarst as a short-term permafrost disturbance, Central Yakutia. Permafr Periglac Process 15(1):81–87. https://doi.org/10.1002/ppp.473

Brown J, Hinkel KM, Nelson F (2000) The circumpolar active layer monitoring (CALM) program: research designs and initial results. Polar Geogr 24(3):166–258

Brutsaert W, Hiyama T (2012) The determination of permafrost thawing trends from long-term streamflow measurements with an application in eastern Siberia. J Geophys Res Atmos 117(D22):D22110. https://doi.org/10.1029/2012JD018344

Bulygina O, Groisman PY, Razuvaev VN, Korshunova NN (2011) Changes in snow cover characteristics over Northern Eurasia since 1966. Environ Res Lett 6:04520. https://doi.org/10.1088/1748-9326/6/4/045204

Burn C (1997) Cryostratigraphy, paleogeography, and climate change during the early Holocene warm interval, western Arctic coast, Canada. Can J Earth Sci 34(7):912–925

Burn C (2000) The thermal regime of a retrogressive thaw slump near Mayo, Yukon territory. Can J Earth Sci 37(7):967–981

Burn C, Friele P (1989) Geomorphology, vegetation succession, soil characteristics and permafrost in retrogressive thaw slumps near Mayo. Yukon Territory Arctic:31–40

Burn CR, Nelson FE (2006) Comment on "A projection of severe near-surface permafrost degradation during the 21st century" by David M. Lawrence and Andrew G. Slater. Geophysical Research Letters 33 (21). https://doi.org/10.1029/2006GL027077

Carstensen J, Weydmann A (2012) Tipping points in the Arctic: eyeballing or statistical significance? Ambio 41:34–43. https://doi.org/10.1007/s13280-011-0223-8

Chadburn S, Burke E, Essery R, Boike J, Langer M, Heikenfeld M, Cox P, Friedlingstein P (2015) An improved representation of physical permafrost dynamics in the JULES land-surface model. Geosci Model Dev 8(5):1493–1508

Christensen TR, Jonasson S, Callaghan TV, Havström M (1995) Spatial variation in high latitude methane flux-a transect across tundra environments in Siberia and the European Arctic. J Geophys Res Oceans D10:21035–21045

Christensen TR, Ekberg A, Ström L, Mastepanov M, Panikov N, Öquist M, Svensson BH, Nykänen H, Martikainen PJ, Oskarsson H (2003) Factors controlling large scale variations in methane emissions from wetlands. Geophys Res Lett 30(7). https://doi.org/10.1029/2002gl016848

Cohen J, Screen JA, Furtado JC, Barlow M, Whittleston D, Coumou D, Francis J, Dethloff K, Entekhabi D, Overland J (2014) Recent Arctic amplification and extreme mid-latitude weather. Nat Geosci 7(9):627

Connon RF, Quinton WL, Craig JR, Hayashi M (2014) Changing hydrologic connectivity due to permafrost thaw in the lower Liard River valley, NWT, Canada. Hydrol Process 28(14):4163–4178

Connon R, Devoie É, Hayashi M, Veness T, Quinton W (2018) The influence of shallow taliks on permafrost thaw and active layer dynamics in subarctic Canada. J Geophys Res Earth 123(2):281–297

Corradi C, Kolle O, Walter K, Zimov SA, Schulze ED (2005) Carbon dioxide and methane exchange of a north-east Siberian tussock tundra. Glob Chang Biol 0(0):051115033519001. https://doi.org/10.1111/j.1365-2486.2005.01023.x

Cory RM, Crump BC, Dobkowski JA, Kling GW (2013) Surface exposure to sunlight stimulates CO_2 release from permafrost soil carbon in the Arctic. Proc Natl Acad Sci 110(9):3429–3434

Costard F, Gautier E, Brunstein D, Hammadi J, Fedorov A, Yang D, Dupeyrat L (2007) Impact of the global warming on the fluvial thermal erosion over the Lena River in Central Siberia. Geophys Res Lett 34(14):L14501. https://doi.org/10.1029/2007GL030212

Denfeld BAFK, Sobczak WV, Mann PJ, Holmes RM (2013) Summer CO_2 evasion from streams and rivers in the Kolyma River basin, North-East Siberia. Polar Res 32(1):19704

Dentener F (2006) Global maps of atmospheric nitrogen deposition, 1860, 1993, and 2050. ORNL DAAC, Oak Ridge. https://doi.org/10.3334/ORNLDAAC/830

Dorrepaal E, Toet S, van Logtestijn RS, Swart E, van de Weg MJ, Callaghan TV, Aerts R (2009) Carbon respiration from subsurface peat accelerated by climate warming in the subarctic. Nature 460(7255):616

Drake TW, Wickland KP, Spencer RG, McKnight DM, Striegl RG (2015) Ancient low–molecular-weight organic acids in permafrost fuel rapid carbon dioxide production upon thaw. Proc Natl Acad Sci 112(45):13946–13951

Dutta K, Schuur E, Neff J, Zimov S (2006) Potential carbon release from permafrost soils of north-eastern Siberia. Glob Chang Biol 12(12):2336–2351

Elberling B, Christiansen HH, Hansen BU (2010) High nitrous oxide production from thawing permafrost. Nat Geosci 3(5):332

Elberling B, Michelsen A, Schädel C, Schuur EA, Christiansen HH, Berg L, Tamstorf MP, Sigsgaard C (2013) Long-term CO_2 production following permafrost thaw. Nat Clim Chang 3(10):890

Elmendorf SC, Henry GH, Hollister RD, Björk RG, Bjorkman AD, Callaghan TV, Collier LS, Cooper EJ, Cornelissen JH, Day TA (2012) Global assessment of experimental climate warming on tundra vegetation: heterogeneity over space and time. Ecol Lett 15(2):164–175

Emmerton CA, Lesack LF, Marsh P (2007) Lake abundance, potential water storage, and habitat distribution in the Mackenzie River Delta, western Canadian Arctic. Water Resour Res 43(5):W05419. https://doi.org/10.1029/2006WR005139

Eugster W, Rouse WR, Pielke RA Sr, Mcfadden JP, Baldocchi DD, Kittel TG, Chapin FS, Liston GE, Vidale PL, Vaganov E (2000) Land–atmosphere energy exchange in Arctic tundra and boreal forest: available data and feedbacks to climate. Glob Chang Biol 6(S1):84–115

Fedorov AN, Konstantinov PY (2009) Response of permafrost landscapes of central Yakutia to current changes of climate, and anthropogenic impacts. Geogr Nat Resour 30(2):146–150

Fedorov A, Gavriliev P, Konstantinov P, Hiyama T, Iijima Y, Iwahana G (2014) Estimating the water balance of a thermokarst lake in the middle of the Lena River basin, eastern Siberia. Ecohydrology 7(2):188–196

Fedotov A, Phedorin M, Suvorov A, Melgunov M, Khodzher T (2012) Permafrost thawing inferred from Arctic lake sediment of the Taimyr peninsula, East Siberia, Russia. Int J Environ Stud 69(1):7–19

Feng X, Vonk JE, Van Dongen BE, Gustafsson Ö, Semiletov IP, Dudarev OV, Wang Z, Montluçon DB, Wacker L, Eglinton TI (2013) Differential mobilization of terrestrial carbon pools in Eurasian Arctic river basins. Proc Natl Acad Sci 110(35):14168–14173

Fisher JP, Estop-Aragonés C, Thierry A, Charman DJ, Wolfe SA, Hartley IP, Murton JB, Williams M, Phoenix GK (2016) The influence of vegetation and soil characteristics on active-layer thickness of permafrost soils in boreal forest. Glob Chang Biol 22(9):3127–3140

Fortier D, Allard M, Shur Y (2007) Observation of rapid drainage system development by thermal erosion of ice wedges on Bylot Island, Canadian Arctic archipelago. Permafr Periglac Process 18(3):229–243

Frank D, Reichstein M, Bahn M, Thonicke K, Frank D, Mahecha MD, Smith P, Van der Velde M, Vicca S, Babst F (2015) Effects of climate extremes on the terrestrial carbon cycle: concepts, processes and potential future impacts. Glob Chang Biol 21(8):2861–2880

French HM (2018) The periglacial environment. Wiley, Oxford, 515 p

French H, Shur Y (2010) The principles of cryostratigraphy. Earth Sci Rev 101(3–4):190–206

Frey KE, McClelland JW (2009) Impacts of permafrost degradation on arctic river biogeochemistry. Hydrol Process 23(1):169–182

Frey KE, Smith LC (2005) Amplified carbon release from vast west Siberian peatlands by 2100. Geophys Res Lett 32(9):L09401. https://doi.org/10.1029/2004GL022025

Frey KE, McClelland JW, Holmes RM, Smith LC (2007a) Impacts of climate warming and permafrost thaw on the riverine transport of nitrogen and phosphorus to the Kara Sea. J Geophys Res Biogeosci 112(G4):G04S58. https://doi.org/10.1029/2006JG000369

Frey KE, Siegel DI, Smith LC (2007b) Geochemistry of west Siberian streams and their potential response to permafrost degradation. Water Resour Res 43(3):W03406. https://doi.org/10.1029/2006WR004902

Froese DG, Westgate JA, Reyes AV, Enkin RJ, Preece SJ (2008) Ancient permafrost and a future, warmer Arctic. Science 321(5896):1648–1648

Göckede M, Kittler F, Kwon MJ, Burjack I, Heimann M, Kolle O, Zimov N, Zimov S (2017) Shifted energy fluxes, increased Bowen ratios, and reduced thaw depths linked with drainage-induced changes in permafrost ecosystem structure. Cryosphere 11(6):2975–2996

Gooseff MN, Balser A, Bowden WB, Jones JB (2009) Effects of hillslope thermokarst in northern Alaska. EOS Trans Am Geophys Union 90(4):29–30

Groisman PY, Sherstyukov BG, Razuvaev VN, Knight RW, Enloe JG, Stroumentova NS, Whitfield PH, Førland E, Hannsen-Bauer I, Tuomenvirta H (2007) Potential forest fire danger over northern Eurasia: changes during the 20th century. Glob Planet Chang 56(3–4):371–386

Grosse G, Romanovsky V, Walter K, Morgenstern A, Lantuit H, Zimov S (2008) Distribution of thermokarst lakes and ponds at three yedoma sites in Siberia. In: 9th international conference on permafrost, Fairbanks, 2008. Proceedings 9th international conference on Permafrost, pp 551–556

Grosse G, Harden J, Turetsky M, McGuire AD, Camill P, Tarnocai C, Frolking S, Schuur EAG, Jorgenson T, Marchenko S, Romanovsky V, Wickland KP, French N, Waldrop M, Bourgeau-Chavez L, Striegl RG (2011) Vulnerability of high-latitude soil organic carbon in North America to disturbance. J Geophys Res 116. https://doi.org/10.1029/2010jg001507

Grosse G, Jones B, Arp C (2013) Thermokarst lakes, drainage, and drained basins. In: Shroder J, Giardino R, Harbor J (eds) Glacial and periglacial geomorphology, Treatise on geomorphology, vol 8, pp 325–353. https://doi.org/10.1016/b978-0-12-374739-6.00216-5

Guo L, Macdonald RW (2006) Source and transport of terrigenous organic matter in the upper Yukon River: evidence from isotope ($\delta^{13}C$, $\delta^{14}C$, and $\delta^{15}N$) composition of dissolved, colloidal, and particulate phases. Glob Biogeochem Cycles 20(2):GB2011. https://doi.org/10.1029/2005GB002593

Guo L, Ping CL, Macdonald RW (2007) Mobilization pathways of organic carbon from permafrost to arctic rivers in a changing climate. Geophys Res Lett 34(13):L13603. https://doi.org/10.1029/2007GL030689

Gustafsson Ö, van Dongen BE, Vonk JE, Dudarev OV, Semiletov IP (2011) Widespread release of old carbon across the Siberian Arctic echoed by its large rivers. Biogeosciences 8(6):1737–1743. https://doi.org/10.5194/bg-8-1737-2011

Hanssen RF (2001) Radar interferometry: data interpretation and error analysis, vol 2. Springer, Dordrecht, 308 p

Harsch MA, Hulme PE, McGlone MS, Duncan RP (2009) Are treelines advancing? A global meta-analysis of treeline response to climate warming. Ecol Lett 12(10):1040–1049

Hicks Pries CE, Schuur EA, Crummer KG (2013a) Thawing permafrost increases old soil and autotrophic respiration in tundra: partitioning ecosystem respiration using delta(13) C and (14) C. Glob Chang Biol 19(2):649–661. https://doi.org/10.1111/gcb.12058

Hicks Pries CE, Schuur E, Vogel JG, Natali SM (2013b) Moisture drives surface decomposition in thawing tundra. J Geophys Res Biogeo 118(3):1133–1143

Hicks Pries CE, van Logtestijn RS, Schuur EA, Natali SM, Cornelissen JH, Aerts R, Dorrepaal E (2015) Decadal warming causes a consistent and persistent shift from heterotrophic to autotrophic respiration in contrasting permafrost ecosystems. Glob Chang Biol 21(12):4508–4519. https://doi.org/10.1111/gcb.13032

Hilton RG, Galy V, Gaillardet J, Dellinger M, Bryant C, O'regan M, Gröcke DR, Coxall H, Bouchez J, Calmels D (2015) Erosion of organic carbon in the Arctic as a geological carbon dioxide sink. Nature 524(7563):84

Hinkel KM, Hurd JK Jr (2006) Permafrost destabilization and thermokarst following snow fence installation, Barrow, Alaska, USA. Arct Antarct Alp Res 38(4):530–539

Hinkel KM, Eisner WR, Bockheim JG, Nelson FE, Peterson KM, Dai X (2003) Spatial extent, age, and carbon stocks in drained thaw lake basins on the Barrow Peninsula, Alaska. Arct Antarct Alp Res 35(3):291–300

Hinzman LD, Bettez ND, Bolton WR, Chapin FS, Dyurgerov MB, Fastie CL, Griffith B, Hollister RD, Hope A, Huntington HP (2005) Evidence and implications of recent climate change in northern Alaska and other arctic regions. Clim Chang 72(3):251–298

Hinzman LD, Deal CJ, McGuire AD, Mernild SH, Polyakov IV, Walsh JE (2013) Trajectory of the Arctic as an integrated system. Ecol Appl 23(8):1837–1868

Hollesen J, Elberling B, Jansson P-E (2011) Future active layer dynamics and carbon dioxide production from thawing permafrost layers in Northeast Greenland. Glob Chang Biol 17(2):911–926

Holmes R, McClelland JW, Raymond PA, Frazer BB, Peterson BJ, Stieglitz M (2008) Lability of DOC transported by Alaskan rivers to the Arctic Ocean. Geophys Res Lett 35(3):L03402. https://doi.org/10.1029/2007GL032837

Holmes RM, McClelland JW, Peterson BJ, Tank SE, Bulygina E, Eglinton TI, Gordeev VV, Gurtovaya TY, Raymond PA, Repeta DJ (2012) Seasonal and annual fluxes of nutrients and organic matter from large rivers to the Arctic Ocean and surrounding seas. Estuar Coasts 35(2):369–382

Holmes RM, Coe MT, Fiske GJ, Gurtovaya T, McClelland JW, Shiklomanov AI, Spencer RG, Tank SE, Zhulidov AV (2013) Climate change impacts on the hydrology and biogeochemistry of Arctic rivers. In: Goldman, CR, Kumagai, M, Robarts, RD (eds) Climatic change and global warming of inland waters: impacts and mitigation for ecosystems and societies. Wiley, Somerset, pp 1–26. doi:https://doi.org/10.1002/9781118470596

Holmes R, Shiklomanov, AI, Suslova, A, Tretiakov, M, McClelland, JW, Spencer, RGM, Tank, SE (2018) River discharge. Arctic report card 2018. NOAA. https://arctic.noaa.gov/Report-Card/Report-Card-2018

Hopkins D, Kidd J (1988) Thaw lake sediments and sedimentary environments. In: Proceedings of the 5th international permafrost conference, pp 790–795

IPCC (2013) Climate change 2013: the physical science basis. Contribution of working group I to the fifth assessment report of the intergovernmental panel on climate change. Cambridge University Press. https://doi.org/10.1017/CBO9781107415324

Iijima Y, Fedorov AN, Park H, Suzuki K, Yabuki H, Maximov TC, Ohata T (2010) Abrupt increases in soil temperatures following increased precipitation in a permafrost region, Central Lena River basin, Russia. Permafr Periglac Process 21(1):30–41. https://doi.org/10.1002/ppp.662

Iijima Y, Ohta T, Kotani A, Fedorov AN, Kodama Y, Maximov TC (2014) Sap flow changes in relation to permafrost degradation under increasing precipitation in an eastern Siberian larch forest. Ecohydrology 7(2):177–187

Iijima Y, Nakamura T, Park H, Tachibana Y, Fedorov AN (2016) Enhancement of Arctic storm activity in relation to permafrost degradation in eastern Siberia. Int J Climatol 36(13):4265–4275

Iwahana G, Machimura T, Kobayashi Y, Fedorov AN, Konstantinov PY, Fukuda M (2005) Influence of forest clear-cutting on the thermal and hydrological regime of the active layer near Yakutsk, eastern Siberia. J Geophys Res Biogeo 110(G2):G02004. https://doi.org/10.1029/2005JG000039

Iwahana G, Takano S, Petrov RE, Tei S, Shingubara R, Maximov TC, Fedorov AN, Desyatkin AR, Nikolaev AN, Desyatkin RV (2014) Geocryological characteristics of the upper permafrost in a tundra-forest transition of the Indigirka River Valley, Russia. Pol Sci 8(2):96–113

Jeffries MO, Zhang T, Frey K, Kozlenko N (1999) Estimating late-winter heat flow to the atmosphere from the lake-dominated Alaskan North Slope. J Glaciol 45(150):315–324

Jepsen SM, Voss CI, Walvoord MA, Minsley BJ, Rover J (2013) Linkages between lake shrinkage/expansion and sublacustrine permafrost distribution determined from remote sensing of interior Alaska, USA. Geophys Res Lett 40(5):882–887

Johansson M, Callaghan TV, Bosiö J, Åkerman HJ, Jackowicz-Korczynski M, Christensen TR (2013) Rapid responses of permafrost and vegetation to experimentally increased snow cover in subarctic Sweden. Environ Res Lett 8(3):035025. https://doi.org/10.1088/1748-9326/8/3/035025

Jones J Jr, Petrone KC, Finlay JC, Hinzman LD, Bolton WR (2005) Nitrogen loss from watersheds of interior Alaska underlain with discontinuous permafrost. Geophys Res Lett 32(2):L02401. https://doi.org/10.1029/2004GL021734

Jones BM, Grosse G, Arp C, Jones M, Walter Anthony K, Romanovsky V (2011) Modern thermokarst lake dynamics in the continuous permafrost zone, northern Seward peninsula, Alaska. J Geophys Res Biogeosci 116(G2):G00M03. https://doi.org/10.1029/2011JG001666

Jones MC, Grosse G, Jones BM, Walter Anthony K (2012) Peat accumulation in drained thermokarst lake basins in continuous, ice-rich permafrost, northern Seward Peninsula, Alaska. J Geophys Res Biogeo 117(G2)

Jorgenson MT, Shur Y (2007) Evolution of lakes and basins in northern Alaska and discussion of the thaw lake cycle. J Geophys Res Earth Surf 112(F2):F02S17. https://doi.org/10.1029/2006JF000531

Jorgenson MT, Racine CH, Walters JC, Osterkamp TE (2001) Permafrost degradation and ecological changes associated with a warmingclimate in Central Alaska. Clim Chang 48(4):551–579

Jorgenson MT, Shur YL, Pullman ER (2006) Abrupt increase in permafrost degradation in Arctic Alaska. Geophys Res Lett 33(2):L02503. https://doi.org/10.1029/2005gl024960

Jorgenson MT, Romanovsky V, Harden J, Shur Y, O'Donnell J, Schuur EA, Kanevskiy M, Marchenko S (2010) Resilience and vulnerability of permafrost to climate change. Can J For Res 40(7):1219–1236

Jorgenson MT, Harden J, Kanevskiy M, O'Donnell J, Wickland K, Ewing S, Manies K, Zhuang Q, Shur Y, Striegl R (2013) Reorganization of vegetation, hydrology and soil carbon after permafrost degradation across heterogeneous boreal landscapes. Environ Res Lett 8(3):035017

Jorgenson NT, Kanevsky M, Shur Y, Moskalenko N, Brown DRN, Wickland K, Striegl R, Koch J (2015) Role of ground ice dynamics and ecological feedbacks in recent ice wedge degradation and stabilization. J Geophys Res Earth Surf 120:2280–2297. https://doi.org/10.1002/2015JF003602

Karlsson J, Lyon S, Destouni G (2014) Temporal behavior of Lake size-distribution in a thawing permafrost landscape in Northwestern Siberia. Remote Sens 6(1):621–636. https://doi.org/10.3390/rs6010621

Katamura F, Fukuda M, Bosikov NP, Desyatkin RV, Nakamura T, Moriizumi J (2006) Thermokarst formation and vegetation dynamics inferred from a palynological study in Central Yakutia, Eastern Siberia, Russia. Arct Antarct Alp Res 38(4):561–570

Kattsov VM, Walsh JE, Chapman WL, Govorkova VA, Pavlova TV, Zhang X (2007) Simulation and projection of Arctic freshwater budget components by the IPCC AR4 global climate models. J Hydrometeorol 8(3):571–589

Kessler M, Plug LJ, Walter Anthony K (2012) Simulating the decadal-to millennial-scale dynamics of morphology and sequestered carbon mobilization of two thermokarst lakes in NW Alaska. J Geophys Res Biogeosci 117(G2):G00M06. https://doi.org/10.1029/2011JG001796

Keuper F, Van Bodegom PM, Dorrepaal E, Weedon JT, Van Hal J, Van Logtestijn RS, Aerts R (2012a) A frozen feast: thawing permafrost increases plant-available nitrogen in subarctic peatlands. Glob Chang Biol 18(6):1998–2007

Keuper F, Parmentier FJ, Blok D, van Bodegom PM, Dorrepaal E, van Hal JR, van Logtestijn RS, Aerts R (2012b) Tundra in the rain: differential vegetation responses to three years of experimentally doubled summer precipitation in Siberian shrub and Swedish bog tundra. Ambio 41(Suppl 3):269–280. https://doi.org/10.1007/s13280-012-0305-2

Kharin VV, Zwiers F, Zhang X, Wehner M (2013) Changes in temperature and precipitation extremes in the CMIP5 ensemble. Clim Chang 119(2):345–357

Khvorostyanov D, Krinner G, Ciais P, Heimann M, Zimov S (2008a) Vulnerability of permafrost carbon to global warming. Part I: model description and role of heat generated by organic matter decomposition. Tellus B: Chem Phys Meteorol 60(2):250–264

Khvorostyanov D, Ciais P, Krinner G, Zimov S, Corradi C, Guggenberger G (2008b) Vulnerability of permafrost carbon to global warming. Part II: sensitivity of permafrost carbon stock to global warming. Tellus B Chem Phys Meteorol 60(2):265–275

Kirpotin SN, Polishchuk Y, Bryksina N (2009) Abrupt changes of thermokarst lakes in Western Siberia: impacts of climatic warming on permafrost melting. Int J Environ Stud 66(4):423–431. https://doi.org/10.1080/00207230902758287

Kirpotin S, Polishchuk Y, Bryksina N, Sugaipova A, Kouraev A, Zakharova E, Pokrovsky OS, Shirokova L, Kolmakova M, Manassypov R, Dupre B (2011) West Siberian palsa peatlands: distribution, typology, cyclic development, present day climate-driven changes, seasonal hydrology and impact on CO_2 cycle. Int J Environ Stud 68(5):603–623. https://doi.org/10.1080/00207233.2011.593901

Kittler F, Heimann M, Kolle O, Zimov N, Zimov S, Göckede M (2017) Long-term drainage reduces CO_2 uptake and CH_4 emissions in a Siberian permafrost ecosystem. Glob Biogeochem Cycles 31:1704–1717. https://doi.org/10.1002/2017GB005774

Knoblauch C, Beer C, Sosnin A, Wagner D, Pfeiffer EM (2013) Predicting long-term carbon mineralization and trace gas production from thawing permafrost of Northeast Siberia. Glob Chang Biol 19(4):1160–1172

Knoblauch C, Beer C, Liebner S, Grigoriev MN, Pfeiffer E-M (2018) Methane production as key to the greenhouse gas budget of thawing permafrost. Nat Clim Chang 8(4):309

Koch JC, Gurney K, Wipfli MS (2014) Morphology-dependent water budgets and nutrient fluxes in arctic thaw ponds. Permafr Periglac Process 25(2):79–93

Koch J, Jorgenson M, Wickland K, Kanevskiy M, Striegl R (2018) Ice wedge degradation and stabilization impacts water budgets and nutrient cycling in arctic trough ponds. J Geophys Res Biogeo 123:2604–2616. https://doi.org/10.1029/2018JG004528

Kokelj S, Jenkins R, Milburn D, Burn C, Snow N (2005) The influence of thermokarst disturbance on the water quality of small upland lakes, Mackenzie Delta region, Northwest Territories, Canada. Permafr Periglac Process 16(4):343–353

Kokelj SV, Lantz TC, Kanigan J, Smith S, Coutts R (2009) Origin and polycyclic behaviour of tundra thaw slumps, Mackenzie Delta region, Northwest Territories, Canada. Permafr Periglac Process 20(2):173–184

Kokelj SV, Lantz TC, Tunnicliffe J, Segal R, Lacelle D (2017) Climate-driven thaw of permafrost preserved glacial landscapes, northwestern Canada. Geology 45(4):371–374

Kravtsova V, Bystrova A (2009) Changes in thermokarst lake sizes in different regions of Russia for the last 30 years (Изменение размеров термокарстовых озер в различных районах россии за последние 30 лет). Kriosfera Zemli (Earth Cryosphere) 13:16–26

Kurylyk BL, Hayashi M, Quinton WL, McKenzie JM, Voss CI (2016) Influence of vertical and lateral heat transfer on permafrost thaw, peatland landscape transition, and groundwater flow. Water Resour Res 52(2):1286–1305

Kwon MJ, Heimann M, Kolle O, Luus KA, Schuur EAG, Zimov N, Zimov SA, Göckede M (2016) Long-term drainage reduces CO_2 uptake and increases CO_2 emission on a Siberian floodplain due to shifts in vegetation community and soil thermal characteristics. Biogeosciences 13(14):4219–4235. https://doi.org/10.5194/bg-13-4219-2016

Kwon MJ, Beulig F, Ilie I, Wildner M, Kusel K, Merbold L, Mahecha MD, Zimov N, Zimov SA, Heimann M, Schuur EAG, Kostka JE, Kolle O, Hilke I, Gockede M (2017) Plants, microorganisms, and soil temperatures contribute to a decrease in methane fluxes on a drained Arctic floodplain. Glob Chang Biol 23(6):2396–2412. https://doi.org/10.1111/gcb.13558

Labrecque S, Lacelle D, Duguay CR, Lauriol B, Hawkings J (2009) Contemporary (1951–2001) evolution of lakes in the Old Crow Basin, Northern Yukon, Canada: Remote sensing, numerical modeling, and stable isotope analysis. Arctic 62(2):225–238

Lacelle D, Bjornson J, Lauriol B (2010) Climatic and geomorphic factors affecting contemporary (1950–2004) activity of retrogressive thaw slumps on the Aklavik Plateau, Richardson Mountains, NWT, Canada. Permafr Periglac Process 21(1):1–15

Lamoureux SF Lafrenière MJ (2014) Seasonal fluxes and age of particulate organic carbon exported from Arctic catchments impacted by localized permafrost slope disturbances. Environ Res Lett 9(5):045002. https://doi.org/10.1088/1748-9326/9/4/045002

Langer M, Westermann S, Muster S, Piel K, Boike J (2011) The surface energy balance of a polygonal tundra site in northern Siberia-part 1: Spring to fall. Cryosphere 5(1):151

Langer M, Westermann S, Walter Anthony K, Wischnewski K, Boike J (2015) Frozen ponds: production and storage of methane during the Arctic winter in a lowland tundra landscape in northern Siberia, Lena River delta. Biogeosciences 12(4):977–990. https://doi.org/10.5194/bg-12-977-2015

Lantuit H, Pollard W (2008) Fifty years of coastal erosion and retrogressive thaw slump activity on Herschel Island, southern Beaufort Sea, Yukon territory, Canada. Geomorphology 95(1–2):84–102

Lantz TC, Kokelj SV (2008) Increasing rates of retrogressive thaw slump activity in the Mackenzie Delta region, NWT, Canada. Geophys Res Lett 35(6):L06502. https://doi.org/10.1029/2007GL032433

Lara MJ, McGuire AD, Euskirchen ES, Tweedie CE, Hinkel KM, Skurikhin AN, Romanovsky VE, Grosse G, Bolton WR, Genet H (2015) Polygonal tundra geomorphological change in response to warming alters future CO_2 and CH_4 flux on the Barrow Peninsula. Glob Chang Biol 21(4):1634–1651. https://doi.org/10.1111/gcb.12757

Laurion I, Vincent WF, MacIntyre S, Retamal L, Dupont C, Francus P, Pienitz R (2010) Variability in greenhouse gas emissions from permafrost thaw ponds. Limnol Oceanogr 55(1):115–133

Lawrence DM, Slater AG, Tomas RA, Holland MM, Deser C (2008) Accelerated Arctic land warming and permafrost degradation during rapid sea ice loss. Geophys Res Lett 35(11):L11506. https://doi.org/10.1029/2008GL033985

Leibman M, Kizakov A, Sulerzhitsky L, Zaretskaia N (2003) Dynamics of landslide slopes and their development on Yamal Peninsula. In: Permafrost (ed) Proceedings of the 8th international conference on permafrost. Swets and Zeitlinger, Lisse, pp 651–656

Leibman M, Khomutov A, Kizyakov A (2014) Cryogenic landslides in the West-Siberian plain of Russia: classification, mechanisms, and landforms. In: Landslides in cold regions in the context of climate change. Springer, Cham, pp 143–162

Lenz J, Wetterich S, Jones BM, Meyer H, Bobrov A, Grosse G (2016) Evidence of multiple thermokarst lake generations from an 11 800-year-old permafrost core on the northern S eward Peninsula, Alaska. Boreas 45(4):584–603

Lesack LF, Marsh P, Hicks FE, Forbes DL (2013) Timing, duration, and magnitude of peak annual water-levels during ice breakup in the Mackenzie Delta and the role of river discharge. Water Resour Res 49(12):8234–8249

Lesack LF, Marsh P, Hicks FE, Forbes DL (2014) Local spring warming drives earlier river-ice breakup in a large Arctic delta. Geophys Res Lett 41(5):1560–1567

Lewkowicz AG (1987) Headwall retreat of ground-ice slumps, Banks Island, Northwest Territories. Can J Earth Sci 24(6):1077–1085

Lewkowicz AG, Harris C (2005) Frequency and magnitude of active-layer detachment failures in discontinuous and continuous permafrost, northern Canada. Permafr Periglac Process 16(1):115–130

Lewkowicz AG, Way RG (2019) Extremes of summer climate trigger thousands of thermokarst landslides in a High Arctic environment. Nat Commun 10(1):1329. https://doi.org/10.1038/s41467-019-09314-7

Li B, Heijmans MM, Blok D, Wang P, Karsanaev SV, Maximov TC, van Huissteden J, Berendse F (2017a) Thaw pond development and initial vegetation succession in experimental plots at a Siberian lowland tundra site. Plant Soil 420(1–2):147–162

Li B, Heijmans, MMPD, Limpens J, Gallagher A, Van Huissteden J, Maximov TC, Berendse F (2017b) Peat moss carpet reduces methane emission and facilitates shrub expansion in Siberian Arctic tundra. In: A clash of plants. Vegetation succession and its interaction with permafrost dynamics in the Arctic lowland tundra. Thesis, Wageningen University, Wageningen, pp 26–42

Liebner S, Zeyer J, Wagner D, Schubert C, Pfeiffer EM, Knoblauch C (2011) Methane oxidation associated with submerged brown mosses reduces methane emissions from Siberian polygonal tundra. J Ecol 99(4):914–922

Liljedahl A, Hinzman L, Harazono Y, Zona D, Tweedie C, Hollister RD, Engstrom R, Oechel W (2011) Nonlinear controls on evapotranspiration in arctic coastal wetlands. Biogeosciences 8(11):3375–3389

Liljedahl AK, Boike J, Daanen RP, Fedorov AN, Frost GV, Grosse G, Hinzman LD, Iijma Y, Jorgenson JC, Matveyeva N, Necsoiu M, Raynolds MK, Romanovsky VE, Schulla J, Tape KD, Walker DA, Wilson CJ, Yabuki H, Zona D (2016) Pan-Arctic ice-wedge degradation in warming permafrost and its influence on tundra hydrology. Nat Geosci 9(4):312–318. https://doi.org/10.1038/ngeo2674

Lipson DA, Zona D, Raab TK, Bozzolo F, Mauritz M, Oechel W (2012) Water-table height and microtopography control biogeochemical cycling in an Arctic coastal tundra ecosystem. Biogeosciences 9(1):577–591

Liu L, Zhang T, Wahr J (2010) InSAR measurements of surface deformation over permafrost on the North slope of Alaska. J Geophys Res Earth 115(F3):F03023. https://doi.org/10.1029/2009JF001547

Liu L, Schaefer K, Zhang T, Wahr J (2012) Estimating 1992–2000 average active layer thickness on the Alaskan north slope from remotely sensed surface subsidence. J Geophys Res Earth 117(F1):F01005. https://doi.org/10.1029/2011JF002041

Loranty MM, Abbott BW, Blok D, Douglas TA, Epstein HE, Forbes BC, Jones BM, Kholodov AL, Kropp H, Malhotra A (2018) Reviews and syntheses: changing ecosystem influences on soil thermal regimes in northern high-latitude permafrost regions. Biogeosciences 15:5287–5313. https://doi.org/10.5194/bg-15-5287-2018

Lougheed VL, Butler MG, McEwen DC, Hobbie JE (2011) Changes in tundra pond limnology: re-sampling Alaskan ponds after 40 years. Ambio 40(6):589

Luke C, Cox P (2011) Soil carbon and climate change: from the Jenkinson effect to the compost-bomb instability. Eur J Soil Sci 62(1):5–12

Lund M, Falk JM, Friborg T, Mbufong HN, Sigsgaard C, Soegaard H, Tamstorf MP (2012) Trends in CO_2 exchange in a high Arctic tundra heath, 2000–2010. J Geophys Res Biogeo 117(G2):G02001. https://doi.org/10.1029/2011JG001901

Lupascu M, Welker J, Seibt U, Maseyk K, Xu X, Czimczik C (2014) High Arctic wetting reduces permafrost carbon feedbacks to climate warming. Nat Clim Chang 4(1):51

Mann P, Eglinton TI, McIntyre CP, Zimov N, Davydova A, Vonk JE, Holmes RM, Spencer RG (2015) Utilization of ancient permafrost carbon in headwaters of Arctic fluvial networks. Nat Commun 6:7856. https://doi.org/10.1038/ncomms8856

Mann PJ, Spencer RG, Hernes PJ, Six J, Aiken GR, Tank SE, McClelland JW, Butler KD, Dyda RY, Holmes RM (2016) Pan-arctic trends in terrestrial dissolved organic matter from optical measurements. Front Earth Sci 4:25

Marsh P, Russell M, Pohl S, Haywood H, Onclin C (2009) Changes in thaw lake drainage in the Western Canadian Arctic from 1950 to 2000. Hydrol Proc Intl J 23(1):145–158

Marushchak M, Pitkämäki A, Koponen H, Biasi C, Seppälä M, Martikainen P (2011) Hot spots for nitrous oxide emissions found in different types of permafrost peatlands. Glob Chang Biol 17(8):2601–2614

Matsuoka N (2001) Solifluction rates, processes and landforms: a global review. Earth Sci Rev 55(1–2):107–134

Mazhitova G, Malkova G, Chestnykh O (2008) Zamolodchikov D recent decade thaw-depth dynamics in the European Russian Arctic based on the Circumpolar Active Layer Monitoring (CALM) data. In: Proceedings of the ninth international conference on Permafrost, pp 1155–1160

McClelland JW, Holmes RM, Peterson BJ, Stieglitz M (2004) Increasing river discharge in the Eurasian Arctic: consideration of dams, permafrost thaw, and fires as potential agents of change. J Geophys Res Atmos 109(D18):D18102. https://doi.org/10.1029/2004JD004583

McClelland J, Stieglitz M, Pan F, Holmes RM, Peterson BJ (2007) Recent changes in nitrate and dissolved organic carbon export from the upper Kuparuk River, North Slope, Alaska. J Geophys Res Biogeosci 112(G4):G04S60. https://doi.org/10.1029/2006JG000371

McClelland JW, Holmes RM, Peterson BJ, Raymond PA, Striegl R, Zhulidov AV, Zimov SA, Zimov N, Tank SE, Spencer RG (2016) Particulate organic carbon and nitrogen export from major Arctic rivers. Glob Biogeochem Cycles 30(5):629–643

McGuire AD, Lawrence DM, Koven C, Clein JS, Burke E, Chen G, Jafarov E, MacDougall AH, Marchenko S, Nicolsky D (2018) Dependence of the evolution of carbon dynamics in the northern permafrost region on the trajectory of climate change. Proceedings of the National Academy of Sciences:201719903

Merbold L, Kutsch WL, Corradi C, Kolle O, Rebmann C, Stoy PC, Zimov SA, Schulze ED (2009) Artificial drainage and associated carbon fluxes (CO_2/CH_4) in a tundra ecosystem. Glob Chang Biol 15(11):2599–2614

Mesquita P, Wrona FJ, Prowse TD (2010) Effects of retrogressive permafrost thaw slumping on sediment chemistry and submerged macrophytes in Arctic tundra lakes. Freshw Biol 55(11):2347–2358. https://doi.org/10.1111/j.1365-2427.2010.02450.x

Minke M, Donner N, Karpov N, de Klerk P, Joosten H (2009) Patterns in vegetation composition, surface height and thaw depth in polygon mires in the Yakutian Arctic (NE Siberia): a microtopographical characterisation of the active layer. Permafr Periglac Process 20(4):357–368

Morgenstern A, Grosse G, Günther F, Fedorova I, Schirrmeister L (2011) Spatial analyses of thermokarst lakes and basins in Yedoma landscapes of the Lena Delta. Cryosphere 5(4):849–867. https://doi.org/10.5194/tc-5-849-2011

Morgenstern A, Ulrich M, Günther F, Roessler S, Fedorova IV, Rudaya NA, Wetterich S, Boike J, Schirrmeister L (2013) Evolution of thermokarst in east Siberian ice-rich permafrost: a case study. Geomorphology 201:363–379. https://doi.org/10.1016/j.geomorph.2013.07.011

Murton JB (1996) Thermokarst-lake-basin sediments, Tuktoyaktuk coastlands, western arctic Canada. Sedimentology 43(4):737–760

Murton JB (2001) Thermokarst sediments and sedimentary structures, Tuktoyaktuk coastlands, western Arctic Canada. Glob Planet Chang 28(1–4):175–192

Muskett RR, Romanovsky VE (2009) Groundwater storage changes in arctic permafrost watersheds from GRACE and in situ measurements. Environ Res Lett 4(4):045009

Myers-Smith IH, Forbes BC, Wilmking M, Hallinger M, Lantz T, Blok D, Tape KD, Macias-Fauria M, Sass-Klaassen U, Lévesque E (2011) Shrub expansion in tundra ecosystems: dynamics, impacts and research priorities. Environ Res Lett 6(4):045509

Nakagawa F, Yoshida N, Nojiri Y, Makarov V (2002) Production of methane from alasses in eastern Siberia: implications from its ^{14}C and stable isotopic compositions. Glob Biogeochem Cycles 16(3). https://doi.org/10.1029/2000GB001384

Natali SM, Schuur EA, Trucco C, Pries CEH, Crummer KG, Lopez AFB (2011) Effects of experimental warming of air, soil and permafrost on carbon balance in Alaskan tundra. Glob Chang Biol 17(3):1394–1407

Natali SM, Schuur EA, Webb EE, Pries CEH, Crummer KG (2014) Permafrost degradation stimulates carbon loss from experimentally warmed tundra. Ecology 95(3):602–608

Natali SM, Schuur EAG, Mauritz M, Schade JD, Celis G, Crummer KG, Johnston C, Krapek J, Pegoraro E, Salmon VG, Webb EE (2015) Permafrost thaw and soil moisture driving CO_2 and CH_4 release from upland tundra. J Geophys Res Biogeo 120(3):525–537. https://doi.org/10.1002/2014jg002872

Nauta AL, Heijmans MMPD, Blok D, Limpens J, Elberling B, Gallagher A, Li B, Petrov RE, Maximov TC, van Huissteden J, Berendse F (2014) Permafrost collapse after shrub removal shifts tundra ecosystem to a methane source. Nat Clim Chang 5(1):67–70. https://doi.org/10.1038/nclimate2446

Neff JC FJ, Zimov SA, Davydov SP, Carrasco JJ, Schuur EA, Davydova AI (2006) Seasonal changes in the age and structure of dissolved organic carbon in Siberian rivers and streams. Geophys Res Lett 33(23):L23401. https://doi.org/10.1029/2006GL028222

Nelson F, Hinkel K, Shiklomanov N, Mueller G, Miller L, Walker D (1998) Active-layer thickness in north Central Alaska: systematic sampling, scale, and spatial autocorrelation. J Geophys Res Atmos 103(D22):28963–28973

Nitze I, Grosse G, Jones BM, Romanovsky VE, Boike J (2018) Remote sensing quantifies widespread abundance of permafrost region disturbances across the Arctic and Subarctic. Nat Commun 9(1):5423. https://doi.org/10.1038/s41467-018-07663-3

Nowinski NS, Taneva L, Trumbore SE, Welker JM (2010) Decomposition of old organic matter as a result of deeper active layers in a snow depth manipulation experiment. Oecologia 163(3):785–792

O'Donnell JA, Aiken GR, Walvoord MA, Butler KD (2012a) Dissolved organic matter composition of winter flow in the Yukon River basin: implications of permafrost thaw and increased groundwater discharge. Glob Biogeochem Cycles 26(4):GB0E06. https://doi.org/10.1029/2012GB004341

O'Donnell JA, Jorgenson MT, Harden JW, McGuire AD, Kanevskiy MZ, Wickland KP (2012b) The effects of permafrost thaw on soil hydrologic, thermal, and carbon dynamics in an Alaskan peatland. Ecosystems 15(2):213–229

Oberbauer SF, Tweedie CE, Welker JM, Fahnestock JT, Henry GH, Webber PJ, Hollister RD, Walker MD, Kuchy A, Elmore E (2007) Tundra CO_2 fluxes in response to experimental warming across latitudinal and moisture gradients. Ecol Monogr 77(2):221–238

Osterkamp T (2005) The recent warming of permafrost in Alaska. Glob Planet Chang 49(3):187–202

Overeem I, Syvitsky J (2010) Sedimentaryfluxes and budgets in changing cold environments: quantitative analysis and scaling issues. Geogr Ann Ser B 92(2):285–296

Park H, Yabuki H, Ohata T (2012) Analysis of satellite and model datasets for variability and trends in Arctic snow extent and depth, 1948–2006. Pol Sci 6(1):23–37

Park H, Walsh J, Fedorov A, Sherstiukov A, Iijima Y, Ohata T (2013) The influence of climate and hydrological variables on opposite anomaly in active-layer thickness between Eurasian and north American watersheds. Cryosphere 7(2):631–645

Park H, Yoshikawa Y, Oshima K, Kim Y, Ngo-Duc T, Kimball JS, Yang D (2016) Quantification of warming climate-induced changes in terrestrial Arctic River ice thickness and phenology. J Clim 29(5):1733–1754

Parmentier F, Van Der Molen M, Van Huissteden J, Karsanaev S, Kononov A, Suzdalov D, Maximov T, Dolman A (2011) Longer growing seasons do not increase net carbon uptake in the northeastern Siberian tundra. J Geophys Res Biogeo 116(G4):G04013. https://doi.org/10.1029/2011JG001653

Payette S, Delwaide A, Caccianiga M, Beauchemin M (2004) Accelerated thawing of subarctic peatland permafrost over the last 50 years. Geophys Res Lett 31(18):L18208. https://doi.org/10.1029/2004GL020358

Paytan A, Lecher AL, Dimova N, Sparrow KJ, Kodovska FG-T, Murray J, Tulaczyk S, Kessler JD (2015) Methane transport from the active layer to lakes in the Arctic using Toolik Lake, Alaska, as a case study. Proc Natl Acad Sci 112(12):3636–3640

Pizano C, Barón AF, Schuur EA, Crummer KG, Mack MC (2014) Effects of thermo-erosional disturbance on surface soil carbon and nitrogen dynamics in upland arctic tundra. Environ Res Lett 9(7):075006

Plug LJ, West JJ (2009) Thaw lake expansion in a two-dimensional coupled model of heat transfer, thaw subsidence, and mass movement. J Geophys Res Earth 114(F1):F01002. https://doi.org/10.1029/2006JF000740

Plug LJ, Walls C, Scott B (2008) Tundra lake changes from 1978 to 2001 on the Tuktoyaktuk peninsula, western Canadian Arctic. Geophys Res Lett 35(3):L03502. https://doi.org/10.1029/2007GL032303

Pokrovsky OS, Shirokova LS, Kirpotin SN, Kulizhsky SP, Vorobiev SN (2013) Impact of western Siberia heat wave 2012 on greenhouse gases and trace metal concentration in thaw lakes of discontinuous permafrost zone. Biogeosciences 10(8):5349–5365

Prowse T, Alfredsen K, Beltaos S, Bonsal BR, Bowden WB, Duguay CR, Korhola A, McNamara J, Vincent WF, Vuglinsky V (2011) Effects of changes in arctic lake and river ice. Ambio 40(1):63–74

Rawlins MA, Steele M, Holland MM, Adam JC, Cherry JE, Francis JA, Groisman PY, Hinzman LD, Huntington TG, Kane DL (2010) Analysis of the Arctic system for freshwater cycle intensification: observations and expectations. J Clim 23(21):5715–5737

Raymond PA, McClelland J, Holmes R, Zhulidov A, Mull K, Peterson B, Striegl R, Aiken G, Gurtovaya T (2007) Flux and age of dissolved organic carbon exported to the Arctic Ocean: a carbon isotopic study of the five largest arctic rivers. Glob Biogeochem Cycles 21(4):GB4011. https://doi.org/10.1029/2007GB002934

Repo E, Huttunen J, Naumov A, Chichulin A, Lapshina E, Bleuten W, Martikainen P (2007) Release of CO_2 and CH_4 from small wetland lakes in western Siberia. Tellus B Chem Phys Meteorol 59(5):788–796

Repo ME, Susiluoto S, Lind SE, Jokinen S, Elsakov V, Biasi C, Virtanen T, Martikainen PJ (2009) Large N_2O emissions from cryoturbated peat soil in tundra. Nat Geosci 2(3):189

Reyes FR, Lougheed VL (2015) Rapid nutrient release from permafrost thaw in arctic aquatic ecosystems. Arct Antarct Alp Res 47(1):35–48

Riordan B, Verbyla D, McGuire AD (2006) Shrinking ponds in subarctic Alaska based on 1950–2002 remotely sensed images. J Geophys Res Biogeo 111(G4):G04002. https://doi.org/10.1029/2005JG000150

Romanovsky V, Osterkamp T (1997) Thawing of the active layer on the coastal plain of the Alaskan Arctic. Permafr Periglac Process 8(1):1–22

Rowland J, Travis B, Wilson C (2011) The role of advective heat transport in talik development beneath lakes and ponds in discontinuous permafrost. Geophys Res Lett 38(17):L17504. https://doi.org/10.1029/2011GL048497

Ryckeboer J, Mergaert J, Vaes K, Klammer S, De Clercq D, Coosemans J, Insam H, Swings J (2003) A survey of bacteria and fungi occurring during composting and self-heating processes. Ann Microbiol 53(4):349–410

Sachs T, Giebels M, Boike J, Kutzbach L (2010) Environmental controls on CH_4 emission from polygonal tundra on the microsite scale in the Lena river delta, Siberia. Glob Chang Biol 16(11):3096–3110

Sánchez-García L, Alling V, Pugach S, Vonk J, Van Dongen B, Humborg C, Dudarev O, Semiletov I, Gustafsson Ö (2011) Inventories and behavior of particulate organic carbon in the Laptev and East Siberian seas. Glob Biogeochem Cycles 25(2)

Sannel A, Kuhry P (2011) Warming-induced destabilization of peat plateau/thermokarst lake complexes. J Geophys Res Biogeo 116(G3):G03035. https://doi.org/10.1029/2010JG001635

Sazonova T, Romanovsky V (2003) A model for regional-scale estimation of temporal and spatial variability of active layer thickness and mean annual ground temperatures. Permafr Periglac Process 14(2):125–139

Schädel C, Bader MK-F, Schuur EA, Biasi C, Bracho R, Čapek P, De Baets S, Diáková K, Ernakovich J, Estop-Aragones C (2016) Potential carbon emissions dominated by carbon dioxide from thawed permafrost soils. Nat Clim Chang 6(10):950

Scheffer M, Bascompte J, Brock WA, Brovkin V, Carpenter SR, Dakos V, Held H, Van Nes EH, Rietkerk M, Sugihara G (2009) Early-warning signals for critical transitions. Nature 461(7260):53

Schleusner P, Biskaborn BK, Kienast F, Wolter J, Subetto D, Diekmann B (2015) Basin evolution and palaeoenvironmental variability of the thermokarst lake E l'gene-K yuele, Arctic Siberia. Boreas 44(1):216–229

Schreiner K, Bianchi TS, Rosenheim BE (2014) Evidence for permafrost thaw and transport from an Alaskan North slope watershed. Geophys Res Lett 41(9):3117–3126. https://doi.org/10.1002/2014GL059514

Schuur EA, Crummer KG, Vogel JG, Mack MC (2007) Plant species composition and productivity following permafrost thaw and thermokarst in Alaskan tundra. Ecosystems 10(2):280–292

Schuur EA, Bockheim J, Canadell JG, Euskirchen E, Field CB, Goryachkin SV, Hagemann S, Kuhry P, Lafleur PM, Lee H (2008) Vulnerability of permafrost carbon to climate change: implications for the global carbon cycle. Bioscience 58(8):701–714

Schuur EA, Vogel JG, Crummer KG, Lee H, Sickman JO, Osterkamp T (2009) The effect of permafrost thaw on old carbon release and net carbon exchange from tundra. Nature 459(7246):556

Schuur EA, McGuire AD, Schadel C, Grosse G, Harden JW, Hayes DJ, Hugelius G, Koven CD, Kuhry P, Lawrence DM, Natali SM, Olefeldt D, Romanovsky VE, Schaefer K, Turetsky MR, Treat CC, Vonk JE (2015) Climate change and the permafrost carbon feedback. Nature 520(7546):171–179. https://doi.org/10.1038/nature14338

Semiletov IP, Pipko II, Shakhova NE, Dudarev OV, Pugach SP, Charkin AN, McRoy C, Kosmach D, Gustafsson Ö (2011) Carbon transport by the Lena River from its headwaters to the Arctic Ocean, with emphasis on fluvial input of terrestrial particulate organic carbon vs. carbon transport by coastal erosion. Biogeosciences 8(9):2407–2426

Semiletov IP, Shakhova NE, Sergienko VI, Pipko II, Dudarev OV (2012) On carbon transport and fate in the east Siberian Arctic land–shelf–atmosphere system. Environ Res Lett 7(1):015201

Seppälä M (2011) Synthesis of studies of palsa formation underlining the importance of local environmental and physical characteristics. Quat Res 75(2):366–370

Shiklomanov A, Lammers R (2009) Record Russian river discharge in 2007 and the limits of analysis. Environ Res Lett 4(4):045015

Shiklomanov AI, Lammers RB, Rawlins M, Smith L, Pavelsky T (2007) Temporal and spatial variations in maximum river discharge from a new Russian data set. Journal of geophysical research: Biogeosciences 112(G4):G04S53. https://doi.org/10.1029/2006JG000352

Shiklomanov NI, Streletskiy DA, Nelson FE, Hollister RD, Romanovsky VE, Tweedie CE, Bockheim JG, Brown J (2010) Decadal variations of active-layer thickness in moisture-controlled landscapes, Barrow, Alaska. J Geophys Res Biogeosci 115(G4):G00I04. https://doi.org/10.1029/2009JG001248

Shkolnik I, Pavlova T, Efimov S, Zhuravlev S (2017) Future changes in peak river flows across northern Eurasia as inferred from an ensemble of regional climate projections under the IPCC RCP8.5 scenario. Climate Dynam https://doi.org/10.1007/s00382-017-3600-6

Shur Y, Jorgenson M (2007) Patterns of permafrost formation and degradation in relation to climate and ecosystems. Permafr Periglac Process 18(1):7–19

Sistla SA, Moore JC, Simpson RT, Gough L, Shaver GR, Schimel JP (2013) Long-term warming restructures Arctic tundra without changing net soil carbon storage. Nature 497(7451):615

Sjögersten S, Caul S, Daniell T, Jurd A, O'Sullivan O, Stapleton C, Titman JJ (2016) Organic matter chemistry controls greenhouse gas emissions from permafrost peatlands. Soil Biol Biochem 98:42–53

Smith LC, Sheng Y, MacDonald G, Hinzman L (2005) Disappearing arctic lakes. Science 308(5727):1429–1429

Soudzilovskaia NA, Bodegom PM, Cornelissen JH (2013) Dominant bryophyte control over high-latitude soil temperature fluctuations predicted by heat transfer traits, field moisture regime and laws of thermal insulation. Funct Ecol 27(6):1442–1454

Squires MM, Lesack LF, Hecky RE, Guildford SJ, Ramlal P, Higgins SN (2009) Primary production and carbon dioxide metabolic balance of a lake-rich arctic river floodplain: partitioning of phytoplankton, epipelon, macrophyte, and epiphyton production among lakes on the Mackenzie Delta. Ecosystems 12(5):853–872

St. Jacques JM, Sauchyn DJ (2009) Increasing winter baseflow and mean annual streamflow from possible permafrost thawing in the Northwest Territories, Canada. Geophys Res Lett 36(1):L01401. https://doi.org/10.1029/2008GL035822

Stieglitz M, Déry S, Romanovsky V, Osterkamp T (2003) The role of snow cover in the warming of arctic permafrost. Geophys Res Lett 30(13):1721. https://doi.org/10.1029/2003GL017337

Striegl RGDM, McDonald CP, Rover JR, Stets EG (2012) Carbon dioxide and methane emissions from the Yukon River system. Global Biogeochem Cycles 24(4):GB0E05. https://doi.org/10.1029/2012GB004306

Ström L, Ekberg A, Mastepanov M, Røjle Christensen T (2003) The effect of vascular plants on carbon turnover and methane emissions from a tundra wetland. Glob Chang Biol 9(8):1185–1192

Sturm M, Racine C, Tape K (2001) Climate change: increasing shrub abundance in the Arctic. Nature 411(6837):546

Sturm M, Douglas T, Racine C, Liston GE (2005) Changing snow and shrub conditions affect albedo with global implications. J Geophys Res Biogeo 110(G1):G01004. https://doi.org/10.1029/2005JG000013

Sugimoto A, Naito D, Yanagisawa N, Ichiyanagi K, Kurita N, Kubota J, Kotake T, Ohata T, Maximov T, Fedorov A (2003) Characteristics of soil moisture in permafrost observed in East Siberian taiga with stable isotopes of water. Hydrol Process 17(6):1073–1092

Tape KD, Verbyla D, Welker JM (2011) Twentieth century erosion in Arctic Alaska foothills: the influence of shrubs, runoff, and permafrost. J Geophys Res Biogeo 116(G4):G04024. https://doi.org/10.1029/2011JG001795

Teltewskoi A, Beermann F, Beil I, Bobrov A, De Klerk P, Lorenz S, Lüder A, Michaelis D, Joosten H (2016) 4000 years of changing wetness in a permafrost polygon Peatland (Kytalyk, NE Siberia): a comparative high-resolution multi-proxy study. Permafr Periglac Process 27(1):76–95

Trucco C, Schuur EAG, Natali SM, Belshe EF, Bracho R, Vogel J (2012) Seven-year trends of CO_2 exchange in a tundra ecosystem affected by long-term permafrost thaw. J Geophys Res Biogeo 117(G2):G02031. https://doi.org/10.1029/2011jg001907

Turetsky M, Wieder R, Vitt D, Evans R, Scott K (2007) The disappearance of relict permafrost in boreal North America: effects on peatland carbon storage and fluxes. Glob Chang Biol 13(9):1922–1934

Turetsky M, Treat C, Waldrop M, Waddington J, Harden J, McGuire A (2008) Short-term response of methane fluxes and methanogen activity to water table and soil warming manipulations in an Alaskan peatland. J Geophys Res Biogeosci 113(G3):G00A10. https://doi.org/10.1029/2007JG000496

Ukraintseva N, Leibman M, Streletskaya I, Mikhaylova T (2014) Geochemistry of plant-soil-permafrost system on landslide-affected slopes, Yamal, Russia as an indicator of landslide age. In: Landslides in cold regions in the context of climate change. Springer, Cham, pp 107–131

Van der Molen M, Van Huissteden J, Parmentier F, Petrescu A, Dolman A, Maximov T, Kononov A, Karsanaev S, Suzdalov D (2007) The growing season greenhouse gas balance of a continental tundra site in the Indigirka lowlands, NE Siberia. Biogeosciences 4(6):985–1003. https://doi.org/10.5194/bg-4-985-2007

Van Hardenbroek M, Lotter AF, Bastviken D, Duc N, Heiri O (2012) Relationship between $\delta^{13}C$ of chironomid remains and methane flux in Swedish lakes. Freshw Biol 57(1):166–177

Van Huissteden J (1990) Tundra rivers of the last glacial: sedimentation and geomorphological processes during the Middle Pleniglacial in Twente, Eastern Netherlands. In: Van Amerom H (ed) Mededelingen Rijks Geologische Dienst, vol 44–3. Rijks Geologische Dienst, pp 1–138

Van Huissteden J, Dolman AJ (2012) Soil carbon in the Arctic and the permafrost carbon feedback. Curr Opin Environ Sustain 4(5):545–551. https://doi.org/10.1016/j.cosust.2012.09.008

Van Huissteden J, Maximov TC, Dolman AJ (2005) High methane flux from an arctic floodplain (Indigirka lowlands, eastern Siberia). J Geophys Res Biogeo 110(G2):G02002. https://doi.org/10.1029/2005jg000010

Van Huissteden J, Maximov TC, Dolman AJ (2009) Correction to "High methane flux from an arctic floodplain (Indigirka lowlands, eastern Siberia). J Geophys Res Biogeosci 114(G2):n/a–n/a. https://doi.org/10.1029/2009jg001040

Van Huissteden J, Berrittella C, Parmentier F, Mi Y, Maximov T, Dolman A (2011) Methane emissions from permafrost thaw lakes limited by lake drainage. Nat Clim Chang 1(2):119–123

Van Huissteden J, Vandenberghe J, Gibbard PL, Lewin J (2013) Permafrost and Periglacial features | Periglacial fluvial sediments and forms. In: Elias SA (ed) Encyclopedia of the quaternary, vol 3. Elsevier, Amsterdam, pp 490–499. https://doi.org/10.1016/b978-0-444-53643-3.00108-4

Vandenberghe J (2003) Climate forcing of fluvial system development: an evolution of ideas. Quat Sci Rev 22(20):2053–2060

Vandenberghe J (2008) The fluvial cycle at cold–warm–cold transitions in lowland regions: a refinement of theory. Geomorphology 98(3–4):275–284

Verville J, Hobbie S, Chapin FS, Hooper D (1998) Response of tundra CH_4 and CO_2 flux tomanipulation of temperature and vegetation. Biogeochemistry 41(3):215–235

Vihma T, Screen J, Tjernström M, Newton B, Zhang X, Popova V, Deser C, Holland M, Prowse T (2016) The atmospheric role in the Arctic water cycle: a review on processes, past and future changes, and their impacts. J Geophys Res Biogeo 121(3):586–620

Vonk JE, Gustafsson Ö (2013) Permafrost-carbon complexities. Nat Geosci 6(9):675–676. https://doi.org/10.1038/ngeo1937

Vonk JE, Sanchez-Garcia L, Semiletov I, Dudarev O, Eglinton T, Andersson A, Gustafsson Ö (2010) Molecular and radiocarbon constraints on sources and degradation of terrestrial organic carbon along the Kolyma paleoriver transect, east Siberian Sea. Biogeosciences 7(10):3153–3166

Vonk JE, Mann PJ, Davydov S, Davydova A, Spencer RGM, Schade J, Sobczak WV, Zimov N, Zimov S, Bulygina E, Eglinton TI, Holmes RM (2013) High biolability of ancient permafrost carbon upon thaw. Geophys Res Lett 40(11):2689–2693. https://doi.org/10.1002/grl.50348

Vonk JE, Semiletov IP, Dudarev OV, Eglinton TI, Andersson A, Shakhova N, Charkin A, Heim B, Gustafsson Ö (2014) Preferential burial of permafrost-derived organic carbon in Siberian-Arctic shelf waters. J Geophys Res Oceans 119(12):8410–8421

Vonk JE, Tank SE, Bowden WB, Laurion I, Vincent WF, Alekseychik P, Amyot M, Billet M, Canario J, Cory RM (2015) Reviews and syntheses: effects of permafrost thaw on Arctic aquatic ecosystems. Biogeosciences 12(23):7129–7167

Walker D, Jia G, Epstein H, Raynolds M, Chapin Iii F, Copass C, Hinzman L, Knudson J, Maier H, Michaelson G (2003) Vegetation-soil-thaw-depth relationships along a low-arctic bioclimate gradient, Alaska: synthesis of information from the ATLAS studies. Permafr Periglac Process 14(2):103–123

Walter Anthony KM, Anthony P, Grosse G, Chanton J (2012) Geologic methane seeps along boundaries of Arctic permafrost thaw and melting glaciers. Nat Geosci 5(6):419–426. https://doi.org/10.1038/ngeo1480

Walter Anthony KM, Zimov SA, Grosse G, Jones MC, Anthony PM, Chapin FS 3rd, Finlay JC, Mack MC, Davydov S, Frenzel P, Frolking S (2014) A shift of thermokarst lakes from carbon sources to sinks during the Holocene epoch. Nature 511(7510):452–456. https://doi.org/10.1038/nature13560

Walter Anthony K, Daanen R, Anthony P, Schneider von Deimling T, Ping C-L, Chanton JP, Grosse G (2016) Methane emissions proportional to permafrost carbon thawed in Arctic lakes since the 1950s. Nat Geosci 9:679–682

Walter Anthony K, von Deimling TS, Nitze I, Frolking S, Emond A, Daanen R, Anthony P, Lindgren P, Jones B, Grosse G (2018) 21st-century modeled permafrost carbon emissions accelerated by abrupt thaw beneath lakes. Nat Commun 9(1):3262

Walter KM, Zimov SA, Chanton JP, Verbyla D, Chapin FS 3rd (2006) Methane bubbling from Siberian thaw lakes as a positive feedback to climate warming. Nature 443(7107):71–75. https://doi.org/10.1038/nature05040

Walter KM, Smith LC, Chapin FS, 3rd (2007a) methane bubbling from northern lakes: present and future contributions to the global methane budget. Philos Trans A Math Phys Eng Sci 365 (1856):1657–1676. doi:https://doi.org/10.1098/rsta.2007.2036

Walter KM, Edwards ME, Grosse G, Zimov SA, Chapin FS 3rd (2007b) Thermokarst lakes as a source of atmospheric CH_4 during the last deglaciation. Science 318(5850):633–636. https://doi.org/10.1126/science.1142924

Walter K, Chanton J, Chapin F, Schuur E, Zimov S (2008) Methane production and bubble emissions from arctic lakes: isotopic implications for source pathways and ages. J Geophys Res Biogeosci 113(G3):G00A08. https://doi.org/10.1029/2007JG000569

Walvoord MA, Kurylyk BL (2016) Hydrologic impacts of thawing permafrost – a review. Vadose Zone J 15 (6):0. doi:https://doi.org/10.2136/vzj2016.01.0010

Walvoord MA, Striegl RG (2007) Increased groundwater to stream discharge from permafrost thawing in the Yukon River basin: potential impacts on lateral export of carbon and nitrogen. Geophys Res Lett 34(12):L12402. https://doi.org/10.1029/2007GL030216

Wang P, Heijmans MM, Mommer L, van Ruijven J, Maximov TC, Berendse F (2016) Belowground plant biomass allocation in tundra ecosystems and its relationship with temperature. Environ Res Lett 11(5):055003

Wickland K, Aiken GR, Butler K, Dornblaser MM, Spencer RGM, Striegl RG (2012) Biodegradability of dissolved organic carbon in the Yukon River and its tributaries: Seasonality and importance of inorganic nitrogen. Glob Biogeochem Cycles 26(4):GB0E03. https://doi.org/10.1029/2012GB004342

Wickland KP, Waldrop MP, Aiken GR, Koch JC, Jorgenson MT, Striegl RG (2018) Dissolved organic carbon and nitrogen release from boreal Holocene permafrost and seasonally frozen soils of Alaska. Environ Res Lett 13(6):065011

Wik M, Thornton BF, Bastviken D, MacIntyre S, Varner RK, Crill PM (2014) Energy input is primary controller of methane bubbling in subarctic lakes. Geophys Res Lett 41(2):555–560

Wik M, Varner RK, Anthony KW, MacIntyre S, Bastviken D (2016) Climate-sensitive northern lakes and ponds are critical components of methane release. Nat Geosci 9(2):99

Wik M, Johnson JE, Crill PM, DeStasio JP, Erickson L, Halloran MJ, Fahnestock MF, Crawford MK, Phillips SC, Varner RK (2018) Sediment characteristics and methane ebullition in three subarctic lakes. J Geophys Res Biogeo 123(8):2399–2411

Wild B, Andersson A, Bröder L, Vonk J, Hugelius G, McClelland JW, Song W, Raymond PA, Gustafsson Ö (2019) Rivers across the Siberian Arctic unearth the patterns of carbon release from thawing permafrost. Proc Natl Acad Sci 116(21):10280–10285

Williams PJ, Smith MW (1991) The frozen earth. Fundamentals of geocryology. Cambridge University Press, Cambridge, 306 p

Woo M-K (2012) Permafrost hydrology. Springer, 563 p

Wrona FJ, Johansson M, Culp JM, Jenkins A, Mård J, Myers-Smith IH, Prowse TD, Vincent WF, Wookey PA (2016) Transitions in Arctic ecosystems: ecological implications of a changing hydrological regime. J Geophys Res Biogeo 121(3):650–674

Wu P, Wood R, Stott P (2005) Human influence on increasing Arctic river discharges. Geophys Res Lett 32(2):L02703. https://doi.org/10.1029/2004GL021570

Ye H, Ladochy S, Yang D, Zhang T, Zhang X, Ellison M (2004) The impact of climatic conditions on seasonal river discharges in Siberia. J Hydrometeorol 5(2):286–295

Yi Y, Kimball JS, Chen R, Moghaddam M, Reichle RH, Mishra U, Zona D, Oechel WC (2018) Characterizing permafrost active layer dynamics and sensitivity to landscape spatial heterogeneity in Alaska. Cryosphere 12(1):145–161. https://doi.org/10.5194/bg-15-5287-2018

Yoshikawa K, Hinzman LD (2003) Shrinking thermokarst ponds and groundwater dynamics in discontinuous permafrost near council, Alaska. Permafr Periglac Process 14(2):151–160

Zakharova EA, Kouraev AV, Stephane G, Franck G, Desyatkin RV, Desyatkin AR (2018) Recent dynamics of hydro-ecosystems in thermokarst depressions in Central Siberia from satellite and in situ observations: importance for agriculture and human life. Sci Total Environ 615:1290–1304

Zhang T, Frauenfeld OW, Serreze MC, Etringer A, Oelke C, McCreight J, Barry RG, Gilichinsky D, Yang D, Ye H (2005) Spatial and temporal variability in active layer thickness over the Russian Arctic drainage basin. J Geophys Res Atmos 110(D16):D16101. https://doi.org/10.1029/2004JD005642

Zhang W, Miller P, Jansson C, Samuelsson P, Mao J, Smith B (2018) Self-amplifying feedbacks accelerate greening and warming of the Arctic. Geophys Res Lett 45:7102–7111. https://doi.org/10.1029/2018GL077830

Zimov S, Semiletov I, Daviodov S, Voropaev YV, Prosyannikov S, Wong C, Chan Y-H (1993) Wintertime CO_2 emission from soils of Northeastern Siberia. Arctic 46(3):197–204

Zimov S, Voropaev YV, Semiletov I, Davidov S, Prosiannikov S, Chapin FS, Chapin M, Trumbore S, Tyler S (1997) North Siberian lakes: a methane source fueled by Pleistocene carbon. Science 277(5327):800–802

Zimov SA, Davydov SP, Zimova GM, Davydova AI, Schuur EAG, Dutta K, Chapin FS (2006) Permafrost carbon: stock and decomposability of a globally significant carbon pool. Geophys Res Lett 33(20). https://doi.org/10.1029/2006gl027484

Zinder SH (1990) Conversion of acetic acid to methane by thermophiles. FEMS Microbiol Rev 6(2–3):125–137

Zona D, Oechel W, Kochendorfer J, Paw U, Salyuk A, Olivas P, Oberbauer S, Lipson D (2009) Methane fluxes during the initiation of a large-scale water table manipulation experiment in the Alaskan Arctic tundra. Glob Biogeochem Cycles 23(2):GB2013. https://doi.org/10.1029/2009GB003487

Zona D, Lipson DA, Paw U, Kyaw T, Oberbauer SF, Olivas P, Gioli B, Oechel WC (2012) Increased CO_2 loss from vegetated drained lake tundra ecosystems due to flooding. Glob Biogeochem Cycles 26(2):GB2004. https://doi.org/10.1029/2011GB004037

Zona D, Gioli B, Commane R, Lindaas J, Wofsy SC, Miller CE, Dinardo SJ, Dengel S, Sweeney C, Karion A, Chang RY, Henderson JM, Murphy PC, Goodrich JP, Moreaux V, Liljedahl A, Watts JD, Kimball JS, Lipson DA, Oechel WC (2016) Cold season emissions dominate the Arctic tundra methane budget. Proc Natl Acad Sci U S A 113(1):40–45. https://doi.org/10.1073/pnas.1516017113

Chapter 6
Vegetation Change

Abstract Vegetation in permafrost areas is already changing as a result of climate change. Since vegetation strongly affects the heat balance of permafrost soils and the amount of ecosystem carbon in soils and biomass, this is an important aspect of the permafrost carbon cycle. Zonal vegetation change is currently showing mostly as shrub expansion in the southern parts of the tundra biome, while northward migration of the tree line is lagging behind. Satellite remote sensing has recorded this 'Arctic greening', but it is not uniform throughout the Arctic – also 'browning' occurs regionally for various reasons; amongst these are the effects of regional climate variation and effects of permafrost thaw. A larger vegetation biomass does not necessarily result in a larger soil carbon content; in some cases a decrease or depth shifts of soil carbon have been reported. Wildfires that may induce permafrost thaw are expected to increase in case of drier and warmer climates; this includes tundra fires.

Permafrost thaw and decomposition of thawed organic matter liberates nutrients, but it depends on plant traits to what extent the additional nutrients increase the vegetation and biomass. Abrupt permafrost thaw results in severe ecosystem disturbance. Ecosystems are resilient to a certain amount of disturbance, although these disturbances results in large soil carbon losses that only can be regained in time scales of centuries to millennia, and in particular the 'certain amount of disturbance' needs assessment. Superposed on the effects of climate change are other anthropogenic disturbances of ecosystems that affect permafrost.

Keywords Vegetation change · Tree line · Shrub expansion · Arctic greening · Arctic browning · Fire · Nutrients · Ecosystem resilience · Anthropogenic disturbance

The tight connections between permafrost, soil carbon and vegetation already have been touched upon on in the previous chapters. The focus of this chapter is the extent to which vegetation and ecosystems influence the permafrost carbon cycle directly, and in turn, how permafrost thaw interacts with vegetation. This include resilience of vegetation and soil carbon to disturbances caused by permafrost thaw.

Organisms that live near the limits of their environmental range are particularly sensitive to changes in their environment. For permafrost ecosystems and vegeta-

J. van Huissteden, *Thawing Permafrost*,
https://doi.org/10.1007/978-3-030-31379-1_6

tion, the effect of climate change will be felt as higher air temperatures, a longer growing season and changed precipitation patterns. A higher atmospheric CO_2 concentration increases primary production and decrease water use, although this may be restricted by nutrient availability. Globally, CO_2 fertilisation is responsible for about 70% of global greening recorded by satellites. However, from experimental data on free air CO_2 enrichment in various ecosystems it is not clear that higher CO_2 concentrations lead to more ecosystem carbon storage (Norby and Zak 2011); higher above-ground ecosystem production may be accompanied by faster below-ground carbon cycling. For northern forests and tundra, temperature is the most important factor that drives plant production (Zhu et al. 2016a). The vegetation itself is already adapting to a warmer climate in a variety of ways, including shifts in species composition of the vegetation (Soja et al. 2007; Myers-Smith et al. 2011).

Climate change will cause shifts in vegetation zones. Besides these zonal changes, also local shifts in vegetation occur as an effect of permafrost thaw, causing changes in soil temperature, active layer thickness, soil moisture and nutrient conditions. These changes feed back on the carbon cycle. Furthermore, human activities also have an increasing impact on vegetation of permafrost regions.

As seen in Chaps. 2 and 4, the following changes may be expected to occur with vegetation change (Fig. 6.1):

- Surface energy balance: radiation balance, sensible and latent heat exchange;
- Heat conduction into the ground: insulation by vegetation, snow cover thickness;

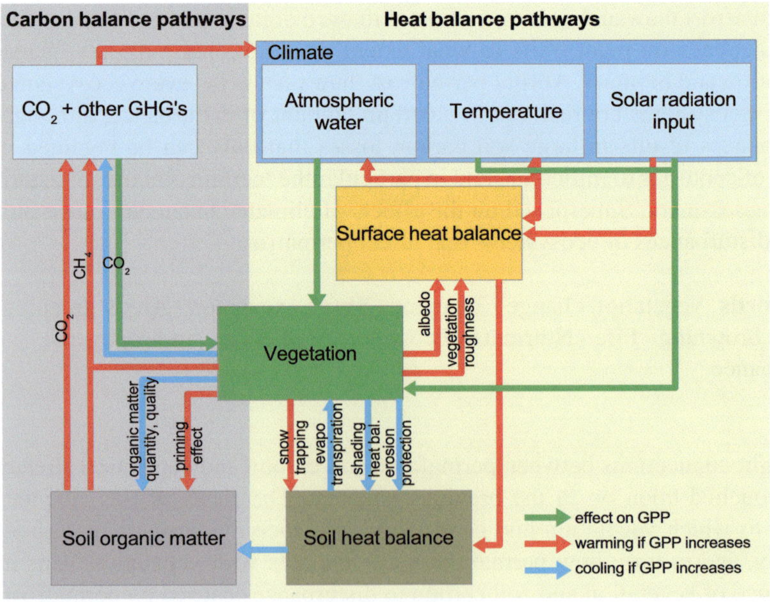

Fig. 6.1 Interactions of vegetation with climate, heat balance and carbon cycle. The vegetation determines carbon input into the soil, and in wetlands CH_4 gas exchange. Vegetation has a mainly soil cooling effect via the soil heat and water balance

- Primary production: quantity of organic matter input to the soil;
- Root system changes: depth of organic matter input to the soil;
- Vegetation composition change: quality of organic matter input;
- Vegetation-driven changes in decomposition rates;
- Change of CH_4 production and emission pathways from wet soils, ponds and lakes.;

The 'ecosystem protection' of permafrost (Shur and Jorgenson 2007; Chap. 5) depends on vegetation cover. Increased primary production by higher CO_2 concentration, warmer and longer growing seasons and nutrient availability may result in increase of organic matter input to the soil (e.g. Parmentier et al. 2011; Hicks Pries et al. 2013), but soil environmental factors ultimately determine if more organic matter is preserved (next sections). Various authors have mentioned the effect of vegetation on litter quality and lability of fresh organic matter (e.g. Aerts 2006; Myers-Smith et al. 2011). Increased input of labile organic matter affects also the decomposition of older soil carbon by the priming effect (Chap. 4; Kuzyakov 2010; Bernal et al. 2016). Local changes of wetland vegetation driven by permafrost thaw affect CH_4 production and emission pathways from the soil (Turetsky et al. 2014). At the same time these ecosystem changes also feed back on climate change with their effects on the surface heat balance and carbon balance, with a general trend of amplification of climate warming (Bonan 2008; Chapin et al. 2010).

Arctic plant species display a wide range of traits that affect biogeochemical feedbacks to their environment: woody or herbaceous, evergreen or deciduous, aerenchymous or non-aerenchymous, mycorrhizal or non-mycorrhizal, vascular or non-vascular plants. Relatively small changes in species composition therefore may significantly alter these biogeochemical feedbacks (e.g. Andresen et al. 2017). Furthermore, plant species react in different ways on environmental change, such as changes in climate, moisture and nutrient availability in the soil. Wookey et al. (2009) give a useful introduction to plant traits that influence these interactions. Several model predictions suggest an increasing carbon storage in biomass over permafrost areas in the coming century. However, factors that could decrease carbon biomass such as water stress, nutrient limitation, disturbance by permafrost collapse, herbivory, fire or human activities are not yet adequately included in models (Abbott et al. 2016).

Another aspect of adaptation of vegetation to climate change is plant trait plasticity and evolution. Plasticity is change of the plant phenotype, within the genetic variability of the species, resulting in an adaptation of plant traits to changing environmental conditions. Adaptations in leaf size and shape, leaf chemistry, stomata density and size, bud-break and flowering time, seed and seedling traits, root to shoot ratio, root length and metabolic rate have been reported. Adaptation by plasticity are likely to affect the predicted rates of ecosystem carbon exchange, with largely unknown consequences for future atmospheric CO_2 concentrations (Nicotra et al. 2010). Plasticity varies between species and determines to what extent they can adapt to rapidly changing environmental conditions. Most adaptations of plants are expected to be driven by plasticity, but evolutionary changes cannot be excluded (Berteaux et al. 2004).

The concepts of vulnerability and resilience are important to understand the ecosystem and carbon cycle effects of vegetation change (Chapin et al. 2010). Vulnerability is the degree to which a system is likely to change as a result of an environmental disturbance, e.g. climate change or permafrost thaw. Resilience is the capacity of the system to maintain or restore function and structure when confronted with disturbances. In particular the question to what extent net carbon uptake in ecosystem biomass will compensate for losses of carbon from permafrost soils is an important topic in this chapter.

6.1 Zonal Vegetation Shifts

For zonal vegetation shifts, we tend to think in the types of ecosystems and biomes that we know today. However, it is not certain to what extent they will survive a warming climate at all. Future ecosystems on permafrost may be novel, possibly with no current eco-climatological analog (Macias-Fauria et al. 2012; Loranty et al. 2018b). For instance, the climatic changes at the termination of the Last Glacial made an entire zonal biome almost disappear: the mammoth steppe, the cold steppe zone dominated by forbs and grasses, populated with large herbivores, and depositional environment of Yedoma deposits (Zimov et al. 2012).

In general, high-latitude vegetation is limited by low temperatures, and a change of climate towards higher soil and air temperatures should increase vegetation productivity and results in a northward shift of species and even biomes. Evidence for vegetation biomass increase throughout the tundra zone is derived from vegetation indices from remote sensing. Over most of the Arctic tundra, NDVI (normalized difference vegetation index) has increased, in line with increasing shrub and other vegetation biomass, as Bhatt et al. (2010) conclude from a large NDVI dataset from the NOAA Advanced Very High Resolution Radiometer (AVHRR) sensor over 1982 to 2008.

However, the response of zonal vegetation change to climate change is complicated by multiple relations of species with climate.

- Climatic factors are the growing season length, temperature, snow depth and timing of snow melt, precipitation, and incoming solar radiation. Because of varying response of species to these factors, migration will differ among species.
- Edaphic factors (related to soils) limit rates of vegetation change, in particular in the High Arctic where soil nutrients may be limiting. Limited nutrient supply delays the encroachment of species that profit from new climatic conditions. The release of nutrients from thawing permafrost and faster decomposition of soil organic matter is therefore an important factor in shrub expansion (Weintraub and Schimel 2005; Myers-Smith et al. 2011). Nitrogen fixation by bacteria and cyanobacteria also influence migration rates (Wookey et al. 2009).

- Plant dispersal mechanisms affect migration rate of species and may retard migration (e.g. McGuire et al. 2002). For instance, many tree species have relatively short seed dispersal distances and it takes time to grow a tree from a seedling. Landscape barriers (e.g. a mountain chain or extensive wetlands) also can retard migration (MacDonald et al. 2007; Frost and Epstein 2014).
- Interactions of plants with the ecosystem modify climate change effects, e.g. competition between species, grazing by herbivores, pathogens and presence of soil mycorrhizae.
- Disturbances by thawing permafrost, fire and human activities increasingly affect the vegetation (Myers-Smith et al. 2011, 2015; Lara et al. 2018). However, disturbances can work in two directions: in the case of trees and shrubs, permafrost thaw can negatively affect growing conditions by subsidence, waterlogging of soils and erosion, or create favourable locations for new establishment by removal of competing species, increased nutrient supply and exposing mineral soil (Fig. 6.2; Myers-Smith et al. 2011; Frost and Epstein 2014).

6.1.1 Present Climate-Related Vegetation Change

In all boreal forest regions, warming has occurred, and is expected to continue, with 3–7 °C in Central Alaska and 4–5 °C in the Canadian boreal forest zone by 2100 (Chapin et al. 2010; Price et al. 2013; Verbyla 2011). With a globally averaged projection of a warming of 4 °C by the end of this century, boreal forest regions could experience temperature increases from 4° to 11 °C, with only a modest increase in precipitation, resulting in a major transformation of the forest (Gauthier et al. 2015). At the southern limits of boreal forests, temperate trees are invading (Kullman 2002; Fisichelli et al. 2014). However, the climatic trends do not necessarily increase green biomass; a conspicuous browning trend is visible in remote sensing-derived NDVI data (Verbyla 2011; see below).

Fig. 6.2 Contrasting effect of permafrost thaw on shrubs in Siberian tundra. (**a**) Partly stabilized thaw slumps on a west-facing thaw lake bank colonized by fast growing *Salix* shrubs. (**b**) Drowning *Betula nana* shrubs in a thaw pond. Kytalyk reserve, Indigirka lowlands. (Photos by the author)

The boreal forest is dominated by a small number of tree species, predominantly evergreen conifers (*Picea, Pinus, Abies*), deciduous conifers of the genus *Larix*, and deciduous broadleaf genera (*Betula, Populus, Salix)*. Besides the areal expansion of the boreal forest, also changes in species composition are expected. Northern tree expansion is limited by cool and short growing seasons. In particular temperatures during the early part of the growing season are critical. Cold soils also restrict the availability of nutrients by slowing decomposition (MacDonald et al. 2007; Price et al. 2013). In northern Eurasia, the treeline zone corresponds to average July temperatures of 12.5–10.0 °C and a growing degree day number of 800 GDD (Kaplan et al. 2003; see Sect. 4.2.1). In general, the responses of boreal forest to recent climate change has been more heterogeneous than those in tundra ecosystems (Loranty et al. 2018b). The causes of this varying response are species differences in physiology, effects of temperature-induced drought, disturbance regimes from fire and herbivory, and vegetation successional dynamics and migration rates.

At the northern fringe of the boreal forest, biome shifts are modest. Most studies pertain to mountain slopes, where the forest-tundra transition is compressed in a short horizontal distance, whereas it is much wider in lowlands. During the Holocene, climate oscillations in the order of ±2 °C have occurred at high latitudes (Price et al. 2013). At the time of the Holocene thermal maximum (between 10–3 kyr) and the Medieval Warm Period the treeline was located further north than at present in Russia (Bigelow et al. 2003; MacDonald et al. 2007). The current treeline has not yet expanded back to that position. The maximum Postglacial migration rate of trees in Northern Canada was 10 km per century; this indicates that tree migration may not be able to keep up with present climate change (Aitken et al. 2008; Price et al. 2013). Moreover, the response of the treeline depends also on other factors than climate change, as listed above (MacDonald et al. 2007).

In Russia and Siberia, evidence of recruitment of new trees and increased productivity is seen near the treeline, shown by a comparison of military intelligence satellite images collected between 1965 and 1969 and modern high resolution images. Tree line advance is most conspicuous in uplands; in some lowland areas little change is seen, and in one case in Eastern Siberia treeline retreated, caused by permafrost degradation (Frost and Epstein 2014; Fig. 6.3). The expansion of tall shrubs and trees is correlated with an increase in MAAT, annual and winter precipitation. The treeline expansion in lowlands is modest and stays south of the treeline before the Little Ice Age. Evidence from tree rings suggests that the migration of the treeline lags decennia on climate warming. However, in lowlands the presence of wetlands retards treeline expansion. In wet terrain with ice wedge polygons Larch trees establish preferentially on relatively dry polygon rims (Frost and Epstein 2014).

In Northern Siberia, presently deciduous Larix species dominate, but future projections and present observations indicate an increase in evergreen conifers, which is attributed to warmer winters (ACIA 2004; Kharuk et al. 2007). Zhang et al. (2011) have demonstrated with a dynamic vegetation model, that Siberian Larch and permafrost soils are a closely coupled system; permafrost thaw is a water source for the trees in summer. Warming of the climate by about 2 °C would result in a replacement of Larch by evergreen coniferous trees.

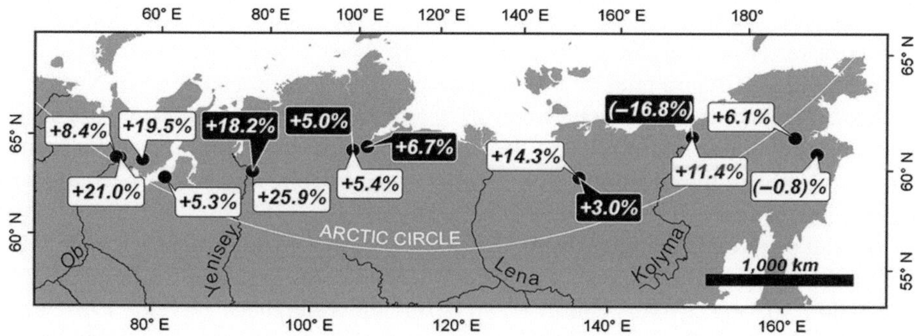

Fig. 6.3 Summary of net changes in tall shrub (white boxes) and larch (black boxes) cover at 11 tundra ecotones in northern Siberia. Negative values are given in parentheses. (Republished with permission of John Wiley and Sons Inc., from 'Tall shrub and tree expansion in Siberian tundra ecotones since the 1960s', by Frost and Epstein (2014), Global Change Biology 20 © 2014; permission conveyed through Copyright Clearance Center, Inc.)

In several regions, tree-ring studies have found a decrease of forest productivity due to drought (Loranty et al. 2018b). Within the North American boreal forest zone, clear regional differences exist in the effects of moisture availability. In Western Canada and Central Alaska, the incidence of drought has increased, restricting tree growth. By contrast, in Eastern Canada moister condition prevail, and are expected to occur in the future (Chapin et al. 2010; Price et al. 2013). Where moisture is not limiting, forest productivity has increased in Canada (Price et al. 2011, 2013). Soil moisture limitation is presently the primary factor limiting production of Alaskan boreal trees (Yarie and Van Cleve 2010). On the other hand, moisture increase also may decrease productivity, as demonstrated by Iijima et al. (2010, 2014) for the Larch forest of the taiga in Central Siberia, where a number of wet years led to decline of forest growth.

Insect pest outbreaks have also severe impacts on tree productivity in the boreal forest, e.g. the mountain pine beetle (*Dendroctonus ponderosae)* in Canada and the winter moth (*Operophtera brumata)* in the mountain birch zone in Scandinavia (e.g. Kurz et al. 2008). In general, many insect species will benefit from higher temperatures, but it is uncertain how climate change affects the relevant population processes of survival, reproduction, and dispersal, and how future ecological relations evolve (Price et al. 2013). Nevertheless, the combination of climate stress and insect outbreaks is already affecting forest productivity and have severe effects in southern parts of the boreal forest zone (Verbyla 2011).

Forest fires are expected to increase in the boreal forest. Climatic factors driving potential increase of forest fires are higher temperatures and evaporation. In 2010 and 2012, severe drought has led to many boreal peatland fires in Russia (Gorchakov et al. 2014). The greenhouse gas emission from fires are substantial, in particular from the boreal forest zone where extensive peat lands also can be ignited (Turetsky et al. 2011a, b). Wildfires will be discussed more extensively below.

Under less continental climate conditions in Canada and Sweden treeline migration is faster (Kullman 2002; Payette 2007). In Alaska, treeline advance also has been observed (Chapin et al. 2005). In a mountain vegetation transect in Northern Finland, a shift of the treeline between 1976 and 2014 was absent, despite a 2–3 °C warming. Beyond the treeline, significant changes of plant communities were found with expansion of dwarf shrubs. The stability of the treeline might have been caused by reindeer grazing (Vuorinen et al. 2017).

Instead of treeline change, shrub expansion (or shrubification) is seen widely in the tundra north of the treeline, mapped with remote sensing, tree-ring studies and vegetation surveys (Myers-Smith et al. 2011 and references therein; Frost and Epstein 2014). Shrub expansion can react more rapidly to climate changes, because shrubs are already present in the tundra and need no significant migration (Chapin et al. 2005). The shrubification may manifest itself as increased infilling of open patches with shrubs, increased growth in size, and migration (Myers-Smith et al. 2011; Fig. 6.4). The shrubs vary in growth forms: tall multi-stemmed shrubs, erect dwarf shrubs (0.1–0.4 m) and prostrate dwarf shrubs (<0.1 m) of which most of the branches lie upon the ground surface. The expanding shrub species are the taller species and vary regionally. Expanding species are *Empetrum nigrum* and *Juniperus nana* in Scandinavia, *Alnus viridis* in North America and West Siberia, *Betula nana* and *Salix* species throughout the Arctic.

Climatic factors that contribute to shrubification are higher temperatures, longer growing season, incoming solar radiation, change in snow cover and precipitation (Myers-Smith et al. 2011). Furthermore, changes in soil moisture, nutrients, active layer thickness, and disturbance by permafrost thaw contribute (Vogel et al. 2009). A feedback loop can be conceived, involving nutrient liberation by faster soil

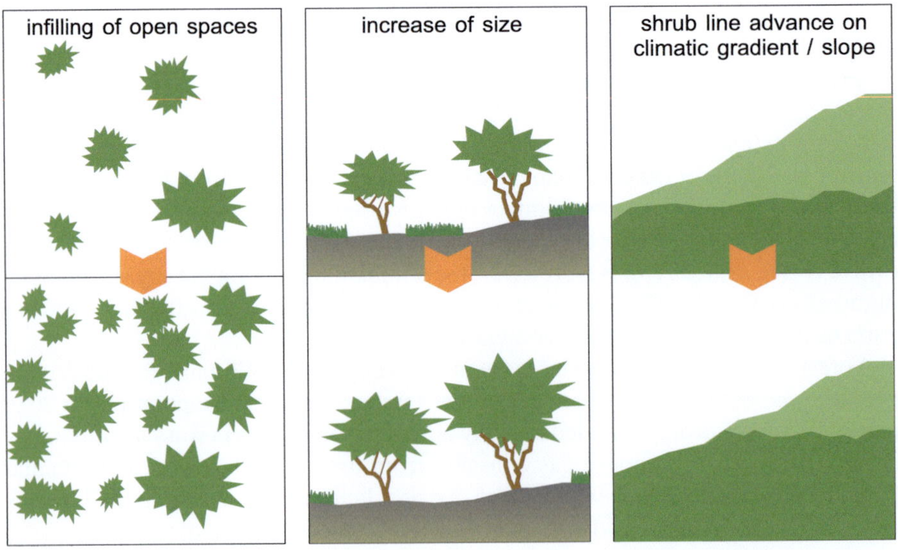

Fig. 6.4 Three general categories of shrub expansion, after Myers-Smith et al. (2011)

organic matter decomposition by warming of the soil, faster growth of shrubs by higher N availability, and winter soil warming by a thicker snow cover induced by shrub expansion (see below; Weintraub and Schimel 2005). However, most shrubs have shallow root systems and may profit less from N availability from warming soils than other plant functional types, and soil shading may reduce soil warming in summer (Wang et al. 2018, see below). Another edaphic factor that facilitates shrub establishment is the exposure of mineral soil, either by cryoturbation or by abrupt permafrost thaw. Frost et al. (2013) describe preferential establishment of Alder shrubs on bare spots in frost boils. Cryogenic landslides, if stabilized, are known to facilitate establishment of Willow shrubs (Leibman et al. 2003, 2014).

Dendrological evidence suggests that summer temperature is the most important driver of shrub growth (Myers-Smith et al. 2015). The climate sensitivity of shrub growth varies regionally, it is largest for the Northwest Siberian Arctic and Northern Europe. It is also larger for taller shrubs, for shrubs growing at the northern limit of their range, and for shrubs growing in moist sites. From a tree-ring study in Northern Siberia, Blok et al. (2011a) concluded that in particular the early summer air temperature is an important driver for growth of *Salix pulchra* and *Betula nana* shrubs. No effect of winter precipitation (snow cover) was found, but for *Betula nana* the summer precipitation in combination with summer warmth contributed also to establishment of new plants (Li et al. 2016).

Further North, barrens are greening, although not uniformly (see next section). The amount of bare ground decreases in vegetation test plots analysed by Elmendorf et al. (Elmendorf et al. 2012). In general, more species-rich and higher biomass plant communities are found at sites where moisture supply is higher than elsewhere in polar deserts (Bliss et al. 1994; Kade et al. 2005; Fig. 6.5). This suggests that increasing precipitation also may enhance plant growth. However, experimentally derived effects are weak. A list of experimental studies of precipitation addition by Keuper et al. (2012a) shows no response for most experiments. Her own experiments

Fig. 6.5 Spring-fed vegetation of mosses in an otherwise dry polar desert environment. Sprengisandur, Iceland. (Photo by the author)

in Siberian dry tundra heath (dominated by *Betula nana*) and palsa vegetation (dominated by *Empetrum hermaphroditum*) in Northern Sweden did not show effects on total biomass, while 200 mm year^{-1} extra precipitation was added. For the Siberian site there was a response of ±30% length growth of shrub branches, but this was minor contribution to the total biomass. Warming of polar desert vegetation in Svalbard by open top tents, and a precipitation supplement of 50% of the average local precipitation had minimal influence in experiments of Robinson et al. (1998).

6.1.2 Arctic Greening and Browning

Trends of increasing biomass are recorded by satellite-derived time series of NDVI. NDVI shows sharp fluctuations, marked by several consecutive years of reversal (e.g. 2012–2015) – 'Arctic browning' as opposite to 'greening' (Bhatt et al. 2013; Epstein et al. 2015, 2018; Fig. 6.6). Slow-down or reversals of the NDVI trend are not uniform throughout the Arctic. Bhatt et al. (2013) attribute decreasing NDVI trends to large-scale circulation changes that lead to lower summer temperatures over Eurasia through increased summer cloud cover, and more cloud-free skies in Eastern North America. This is corroborated by sea level pressure changes suggesting a negative phase of the Arctic Oscillation since the late 1990's.

The overall trend in greenness is positive throughout most of the Arctic (Epstein et al. 2018; Figs. 6.6 and 6.7). Regions with marked increase in greenness are the North Slope of Alaska, southern zones of the Canadian tundra and Eastern Siberia. Regions of marked browning are found in Western Alaska, the Canadian High Arctic, and the Northern and Northwestern Siberian tundra. NDVI greenness

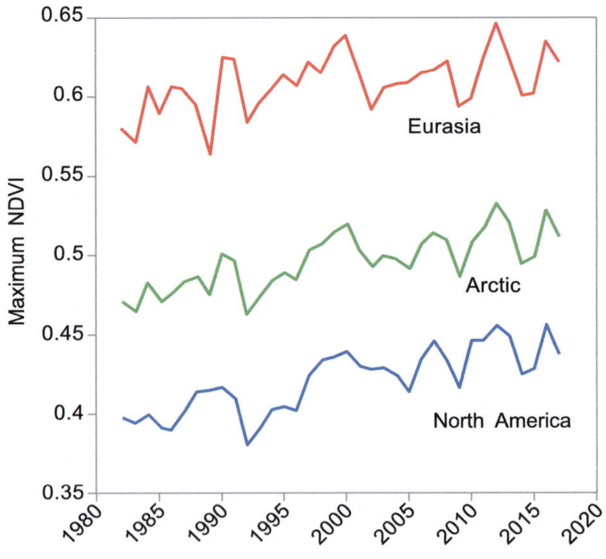

Fig. 6.6 Maximum NDVI trends for Eurasia (top), Arctic as a whole (middle) and North America (bottom). (From Epstein et al. 2018 (NOAA Arctic Report Card 2018); https://arctic.noaa.gov/Report-Card/Report-Card-2018)

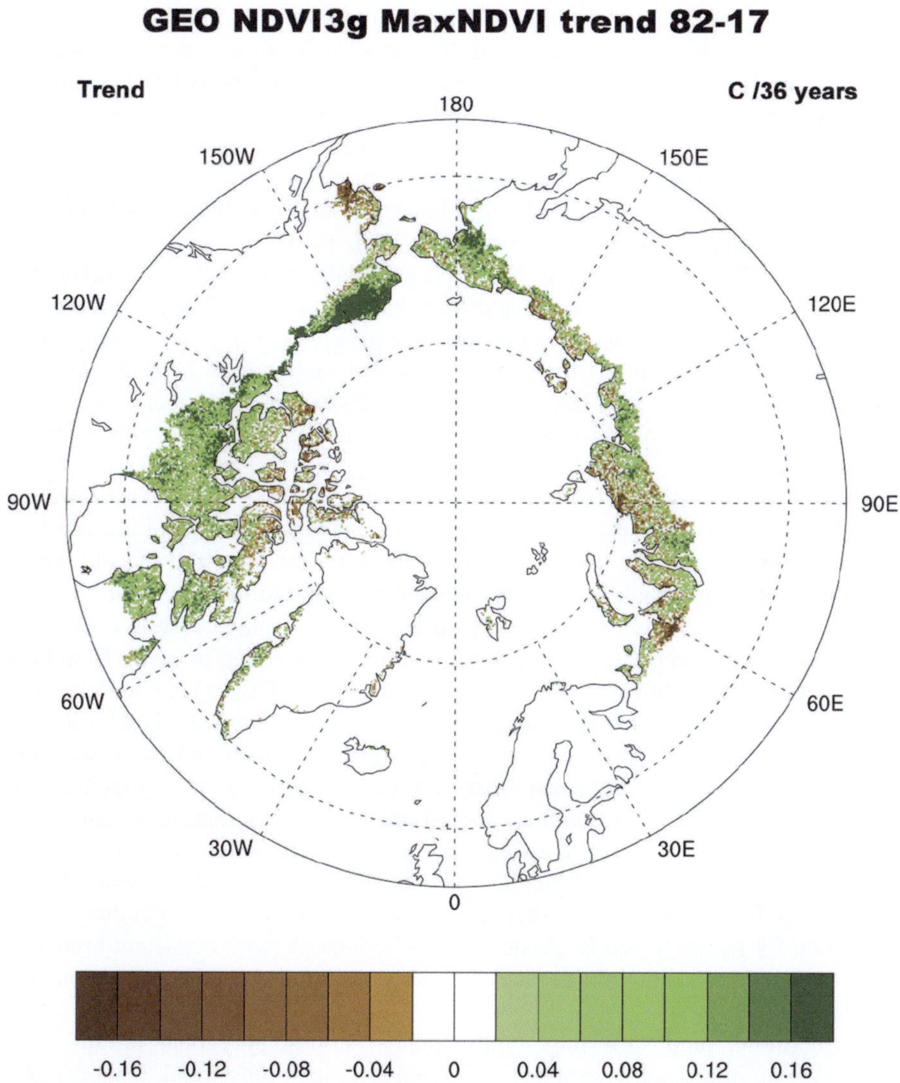

Fig. 6.7 Magnitude of the overall trend in maximum NDVI for the period 1982–2017. (From Epstein et al. 2018 (NOAA Arctic Report Card 2018); https://arctic.noaa.gov/Report-Card/Report-Card-2018)

increase in northernmost tundra zones is relatively modest compared to the summer warmth increase, while in more southerly tundra zones the NDVI increase is stronger (Reichle et al. 2018). Bhatt et al. (2010) correlate the general trend in NDVI (maximum NDVI and integrated NDVI over the summer) with sea-ice retreat. In particular summer sea ice concentration has decreased throughout the Arctic. A decrease in the spring and summer sea ice concentration increases summer warmth in the nearby land areas. The summer warmth index (SWI, the sum of the monthly mean temperatures above freezing in °C per month) over most tundra areas has increased. In turn, vegetation biomass expressed as NDVI has increased (Fig. 6.7), in line with the conclusions of Myers-Smith et al. (2015). Exceptions on the positive trend of NDVI are land regions around the Bering and Chukchi sea, that show a negative NDVI trend – browning. Here spring sea ice has increased during the studied period (Bhatt et al. 2010) or a larger amount of snowfall has resulted in longer duration of spring snow cover (Epstein et al. 2015).

Negative NDVI trends despite a warming climate is a puzzling phenomenon. Although NDVI gives a first order approximation of biomass changes, it does not give precise information on the actual changes in ecosystems. Satellite-based biomass estimates do not allow a clear separation of productivity between vascular plants and mosses, which is important in Arctic environments (Street et al. 2012). To consider the causes of Arctic browning in more detail, information on the ground is essential.

Phoenix and Bjerke (2016) list a number of mechanisms for Arctic browning. First, there are climate variations and extreme events that reduce plant growth: longer lasting snow cover in spring, caused by large winter snow fall or cold spring conditions; summer cooling, e.g. by persistent cloudy summer conditions, and extreme weather events in winter. Extreme winter events may cause low temperature stress, for instance due to a lack of protective snow cover, or result from high winter temperature extremes, for instance rain events leading to ice encasement of plants, and extreme winter warming leading to too early bud bursts (Bokhorst et al. 2011). Second, other browning events are caused by disturbances: tundra fire increase (Sect. 6.1.4); abrupt permafrost thaw, e.g. pond formation, cryogenic landslides, erosion (Chap. 5; Nitze et al. 2018, Lara et al. 2018), and insect outbreaks. Many of these natural events are localized or short-term disturbances rather than regional phenomena and are difficult to detect using satellite remote sensing (Nitze et al. 2018). For most natural events, the ecosystem may recover depending on the time scale for recovery and frequency of these events. A more persistent browning results from increasing human activities such as mining, industrialisation and settlement.

Furthermore, landforms (e.g. tundra ponds, slopes, polygonal terrain, river plains) are an important determinant of greening or browning trend, demonstrated by machine learning methods applied to a large remote sensing data set of the Alaskan North Slope (Lara et al. 2018). Therefore, understanding of how climate interacts with landforms is necessary to understand the future impact of climate change on vegetation. Future climate trajectories add another source of uncertainty. With this machine learning approach, also the future development was tested with different climate scenarios and climate model outputs. While some models predict considerable greening, two models predicted widespread browning under future climate conditions.

6.1.3 Feedbacks on Climate and Soil Temperature

The effects of vegetation on the surface heat balance have been discussed in Chap. 2. Table 2.1 lists the differences in heat balance characteristics from a number of ecosystem types on permafrost. In general, taller vegetation captures more energy from incoming solar radiation. The winter / early spring albedo decreases sharply from about 0.8 in tundra to 0.6–0.2 when tall shrubs and trees are present; also, the summer albedo is lower for taller vegetation. Within the tundra, small differences in shrubs protruding through the snow can generate an albedo decrease of 30% (Sturm et al. 2005).

The aerodynamic resistance is lower for tall canopies, facilitating more turbulent and latent heat exchange with the atmosphere. The sensible heat flux Q_H is larger with taller vegetation, while the fraction of net incoming radiation that is available for heating the ground decreases with vegetation height; from 0.12–0.22 for tundra, to 0.05–0.1 for tall shrubs and trees. Changes of the latent heat flux Q_E are less strongly linked to vegetation structure, since evapotranspiration also depends strongly on variation in precipitation and soil moisture (Boike et al. 2008; Lafleur and Humphreys 2018).

Despite the increased capture of radiative energy by tall vegetations by their lower albedo, the net effect is that less incoming radiation is available for warming the ground. Instead, the radiative energy is used for warming the vegetation canopy where it contributes to a higher sensible heat flux; the larger incoming radiative energy is mostly dissipated to the atmosphere. A cooler microclimate is created below the vegetation in summer. Less incoming radiation reaches the ground surface – the 'shading effect' (Loranty et al. 2018b). Cutting experiments have shown the importance of this shading effect, even for low stature *Betula nana* shrubs. In a statistical analysis of factors that influence the ALT, the tree canopy LAI was the most important factor with a negative influence on ALT, next to moss layer thickness (Fisher et al. 2016). Conversely, shrub establishment helps establishment of new permafrost (Briggs et al. 2014).

The effect of shrubs on active layer thickness and soil temperature is best demonstrated by a shrub removal experiment in which *Betula nana* shrubs were removed from an experimental plot in the tundra of Northeastern Siberia, described by Blok et al. (2010) and Nauta et al. (2014). The removal of the dense shrub canopy resulted in a rapid local ALT increase and surface subsidence (Fig. 6.8). After removal, the ALT increased by 9% during the next growing season. Most of the ALT increase was achieved during eight warm days in the summer. The differences in ALT between vegetation removal and control plots were also demonstrated by differences in the ground heat flux. Measurements of the radiation balance and ground heat flux indicated that the ALT increase was caused by a loss of soil shading by the shrub cover.

In the following years the difference in ALT between control and shrub removal plots increased further, from a difference of 5 cm on average in 2008 to 15 cm in 2012 (control plot ALT average 25.8 cm, experiment average 40.8 cm). By 2010,

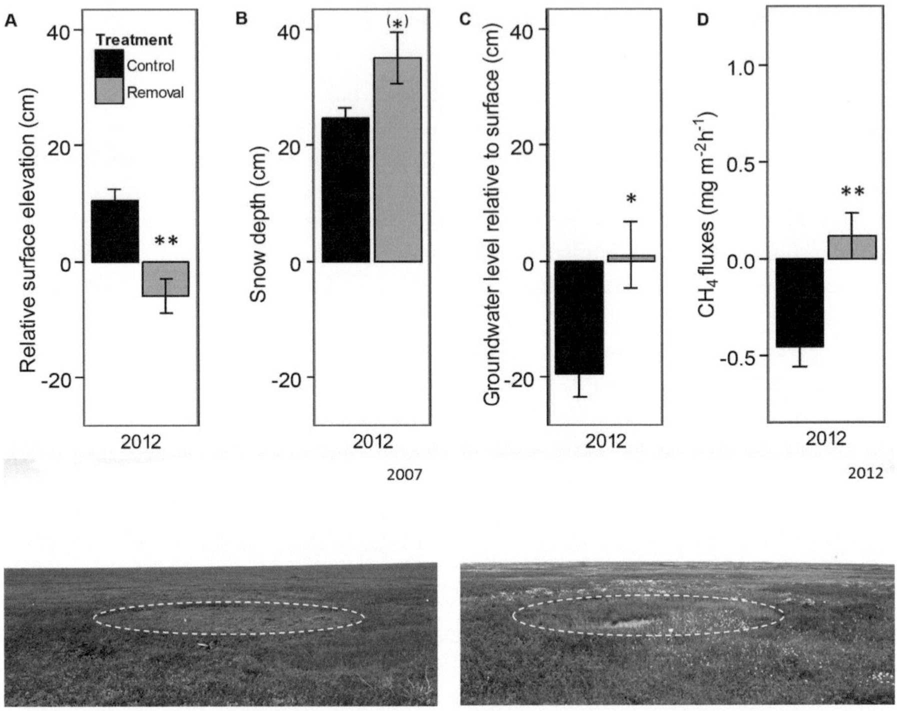

Fig. 6.8 Results of a vegetation removal experiment at Kytalyk Research Station, Indigirka Lowlands, northeastern Siberia. Above: Relative surface elevation, snow depth, groundwater level and methane flux in control and Betula nana removal plots in 2012, 5 years after removal of shrubs. (**a**–**d**) Data are mean values ± standard deviation for 5 plots, for relative surface elevation (**a**), snow depth in April (**b**), groundwater level in mid-July (**c**) and methane flux in early August (**d**). The legend in a applies to all panels. (∗), ∗ and ∗∗ indicate significant differences between the two treatments ($p < 0.10$, $p < 0.05$, $p < 0.01$, respectively). Positive methane (CH_4) fluxes indicate net emissions of methane to the atmosphere; negative values indicate net uptake of methane from the atmosphere. Below: one of the plots just after removal of shrubs, and in summer of 2012, showing presence of standing water. (From Nauta et al. (2014); photos provided by Dr. M.M.P.D. Heijmans. © Springer Nature AG 2014)

the surface of the removal plots began to subside, resulting in a lower surface and higher water table, and in most removal plots temporary or permanent pools developed. The surface subsidence also resulted in an about 10 cm thicker snow cover over the removal plots. The increased soil moisture – up to saturation – increased the thermal conductivity of the soil. Moreover, the soil moisture changes started to feed back on the vegetation; the vegetation became dominated by wetland graminoids. Soil subsidence continued at a pace of 1 cm $year^{-1}$ (Li et al. 2017a). The water-saturated soil resulted in emission of CH_4 which pushed the local ecosystem from a dry heath absorbing CH_4, into a pond emitting CH_4 (Nauta et al. 2014).

Two conclusions can be drawn from this experiment: first, expansion of shrubs in tundra ecosystems by climate warming retards ALT increase, or even decreases

ALT, by decreasing the ground heat flux in summer. Second, the opposite – disruption of a dense vegetation cover – may trigger ALT increase and surface subsidence. It demonstrates the presence of 'climate-driven, ecosystem protected' permafrost of Shur and Jorgenson (2007). However, it is questionable if an ALT decrease is a permanent effect of shrub expansion. The combination of larger sensible heat transfer to the air by taller vegetation and its lower albedo in spring and summer contributes to warming of the atmosphere. Furthermore, also other climate warming effects as earlier snow melt will ultimately increase soil temperature (Euskirchen et al. 2007).

A cut-over experiment in the Siberian boreal forest on continuous permafrost showed a more moderate response of the ALT (Iwahana et al. 2005). An increase in ALT by 14 cm (undisturbed ALT ±105 cm) occurred only during the first year, accompanied by an increase in soil moisture due to reduced evapotranspiration, and a larger active layer ice volume the next winter. The next 2 years the ALT remained stable, although the ground heat flux increased. The stabilization of the ALT was attributed to the latent heat required for thawing the larger amount of ground ice, acting as a buffer for permafrost degradation.

Shrubs tend to trap more snow than open tundra (Sturm et al. 2001; Liston and Sturm 2002; Lafleur and Humphreys 2018). The effect is to increase winter soil temperatures. Together with the shading effects in summer (cooling) the net effect is a smaller amplitude of the seasonal soil temperature range. In a snow cover – shrub manipulation experiment by Myers-Smith and Hik (2013), soil temperatures were 4–5 °C warmer in January, and 2 °C cooler in July under shrub cover. An increase in shrub cover therefore is likely to result in more winter warming by a thicker snow cover, compared to adjacent areas which are more open.

In the boreal forest, the snow is intercepted by the tree canopies, and canopy type (evergreen vs. deciduous) and tree density determine the snow depth and distribution on the ground (Rasmus et al. 2011). Denser forest results in a thinner forest floor snow cover, and individual trees also decrease snow depth below their canopy. The snow cover thickness also influences the forest floor vegetation; dwarf shrubs such as *Vaccinium* preferentially grow in thicker snow; some mosses prefer a thinner snow cover. Tree density changes also influence the understory vegetation by competition for light and nutrients. Loranty et al. (2018a) found that lichen cover in North Siberian larch forest was inversely related to crown density; the lichen cover in turn affected soil temperatures by its insulating properties. Change of the forest density in the future likely will affect snow cover depth and the undergrowth.

The presence of a layer of mosses and shrubs has a strong effect on the penetration of summer thaw into the soil because of the insulating properties of mosses already mentioned in Chap. 2 (e.g. Blok et al. 2011b; Turetsky et al. 2012; Soudzilovskaia et al. 2013). Fisher et al. (2016) found an exponential relationship between moss cover thickness and ALT in boreal forests; small changes in a thin moss layer had a larger effect than changes in a thick moss layer. The insulating effect of a moss layer on ALT is stronger than that of the thickness of the organic layer. For *Sphagnum* mosses, the productivity may be increased with thicker winter snow cover and higher summer temperatures as shown experimentally by Dorrepaal et al. (2004). Meta-analyses of warming experiments in the tundra biome (Walker et al. 2006, Elmendorf et al. Elmendorf et al. 2012) show that shrubs and graminoids profit from warming with taller growth and increase of cover percentages, but mosses and lichens do not profit from warming.

The most important factor in a decline of mosses and lichens is competition for light with taller growing vascular plants (Cornelissen et al. 2001). Besides the effect of moss and lichen ground cover on soil thermal conductivity, lichens also may play a role in the nutrient cycle by fixing N (Sect. 4.5.5) and are staple food for grazers such as reindeer (Wookey et al. 2009). However, the sensitivity of mosses to taller vascular plant growth depends on species. *Sphagnum* species are more resilient (Keuper et al. 2011; Turetsky et al. 2012).

It is difficult to gouge what the net effect of all ecosystem changes in boreal forest and tundra is on the ground heat balance, because of the many opposing effects (Loranty et al. 2018b). For instance, the soil shading effect of shrub cover in the tundra which reduces summer warming of the active layer may be offset by moss and lichen cover decline, as suggested by Walker et al. (2006). A decreased moss and lichen cover decreases the insulating effect of the soil cover, and decreases the albedo, both with a warming effect.

Treeline change and changes in forest composition represent a vegetation-driven warming feedback on climate change. A northward movement of the treeline, which replaces tundra with a high albedo with a relatively low albedo tall shrub and forest vegetation, causes local and regional warming (Bonan et al. 1995; MacDonald et al. 2007). The difference in heating between some boreal and tundra sites may be as high as 50 W m^{-2} (Pielke and Vidale 1995). Beringer et al. (2005) found differences between tundra and forest of 21 W m^{-2} at the Alaskan forest – tundra transition. Species change within the forest may have a similar warming effect; the replacement of deciduous Larch with evergreen conifers decreases the spring albedo (MacDonald et al. 2007). However, an analysis of summer warming in Alaska concludes that the contribution of vegetation change to warming is an order of magnitude smaller compared to the effect of lengthening of the snow-free season, based on satellite observations of albedo and surface temperature and modelling (Chapin et al. 2005; Euskirchen et al. 2007). Nevertheless, Chapin et al. (2005) expect that vegetation change will increase disproportionally in the future, while the effect of earlier snowmelt declines.

6.1.4 Carbon Balance Effects of Vegetation Change

Carbon balance effects of vegetation shifts consist of changes in above-ground biomass and soil carbon storage. It seems logical to assume that with a larger biomass, also the below-ground carbon store will increase. A larger above-ground biomass should result in a larger carbon input into the soil. However, the fate of that additional carbon depends on the conditions for decomposition (Sect. 4.5) which also change. These conditions can change the magnitude of ecosystem carbon uptake from the atmosphere, and even result in net carbon emission. The factors that influence soil carbon decomposition are altered by changed soil temperature, and changes in the quality of soil carbon input (Wookey et al. 2009; Malhotra et al. 2018). Several examples below demonstrate that greater above-ground biomass carbon does not necessarily result in more below-ground carbon, or even decrease of below-ground carbon storage.

Biomass changes in the Arctic have been derived from large scale remote sensing studies using vegetation indices such as NDVI (Sects. 4.2 and 6.1.1). In general, it is assumed that biomass does not change strongly between years due to weather conditions, and trends are not strongly influenced by these differences. However, in High Arctic environments, the year-to-year variability between biomass may be large, because the plants react more strongly on differences in growing season temperature and incoming solar radiation. In High Arctic Svalbard, measured annual biomass variations amount 23–46 g m^{-2} (Van Der Wal and Stien 2014). Another problem with biomass estimates from space is that the below-ground biomass is not included, while the below-ground biomass may be as much as 80% of total biomass in high-latitude environments (Blume-Werry et al. 2016).

Comparing remote sensing data with above-ground biomass data, Epstein et al. (2012) concluded that the southernmost tundra zones (C and D) in the circumpolar Arctic vegetation map of Fig. 4.4 gained 20–26% in biomass in nearly 30 years. In particular in Northern Alaska and Siberia, the increases in biomass are substantial, but in some areas of Western Siberia and Western Alaska, biomass showed small declines of 2–4% (Fig. 6.7). Assuming that the above-ground biomass represents an equivalent amount of below-ground biomass, the total increase of tundra biomass would result in an increase of 40 Pg of C – at most 0.2 Pg $year^{-1}$ – in vegetation biomass. This is only biomass increase; not the amount of carbon that is actually sequestered in soils and vegetation, since this also depends on turnover rates of dead plant material. The 40 Pg C is a conservative estimate because of the low proportion of below-ground biomass that is assumed here. This carbon uptake is dwarfed by the anthropogenic carbon emissions, but has a large impact on the tundra environment.

Myneni et al. (2001) inferred a large CO_2 sink in boreal forests from remote sensing studies between 1981 and 1999, amounting to 0.68 ± 0.34 Pg C $year^{-1}$; nearly 70% is located in Eurasian boreal forests. However, there are signs that the carbon sink in boreal forests is weakening. Stephens et al. (2007) found smaller northern sink fluxes, based on vertical CO_2 concentration profiles. Model-based estimates by Piao et al. (2008) show that northern terrestrial ecosystems (north of 25 °C) may be losing carbon by autumn warming. The higher autumn soil temperatures lengthen the active season for the decomposer community in the soil; the resulting carbon loss offsets the increasing spring uptake. Piao et al. (2011) report a decrease of the NDVI in Northern Eurasia between 1997 and 2006, related to decreased precipitation. Ma et al. (2012) have found a decline of the biomass carbon uptake in the southwestern part of the boreal forest zone in Canada, caused by drought. A similar drought related productivity decline has been found in other regions of North America (Chapin et al. 2005; see above).

Nevertheless, regionally the carbon mass in forest vegetation has increased. A repeated survey in Sweden showed increases of tree biomass (19%) in the Alpine Birch forest zone between 1997 and 2010 (Hedenås et al. 2011). Dolman et al. (2012) estimated the net atmosphere to land carbon flux over Russia (including former Soviet republics) at −0.613 Pg C $year^{-1}$ based on eddy covariance data, forest inventory data and global inversions. Forests represent 90% of this sink. The forest area increased, resulting in a flux of −0.26 Pg C $year^{-1}$ from forests in Asian Russia (largely forest on permafrost) and − 0.17 Pg C $year^{-1}$ in European Russia. Caution is needed however, carbon stock changes in boreal forests need not be climate-related but may result also from recovery from forest exploitation in the past (Kauppi et al. 2010). About two thirds of the boreal forest is currently under some form of management, mostly for wood production. Most of the unmanaged forest occurs in the higher northern latitudes, which includes also areas on permafrost (Gauthier et al. 2015).

In shrub tundra, the growing season CO_2 sink measured by eddy covariance is related to the density of shrubs and therefore may be expected to increase with increasing shrub cover (Lafleur and Humphreys 2018). An extension of the growing season also leads to increased gross primary production (*GPP*). However, analysis of multi-year variability of the summer CO_2 sink in Siberian tundra showed that the increase of *GPP* by longer growing seasons was compensated by increasing R_{eco} (Parmentier et al. 2011). In open top chamber experiments in Alaskan tundra, warming generally increased *GPP*, but also R_{eco}. In particular in dry ecosystems this resulted in a net CO_2 source, while for moist and wet ecosystems the sign varied (Oberbauer et al. 2007).

Many estimates of biomass carbon change include only the above-ground biomass, while the below-ground biomass is considerably larger in cold climate ecosystems. The growing season length for roots and shoots differs considerably. In Alpine tundra in Northern Sweden, the growing season for roots extends well past the growing season for above-ground shoots (Blume-Werry et al. 2016). The above-ground growing season finishes end of July to early August, while the below-ground growing season extends to early September, resulting in an approximately 50% longer below-ground growing season. The above-ground plant parts have their growth peak in early summer. However, while the above-ground parts of plants stop growing and become senescent at the end of the summer, fine roots continue to grow. This allows plants to take up more nutrients in an otherwise nutrient-poor soil. Plants profit in this way from higher autumn temperatures in the soil, which are generally higher than the air temperatures. This phenomenon is poorly represented in vegetation and soil carbon models (Blume-Werry et al. 2016).

The soil ecosystem is affected in several ways by changes in vegetation composition, which in turn feed back on soil carbon storage and carbon fluxes. Besides the effects of vegetation on soil temperature and ALT that influence the soil environment (Sect. 5.2), there are a number of factors that are strongly related to plant traits, productivity and carbon allocation (Wookey et al. 2009, Fig. 6.9):

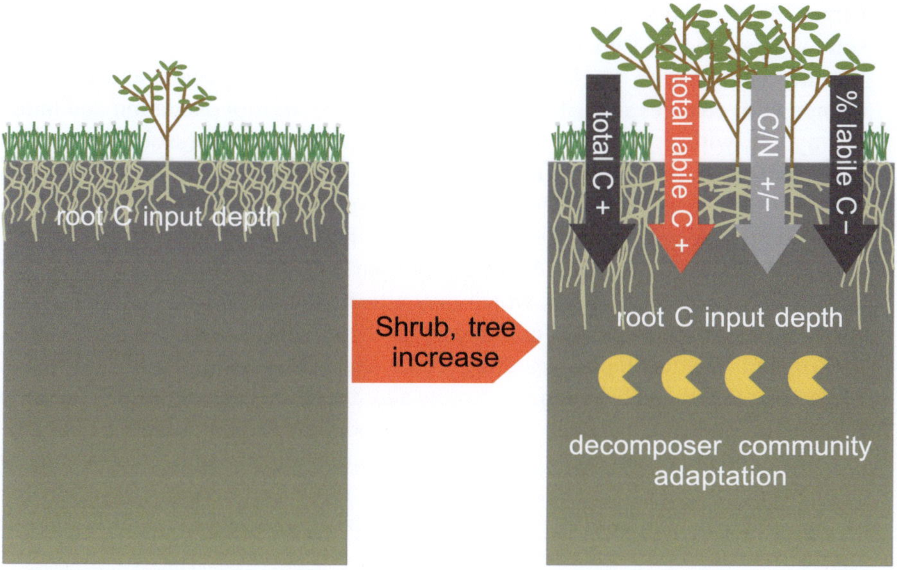

Fig. 6.9 Changes of carbon quantity, quality and carbon allocation with shrub / tree expansion. Black arrows indicate a potential carbon conserving effect, red arrows a potential carbon loss effect

- Changes in the amount of carbon input into the soil from vegetation (e.g. Sistla et al. 2013; Zeh et al. 2019).
- Changes in the nutrient content of the litter input to the soil. Woody material has a lower N content than that of leaves, and evergreen leaves have a generally lower N content than deciduous leaves (Cornelissen et al. 2007a).
- Lability/recalcitrance of the organic matter input the soil. Next to nutrient content, also the organic matter structural components differ in lability (Sect. 4.5). Lignin content is a good indicator for recalcitrance, being relatively recalcitrant material. An increase of the amount of woody biomass will affect decomposition rates negatively. In general, plant biomass composes more slowly in the order forbs > graminoids ≥ deciduous shrubs > evergreen shrubs (Wookey et al. 2009). Cornelissen et al. (2007a) suppose that the ongoing expansion of shrubs with recalcitrant leaf litter increases carbon storage in the soil (however, see Sect. 6.2.2 below). In peat soils, CO_2 emissions are strongly influenced by organic matter biochemistry, influenced in turn by past and present organic matter input from vegetation (Sjögersten et al. 2016). Mosses, in particular *Sphagnum* mosses, decompose slowly (Hobbie et al. 2000). In peatlands, overall decomposition rates are determined by the litter mass ratios of vascular plants and mosses (Dorrepaal et al. 2007).

- Root exudation varies markedly between plant species (Wookey et al. 2009). A change of the depth and amount of input of labile organic compounds may increases overall decomposition rates of stored soil organic matter (Zeh et al. 2019), likely driven by the priming effect (Sect. 4.5.2).

- Effects of litter on soil pH; *Sphagnum* litter adds acidifying components to the soil (Verhoeven and Toth 1995; Limpens et al. 2017).
- Rooting depth varies markedly between different plant species, and also will change with changes in ALT. This changes patterns of root exudation and nutrient uptake within the soil. Dwarf shrubs exploit shallower soil layers than grasses, grasses profit more from an increase in ALT (Wang et al. 2018). Also changes in the seasonal distribution of root development, or root system phenology, can be expected. Changes of the root distribution in the soil brings the depth of carbon allocation to upper/warmer/drier parts of the soil profile or to lower/cooler/wetter parts of the soil, which may affect decomposition rates.
- Adaptations of the microbial community to the quality, amount and location of the organic matter input, resulting from the changes in the root system. Micorrhizal relations between plant roots and microbial population will adapt to the presence of new species. The microbial population also adapts to the biochemical and structural composition of the organic matter input. Organic matter of plants decomposes faster in soils where they grow than in soils where they are absent. This adaptation of the microbial community to organic matter input has been demonstrated also for Arctic species (Wang et al. 2017).

A long-term experiment, a 20 year summer greenhouse warming experiment in Alaskan tundra (Sistla et al. 2013) clearly shows that an increase of above-ground biomass does not increase the amount of soil carbon, only its distribution. Soil respiration increased in the first years of the experiment but declined afterwards. Soil carbon measurements after 20 years showed that total soil carbon did not change, but increased in the deeper mineral soil and decreased on the organic topsoil horizons. This was combined with a 67% increase in biomass, with a near doubling of *Betula nana* shrub cover, at the expense of mosses and lichens, and a larger carbon input from vegetation litter and root growth into the soil. At the same time there is a pronounced interaction of the vegetation with the heat balance. The shrub cover moderated an increase of soil temperatures by the effect soil shading in summer. In the early years of the experiment, the topsoil warmed in summer by 2 °C, in later years soil warming in summer was negligible. Winter warming of about 2 °C occurred as a result of increased snow cover, trapped by the higher vegetation.

Another illustrative example is seen in the mountain Birch zone at the tree line in Northern Scandinavia, which has increased its productivity (Hedenås et al. 2011). Permafrost is absent in the mountain Birch zone while in the tundra discontinuous permafrost occurs. During the Holocene thermal maximum, the Birch zone has extended further upslope, and is now regaining ground again. The understorey of the birch forest is composed of the same plants as those in the tundra. However, the total ecosystem carbon stock in the tundra heath is larger than that of the adjacent birch forest, while the birch forest has a larger above-ground stock (Fig. 6.10). Based on ^{14}C analysis of soil C and respired CO_2 from vegetated and unvegetated plots in both vegetation types, Hartley et al. (2012) found evidence for a strong priming effect (Sect. 4.5.2). This means that the decomposition of old soil carbon in the birch forest soil is enhanced by the addition of recent, labile photosynthates

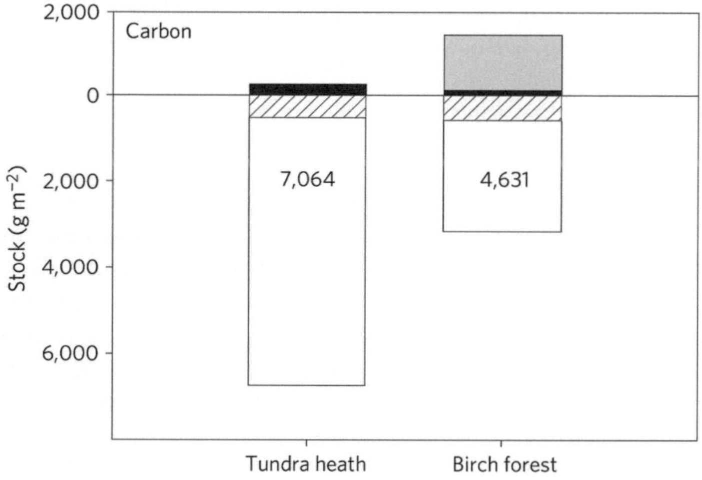

Fig. 6.10 Carbon stocks of tundra heath and mountain birch forest. The stocks are divided into above-ground tree biomass (grey fill), above-ground shrub biomass (black fill), roots (diagonal hash fill) and soils down to the underlying parent material (white fill). Above-ground stocks are shown above the x axis; below-ground stocks are shown below the x axis. The total ecosystem stock values (g m^{-2}) are provided within each soil bar. All differences between the two ecosystems (except root C stocks) were statistically significant ($P < 0.001$). (From Hartley et al. 2012, Copyright © 2012, Springer Nature)

excreted by the root system of the birches. An alternative explanation could be a lower soil moisture induced by higher evapotranspiration of the trees, but there were no clear soil moisture differences. Winter fluxes were not included in this experiment. It is therefore quite well possible that a thicker snow cover and warmer winter soil results in higher carbon loss in the birch forest compared to the tundra heath. Nevertheless, the priming effect should have contributed significantly to the soil carbon decomposition in the birch forest.

An upwards shift of carbon allocation by a change from deeper rooting sedges towards shrubs with a shallower root system also could decrease the below-ground carbon storage. In a long-term fertilization experiment in Alaska, described by Mack et al. (2004) and Sullivan et al. (2007), almost 2 kg m^{-2} was lost in 20 years from the ecosystem; an increase in biomass carbon did not compensate for below-ground losses. Sullivan et al. (2007) estimated that the shift from a deep-rooting system to a shallow rooting system could account for 20–39% of the soil carbon loss; there was 50% less root biomass in the deeper soil layers than in unfertilized plots. Other probable causes for the carbon loss are higher decomposition rates in the topsoil due to higher temperature, and stimulation of decomposition by higher nutrient availability (Mack et al. 2004).

Another source of information is the soil carbon store in current vegetation zones. Mapping of soil organic matter stores in Siberian subarctic tundra and Larch/Pinus/Birch forest on continuous permafrost in eastern Siberia, shows less soil carbon in the forest, 23.4 kg C m^{-2} for the boreal forest and 27.9 kg C m^{-2} (down to 1 m depth) for the tundra (both for non-wetland terrain classes and including fine roots; Siewert et al. 2015; Fig. 4.19). Similar data from Palmtag et al. (2015) show less carbon in Larch woodland soils (17.3 kg C m^{-2}) than in tundra (23.7 kg C m^{-2}). Differences between tussock tundra and shrub tundra vegetation classes are insignificant.

Given these data, and the observations of Sistla et al. (2013), Mack et al. (2004) and Hartley et al. (2012) above, it is unlikely that a transfer of tundra to shrubland or boreal forest on well drained areas will substantially increase the soil organic matter store, and a decrease could occur as well. The climate-driven increase in biomass is a shift from a low above-ground biomass carbon store combined with slow below-ground carbon turnover, towards more above-ground biomass carbon and faster below-ground carbon turnover. The data from Hartley et al. (2012) suggest even a net ecosystem carbon decrease with forest expansion, by enhanced soil respiration (Fig. 6.10).

Ecosystem carbon changes in the High Arctic are equally uncertain. Palmtag et al. (2015) note a large difference in soil carbon content between barren (average 2.7 kg C m^{-2}), Arctic heath (15.9 kg C m^{-2}) in Northeast Greenland and Subarctic tundra in Northeast Siberia (23.7 kg C m^{-2}). Warmer (and wetter) climate in the High Arctic potentially may increase soil carbon stocks by increase of vegetation productivity. However, Epstein et al. (2012) found only moderate changes of the vegetation biomass in High Arctic barrens (0.41 g m^{-2} $year^{-1}$) compared to tundra vegetations (1.68–4.61 g m^{-2} $year^{-1}$). The strongest increase occurs in wet, non-tussock sedge, dwarf-shrub, and moss tundra. Restrictions in biomass increase in barrens are soil moisture or nutrient limitation. Changes in soil carbon decomposition rates in high-arctic semi-desert ecosystems may be strongly dependent on water availability; warming alone decreases the C sink strength, while a combination of warming and wetting increases the ecosystem C sink (Lupascu et al. 2014).

Wetlands in the tundra and boreal forest are environments where increased vegetation biomass may result more easily in below-ground carbon storage in peats, because of the reducing soil conditions. Epstein et al. (2012) see a relatively high increase in biomass in tundra wetlands (>2 g m^{-2} $year^{-1}$). The most active peat formers in the peatland, *Sphagnum* mosses, need not be affected by competition with vascular plants; *Sphagnum* mosses have the ability of competing with length growth with vascular plants (Keuper et al. 2011). Moreover, *Sphagnum* is a strong competitor for nutrients with shrubs, and regulates soil moisture. Provided that permafrost degradation, severe drought or fire does not disrupt *Sphagnum* ecosystems, these may be quite resilient to climate change and continue to sequester carbon (Sect. 6.2.3).

Tarnocai (2006) analysed the potential effects of climate change on Canadian peatlands. In the Mid and High Arctic wetlands, increased primary production by higher temperatures, longer growing seasons, higher moisture availability and higher CO_2 concentration could contribute to increased peat accumulation. These are also the regions where discontinuous and continuous permafrost are at least theoretically less vulnerable to thaw. However, the more southerly peatland regions will be severely affected by permafrost thaw and drought; for coastal peatlands (e.g.

the Hudson Bay Lowlands), also flooding by rising sea level will contribute to peat loss.

Many of the permafrost features at the southern limits of boreal peatlands can be considered as relict, and because of their higher surface elevation and dry conditions at their surface, may limit the carbon uptake by an increase of peatland vegetation productivity in warmer and more CO_2-rich conditions. Degradation of ice-rich palsas and peat plateaus is already proceeding at a high pace and will result in significant carbon loss from peatlands (Sect. 5.3.1, O'Donnell et al. 2012). In the resulting thaw depressions, this is followed by regrowth of plant biomass and accumulation of fresh peat (Turetsky et al. 2007). The vegetation change and carbon balance of permafrost degradation will be discussed in more detail in Sect. 6.2.

6.1.5 Fire

In the late summer of 2007, an unprecedented tundra wildfire, the Anaktuvuk River fire, burned 1039 km^2 tundra on the North Slope of the Brooks Range in Alaska (Mack et al. 2011). The ecosystem carbon loss amounted 2016 ± 435 g C m^{-2}, of which ±72% was soil carbon. The entire burned area released 2.1 Tg C to the atmosphere during the fire, which is at least half of the modelled annual net C sink for the entire Arctic tundra biome (3–4 TgC $year^{-2}$, McGuire et al. 2009).

Wildfire incidence is related to climate, but has in turn a large impact on the climate by the emission of climate-warming greenhouse gases (CO, CO_2, CH_4), heat absorbing black carbon (soot) particles, and aerosols that have a cooling effect. Black carbon, when deposited on snow, decreases the albedo. In permafrost regions, fires can trigger thaw of permafrost (Kasischke et al. 2005; Van der Werf et al. 2010, 2017).

Generally, fire frequency and extent decrease sharply going from the boreal forest towards the tundra (Nitze et al. 2018). However, wildfire is becoming more frequent in tundras. Fires are a common feature for many northern ecosystems, also without human intervention. Natural fire frequencies are related to climate change (Kuhry 1994) and change also as a result of human activities (Flannigan et al. 2005, 2009; Kasischke et al. 2010). Fires are ignited by humans for various accidental or purposeful reasons. Lightning is a natural source of fire ignition. Lightning-ignited fires can burn large areas because they also occur in remote areas. In Canada, lightning-ignited fires contribute to 85% of the burned area, while representing only 35% of the fires reported in Canada (Podur et al. 2003; Kasischke et al. 2010).

Fire activity is influenced by weather conditions, amount of fuel in the shape of dead or living biomass, its mode of ignition and human fire control. The impact of fire on forest or tundra ecosystems is determined by fire frequency, fire size, intensity (amount of energy released by the fire), the season in which it occurs, and fire type. Fire types are tree crown, surface, and ground or peat fires, which are controlled by the intensity and fuel characteristics, that are in turn determined by vegetation structure and moisture (Flannigan et al. 2000).

Fire interval records are derived from historical records, paleo-records of charcoal layers in lake sediments, and surveys of past fire damage in forests and by remote sensing. Fire intervals show a wide variation from decades to millennia (Kuhry 1994; Katamura et al. 2009; Higuera et al. 2011; Kharuk et al. 2011; Jones et al. 2013; Hu et al. 2015). Boreal fire intervals typically range from ca. 50 to 500 years. Kharuk et al. (2011) report fire intervals in Siberian taiga forest of 130 to 350 year with a 200 ± 50 mean. Fire intervals at high latitudes are longer than at lower latitudes. Charcoal particles in tundra peats indicate that tundra fires also occurred in the past (Teltewskoi et al. 2016). Fire intervals in the Low Arctic shrub tundra are in the order of 140–480 year (Higuera et al. 2011). Further north, biomass is lower, and fire rarely occurs.

In particular peat fires emit high amounts of CH_4; in boreal regions with peatlands a large amount of carbon emission emanates from burning peat. The released amount of C is difficult to estimate because of uncertainties in burn depths (Van der Werf et al. 2010). Between 1997 and 2004, the average annual global fire emissions were 2.5 Pg C year^{-1}, of which boreal regions contributed about 9%. Boreal North America emitted between 1997 and 2004 44 ± 34 Tg C year^{-1}, Boreal Eurasia 188 ± 133 Tg C year^{-1} (Van der Werf et al. 2010; Fig. 6.11). Some of the burnt organic matter is also preserved in soils. Charcoal is an ubiquitous component in permafrost soils in boreal forests and is considered as recalcitrant organic matter in soils. It accumulates in soils and sediments (Guggenberger et al. 2008; Katamura et al. 2009). Besides the emissions, the change of vegetation and soil surface by fire alters the surface energy balance, initiating permafrost thaw (Shur and Jorgenson 2007; Jones et al. 2015).

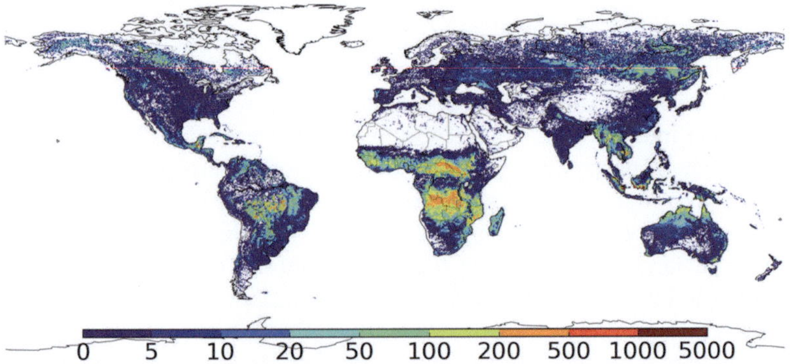

Fig. 6.11 Mean annual fire emissions (g Cm^{-2} year^{-1}) averaged over 1997–2004. Model results by Van der Werf et al. (2017). Earth System Science Data 9, https://doi.org/10.5194/essd-9-697-2017, © Author(s) 2017, published under Creative Commons License 3.0 https://creativecommons.org/licenses/by/3.0/, image provided courteously by the authors. Image provided courteously by Prof. Van der Werf

An important post-fire indirect effect is increase of the albedo of the snow cover in winter and spring, by partial destruction of tree or shrub cover. In the first years after the fire, the recovery of the vegetation by low vegetation (grasses and forbs) decreases the summer albedo again, relative to unburnt forest and the charred burn floor. The result is a local cooling effect on the climate. Randerson et al. (2006) calculated the net radiative effect of a forest fire in Alaska. They concluded that the radiative effect of higher albedo and carbon uptake in the regenerating vegetation in de decades after the fire, was higher than the radiative effect of the fire emissions and black carbon deposition on snow. Averaged over an 80 year period (the time it takes for the forest to regenerate), the net radiative effect was climate cooling: -2.3 ± 2.2 W m^{-2} year^{-1}. Averaged over long periods therefore, the net effect of a normal fire cycle in upland boreal forest needs not be a contribution to warming of the climate. However, the net radiative cooling effect is marginal, and the effects on permafrost degradation and related greenhouse gas emissions, the large emission from burning peatland (Turetsky et al. 2011a, b), and recent acceleration of the fire cycle may offset the results of Randerson et al. (2006) towards net warming.

The type, severity, extent and recurrence interval of fires strongly differs with weather and ecosystem characteristics.

- The amount and distribution of fuel, which is determined by the ecosystem. This is determined by the biomass, its quality (living or dead biomass), the ecosystem structure (the amount of biomass in tree crowns, in undergrowth, mosses, and organic soils), and the size of biomass elements (tree boles which burn slowly and incomplete or thin branches and needles, which burn completely). In evergreen needle leaf forests, more biomass is present in tree crowns resulting in crown fires, while in deciduous Larch forests a part of this biomass is transferred to the ground surface each year, which favours ground fires (Kasischke et al. 2005; Forkel et al. 2012).
- The soil, which consist of varying thickness of organic layers or peat deposits. It determines how much soil carbon may be lost and whether smouldering peat fires may contribute to high CH_4 and CO emissions. For the forest fire science community the organic soil horizons, with their intricate ecosystem, are known simply as 'duff'.
- Permafrost protects peats against deep burning to some extent, by higher soil moisture and the latent heat effect of thawing ground ice (Kasischke et al. 2010). Depending on the soil moisture conditions and nature of the organic surface layer up to 30 cm of organic material can be burned (Kasischke et al. 2005; Jones et al. 2015). Soil carbon loss in forest burns in Alaska in ice-rich, yedoma-like uplands was negligible after fire, while soil on rocky uplands showed loss of permafrost, increased subsurface drainage, drier soil conditions and significant soil carbon loss (Jorgenson et al. 2013).
- Fuel exhaustion after a fire and slow regrowth of vegetation – new fuel stores – determine the fire return interval of ecosystems (e.g. Hu et al. 2015). Fire frequency records therefore show large differences between eco-regions.

- The weather conditions determine the incidence of lightning that ignite fires, the moisture conditions of the fuel, and winds that may spread the fire over large distances. Low moisture and high winds cause severe, high energy fires which can burn deeply into the soil and spread over a large distance (Flannigan et al. 2005; Turetsky et al. 2012; Forkel et al. 2012; Hu et al. 2015). Wet soil conditions restrict burnt area (Bartsch et al. 2009). In the Siberian taiga, the relation of soil moisture and fire with precipitation and evaporation proves not straightforward; it may include effect of dry periods in the previous year, carrying over low soil moisture conditions to the next year (Forkel et al. 2012).
- Fire regimes display a seasonality. In Siberian Larch forest, most fires start early in the growing season, when vegetation is withdrawing water from the soil at a high rate and the soil has still not thawed deeply, which limits the water supply (Forkel et al. 2012; Ohta et al. 2008). In tundra environments, standing dead graminoid biomass in early summer is an easily flammable fuel stock.
- Landscape characteristics may promote or discourage fires, e.g. lakes can act as effective fire barriers (Hu et al. 2015); terrain slope orientation also may facilitate spreading of fire (Forkel et al. 2012).

These factors also illustrate the relation of fire incidence and fire carbon emissions with climate. Statistical analysis of climate and fire data in Alaska shows that summer precipitation and temperature explained 90% of the variability of annual burnt area for tundra fires. Increase of fire incidence in the tundra shows a threshold behaviour, with thresholds at approximately 11 °C average summer temperature and 150 mm summer precipitation. Above the temperature limit and below the precipitation limit, the burnt area in tundra rises markedly (Hu et al. 2015; Fig. 6.12). Warmer summer temperatures cause more thunderstorms (e.g. Hu et al. 2015). Balzter et al. (2005) found significant relations of burned forest area in Siberia with the Arctic Oscillation, summer temperatures and precipitation; Jupp et al. (2006) found a relation with early summer rainfall. For Canada, temperature and fuel moisture relates with monthly burned area (Flannigan et al. 2005). Experimental drying of peat soils has been shown to increase carbon loss from burning boreal peat, which may be expected to increase in future warmer climates (Turetsky et al. 2011b). A longer summer season in Alaska and Canada results in a shift to late season fires. In permafrost areas such fires can burn deeper into the soil because of deeper ALT (Flannigan et al. 2005; Kasischke et al. 2010; Turetsky et al. 2011a).

Many publications indicate a recent change in the fire regime in boreal forests: increase in the incidence of fires, severity of fires, size of fires and seasonal shifts (e.g. Kasischke et al. 2005, 2010; Turetsky et al. 2004, 2011a; Kasischke and Turetsky 2006; Tchebakova et al. 2009; Mack et al. 2011) and tundra (Hu et al. 2015; Jones et al. 2015). Higuera et al. (2011) concluded from a record of charcoal layers in lake sediments that tundra fire frequency was higher than based on historical sources. The estimation of fire return intervals based on relatively short historical records is difficult, given the typical fire return intervals in the order of tens of years, and large annual differences in burnt area. Nevertheless, also paleo-records show relations of fire frequencies with climate throughout the Holocene (Kuhry

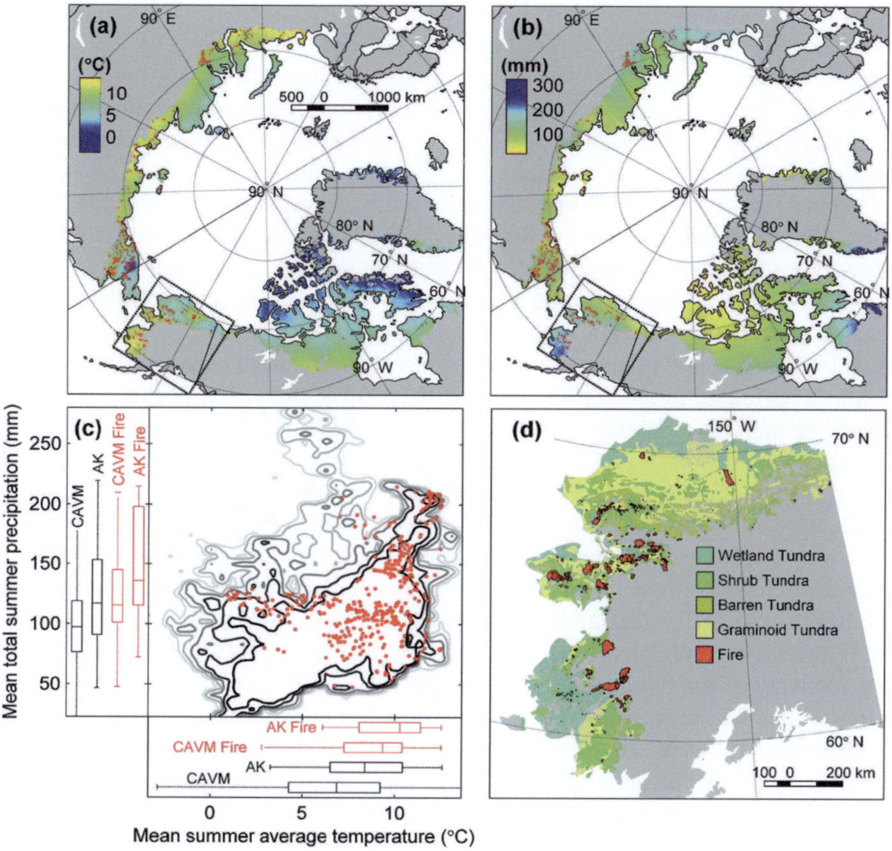

Fig. 6.12 Fire and climate in Arctic tundra, as depicted by Hu et al. 2015, (based on CAVM Team 2003; Walker et al. 2005; Harris et al. 2014). (**a**) Mean summer average temperature and (**b**) mean total summer precipitation (both 2000–2012 CE); (**c**) climate space (as range of mean summer average temperature and mean total summer precipitation) occupied by circumpolar Arctic tundra, and tundra fire occurrences. Lighter (darker) contour lines depict infrequently (frequently) occurring climate space for Arctic tundra; (**d**) spatial distribution of Alaskan tundra fires from 1950–2013 (Alaska Interagency Coordination Center, http://fire.ak.blm.gov, viewed 7 Feb 2014). Red points in panels (**a**), (**b**), and (**c**) represent pixels with nonzero burned-area estimates between 2002 and 2013 by the MCD64A1burned-area-data product, taken from the GFED4 dataset (Giglio et al. 2013). (Republished with permission of The Ecological Society of America, from Frontiers in Ecology and the Environment 13(7), 'Arctic tundra fires: natural variability and responses to climate change' by Hu et al. (2015); © The Ecological Society of America; permission conveyed through Copyright Clearance Center, Inc.)

1994). Fire frequency in Siberian Larch forest was lower during the Little Ice Age (Kharuk et al. 2011).

Based on relations of fire with climate, in particular temperature and moisture, increase of future fire activity is expected for boreal and Arctic areas (Balshi et al. 2009; Flannigan et al. 2005); additionally, fires ignited by human activity are expected to increase (Kasischke et al. 2010). Balshi et al. (2009) modelled an increase of decadal total carbon emissions from fire of 2.5–4.4 times by 2091–2100, relative to the last decade of the twentieth century, depending on greenhouse gas emission / climate scenarios. Uncertainties in future estimates of carbon emissions by fire are the effects of post-fire vegetation succession, changes in soil carbon by permafrost thaw, and the effectivity of human efforts to curb fires.

Vegetation recovery may start soon after a fire, depending on the severity of the burn. Many plant species have traits which are adapted to a stable fire regime and allow quick regeneration from roots if not burnt severely (Johnstone et al. 2010; Forkel et al. 2012). Regeneration also depends on soil type and amount of burnt organic matter. Mineral soils are good seedbeds for establishing new seedlings, a thick layer of charred organic matter not. Boreal forests have been markedly stable despite a regime of frequent fires. Re-establishment of trees may be slow and take a decade or more to accelerate. However, rapid establishment of high density stands thereafter may result in rapid vegetation carbon accumulation to a level similar to that of mature stands (Alexander et al. 2012).

Changes in the fire regime, leading to more frequent fires or more destructive fires, may affect the resilience of ecosystems to fire strongly and result in a change of species composition, e.g. from conifer to deciduous forest (Johnstone 2006; Johnstone et al. 2010). In Southern and Central Siberia, increased fire frequency in the future may shift taiga forests to grasslands (Tchebakova et al. 2009). In the tundra, recovery of the biomass is rapid, assisted by nutrients released from ashes; in 5–10 years surface greenness as determined from satellite data returns to stable levels again (Hu et al. 2015). Tundra fires facilitate shrub establishment (Myers-Smith et al. 2011), and older fire scars are visible from increased shrub growth many decades after the fire (Jones et al. 2013). However, lichens regenerate very slowly after fire (Hu et al. 2015).

The effects of fires on permafrost are multiple and large, governed by changes in surface heat balance, soil temperature and ALT. A + before each paragraph below indicates a soil warming effect, a – soil cooling, and ± uncertainty.

- + The *summer albedo decreases* strongly shortly after the fire, by the presence of black, charred soil and vegetation. The tundra fire described in the first paragraph in this section decreased the albedo by 50–71% (Mack et al. 2011).
- – In post-fire years, vegetation succession gradually will restore the summer albedo. Altered vegetation structure and composition *increases summer albedo* compared to the original vegetation; this may persist for decades (Randerson et al. 2006; Rocha et al. 2012).
- – The winter – early *spring albedo increases* because of less interruption of the snow cover by vegetation (e.g. Randerson et al. 2006).
- + The thickness of the *surface organic layer is reduced* including its insulating properties, resulting in a stronger ground heat flux, both in winter and summer (Liljedahl et al. 2007; Nossov et al. 2013; Jafarov et al. 2013).
- + The *sensible and latent heat fluxes are changed* by a simpler vegetation canopy structure after the fire. The effect differs with the original vegetation structure

(tundra vs. forest) and interacts with albedo change (Chambers et al. 2005). In the tundra, fire has a relatively small effect on the vegetation canopy structure and hence, on the turbulent energy fluxes; the effect of albedo change dominates. In forest, the surface becomes considerably smoother (strong decrease in roughness length). The changes in the turbulent fluxes are therefore more substantial, causing a reduction of the latent heat exchange. The albedo decrease increases the surface temperature, which results in an increase of both longwave radiation emission and sensible heat flux; the evapotranspiration is also decreased by loss of transpiring leaf surface. Altogether, the surface receives more incoming radiation but cannot exchange this radiation effectively with the atmosphere. Surface temperature rises, and more energy is available for the ground heat flux, which is enhanced by increased soil moisture. In general, a decrease of a protective vegetation cover increases ALT (Sect. 2.3).

- ± A reduced shrub canopy is less effective in *snow trapping in winter*, which may result in a thinner snow cover (Liljedahl et al. 2007). However, fires also facilitate the growth of shrubs in tundra environments (Jones et al. 2013).
- ± *Soil moisture may increase* by reduced evapotranspiration (Liljedahl et al. 2007), although this is modified by soil hydrology. The water storage space in the active layer is increased also by increasing thaw depth (Nossov et al. 2013). Permafrost degradation after fire may result in *increased drainage* and decreasing soil moisture (Liljedahl et al. 2007; Jones et al. 2015).

Altogether, an increase in thaw depth and soil temperature is common, that may persist over decades, as a result of the changed surface heat balance (Liljedahl et al. 2007; Jafarov et al. 2013). Nossov et al. (2013) report a doubling of thaw depth, an increase in mean annual soil temperature at the surface (from −0.6 °C to +2.1 °C) and at 1 m depth (−1.7 °C to +0.4 °C), and a two-fold increase in net soil heat input at a burned boreal forest site. The higher soil temperatures may trigger further permafrost degradation, amplifying permafrost carbon loss (Hu et al. 2015). A permafrost collapse pond in Alaskan boreal forest expanded significantly after fire (Myers-Smith et al. (2008). Charcoal layers indicated that this collapse pond also was initiated by an earlier fire (Fig. 6.13).

Jones et al. (2015) followed the effects on permafrost thaw of the Anaktuvuk river fire mentioned above, with LIDAR and INSAR mapping of surface elevation changes, 2 and 7 years after the fire when vegetation restoration was underway. Thaw subsidence was detected in 34% of the burned area studied, compared to less than 1% in similar undisturbed tundra. In the first year after the fire, active layer detachment slides and retrogressive thaw slumps were initiated. These stabilized within 5 years. However, the LIDAR data showed large scale subsidence and ice wedge thaw (Fig. 6.14), in particular over ice-rich Yedoma deposits and severely burnt areas. Seven years after the burn, the ground thermal profile was still adjusting to the heat balance changes at the surface. The changing microtopography was accompanied by gully development. Liljedahl et al. (2007) also report severe gully erosion after a tundra fire.

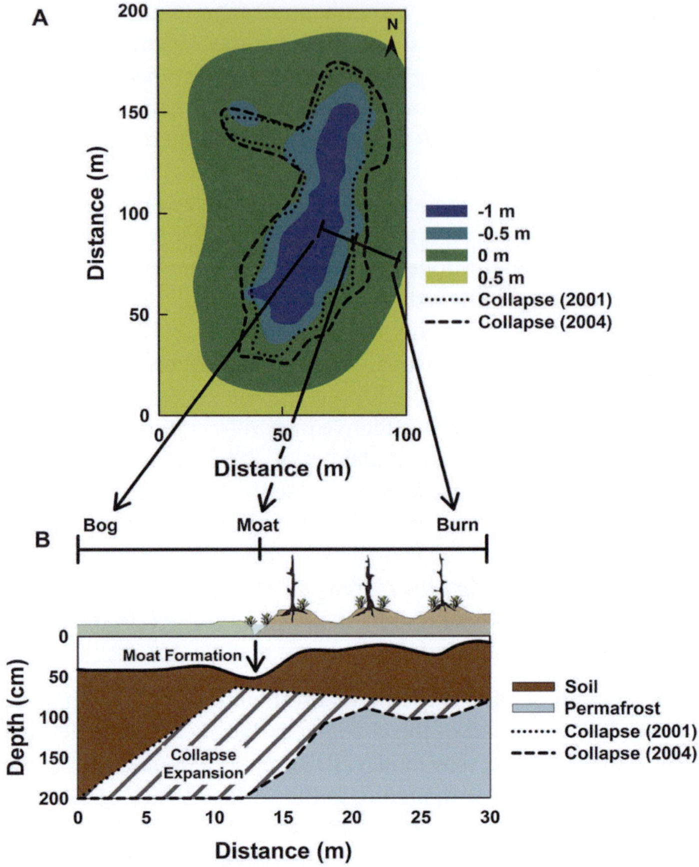

Fig. 6.13 Permafrost collapse expansion in Central Alaskan boreal forest peat since a fire in 2001, illustrating the spatial extent of the collapse (**a**). The extent of the collapse in 2001 is indicated by the dotted line and in 2004 is indicated by the dashed line. (**b**) Cross section of transect from centre of peatland to burn zone illustrating lateral subsidence and moat (waterlogged collapse margin depression) formation. The soil surface topography is indicated by the shades of colour in plot A and by the bold solid line in plot B. Frozen ground is represented by the blue polygon and the ground that has thawed since the fire is indicated by the hatched polygon. (From Myers-Smith et al. 2008. Biogeosciences 5, www.biogeosciences.net/5/1273/2008/, © Author(s) 2008, published under Creative Commons License 3.0 https://creativecommons.org/licenses/by/3.0/)

Change in microbial communities may reduce the soil respiration. Taş et al. (2014) found that the microbial community diversity and potential for respiration was reduced over all soil depths by fire, combined with a decrease in soil moisture and respiration potential. This potentially reduces emission of greenhouse gases. However, incubation data by Sawamoto et al. (2000) and soil respiration data by Takakai et al. (2008) show an increase of microbial respiration and CH_4 uptake in burnt forest soils in Siberia.

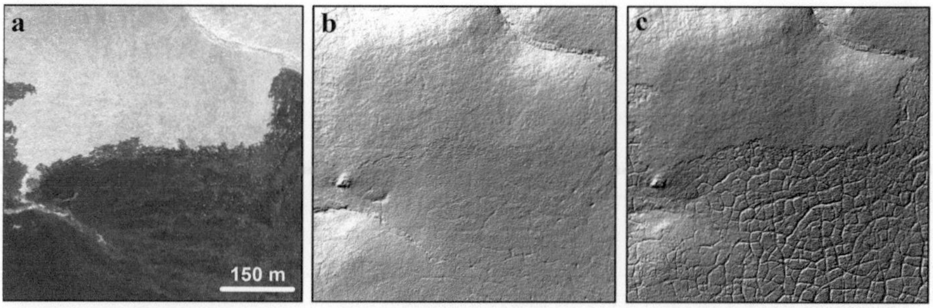

Fig. 6.14 Detection of permafrost thaw subsidence and thermokarst initiation in burned tundra using multi-temporal LiDAR. (**a**) A Quickbird image from 05 July 2008, the year following the fire, showing a portion of the northern extent of the burn area and the distinction between burned (dark) and unburned (light) tundra. Hillshade images of the (**b**) 2009 and (**c**) 2014 1 m resolution LiDAR DTMs showing ice wedge degradation in the burn area. Part a, b, c in Fig. 4 from Jones et al. (2015): Recent Arctic tundra fire initiates widespread thermokarst development. Nature Scientific Reports DOI:https://doi.org/10.1038/srep15865; published under Creative Commons License 4.0 http://creativecommons.org/licenses/by/4.0/

Summarizing, fires have a long-term influence on the carbon cycle of permafrost ecosystems. The most important is the increase in soil temperature and ALT by changes in the surface heat balance. This may increase soil respiration rates. Vegetation succession reduces these post-fire effects over time, but soil thermal anomalies may persist for a long time, and may lead to further permafrost thaw, erosion and carbon release from permafrost soils. As seen in the previous section, climate warming and increase of vegetation biomass shifts ecosystem carbon towards above-ground biomass. This makes ecosystem carbon more vulnerable to wildfires.

6.2 Thawing Permafrost and Vegetation

As seen in the previous chapter, vegetation is strongly affected by the various forms of permafrost degradation. This represents a part of the 'Arctic browning' discussed in Sect. 6.1.1, but permafrost thaw also may contribute to greening by stimulating net primary production by released nutrients. It is a balance between ecosystem disturbance and ecosystem resilience or ecosystem resistance (e.g. Van Huissteden and Dolman 2012; Jorgenson et al. 2013). The future of permafrost ecosystems depends on how much disturbance can be accommodated by their resilience or resistance before large-scale reorganization of ecosystems occur. A resilient ecosystem shows a temporary change by disturbance but returns to its previous state by recovery after some time; a resistant ecosystem resists change without clear change of the system.

In this chapter it is explored what effects permafrost thaw has on ecosystems, how they recover, what the consequences are for the carbon cycle. Much has already been discussed in Chap. 5 where the accent was put on soil temperature and hydrology. Here the ecological effects have focus. As seen in the previous chapter, permafrost thaw can be either gradual, or result in rapid changing soil surface conditions (abrupt thaw) by subsidence and erosion. A number of potential changes that affect vegetation emerge from Chap. 5:

- Exposure of previously frozen carbon and nutrients to decomposition;
- Exposure of relatively unweathered soil parent material by erosion, enhancing nutrient availability;
- Changes in active layer depth, that affect the rooting depth of plants;
- Changes in soil temperature in the root zone;
- Hydrological changes expressed by changes in soil moisture by subsidence (moister soils), improved soil drainage (drier soils) or change in the precipitation/evaporation balance;
- Submergence of vegetation by creation of thaw ponds;
- Removal of the topsoil and eventually deeper parts of the soil profile, including existing vegetation, by erosion.

Whatever the nature of permafrost thaw is, gradual or abrupt, nutrients have a central role. If nutrients are released by enhanced decomposition, strong changes of the vegetation composition and primary production are common (e.g. Van Wijk et al. 2004; Gough and Hobbie 2003).

6.2.1 Effects of Permafrost Thaw on Vegetation: Nutrient Release

For the vegetation, the most important effects of increased soil organic matter decomposition caused by permafrost thaw is the release of nutrients in nutrient-limited ecosystems. The carbon to nitrogen ratios of organic matter in soils may be a factor two or more lower than those in plants. In theory, nutrients made available by decomposition of soil organic matter (SOM) could increase C storage. If the vegetation has an average C/N ratio of 60 and SOM has a C/N ratio of 20 (Sect. 4.5.5), then the release of one unit of N from SOM decomposition could result in the storage of 3 units of C in new vegetation biomass, which is a net gain of 2 units of C at the ecosystem scale (Hobbie et al. 2002). Of course, this holds only if no restrictions apply to full mineralization of N into plant-available compounds, no N is lost from the ecosystems, and plants can take up all available N. In practice, these conditions are seldom met.

The additional nutrients have various sources: surface water (Koch et al. 2014, 2018), nutrients liberated from thawed soil and decomposition of thawed organic matter (Keuper et al. 2012b; Koch et al. 2018) and groundwater (Jorgenson et al. 2001). Incubation experiments by Reyes and Lougheed (2015) show that permafrost soil readily releases mineral nutrients (K, Ca, Fe, Mg, S, Si) at thawing. The release of N from organic matter is driven by soil environment parameters that influence decomposition rate such as microbial activity, temperature and oxidation-reduction potential. Frozen soil carbon from below the permafrost table is an up to seven times larger source of plant available N than the organic carbon from the root zone (Keuper et al. 2012a). Thawing permafrost is therefore a large nutrient source.

Plants vary in nutrient acquisition strategies and conservation of nutrients (Freschet et al. 2010a, 2012a, 2012b). A plant economics spectrum can be discerned, going from species with rapid acquisition of resources, correlating with high specific leaf area, high leaf N and P (low C/N ratio), to species which focus on resource conservation, correlated with high leaf dry matter content, lignin content and high C/N ratio. Freschet et al. (2012a) demonstrated that this spectrum affects all plant organs, from leaves and stems to roots, in subarctic heath vegetation. Thus, potentially, a high leaf decomposability predicts a high root decomposability, although there are differences in decomposability between organs of the same plant (leaves decompose faster than stems and roots). Plants also resorb significant amounts of nutrients from organs before senescence, mostly from leaves, as a nutrient conservation strategy (Freschet et al. 2010b).

The release of nutrients from thawing permafrost is therefore a true game changer in ecosystems. It strongly influences vegetation composition, the quality of the organic matter added to the soil, the biochemistry of soil organic matter, and the composition of microbial communities. Variation of litter/root organic matter quality among plant species is more important than environmental conditions (e.g. temperature) as a driver of root decomposition (Hobbie 1996; Aerts 2006; Iversen et al. 2015). Plants exert a biochemical influence on litter and topsoil decomposer community composition, thereby having an indirect effects on decomposition rates (Freschet et al. 2012b). The decomposer community is adapted to the average quality of the litter produced (Freschet et al. 2012b). If the quality of the litter that is produced departs from the quality to which the decomposer community is adapted, its decomposition rate will be influenced negatively. For example, a higher content of lignin and aromatics correlates with a larger proportion of fungi in the microbial biomass (Sjögersten et al. 2016). For these reasons, a change in the plant community also induces adaption of the microbial decomposer community over time.

Both soil warming and increased nutrient availability may drive an increase of primary production. It is difficult to separate both in experiments. Warming experiments, using a combination of artificially thicker winter snow cover and open top chambers by Natali et al. (2011, 2014, 2015), Sect. 5.2.3) increased the soil temperature by 1.5 °C and resulted in increased soil organic carbon decomposition, which released nutrients, and increased vegetation productivity.

A combined soil warming – nutrient addition experiment by Wang et al. (2017) demonstrates a dominating role of nutrients compared to warming. Warming in this experiment was achieved with heating wires at 15 cm depth in the soil, powered by solar panels; air warming did not occur. The soil warming increased the active layer depth on average with 7 cm and increased the soil temperature in the deeper layers with 0.7 °C. Significant nutrient release, related to higher soil temperatures and thicker ALT was not detected, except for a marginally higher amount of exchangeable phosphorus in the deeper mineral soil. The soil warming alone did not increase the ecosystem above-ground or below-ground biomass. Only sedges profited from warming, increasing their above-ground biomass, and the average root depth of the plant community increased. By contrast, nutrient addition had a much stronger effect. Fertilization increased the above-ground and below-ground biomass of grasses. For shrubs, the below-ground biomass decreased, while their above-ground biomass increased. The biomass of mosses and lichens decreased, as result of the competition for light with the faster growing vascular plants.

The fertilization experiment had a significant cooling effect on the soil. The ALT was reduced by 4 cm and soil temperatures by 0.6–0.9 °C, probably by increased soil shading by the higher vascular plant biomass. The experiment demonstrates also the effect of plasticity and shading of the soil. The sedges adapted their rooting depth to the depth where most nutrients were available, by lateral spreading of roots instead of depth growth. Shrub roots decreased, but this may have been compensated by additional nutrient availability through mycorrhiza. In general, more nutrients result in a lower root to shoot ratio (Chapin 1980; Wang et al. 2017).

In polar desert environments the effect of nutrient supply is also larger than that of warming or precipitation increase, as shown by experiments of Robinson et al. (1998) in Svalbard. Warming by open top tents and a precipitation increase of 50% did have a minimal effect on productivity, while N, P and K addition increased biomass and species composition. Colonization of dry sites in barrens is enhanced by the establishment of cyanobacterial crusts; these provide nutrients by N fixation, and enhance soil moisture and temperature (Cooper 2011).

However, caution is needed. The effects of nutrient release from thawing organic matter on vegetation productivity are filtered by the complicated relations of organic matter decomposition with temperature (Sect. 4.5), the modification of nutrient uptake by plant traits, interaction of vegetation with soil temperature (see above), and thus show variation with ecosystem type and soil conditions. Root traits (root distribution, anatomy, chemistry, mycorrhizal symbiosis) also determine how species react on changes in soil temperature, moisture and nutrient availability. Moreover, the experimental approaches to study the effects of warming, wetting and nutrient supply are generally of short duration and contain artefacts such as stepwise changes and relatively high doses that might not be representative for changes occurring under natural conditions. Adaptations of the root system and microbial community to new conditions take time (Hobbie et al. 2002; Sullivan et al. 2007).

6.2.2 Below-Ground Interaction of Root Systems with Nutrients and Soil Carbon

The effects of vegetation on organic matter input in the soil depends on litter produced at the surface, and the input from the root system, exudates and dead roots (Sect. 6.1.3). In particular the root system is still a black box (Sullivan et al. 2007). In tundra ecosystems the below-ground biomass is generally larger that the above-ground biomass. The high root biomass is often related to limitations in nutrient availability and harsh climatic conditions, where roots are needed for nutrient storage and vegetative propagation. In polar deserts however, the above-ground biomass is higher than the below-ground biomass. The root distribution is extremely shallow; nearly all roots are found in the top 30 cm, because of restrictions of the roots by soil temperature, active layer thickness, soil instability caused by cryoturbation and water saturated conditions at the base of the active layer.

Nutrients are taken up by plants depending on the architecture of the root system (Wang et al. 2016, 2017). Typical wetland species such as sedges have deeper root penetration than other species, their aerenchymous tissue allows growth in anoxic soil; dwarf shrubs such as *Betula nana* have generally shallow root systems. A review paper by Iversen et al. (2015) lists root traits of a large number of tundra plant species. The root depth of various species groups can be characterized as follows (summary from Iversen et al. 2015):

- Grasses: variable, adaptive to nutrient availability (e.g. Wang et al. 2017);
- Sedges (Carex, Eriophorum): in general deeper roots than grasses, up to ~45 cm;
- Deciduous shrubs: shallow – intermediate;
- Evergreen, ericaceous shrubs: mostly shallow, up to 15 cm;
- Forbs: variable, mostly shallow but some species intermediate – deep.

However, the root system is not static. With help of mycorrhizal symbiosis shrubs potentially can extract nutrients from deeper levels. A change in nutrient conditions and soil temperature results in adaptations of plant roots, depending on the plasticity of species (Sect. 6.2.1). In particular non-woody fine roots (<2 mm) are important for the ecosystem carbon and nutrient budgets. They represent the highly active part of the root system: high respiration rate, high nutrient content, short lifespan and the release of water and soluble exudates.

Injection experiments with $^{15}NH_4$ at different depths (5 cm, 15 cm and near the thaw front) in Siberian tundra soils demonstrates the combined effect of root traits and nutrient availability (Wang et al. 2018). For graminoids (grasses and sedges), the deep injection of ^{15}N increased leaf ^{15}N content more than shallow injection. Dwarf shrubs also took up N from deep injection, but to a lesser extent. Similar studies (Keuper et al. 2014; Oulehle et al. 2016; Zhu et al. 2016b) also showed that deeper rooting species take their nutrients from deeper soil layers. This is not confined to graminoids. In a sub-Arctic peatland in Sweden, the deep rooting perennial forb *Rubus chamaemorus* used added N at the thaw front.

This suggests that deeper rooting species, in particular graminoids, would profit more from nutrients liberated from permafrost thaw than shrubs. However, shrub expansion is the general pattern seen in the tundra instead of graminoid expansion (Sect. 6.1). Likely, warming of permafrost soils liberates more nutrients in the topsoil than near the base of the active layer, which remains relatively cool throughout the growing season. Moreover, the thermal effect of shrub growth is a thinner active layer (Blok et al. 2010; Juszak et al. 2016, 2017). This results in a decrease of the deep nutrient source from which grasses profit. Time also may be a factor, related to longer term adaptation of plants. Hobbie et al. (2002) cites nutrient addition experiments in Alaska, which also favoured grass growth in the short term of a few years, whereas on the longer term shrub dominance resulted.

The root depths above give an approximate indication at which depth an expansion of the biomass of a certain species or plant functional type contributes to new soil carbon allocation. For instance, expansion of sedges would shift carbon allocation downwards, expansion of shrubs upwards. A change in the root distribution therefore may result in more organic matter at the depth at which root biomass increases. In general, addition of organic matter at deeper levels in the ALT is more effective because of lower temperatures and more frequent anaerobic conditions that reduce the decomposition rates. On the other hand, exudation of large amounts of labile organic matter by roots may enhance aerobic decomposition rates at shallower depths as result of the priming effect (Sects. 4.5.2 and 6.1.4).

An illustration of this effect is given by the fertilization experiment described in Sect. 6.1.4 (Mack et al. 2004; Sullivan et al. 2007). This changed tussock tundra to shrub tundra, with a shallower root system. The result was a loss of soil carbon due to reduced input of organic matter in the deeper soil where conditions are better for carbon conservation. On the other hand, a deeper rooted system resulting from the expansion of sedges – generally occurring together with waterlogged soil conditions – would shift soil carbon allocation downward, in the wetland case into a zone where decomposition is slow due to anaerobic conditions and lower temperatures (Ström et al. 2005, 2012).

The type of organic matter also interacts with rooting depth and decomposition (Sect. 6.1.3). Root exudates, generally labile, low molecular weight compounds, contribute to substrate for methanogenesis if excreted below the water table in largely anaerobic environments (e.g. Ström et al. 2012; Sect. 4.5.6). If excreted above the water table in shallow soil horizons, exudates could contribute to the priming effect. This enhances decomposition rate by a higher energy availability, and stimulation of enzyme activity by the decomposition of labile organic compounds (Sect. 4.5.1). This effect is best known from temperate environments (Iversen et al. 2015), but the example above from Hartley et al. (2012) in Sect. 6.1.4 shows that it operates in high-latitude soils as well.

Many experiments with nutrients and warming have shown that when the vascular plant cover increases, the biomass of mosses and lichens decreases (e.g. Sistla et al. 2013). Mosses have generally low decomposition rates compared to vascular plants, and mosses facilitate peat accumulation (Hobbie 1996; Cornelissen et al. 2007b; Turetsky et al. 2012). In general *Sphagnum* species decompose slowly because of low pH resulting from a high cation exchange capacity and recalcitrant tissue, although their susceptibility to decomposition differs per species (Limpens et al. 2017). Lichens are in general more easily decomposable than mosses, but again there is large variability among species (Cornelissen et al. 2007b). Sjögersten et al. (2016) found more aromatic compounds in lichen-rich soils, which contribute to slower degradation. Because of the generally slower decomposition of moss litter, decrease of a moss cover decreases also topsoil organic matter. This may be amplified by loss of the cooling effect that mosses have in summer on the topsoil.

6.2.3 Abrupt Thaw and Vegetation

The most drastic changes of vegetation are produced by abrupt thaw phenomena: subsidence and erosion. For erosion features, vegetation has a prominent role in their stabilization. The role of vegetation in thaw subsidence (Sect. 5.3.4) is larger, in particular with respect to CH_4 emission and renewed carbon accumulation. This is the main subject here.

Although ground ice thaw is obviously the most important component in the loss of soil volume, the loss of organic matter by increased peat decomposition also should contribute to subsidence, as is known from drained peatlands (e.g. Schothorst 1977). The initial soil carbon stock before thaw subsidence is generally not known, but is important to estimate the amount of soil carbon loss to estimate its net effect on the carbon cycle. Collapse of palsas and peat plateaus changes the ecosystem from relatively slowly accumulating peat in a relatively dry environment, to an environment which is wetter and peat accumulates faster.

Subsidence resulting in water saturated soils or ponding of water changes the species composition drastically, by eliminating species that are not adapted to water-saturated soils and creating a favourable environment for wetland species (e.g. Turetsky et al. 2007; Li et al. 2017a, b; Fig. 5.10; Sect. 5.3.1). Next, subsidence and associated erosion creates also an irregular micro-relief, where wetter and drier conditions may occur close together (Schuur et al. 2007; Vogel et al. 2009). Often a distinct moat or channel is present along the edges of a permafrost collapse features (Fig. 6.13), with standing water and wetland vegetation of *Carex* and *Eriophorum* sp. (Myers-Smith et al. 2008). This moat results from recent subsidence along the margin, while at larger distance from the margin new peat formation may have elevated the surface again.

However, even in case of significant subsidence, vegetation change needs not be very outspoken, if the existing vegetation already tolerates wet soil conditions. In a 7 year long snow addition experiment on a palsa surface by Johansson et al. (2013), the surface elevation changed on average 24 cm, soil temperature increased by 1.5 °C, soil moisture at the end of the growing season almost doubled and ALT increased by almost 20%. Nevertheless, no significant change in the vegetation composition occurred. The only difference was a more vigorous growth of the dominant species *Eriophorum vaginatum*.

In boreal forest peatlands with permafrost in continental North America, ombrotrophic unfrozen peat surfaces have a vegetation of Black Spruce (*Picea mariana*), ericaceous shrubs (e.g. *Andromeda, Ledum sp.*), and a surface cover of *Sphagnum* species. On elevated palsas and peat plateaus, the soil is drier and tree stands are denser and taller. True mosses (e.g. *Pleurozium schreberi, Hylocomium splendens*) and lichens dominate the surface; peat accumulation rate in these relatively dry environments is low. Permafrost thaw creates depressions with waterlogged and flooded soil, where sedges (*Carex* and *Eriophorum* sp.) and *Sphagnum* species dominate. Trees die off when their root system submerges, or when they lose their support and fall. The edges of the depressions show collapsing, dried peat, dominated by lichens, shrubs and forbs. The wet depressions are treeless, and receive more solar radiation warming the soil, which is further enhanced by the low albedo of water at the surface. In general, the gross primary production is high. O'Donnel et al. (2012) found evidence of fast organic matter decomposition in the permafrost collapse depressions, which provides nutrients to a re-establishing vegetation. The high decomposition is probably related to moisture availability, litter type and higher bacterial activity (Camill et al. 2001, O'Donnel et al. 2012). Over time an ombrotrophic bog vegetation may develop again in the depressions (Beilman et al. 2001; Turetsky et al. 2007; O'Donnel et al. 2012).

In Northern Swedish palsa bogs, the intact palsa surface is dominated by a moist/dry ombrotrophic vegetation of shrubs, mosses and lichen. Collapse by permafrost thaw is shown by a hummocky surface where shrubs, *Eriophorum* and depressions with floating Sphagnum mats alternate. Further thaw results in establishment of a sedge vegetation, developing either into *Sphagnum* bog or sedge fen vegetation over time (Christensen et al. 2004; Malmer et al. 2005). A similar succession is seen in the Siberian tundra. Moderately drained tussock tundra is dominated by *Eriophorum vaginatum* tussocks with grasses, mosses, lichens and shrubs like *Betula nana*, *Vaccinium sp.*; dry tundra is generally dominated by *Betula nana*, mosses and lichens. Thaw of ice-rich permafrost and creation of waterlogged depressions or pools results in death of *Betula nana*, and subsequent colonization by sedges (*Eriophorum, Carex sp*; Fig. 5.10). Over time, *Sphagnum* (usually *S. fuscum*) establishes between the sedges and grows into more or less continuous carpets, eventually followed by re-establishment of shrubs (Osterkamp et al. 2009; Li et al. 2017b).

O'Donnel et al. (2012) have made a reconstruction of the amount of peat that should have been present in permafrost peatland collapse scars of ages ranging from about 1200 years to recent, in Central Alaska. This estimate is based on radiocarbon and tree-ring dated peat cores, and estimation of peat accumulation and decomposition rates with a model. For estimation of peat accumulation rates, an exponential model of peat accumulation and decomposition is applied and fitted to the peat ages in the cores. This provides estimates of the annual input of carbon and decomposition rates for both the forest peat and the newly accumulated peat in the thaw depression. The collapse of the peat plateau surface results in subsidence and inundation of forest peat. To derive the change in the carbon stocks of the forest peat in the forest peat after thaw, a first order decomposition model is used (Sect. 4.5.4). The estimated decomposition constant k of thawed forest peat is two orders of magnitude higher than that of intact forest peat; the decomposition constant for new bog peat is also higher by an order of magnitude (O'Donnel et al. 2012). The higher decomposition rates are attributed to higher soil temperatures in the thaw depressions, and for the thawed forest peat, an increase of the microbial population after thaw.

Figure 6.15 shows the modelled old peat accumulation in the forest peat, the peat decomposed after thaw, and the regrowth of fresh peat, based on the model equations of O'Donnel et al. (2012). Figure 6.15a depicts the carbon accumulation of uninterrupted peat formation on a forested permafrost plateau, Fig. 6.15b the carbon loss after permafrost thaw, followed by renewed bog peat formation with a higher

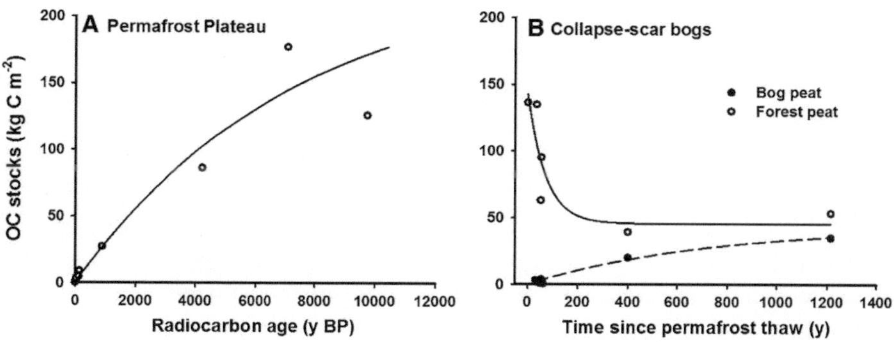

Fig. 6.15 (**a**) Accumulation of peat soil carbon at a permafrost plateau site in Central Alaska as determined by fitting a carbon accumulation model to the carbon stock and radiocarbon age data. (**b**) Comparison of old forest and young bog peat C stocks over time since permafrost thaw at Koyukuk Flats study region. The inundated forest peat layer refers to the organic matter originally accumulated and stored in soils of permafrost plateaus, which ultimately becomes saturated and buried following permafrost thaw. Points at t = 0 reflect carbon stocks from peat plateau profiles. Where t > 0, points reflect collapse-scar bog age. Loss of OC from forest peat is determined by the carbon accumulation model. From O'Donnell et al. 2012. © 2011, Springer Science Business Media

accumulation rate than that of the forest peat. For a collapse scar bog of 60 years old, the net average carbon loss rates (accumulation – loss) range between 178 and 1389 g C m^{-2} $year^{-1}$. For a 1200 year old bog, loss rates range between 1 and 40 g C m^{-2} $year^{-1}$ over the years since thaw. Therefore, the time span required for new peat formation to balance the loss of peat carbon by thaw is more than 1200 years, despite high carbon accumulation rates in the collapse scar bogs.

The organic carbon accumulation rate for young peat in permafrost collapse is about 2.7 times higher than the carbon accumulation of forest peat with permafrost (23 ± 8 vs. 61 ± 2 g C m^{-2} $year^{-1}$; O'Donnel et al. 2012). A similar multiplication factor (2.5) of C accumulation in recent bog peat with sedges and *Sphagnum* over forest peat underlain by permafrost is reported for the southern margin of the Canadian permafrost region by Camill et al. (2001) and Turetsky et al. (2007). The accumulation rate of young bog peat was found to decrease exponentially over time to an approximately constant level of ±120 g C m^{-2} $year^{-1}$.

The data of O'Donnell et al. (2012) and Turetsky et al. (2007) suggest that new peat formation in thaw depressions results in an increase in carbon uptake by nutrient-stimulated increase of gross primary production. However, CO_2 fluxes measured in thaw depressions in Central Alaska (Vogel et al. 2009; Trucco et al. 2012) showed that, although *GPP* increases with permafrost thaw, R_{eco} offsets the higher *GPP*. The R_{eco} carbon losses are in particular high in autumn and winter, and highest on a site which has undergone long (>50 years) and deep thaw, with a relatively high water table. This site was a net source of CO_2 to approximately neutral, while sites with less deep thaw were either CO_2-neutral or a small sink.

Moreover, wet thaw depressions have a significant climate warming effect by the emission of CH_4. The increase of CH_4 emission is induced by:

- Water-saturated, often flooded soil conditions causing anaerobic soil conditions in the depressions created by permafrost thaw (Sect. 4.5.6);
- The establishment of a wetland sedge vegetation, with production of labile organic compounds and deeper penetration of roots;
- This vegetation has aerenchymous roots and stems, which transport CH_4 to the atmosphere, bypassing methanotrophy in the upper soil zone or moss cover (Fig. 4.25).

Below the water table, in the anoxic soil environment, root penetration is aided by the aerenchymous tissue that facilitates oxygen supply to the roots (Sect. 4.5.6). Common cold-climate species that are aerenchymous are the grasses *Arctophila fulva*, *Dupontia fischeri*, many *Carex* species, and *Eriophorum* species (Iversen et al. 2015). This deeper penetration of roots is accompanied by the release of relatively labile organic matter (Ström et al. 2003, 2012; Hodgkins et al. 2014).

The release of labile compounds is species-specific. In the root zone of *Eriophorum scheuchzeri* high acetate production was observed; shading reduced acetate production, showing that this was closely linked to photosynthesis. The acetate production by Eriophorum was about 8 times higher than in the root zone of the wetland grass *Dupontia* growing at the same location (Ström et al. 2003). Smaller differences were found between *Eriophorum* and *Carex* species (Ström and Christensen 2007). The pool of low molecular weight organic compounds in the same fen consisted for 75–83% of organic acids, the remainder consisted of simple carbohydrates and amino acids. The organic acids consisted mainly of acetic acid, the substrate for acetoclastic methanogenesis (Ström et al. 2012).

Plant species therefore control the CH_4 production, next to the microbial fermentation reactions producing organic acids, and soil environmental variables as redox conditions and temperature (Sect. 4.5.3; Olefeldt et al. 2013). Furthermore, positive correlations between plant productivity (*NEE* or *NEP*) and CH_4 fluxes have been found in many studies. These are related to plant biomass and the production of methanogenic substrate by root exudation (Ström and Christensen 2007 and references therein). On the other hand, also the CH_4 oxidation in the rhizosphere varies per species, which influences the plant-transported CH_4 emission (Ström et al. 2005).

In permafrost collapse depressions in the Stordalen mire in Northern Sweden, thaw increases the potential for CH_4 and CO_2 production as shown by incubations (Fig. 6.16), and shifts the methanogenesis from the hydrogenotrophic to the acetoclastic pathway, indicating the presence of more labile compounds. At the same time, the C/N ratio of the peat decreases, reflecting the shifts in plant community, and the pH increases. Analysis of organic matter structural compounds in peat and dissolved organic matter indicates the presence of more labile compounds with progress of thaw (Hodgkins et al. 2014). These changes contribute to higher decomposition rates of the peat.

A significant increase of CH_4 emission has been reported for many wet terrain depressions created by permafrost thaw (e.g. Vitt et al. 2000; Camill et al. 2001; Turetsky et al. 2000, 2002; Christensen et al. 2004; Ström and Christensen 2007;

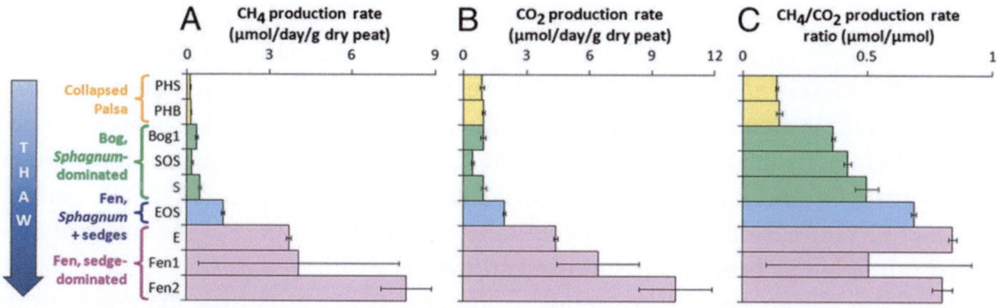

Fig. 6.16 CH_4 and CO_2 production rates in anaerobic incubations (±SE between replicates), which represent relative organic matter lability as potential decomposition rates. From Hodgkins et al. 2014, Changes in peat chemistry associated with permafrost thaw increase greenhouse gas production. Proc Natl Acad Sci USA 111 (16):5819–5824. doi:https://doi.org/10.1073/pnas.1314641111. Republished with permission of PNAS

Turetsky et al. 2007; O'Donnel et al. 2012; Lara et al. 2012; Olefeldt et al. 2013; Andresen et al. 2017). Permafrost thaw in the Stordalen mire palsas in Northern Sweden resulted in CH_4 emissions up to 36 mg CH_4 m^{-2} h^{-1} and increased dark respiration CO_2 emissions, both high enough to transfer the thawed areas into a net source of radiative forcing, despite a higher *NEE*.

Concluding, variation in plant traits with respect to productivity, root exudation and root morphology determine the high CH_4 emission of thaw subsidence depressions and ponds. Vegetation composition further affects the CH_4 emission by symbiotic relations of methanotrophs with mosses (Sect. 4.5.6). This combines with vegetation effects on soil temperature and ALT, to create a high variability of CH_4 emission among vegetation types (e.g. Turetsky et al. 2002; Van Huissteden et al. 2005, 2009; Ström and Christensen 2007; Olefeldt et al. 2013; Li et al. 2017a, b). Changes in soil temperature and microbial population in the thaw ponds contribute to high CO_2 emission also.

Relations of CO_2 and CH_4 exchange and plant communities and micro-relief features such as thaw ponds have been successfully applied to upscaling of measured fluxes to a landscape using satellite images and vegetation mapping data. These methods also have been used also to map changes over time (e.g. Johansson et al. 2006; Lara et al. 2012, 2015; Budishchev et al. 2014; Marushchak et al. 2016; Andresen et al. 2017). The latter authors infer a 60% increase of CH_4 emission from expansion of the aerenchymous wetland species *Carex aquatilis* and *Arctophila fulva* in the Northern Alaskan coastal plain.

6.2.4 Resilience

Sedimentary records in lakes and palaeobotanical records in have shown that disturbances by fire and Holocene climate fluctuations have occurred before in permafrost ecosystems (e.g. Katamura et al. 2009; Teltewskoi et al. 2016). Ecosystems have repeatedly experienced disturbances including thaw, and have a resilience to disturbances, allowing recovery. Moreover, the accumulation of ground ice and frost heave makes permafrost ecosystems vulnerable to abrupt thaw, also without climate change (Kanevskiy et al. 2017). Permafrost ecosystems therefore may show pseudo-cyclic behaviour induced by ground ice thaw followed by renewed ground ice accumulation (Jorgenson et al. 2015). Like thaw may induce carbon losses and changes ecosystem CO_2 and CH_4 emissions, post-thaw vegetation succession will change CO_2 and CH_4 emission rates, and eventually may restore a part of the lost soil carbon, as already has been discussed in the previous section.

Here, this resilience of ecosystems to the effects of permafrost thaw is considered in more detail, in particular the role it may take in mitigating the effects of permafrost thaw. Important questions for understanding future trajectories of change of permafrost ecosystems are: how do ecosystems recover from disturbance, how long it takes and how much of the lost ecosystem carbon is gained back by recovery. However, the fact that ecosystems show a certain amount of resilience does not dismiss the question of how much disturbance permafrost ecosystems can take before also this resilience breaks down.

Li et al. (2017b) describe a succession of sedge (*Eriophorum/Carex)* vegetation to *Sphagnum* carpet in thaw ponds in Northeast Siberia (Fig. 5.10). They compared plots of sedge vegetation, a plot with *Sphagnum* spontaneously developing in a sedge vegetation, a similar plot but with *Sphagnum* removed, and a sedge vegetation plot with added *Sphagnum*. In the sedge vegetation, CH_4 emissions and CO_2 uptake were significantly higher than in the Sphagnum carpet. Where Sphagnum was removed, CH_4 emissions increased, and where Sphagnum was added, CH_4 emissions decreased (Fig. 6.17). This was attributed to the effect of strong methanotrophy in the unsaturated *Sphagnum* layer (Kip et al. 2010; Raghoebarsing et al. 2005). Interestingly, on further development the *Sphagnum* carpet becomes a seedbed for the establishment of shrub seedlings of *Betula nana* and *Salix pulchra*, indicating further terrestrialisation of the ponds (Li et al. 2017b; Magnússon et al. submitted). This might result in a return of negative CH_4 fluxes, which are typical for the dense *Betula nana* vegetation in which the ponds developed.

A similar succession was found in a permafrost collapse pond in Alaska by Myers-Smith et al. (2008) based on palaeobotanical and diatom analysis of peat cores. The collapse pond was initiated about 100 years ago by fire. Initially the surface was colonized by a sedge-dominated peatland with nutrient-rich conditions. Around 1970 a shift towards a Sphagnum-dominated peatland occurred. This indicates that the conditions became drier, more acidic and/or more nutrient poor. Here the succession from sedge lawn to moss peat is attributed to a regional shift in climate to drier conditions rather than internal succession.

Ice wedge thaw ponds tend to be deeper than the shallow ponds described above. In ice wedge thaw ponds in Alaskan tundra, Jorgenson et al. (2015) discern several stages in the degradation and stabilization of wedge ice thaw (Fig. 6.18):

- Intact wedges have narrow, well vegetated troughs.
- Initial degradation is shown by shallow flooding, abundant hydrophytic sedges; the ALT reaches down to the wedge ice.

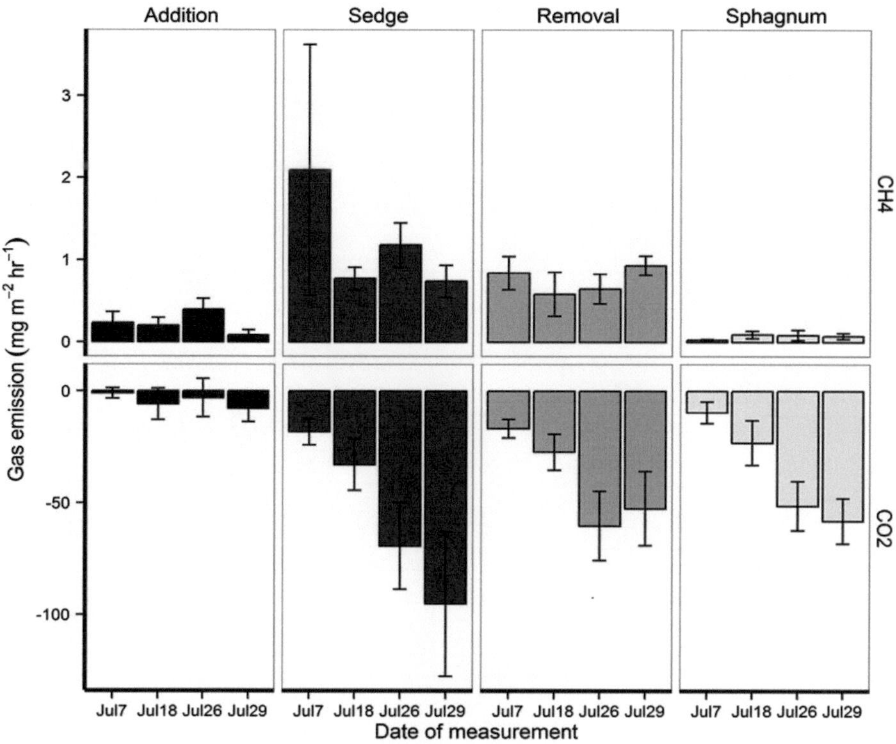

Fig. 6.17 Mean gas fluxes (Above: CH_4 flux; Below: CO_2 flux) of four different experimental plots (Addition: Sphagnum addition in sedge lawn (black), Sedge lawn (dark grey), Removal: Sphagnum removal, leaving sedges (medium grey) and Sphagnum plots (light grey)) on July 8, 18, 26 and 29, 2014. The error bars reflect the standard errors. Negative gas fluxes indicate net absorption/uptake, whereas positive fluxes indicate net emission. (From Li et al. 2017a, b)

- Advanced degradation shows wide, deeply flooded troughs, with aquatic mosses and vascular plants.
- Initial stabilization troughs are shallowly flooded and dominated by sedges.
- In advanced stabilization, newly formed peat emerges above the water table, with a diverse vegetation including initial establishment of shrubs.

In the degradational stages, heat balance and soil thermal feedbacks result in an increasing soil temperature, e.g. by snow accumulation in the troughs above the ice wedges. This culminates in higher winter soil temperatures, and higher summer temperatures in the advanced degradation stage. During stabilization, a layer of sediment develops, derived from peat formation and erosion of the banks of the troughs. In combination with drainage and regrowth of ice lenses, the surface emerges again above the water table. A new ice-rich intermediate layer may be created in this way. In this stabilization stage also the soil temperatures decrease again.

Many ice wedges have undergone earlier phases of thaw, evidenced by the presence of cave ice, created by refreezing of water in eroded ice wedge parts. A cyclical

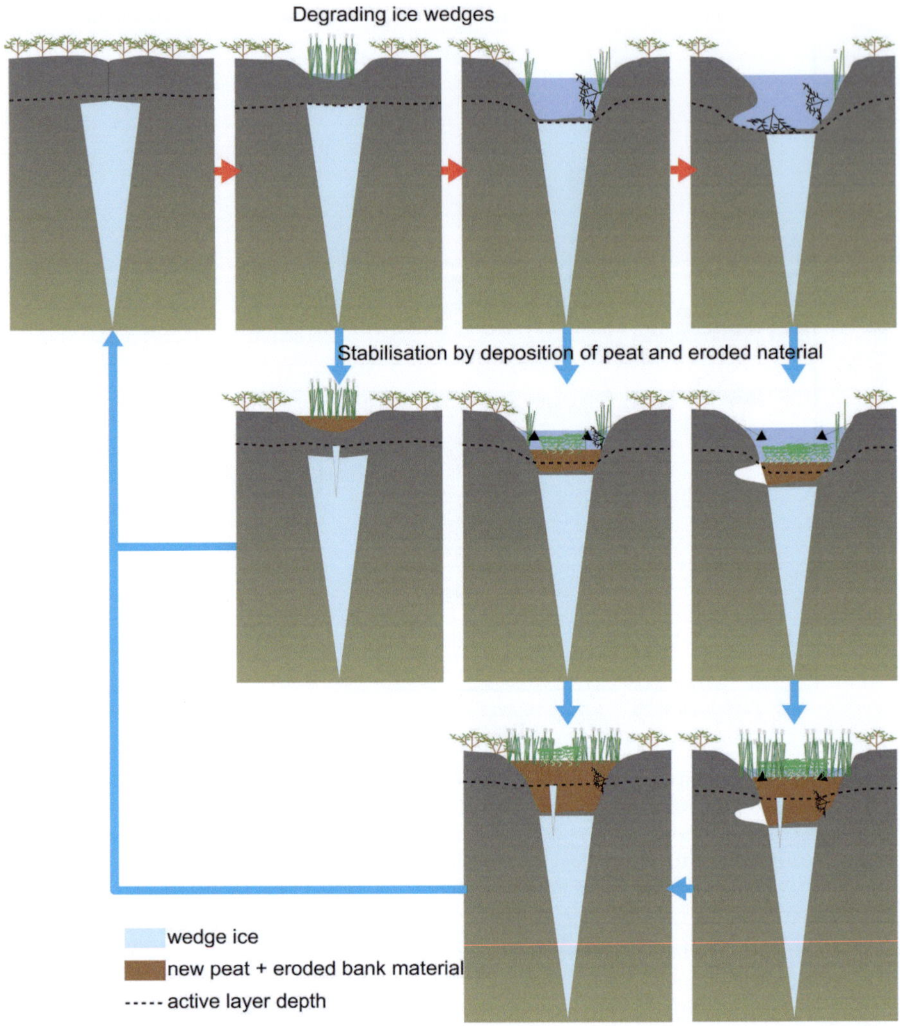

Fig. 6.18 Stages of ice-wedge degradation (red arrows) and stabilization (blue arrows). (Simplified after Jorgenson et al. (2015) and Kanevskiy et al. (2017))

model of ice wedge destabilization and stabilization is proposed by Jorgenson et al. (2015). In this model, vegetation succession is an important component to stabilize ice wedge thaw processes and prevent large scale loss of surface permafrost and carbon. Vegetation succession in the ice wedge troughs is facilitated by infilling with sediment and drainage. Jorgenson et al. (2015) propose that the process of stabilization by vegetation succession prevents extensive ice wedge thaw, which otherwise may proceed to thaw lake development. Without stabilization, an alternative route may be erosion of ice wedges and widespread drainage (Liljedahl et al. 2016), depending on the local relief.

Kanevskiy et al. (2017) have extended this conceptual model including multiple pathways of degradation and stabilization, depending on the severity of degradation. Stabilisation of even deeply thawed ice wedges starts with accumulation of a new layer of sediment above the wedge ice, which consists of material eroded from the banks of troughs and new peat formation. Central to this model is the development of a new ice-rich intermediate layer above ice wedges and renewed ice wedge growth in sediment in the thawed ice wedge troughs. This intermediate layer serves as a buffer for further thaw of the ice wedge ice below.

Even in thaw lakes renewed carbon accumulation occurs in lake sediments. The lakes may turn over time from a source of carbon to a sink (Walter-Anthony et al. 2014). Their study of sedimentary successions in exposed thaw lake sediments in Yedoma deposits in North Siberia quantified carbon accumulation lake sediments. The basal radiocarbon dates of these successions indicate lake formation during the Last Glacial Termination and Early Holocene. The lakes have been filled in with a succession of peat with various admixtures of mineral material. The organic matter is derived in approximately equal amounts from eroded vegetation along the shore, littoral sedge vegetation and benthic brown mosses (*Amblystegiaceae*). During the infilling, partial or complete lake drainage resulted in regrowth of permafrost (Fig. 6.19). Walter-Anthony et al. (2014) derived average Holocene carbon accumulation rates for Siberian thaw lakes, averaging 47 ± 10 g C m^2 $year^1$.

The carbon accumulation rate from the sedimentation in the lake basins is estimated at 47 ± 10 g Cm^{-2} $year^{-1}$, which is 4–5 times higher than for other high-latitude lakes. The organic production is driven by nutrients (N and P) derived from shoreline erosion and weathering; the brown mosses are particularly responsive to P fertilization which may have been released from the surrounding Yedoma deposits. Based on estimates of the CO_2 and CH_4 emitted by thaw lake during formation and infilling by sediment, (Walter-Anthony et al. 2014) estimate that about 5 millennia ago, the carbon sequestration in lake sediments made the thaw lakes from the Last Glacial Termination turn from an radiative source to a sink (Fig. 6.19B).

Erosion – cryogenic landslides and gully erosion – results in removal of vegetation (Sect. 5.3.4). The largest role for vegetation in these cases of abrupt thaw is stabilisation and re-colonization. Generally the freshly exposed mineral soils are rapidly colonized by vegetation, after the intensity of erosion processes decreases. Such a decrease of intensity is most likely related to weather conditions – a cooler or drier summer period that stabilizes thaw and mass-wasting processes. The establishment of the vegetation subsequently plays a stabilizing role by retaining slope material with roots. In landslides, large masses of seedlings of various plants often cover the landslide material that has stopped moving. However, vegetation rafts in landslides (Figs. 3.24 and 3.25) also contribute. During an initial pioneer phase, a vegetation with high grasses and forbs dominates, which is gradually replaced by forest or shrub tundra vegetation (Burn and Friele 1989; Schuur et al. 2007; Ukraintseva et al. 2014; Fig. 6.20). A dense willow shrub cover often remains for decennia after landslide activity.

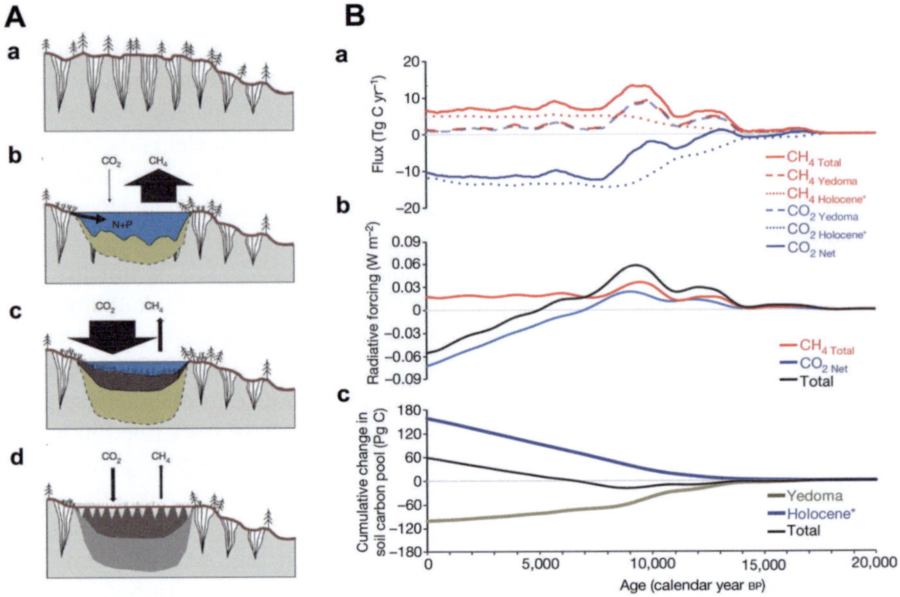

Fig. 6.19 Carbon cycle during the development of deep thaw lakes, from Walter-Anthony et al. (2014). (**A***a*) Yedoma with massive Pleistocene ice wedges. (**A***b*) Thaw lake expansion (thaw bulb shown in yellow; thaw boundary shown as a dotted line) accompanied by Pleistocene-aged CH_4 emissions and release of N and P from yedoma into lakes, stimulating aquatic productivity and CO_2 uptake. (**A***c*) Partially drained lake with atmospheric CO_2 uptake by plants forming thick Holocene organic carbon deposits (brown), exceeding CH_4 emissions from contemporary organic matter decay. (**A***d*) Refreezing of remaining Pleistocene and Holocene carbon in sediments following complete lake drainage (new ice wedges are shown as triangles), with peatland-type CH_4 emissions and CO_2 uptake. The thicknesses of CH_4 and CO_2 arrows are scaled by relative magnitude on a carbon-mass basis. (**B***a*) Carbon flux trajectories for yedoma-region thaw lake basins determined by basal date frequency of lake deposits (see Walter-Anthony et al. 2014)). Negative flux: CO_2-C uptake by peat formation; positive flux: CO_2-C emission from thawed yedoma decay and CH_4-C emissions, positive flux. The solid red line in Ba is the sum of CH_4-C emissions from decay of thawed yedoma deposits (dashes) and younger organic matter in thaw lake basins basins (dots). (**B***b*) Radiative forcing due to atmospheric perturbations in CH_4 and CO_2 concentration for flux trajectories shown in **B***a*. (**B***c*) Cumulative changes in the yedoma-region permafrost carbon pool due to loss of yedoma carbon to the atmosphere by thermokarst-lake formation (negative) and atmospheric carbon uptake and burial by the same lakes (positive). © Springer Nature AG 2014

The re-vegetated sites may contain more carbon in above-ground biomass, as they remain well visible by a taller vegetation (Ukraintseva et al. 2014). However, the restoration of soil carbon is slow. In the thaw slumps in boreal forest of Burn and Friele (1989), vegetation was restored to the original forest composition in 35–50 years, but soil organic horizons remained thin and contained a less than half of the organic matter than the original forest soil. Unfortunately, the carbon losses and gains of cryogenic landslides are poorly quantified, despite their strongly increasing importance as permafrost degradation features (Lewkowicz and Way 2019).

Fig. 6.20 (**a**) Stabilizing vegetation with high forbs on a landslide. (**b**) Stabilized landslide covered with willow bushes, on the right side new landslide activity develops. Kytalyk Reseve, Indigirka Lowlands, NE Siberia. (Photos by the author)

The post-thaw vegetation successions are an expression of the resilience of permafrost ecosystems. This resilience combines three aspects:

- Restoration of vegetation of ecosystems that have been subject to thaw disturbance, and establishment of an ongoing vegetation succession going in the direction of the pre-thaw system (but not necessarily reaching this stadium)
- Stabilization of thaw and protection of ground ice against further thaw, eventually with new ice accumulation. Both sedimentation (organic material, eroded material) and re-establishment of the vegetation cover contribute to this stabilization.
- Restoration of at least part of the lost soil carbon by renewed accumulation of organic matter.

A complete autonomous vegetation succession need not be responsible for the restoration of the carbon stocks. The transfer from sedge marsh to *Sphagnum* bog described by Myers-Smith et al. (2008) has been ascribed to a climate shift. In the stabilization by ice wedge thaw of Jorgenson et al. (2015) and Kanevskiy et al. (2017), sedimentation and drainage influence the vegetation succession. In the case of stabilizing thaw slumps and active layer slides, a cool and dry summer may aid in stabilization. Likely the succession is not completely autonomous: the incidence of a cooler period aids stabilization. Conversely, a continued warming may prevent stabilization.

It is also questionable how long it takes to restore the amount of carbon that was lost by thaw processes – if it ever occurs. In the permafrost collapse sites of O'Donnel et al. (2012) it takes some 1200 years before the carbon loss was more or less compensated. The succession in the tundra ponds of Li et al. (2017a, b) and Magnússon et al. (submitted) is relatively fast, in the order of decades, but the final stage – a complete return of a dense shrub vegetation underlain by ice-rich permafrost – has not been observed. Moreover, the carbon losses and gains have not been quantified in this case. However, despite restoration of carbon stocks, the radiative effect of vegetation succession is important; in this case CH_4 emission is strongly reduced by enhanced methanotrophy in the newly established *Sphagnum* carpets (Li et al. 2017b).

This resilience should be accounted for in models of future carbon emission from permafrost areas to the atmosphere, because it has a not negligible effect on the magnitude permafrost carbon feedback (Jorgenson et al. 2010; Van Huissteden and Dolman 2012; Grosse et al. 2011). Aspects that influence this compensation effect are the length of the recovery time of vegetation and carbon stores of the ecosystem. The longer the duration of recovery, the smaller the effect of resilience (Van Huissteden and Dolman 2012). From the above, it appears that the typical time scales for restoration in cases of abrupt thaw features are:

- Restoration of vegetation, including its carbon sink: decades to centuries
- Restoration of lost carbon stores: order of 1–2 millennia (e.g.collapse scars of O'Donnel et al. O'Donnell et al. 2012)
- Compensation of radiative forcing form carbon losses if CH_4 emission occurred: several millennia (thaw lakes).

Furthermore, resilience has its limits. By high frequency and large area of disturbance impacts, a large part of an area may become disturbed, and thresholds beyond which recovery is impossible may be crossed. Ecosystem resilience is likely to be lost, invoking drastic changes of ecosystems over large areas into another state (Scheffer et al. 2001, 2009). The fast expansion of thaw lakes during the Last Glacial termination, which affected the high-latitude permafrost landscape on a huge scale (Walter et al. 2007) is probably an example. Other possible examples that suggest that large-scale changes are being initiated, may be the recently observed 60-fold increase of thaw slumps on eastern Banks Island (Lewkowicz and Way 2019) and the widespread ice wedge thaw reported by Liljedahl et al. (2016). These disturbances, occurring with a high frequency and large spatial coverage have to potential to alter existing ecosystems on at least regional scales.

6.3 Human Vegetation Disturbance: Industrialisation and Agriculture

Many human activities may cause local vegetation disturbance that can initiate permafrost thaw and subsequent carbon losses. Three main groups of human activities are expanding in area in permafrost areas, industrialisation, transport and agriculture (Nelson et al. 2002; French 2018). The vegetation in permafrost environments is very fragile, and inadvertent damage is easily done and may remain visible for a long time (Fig. 6.21). In turn, human infrastructure itself is increasingly vulnerable to the climate-induced permafrost thaw, with possible further release of greenhouse

Fig. 6.21 Old vehicle track in Siberian tundra. Kytalyk Research Station. (Photo by the author)

gases as a consequence. The goal of this section is to explore effects human activities on the permafrost carbon cycle (besides the effects of anthropogenic global warming).

Industrialisation is expanding in particular on behalf of increasing oil and gas extraction, and other extractive industries. These industries have led to establishment of several large urban centres. Besides mines or well pads, mining activities include a large range of additional infrastructural components: housing facilities, access roads, oil and gas processing facilities, waste treatment facilities such as mine tailing ponds, pipelines and other transport facilities such as ports. In the future, the navigability of the Arctic ocean will increase in summer, which will expand shipping-related facilities along the Arctic coast. With increasing geopolitical tension, military facilities (including exercise grounds) are also likely to grow.

All this infrastructure is vulnerable to thawing permafrost, which may cause severe destabilization and loss of its integrity. In Russia, 70% of infrastructure within current permafrost areas is located in areas with high risks for permafrost thaw. This includes 45% of the Russian oil and gas infrastructure (Hjort et al. 2018).

Permafrost engineering practices developed during the twentieth century ensure stability under present climate conditions if properly designed and maintained (see e.g. French 2018 for a summary of approaches). However, it is not known to what extent structures remain stable under conditions of future climate warming. Climate change effects may not be included in design of structures. For instance, design criteria usually require the use of current climate statistics only. It then depends on the safety margins that have been used in the design, to what extent structures may withstand the effects of climate change (Shiklomanov et al. 2017).

In many cases of structural damage to buildings and infrastructure there is discussion on the distinction between the effects of climate-induced permafrost changes, and socio-economic factors such as improper use, lack of maintenance, age or design and construction flaws. However, a general pattern of climate change emerges from reports on infrastructure damage in Russia. Bearing capacity of pile foundations depends partly on foundation design and partly on permafrost characteristics. Based on climate projections, Shiklomanov et al. (2017) estimate that bearing capacity of pile foundations in Northwest Siberian cities as Salekhard and Anadyr potentially decreases to critical levels by the mid 2020s. In Yakutsk and Norilsk this would be around the 2040s. Analysis of permafrost thaw risks to infrastructure by Hjort et al. (2018) shows that climate scenarios (the RCP pathways) have little influence on the risk level until 2050. On the longer term however, achievement of the goals of the Paris climate agreement would make a clear difference in potential damage.

Oil and gas extraction occurs widespread in permafrost areas: Northern Alaska and Canada, Northern Russia, West Siberia, with the most recent and fastest developments in the Yamal peninsula. Over time, the oil and gas industry has reduced its impact by technological development in well drilling, such as deviated and horizontal drilling. This allows better planning of well locations and concentration of activities in less well pads. However, developments in enhanced oil recovery and unconventional fossil fuels (shale gas and oil) induce a strong increase in activities and infrastructure. Enhanced oil recovery (EOR) requires more water, chemical additives, a larger number of wells and results in an increase of waste water production. This holds even more fore shale oil and gas, which require a dense network of wells and strongly increase water use and waste water production (e.g. Kondash et al. 2017). Shale oil reserves occur in Western Siberia and are partly overlain by discontinuous and continuous permafrost; in older oil fields such as those in Northern Alaska, EOR techniques are increasingly applied. Spills of oil, fuel and other hydrocarbon products occur ubiquitously in relation to all mining and other infrastructure in permafrost areas, and are persistent and difficult to clean up (Filler et al. 2009).

Several Arctic areas are highly important energy suppliers to large population centres further south, e.g. the Yamalo-Nenets region in Northwestern Siberia, which supplies more than one third of the gas pipeline imports of the European Union (Hjort et al. 2018). Structural damage to oil and gas infrastructure increases the risk of environmental pollution including oil spills and fugitive CH_4 leaks (Mazhitova et al. 2004; Jernelöv 2010; Hjort et al. 2018). About 21% of the accidents along West Siberian pipeline network is related to permafrost thaw (Anisimov and Reneva 2006). The risk of fugitive CH_4 emissions from oil and gas infrastructure that is damaged by permafrost thaw adds to the other CH_4 emission sources from permafrost thaw.

Besides damage of climate-induced permafrost thaw to infrastructure, the increasing human activities in permafrost areas also contribute to an increase of permafrost thaw in the areas that are affected by these activities. Perhaps one of the best documented oil fields in terms of its impact on permafrost is the Prudhoe Bay oilfield in Northern Alaska (Raynolds et al. 2014). The authors list a number of impacts on the permafrost surface environment, apart from the direct impacts of building infrastructure, roads, gravel pits, urban infrastructure etc.:

- Ice wedge thaw and other thaw subsidence by removal of vegetation, which may or may not be related to infrastructure. When a direct link to infrastructure is absent, climate change also is a likely explanation.
- Off-road vehicle traffic which disturbs vegetation .
- Flooding resulting from road construction, which blocks surface drainage if improperly constructed; flooding contributes to permafrost thaw (Sect. 5.4.3).
- Snow drifts accumulating alongside roads, which increase soil temperatures.
- Dust deposition on vegetation near roads may kill vegetation, or enhance thaw of the snow cover in spring; both act to warm the soil in summer.

The effects of flooding can be severe and long-lasting. A pipeline construction road constructed in the 1970s blocked normal surface drainage in Alaska. Survey by radar interferometry in 2006–2010 revealed thaw subsidence of 2–8 cm $year^{-1}$ over a large area, equivalent to an ice volume loss of about 1.2×10^7 m^3 $year^{-1}$ (Liu et al. 2015). These effects also hold for other industrial and mining activities.

Other effects of oil and gas industry not mentioned by Raynolds et al. (2014), but manifest in many oil and gas fields are:

- Oil or waste water spills and the effects of cleaning operations (Jernelöv 2010) or spills of mine tailings in case of other mining activities (e.g. Elberling et al. 2003). Given the chemical composition of mining wastes (brines in the case of oil and gas industry, or acid and heavy-metal polluted wastes in case of other mining activities) accidents are highly detrimental to the environment and destructive for vegetation.
- CH_4, other volatile hydrocarbon, and black carbon emissions from flaring by incomplete combustion of flared gas, from gas leaks and accidents (e.g. Karion et al. 2013).
- Fugitive gas emissions from improperly abandoned wells (e.g. Townsend-Small et al. 2016).
- Large scale subsidence, including induced seismicity resulting from reservoir compaction by oil/gas extraction; in flat areas this potentially changes drainage systems (e.g. van Thienen-Visser et al. 2015).

The first three are an operational risk factor for the industry and normally will be prevented as much as possible. However, the effects of permafrost degradation on structural stability of infrastructure sketched above will further aggravate the risks. As indicated by Shiklomanov et al. (2017) there is a significant socio-economic aspect, in the shape of management of infrastructure, that aggravates the risks induced by climate change. In the case of industrial decline (as seen in the early '90s at the breakdown of the Soviet Union) management activities may fall short, in particular with respect to accident management. Large scale subsidence by reservoir compaction is unavoidable for oil and gas fields. In flat, poorly drained areas it is likely to increase poor drainage and flooding, with the risk of increasing permafrost thaw and subsidence and the development or expansion of thaw lakes. Negative effects on permafrost stability might be mitigated by careful water management planning.

Summarizing, the side effects of common industrial activities are an enhancement of localized permafrost thaw by flooding and vegetation disruption (Raynolds et al. 2014). The interaction of human-induced and climate-induced permafrost thaw with mining infrastructure potentially exacerbates CH_4 emissions and other spills (Hjort et al. 2018), and it it highly depends on future environmental management if this can be prevented or not. To my knowledge, an assessment of future risks of CH_4 emissions from oil and gas development of the Arctic still has to be made. In the same vein, there is no integral assessment of carbon emissions from permafrost soils due to enhanced permafrost thaw by man-made disturbance. Such an assessment needs to include socio-economic aspects.

Agriculture in the shape of reindeer herding has been a traditional activity in the Arctic for centuries. In recent decennia also forestry has become important. In particular in southern permafrost areas, agricultural activities have been existing for a long time, and are increasing. For instance, in central Yakutia there are many agricultural communities (mainly livestock), and around the city of Yakutsk also crop production is expanding. Clearance of forest for settlement and agricultural activities in Central Siberia has frequently resulted in deep thaw subsidence. This reduces areas suitable for agriculture and settlement (Fedorov et al. 2016; French 2018; Fig. 5.1).

Agriculture expansion may have as a side effect export of nutrients (NH_3) via the air, and industrial development and traffic contribute to emissions of other reactive N components. On a global scale, this may have an – uncertain – cooling effect on the climate by relieving nutrient constraints on CO_2 uptake in vegetation, but at the same time these components can have an acidifying effect on soils (Erisman et al. 2011). Atmospheric transport of reactive N includes long distance transport from southerly located industrial and agricultural areas. The importance of nutrients in pristine permafrost ecosystems also prompts the question what the effect will be of an increased addition of reactive N deposition from the atmosphere generated by fossil fuel combustion and agricultural activities in higher latitudes; currently this is generally less than 1 kg N ha^{-1} $year^{-1}$ in most of the Arctic (Sect. 4.5.5).

Potentially it may further enhance the primary production of high latitude ecosystems (Hobbie et al. 2002), but many temperate and Arctic ecosystems are not only limited by N, but also by phosphorus. Responses of vegetation to any increase in N deposition, or increase in N availability due to enhanced soil organic matter decomposition, will also depend on the increase of P availability (Chapin et al. 1978; Gordon et al. 2001; Wassen et al. 2005). P delivery by weathering may be enhanced when mineral soil is involved in thaw and reworking, as occurs with erosive permafrost degradation (Hobbie et al. 2002). Next, enhanced nutrient availability does not necessarily result in greater soil C storage (e.g. Mack et al. 2004, and previous sections). On the other hand, increase of vegetation biomass, e.g. shrub expansion may contribute to mitigation of permafrost thaw – at least temporary – by its effect on the soil heat balance.

References

Abbott BW, Jones JB, Schuur EA, Chapin FS III, Bowden WB, Bret-Harte MS, Epstein HE, Flannigan MD, Harms TK, Hollingsworth TN (2016) Biomass offsets little or none of permafrost carbon release from soils, streams, and wildfire: an expert assessment. Environ Res Lett 11(3):034014. https://doi.org/10.1088/1748-9326/11/3/034014

ACIA (2004) Impacts of a warming Arctic. Cambridge, Arctic climate impact assessment, 144 p

Aerts R (2006) The freezer defrosting: global warming and litter decomposition rates in cold biomes. J Ecol 94(4):713–724

Aitken SN, Yeaman S, Holliday JA, Wang T, Curtis-McLane S (2008) Adaptation, migration or extirpation: climate change outcomes for tree populations. Evol Appl 1(1):95–111

Alexander HD, Mack MC, Goetz S, Loranty MM, Beck PS, Earl K, Zimov S, Davydov S, Thompson CC (2012) Carbon accumulation patterns during post-fire succession in cajander larch (Larix cajanderi) forests of Siberia. Ecosystems 15(7):1065–1082

Andresen CG, Lara MJ, Tweedie CE, Lougheed VL (2017) Rising plant-mediated methane emissions from arctic wetlands. Glob Chang Biol 23(3):1128–1139

Anisimov O, Reneva S (2006) Permafrost and changing climate: the Russian perspective. AMBIO J Hum Environ 35(4):169–176

Balshi MS, McGuire AD, Duffy P, Flannigan M, Kicklighter DW, Melillo J (2009) Vulnerability of carbon storage in North American boreal forests to wildfires during the 21st century. Glob Chang Biol 15(6):1491–1510

Balzter H, Gerard FF, George CT, Rowland CS, Jupp TE, McCallum I, Shvidenko A, Nilsson S, Sukhinin A, Onuchin A (2005) Impact of the Arctic oscillation pattern on interannual forest fire variability in Central Siberia. Geophys Res Lett 32(14):L14709. https://doi.org/10.1029/2005GL022526

Bartsch A, Balzter H, George C (2009) The influence of regional surface soil moisture anomalies on forest fires in Siberia observed from satellites. Environ Res Lett 4(4):045021. https://doi.org/10.1088/1748-9326/4/4/045021

Beilman DW, Vitt DH, Halsey LA (2001) Localized permafrost peatlands in western Canada: definition, distributions, and degradation. Arct Antarct Alp Res 33(1):70–77

Beringer J, Chapin FS, Thompson CC, McGuire AD (2005) Surface energy exchanges along a tundra-forest transition and feedbacks to climate. Agric For Meteorol 131(3):143–161

Bernal B, McKinley DC, Hungate BA, White PM, Mozdzer TJ, Megonigal JP (2016) Limits to soil carbon stability; deep, ancient soil carbon decomposition stimulated by new labile organic inputs. Soil Biol Biochem 98:85–94

Berteaux D, Réale D, McAdam AG, Boutin S (2004) Keeping pace with fast climate change: can arctic life count on evolution? Integr Comp Biol 44(2):140–151

Bhatt US, Walker DA, Raynolds MK, Comiso JC, Epstein HE, Jia G, Gens R, Pinzon JE, Tucker CJ, Tweedie CE (2010) Circumpolar Arctic tundra vegetation change is linked to sea ice decline. Earth Interact 14(8):1–20

Bhatt U, Walker D, Raynolds M, Bieniek P, Epstein H, Comiso J, Pinzon J, Tucker C, Polyakov I (2013) Recent declines in warming and vegetation greening trends over pan-Arctic tundra. Remote Sens 5(9):4229–4254

Bigelow NH, Brubaker LB, Edwards ME, Harrison SP, Prentice IC, Anderson PM, Andreev AA, Bartlein PJ, Christensen TR, Cramer W (2003) Climate change and Arctic ecosystems: 1. Vegetation changes north of 55° N between the last glacial maximum, mid-Holocene, and present. J Geophys Res Atmos 108(D19):8170. https://doi.org/10.1029/2002JD002558

Bliss L, Henry G, Svoboda J, Bliss D (1994) Patterns of plant distribution within two polar desert landscapes. Arct Alp Res 26(1):46–55

Blok D, Heijmans MM, Schaepman-Strub G, Kononov A, Maximov T, Berendse F (2010) Shrub expansion may reduce summer permafrost thaw in Siberian tundra. Glob Chang Biol 16(4):1296–1305

Blok D, Sass-Klaassen U, Schaepman-Strub G, Heijmans M, Sauren P, Berendse F (2011a) What are the main climate drivers for shrub growth in Northeastern Siberian tundra? Biogeosciences 8(5):1169–1179

Blok D, Heijmans MMPD, Schaepman-Strub G, van Ruijven J, Parmentier FJW, Maximov TC, Berendse F (2011b) The cooling capacity of mosses: controls on water and energy fluxes in a Siberian tundra site. Ecosystems 14(7):1055–1065. https://doi.org/10.1007/s10021-011-9463-5

Blume-Werry G, Wilson SD, Kreyling J, Milbau A (2016) The hidden season: growing season is 50% longer below than above ground along an arctic elevation gradient. New Phytol 209(3):978–986

Boike J, Wille C, Abnizova A (2008) Climatology and summer energy and water balance of polygonal tundra in the Lena River Delta, Siberia. J Geophys Res 113(G3). https://doi.org/10.1029/2007jg000540

Bokhorst S, Bjerke JW, Street L, Callaghan TV, Phoenix GK (2011) Impacts of multiple extreme winter warming events on sub-Arctic heathland: phenology, reproduction, growth, and CO_2 flux responses. Glob Chang Biol 17(9):2817–2830

Bonan GB (2008) Forests and climate change: forcings, feedbacks, and the climate benefits of forests. Science 320(5882):1444–1449

Bonan GB, Chapin FS, Thompson SL (1995) Boreal forest and tundra ecosystems as components of the climate system. Clim Chang 29(2):145–167

Briggs MA, Walvoord MA, McKenzie JM, Voss CI, Day-Lewis FD, Lane JW (2014) New permafrost is forming around shrinking Arctic lakes, but will it last? Geophys Res Lett 41(5):1585–1592

Budishchev A, Mi Y, van Huissteden J, Belelli-Marchesini L, Schaepman-Strub G, Parmentier FJW, Fratini G, Gallagher A, Maximov TC, Dolman AJ (2014) Evaluation of a plot-scale methane emission model using eddy covariance observations and footprint modelling. Biogeosciences 11(17):4651–4664. https://doi.org/10.5194/bg-11-4651-2014

Burn C, Friele P (1989) Geomorphology, vegetation succession, soil characteristics and permafrost in retrogressive thaw slumps near Mayo, Yukon territory. Arctic 42(1):31–40

Camill P, Lynch JA, Clark JS, Adams JB, Jordan B (2001) Changes in biomass, aboveground net primary production, and peat accumulation following permafrost thaw in the boreal peatlands of Manitoba, Canada. Ecosystems 4(5):461–478

Chambers S, Beringer J, Randerson J, Chapin Iii F (2005) Fire effects on net radiation and energy partitioning: contrasting responses of tundra and boreal forest ecosystems. J Geophys Res Atmos 110(D9):D09106. https://doi.org/10.1029/2004JD005299

Chapin FS III (1980) The mineral nutrition of wild plants. Annu Rev Ecol Syst 11(1):233–260

Chapin F III, Barsdate R, Barel D (1978) Phosphorus cycling in Alaskan coastal tundra: a hypothesis for the regulation of nutrient cycling. Oikos:189–199

Chapin FS, Sturm M, Serreze MC, McFadden J, Key J, Lloyd A, McGuire A, Rupp T, Lynch A, Schimel JP (2005) Role of land-surface changes in Arctic summer warming. Science 310 (5748):657-660

Chapin F, McGuire AD, Ruess RW, Hollingsworth TN, Mack M, Johnstone J, Kasischke E, Euskirchen E, Jones J, Jorgenson M (2010) Resilience of Alaska's boreal forest to climatic change. Can J For Res 40(7):1360–1370

Christensen TR, Johansson T, Åkerman HJ, Mastepanov M, Malmer N, Friborg T, Crill P, Svensson BH (2004) Thawing sub-arctic permafrost: effects on vegetation and methane emissions. Geophys Res Lett 31(4):L04501. https://doi.org/10.1029/2003GL018680

Cooper E (2011) Polar desert vegetation and plant recruitment in Murchisonfjord, Nordaustlandet, Svalbard. Geogr Ann Ser B 93(4):243–252

Cornelissen JHC, Callaghan TV, Alatalo J, Michelsen A, Graglia E, Hartley A, Hik D, Hobbie S, Press M, Robinson CH (2001) Global change and arctic ecosystems: is lichen decline a function of increases in vascular plant biomass? J Ecol 89(6):984–994

Cornelissen JH, Van Bodegom PM, Aerts R, Callaghan TV, Van Logtestijn RS, Alatalo J, Stuart Chapin F, Gerdol R, Gudmundsson J, Gwynn-Jones D (2007a) Global negative vegetation

feedback to climate warming responses of leaf litter decomposition rates in cold biomes. Ecol Lett 10(7):619–627
Cornelissen JH, Lang SI, Soudzilovskaia NA, During HJ (2007b) Comparative cryptogam ecology: a review of bryophyte and lichen traits that drive biogeochemistry. Ann Bot 99(5):987–1001
Dolman A, Shvidenko A, Schepaschenko D, Ciais P, Tchebakova N, Chen T, Van Der Molen M, Belelli Marchesini L, Maximov T, Maksyutov S (2012) An estimate of the terrestrial carbon budget of Russia using inventory-based, eddy covariance and inversion methods. Biogeosciences 9(12):5323–5340
Dorrepaal E, Aerts R, Cornelissen JH, Callaghan TV, Van Logtestijn RS (2004) Summer warming and increased winter snow cover affect *Sphagnum fuscum* growth, structure and production in a sub-arctic bog. Glob Chang Biol 10(1):93–104
Dorrepaal E, Cornelissen JH, Aerts R (2007) Changing leaf litter feedbacks on plant production across contrasting sub-arctic peatland species and growth forms. Oecologia 151(2):251–261
Elberling B, Balić-Žunić T, Edsberg A (2003) Spatial variations and controls of acid mine drainage generation. Environ Geol 43(7):806–813
Elmendorf SC, Henry GH, Hollister RD, Björk RG, Boulanger-Lapointe N, Cooper EJ, Cornelissen JH, Day TA, Dorrepaal E, Elumeeva TG (2012) Plot-scale evidence of tundra vegetation change and links to recent summer warming. Nat Clim Chang 2(6):453
Epstein HE, Raynolds MK, Walker DA, Bhatt US, Tucker CJ, Pinzon JE (2012) Dynamics of aboveground phytomass of the circumpolar Arctic tundra during the past three decades. Environ Res Lett 7(1):015506
Epstein J, Bhatt U, Raynolds MK, Walker DA, Bieniek PA, Tucker CJ, Pinzon J, Myers-Smith IH, Forbes BC, Macias-Fauria M, Boelman NT, Sweet SK (2015) Tundra greenness. NOAA Arctic report card. https://arctic.noaa.gov/portals/7/arcticreportcard/documents/arcticreport-card_full_report2015.pdf
Epstein J, Bhatt U, Raynolds M, Walker D, Forbes B, Phoenix G, Bjerke J, Tommervik H, Karlsen SR, Myneni R, Park T, Goetz S, Jia G (2018) Tundra greenness. NOAA Arctic report card. https://arctic.noaa.gov/Report-Card/Report-Card-2018
Erisman JW, Galloway J, Seitzinger S, Bleeker A, Butterbach-Bahl K (2011) Reactive nitrogen in the environment and its effect on climate change. Curr Opin Environ Sustain 3(5):281–290
Euskirchen E, McGuire A, Chapin F III (2007) Energy feedbacks of northern high-latitude ecosystems to the climate system due to reduced snow cover during 20th century warming. Glob Chang Biol 13(11):2425–2438
Fedorov A, Konstantinov P, Bashirin N, Desyatkin R, Iijima Y, Park H, Ulrich M, Sejourne A, Cosard F, Grenier C, Hiyama T, Iwahana G (2016) Climate warming and human-induced thermokarst activity in Central Yakutia and its social response. In: Günther F, Morgenstern A (eds) Tenth International Conference on Permafrost (XI ICOP), 20–24 June 2016, Potsdam, Germany, pp 1200–1201. https://doi.org/10.2312/GFZ.LIS.2016.001
Filler DM, Van Stempvoort DR, Leigh MB (2009) Remediation of frozen ground contaminated with petroleum hydrocarbons: feasibility and limits. In: Margesin R (ed) Permafrost soils, Soil biology, vol 16. Springer, Heidelberg, pp 279–301
Fisher JP, Estop-Aragonés C, Thierry A, Charman DJ, Wolfe SA, Hartley IP, Murton JB, Williams M, Phoenix GK (2016) The influence of vegetation and soil characteristics on active-layer thickness of permafrost soils in boreal forest. Glob Chang Biol 22(9):3127–3140
Fisichelli NA, Frelich LE, Reich PB (2014) Temperate tree expansion into adjacent boreal forest patches facilitated by warmer temperatures. Ecography 37(2):152–161
Flannigan MD, Stocks BJ, Wotton BM (2000) Climate change and forest fires. Sci Total Environ 262(3):221–229
Flannigan MD, Logan KA, Amiro BD, Skinner WR, Stocks B (2005) Future area burned in Canada. Clim Chang 72(1–2):1–16
Flannigan M, Stocks B, Turetsky M, Wotton M (2009) Impacts of climate change on fire activity and fire management in the circumboreal forest. Glob Chang Biol 15(3):549–560

Forkel M, Thonicke K, Beer C, Cramer W, Bartalev S, Schmullius C (2012) Extreme fire events are related to previous-year surface moisture conditions in permafrost-underlain larch forests of Siberia. Environ Res Lett 7(4):044021
French HM (2018) The periglacial environment. John Wiley & Sons ltd, Oxford 515 p
Freschet GT, Cornelissen JH, Van Logtestijn RS, Aerts R (2010a) Evidence of the 'plant economics spectrum'in a subarctic flora. J Ecol 98(2):362–373
Freschet GT, Cornelissen JH, van Logtestijn RS, Aerts R (2010b) Substantial nutrient resorption from leaves, stems and roots in a subarctic flora: what is the link with other resource economics traits? New Phytol 186(4):879–889
Freschet GT, Aerts R, Cornelissen JH (2012a) A plant economics spectrum of litter decomposability. Funct Ecol 26(1):56–65
Freschet GT, Aerts R, Cornelissen JH (2012b) Multiple mechanisms for trait effects on litter decomposition: moving beyond home-field advantage with a new hypothesis. J Ecol 100(3):619–630
Frost GV, Epstein HE (2014) Tall shrub and tree expansion in Siberian tundra ecotones since the 1960s. Glob Chang Biol 20(4):1264–1277
Frost GV, Epstein HE, Walker DA, Matyshak G, Ermokhina K (2013) Patterned-ground facilitates shrub expansion in low Arctic tundra. Environ Res Lett 8(1):015035
Gauthier S, Bernier P, Kuuluvainen T, Shvidenko A, Schepaschenko D (2015) Boreal forest health and global change. Science 349(6250):819–822
Giglio L, Randerson JT, van der Werf GR (2013) Analysis of daily, monthly, and annual burned area using the fourth-generation global fire emissions database (GFED4). J Geophys Res Biogeo 118(1):317–328
Gorchakov G, Sitnov S, Sviridenkov M, Semoutnikova E, Emilenko A, Isakov A, Kopeikin V, Karpov A, Gorchakova I, Verichev K (2014) Satellite and ground-based monitoring of smoke in the atmosphere during the summer wildfires in European Russia in 2010 and Siberia in 2012. Int J Remote Sens 35(15):5698–5721
Gordon C, Wynn J, Woodin SJ (2001) Impacts of increased nitrogen supply on high Arctic heath: the importance of bryophytes and phosphorus availability. New Phytol 149(3):461–471
Gough L, Hobbie SE (2003) Responses of moist non-acidic arctic tundra to altered environment: productivity, biomass, and species richness. Oikos 103(1):204–216
Grosse G, Harden J, Turetsky M, McGuire AD, Camill P, Tarnocai C, Frolking S, Schuur EAG, Jorgenson T, Marchenko S, Romanovsky V, Wickland KP, French N, Waldrop M, Bourgeau-Chavez L, Striegl RG (2011) Vulnerability of high-latitude soil organic carbon in North America to disturbance. J Geophys Res 116. https://doi.org/10.1029/2010jg001507
Guggenberger G, Rodionov A, Shibistova O, Grabe M, Kasansky OA, Fuchs H, Mikheyeva N, Zrazhevskaya G, Flessa H (2008) Storage and mobility of black carbon in permafrost soils of the forest tundra ecotone in northern Siberia. Glob Chang Biol 14(6):1367–1381
Harris I, Jones PD, Osborn TJ, Lister DH (2014) Updated high-resolution grids of monthly climatic observations–the CRU TS3. 10 dataset. Int J Climatol 34(3):623–642
Hartley IP, Garnett MH, Sommerkorn M, Hopkins DW, Fletcher BJ, Sloan VL, Phoenix GK, Wookey PA (2012) A potential loss of carbon associated with greater plant growth in the European Arctic. Nat Clim Chang 2(12):875
Hedenås H, Olsson H, Jonasson C, Bergstedt J, Dahlberg U, Callaghan TV (2011) Changes in tree growth, biomass and vegetation over a 13-year period in the Swedish sub-Arctic. Ambio 40(6):672
Hicks Pries CE, Schuur EA, Crummer KG (2013) Thawing permafrost increases old soil and autotrophic respiration in tundra: partitioning ecosystem respiration using $\delta^{13}C$ and ^{14}C. Glob Chang Biol 19(2):649–661. https://doi.org/10.1111/gcb.12058
Higuera PE, Chipman ML, Barnes JL, Urban MA, Hu FS (2011) Variability of tundra fire regimes in Arctic Alaska: millennial-scale patterns and ecological implications. Ecol Appl 21(8):3211–3226

Hjort J, Karjalainen O, Aalto J, Westermann S, Romanovsky VE, Nelson FE, Etzelmüller B, Luoto M (2018) Degrading permafrost puts Arctic infrastructure at risk by mid-century. Nat Commun 9(1):5147

Hobbie SE (1996) Temperature and plant species control over litter decomposition in Alaskan tundra. Ecol Monogr 66(4):503–522

Hobbie SE, Schimel JP, Trumbore SE, Randerson JR (2000) Controls over carbon storage and turnover in high-latitude soils. Glob Chang Biol 6(S1):196–210

Hobbie SE, Nadelhoffer KJ, Högberg P (2002) A synthesis: the role of nutrients as constraints on carbon balances in boreal and arctic regions. Plant Soil 242(1):163–170

Hodgkins SB, Tfaily MM, McCalley CK, Logan TA, Crill PM, Saleska SR, Rich VI, Chanton JP (2014) Changes in peat chemistry associated with permafrost thaw increase greenhouse gas production. Proc Natl Acad Sci U S A 111(16):5819–5824. https://doi.org/10.1073/pnas.1314641111

Hu FS, Higuera PE, Duffy P, Chipman ML, Rocha AV, Young AM, Kelly R, Dietze MC (2015) Arctic tundra fires: natural variability and responses to climate change. Front Ecol Environ 13(7):369–377

Iijima Y, Fedorov AN, Park H, Suzuki K, Yabuki H, Maximov TC, Ohata T (2010) Abrupt increases in soil temperatures following increased precipitation in a permafrost region, Central Lena River basin, Russia. Permafr Periglac Process 21(1):30–41. https://doi.org/10.1002/ppp.662

Iijima Y, Ohta T, Kotani A, Fedorov AN, Kodama Y, Maximov TC (2014) Sap flow changes in relation to permafrost degradation under increasing precipitation in an eastern Siberian larch forest. Ecohydrology 7(2):177–187

Iversen CM, Sloan VL, Sullivan PF, Euskirchen ES, McGuire AD, Norby RJ, Walker AP, Warren JM, Wullschleger SD (2015) The unseen iceberg: plant roots in arctic tundra. New Phytol 205(1):34–58

Iwahana G, Machimura T, Kobayashi Y, Fedorov AN, Konstantinov PY, Fukuda M (2005) Influence of forest clear-cutting on the thermal and hydrological regime of the active layer near Yakutsk, eastern Siberia. J Geophys Res Biogeo 110(G2):G02004. https://doi.org/10.1029/2005JG000039

Jafarov EE, Romanovsky VE, Genet H, McGuire AD, Marchenko SS (2013) The effects of fire on the thermal stability of permafrost in lowland and upland black spruce forests of interior Alaska in a changing climate. Environ Res Lett 8(3):035030

Jernelöv A (2010) The threats from oil spills: now, then, and in the future. Ambio 39(5–6):353–366

Johansson T, Malmer N, Crill PM, Friborg T, Aakerman JH, Mastepanov M, Christensen TR (2006) Decadal vegetation changes in a northern peatland, greenhouse gas fluxes and net radiative forcing. Glob Chang Biol 12(12):2352–2369

Johansson M, Callaghan TV, Bosiö J, Åkerman HJ, Jackowicz-Korczynski M, Christensen TR (2013) Rapid responses of permafrost and vegetation to experimentally increased snow cover in sub-arctic Sweden. Environ Res Lett 8(3):035025. https://doi.org/10.1088/1748-9326/8/3/035025

Johnstone JF (2006) Response of boreal plant communities to variations in previous fire-free interval. Int J Wildland Fire 15(4):497–508

Johnstone JF, Chapin FS, Hollingsworth TN, Mack MC, Romanovsky V, Turetsky M (2010) Fire, climate change, and forest resilience in interior Alaska. Can J For Res 40(7):1302–1312

Jones BM, Breen AL, Gaglioti BV, Mann DH, Rocha AV, Grosse G, Arp CD, Kunz ML, Walker DA (2013) Identification of unrecognized tundra fire events on the north slope of Alaska. J Geophys Res Biogeo 118(3):1334–1344

Jones BM, Grosse G, Arp CD, Miller E, Liu L, Hayes DJ, Larsen CF (2015) Recent Arctic tundra fire initiates widespread thermokarst development. Sci Rep 5:15865

Jorgenson MT, Racine CH, Walters JC, Osterkamp TE (2001) Permafrost degradation and ecological changes associated with a warmingclimate in Central Alaska. Clim Chang 48(4):551–579

Jorgenson MT, Romanovsky V, Harden J, Shur Y, O'Donnell J, Schuur EA, Kanevskiy M, Marchenko S (2010) Resilience and vulnerability of permafrost to climate change. Can J For Res 40(7):1219–1236

Jorgenson MT, Harden J, Kanevskiy M, O'Donnell J, Wickland K, Ewing S, Manies K, Zhuang Q, Shur Y, Striegl R (2013) Reorganization of vegetation, hydrology and soil carbon after permafrost degradation across heterogeneous boreal landscapes. Environ Res Lett 8(3):035017

Jorgenson M, Kanevskiy M, Shur Y, Moskalenko N, Brown D, Wickland K, Striegl R, Koch J (2015) Role of ground ice dynamics and ecological feedbacks in recent ice wedge degradation and stabilization. J Geophys Res Earth 120(11):2280–2297. https://doi.org/10.1002/2015JF003602

Jupp TE, Taylor CM, Balzter H, George CT (2006) A statistical model linking Siberian forest fire scars with early summer rainfall anomalies. Geophys Res Lett 33(14)

Juszak I, Eugster W, Heijmans MM, Schaepman-Strub G (2016) Contrasting radiation and soil heat fluxes in Arctic shrub and wet sedge tundra. Biogeosciences 13(13):4049–4064

Juszak I, Iturrate-Garcia M, Gastellu-Etchegorry J-P, Schaepman ME, Maximov TC, Schaepman-Strub G (2017) Drivers of shortwave radiation fluxes in Arctic tundra across scales. Remote Sens Environ 193:86–102

Kade A, Walker DA, Raynolds MK (2005) Plant communities and soils in cryoturbated tundra along a bioclimate gradient in the low Arctic, Alaska. Phytocoenologia 35(4):761–820

Kanevskiy M, Shur Y, Jorgenson T, Brown DNR, Moskalenko N, Brown J, Walker DA, Raynolds MK, Buchhorn M (2017) Degradation and stabilization of ice wedges: implications for assessing risk of thermokarst in northern Alaska. Geomorphology 297:20–42

Kaplan JO, Bigelow NH, Prentice IC, Harrison SP, Bartlein PJ, Christensen TR, Cramer W, Matveyeva NV, McGuire AD, Murray DF (2003) Climate change and Arctic ecosystems: 2. Modeling, paleodata-model comparisons, and future projections. J Geophys Res Atmos 108 (D19)

Karion A, Sweeney C, Pétron G, Frost G, Michael Hardesty R, Kofler J, Miller BR, Newberger T, Wolter S, Banta R (2013) Methane emissions estimate from airborne measurements over a western United States natural gas field. Geophys Res Lett 40(16):4393–4397

Kasischke ES, Turetsky MR (2006) Recent changes in the fire regime across the north American boreal region—spatial and temporal patterns of burning across Canada and Alaska. Geophys Res Lett 33(9):L09703. https://doi.org/10.1029/2006GL025677

Kasischke ES, Hyer EJ, Novelli PC, Bruhwiler LP, French NH, Sukhinin AI, Hewson JH, Stocks BJ (2005) Influences of boreal fire emissions on northern hemisphere atmospheric carbon and carbon monoxide. Glob Biogeochem Cycles 19(1):GB1012. https://doi.org/10.1029/2004GB002300

Kasischke ES, Verbyla DL, Rupp TS, McGuire AD, Murphy KA, Jandt R, Barnes JL, Hoy EE, Duffy PA, Calef M (2010) Alaska's changing fire regime—implications for the vulnerability of its boreal forests. Can J For Res 40(7):1313–1324

Katamura F, Fukuda M, Bosikov NP, Desyatkin RV (2009) Charcoal records from thermokarst deposits in central Yakutia, eastern Siberia: implications for forest fire history and thermokarst development. Quat Res 71(1):36–40

Kauppi PE, Rautiainen A, Korhonen KT, Lehtonen A, Liski J, Nöjd P, Tuominen S, Haakana M, Virtanen T (2010) Changing stock of biomass carbon in a boreal forest over 93 years. For Ecol Manag 259(7):1239–1244

Keuper F, Dorrepaal E, Van Bodegom PM, Aerts R, Van Logtestijn RS, Callaghan TV, Cornelissen JH (2011) A race for space? How Sphagnum fuscum stabilizes vegetation composition during long-term climate manipulations. Glob Chang Biol 17(6):2162–2171

Keuper F, Parmentier FJ, Blok D, van Bodegom PM, Dorrepaal E, van Hal JR, van Logtestijn RS, Aerts R (2012a) Tundra in the rain: differential vegetation responses to three years of experimentally doubled summer precipitation in Siberian shrub and Swedish bog tundra. Ambio 41(Suppl 3):269–280. https://doi.org/10.1007/s13280-012-0305-2

Keuper F, Van Bodegom PM, Dorrepaal E, Weedon JT, Van Hal J, Van Logtestijn RS, Aerts R (2012b) A frozen feast: thawing permafrost increases plant-available nitrogen in subarctic peatlands. Glob Chang Biol 18(6):1998–2007

Keuper F, Dorrepaal E, Bodegom P, Logtestijn R, Venhuizen G, Hal J, Aerts R (2014) Foraging the thaw front: increased nutrient uptake at the permafrost surface enhances biomass production of deep-rooting subarctic peatland species. In: Keuper F Direct and indirect effects of climatic changes on vegetation productivity and species composition of permafrost peatlands. Thesis, Department of Ecological Sciences, Faculty of Science, VU University VU University, Amsterdam, pp 67–84

Kharuk V, Ranson K, Dvinskya M (2007) Evidence of evergreen conifer invasion into larch dominated forests during recent decades in Central Siberia. Eurasian J Forest Res 10(2):163–171

Kharuk VI, Ranson KJ, Dvinskaya ML, Im ST (2011) Wildfires in northern Siberian larch dominated communities. Environ Res Lett 6(4):045208

Kip N, van Winden JF, Pan Y, Bodrossy L, Reichart G-J, Smolders AJP, Jetten MSM, Damsté JSS, Op den Camp HJM (2010) Global prevalence of methane oxidation by symbiotic bacteria in peat-moss ecosystems. Nat Geosci 3(9):617–621. https://doi.org/10.1038/ngeo939

Koch JC, Gurney K, Wipfli MS (2014) Morphology-dependent water budgets and nutrient fluxes in arctic thaw ponds. Permafr Periglac Process 25(2):79–93

Koch JC, Jorgenson M, Wickland K, Kanevskiy M, Striegl R (2018) Ice wedge degradation and stabilization impacts water budgets and nutrient cycling in arctic trough ponds. J Geophys Res Biogeo 123:2604–2616. https://doi.org/10.1029/2018JG004528

Kondash AJ, Albright E, Vengosh A (2017) Quantity of flowback and produced waters from unconventional oil and gas exploration. Sci Total Environ 574:314–321

Kuhry P (1994) The role of fire in the development of Sphagnum-dominated peatlands in western boreal Canada. Journal of Ecology:899–910

Kullman L (2002) Rapid recent range-margin rise of tree and shrub species in the Swedish Scandes. J Ecol 90(1):68–77

Kurz WA, Dymond C, Stinson G, Rampley G, Neilson E, Carroll A, Ebata T, Safranyik L (2008) Mountain pine beetle and forest carbon feedback to climate change. Nature 452(7190):987

Kuzyakov Y (2010) Priming effects: interactions between living and dead organic matter. Soil Biol Biochem 42(9):1363–1371

Lafleur PM, Humphreys ER (2018) Tundra shrub effects on growing season energy and carbon dioxide exchange. Environ Res Lett 13(5):055001

Lara M, Villarreal S, Johnson D, Hollister R, Webber P, Tweedie C (2012) Estimated change in tundra ecosystem function near Barrow, Alaska between 1972 and 2010. Environ Res Lett 7(1):015507

Lara MJ, McGuire AD, Euskirchen ES, Tweedie CE, Hinkel KM, Skurikhin AN, Romanovsky VE, Grosse G, Bolton WR, Genet H (2015) Polygonal tundra geomorphological change in response to warming alters future CO_2 and CH_4 flux on the Barrow Peninsula. Glob Chang Biol 21(4):1634–1651. https://doi.org/10.1111/gcb.12757

Lara MJ, Nitze I, Grosse G, Martin P, McGuire AD (2018) Reduced arctic tundra productivity linked with landform and climate change interactions. Sci Rep 8(1):2345

Leibman M, Kizakov A, Sulerzhitsky L, Zaretskaia N (2003) Dynamics of landslide slopes and their development on Yamal Peninsula. In: Permafrost. Proceedings of the 8th international conference on permafrost. Swets and Zeitlinger, Lisse, 2003. pp 651–656

Leibman M, Khomutov A, Kizyakov A (2014) Cryogenic landslides in the West-Siberian plain of Russia: classification, mechanisms, and landforms. In: Landslides in cold regions in the context of climate change. Springer, Cham, pp 143–162

Lewkowicz AG, Way RG (2019) Extremes of summer climate trigger thousands of thermokarst landslides in a high Arctic environment. Nat Commun 10(1):1329. https://doi.org/10.1038/s41467-019-09314-7

Li B, Heijmans MM, Berendse F, Blok D, Maximov T, Sass-Klaassen U (2016) The role of summer precipitation and summer temperature in establishment and growth of dwarf shrub Betula nana in northeast Siberian tundra. Polar Biol 39(7):1245–1255

Li B, Heijmans MM, Blok D, Wang P, Karsanaev SV, Maximov TC, van Huissteden J, Berendse F (2017a) Thaw pond development and initial vegetation succession in experimental plots at a Siberian lowland tundra site. Plant Soil 420(1–2):147–162

Li B, Heijmans MMPD, Limpens J, Gallagher A, Van Huissteden J, Maximov TC, Berendse F (2017b) Peat moss carpet reduces methane emission and facilitates shrub expansion in Siberian Arctic tundra. In: Li B (ed) A clash of plants. Vegetation succession and its interaction with permafrost dynamics in the Arctic lowland tundra. PhD thesis, Wageningen, The Netherlands, pp 26–42

Liljedahl A, Hinzman L, Busey R, Yoshikawa K (2007) Physical short-term changes after a tussock tundra fire, Seward Peninsula, Alaska. Journal of Geophysical Research: Earth Surface 112 (F2):F02S07. https://doi.org/10.1029/2006JF000554

Liljedahl AK, Boike J, Daanen RP, Fedorov AN, Frost GV, Grosse G, Hinzman LD, Iijma Y, Jorgenson JC, Matveyeva N, Necsoiu M, Raynolds MK, Romanovsky VE, Schulla J, Tape KD, Walker DA, Wilson CJ, Yabuki H, Zona D (2016) Pan-Arctic ice-wedge degradation in warming permafrost and its influence on tundra hydrology. Nat Geosci 9(4):312–318. https://doi.org/10.1038/ngeo2674

Limpens J, Bohlin E, Nilsson MB (2017) Phylogenetic or environmental control on the elemental and organo-chemical composition of Sphagnum mosses? Plant Soil 417(1–2):69–85

Liston GE, Sturm M (2002) Winter precipitation patterns in arctic Alaska determined from a blowing-snow model and snow-depth observations. J Hydrometeorol 3(6):646–659

Liu L, Schaefer K, Chen A, Gusmeroli A, Zebker H, Zhang T (2015) Remote sensing measurements of thermokarst subsidence using InSAR. J Geophys Res Earth 120(9):1935–1948

Loranty MM, Berner LT, Taber ED, Kropp H, Natali SM, Alexander HD, Davydov SP, Zimov NS (2018a) Understory vegetation mediates permafrost active layer dynamics and carbon dioxide fluxes in open-canopy larch forests of northeastern Siberia. PLoS One 13(3):e0194014

Loranty MM, Abbott BW, Blok D, Douglas TA, Epstein HE, Forbes BC, Jones BM, Kholodov AL, Kropp H, Malhotra A (2018b) Reviews and syntheses: changing ecosystem influences on soil thermal regimes in northern high-latitude permafrost regions. Biogeosciences 15:5287–5313. https://doi.org/10.5194/bg-15-5287-2018

Lupascu M, Welker J, Seibt U, Maseyk K, Xu X, Czimczik C (2014) High Arctic wetting reduces permafrost carbon feedbacks to climate warming. Nat Clim Chang 4(1):51

Ma Z, Peng C, Zhu Q, Chen H, Yu G, Li W, Zhou X, Wang W, Zhang W (2012) Regional drought-induced reduction in the biomass carbon sink of Canada's boreal forests. Proc Natl Acad Sci 109(7):2423–2427

MacDonald G, Kremenetski K, Beilman D (2007) Climate change and the northern Russian treeline zone. Philos Trans R Soc B Biol Sci 363(1501):2283–2299

Macias-Fauria M, Forbes BC, Zetterberg P, Kumpula T (2012) Eurasian Arctic greening reveals teleconnections and the potential for structurally novel ecosystems. Nat Clim Chang 2(8):613

Mack MC, Schuur EA, Bret-Harte MS, Shaver GR, Chapin FS III (2004) Ecosystem carbon storage in arctic tundra reduced by long-term nutrient fertilization. Nature 431(7007):440

Mack MC, Bret-Harte MS, Hollingsworth TN, Jandt RR, Schuur EA, Shaver GR, Verbyla DL (2011) Carbon loss from an unprecedented Arctic tundra wildfire. Nature 475(7357):489

Magnússon R, Rotbarth R, Sass-Klaassen U, Heijmans MMPD, Limpens J, Kleijn D, Van Huissteden J, Maximov TC (submitted) Coupled vegetation succession and permafrost dynamics in thaw ponds in the Siberian lowland tundra. Ecology

Malhotra A, Moore TR, Limpens J, Roulet NT (2018) Post-thaw variability in litter decomposition best explained by microtopography at an ice-rich permafrost peatland. Arctic, Antarctic, and Alpine Res 50 (1):e1415622

Malmer N, Johansson T, Olsrud M, Christensen TR (2005) Vegetation, climatic changes and net carbon sequestration in a North-Scandinavian subarctic mire over 30 years. Glob Chang Biol 11(11):1895–1909

Marushchak M, Friborg T, Biasi C, Herbst M, Johansson T, Kiepe I, Liimatainen M, Lind S, Martikainen P, Virtanen T (2016) Methane dynamics in the subarctic tundra: combining stable isotope analyses, plot-and ecosystem-scale flux measurements. Biogeosciences 13(2):597–608

Mazhitova G, Karstkarel N, Oberman N, Romanovsky V, Kuhry P (2004) Permafrost and infrastructure in the Usa Basin (Northeast European Russia): possible impacts of global warming. AMBIO J Hum Environ 33(6):289–295

McGuire A, Wirth C, Apps M, Beringer J, Clein J, Epstein H, Kicklighter D, Bhatti J, Chapin Iii F, De Groot B (2002) Environmental variation, vegetation distribution, carbon dynamics and water/energy exchange at high latitudes. J Veg Sci 13(3):301–314

McGuire AD, Anderson LG, Christensen TR, Dallimore S, Guo L, Hayes DJ, Heimann M, Lorenson TD, Macdonald RW, Roulet N (2009) Sensitivity of the carbon cycle in the Arctic to climate change. Ecol Monogr 79(4):523–555

Myers-Smith IH, Hik DS (2013) Shrub canopies influence soil temperatures but not nutrient dynamics: an experimental test of tundra snow–shrub interactions. Ecol Evol 3(11):3683–3700

Myers-Smith I, Harden JW, Wilmking M, Fuller CC, McGuire AD, Chapin FS III (2008) Wetland succession in a permafrost collapse: interactions between fire and thermokarst. Biogeosciences 5(5):1273–1286

Myers-Smith IH, Forbes BC, Wilmking M, Hallinger M, Lantz T, Blok D, Tape KD, Macias-Fauria M, Sass-Klaassen U, Lévesque E (2011) Shrub expansion in tundra ecosystems: dynamics, impacts and research priorities. Environ Res Lett 6(4):045509

Myers-Smith IH, Elmendorf SC, Beck PS, Wilmking M, Hallinger M, Blok D, Tape KD, Rayback SA, Macias-Fauria M, Forbes BC (2015) Climate sensitivity of shrub growth across the tundra biome. Nat Clim Chang 5(9):887

Myneni R, Dong J, Tucker C, Kaufmann R, Kauppi P, Liski J, Zhou L, Alexeyev V, Hughes M (2001) A large carbon sink in the woody biomass of northern forests. Proc Natl Acad Sci 98(26):14784–14789

Natali SM, Schuur EA, Trucco C, Pries CEH, Crummer KG, Lopez AFB (2011) Effects of experimental warming of air, soil and permafrost on carbon balance in Alaskan tundra. Glob Chang Biol 17(3):1394–1407

Natali SM, Schuur EA, Webb EE, Pries CEH, Crummer KG (2014) Permafrost degradation stimulates carbon loss from experimentally warmed tundra. Ecology 95(3):602–608

Natali SM, Schuur EAG, Mauritz M, Schade JD, Celis G, Crummer KG, Johnston C, Krapek J, Pegoraro E, Salmon VG, Webb EE (2015) Permafrost thaw and soil moisture driving CO_2 and CH_4 release from upland tundra. J Geophys Res Biogeo 120(3):525–537. https://doi.org/10.1002/2014jg002872

Nauta AL, Heijmans MMPD, Blok D, Limpens J, Elberling B, Gallagher A, Li B, Petrov RE, Maximov TC, van Huissteden J, Berendse F (2014) Permafrost collapse after shrub removal shifts tundra ecosystem to a methane source. Nat Clim Chang 5(1):67–70. https://doi.org/10.1038/nclimate2446

Nelson FE, Anisimov OA, Shiklomanov NI (2002) Climate change and hazard zonation in the circum-Arctic permafrost regions. Nat Hazards 26(3):203–225

Nicotra AB, Atkin OK, Bonser SP, Davidson AM, Finnegan E, Mathesius U, Poot P, Purugganan MD, Richards CL, Valladares F (2010) Plant phenotypic plasticity in a changing climate. Trends Plant Sci 15(12):684–692

Nitze I, Grosse G, Jones BM, Romanovsky VE, Boike J (2018) Remote sensing quantifies widespread abundance of permafrost region disturbances across the Arctic and subarctic. Nat Commun 9(1):5423. https://doi.org/10.1038/s41467-018-07663-3

Norby RJ, Zak DR (2011) Ecological lessons from free-air CO_2 enrichment (FACE) experiments. Annu Rev Ecol Evol Syst 42:181–203

Nossov DR, Jorgenson MT, Kielland K, Kanevskiy MZ (2013) Edaphic and microclimatic controls over permafrost response to fire in interior Alaska. Environ Res Lett 8(3):035013
O'Donnell JA, Jorgenson MT, Harden JW, McGuire AD, Kanevskiy MZ, Wickland KP (2012) The effects of permafrost thaw on soil hydrologic, thermal, and carbon dynamics in an Alaskan peatland. Ecosystems 15(2):213–229
Oberbauer SF, Tweedie CE, Welker JM, Fahnestock JT, Henry GH, Webber PJ, Hollister RD, Walker MD, Kuchy A, Elmore E (2007) Tundra CO_2 fluxes in response to experimental warming across latitudinal and moisture gradients. Ecol Monogr 77(2):221–238
Ohta T, Maximov TC, Dolman AJ, Nakai T, van der Molen MK, Kononov AV, Maximov AP, Hiyama T, Iijima Y, Moors EJ (2008) Interannual variation of water balance and summer evapotranspiration in an eastern Siberian larch forest over a 7-year period (1998–2006). Agric For Meteorol 148(12):1941–1953
Olefeldt D, Turetsky MR, Crill PM, McGuire AD (2013) Environmental and physical controls on northern terrestrial methane emissions across permafrost zones. Glob Chang Biol 19(2):589–603
Osterkamp T, Jorgenson M, Schuur E, Shur Y, Kanevskiy M, Vogel J, Tumskoy V (2009) Physical and ecological changes associated with warming permafrost and thermokarst in interior Alaska. Permafr Periglac Process 20(3):235–256
Oulehle F, Rowe E, Myška O, Chuman T, Evans C (2016) Plant functional type affects nitrogen use efficiency in high-Arctic tundra. Soil Biol Biochem 94:19–28
Palmtag J, Hugelius G, Lashchinskiy N, Tamstorf MP, Richter A, Elberling B, Kuhry P (2015) Storage, landscape distribution, and burial history of soil organic matter in contrasting areas of continuous permafrost. Arct Antarct Alp Res 47(1):71–88
Parmentier F, Van Der Molen M, Van Huissteden J, Karsanaev S, Kononov A, Suzdalov D, Maximov T, Dolman A (2011) Longer growing seasons do not increase net carbon uptake in the northeastern Siberian tundra. J Geophys Res Biogeo 116(G4):G04013. https://doi.org/10.1029/2011JG001653
Payette S (2007) Contrasted dynamics of northern Labrador tree lines caused by climate change and migrational lag. Ecology 88(3):770–780
Phoenix GK, Bjerke JW (2016) Arctic browning: extreme events and trends reversing arctic greening. Glob Chang Biol 22(9):2960–2962
Piao S, Ciais P, Friedlingstein P, Peylin P, Reichstein M, Luyssaert S, Margolis H, Fang J, Barr A, Chen A (2008) Net carbon dioxide losses of northern ecosystems in response to autumn warming. Nature 451(7174):49
Piao S, Wang X, Ciais P, Zhu B, Wang T, Liu J (2011) Changes in satellite-derived vegetation growth trend in temperate and boreal Eurasia from 1982 to 2006. Glob Chang Biol 17(10):3228–3239
Pielke RA, Vidale PL (1995) The boreal forest and the polar front. J Geophys Res Atmos 100(D12):25755–25758
Podur J, Martell DL, Csillag F (2003) Spatial patterns of lightning-caused forest fires in Ontario, 1976–1998. Ecol Model 164(1):1–20
Price D, McKenney D, Joyce L, Siltanen R, Papadopol P, Lawrence K (2011) High resolution interpolation of IPCC AR4 GCM climate scenarios for Canada. Nat Resour Can, Can for Serv North for Cent, Edmonton, Alta Inf rep NOR-X-421
Price DT, Alfaro R, Brown K, Flannigan M, Fleming R, Hogg E, Girardin M, Lakusta T, Johnston M, McKenney D (2013) Anticipating the consequences of climate change for Canada's boreal forest ecosystems. Environ Rev 21(4):322–365
Raghoebarsing AA, Smolders AJ, Schmid MC, Rijpstra WIC, Wolters-Arts M, Derksen J, Jetten MS, Schouten S, Damsté JSS, Lamers LP (2005) Methanotrophic symbionts provide carbon for photosynthesis in peat bogs. Nature 436(7054):1153
Randerson JT, Liu H, Flanner MG, Chambers SD, Jin Y, Hess PG, Pfister G, Mack M, Treseder K, Welp L (2006) The impact of boreal forest fire on climate warming. Science 314(5802):1130–1132
Rasmus S, Lundell R, Saarinen T (2011) Interactions between snow, canopy, and vegetation in a boreal coniferous forest. Plant Ecol Diver 4(1):55–65

Raynolds MK, Walker DA, Ambrosius KJ, Brown J, Everett KR, Kanevskiy M, Kofinas GP, Romanovsky VE, Shur Y, Webber PJ (2014) Cumulative geoecological effects of 62 years of infrastructure and climate change in ice-rich permafrost landscapes, Prudhoe Bay oilfield, Alaska. Glob Chang Biol 20(4):1211–1224. https://doi.org/10.1111/gcb.12500

Reichle L, Epstein H, Bhatt U, Raynolds M, Walker D (2018) Spatial heterogeneity of the temporal dynamics of Arctic tundra vegetation. Geophys Res Lett 45(17):9206–9215

Reyes FR, Lougheed VL (2015) Rapid nutrient release from permafrost thaw in arctic aquatic ecosystems. Arct Antarct Alp Res 47(1):35–48

Robinson CH, Wookey PA, Lee JA, Callaghan TV, Press MC (1998) Plant community responses to simulated environmental change at a high arctic polar semi-desert. Ecology 79(3):856–866

Rocha AV, Loranty MM, Higuera PE, Mack MC, Hu FS, Jones BM, Breen AL, Rastetter EB, Goetz SJ, Shaver GR (2012) The footprint of Alaskan tundra fires during the past half-century: implications for surface properties and radiative forcing. Environ Res Lett 7(4):044039

Sawamoto T, Hatano R, Yajima T, Takahashi K, Isaev A (2000) Soil respiration in Siberian taiga ecosystems with different histories of forest fire. Soil Sci Plant Nutr 46(1):31–42

Scheffer M, Carpenter S, Foley JA, Folke C, Walkerk B (2001) Catastrophic shifts in ecosystems. Nature 413:591–596

Scheffer M, Bascompte J, Brock WA, Brovkin V, Carpenter SR, Dakos V, Held H, Van Nes EH, Rietkerk M, Sugihara G (2009) Early-warning signals for critical transitions. Nature 461(7260):53

Schothorst C (1977) Subsidence of low moor peat soils in the western Netherlands. Geoderma 17(4):265–291

Schuur EA, Crummer KG, Vogel JG, Mack MC (2007) Plant species composition and productivity following permafrost thaw and thermokarst in Alaskan tundra. Ecosystems 10(2):280–292

Shiklomanov NI, Streletskiy DA, Swales TB, Kokorev VA (2017) Climate change and stability of urban infrastructure in Russian permafrost regions: prognostic assessment based on GCM climate projections. Geogr Rev 107(1):125–142

Shur Y, Jorgenson M (2007) Patterns of permafrost formation and degradation in relation to climate and ecosystems. Permafr Periglac Process 18(1):7–19

Siewert MB, Hanisch J, Weiss N, Kuhry P, Maximov TC, Hugelius G (2015) Comparing carbon storage of Siberian tundra and taiga permafrost ecosystems at very high spatial resolution. J Geophys Res Biogeo 120(10):1973–1994

Sistla SA, Moore JC, Simpson RT, Gough L, Shaver GR, Schimel JP (2013) Long-term warming restructures Arctic tundra without changing net soil carbon storage. Nature 497(7451):615

Sjögersten S, Caul S, Daniell T, Jurd A, O'Sullivan O, Stapleton C, Titman JJ (2016) Organic matter chemistry controls greenhouse gas emissions from permafrost peatlands. Soil Biol Biochem 98:42–53

Soja AJ, Tchebakova NM, French NH, Flannigan MD, Shugart HH, Stocks BJ, Sukhinin AI, Parfenova E, Chapin FS III, Stackhouse PW Jr (2007) Climate-induced boreal forest change: predictions versus current observations. Glob Planet Chang 56(3–4):274–296

Soudzilovskaia NA, Bodegom PM, Cornelissen JH (2013) Dominant bryophyte control over high-latitude soil temperature fluctuations predicted by heat transfer traits, field moisture regime and laws of thermal insulation. Funct Ecol 27(6):1442–1454

Stephens BB, Gurney KR, Tans PP, Sweeney C, Peters W, Bruhwiler L, Ciais P, Ramonet M, Bousquet P, Nakazawa T (2007) Weak northern and strong tropical land carbon uptake from vertical profiles of atmospheric CO_2. Science 316(5832):1732–1735

Street LE, Stoy PC, Sommerkorn M, Fletcher BJ, Sloan VL, Hill TC, Williams M (2012) Seasonal bryophyte productivity in the sub-Arctic: a comparison with vascular plants. Funct Ecol 26(2):365–378

Ström L, Christensen TR (2007) Below ground carbon turnover and greenhouse gas exchanges in a sub-arctic wetland. Soil Biol Biochem 39(7):1689–1698

Ström L, Ekberg A, Mastepanov M, Røjle Christensen T (2003) The effect of vascular plants on carbon turnover and methane emissions from a tundra wetland. Glob Chang Biol 9(8):1185–1192

Ström L, Mastepanov M, Christensen TR (2005) Species-specific effects of vascular plants on carbon turnover and methane emissions from wetlands. Biogeochemistry 75(1):65–82

Ström L, Tagesson T, Mastepanov M, Christensen TR (2012) Presence of Eriophorum scheuchzeri enhances substrate availability and methane emission in an Arctic wetland. Soil Biol Biochem 45:61–70

Sturm M, Holmgren J, McFadden JP, Liston GE, Chapin FS III, Racine CH (2001) Snow–shrub interactions in Arctic tundra: a hypothesis with climatic implications. J Clim 14(3):336–344

Sturm M, Douglas T, Racine C, Liston GE (2005) Changing snow and shrub conditions affect albedo with global implications. J Geophys Res Biogeo 110(G1):G01004. https://doi.org/10.1029/2005JG000013

Sullivan PF, Sommerkorn M, Rueth HM, Nadelhoffer KJ, Shaver GR, Welker JM (2007) Climate and species affect fine root production with long-term fertilization in acidic tussock tundra near Toolik Lake, Alaska. Oecologia 153(3):643–652

Takakai F, Desyatkin AR, Lopez CL, Fedorov AN, Desyatkin RV, Hatano R (2008) Influence of forest disturbance on CO_2, CH_4 and N_2O fluxes from larch forest soil in the permafrost taiga region of eastern Siberia. Soil Sci Plant Nutr 54(6):938–949

Tarnocai C (2006) The effect of climate change on carbon in Canadian peatlands. Glob Planet Chang 53(4):222–232

Taş N, Prestat E, McFarland JW, Wickland KP, Knight R, Berhe AA, Jorgenson T, Waldrop MP, Jansson JK (2014) Impact of fire on active layer and permafrost microbial communities and metagenomes in an upland Alaskan boreal forest. ISME J 8(9):1904

Tchebakova N, Parfenova E, Soja A (2009) The effects of climate, permafrost and fire on vegetation change in Siberia in a changing climate. Environ Res Lett 4(4):045013

Team CCAVM (2003) Circumpolar Arctic vegetation map. Conservation of Arctic Flora and Fauna (CAFF) Map No 1. Anchorage, AK. US Fish and Wildlife Service. www.geobotany.uaf.edu/cavm

Teltewskoi A, Beermann F, Beil I, Bobrov A, De Klerk P, Lorenz S, Lüder A, Michaelis D, Joosten H (2016) 4000 years of changing wetness in a permafrost polygon Peatland (Kytalyk, NE Siberia): a comparative high-resolution multi-proxy study. Permafr Periglac Process 27(1):76–95

Townsend-Small A, Ferrara TW, Lyon DR, Fries AE, Lamb BK (2016) Emissions of coalbed and natural gas methane from abandoned oil and gas wells in the United States. Geophys Res Lett 43(5):2283–2290

Trucco C, Schuur EAG, Natali SM, Belshe EF, Bracho R, Vogel J (2012) Seven-year trends of CO_2 exchange in a tundra ecosystem affected by long-term permafrost thaw. J Geophys Res Biogeo 117(G2):G02031. https://doi.org/10.1029/2011jg001907

Turetsky MR, Wieder RK, Williams CJ, Vitt DH (2000) Organic matter accumulation, peat chemistry, and permafrost melting in peatlands of boreal Alberta. Ecoscience 7(3):115–122

Turetsky MR, Wieder RK, Vitt DH (2002) Boreal peatland C fluxes under varying permafrost regimes. Soil Biol Biochem 34(7):907–912

Turetsky MR, Amiro BD, Bosch E, Bhatti JS (2004) Historical burn area in western Canadian peatlands and its relationship to fire weather indices. Glob Biogeochem Cycles 18(4):GB4014. https://doi.org/10.1029/2004GB002222

Turetsky M, Wieder R, Vitt D, Evans R, Scott K (2007) The disappearance of relict permafrost in boreal North America: effects on peatland carbon storage and fluxes. Glob Chang Biol 13(9):1922–1934

Turetsky MR, Kane ES, Harden JW, Ottmar RD, Manies KL, Hoy E, Kasischke ES (2011a) Recent acceleration of biomass burning and carbon losses in Alaskan forests and peatlands. Nat Geosci 4(1):27

Turetsky M, Donahue W, Benscoter B (2011b) Experimental drying intensifies burning and carbon losses in a northern peatland. Nat Commun 2:514

Turetsky MR, Bond-Lamberty B, Euskirchen E, Talbot J, Frolking S, McGuire AD, Tuittila ES (2012) The resilience and functional role of moss in boreal and arctic ecosystems. New Phytol 196(1):49–67

Turetsky MR, Kotowska A, Bubier J, Dise NB, Crill P, Hornibrook ER, Minkkinen K, Moore TR, Myers-Smith IH, Nykanen H, Olefeldt D, Rinne J, Saarnio S, Shurpali N, Tuittila ES, Waddington JM, White JR, Wickland KP, Wilmking M (2014) A synthesis of methane emissions from 71 northern, temperate, and subtropical wetlands. Glob Chang Biol 20(7):2183–2197. https://doi.org/10.1111/gcb.12580

Ukraintseva N, Leibman M, Streletskaya I, Mikhaylova T (2014) Geochemistry of plant-soil-permafrost system on landslide-affected slopes, Yamal, Russia as an indicator of landslide age. In: Landslides in cold regions in the context of climate change. Springer, Cham, pp 107–131

Van Der Wal R, Stien A (2014) High-arctic plants like it hot: a long-term investigation of between-year variability in plant biomass. Ecology 95(12):3414–3427

Van der Werf GR, Randerson JT, Giglio L, Collatz G, Mu M, Kasibhatla PS, Morton DC, DeFries R, Yv J, van Leeuwen TT (2010) Global fire emissions and the contribution of deforestation, savanna, forest, agricultural, and peat fires (1997–2009). Atmos Chem Phys 10(23):11707–11735

Van Der Werf GR, Randerson JT, Giglio L, Van Leeuwen TT, Chen Y, Rogers BM, Mu M, Van Marle MJ, Morton DC, Collatz GJ (2017) Global fire emissions estimates during 1997-2016. Earth System Science Data 9:697–720. https://doi.org/10.5194/essd-9-697-2017

Van Huissteden J, Dolman AJ (2012) Soil carbon in the Arctic and the permafrost carbon feedback. Curr Opin Environ Sustain 4(5):545–551. https://doi.org/10.1016/j.cosust.2012.09.008

Van Huissteden J, Maximov TC, Dolman AJ (2005) High methane flux from an arctic floodplain (Indigirka lowlands, eastern Siberia). J Geophys Res Biogeo 110(G2):G02002. https://doi.org/10.1029/2005jg000010

Van Huissteden J, Maximov TC, Dolman AJ (2009) Correction to "High methane flux from an arctic floodplain (Indigirka lowlands, eastern Siberia). J Geophys Res Biogeosci 114 (G2):n/a–n/a. doi:https://doi.org/10.1029/2009jg001040

van Thienen-Visser K, Pruiksma J, Breunese J (2015) Compaction and subsidence of the Groningen gas field in the Netherlands. Proceedings of the International Association of Hydrological Sciences 372:367–373

Van Wijk M, Clemmensen KE, Shaver G, Williams M, Callaghan T, Chapin F III, Cornelissen J, Gough L, Hobbie S, Jonasson S (2004) Long-term ecosystem level experiments at Toolik Lake, Alaska, and at Abisko, Northern Sweden: generalizations and differences in ecosystem and plant type responses to global change. Glob Chang Biol 10(1):105–123

Verbyla D (2011) Browning boreal forests of western North America. Environ Res Lett 6(4):041003

Verhoeven J, Toth E (1995) Decomposition of Carex and Sphagnum litter in fens: effect of litter quality and inhibition by living tissue homogenates. Soil Biol Biochem 27(3):271–275

Vitt D, Halsey LA, Zoltai SC (2000) The changing landscape of Canada's western boreal forest: the current dynamics of permafrost. Can J For Res 30:283–287

Vogel J, Schuur EA, Trucco C, Lee H (2009) Response of CO_2 exchange in a tussock tundra ecosystem to permafrost thaw and thermokarst development. J Geophys Res Biogeo 114(G4)

Vuorinen KE, Oksanen L, Oksanen T, Pyykönen A, Olofsson J, Virtanen R (2017) Open tundra persist, but arctic features decline—vegetation changes in the warming Fennoscandian tundra. Glob Chang Biol 23(9):3794–3807

Walker DA, Raynolds MK, Daniëls FJ, Einarsson E, Elvebakk A, Gould WA, Katenin AE, Kholod SS, Markon CJ, Melnikov ES (2005) The circumpolar Arctic vegetation map. J Veg Sci 16(3):267–282

Walker MD, Wahren CH, Hollister RD, Henry GH, Ahlquist LE, Alatalo JM, Bret-Harte MS, Calef MP, Callaghan TV, Carroll AB (2006) Plant community responses to experimental warming across the tundra biome. Proc Natl Acad Sci 103(5):1342–1346

Walter K, Edwards M, Grosse G, Zimov S, Chapin F (2007) Thermokarst lakes as a source of atmospheric CH_4 during the last deglaciation. Science 318(5850):633–636

Walter-Anthony KM, Zimov SA, Grosse G, Jones MC, Anthony PM, Chapin FS 3rd, Finlay JC, Mack MC, Davydov S, Frenzel P, Frolking S (2014) A shift of thermokarst lakes from carbon sources to sinks during the Holocene epoch. Nature 511(7510):452–456. https://doi.org/10.1038/nature13560

Wang P, Heijmans MM, Mommer L, van Ruijven J, Maximov TC, Berendse F (2016) Belowground plant biomass allocation in tundra ecosystems and its relationship with temperature. Environ Res Lett 11(5):055003

Wang P, van Ruijven J, Heijmans MM, Berendse F, Maksimov A, Maximov T, Mommer L (2017) Short-term root and leaf decomposition of two dominant plant species in a Siberian tundra. Pedobiologia 65:68–76

Wang P, Limpens J, Nauta A, van Huissteden C, Quirina van Rijssel S, Mommer L, de Kroon H, Maximov TC, Heijmans MM (2018) Depth-based differentiation in nitrogen uptake between graminoids and shrubs in an Arctic tundra plant community. J Veg Sci 29(1):34–41

Wassen MJ, Venterink HO, Lapshina ED, Tanneberger F (2005) Endangered plants persist under phosphorus limitation. Nature 437(7058):547

Weintraub MN, Schimel JP (2005) Nitrogen cycling and the spread of shrubs control changes in the carbon balance of arctic tundra ecosystems. Bioscience 55(5):408–415

Wookey PA, Aerts R, Bardgett RD, Baptist F, Bråthen KA, Cornelissen JH, Gough L, Hartley IP, Hopkins DW, Lavorel S (2009) Ecosystem feedbacks and cascade processes: understanding their role in the responses of Arctic and alpine ecosystems to environmental change. Glob Chang Biol 15(5):1153–1172

Yarie J, Van Cleve K (2010) Long-term monitoring of climatic and nutritional affects on tree growth in interior Alaska. Can J For Res 40(7):1325–1335

Zeh L, Limpens J, Erhagen B, Bragazza L, Kalbitz K (2019) Plant functional types and temperature control carbon input via roots in peatland soils. Plant Soil:1–20

Zhang N, Yasunari T, Ohta T (2011) Dynamics of the larch taiga–permafrost coupled system in Siberia under climate change. Environ Res Lett 6(2):024003

Zhu Z, Piao S, Myneni RB, Huang M, Zeng Z, Canadell JG, Ciais P, Sitch S, Friedlingstein P, Arneth A (2016a) Greening of the earth and its drivers. Nat Clim Chang 6(8):791

Zhu Q, Iversen CM, Riley WJ, Slette IJ, Vander Stel HM (2016b) Root traits explain observed tundra vegetation nitrogen uptake patterns: implications for trait-based land models. J Geophys Res Biogeo 121(12):3101–3112

Zimov S, Zimov N, Tikhonov A, Chapin F III (2012) Mammoth steppe: a high-productivity phenomenon. Quat Sci Rev 57:26–45

Chapter 7
Methane

Abstract Methane is the most widely known greenhouse gas emitted from permafrost areas. The major sources are: wetland ecosystem emission, emission from thawed old carbon from permafrost, and deeper sources, including intra-permafrost and below-permafrost free gas reservoirs and gas hydrates. The hydrate source occurs mainly at the base of the permafrost at depths below ~200 m and is unlikely to be disturbed by permafrost warming on a time scale of centuries. However, shallower hydrates have been reported from boreholes. Free gas reservoirs are also known from taliks. Permafrost acts as a 'lid' on deeper methane sources, but with permafrost warming this lid will be increasingly leaky, in particular in areas of thin continuous and discontinuous permafrost that show rapid warming. One of these areas is Northwestern Siberia, from which a new phenomenon has been reported: explosive crater formation with emission of CH_4. The chapter concludes with a perspective on CH_4 emission in comparison with other greenhouse gases. CO_2 dominates emission from thawing permafrost. However, CH_4 emission from shallow intrapermafrost gas reservoirs is poorly quantified, and ecosystem CH_4 emission can react fast on changes in the Arctic climate system, such as sea ice decline. Next, N_2O may become an increasingly important greenhouse gas, related to increasing erosion phenomena.

'Methane from permafrost' has become a climate change icon, but a rather imprecise and unspecified one. Three sources of CH_4 from permafrost can be discerned:

- Ecosystem CH_4 emission, as discussed in Sects. 4.5.6 and 6.2.3. This is derived from recently produced, labile organic matter and can be considered as a part of the ecosystem respiration. This does not exclude that some old soil carbon is respired as well, in particular in permafrost thaw ponds, but the CH_4 produced would have a largely recent radiocarbon age. This is a source that is not specifically depending on permafrost thaw for the production and release to the atmosphere of CH_4, but it is related to the presence of wetlands. However, permafrost thaw strongly influences wetland ecology and wetland area and therefore also wetland CH_4 emission.
- Old carbon CH_4, which is produced from old organic matter that is thawed out of older sedimentary deposits, mostly Early Holocene peat and alluvial deposits, or Pleistocene Yedoma deposits. It is mostly produced by thaw below thaw lakes

J. van Huissteden, *Thawing Permafrost*,
https://doi.org/10.1007/978-3-030-31379-1_7

and ponds. Although it may be mixed with CH_4 derived from recent photosynthetic products, it has radiocarbon ages indicating a much older age than the ecosystem CH_4 emission (Walter-Anthony et al. 2016). See Sect. 5.3.3. Both production and release of CH_4 are related to permafrost thaw.
- Deep CH4 sources: gas hydrates at the base of permafrost or at higher levels, gas trapped in deeper permafrost levels at depths of tens of metres and deeper, and methane seeps from deeper gas reservoirs (Walter-Anthony et al. 2012). CH_4 from these sources may be very old and may have infinite radiocarbon ages (> ±70 kyr). Compared to the two sources of CH_4 above, deep CH_4 sources do not represent new production of CH_4 from permafrost organic matter, but release of stored CH_4 that is initiated or enhanced by permafrost thaw.

This third source is the main subject of this chapter. The other two have been discussed in previous chapters in close relation to soil organic matter turnover and thaw lake formation (Chaps. 4 and 5) and vegetation succession (Chap. 6). However, there is no sharp distinction between these sources. The CH_4 originating from these sources can only be distinguished by analysis of the stable isotopes of C and H, and by ^{14}C dating of the gas (e.g. Walter et al. 2008). Thaw lake CH_4 emissions will contain a certain amount of recently produced ecosystem CH_4, while deep CH_4 also may be included. Ecosystem CH_4 emission will contain some CH_4 derived from older soil carbon. Furthermore, in this chapter the importance of the various CH_4 sources in relation climate change is discussed, and put in the perspective of other greenhouse gases emanating from permafrost soils and ecosystems.

7.1 Deep CH_4 Sources

Like the ecosystem and old carbon CH_4 sources, most of the deep CH_4 sources are of microbial origin. However, the deep CH_4 source could include thermogenic CH_4, produced by subsurface heating (temperatures above 110 °C) of sedimentary organic matter, generally in hydrocarbon source rocks. In the Earth's mantle and crust, abiotic CH_4 is also formed by a range of processes, which do not involve organic matter: high temperature reactions within the mantle (juvenile CH_4), and various water – rock – gas interactions at lower temperatures, e.g. serpentinisation (hydration forming serpentine group minerals) of ferromagnesian minerals as pyroxene and olivine (Etiope and Sherwood Lollar 2013).

Microbial (biogenic) and thermogenic CH_4 or abiotic CH_4 can be distinguished from each other based on their content of the heavier C and H stable isotopes in CH_4, ^{13}C and Deuterium (D). Figure 7.1 shows isotopic composition of CH_4 from various origin. Microbes producing CH_4 discriminate against the heavier isotopes of C and H, and microbial CH_4 is therefore isotopically lighter than thermogenic CH_4. Furthermore, CH_4 produced from old organic carbon tends be produced by the hydrogenotrophic reaction (Sect. 4.5.3; Walter et al. 2008). This is isotopically lighter than CH_4 produced by acetoclastic reactions; $\delta^{13}C$ values of hydrogenotrophic

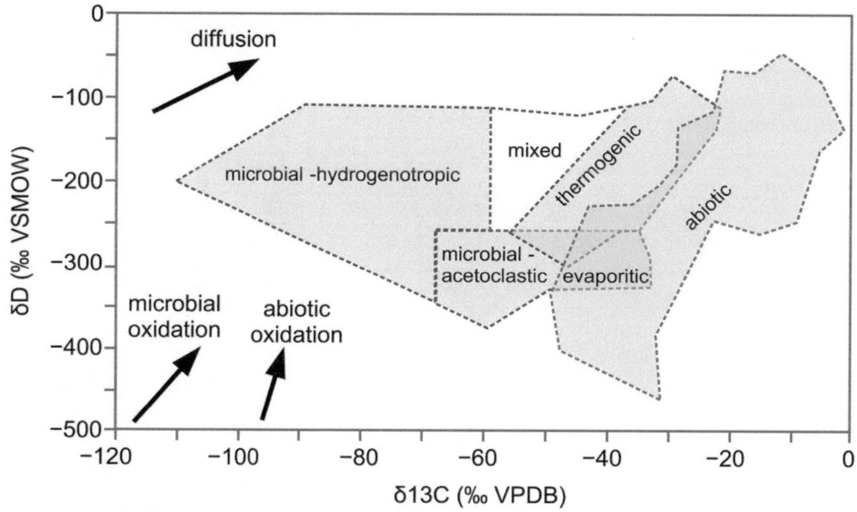

Fig. 7.1 Diagram for classification of methane sources based on $\delta^{13}C$-CH_4 and δD-CH_4 isotope information. Based on Etiope and Sherwood Lollar (2013). Values are referenced to standards VPDB and VSMOW. The diagram includes including biotic CH_4 observed in subsoil petroleum systems. Microbial-evaporitic pertains to microbial CH_4 generated in hypersaline ecosystems. Abiotic CH_4 is derived from various subsurface sources that do not involve organic carbon: reactions in the mantle and magmatic processes, and low-temperature reactions of gas-water-rock reactions. The arrows indicate the direction of shift in fractionation caused by microbial oxidation, abiotic oxidation, and diffusion. See original publication for data sources of this graph

CH_4 can be as low as 110‰. The CH_4 source also may be traced by admixtures of non-methane hydrocarbons such as ethane, and noble gases (Hausmann et al. 2016; Etiope and Sherwood Lollar 2013). Microbial oxidation of CH_4 during transport from the source modifies the isotopic composition, resulting in an isotopically heavier composition of the remaining CH_4 (e.g. Sugimoto and Wada 1993; Whiticar 1999; Fig. 7.1).

A specific type of subsurface form of CH_4 that occurs in permafrost areas is methane hydrate. Gas hydrates (also known as clathrates) form under high pressure and low temperature, when CH_4 (and other gases) are locked in a cage of water molecules (Fig. 7.2). The configuration of the water cage depends the type of gas that is included. For the most common methane hydrate structure, one m^3 of hydrate contains 164 m^3 of CH_4 at standard atmospheric temperature and pressure (Kvenfolden 1988). Three conditions have to be fulfilled for the formation of methane hydrate in sedimentary rocks. The dissolved CH_4 concentration in pore water has to exceed the saturation value for hydrate formation, and hydrate has to be stable given the local temperature and pressure conditions. At atmospheric pressures, methane hydrate is never stable, but depending on the pressure, it may be stable at temperatures above zero (Fig. 7.3).

The saturation of CH_4 will be sufficiently high in subsurface zones where CH_4 is present as a result of microbial reactions, or where it is trapped after upward

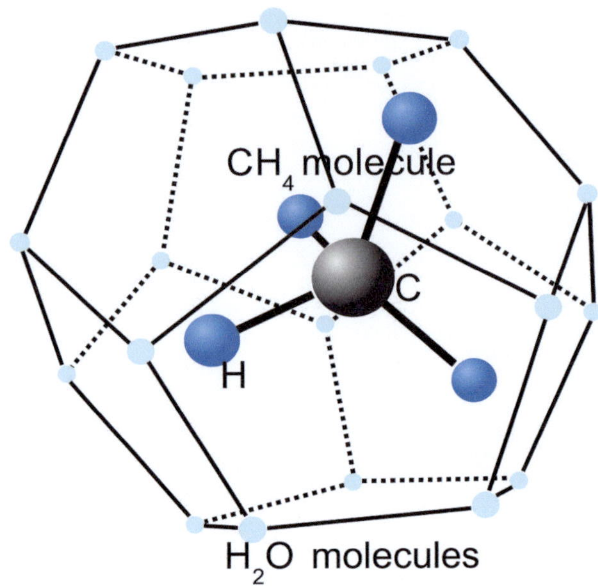

Fig. 7.2 Methane hydrate: CH_4 molecule within a clathrate cage formed by water molecules. (Redrawn from Beauchamp 2004)

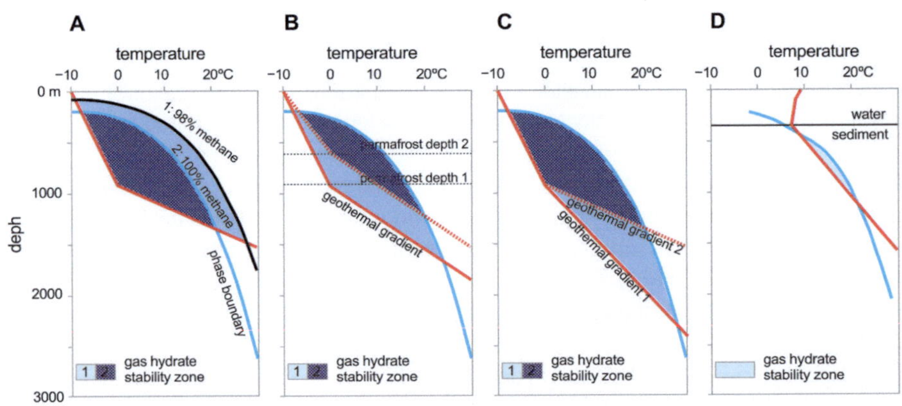

Fig. 7.3 The gas hydrate stability zone associated with terrestrial permafrost. The curved lines represent the hydrate – gas phase boundary, the straight lines the geothermal gradient within the permafrost (as affected by the presence of ice) and below the permafrost. (**a**) the effect of gas composition on the gas hydrate stability zone (pure CH_4 and 98% CH_4, 2% other hydrocarbons). (**b**) the effect of permafrost depth, which shifts the geothermal gradient. (**c**) the effect of the geothermal gradient below the permafrost. (**d**) the hydrate stability zone in shallow offshore regions. The pore-pressure gradient in (**a–c**) is 9.795 kPa/m. (After O'Connor et al. 2010, Kvenfolden and Lorensen 2001, and Collett et al. 2011)

migration of the gas. The pressure at depth in unfrozen sediments is determined by the hydrostatic pressure, which is the weight of the water column if there are no confining layers that act as a seal to water flow. Deep, continuous permafrost is a confining layer, therefore the weight of the solids, the lithostatic pressure, also has to be taken into account. The temperature gradient in the permafrost is determined by the geothermal heat gradient and the thermal properties of the frozen and unfrozen strata (Fig. 7.3). Both pressure and temperature increase with depth, resulting in a limited region of temperature and pressure conditions where hydrates are stable (Fig. 7.3).

In Fig. 7.3a, the curved right side line marked ‘phase boundary’ gives the phase transition from hydrate to free gas based on (hydrostatic) pressure and temperature at a given depth. The left hand (red) line is the temperature gradient. The area between those lines represents the depth and temperature conditions where hydrate is stable – the hydrate stability zone. For permafrost, at depths below ~225 m, hydrate can be expected. The hydrate stability zone stretches well below the base of the permafrost. In marine environments, hydrate formation can be expected below a water column of slightly less than 200 m. At those depths the water temperatures are also low, in particular in Arctic seas. In the water column, CH_4 concentration is usually too low for hydrate formation, but within bottom sediment hydrate may form. Gas hydrates also have been found in deep lakes such as Lake Baikal. Figure 7.3d shows the hydrate stability zone for shallow seas. The actual base of the hydrate stability zones depends on the local geothermal gradient; this varies considerably from region to region (O’Connor et al. 2010; Collett et al. 2010, 2011). Furthermore, the gas composition (presence of additional higher alkanes) affects stability. Figure 7.3 which includes the effects of composition and geothermal gradient on the gas hydrate stability zone. Climatic conditions that are favourable for formation of both hydrates and permafrost have existed in the Arctic at least since the end of the Pliocene (Collett et al. 2010).

Because of the configuration of the hydrate stability zone, most hydrate can be expected near the base of the permafrost and below it. CH_4 escaping from deeper seated gas reservoirs in sedimentary basins will seep upwards and collect near the base of the permafrost which acts as a stratigraphical trap. Hydrate collects in layers, nodules and in pores, preferably in porous sedimentary units such as sands and gravels (Kvenfolden 1988). For instance in the Kuparuk oil field on the Alaskan North coast, hydrate layers are found around 500 m depth just above the base of the permafrost, and below the permafrost down to about 700 m (Kvenfolden 1988). The hydrate occurs as nodules, veins, and layers in sediments (Fig. 7.4). However, not only in sedimentary basins hydrate occurs: Stotler et al. (2010) report methane hydrate from crystalline shield rocks in Canada. This CH_4 was probably from very ancient thermogenic or abiotic origin.

The upper boundary of the gas hydrate stability zone of about 225 m does not exclude that some methane hydrate can exists at shallower depths in permafrost. First, hydrate can persist as a metastable compound, several degrees above its normal thermodynamic dissolution temperature. There is an energy barrier for the formation of bubbles of CH_4, the dissolution of hydrates appears to be controlled by diffusion (Buffett and Zatsepina 1999; Rehder et al. 2004). Second, if CH_4 is mixed with some higher alkanes of thermogenic origin, the stability zone can be extended up to 100 m (Ruppel and Kessler 2017). Third, local pressure conditions, deviating

Fig. 7.4 Methane hydrate in a drill core in gravel deposits at Mallik, Mackenzie Delta region. (Courtesy of USGS, source https://www.usgs.gov/media/images/mallik-gas-hydrate-sample, accessed April 2019)

from the hydrostatic pressure of Fig. 7.3, may allow the existence of CH_4 at shallower depth.

Evidence for the presence of CH_4 hydrates at shallow depths, up to 25–30 m below the surface is reported by Rivkina et al. (2001, 2007). It is assumed that within the pore spaces of the permafrost high pressures may be created during freezing, allowing hydrate formation, High water pressures indeed may develop in freezing taliks (Mackay 1998; Sect. 3.4.2). $\delta^{13}C$ data of CH_4 from these cores demonstrate a microbial origin of the CH_4. Active methanogenic cultures of both acetoclastic and hydrogenotrophic Archaea were obtained as well from boreholes (Rivkina et al. 2007). In West Siberia, possible hydrates or pressurized free gas at shallow depths result in blowout risk for oil and gas drilling (Yakushev and Chuvilin 2000). However, the amount of hydrates at shallow depths is likely to be very minor, compared to hydrates with the normal gas hydrate stability zone in permafrost; abnormal pressure conditions need to be invoked for their existence (Ruppel and Kessler 2017).

In marine environments, hydrates occur in a stability zone in sediments below the surface; in deep ocean environments the top of the hydrate stability zone occurs at the surface. The base of the hydrate stability zone, where temperature is too high for stable hydrates, is marked by a specific seismic reflector known as the Bottom Simulating Reflector (BSR), which marks the contrast between solid hydrate and free gas in the pores beneath it.

The Arctic ocean is a special case, because the occurrence of extensive areas of subsea permafrost, which originated during low sea level stands during the ice ages (Figs. 1.1 and 1.22). On the East Siberian shelf, continuous permafrost is preserved down to water depths of 60 m; at greater depths, permafrost is discontinuous or patchy. The permafrost is thickest in shallow water areas, where it approaches the thickness of subaerial permafrost. Below the submarine extensions of the present river valleys draining to the Arctic Ocean, permafrost is thinner and open taliks may be present (Romanovskii et al. 2004, 2005). Thaw lake formation started before the marine transgression; the formation of thaw lakes had a prominent role in the landward transgression of the sea. Thaw lakes were transferred into lagoons, and subsequent ground ice thaw and wave erosion contributed strongly to shoreline erosion. However, Romanovskii et al. (2004) infer that the taliks below these lakes are still closed taliks. The subsea permafrost degrades very slowly because the temperature of bottom water above the sea floor tends to be below zero, in the range of −0.5 to −1.8 °C (Romanovskii et al. 2005). The subsea permafrost is situated within the gas hydrate stability zone, and just as beneath terrestrial permafrost, occurrences of hydrate are present in similar kind of patchy distribution. Modelling of the evolution of subsea permafrost and the gas hydrate stability zone has shown that the hydrate stability zone is stable, or even expanding due to a rising sea level and sedimentation (Romanovskii et al. 2005; Ruppel and Kessler 2017).

The total amount of hydrates in permafrost is poorly known. Marine seafloor gas hydrates are estimated to range between 170 and 1000 Gt CH_4-C (Archer et al. 2009; O'Connor et al. 2010). The amount below and in terrestrial permafrost is considerably less than the quantity of marine hydrates (Brook et al. 2008; O'Connor et al. 2010). Estimates depend on modelling of the conditions for gas hydrate stability, and what is known from exploration drilling. Collett et al. (2010) list several gas hydrate fields on the Alaskan North Slope, northern Canada and northern Eurasia. Estimates of the total amount of terrestrial gas hydrates by various authors are listed by Collett et al. (2010) and range between 6.9 and 16,728 Gt C – varying over four orders of magnitude. Earlier estimates are the highest, since it was often assumed dat most of the pore space in the gas hydrate stability zone was filled with hydrate. However, based on more recent drilling data, estimates have decreased considerably (Bosswell and Collett 2011). Based on Bosswell and Collett (2011), Ruppel and Kessler (2017) arrive at an estimate of the global hydrate reservoir of 1800 Gt CH_4-C, with 20 Gt CH_4-C in terrestrial permafrost in the Northern Hemisphere.

CH_4 as free gas is common in permafrost. This may be CH_4 generated as ecosystem CH_4 during deposition of sediments, e.g. in lacustrine, fluvial or deltaic deposits, and trapped as bubbles in permafrost. However, also CH_4 formation within permafrost from buried organic matter may contribute. Various methanogenic Archeae were found in a borehole at 22.5 m depth in the Kolyma lowlands, in a layer with abundant organic material, including novel species (Shcherbakova et al. 2016). Tuorto et al. (2014) showed that even replication occurs at subzero temperatures, showing that some strains not only persist on a subsistence level. Other favourable environments for methanogens and other bacterial life are cryopegs (Gilichinsky et al. 2003).

Brouchkov and Fukuda (2002) found CH_4 concentrations as high as 6000 ppm in frozen soil and ice wedges, and found low rate CH_4 production in deeper soil material at subzero temperatures. Rivkina et al. (2001, 2007) measured CH_4 concentrations in drill cores up to 50 m depth in the East Siberian coastal plain west of the Kolyma River. These cores penetrated Holocene thaw lake and alluvial sediments, the Yedoma, and Plio-Pleistocene marine and terrestrial sands and silts. The CH_4 concentrations were highest in Holocene thaw lake sediments (up to 20 ml/kg) and in Middle Pleistocene – Pliocene marine deposits (up to 50 ml/kg), but very low in Yedoma deposits. The dry steppe environment under which the Yedoma deposits accumulated and were syngenetically frozen, were unfavourable for ecosystem CH_4 formation, while epigentically frozen sedimentary environments accumulated ecosystem CH_4 in a non-frozen state before freezing. The CH_4 concentration did not show any systematic trends with depth, that could suggest seepage from a deep subsurface source. In the fine-grained ice-rich deposits there is no evidence for significant vertical migration of CH_4.

However, significant migration and accumulation has been shown to occur. Pore water expulsion by freezing of lake taliks is a mechanism that contributes to migration to form CH_4 accumulations in course grained reservoirs, and it also may contribute to high pressures, eventually enabling hydrate formation. Kraev et al. (2013) report on gas accumulations in coarse-grained fluvial deposits at 23 m depth in the Kolyma lowlands, which contained CH_4 with a considerable younger age (± 12,000 year) than the deposits in which it was found (± 31,000 kyr). This gas should have been migrated into the reservoir. Isotopic characterisation and flux measurements of CH_4 emitted from boreholes in Northwest and Northeast Siberia by Kraev et al. (2019) showed that gas is trapped in particular in epigenetically frozen deposits that were saturated with CH_4 before freezing. Essentially this is trapped ecosystem CH_4. High emissions were found in a borehole in an active pingo, associated with upwelling water; an inactive pingo did not show high amounts of gas.

The proposed migration mechanism is expulsion of water and CH_4 from overlying deposits by epigenetic freezing of a talik; a theoretical model is published by Kraev et al. (2013). This mechanism depends on the freezing rate of the talik and soil/sediment properties and is most effective in coarser-grained material and at high freezing rates. Depending on the sediment properties, there is a threshold freezing rate that forces water to move to the freezing front to form ice lenses with inclusion of trapped CH_4, or results in water and CH_4 expulsion (Sect. 3.2). Epigenetic refreezing of taliks underneath former thaw lakes is common in lowlands with continuous permafrost, as shown by the many closed system pingos (Sect. 3.4.2). Figure 7.5 shows a conceptual model of CH_4 accumulation based on a

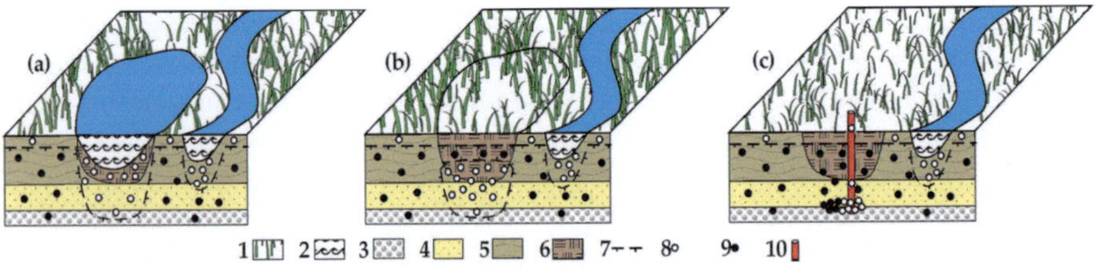

Fig. 7.5 Paleo-environmental reconstruction of a sedimentary succession at the Kolyma floodplain, with redistribution of CH_4 by epigenetic freezing. (**a**) Soils of the thaw bulb (talik) under a floodplain lake with potential methane production; (**b**) Freezing of the methane-enriched soils of the thaw bulb after the lake drainage; (**c**) CH_4 trapped in the pores and cavities of permafrost after complete freezing of the thaw bulb. Key: 1 – floodplain wetland; 2 – warm riverine and lacustrine waters; 3 – riverine alluvium, gravel layer; 4 – riverine alluvium, sand layer; 5 – floodplain alluvium, silt layer; 6 –lacustrine deposits, peaty silts; 7 – permafrost table; 8 – free methane; 9 – trapped methane; 10 – borehole. (From Kraev et al. (2017); figure courteously provided by Dr. Kraev)

paleo-environmental reconstruction of fluvial deposits in coarse-grained subsurface layers according to Kraev et al. (2017).

Migration of gas from deeper reservoirs also contributes to emissions in permafrost areas (Walter-Anthony et al. 2012; Kohnert et al. 2017). It is a common feature in sedimentary basins with hydrocarbon source rocks and in geothermal areas (Etiope and Klusman 2002). Microseepage is the slow, continuous loss of CH_4 from hydrocarbon reservoir and source rocks. Low rates of seepage are subject to oxidation in surface soils. Macroseepage consists of concentrated hot spots of high rate CH_4 emission, often with a surface expression as mud volcanoes, where mud and gases under pressure are emitted in volcano-like structures. Permeable faults and fractures in rocks contribute to seepage. Micro and macro-seepage occur both subaerially, as well as at the sea floor. Macroseepage is often associated with heat flow and seepage of highly saline water (Etiope and Klusman 2002), and may penetrate permafrost. An example of CH_4 seepage associated with brines in continuous permafrost is reported by Niederberger et al. (2010) from Axel Heiberg Island in the Canadian High Arctic. These springs are cold (temperatures between −5 and +10 °C) and moderately saline to hypersaline (up to 24% salts).

Figure 7.6 gives a summary of deep permafrost CH_4 sources and their origin discussed above. Emission of this CH_4 from geological origin is strongly influenced by the state of the cryosphere, which includes permafrost as well as glaciers. Walter-Anthony et al. (2012) characterize this as the cryospheric cap, that traps gas leaking from deep geologic reservoirs by the presence of permafrost or glacier ice. In their terminology, subcap CH_4 consists of sources of CH_4 from below the permafrost or

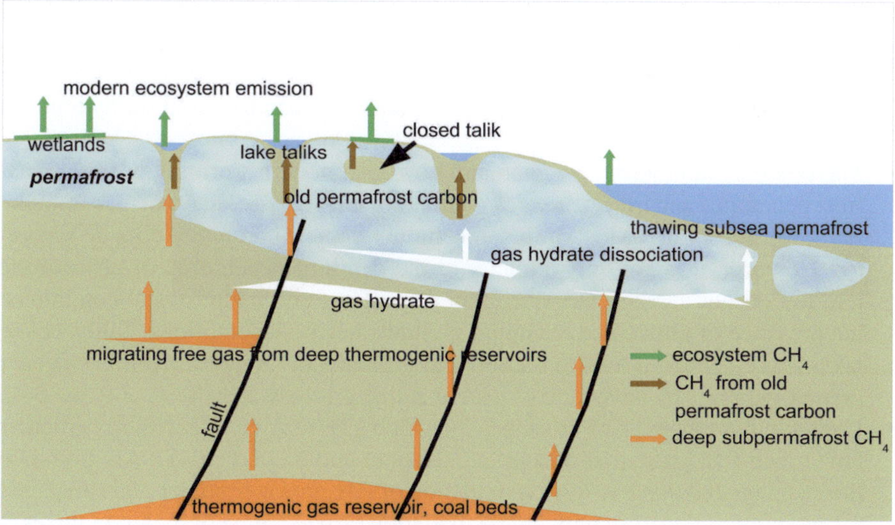

Fig. 7.6 Deep CH_4 sources in permafrost, vs. modern microbial ecosystem sources and sources from old permafrost carbon. Fault lines are potential conduits for gas seepage from the deep subsurface. (Simplified after Ruppel and Kessler 2017)

glacier base, including hydrates, or fossil carbon, e.g. coal beds. Satellite image mapping of early winter ice cover on lakes reveals areas where large hotspot emissions occur, since lakes do not freeze readily at such locations. Based on mapping of emission hotspots in lakes in Alaska and Greenland and chemical/isotope analysis of emitted gas, Walter-Anthony et al. (2012) discriminate the following states of the cryospheric cap related to permafrost:

- Continuous permafrost regions where deep taliks below rivers and lakes increase permafrost permeability locally (Northern Alaska, West Siberia);
- Formerly continuous permafrost areas but now discontinuous, where geologic seeps are relatively rare, and if present, are determined by fault tectonics rather than permafrost presence;
- Former glacier-covered areas during the Last Glacial Maximum (South-Central Alaska) where unloading of ice changes subsurface stress conditions and activates faults and rock fractures, that increase the permeability for gas seepage.

7.2 Climate Change Related Release of Deep Permafrost CH_4

How vulnerable are deep CH_4 sources to climate change and how large is the feedback on climate warming they can produce? The uncertainties and knowledge gaps are large:

- As seen above, there are a variety of sources ranging from methane hydrate at the deep base of the permafrost up to depths of 225 m, and shallower pockets of gaseous CH_4 and possibly hydrates above this depth. Moreover, the size of these stores is poorly known (previous section). This holds in particular for shallower gas trapped in permafrost; most research attention has focussed on hydrates at the permafrost base.
- There is uncertainty on how deep and how fast modern warming of the permafrost penetrates and may generate conduits for CH_4 release, in particular from the shallower reservoirs. Most of this uncertainty depends on future climate scenarios. However, it is not only a matter of the depth of penetration of of warming into the deeper subsoil, but also local permafrost and hydrological conditions: the presence of taliks, and changing groundwater flow conditions. Taliks below lakes reach depths of tens of metres, and groundwater flow can facilitate deeper permafrost thaw by advection of heat or saline groundwater (Sects. 3.9 and 5.4).
- Uncertainty in the type of processes by which CH_4 may be released, in particular how much CH_4 oxidation occurs (e.g. Ruppel and Kessler 2017). Are high rate fluxes generated that can bypass oxidation of CH_4 in water bodies and soils, or is release slow and diffuse with ample oxidation? An example of high rate fluxes is seen in ebullition hot spots in thaw lakes (e.g. Walter et al. 2008).

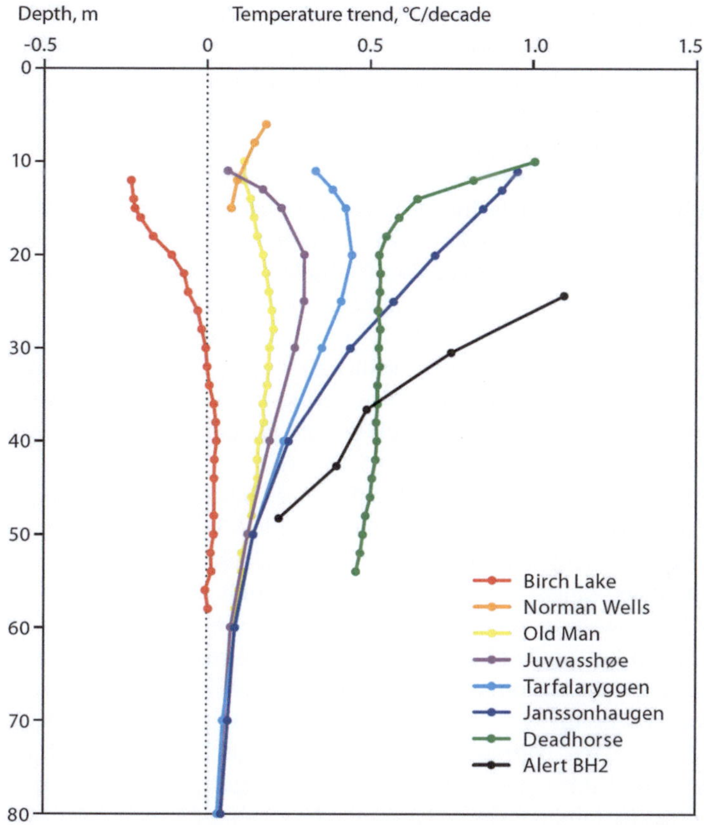

Fig. 7.7 Observed linear trends in ground temperature between 2000 and 2014 as a function of depth, for the three Nordic key-monitoring sites at Svalbard (Janssonhaugen) and in Scandinavia (Tarfalaryggen and Juvvasshøe), the Canadian sites Alert BH2 and Norman Wells, and the Alaskan sites Deadhorse, Old Man, and Birch Lake (From AMAP (2017) Fig. 4.3; Republished with permission from Arctic Monitoring and Assessment Programme)

Deep permafrost warming at depths >20 m is shown by permafrost boreholes throughout the Arctic, with trends in the order of 0.21–0.82 °C per decade (Fig. 1.28; AMAP 2017). In Fig. 7.7, temperature trends at depth are shown for a few deeper permafrost boreholes. Most sites show significant warming, except for Central Alaska, where thinner snow cover and a lack of air temperature increase contributed to a cooling trend at the surface. Continuous permafrost sites with strongly increasing surface temperatures in the High Arctic show the highest temperature trends. Warming trends are even seen at depths of 50 m and deeper. Current warming trends at depth are well below 0.1 °C $year^{-1}$, but may increase in the future. The depth range of significant warming of 20 m or more is within the reach of shallower gas pockets that have been reported, but this penetration depth remains well above the range where the top of the gas hydrate stability zone is located at about 225 m. The large bulk of hydrates in the gas hydrate stability zone at the base of the continental permafrost has been stable on the time scale of millennia during interglacial periods of the Pleistocene, and for release under present climate change a lag time in the order of centuries is expected (Ruppel 2011).

On the other hand, the present rate of warming is high. Continued climate warming will also raise the temperature in the gas hydrate stability zone, resulting in a slow decrease of thickness of the gas hydrate stability zone from above and below, with dissolution of hydrates forming free gas that can migrate upwards. It then depends on local geology and permafrost configuration – the presence of confining layers that may form stratigraphical traps, faults, taliks and cryopegs – how much gas is trapped before it reaches the atmosphere (Ruppel and Kessler 2017). It is therefore uncertain what the fate of CH_4 from deep permafrost hydrates is – trapped below stratigraphical barriers in the subsurface, gradual leakage towards the surface including partial oxidation, or release by gas seeps. Oxidation is restricted to the topsoil, unless the released CH_4 passes sulfate containing marine deposits where anaerobic oxidation may occur (Sect. 4.5.3).

Anomalous warming is associated with water bodies – talik development below lakes and rivers reach several tens of metres; up to 95 m has been reported (Sects. 3.9 and 5.3). Taliks may therefore intercept CH_4 released from hydrate dissociation. Hot spot methane emissions in lakes that may point to activation of deeper CH_4 reservoirs by thaw have been reported by Walter et al. (2006, 2007a, 2008). The high CH_4 age of hotspot bubble emissions in lakes already indicate that this CH_4 is not produced by lake bottom ecosystem emission, but is produced from old permafrost carbon that has thawed (Walter et al. 2008). However, although the decomposition of old permafrost carbon is the most likely source for even high flux hotspot emissions in thaw lakes, the activation of shallow gas pockets or CH_4 rising from deeper sources by talik development cannot be excluded.

The depth of a talik beneath a lake or river depends on bottom water temperature, the thermal conductivity of the talik and permafrost, and the age of the water body (Ling and Zhang 2003; West and Plug 2008). Climate warming is expected to reduce the period of ice cover in lakes, increasing the energy input into the lake water, warming surface waters and changing lake water stratification (Wik et al. 2016; Woolway and Merchant 2019); for high-latitude lakes this also will affect lake bottom temperatures, resulting in deeper taliks over time. However, deeper talik development or development of new lake taliks is a process occurring on time scale of centuries rather than decennia. Therefore, a significant increase of the release of deep hydrate CH_4 is not to be expected on a time scale of decennia. Another problem is that emission of CH_4 from hydrates will be difficult to detect because hydrate dissociation leaves no specific isotopic signature; only a careful comparison of the composition of hydrates and that of emitted gas could confirm hydrate dissociation (Ruppel and Kessler 2017).

In discontinuous permafrost, transport pathways of deep CH_4 sources towards the surface are more likely, and already active (Walter Anthony et al. 2012). Airborne mapping of CH_4 fluxes over the Mackenzie Delta region show large differences in CH_4 fluxes between discontinuous permafrost in the Mackenzie Delta and areas to the east of the delta where permafrost is continuous (Kohnert et al. 2017; Fig. 7.8).

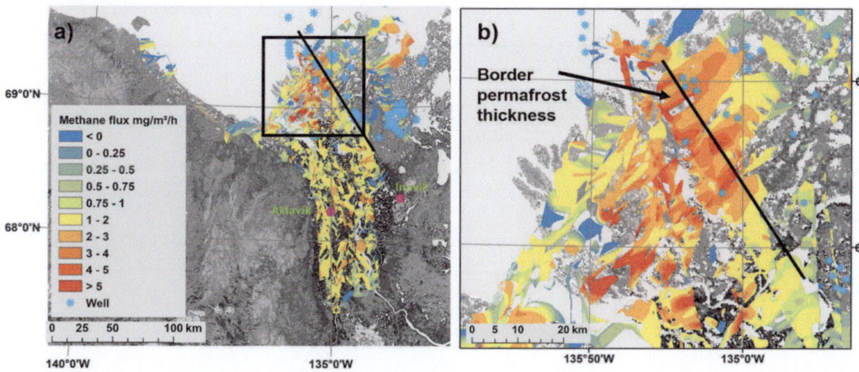

Fig. 7.8 CH_4 flux map derived from airborne CH_4 flux measurements. (**a**) CH_4 flux topography for 2 years of flight campaign (2012 and 2013) containing data with a standard error < 30%. Fluxes >5.0 mg m^{-2} h^{-1} are considered to be of geologic origin. Blue asterisks show the location of oil or gas wells. (**b**) Magnification of the area with the highest CH_4 fluxes marked with the black square in (**a**). Legend as in (**a**). West of the black line the permafrost is discontinuous and thin, east of it continuous and thick. (Figure courteously provided by Dr. K. Kohnert. From Kohnert et al. (2017): Strong geologic methane emissions from discontinuous terrestrial permafrost in the Mackenzie Delta, Canada, Scientific Reports volume 7, DOI:https://doi.org/10.1038/s41598-017-05783-2; Creative Commons License 4.0, https://creativecommons.org/licenses/by/4.0/)

These data strongly suggest that new emission pathways for geologic sources of CH_4 will be opened when continuous permafrost gives way to discontinuous permafrost. Since geologic CH_4 emission is often associated with anomalous heat flow and seepage of highly saline water (Etiope and Klusman 2002), this could also amplify permafrost thaw.

Future projections of permafrost area based on the IPCC CMIP5 Earth System models show a very large range, depending on the future greenhouse gas concentration pathways (Chap. 8; Fig. 7.9). For the high greenhouse gas emission scenario RCP 8.5, climate-sustainable permafrost would be reduced to the Canadian High Arctic and northernmost Siberia; this would include both discontinuous and continuous permafrost. Chadburn et al. (2017b) infer from observations on present permafrost extent a sensitivity of permafrost area to global warming of 4 million km^2 $°C^{-1}$. A retreat of the permafrost area will include a conversion of continuous into increasingly discontinuous permafrost, in particular in areas where the continuous permafrost is relatively thin (Sect. 5.4.1), although relict permafrost will remain at depth. Walter-Anthony et al. (2012) mention specifically West Siberia as an area where continuous permafrost is relatively thin, and vulnerable to an increase of geologic emissions.

Based on modelling of permafrost warming and thaw lake expansion, Schneider von Deimling et al. (2015) arrive at an estimate of a maximum emission of 50 Tg CH_4 $year^{-1}$ for the Yedoma deposits by the mid of the next century. These are the emissions from thawing old permafrost carbon; the subcap geologic emissions of Walter-Anthony et al. (2012) are not included in this estimate. Likewise, the effects

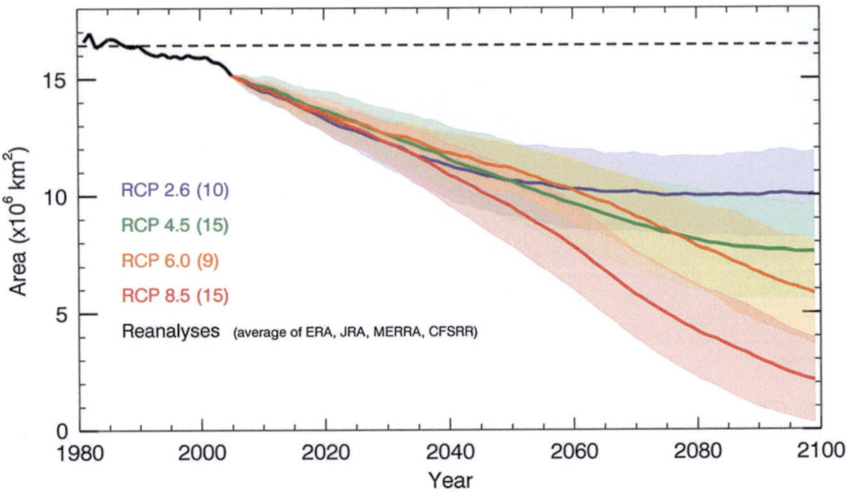

Fig. 7.9 Projected change in sustainable (continuous and discontinuous) permafrost area based on climate change with respect to present, based on the IPCC CMIP5 Earth System models. Permafrost area is calculated from model output using the surface frost index of Nelson and Outcalt (1987). Shaded areas represent one standard deviation across the CMIP5 models, and the dashed black line is the present-day total area of continuous and discontinuous permafrost. The continuous black line is based on reanalysis data of historic weather. The RCP's (denoted with different colours) are the IPCC greenhouse gas concentration pathways used in the models; the numbers in brackets are the number of models contributing to each RCP estimate. (Republished with permission of American Meteorological Society from 'Diagnosing present and future permafrost from climate models' by Slater and Lawrence (2013), Journal of Climate, Vol. 26-–15, 2013; permission conveyed through Copyright Clearance Center, Inc.)

of potential microbial self-heating of permafrost carbon (Khvorostyanov et al. 2008a, b) have not been included. As discussed in Sect. 5.2.4, this effect has only been demonstrated in models, and needs a high rate of warming.

Because of repeated submergence during interglacials of the Pleistocene, also the subsea permafrost permafrost on the Arctic Ocean shelf is relatively thin compared to the thickness of unglaciated terrestrial areas. Land permafrost is 8–12 °C colder than seafloor permafrost in the same region (Shakhova et al. 2014). Shakhova et al. (2010a, b, 2014) have shown that sea water above the East Siberian Arctic Shelf (ESAS hereafter) has an elevated dissolved CH_4 concentration, and that also the atmospheric mixing ratio of CH_4 above the sea is anomalously high; 50% of the surface water is supersaturated with CH_4 with respect to the atmosphere, and 80% of the bottom water, indicating the sea floor as a CH_4 source (Fig. 7.10). The high CH_4 flux is caused by seeps of dissolving hydrates and geologic emissions through taliks in the subsea permafrost. High resolution seismic data of the sea floor shows presence of seeps; in addition, many bubble plumes from seeps were detected with sonar.

Shakhova et al. (2014) estimated that 100–630 mg CH_4 m^{-2} d^{-1} enters the water column by bubble plumes from partially thawed permafrost. Rising from deeper

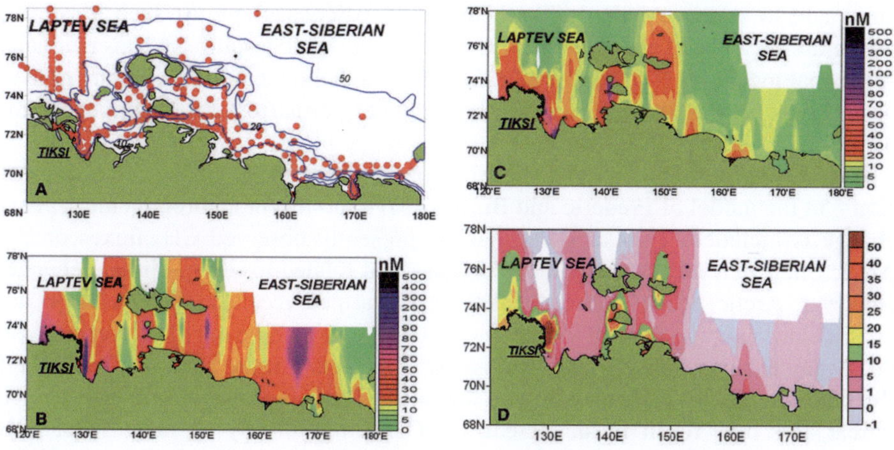

Fig. 7.10 Summertime observations of dissolved CH4 in the water of East Siberian Arctic Ocean Shelf, from Shakhova et al. (2010b). (**a**). Positions of oceanographic stations in the eastern Laptev Sea and East Siberian Sea; bathymetry lines for 10, 20, and 50 m depth are shown in blue. (**b**). Dissolved CH_4 in bottom water. (**c**). Dissolved CH4 in surface water. (**d**). Fluxes of CH_4 venting to the atmosphere over the East Siberian Arctic Ocean Shelf. (Republished with permission of AAAS from 'Extensive methane venting to the atmosphere from sediments of the East Siberian Arctic Shelf' by Shakhova et al. (2010b), Science, Vol. 327, Issue 5970, 2010; permission conveyed through Copyright Clearance Center, Inc.)

waters, bubbles may dissolve in the water column; gas exchange will decrease the CH_4 concentration in the bubbles, and dissolved CH_4 is subject to oxidation (Shakhova et al. 2010a). Since the water depth over the ESAS is relatively shallow (< 50 m), in the summer bubbles could reach the surface rapidly without much oxidation losses. Moreover, low temperatures limit dissolution and oxidation. In particular during storms, the exchange of CH_4 from sea water to the atmosphere is high. However, Thornton et al. (2016) report turbulence-driven CH_4 fluxes of 3.0–3.8 mg CH_4 m^{-2} d^{-1} for open water in the Laptev and East Siberian seas, indicating losses in the water column. Also, storage below ice in winter results in losses.

Present-day microbial sources in Holocene seafloor sediments, or Pleistocene carbon in taliks or the water column cannot explain the high water CH_4 concentrations in the ESAS. The sources of the CH_4 are assumed to be geologic sources (below the ESAS, tectonic faults and rift zones with horsts and grabens occur) and destabilisation of methane hydrates. Another source is submarine groundwater discharge (Lecher et al. 2016). Shakhova et al. (2010b, 2014) suppose open taliks going through the subsea permafrost as important conduits. These taliks are related to paleo-river channels across the shelf during Pleistocene low sealevel stands, and to the creation of thaw lakes at the Last Glacial Termination (Romanovskii et al. 2004, 2005). Frederic and Buffett (2014) have modelled heat transfer from sea water during the Holocene into the Arctic shelf seas. They show that, despite the near-zero bottom water temperatures, this heat transfer has increased the permafrost temperature and reduced the ice content of the permafrost sufficiently to create taliks and to widen existing taliks below river paleochannels. At such locations, temperatures will rise enough to dissociate methane hydrates into gaseous methane. A thermal model by Nicolsky et al. (2012) shows that open talik development may occur below former thaw lakes, and show thinner permafrost below grabens compared to horsts.

This hydrate destabilisation is the effect of Holocene sea level rise and transgression – in the model of Frederic and Buffett (2014) no recent anthropogenic climate change is included. This suggests that the presently observed CH_4 fluxes on the Arctic Shelf may be an autonomous process, not related to recent climate change. However, Arctic ocean surface waters have warmed recently (Dmitrenko et al. 2011). Until now this has led to only modest deepening of the top of the permafrost, but it may enhance subsea permafrost thaw in shallow coastal parts of Arctic Shelf (Shakhova et al. 2013). Depending on the rate of future Arctic warming and sea ice retreat, this may result in an increase of CH_4 emission, by an increase of CH_4 exchange of Arctic shelf seas to the atmosphere, and increasing perforation of the permafrost lid on geologic sources and hydrates. On the other hand, this has to be compared to rates of CO_2 uptake by the ocean by possibly enhanced nutrient upwelling (Pohlman et al. 2017).

Other reports of the effect of hydrate destabilisation below subsea permafrost are given by Paull et al. (2007), Portnov et al. (2013, 2014) and Serov et al. (2015) from respectively the Beaufort Sea and the Kara Sea shelves. Paull et al. (2007) and Serov et al. (2015) describe pingo-like features created by the pressure of dissociating gas hydrates (see also next section).

7.3 Cryovolcanism: Gas Emission Craters

In the summer of 2014 a hitherto unknown phenomenon in permafrost landscapes was discovered: an approximately 30 m wide crater, at least 50 m deep, surrounded by a ring wall of ejected blocks of host sediments and active layer blocks (Fig. 7.11). Leibman et al. (2014) were the first to give a scientific account. The crater is situated on the Yamal peninsula in the continuous permafrost zone, in a hilly tundra landscape. High CH_4 and H_2S concentrations were measured inside the crater, suggesting a gas outburst. In the Yamal region, where one of the major Russian gas fields is situated, frequent gas blowouts during drilling occur (Yakushev and Chuvilin 2000). An origin as meteorite impact, or an anthropogenic origin as a result of drilling operations in the Bovanenko gas field (given the distance to drill sites) can be excluded (Olenchenko et al. 2015). In the crater wall peaty tundra soil, frozen marine silts and clays, and layers of tabular ground ice were exposed. The ejecta, consisting of the same materials were found up to 120 m from the crater, indicating formation by an explosive process. Based on analysis of satellite images preceding the discovery, the crater should have formed in the late autumn of 2013. The regional permafrost thickness varies between 100 and 280 m, with permafrost temperature from −2 to −7 °C (Kizyakov et al. 2017).

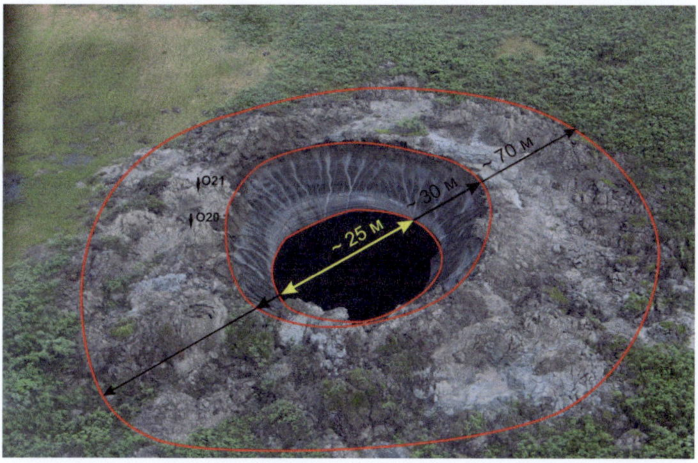

Fig. 7.11 Ring wall, upper funnel, and inner hole size of the first described crater on Yamal Peninsula based GPS survey by Leibman et al. (2014). O20 and O21 are measurement points. (Photo by A.V. Hanoverian; image courteously provided by Dr. A.I Kyziakov and Dr. M.O. Leibman, from Leibman et al. 2014)

More crater discoveries followed (Fig. 7.12; Kizyakov et al. 2017, 2018). Two craters were mapped in detail with high resolution satellite image stereo-pairs. The first discovered crater in Central Yamal (GEC-1 in Fig. 7.12) showed presence of a very clear mound of 5–6 m high before the crater formed, as seen on an image from the previous spring (Fig. 7.13). This was also observed at the second crater on the Gydan peninsula (AntGEC), although that bulge was lower (about 2 m). The crater on the Gydan peninsula had a smaller funnel diameter (10–13 m vs. 15–16 m for crater 1), was less deep (15 m vs > 50 m), and did not show a clear ring wall of ejecta, although ejected material was found abundantly around the crater (Kizyakov et al. 2017). Ejected blocks did form impact depressions, sometimes more than 10 m wide. At the GEC-2 crater, an ejected block of frozen sediment was found at 293 m from the crater; the width of the impact zones around the craters vary from site tot site (Kizyakov et al. 2018). The mound development indicates buildup of subsurface pressure before crater formation. The ejected blocks attest of the explosive and hazardous nature of crater formation, being a threat to local inhabitants and infrastructure (Kizyakov et al. 2018).

Geophysical surveys by Olenchenko et al. (2015) of the first Yamal crater give insight into the geological structures that may be linked to crater formation. The subsurface geology consists of unconsolidated fine-grained Paleogene–Quaternary sediments down to a depth of 500 m. Faults in the subsurface, related to rift tectonics, are common. The upper permafrost consists of very ice-rich deposits, including ice wedges, tabular ground ice bodies and pingos. The crater is situated within the boundaries of a former thaw lake. The geo-electric profiles by Olenchenko et al. (2015) show a complex distribution of ground ice. The crater is located at the boundary of a thick ice-rich unit (Fig. 3.39). A geomagnetic survey indicated the presence

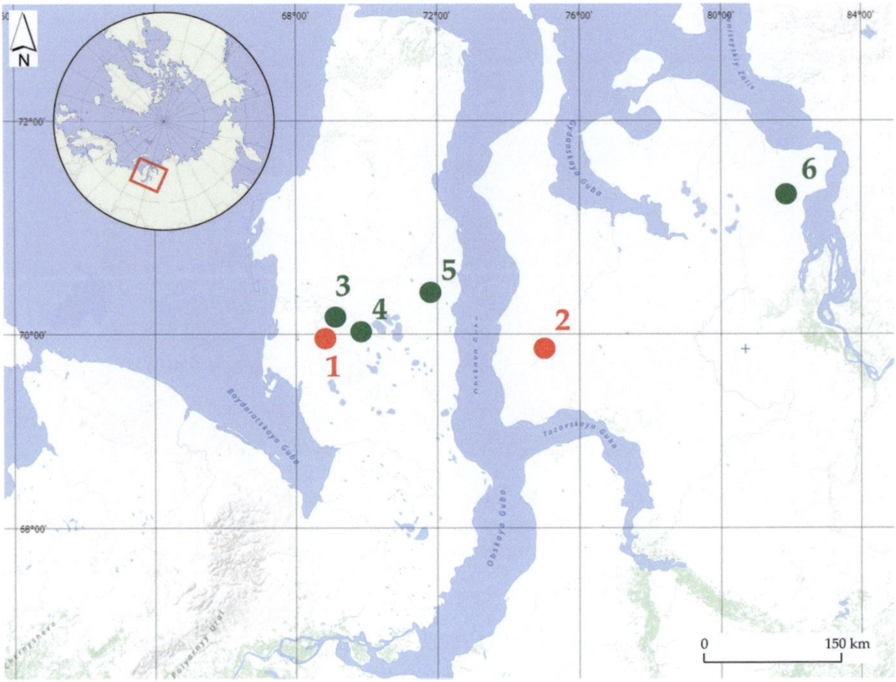

Fig. 7.12 Location map of gas-emission craters known in 2017; from Kizyakov et al. (2018) https://www.mdpi.com/2072-4292/9/10/1023. Map courteously provided by Dr. A.I. Kyziakov © 2017 by the authors. Licensee MDPI, Basel, Switzerland. Creative Commons Attribution (CC BY) license (http://creativecommons.org/licenses/by/4.0/)

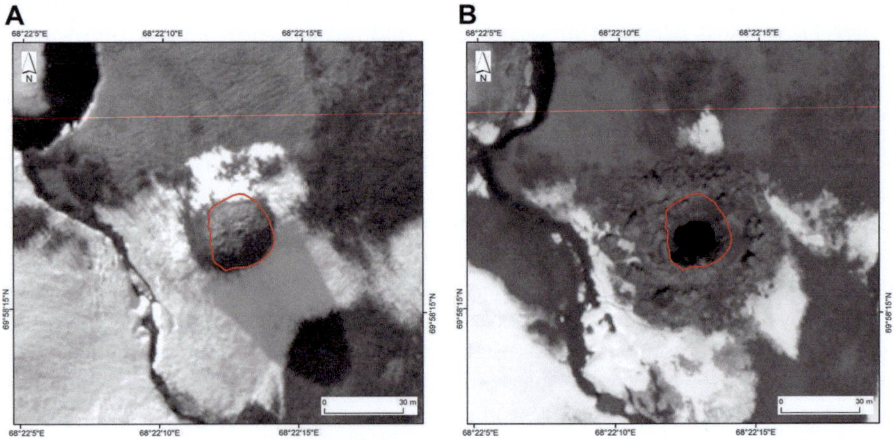

Fig. 7.13 (**a**). WorldView-1 image dated 09 June 2013 (before the crater formation) of crater 1 in Fig. 7.12; (**b**). WorldView-1 image dated 15 June 2014 (after the crater formation). The red line represents the crater edge as of 15 June 2014. (From Kizyakov et al. (2017). Figure courteously provided by Dr. A.I Kyziakov. WorldView © 2019 MAXAR; processed image and data © 2017 by the authors. Licensee MDPI, Basel, Switzerland. Creative Commons Attribution (CC BY) license (http://creativecommons.org/licenses/by/4.0/))

of faults crossing each other near the crater location. A TEM (Transient Electro Magnetic) survey indicated an upwards bulge in the base of the permafrost and the presence of an anomalous high resistivity layer between 60–80 m depth, which was interpreted as methane hydrate bearing sand. Another geophysical section indicates the presence of a refreezing lake talik (Olenchenko, pers. comm.).

Based on these data, Olenchenko et al. (2015) inferred two possible origins of the crater: upward migration of gas from deep gas reservoirs and accumulation in a talik of the paleo-lake, or overpressure developed by decomposition of gas hydrates and upwards migration of this gas. Warming of the permafrost is enhanced by a series of anomalously warm summers and may have contributed to gas migration (Leibman et al. 2014).

Buldovicz et al. (2018) propose a large role for CO_2 during the crater outburst. Besides CH_4, CO_2 also forms in taliks, and by contrast to CH_4, CO_2 is highly soluble in water. Depressurisation of dissolved CO_2 results in a rapid volume expansion and bubble formation as happens at the opening of a champaign bottle; this process may have contributed to crater formation. Buldovicz et al. (2018) modelled the pressures that may be created at talik freezing, and conclude that the generated gas pressure by itself is insufficient to to break the overlying frozen soil cap. They invoke thermal contraction cracking in autumn as a mechanism triggering the first explosive release of gas. Initially, the gas escapes through the cracks, provoking depressurization (the pneumatic stage of a few minutes), followed by a hydraulic stage, which is the fast release of pressurized CO_2 from water, that may last a few hours (Fig. 7.14). To

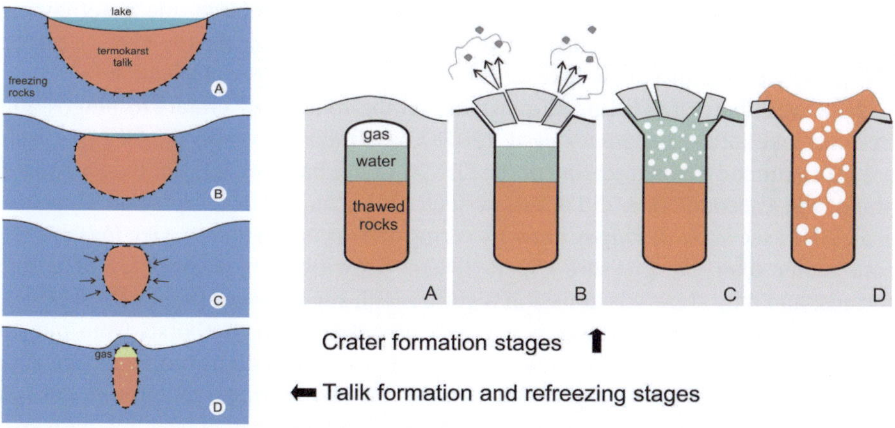

Fig. 7.14 Gas emitting rater formation after Buldovicz et al. (2018). Left: Stages in talik formation refreezing with pingo formation. A: Talik under the lake; B: lateral freezing of the talik upon lake shrinking; C: formation of a freezing closed talik; D: growth of a frost mound (pingo). Right: stages in the cryogenic eruption. A: Pre-explosion; B: pneumatic stage with first gas pressure release; C: hydraulic stage with depressurizing of dissolved CO2, and D: phreatic stage. Combination of Figs. 2 and 3 in Buldovicz et al. (2018) Cryovolcanism on the Earth: Origin of a spectacular crater in the Yamal peninsula (Russia); Scientific Reports 8(1); available under Creative Commons License 4.0, https://creativecommons.org/licenses/by/4.0/

explain the large depth of the crater, it is assumed that at the final stage of freezing, the talik takes a approximate cylindrical shape.

Isotopic data suggest a microbiological origin of the CH_4 emitted from the craters (Buldovicz et al. 2018), being depleted in ^{13}C. However, the CH_4 analysis of Buldovicz et al. (2018) has been done on samples dating at least 1 year after crater formation, when microbial CH_4 production should have started in the crater lake (Savvichev et al. 2018). Measurements of radioactivity near the crater do not show anomalies that might indicate a contribution of a deep gas source (Leibman et al. 2014; Olenchenko et al. 2015).

In a few years, the craters transform into small lakes with an irregular outline by caving in of the crater wall (Babkina et al. 2017). By erosion of the wall, the surrounding vegetation and soil carbon is deposited in the lake (Kizyakov et al. 2017; Savvichev et al. 2018). Compared to nearby lakes that have no crater origin, the crater lakes have a high DOC concentration, a high amount of dissolved ions derived from marine and continental deposits in the crater walls, and high dissolved CH_4. However, the rate of anaerobic microbial conversion of organic matter (methanogenesis and sulfate reduction) was low in the former craters, compared to non-crater lakes. Also, the diversity of the microbial community was lower, indicating the young ecosystem developing in the lake bottom sediment (Savvichev et al. 2018).

The discovery of craters generates widespread speculation about their origin, and their relation to other cryogenic processes such pingos and thaw lakes. Buldovicz et al. (2018) classify the craters as cryovolcanism, volcano-like features that emit volatiles and water, also occurring on other planets and moons in the Solar System. Leibman et al. (2014) and Olenchenko et al. (2015) infer that destabilization of gas hydrates may be involved. Buldovicz et al. (2018) assign a large role to refreezing of a talik beneath a drained lake basin and pingo formation for the GEC-1 crater. The presence of a refreezing lake talik is supported by the geophysical data.

However, there are still open questions on the relation of craters to pingos and freezing lake taliks. Buldovicz et al. (2018) base their theory on the GEC-1 crater only. The mound before creation of the GEC-1 crater has grown faster than a normal pingo, and a second crater did not show such a clear mound development (Kizyakov et al. 2017; see above). Pingos grow by spring-fed accumulation of a ice massive ice core rather than gas pressure (Sect. 3.4.2). Growing pingos are sites of active springs, and may show gas emission together with spring water (Kraev et al. 2019). However, normal pingo growth is a slower process and gas outbursts from pingos have not been reported before. Other observed crater sites do not show a clear link with a drained lake basin, and crater formation is not restricted to autumn, but has also been reported in summer (Kizyakov et al. 2018). There may be a link with erosion. Sites depicted by Kizyakov et al. (2017, 2018) show evidence of relatively recent erosion (e.g. cryogenic landslides near GEC-2 and gully erosion near AntGEC). Besides higher temperatures, also increasing removal of soil and ice by cryogenic landslides and water erosion may have decreased overburden pressure on gas reservoirs, and allowing faster warming of permafrost.

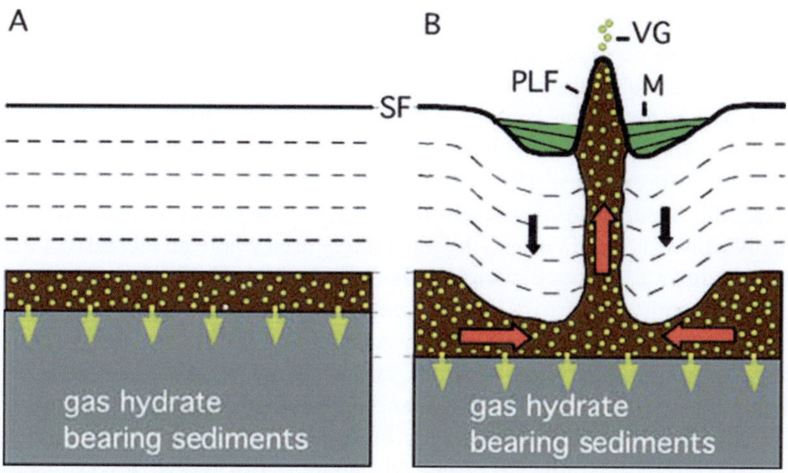

Fig. 7.15 Pingo-like feature (PLF) on the bottom of the Beaufort Sea (Kaglulik PLF), Nortern Canada. (**a**, **b**): Schematic drawing outlining PLF and moat formation (M) associated with gas hydrate decomposition. (**a**): Cross-section of the permafrost-bearing Arctic seafloor (SF) after being transgressed by Arctic Ocean water. As the subsurface warms, the top of the gas hydrate stability zone will move downward. Warming results in gas hydrate decomposition in a gradually thickening zone (brown), releasing gaseous methane into the sediments (yellow). Bubble formation associated with this phase change will create over-pressured conditions. (**b**): Presumed material flow (red arrows) both laterally and vertically in response to overpressure. Displaced sediments rise upward to form the PLF and allow the gas to vent (VG). As the pressure is dissipated through both the transfer of solids and degassing, subsidence in the area immediately surrounding the PLF (black arrows) creates the moat. (Republished with permission of John Wiley and Sons Inc., from Paull et al. (2007): Origin of pingo-like features on the Beaufort Sea shelf and their possible relationship to decomposing methane gas hydrates. Geophysical Research Letters 34 (1); permission conveyed through Copyright Clearance Center, Inc.)

Related features found on subsea permafrost are pingo-like features (PLF) observed on the shelf seafloor of the Beaufort Sea and the Kara Sea offshore of the Yamal Peninsula (Paull et al. 2007; Serov et al. 2015). These have been formed after the Holocene marine transgression. The feature described by Paull et al. (2007) shows venting of old, microbial CH_4. It has apparently formed by upwards extrusion of Pleistocene aged sediments with a high ground ice content (Fig. 7.15). A moat, filled with recent marine sediment around the PLF indicates depletion and collapse of a probable hydrate reservoir around it. From the two PLF's in the Kara sea described by Serov et al. (2015), one shows high CH_4 concentrations in sediment cores. Based on the gas composition, it is assumed that the gas originates from permafrost-related gas hydrates.

Whatever the source of the emitted gas is, the appearance of gas emitting craters in subaerial permafrost is a new phenomenon, not known from permafrost literature predating the Yamal craters, despite at least a century of increasingly intensive permafrost research. Deeper warming of permafrost may serve as an explanation for this new phenomenon (Leibman et al. 2014). If the craters originate from pressurization of closed taliks by refreezing as hypothesized by Buldovicz et al. (2018), a climate warming origin is not needed to explain crater formation. Talik refreezing is an ubiquitous process throughout continuous permafrost areas. However, crater formation has not been reported before in this context. The first hypothesis would agree well with the increasingly leaky cryospheric lid on geologic CH_4 emissions (previous section, Walter-Anthony et al. 2012).

So far, the occurrence of gas emitting craters appears to be restricted to West Siberia. Northwest Siberia is also the most likely area where an increase of this type of cryovolcanism by climate warming should be expected. Factors are continuous, but relatively thin permafrost, more intense climatic warming than elsewhere in Siberia, a geology with deeper gas reservoirs that is favourable for geologic CH_4 seeps and accumulation of gas hydrates, and potential presence of pressurized gas reservoirs related to thaw lake drainage and talik refreezing. It is as yet unknown to what extent this phenomenon may occur elsewhere; time will learn. In any case, it shows that relatively shallow gas reservoirs, and possibly hydrate reservoirs, may be destabilized by warming of the permafrost. The potential amount of CH_4 that may be released in this way is presently unquantified.

7.4 CH_4 Emissions in Perspective: Ecosystem Emissons, CO_2 and N_2O

Although 'methane from permafrost' has attracted widespread media attention, it is not the only greenhouse gas emitted by warming permafrost – and probably even not the most important one, at least on the short term of decades. Media attention also focusses mostly on the most spectacular CH_4 sources such as bubble plumes in lakes and at sea, and the craters discussed in the previous section.

A recent review (Schuur et al. 2015) concluded that greenhouse gas emissions from permafrost soils likely will not cause abrupt climate change. Based on incubation experiments it is argued that most of the carbon emissions from permafrost is CO_2 emission, because of the much faster aerobic decomposition compared to anaerobic decomposition. Moreover, longer term incubations have shown that decomposition decreases over time because of depletion of the most labile carbon pool. How thaw manifests itself geomorphologically – by formation of lakes and ponds with anaerobic conditions, or as erosive features where permafrost carbon is subject to aerobic conditions, determines the partitioning of the permafrost carbon loss in CO_2 and CH_4 (Chap. 5). Models reviewed by Schuur et al. (2015) predict variable carbon loss rates. However, most of the model-predicted carbon loss occurs after 2100, which leads to the conclusion that on a time scale of decennia very high abrupt emissions are unlikely. Nevertheless, abrupt thaw processes, such as erosion and thaw pond development are not represented in models. As seen in Chap. 5, in particular permafrost soil erosion processes are increasing at an alarming rate in some regions (e.g. Abbott and Jones 2015; Lewkowicz and Way 2019). There is high uncertainty in the model-predicted emissions and surprises cannot be ruled out (Schuur et al. 2015).

The dominance of CO_2 emissions does not dismiss the importance of CH_4. First, its radiative contribution to climate change has been upgraded. The 100 year global warming potential of CH_4 was revised upwards from 23 (Houghton et al. 2001) to 34 (Stocker et al. 2013) in the IPCC reports. Second, a renewed rise of atmospheric CH_4 concentration has started in 2007, continuing until 2019. The factors that contributed to this renewed rise remain unclear (Kirschke et al. 2013; Nisbett et al. 2019), but wetlands may be included as a source since the $\delta^{13}C$ ratio of atmospheric CH_4 is decreasing, suggesting a biogenic contribution. This could be permafrost wetland sources, agricultural sources or lower latitude wetland sources. Furthermore, a decrease of $\delta^{13}C$ also could result from a decrease of the atmospheric sink of CH_4 (Nisbett et al. 2019). Worden et al. (2017) infer a net emission increase for both fossil fuel (±50–75% of total increase) and biogenic sources (including wetlands and agriculture) to reconcile isotope and atmospheric chemistry data with strength of sources. In the extensive analysis by Nisbett et al. (2019), increasing CH_4 emission from Arctic sources is included as a potential, but still uncertain option.

A potential surprise is the unquantified amount of CH_4 stored in the upper tens of metres of permafrost and its release mechanisms, discussed in the previous sections. Next, it is hardly known to what extent the dwindling of the cryospheric cap on geologic emissions of CH_4 will add to the high-latitude CH_4 emissions on a decadal to century scale (Walter-Anthony et al. 2012). Ecosystem emissions are also highly sensitive to changes in temperature, vegetation and hydrology. Even a simple change in vegetation can increase the ecosystem emissions (Andresen et al. 2017). However, ecosystem emissions also may decrease by decline of wetlands and drier conditions as seen in previous chapters.

A factor that influences CH_4 emission in more than one way is the rate of sea ice decline (Parmentier et al. 2013).

- First, sea ice decline results in increasing warming of the Arctic ocean, including an increase in shallow water temperatures over the shelves with subsea permafrost. Although large scale destabilization of methane hydrates at the base of the subsea permafrost is not likely at a time scale of decennia, localized expansion of taliks may increase emission of CH_4 (see above).
- Second, terrestrial warming, increase of winter precipitation, warmer autumns and earlier springs likely enhances ecosystem CH_4 emission in the wetland-rich coastal plains along the Eurasian and North American Arctic coasts by wetter and warmer soil conditions (Parmentier et al. 2013, 2015). This ecosystem CH_4 emission reacts to climate change on much shorter time scales than subsea and subcap emissions, as shown by increased emission of CH_4 with a wetland isotopic signature from the Arctic during warm summers (Dlugokencky et al. 2009). The highest terrestrial warming is recorded near the Arctic sea board in regions where most sea ice is lost, which affects most of the tundra biome. Most of the warming occurs in autumn, which may increase the high autumn – early winter emissions reported in Sect. 4.6.2.

Parmentier et al. (2015) showed that a significant increase of ecosystem emission rates already occurs in the extensive coastal plain wetlands of the Arctic Ocean in reaction to sea ice decline. There are insufficient long term records of CH_4 emission from ecosystems, but the data that exist, can be used to validate modelled emissions over longer time scales. Parmentier et al. (2015) found clear spatial correlations between sea ice loss and increasing CH_4 flux on nearby land areas. The relation between CH_4 fluxes and sea ice was estimated using three different soil CH_4 flux models and reanalysis data. All models showed the same trends. On average, the simulated emissions for 2005–2010 were 1.7 Tg CH_4 year^{-1} higher compared to 1981–1990. There were clear negative correlations of CH_4 emission with decrease of sea ice within a radius of 2000 km of each model grid cell, indicating that sea ice decline is connected with increase of CH_4 emissions on nearby wetland ecosystems (Fig. 7.16). The correlations are strongest in spring and autumn.

The increase of CH_4 is largely temperature-driven, although snow cover thickness also could enhance winter emissions (winter emissions are not well represented in CH_4 models). Slightly positive correlations (less CH_4 emission) were found over the Canadian Arctic Archipelago. However, this is a High Arctic area with relatively few wetlands, where uptake of CH_4 (also temperature dependent) may dominate (Juncher Jørgenson et al. 2014). This demonstrates again the importance of ecosystem emission processes and the uncertainties therein, caused by the complexities of the interaction between permafrost thaw and soil temperature, wetland hydrology, labile carbon supply and vegetation sketched in Chaps. 4, 5 and 6.

The third important greenhouse gas that originates from permafrost ecosystems is nitrous oxide, N_2O, which is a greenhouse gas that is usually linked to heavily fertilized agricultural soils, but recently has been shown to have hotspot sources in the Arctic. The sources and mechanisms behind high-arctic N_2O emissions have been discussed in Sects. 4.5.3 and 4.5.5. Hitherto reported sources are characterised by peat soils which have no vegetation cover as a result form erosional processes or cryoturbation (Repo et al. 2009; Marushchak et al. 2011; Abbott and Jones 2015). Arctic peat soils generally have a high N concentration; N compounds produced by mineralization of soil organic matter are not taken up by vegetation, and N_2O may be produced by nitrification or denitrification. N_2O fluxes from soils are notoriously difficult to measure; they are of an episodic nature, and the low concentrations involved require high precision and resources demanding instrumentation, in particular for eddy covariance (Kroon et al. 2007). For this reason, and the low expectancy of high emissions, there are hardly observations in the Arctic of N_2O emissions.

A recent survey using airborne eddy covariance observations from a low flying small aircraft over the Alaskan Arctic coastal plain showed higher N_2O emissions than expected (Wilkerson et al. 2019). The observed daily mean differs significantly from zero emission, and lies within a 90% confidence interval between 2.2 and 4.7 mg N_2O m^{-2} d^{-1}. The emissions show a spotty pattern, characteristic of locally high emissions related to ecosystem conditions.

It is unknown to what extent there is an increasing trend of N_2O emissions, but several lines of evidence from known sources and processes make this likely (Wilkerson et al. 2019).

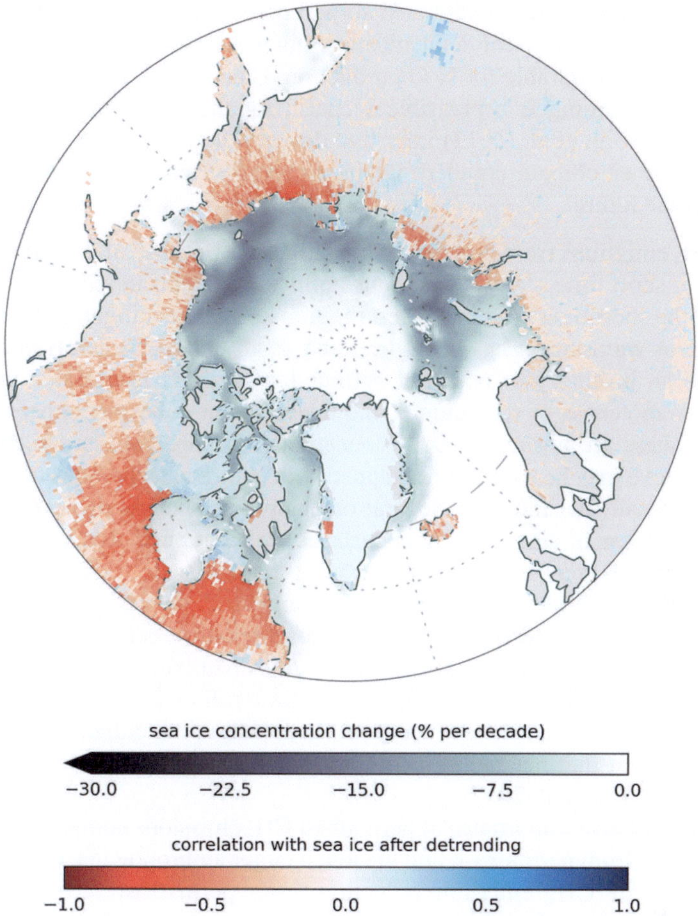

Fig. 7.16 Correlations between 1981 and 2010 May–October terrestrial methane emissions and sea ice concentration. Red to blue colours depict the r value of the correlation between modelled methaneemissions and sea ice concentration within 2000 km, averaged for three models. The linear trend in sea ice concentration is shown to indicate areas of high retreat. Note that high correlations do not necessarily equal high emissions. The correlated time series have been de-trended to minimize spurious correlations. Republished with permission of John Wiley and Sons Inc., from Parmentier et al. (2015): Rising methane emissions from northern wetlands associated with sea ice decline. Geophysical Research Letters 42 (17); permission conveyed through Copyright Clearance Center, Inc.

- First, disturbance of peaty soils by permafrost erosion is known to create environments that are favourable of N_2O emissions (Marushchack et al. 2011; Abbott and Jones 2015; Sect. 4.5.5).
- Second, experimental thawing, drainage and rewetting of permafrost soils (Elberling et al. 2010) and in situ open top chamber warming (Voigt et al. 2017) have shown to result in N_2O emission.

- Third, continuously water-saturated soils suppress N_2O formation by creating a too low redox potential for denitrification, but water table changes by drying and rewetting are favourable for N_2O production. Enhanced drainage, evapotranspiration and warming is a plausible scenario for the future in permafrost soils (Sect. 5.4). Voigt et al. (2017) infer that drought, heat and vegetation stress in a future warmer climate could result in higher N_2O emission from permafrost peats in the future.

Can CH_4 emissions from permafrost be mitigated? Because of the climate impact of CH_4 over short time scales, the serious political problems in reducing CO_2 emissions, and the short time span that is left for climate change mitigation to stay below safe limits of warming, it has been proposed as an option. Stolaroff et al. (2012) have made an inventory of potential options for Arctic CH_4 emission mitigation. Given the remoteness of the sources, the huge area and the low concentration of emissions, there are not many. Only concentrated, trapped sources such as those discussed above may be successfully tapped with common techniques that are being used in the fossil fuel industry. Several other options such as dispersal of bubble plumes or increase of albedo of water surfaces are mentioned also.

These methods may be only applicable on a small scale. However, if applied, these methods require access to Arctic ecosystem areas on a large scale. As shown in Sect. 6.3, this brings the risk of disturbance of sensitive ecosystems, resulting in enhancement of greenhouse gas emissions by this disturbance, instead of reduction of emissions. Other options not mentioned by Stolaroff et al. (2012), such as drainage of permafrost wetlands, may reduce CH_4 emission, but enhance CO_2 emission to such an extent that net radiative forcing is increased (Sect. 5.4.6; Kwon et al. 2016, 2017; Kittler et al. 2017).

Given the relative importance of permafrost CH_4 emissions compared to CO_2 and N_2O emissions from permafrost, and the much larger anthropogenic CH_4 emissions, the relevance of CH_4 emission reduction from permafrost areas is doubtful. Moreover, given the vulnerability of permafrost ecosystems to human disturbance, and the resilience of ecosystems, as discussed in the previous chapter, this type of technological solutions are likely to do more harm than good. Instead, it is important to minimize human disturbance of permafrost ecosystems, and to foster their resilience with adequate protection.

References

Abbott BW, Jones JB (2015) Permafrost collapse alters soil carbon stocks, respiration, CH_4, and N_2O in upland tundra. Glob Chang Biol 21(12):4570–4587

AMAP (2017) Snow, Water, Ice and Permafrost in the Arctic (SWIPA) 2017. Arctic Monitoring and Assessment Programme (AMAP). Oslo, Norway, 269 pp

Andresen CG, Lara MJ, Tweedie CE, Lougheed VL (2017) Rising plant-mediated methane emissions from arctic wetlands. Glob Chang Biol 23(3):1128–1139

Archer D, Buffett B, Brovkin V (2009) Ocean methane hydrates as a slow tipping point in the global carbon cycle. Proc Natl Acad Sci 106(49):20596–20601

Babkina E, Khomutov A, Leibman M, Dvornikov Y, Kizyakov A, Babkin E (2017) Paragenesis of thermal denudation with gas-emission crater and lake formation, Yamal Peninsula, Russia. In: EGU General Assembly Conference Abstracts, 2017. p 6026

Beauchamp B (2004) Natural gas hydrates: myths, facts and issues. Compt Rendus Geosci 336(9):751–765

Boswell R, Collett TS (2011) Current perspectives on gas hydrate resources. Energy Environ Sci 4(4):1206–1215

Brook E, Archer D, Frolking S, Lawrence D (2008) Potential for abrupt changes in atmospheric methane. U.S. Climate Change Science Program Synthesis and Assessment Product 3.4: Abrupt Climate Change. U.S. Climate Change Science Program and the Subcommittee on Global Change Research, U.S. Geological Survey, Reston VA

Brouchkov A, Fukuda M (2002) Preliminary measurements on methane content in permafrost, central Yakutia, and some experimental data. Permafr Periglac Process 13(3):187–197. https://doi.org/10.1002/ppp.422

Buffett BA, Zatsepina OY (1999) Metastability of gas hydrate. Geophys Res Lett 26(19):2981–2984

Buldovicz SN, Khilimonyuk VZ, Bychkov AY, Ospennikov EN, Vorobyev SA, Gunar AY, Gorshkov EI, Chuvilin EM, Cherbunina MY, Kotov PI (2018) Cryovolcanism on the Earth: origin of a spectacular crater in the Yamal peninsula (Russia). Sci Rep 8(1):13534

Chadburn SE, Burke E, Cox P, Friedlingstein P, Hugelius G, Westermann S (2017b) An observation-based constraint on permafrost loss as a function of global warming. Nat Clim Chang 7(5):340

Collett TS, Johnson A, Knapp CC, Boswell R (2010) Natural gas hydrates: energy resource potential and associated geologic hazards, AAPG Memoir 89, vol 89. AAPG

Collett TS, Lee MW, Agena WF, Miller JJ, Lewis KA, Zyrianova MV, Boswell R, Inks TL (2011) Permafrost-associated natural gas hydrate occurrences on the Alaska north slope. Mar Pet Geol 28(2):279–294

Dlugokencky E, Bruhwiler L, White J, Emmons L, Novelli PC, Montzka SA, Masarie KA, Lang PM, Crotwell A, Miller JB (2009) Observational constraints on recent increases in the atmospheric CH_4 burden. Geophys Res Lett 36(18):L18803. https://doi.org/10.1029/2009GL03978

Dmitrenko IA, Kirillov SA, Tremblay LB, Kassens H, Anisimov OA, Lavrov SA, Razumov SO, Grigoriev MN (2011) Recent changes in shelf hydrography in the Siberian Arctic: potential for subsea permafrost instability. J Geophys Res Oceans 116(C10):C10027. https://doi.org/10.1029/2011JC007218

Elberling B, Christiansen HH, Hansen BU (2010) High nitrous oxide production from thawing permafrost. Nat Geosci 3(5):332

Etiope G, Klusman RW (2002) Geologic emissions of methane to the atmosphere. Chemosphere 49(8):777–789

Etiope G, Sherwood Lollar B (2013) Abiotic methane on earth. Rev Geophys 51(2):276–299

Frederick J, Buffett B (2014) Taliks in relict submarine permafrost and methane hydrate deposits: pathways for gas escape under present and future conditions. J Geophys Res Earth 119(2):106–122

Gilichinsky D, Rivkina E, Shcherbakova V, Laurinavichuis K, Tiedje J (2003) Supercooled water brines within permafrost—an unknown ecological niche for microorganisms: a model for astrobiology. Astrobiology 3(2):331–341

Hausmann P, Sussmann R, Smale D (2016) Contribution of oil and natural gas production to renewed increase in atmospheric methane (2007–2014): top–down estimate from ethane and methane column observations. Atmos Chem Phys 16(5):3227–3244

Houghton JT, Ding Y, Griggs DJ, Noguer M, van der Linden PJ, Dai X, Maskell K, Johnson C (2001) Climate change 2001: the scientific basis. The Press Syndicate of the University of Cambridge, Cambridge

Juncher Jørgensen C, Lund Johansen KM, Westergaard-Nielsen A, Elberling B (2014) Net regional methane sink in high Arctic soils of Northeast Greenland. Nat Geosci 8(1):20–23. https://doi.org/10.1038/ngeo2305

Khvorostyanov D, Krinner G, Ciais P, Heimann M, Zimov S (2008a) Vulnerability of permafrost carbon to global warming. Part I: model description and role of heat generated by organic matter decomposition. Tellus B Chem Phys Meteorol 60(2):250–264

Khvorostyanov D, Ciais P, Krinner G, Zimov S, Corradi C, Guggenberger G (2008b) Vulnerability of permafrost carbon to global warming. Part II: sensitivity of permafrost carbon stock to global warming. Tellus B Chem Phys Meteorol 60(2):265–275

Kirschke S, Bousquet P, Ciais P, Saunois M, Canadell JG, Dlugokencky EJ, Bergamaschi P, Bergmann D, Blake DR, Bruhwiler L (2013) Three decades of global methane sources and sinks. Nat Geosci 6(10):813

Kittler F, Heimann M, Kolle O, Zimov N, Zimov S, Göckede M (2017) Long-term drainage reduces CO_2 uptake and CH_4 emissions in a Siberian permafrost ecosystem. Glob Biogeochem Cycles 31:1704–1717. https://doi.org/10.1002/2017GB005774

Kizyakov A, Zimin M, Sonyushkin A, Dvornikov Y, Khomutov A, Leibman M (2017) Comparison of gas emission crater Geomorphodynamics on Yamal and Gydan peninsulas (Russia), based on repeat very-high-resolution stereopairs. Remote Sens 9(10):1023

Kizyakov A, Khomutov A, Zimin M, Khairullin R, Babkina E, Dvornikov Y, Leibman M (2018) Microrelief associated with gas emission craters: remote-sensing and field-based study. Remote Sens 10(5):677

Kohnert K, Serafimovich A, Metzger S, Hartmann J, Sachs T (2017) Strong geologic methane emissions from discontinuous terrestrial permafrost in the Mackenzie Delta, Canada. Sci Rep 7(1):5828

Kraev GN, Schultze ED, Rivkina EM (2013) Cryogenesis as a factor of methane distribution in layers of permafrost. Dokl Earth Sci 451(2):882–885. https://doi.org/10.1134/s1028334x13080291

Kraev G, Schulze E-D, Yurova A, Kholodov A, Chuvilin E, Rivkina E (2017) Cryogenic displacement and accumulation of biogenic methane in frozen soils. Atmosphere 8(6):105

Kraev G, Rivkina E, Vishnivetskaya T, Belonosov A, van Huissteden J, Kholodov A, Smirnov A, Kudryavtsev A, Teshebaeva K, Zamolodchikov D (2019) Methane in gas shows from boreholes in epigenetic permafrost of Siberian Arctic. Geosciences 9(2):67

Kroon P, Hensen A, Jonker H, Zahniser M, Veen WVT, Vermeulen A (2007) Suitability of quantum cascade laser spectroscopy for CH_4 and N_2O eddy covariance flux measurements. Biogeosciences 4(5):715–728

Kvenvolden KA (1988) Methane hydrates and global climate. Glob Biogeochem Cycles 2(3):221–229

Kvenvolden KA, Lorenson TD (2001) The global occurrence of natural gas hydrate. Nat Gas Hydrates Occurrence, Distrib Detect 124:3–18

Kwon MJ, Heimann M, Kolle O, Luus KA, Schuur EAG, Zimov N, Zimov SA, Göckede M (2016) Long-term drainage reduces CO_2 uptake and increases CO_2 emission on a Siberian floodplain due to shifts in vegetation community and soil thermal characteristics. Biogeosciences 13(14):4219–4235. https://doi.org/10.5194/bg-13-4219-2016

Kwon MJ, Beulig F, Ilie I, Wildner M, Kusel K, Merbold L, Mahecha MD, Zimov N, Zimov SA, Heimann M, Schuur EAG, Kostka JE, Kolle O, Hilke I, Göckede M (2017) Plants, microorganisms, and soil temperatures contribute to a decrease in methane fluxes on a drained Arctic floodplain. Glob Chang Biol 23(6):2396–2412. https://doi.org/10.1111/gcb.13558

Lecher AL, Kessler J, Sparrow K, Kodovska FGT, Dimova N, Murray J, Tulaczyk S, Paytan A (2016) Methane transport through submarine groundwater discharge to the North Pacific and Arctic Ocean at two Alaskan sites. Limnol Oceanogr 61(S1):S344–S355

Leibman MO, Kizyakov AI, Plekhanov AV, Streletskaya ID (2014) New permafrost feature–deep crater in central Yamal (West Siberia, Russia) as a response to local climate fluctuations. Environ Sustain 4:68–80

Lewkowicz AG, Way RG (2019) Extremes of summer climate trigger thousands of thermokarst landslides in a High Arctic environment. Nat Commun 10(1):1329. https://doi.org/10.1038/s41467-019-09314-7
Ling F, Zhang T (2003) Numerical simulation of permafrost thermal regime and talik development under shallow thaw lakes on the Alaskan Arctic Coastal Plain. J Geophys Res 108(D16):4511. https://doi.org/10.1029/2002JD003014
Mackay JR (1998) Pingo growth and collapse, Tuktoyaktuk Peninsula area, western Arctic coast, Canada: a long-term field study. Géog Phys Quatern 52(3):271–323
Marushchak M, Pitkämäki A, Koponen H, Biasi C, Seppälä M, Martikainen P (2011) Hot spots for nitrous oxide emissions found in different types of permafrost peatlands. Glob Chang Biol 17(8):2601–2614
Nelson FE, Outcalt SI (1987) A computational method for prediction and regionalization of permafrost. Arct Alp Res 19(3):279–288
Nicolsky D, Romanovsky V, Romanovskii N, Kholodov A, Shakhova N, Semiletov I (2012) Modeling sub-sea permafrost in the East Siberian Arctic Shelf: the Laptev Sea region. J Geophys Res Earth 117(F3):F03028. https://doi.org/10.1029/2012JF002358
Niederberger TD, Perreault NN, Tille S, Lollar BS, Lacrampe-Couloume G, Andersen D, Greer CW, Pollard W, Whyte LG (2010) Microbial characterization of a subzero, hypersaline methane seep in the Canadian High Arctic. ISME J 4(10):1326
Nisbet E, Manning M, Dlugokencky E, Fisher R, Lowry D, Michel S, Myhre CL, Platt SM, Allen G, Bousquet P (2019) Very strong atmospheric methane growth in the 4 years 2014–2017: implications for the Paris Agreement. Glob Biogeochem Cycles 33(3):318–342
O'Connor FM, Boucher O, Gedney N, Jones CD, Folberth GA, Coppell R, Friedlingstein P, Collins WJ, Chappellaz J, Ridley J, Johnson CE (2010) Possible role of wetlands, permafrost, and methane hydrates in the methane cycle under future climate change: a review. Rev Geophys 48(4). https://doi.org/10.1029/2010rg000326
Olenchenko V, Sinitsky A, Antonov E, Eltsov I, Kushnarenko O, Plotnikov A, Potapov V, Epov M (2015) Results of geophysical surveys of the area of "Yamal Crater", the new geological structure. Methods 19(4):84–95
Parmentier F-JW, Christensen TR, Sørensen LL, Rysgaard S, McGuire AD, Miller PA, Walker DA (2013) The impact of lower sea-ice extent on Arctic greenhouse-gas exchange. Nat Clim Chang 3(3):195–202. https://doi.org/10.1038/nclimate1784
Parmentier FW, Zhang W, Mi Y, Zhu X, van Huissteden J, Hayes DJ, Zhuang Q, Christensen TR, McGuire AD (2015) Rising methane emissions from northern wetlands associated with sea ice decline. Geophys Res Lett 42(17):7214–7222. https://doi.org/10.1002/2015GL065013
Paull CK, Ussler W, Dallimore SR, Blasco SM, Lorenson TD, Melling H, Medioli BE, Nixon FM, McLaughlin FA (2007) Origin of pingo-like features on the Beaufort Sea shelf and their possible relationship to decomposing methane gas hydrates. Geophys Res Lett 34(1):L01603. https://doi.org/10.1029/2006GL027977
Pohlman JW, Greinert J, Ruppel C, Silyakova A, Vielstädte L, Casso M, Mienert J, Bünz S (2017) Enhanced CO2 uptake at a shallow Arctic Ocean seep field overwhelms the positive warming potential of emitted methane. Proc Natl Acad Sci 114(21):5355–5360
Portnov A, Smith AJ, Mienert J, Cherkashov G, Rekant P, Semenov P, Serov P, Vanshtein B (2013) Offshore permafrost decay and massive seabed methane escape in water depths> 20 m at the South Kara Sea shelf. Geophys Res Lett 40(15):3962–3967
Portnov A, Mienert J, Serov P (2014) Modeling the evolution of climate-sensitive Arctic subsea permafrost in regions of extensive gas expulsion at the West Yamal shelf. J Geophys Res Biogeo 119(11):2082–2094
Rehder G, Kirby SH, Durham WB, Stern LA, Peltzer ET, Pinkston J, Brewer PG (2004) Dissolution rates of pure methane hydrate and carbon-dioxide hydrate in undersaturated seawater at 1000-m depth. Geochim Cosmochim Acta 68(2):285–292
Repo ME, Susiluoto S, Lind SE, Jokinen S, Elsakov V, Biasi C, Virtanen T, Martikainen PJ (2009) Large N_2O emissions from cryoturbated peat soil in tundra. Nat Geosci 2(3):189

Rivkina E, Gilichinsky DA, McKay C, Dallimore S (2001) Methane distribution in permafrost: evidence for an interpore pressure methane hydrate. In: Permafrost response on economic development, environmental security and natural resources. Springer, Dordrecht, pp 487–496

Rivkina E, Shcherbakova V, Laurinavichius K, Petrovskaya L, Krivushin K, Kraev G, Pecheritsina S, Gilichinsky D (2007) Biogeochemistry of methane and methanogenic archaea in permafrost. FEMS Microbiol Ecol 61(1):1–15. https://doi.org/10.1111/j.1574-6941.2007.00315.x

Romanovskii N, Hubberten H-W, Gavrilov A, Tumskoy V, Kholodov A (2004) Permafrost of the east Siberian Arctic shelf and coastal lowlands. Quat Sci Rev 23(11):1359–1369

Romanovskii N, Hubberten H-W, Gavrilov A, Eliseeva A, Tipenko G (2005) Offshore permafrost and gas hydrate stability zone on the shelf of East Siberian Seas. Geo-Mar Lett 25(2–3):167–182

Ruppel CD (2011) Methane hydrates and contemporary climate change. Nat Educ Knowl 2(12):12

Ruppel CD, Kessler JD (2017) The interaction of climate change and methane hydrates. Rev Geophys 55(1):126–168

Savvichev A, Leibman M, Kadnikov V, Kallistova A, Pimenov N, Ravin N, Dvornikov Y, Khomutov A (2018) Microbiological study of Yamal Lakes: a key to understanding the evolution of gas emission craters. Geosciences 8(12):478

Schneider von Deimling T, Grosse G, Strauss J, Schirrmeister L, Morgenstern A, Schaphoff S, Meinshausen M, Boike J (2015) Observation-based modelling of permafrost carbon fluxes with accounting for deep carbon deposits and thermokarst activity. Biogeosciences 12(11):3469–3488

Schuur EA, McGuire AD, Schadel C, Grosse G, Harden JW, Hayes DJ, Hugelius G, Koven CD, Kuhry P, Lawrence DM, Natali SM, Olefeldt D, Romanovsky VE, Schaefer K, Turetsky MR, Treat CC, Vonk JE (2015) Climate change and the permafrost carbon feedback. Nature 520(7546):171–179. https://doi.org/10.1038/nature14338

Serov P, Portnov A, Mienert J, Semenov P, Ilatovskaya P (2015) Methane release from pingo-like features across the South Kara Sea shelf, an area of thawing offshore permafrost. J Geophys Res Earth 120(8):1515–1529

Shakhova N, Semiletov I, Leifer I, Salyuk A, Rekant P, Kosmach D (2010a) Geochemical and geophysical evidence of methane release over the East Siberian Arctic Shelf. J Geophys Res Oceans 115(C8):C08007. https://doi.org/10.1029/2009JC005602

Shakhova N, Semiletov I, Salyuk A, Yusupov V, Kosmach D, Gustafsson Ö (2010b) Extensive methane venting to the atmosphere from sediments of the East Siberian Arctic Shelf. Science 327(5970):1246–1250

Shakhova N, Semiletov I, Leifer I, Sergienko V, Salyuk A, Kosmach D, Chernykh D, Stubbs C, Nicolsky D, Tumskoy V (2014) Ebullition and storm-induced methane release from the East Siberian Arctic Shelf. Nat Geosci 7(1):64

Shcherbakova V, Yoshimura Y, Ryzhmanova Y, Taguchi Y, Segawa T, Oshurkova V, Rivkina E (2016) Archaeal communities of Arctic methane-containing permafrost. FEMS Microbiol Ecol 92(10)

Slater AG, Lawrence DM (2013) Diagnosing present and future permafrost from climate models. J Clim 26(15):5608–5623

Stocker TF, Qin D, Plattner G-K, Tignor M, Allen SK, Boschung J, Nauels A, Xia Y, Bex V, Midgley PM (2013) Climate change 2013: the physical science basis. Working group I contribution to the fifth assessment report of the Intergovernmental Panel on climate change 2013. URL https://www.google.com/books

Stolaroff JK, Bhattacharyya S, Smith CA, Bourcier WL, Cameron-Smith PJ, Aines RD (2012) Review of methane mitigation technologies with application to rapid release of methane from the Arctic. Environ Sci Technol 46(12):6455–6469

Stotler RL, Frape SK, Ahonen L, Clark I, Greene S, Hobbs M, Johnson E, Lemieux J-M, Peltier R, Pratt L (2010) Origin and stability of a permafrost methane hydrate occurrence in the Canadian Shield. Earth Planet Sci Lett 296(3–4):384–394

Sugimoto A, Wada E (1993) Carbon isotopic composition of bacterial methane in a soil incubation experiment: contributions of acetate and CO2H2. Geochim Cosmochim Acta 57(16):4015–4027

Thornton BF, Geibel MC, Crill PM, Humborg C, Mörth CM (2016) Methane fluxes from the sea to the atmosphere across the Siberian shelf seas. Geophys Res Lett 43(11):5869–5877

Tuorto SJ, Darias P, McGuinness LR, Panikov N, Zhang T, Häggblom MM, Kerkhof LJ (2014) Bacterial genome replication at subzero temperatures in permafrost. ISME J 8(1):139

Voigt C, Lamprecht RE, Marushchak ME, Lind SE, Novakovskiy A, Aurela M, Martikainen PJ, Biasi C (2017) Warming of subarctic tundra increases emissions of all three important greenhouse gases–carbon dioxide, methane, and nitrous oxide. Glob Chang Biol 23(8):3121–3138

Walter Anthony KM, Anthony P, Grosse G, Chanton J (2012) Geologic methane seeps along boundaries of Arctic permafrost thaw and melting glaciers. Nat Geosci 5(6):419–426. https://doi.org/10.1038/ngeo1480

Walter Anthony K, Daanen R, Anthony P, Schneider von Deimling T, Ping C-L, Chanton JP, Grosse G (2016) Methane emissions proportional to permafrost carbon thawed in Arctic lakes since the 1950s. Nat Geosci 9:679–682

Walter KM, Zimov SA, Chanton JP, Verbyla D, Chapin FS 3rd (2006) Methane bubbling from Siberian thaw lakes as a positive feedback to climate warming. Nature 443(7107):71–75. https://doi.org/10.1038/nature05040

Walter KM, Smith LC, Chapin FS 3rd (2007a) A methane bubbling from northern lakes: present and future contributions to the global methane budget. Philos Trans A Math Phys Eng Sci 365(1856):1657–1676. https://doi.org/10.1098/rsta.2007.2036

Walter K, Chanton J, Chapin F, Schuur E, Zimov S (2008) Methane production and bubble emissions from arctic lakes: isotopic implications for source pathways and ages. J Geophys Res Biogeo 113(G3):G00A08. https://doi.org/10.1029/2007JG000569

West J, Plug LJ (2008) Time-dependent morphology of thaw lakes and taliks in deep and shallow ground ice. J Geophys Res Earth 113(F1):F01009. https://doi.org/10.1029/2006JF000696

Whiticar M (1999) Carbon and hydrogen isotope systematics of bacterial formation and oxidation of methane. Chem Geol 161:291–314

Wik M, Varner RK, Anthony KW, MacIntyre S, Bastviken D (2016) Climate-sensitive northern lakes and ponds are critical components of methane release. Nat Geosci 9(2):99

Wilkerson J, Dobosy R, Sayres DS, Healy C, Dumas E, Baker B, Anderson JG (2019) Permafrost nitrous oxide emissions observed on a landscape scale using the airborne eddy-covariance method. Atmos Chem Phys 19(7):4257–4268

Woolway RI, Merchant CJ (2019) Worldwide alteration of lake mixing regimes in response to climate change. Nat Geosci 12:271–276. https://doi.org/10.1038/s41561-019-0322-x

Worden JR, Bloom AA, Pandey S, Jiang Z, Worden HM, Walker TW, Houweling S, Röckmann T (2017) Reduced biomass burning emissions reconcile conflicting estimates of the post-2006 atmospheric methane budget. Nat Commun 8(1):2227

Yakushev V, Chuvilin E (2000) Natural gas and gas hydrate accumulations within permafrost in Russia. Cold Reg Sci Technol 31(3):189–197

Chapter 8
Models: Forecasting the Present and Future of Permafrost

Abstract Models are indispensable tools for projecting future permafrost extent and carbon cycle. In this chapter the current state of the art in modelling is compared with the permafrost thaw and carbon cycle processes discussed in the previous chapters. Land surface models that are applied in Earth System/climate models are still missing crucial processes for predicting the future permafrost carbon feedback, show large discrepancies in hindcasting present permafrost extent and carbon and large variation in future permafrost prediction. However, improvement of permafrost-relevant soil physics is promising. Large-scale modelling of dedicated permafrost and permafrost carbon models has shown interesting results, in particular in modelling thaw of ice-rich permafrost.

This chapter concludes with an outlook to the future of permafrost. The most recently published papers demonstrate that in some regions abrupt permafrost thaw is accelerating at an alarming pace. Compared to permafrost decline during the last glacial termination, near future permafrost area is expected to decline at least 2.4 times faster.

Models are indispensable for forecasting future effects of climate change on the permafrost carbon cycle and greenhouse gas emissions. The sparseness of the observation network of the permafrost carbon cycle also necessitates the use of models for hind-casting its present and past state. In the recent decade, large progress has been made in predictive modelling of key aspects of the permafrost and permafrost carbon cycle (AMAP 2017). A wide range of models is relevant, with different scope, spatial scales and various levels of integration of processes:

- Earth system models that integrate the climate system, atmosphere, oceans and terrestrial environments (Fig. 8.1);
- Land surface models to describe the exchange of matter and energy between soils, vegetation and atmosphere, usually as part of the Earth system models;
- Vegetation models, that model the evolution of vegetation and biomass carbon related to climate;
- Soil carbon and nutrient models to track the flow of carbon and nutrients in soils;

J. van Huissteden, *Thawing Permafrost*,
https://doi.org/10.1007/978-3-030-31379-1_8

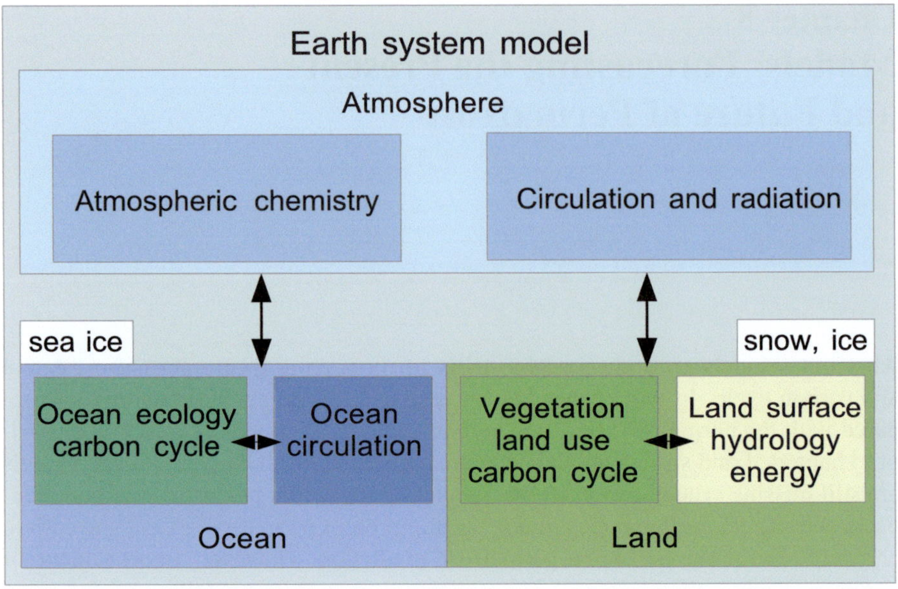

Fig. 8.1 Earth system model (coupled climate model) with its component models. (After Gettelman and Rood 2016)

- Permafrost models, modelling the extent, temperature and active layer thickness of permafrost;
- Hydrological and hydrochemical models for assessing groundwater flow, river runoff and dissolved substances in water;
- Models of geomorphological and sedimentary processes, for quantifying erosion, subsidence, sedimentation.

These models may be used as stand-alone models or combined models. Earth system and land surface models have the highest level of integration, and contain vegetation, soil and hydrological models. At that level of integration, details are lost: a soil model integrated in a land surface model will contain less process and soil detail than a dedicated soil process model. In this chapter, these models will not be discussed in detail, which would require a series of books by itself. Instead, it will be attempted to link existing modelling practices to the processes discussed in the previous chapters, with the focus on permafrost-related processes that are underrepresented in current models. Such an underrepresentation may have good reasons: obstacles in mathematical description, parameter availability or lack of validation data for the process at the desired scale level.

Model spatial scale and complexity are important issues. Models in the Earth and Environmental sciences represent the Earth's surface, the atmosphere, soil, oceans in three dimensions. This needs discretisation of space in a three-dimensional grid, with grid cells or volume elements describing phenomena that are essentially continuous (Gettelman and Rood 2016). The size of these grid cells determines in how much detail processes can be modelled – and the computing time that is needed.

Ideally, a model should incorporate all the relevant processes, given the system it describes. However, much process detail makes a model more complex, with more interacting subsystems, and more computer code (with the risk of bugs). A more complex model also needs more input as initial conditions or process parameters, which may introduce additional uncertainty. For these reasons, complexity is not always better than simplicity. A simplified model can tell more about the basic behaviour of a system than a complex model that is very demanding on uncertain input. For instance, a complex soil carbon model includes biochemical processes of carbon and nutrient transfer, soil moisture, gas diffusion and root system processes. However, to run the model, a large number of parameters and initial conditions of all these components should be known. Many of these parameters are poorly quantified, or not quantified for cold climate soils or even not quantified by measurements at all. Running the model requires assumptions on parameter values. The scale of the model may be that of a single pedon (the smallest unit in soil classification) – which is not a very useful scale for making predictions of the soil carbon evolution at regional or continental scales, and is likely to miss lateral interaction between other pedons. To reliably model the permafrost soil carbon cycle, this soil model should be integrated with at least a soil thermal model that predicts soil freezing, thawing and the active layer depth, and it should be integrated with a vegetation model to get the input side of the carbon balance right.

Uncertainty in the outcome of models is a prime concern when we try to predict future changes with models. It is a subject with many mathematical, philosophical and societal questions, and without a common agreement on terminology (e.g. Walker et al. 2003; Beven 2007), which cannot be discussed in depth here. To guide the discussion, there are three basic sources of uncertainty in models (Gettelman and Rood 2016):

- Model uncertainty: the question whether the model is an adequate representation of the system being modelled. This is the question whether the model gives the correct answer to our questions, if the input of the model is correct (Walker et al. 2003). There are two types of model uncertainty:
  - Technical model uncertainty, ranging from hidden flaws in computer code or processors, to design errors in algorithms.
  - Structural model uncertainty, which is the question whether the models represent the modelled system correctly. In the process of building models, there are many choices to be made on equations that describe processes that are part of the system. For instance, empirical equations that are used for calculating thermal conductivity of soils from soil constituents (Sect. 2.5). If, for a permafrost model, an equation is used that does not include ice content, the model will have difficulty producing the right soil temperatures, whatever the quality of the input is. However, the choices that have been made in building a model need not be flawed, but simply a consequence of incomplete knowledge of the system, or well-reasoned choices based on practical needs, e.g. spatial scale

of the model, computational efficiency and parameter availability. The more complex the models are, the more likely structural differences occur, and there may be no single best model.

- Uncertainty in initial conditions and parameter uncertainty: the current state of the system at the starting point of the model, and uncertainty in the parameters of the processes. A particular source of uncertainty in modelling Arctic systems is the low density of observations in the northern parts of the continents (Chap. 1). Uncertainties in the initial conditions may have a large effect. Many Earth system components are known to be very sensitive to the initial conditions, by the non-linear character of the equations used to model the system (e.g. Kaplan and Glass 2012).
- Scenario uncertainty is uncertainty in the future track of model input, e.g. for climate models the greenhouse gas concentrations in the atmosphere, which are determined by political, social, economic and technological developments in human society, and are not included as processes in the climate model. The effect of various greenhouse gas emission scenarios on high-latitude climates already has been mentioned in Chap. 1.

Modelling toolboxes have been developed to deal with initial condition and parameter uncertainty. This usually includes running a large number of model simulations with different sets of randomly chosen parameters, within a plausible range of parameter values (e.g. Beven 2007). To evaluate model uncertainty, it is a good practice to run different models and compare the range of results, as is done with climate models in the IPCC reports. Model intercomparisons, in which different models are run using the same input data set and the outcomes compared with observations, are a key aspect of the reduction in model uncertainty. It is not surprising that in model inter-comparisons considerable differences between model outcomes arise (Melton et al. 2013; McGuire et al. 2016). Model intercomparisons are very helpful in improvement of models.

A second problem is the variety of scale levels on which environmental problems can be addressed. Complexity of models and inclusion of all relevant processes at all scale levels is not feasible, and well-argued choices have to be made for simplification when going from smaller scale to large scale. Model coupling on various levels is an integral part of Earth System models (Gettelman and Rood 2016). This model coupling lets the sub-models interact with each other – for instance, the vegetation in the land surface model is changed by a changing climate and atmospheric CO_2 concentration, and in turn, the changed vegetation interacts with the water vapour and energy fluxes which are input to the atmosphere. An Earth System model includes a land surface model, which in turn also includes a soil model and a vegetation model. However, the land surface models use grid cells that have sides in the order of tens to hundreds of kilometres, while vegetation and soils vary on much smaller scales. A soil model which serves to calculate soil carbon fluxes as part of the land surface model will therefore use input that is averaged over a large area, and be less complex than a specialized stand-alone soil model.

Basic choices in modelling also include numerical approaches. On the one side are deterministic models that describe processes in terms of deterministic equations, which are numerically solved to generate only one possible model outcome based on a given parameter input. An example may be a model that describes the occurrence of slope failure in a cryogenic landslide based on equations containing slope, soil cohesion, thawed or frozen state and water content. On the other side are stochastic models that contain stochastic processes, which attach a probability to each possible outcome, described by a probability distribution. The slope failure example also could be described by a probability distribution in which a probability distribution of failure is used, given a certain soil material and water content.

A third approach in the spectrum are data-driven models, recently boosted by developments in artificial intelligence. These start with simple regression equations derived from data, without much underlying theory about processes. Neural networks, which imitate connections between nerve cells in the brain, have been around for some time and are helpful to detect patterns in large data sets. These artificial intelligence methods have expanded in the last decennia with increasing computing power and a variety of learning algorithms, and developed into a branch of computer science known as machine learning. An example of an application in permafrost science is presented by Lara et al. 2018.

8.1 Land Surface Models

In Earth System models, the land surface model captures the type of processes that are described in Chaps. 2 and 4. Land surface models include the following processes, listed with increasing sophistication (Bonan 2016; Gettelman and Rood 2016):

- A basic land surface model includes the processes of exchange of water and energy at the Earth surface, including a simple representation of the soil water availability.
- An extended model also includes soil and vegetation, including soil water, soil carbon and nutrients, to model the water vapour exchange (evaporation and transpiration), soil hydrology and changes in net ecosystem exchange of CO_2. The vegetation is usually described as a series of plant functional types.
- A more sophisticated land surface model will include dynamic vegetation, interacting with modelled changes in climate. In this way feedbacks between vegetation and climate can be modelled; vegetation change also affect energy and water vapour exchange.
- For modelling climate change effects in the north, the cryosphere needs to be included: glaciers, ice and snow. Because of the effects of snow cover on albedo, the snow cover sub-models in land surface models may be quite detailed, with a snow cover consisting of various model layers.

- Increasingly, also human activities are incorporated in land surface models since agricultural and urban areas also have a large effect on the exchange of energy, water and carbon at the surface.

In Earth system models, the number of tasks for land surface sub-models is high and diverse. Regionally relevant details – for instance details on permafrost – may be lost. Moreover, it is a huge effort to keep global land surface models up to date with the most recent knowledge on rapidly evolving subjects as permafrost research. Take alone the modelling of soil temperature and active layer thickness (Sect. 2.5, Fig. 5.5). This needs information on:

- soil properties including the thickness of the organic top layer and down to a sufficient depth, in the order of 10 m or more (input data);
- soil water phase change needs to be included (depends on model structure);
- soil water content (from a hydrology sub-model);
- heat balance at the surface (heat and water exchange sub-model, that needs in turn vegetation data or vegetation modelling);
- snow cover (climate model + snow sub-model);
- vegetation (with its layered structure – trees, shrubs and moss cover).

McGuire et al. (2016) published a model intercomparison of 14 land surface models, several of which are included in Earth System models. Included is an inventory of processes of permafrost dynamics and soil carbon that are included in these models (Table 8.1). Some of the models do not include phase transition of water in the soil which is a necessary condition for correct modelling of permafrost tempera-

Table 8.1 Soil thermal processes and soil carbon processes included in land surface models based on a model intercomparison by McGuire et al. (2016)

Process	Nr of models
Soil depth ≥10 m (sufficient to capture seasonal soil temperature variation)	12
Moss layer insulation	3
Organic topsoil insulation included	7
Snow layer insulation included	All
Snow layer insulation with more than one snow layer	9
Soil water phase change included	9
Soil C storage explicitly with depth variation included	5
Moss layer soil C storage included	None
Litter horizon C included	8
Organic horizon C included	3
Ecosystem CH_4-C losses included	5
Fire and other disturbance emissions included	3 (fire) 1 (other)
Lateral C export incuded (LU = Land Use, D = DOC)	3 (LU) 1 (DOC)

The rows with a greyed background are based on 12 models, the other rows on 14 models tures, and most models do not consider the insulation of the moss cover and organic top layer. The models were tested with climate re-analysis data from 1960 to 2009. All models simulated a decrease of the permafrost area over that period (permafrost defined as frozen ground within 3 m depth), largely due to rising air temperature. However, the rate of decrease varied over two orders of magnitude, 200–58,800 km^2 $year^{-2}$. The model sensitivity to precipitation change varied, from conserving permafrost to increasing permafrost loss.

Similar large ranges between model predictions of permafrost area were found in the Coupled Model Inter-comparison Project phase 5 (CMIP5) for the IPCC 5th Assessment report (IPCC 2013; Fig. 7.9). Slater and Lawrence (2013) and Koven et al. (2013) analysed the performance of the land surface part of these models. Instead of the land surface model output, they analysed the permafrost area separately by using permafrost models. The models show a large range of sensitivities of the modelled area of permafrost to climate change: 0.8–2.3 million km^2 $°C^{-1}$, and a wide range of permafrost area loss by the year 2100 depending on the greenhouse emission scenarios: 15–87% under RCP4.5 and 30–99% under the high emission scenarios RCP8.5. Under the RCP8.5 scenario, most models predict that permafrost will be withdrawn towards the High Arctic areas of Canada and northernmost parts of Siberia (Fig. 8.2).

The permafrost presence in Fig. 8.2 is based on a surface frost index (Sect. 2.5.3), in turn based on modelled surface temperatures corrected for modelled snow depth. This approach assumes climate equilibrium, and results in a climate suitability for permafrost, not the actual presence of permafrost. In areas where permafrost disappears, permafrost still may be present as relict or ecosystem-protected permafrost. Modelling of permafrost for a climate equilibrium (stable climate conditions) to which the permafrost is adjusted, results in a smaller permafrost extent than when permafrost extent is modelled with transient climate change. This is the result of a lag time with which permafrost reacts on climate change; penetration of higher temperatures in the soil and thawing takes time. The equilibrium permafrost area is typically 25–38% smaller than that of transient simulations (Slater and Lawrence 2013).

Are these widely diverging model results alarming? Considering the need for forecasting of the permafrost carbon feedback, yes. The land surface models are not designed to model permafrost in the first place, but designed to provide lower boundary conditions for the climate models. However, the effects of permafrost on soil temperature and hydrology are not negligible, a correct permafrost extent is highly desirable for climate modelling. On the other hand, given the complexity of permafrost and ecosystems, this model structure uncertainty is not surprising at all.

Approaches integrating observations and models more closely may decrease the range of model outcomes. Chadburn et al. (2017b) constrain modelled permafrost extent of the the CMIP5 models by using observations on the relation between climate (air temperature) and present permafrost extent. The CMIP5 models show a widely diverging permafrost extent for simulated present and future permafrost extent. By taking modelled global temperature, corrected for Arctic amplification of global temperature increase, and combining it with the observed climate-permafrost

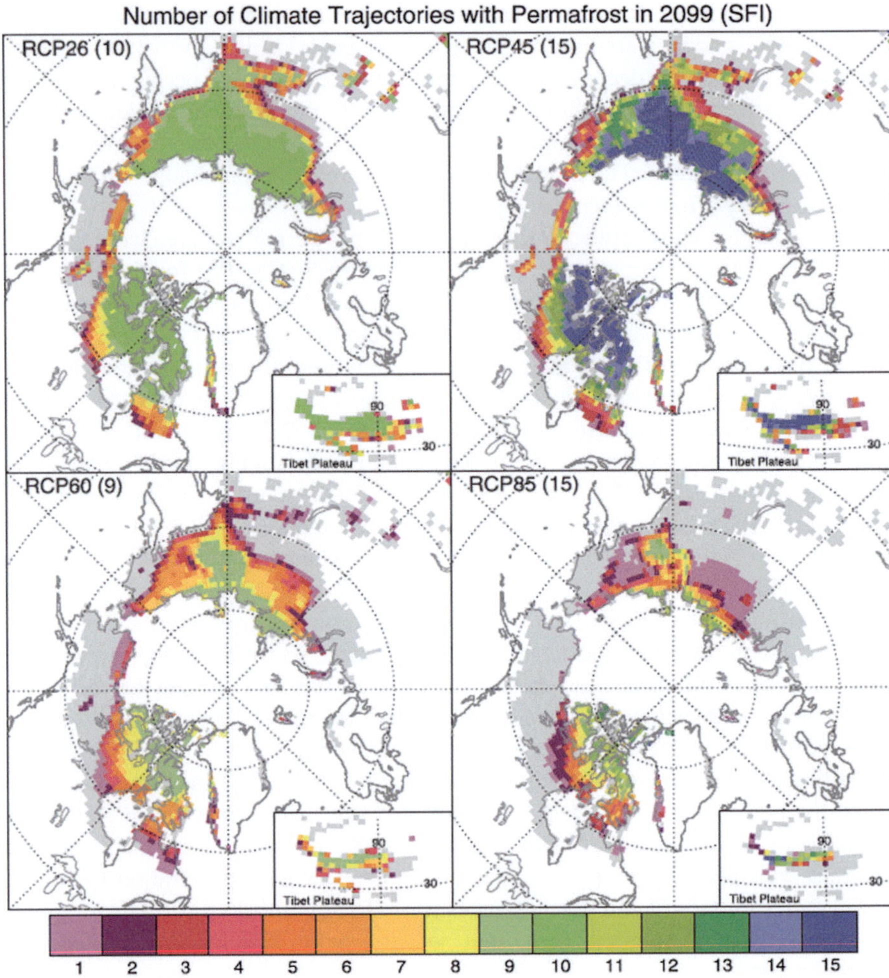

Fig. 8.2 The number of climate model runs with sustainable permafrost in 2099 based on projected change in climate from present-day reanalyses for each IPCC RCP greenhouse gas emission scenario, according to the surface frost index method. The total number of models available for each RCP is given in brackets in each panel. Present-day observed continuous and discontinuous permafrost is shown with grey shading. (Republished with permission of American Meteorological Society from 'Diagnosing present and future permafrost from climate models' by Slater and Lawrence (2013), Journal of Climate, Vol. 26–15, 2013; permission conveyed through Copyright Clearance Center, Inc. © 2013 American Meteorological Society)

relation, a climate sensitivity of permafrost could be derived. This climate sensitivity is independent of the models and the climate scenarios. The resulting sensitivity of climate equilibrium permafrost area to global temperature is 4.0 million km^2 °C^{-1} (+1.0–1.1 million km^2, 1σ confidence interval), which proved 20% higher than previous studies. Based on this climate sensitivity. a stabilisation of global climate

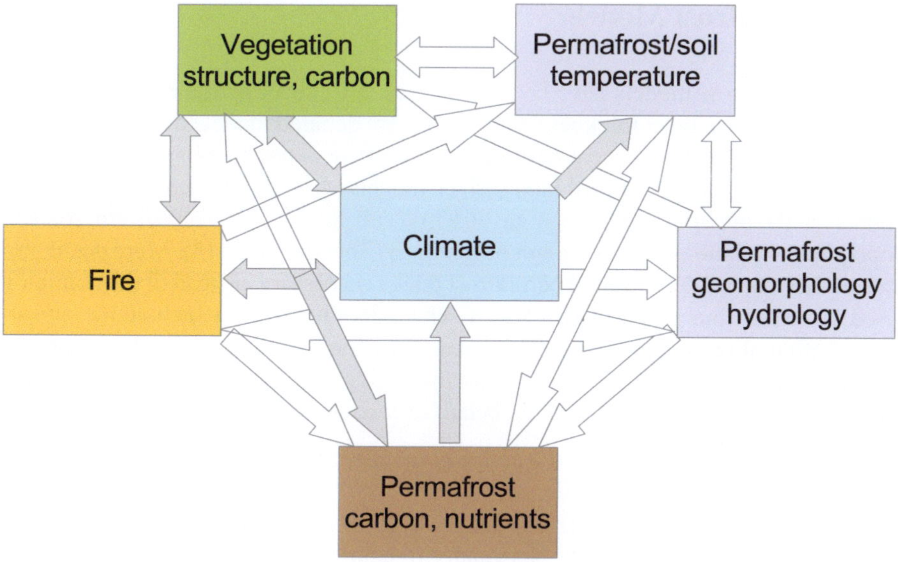

Fig. 8.3 Schematic of land surface subsystems that transfer the effects of climate change to permafrost and the permafrost carbon cycle. The grey arrows indicate relations that are usually addressed in land surface models, white arrows the relations that are not or poorly represented in land surface models. (Based on McGuire et al. 2016)

warming according to the Paris Agreement goals at 2 °C would result in a permafrost loss of eight million km^2, a reduction of approximately 40%.

Improvement of soil temperature and active layer modelling is certainly feasible. The effect of improvement of the modelling of the moss and organic layer effect, a higher resolution soil column and more detailed simulation of snow cover is illustrated for the JULES model by Chadburn et al. (2015a, b). Compared to site observations, the deviations of soil temperatures and active layer thickness were strongly reduced. For the active layer, first 1 m too deep in the model simulations, the model-data discrepancy was reduced to <0.1 m from the observations. Incorporation of other processes related to thaw of ground ice should be possible as well. For improvement of permafrost models, not only vegetation, soil and climate should be interactively coupled, but also the disturbance regimes of fire and geomorphological/hydrological changes of permafrost soils. Figure 8.3 shows the subsystems of the land surface in permafrost areas which could be improved for modelling of climate-permafrost interactions.

8.2 Permafrost Models

Several permafrost models have been developed that model permafrost extent, temperature and active layer thickness, with both more detailed model structure and more detailed input of soil and ecosystem information than is possible in the present generation of clime model-coupled land surface models (AMAP 2017). The land surface models of the previous section operate on a grid cell size in which ecosystem, soil and topography variations simply cannot be resolved (Zhang et al. 2013). More detail also means that a better definition of permafrost thaw is necessary, instead of the definition used in the previous section (ALT >3 m), which does not tell much about the subsurface configuration of permafrost. This subsurface configuration is very important for understanding the effects of thaw on hydrology and gas migration. Zhang et al. (2013) defined stages of permafrost thaw on behalf of modelling as:

- The gradual-thawing stage: the active layer deepens gradually with yearly small fluctuations;
- The increased-thawing stage: the active-layer deepens at a faster rate but without formation of taliks below the seasonally freezing ALT;
- The frequent-talik stage: supra-permafrost taliks form but disappear in cold years, and the depth of permafrost table varies strongly due to the formation and disappearance of taliks;
- The isothermal-permafrost stage: taliks become increasingly thicker and persist regardless of yearly variation in mean annual air temperature;
- The permafrost-free stage: permafrost has disappeared completely, also at depth;
- A criterion not included by Zhang et al. (2013) should be thaw subsidence/erosion/lake formation, for reasons discussed below.

With respect to the permafrost-free stage, a distinction could be made between areas of thin continuous or discontinuous permafrost and areas of thick continuous permafrost, where permafrost may persist for millennia at depth as relict permafrost (Chap. 1).

The use of remote sensing data is helpful for improving environmental input for permafrost models. Langer et al. (2013) have shown the feasibility of permafrost modelling based on remote sensing products. The largest uncertainty arises from unknown thermal properties of snow. Zhang et al. (2013) applied a one-dimensional soil temperature model (NEST, Northern Ecosystem Soil Temperature model) at a very high spatial resolution (10 m), combined with remote sensing input. Zhang et al. (2013) estimate soil and drainage characteristics from ecosystem types, as determined from remote sensing material. Leaf area index (LAI) is also derived from remote sensing; further input consists of topographic information derived from digital elevation models (slope, aspect, viewshed, to determine modification of insolation by topography). Climate input is derived from two climate models. The results of this high resolution model show good agreement with observations of ALT for most ecosystem types (an exception are high alpine mountain slopes). The effect of spatial resolution has been tested by taking larger grid cells, and running the model for the dominant land cover type of each cell – which biased the model outcome towards this dominant land cover. The climate data used for the model do not modify the vegetation. However, model sensitivity tests show that an increasing LAI would reduce active layer thickness by the shading effect. This shows the need for interactive climate-vegetation modelling for permafrost.

A model that includes basic geomorphological and hydrological change of permafrost is CryoGrid 3 (Westermann et al. 2016). CryoGrid 3 includes land-atmosphere coupling and a soil thermal model with conductive heat transport and freeze-thaw processes. The model has an extension called Xice that deals with the effects of excess ice in permafrost soils. It cannot generate excess ice, but it simulates its geomorphological effect upon thaw. When excess ice thaws, it creates subsidence or thaw pond formation, depending on drainage conditions. Subsidence occurs when water from excess ice can drain freely. The model captures observed values of active layer thickness and thaw lake bottom temperature with minor deviations. Forecast runs up to the year 2100 for three different types of cryostratigraphy representative for the Lena Delta region, two drainage scenarios (drainage and no drainage), and two RCP scenarios (RCP 8.5 shown in Fig. 8.4). The cryostratigraphies are permafrost without excess ice, permafrost with layers of medium excess ice, and high ice content Yedoma deposits.

The modelling results show interesting effects of ground ice on permafrost thaw. For the cryostratigraphy without excess ice, permafrost is stable until 2100 with the moderate greenhouse gas emission scenario RCP 4.5. For the high emission RCP 8.5 a supra-permafrost talik develops after 2060. Lakes and subsidence do not develop with this cryostratigraphy. However, the ice-rich cryostratigraphies show subsidence or thaw lake development, depending on the drainage scenario. The ice complex with high amounts of excess ice is most vulnerable. For the RCP 4.5 and 8.5 scenarios, subsidence or lake formation starts in the ice complex at about 2020, and progresses deeper for RCP 8.5. The development of lakes in the 'no drainage' scenario results in deep talik formation, for the drainage scenario it results in subsidence down to 10 m depth, which suggests that deep subsidence and associated erosion could affect ice-rich permafrost in the near future (Westermann et al. 2016).

Two effects are particularly interesting in the CryoGrid output. First, in the moderate warming RCP 4.5 scenario, there is a stabilisation of subsidence after initial rapid development, by a buffer effect of the development of a layer of compacted mineral material or sediment. Second, with the RCP 8.5 scenario + ice-complex + drainage, no talik develops above the permafrost. This is caused by the cooling effect of latent heat of thawing of large amounts of ground ice, combined with volume loss by ground ice thaw. Note that this result will not classify as permafrost thaw according to the criterion used in the previous section (ALT >3 m) or the five stages of thaw of Zhang et al. (2013). This is the reason why subsidence should be included as a criterion for permafrost thaw. For permafrost with large amounts of excess ice, the CryoGrid 3 model also suggests a significant threshold behaviour. Once layers with excess ice start to thaw, rapid subsidence is initiated by irreversible loss of ground ice, which brings progressively deeper layers of excess ice into the active layer. If meltwater can drain from the surface, deep subsidence results. In real

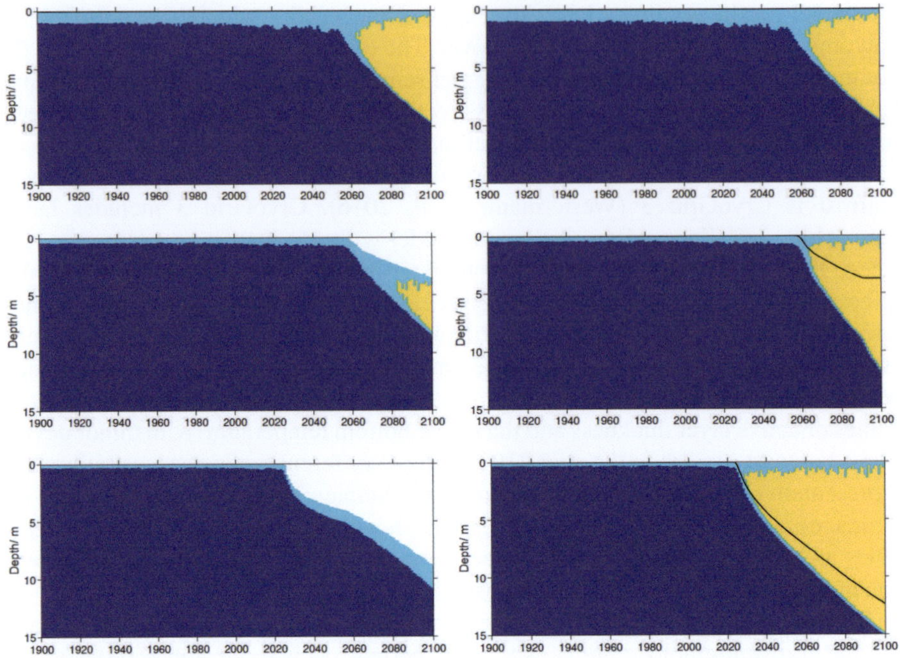

Fig. 8.4 Modelled annual freeze–thaw state for three different cryostratigraphies (no excess ice, medium excess ice, high excess ice yedoma/ice complex) and the drainage scenarios subsidence (left side graphs) and no drainage/lake development (right side graphs), as simulated by CryoGrid 3 Xice forced by CRU-NCEP/CCSM4 climate model output and the future greenhouse gas emission scenario RCP 8.5. The drainage scenarios show deep subsidence for ice-rich permafrost and talik development for low-medium excess ice. The no drainage scenarios indicate lake development and deep talik development. (From 'Simulating the thermal regime and thaw processes of ice-rich permafrost ground with the land-surface model CryoGrid 3' by Westermann et al. (2016) in Geoscientific Model Devel. 9 https://doi.org/10.5194/gmd-9-523-2016; © The authors; published under Creative Commons License 4.0 https://creativecommons.org/licenses/by/4.0/)

permafrost terrain this will be uneven because of the uneven subsurface distribution of ground ice. The channeling of water in these subsidence features may cause considerable further erosion (Westermann et al. 2016).

Nicolsky et al. (2017) developed a high resolution model for the Alaskan North Slope, building on a model by Jafarov et al. (2012), and using the remote-sensing derived ecosystem classification of Zhang et al. (2013). In particular the modelling of snow cover properties is improved in the model, with multiple snow layers and a density model. Soil thermal properties are estimated from borehole temperature data. These improvements significantly contribute to a better model performance if validated on present-day permafrost. For future permafrost, it indicates double active layer thickness by 2050, and widespread supra-permafrost talik development by the end of the century, if the RCP 8.5 scenario is realised.

These examples show that significant improvements of permafrost modelling are possible, at least for stand-alone permafrost models. Specifically, the use of remote sensing data, surface heat balance modelling, improved snow cover sub-models and inclusion of subsidence processes result in better predictions of future permafrost extent. What still is missing in these models, is vegetation that changes interactively with climate change input, which is important for surface heat balance (soil shading) and ground thermal conductivity effects of a surface moss layer (Anisimov and Sherstiukov 2016). All models show – for high greenhouse gas emission scenarios – a far northern retreat of surface permafrost, conformal to the large-scale land surface models in the previous section. The possibility of threshold behaviour with abrupt thaw collapse for ice-rich permafrost (Westermann et al. 2016) should be further explored in models. Other critical thresholds may result from major vegetation shifts (AMAP 2017).

8.3 The Carbon Cycle in Models

One of the essential questions is, how much of the huge permafrost carbon reservoir will enter the atmosphere under the various climate change scenarios, and at what time scale. For that question we first need to know how much permafrost will thaw – as seen above, model forecasts still need improvement. The second part of the answer is, how much soil carbon will be affected by permafrost thaw, how much of that carbon will be decomposed and released into the atmosphere as CO_2 and CH_4, and how much CO_2 is fixed in increasing biomass. For the latter question, also modelling of the nutrient cycle is crucial. More recently, also the third major greenhouse gas – N_2O – is shown to be a greenhouse gas released by permafrost disturbances, with unexpectedly high and potentially increasing emission levels (Sect. 7.4).

Like the differences in future permafrost extent (Sect. 7.1), the differences between the soil and ecosystem carbon balance produced by the land surface models tested by McGuire et al. (2016) diverge widely. Table 8.1 illustrates that most models do not include all soil carbon reservoirs that are crucial for permafrost soils. All models simulated an increase in both vegetation and soil carbon storage, but with more than an order of magnitude difference in the modelled rate of increase. The range of changes in vegetation carbon storage are 24–526 Tg C year^{-1} and in soil carbon 42–637 Tg C year^{-1} across the permafrost region between 1960 and 2009 (Fig. 8.5). Most of the vegetation carbon change resulted in the models from CO_2 fertilization; second was temperature, with some models indicating a negative effect of temperature on vegetation carbon storage.

With the incomplete implementation of soil carbon storage combined with the differences in modelled active layer depths, a large spread of model results is no surprise. The modelling of soil carbon decomposition depends on details such as the soil horizons considered, the various soil carbon pools included (e.g. litter, roots, labile and slowly decomposing pools), the temperature sensitivity of decomposition, and soil moisture and water saturation implementation in the models. Typically, in global land surface models these are simplified for global scale modelling and not specifically applicable to permafrost (Koven et al. 2015a, b). Lateral carbon losses

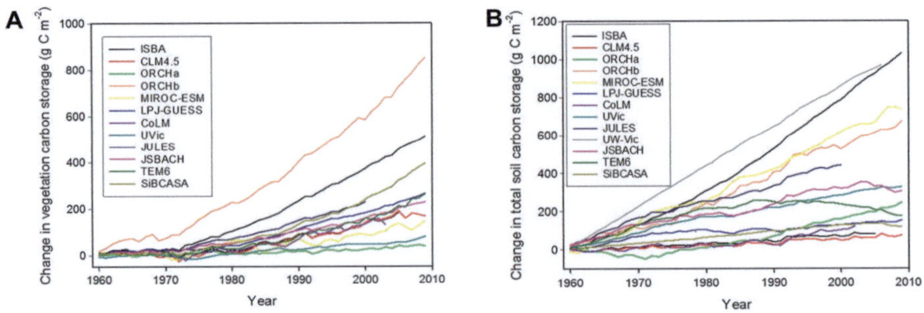

Fig. 8.5 (**a**) Simulated changes in vegetation carbon storage between 1960 and 2009 among 12 land surface models. (**b**) Simulated changes in soil carbon storage between 1960 and 2009 among the same models. (Republished with permission of John Wiley and Sons Inc., from McGuire et al. (2016): Variability in the sensitivity among model simulations of permafrost and carbon dynamics in the permafrost region between 1960 and 2009 Global biogeochemical cycles 30, doi: https://doi.org/10.1002/2016GB005405; permission conveyed through Copyright Clearance Center, Inc)

by transport in rivers, CH_4 emission and disturbances such as fire are included in only a few of the models (often in a simplistic fashion); those that represent these carbon losses had significantly lower increase of carbon storage. The representation of lakes and wetlands differs markedly among the models (McGuire et al. 2016).

Criticism has been voiced on the strong effect of CO_2 fertilisation that is produced by most land surface models. An expert survey by Abbott et al. (2016) suggests that vegetation biomass in the Arctic could decrease by water stress and disturbance rather than increase. There is little field evidence that CO_2 fertilisation results in long-term carbon accumulation in high-latitude ecosystems, because of nutrient limitations and water availability. Nutrient limitation is incompletely included or absent in many models. On the other hand, nutrients liberated from thawing permafrost carbon may enhance vegetation productivity (Sect. 6.2). Field observations have shown that a longer growing season does not lead to increased carbon storage, as also soil respiration increases at the same time (Parmentier et al. 2011; Sect. 6.1.4).

Soil temperature change and active layer change is only part of the story of warming permafrost soils (Chap. 5). Abrupt thaw features resulting from ice-rich permafrost, such as erosion, pond and lake formation are very important agents of carbon loss (Van Huissteden and Dolman 2012). Their importance is rapidly increasing in permafrost areas (Liljedahl et al. 2016; Lewkowicz and Way 2019; Turetsky et al. 2019). These geomorphological processes are not included (and likely are difficult to include on a large scale other than as stochastic processes, Van Huissteden et al. 2011). In Chap. 4, the environmental factors that affect soil carbon decomposition have been discussed. These include uncertain temperature sensitivity, lability, and the proportion of aerobically and anaerobically decomposed soil carbon.

Another source of uncertainty is the initial estimate of permafrost carbon stocks in large scale land surface models. In Sect. 4.4, the size and vertical distribution of the permafrost carbon store has been portrayed. Because of the limited and geo-

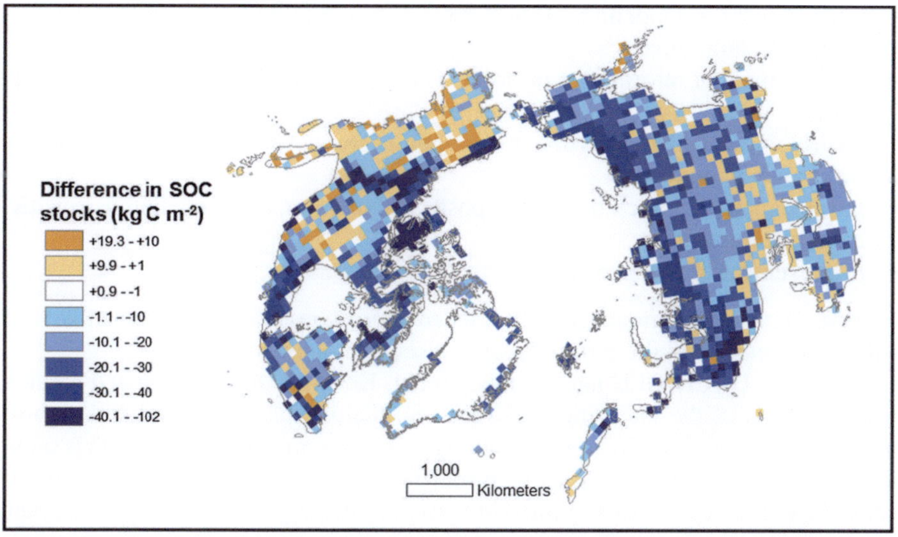

Fig. 8.6 Inconsistencies in soil carbon stocks (to 1 m depth) between observation-based (Northern Circumpolar Soil Carbon Database; www.bbcc.su.se/data/ncscd/) and baseline Earth Surface Model estimates calculated from the mean values in four CMIP5 models (BCC-CSM1.1, CanESM2, MIROC-ESM, and GFDL-ESM2M). Brown (+ sign) identifies over-predictions by models compared to observation-based data; white shows tentative agreement; and blue (− sign) shows model under-predictions. (From Mishra et al. (2013): Empirical estimates to reduce modeling uncertainties of soil organic carbon in permafrost regions: a review of recent progress and remaining challenges, Environ. Res. Lett. 8 (2013), doi:https://doi.org/10.1088/1748-9326/8/3/035020; published under Creative Commons License 3.0, https://creativecommons.org/licenses/by/3.0/)

graphically unevenly spreading of the data, this carbon stock still has large uncertainties. In Earth System models, the soil carbon stocks are initialized by running repeating cycles of the initial climate condition until heterotrophic respiration is balanced by net primary productivity, assuming that the soil carbon stock is then at equilibrium with the climate. This modelled equilibrium carbon stock severely underestimates the permafrost carbon stocks in most permafrost ares (Mishra et al. 2013; Fig. 8.6). Model simulations (CMIP5 models) of the soil carbon stock for the high northern latitudes range between 60 and 820 Pg C, compared to 380–620 Pg C (95% confidence interval, upper metre of soil) for the Northern Circumpolar Soil Carbon Database of Tarnocai et al. (2009) (Todd-Brown et al. 2013).

Summarizing, the main needs for reducing uncertainties in modelling the future permafrost carbon feedback are:

- Better quantification of initial amounts of soil carbon;
- Climate-interactive vegetation modelling to improve both carbon input and effects of climate change on the energy balance;
- The effect of CO_2 fertilization, warming and nutrient dynamics on vegetation biomass;
- Nutrient dynamics changes by decomposition of permafrost carbon;

- Improvement of modelling of disturbances as abrupt thaw, erosion, subsidence, pond formation, fire;
- Improvement of information on soil ice content to facilitate modelling of abrupt thaw;
- Hydrological change affecting soil carbon decomposition environment and lateral transport of carbon;
- Improvement of soil carbon decomposition dynamics, e.g. temperature sensitivity of decomposition;
- Improvement of future permafrost thaw predictions.

Several modelling groups have developed approaches to improve permafrost and soil carbon models in these respects as is discussed below.

Using the ORCHIDEE land surface model, Koven et al. (2009) tested a better representation of the interaction between soil thermal properties and soil organic matter, and added diffusive vertical mixing of organic matter in the soil. This mimics the cryoturbation processes which transfers soil carbon to deeper levels in the active layer (Sect. 3.5). This resulted in a better model approximation of the amount of organic matter in present-day permafrost soils. However, this does not yet include deeper soil carbon inherited from past sedimentation processes, which will need improved geological quantification of carbon stored in sedimentary environments.

Climate-interactive vegetation modelling improves modelling of ecosystem carbon gains and losses, and the feedback of vegetation on local climate. Euskirchen et al. (2009) used the biogeochemistry model TEM (Terrestrial Ecosystem Model) coupled with a dynamic vegetation model, in which plants (defined as plant functional types, PFT) compete for light and nutrients. Driven by the IPCC AR4 climate model simulations (IPCC 2007), all plant functional types showed increases in net primary productivity for the model domain, Northern Alaska. However, also the heterotrophic respiration increased, resulting in a decrease or no change of the net ecosystem carbon uptake. The above-ground biomass increased, in particular for shrubs and spruce forest, which also decreased the albedo of the land surface. Together with an shorter snow season this resulted in a larger regional heat uptake by 2100.

Lawrence and Riley (2015a) experimented with a land surface model (CLM4.5) that included soil N dynamics and evaluated the effects of decomposition of shallow versus deeper soil organic matter for the high emission RCP 8.5 scenario. Permafrost thaw progresses fastest between the years 2020 and 2120. The model shows indeed a synergy between CO_2 fertilisation and reduced nutrient limitation, which results in an increase of the vegetation biomass. The nutrients are mostly derived from shallow soil carbon decomposition. However, the increase in vegetation carbon stock levels off after 2100, due to higher C losses caused by an increase in modelled boreal forest fire frequency. Furthermore, the vegetation increase does not compensate for soil C losses. The model is highly sensitive to assumptions on the rate of decomposition of carbon at deeper levels in the thawed soils, which depend on uncertainties with respect to soil environmental effects on decomposition rates –

microbial activity, moisture, oxygen supply. This results in a large range of estimates of the net total losses of carbon: 21–164 Pg C by 2300.

An entirely different approach is not to expand the models, but incorporation of soil carbon stock data into the model, arriving at a data-constrained approach as discussed in the previous section (Burke et al. 2012, 2013; Harden et al. 2012; Koven et al. 2015b). With the Community Climate System Model (CCSM4), Harden et al. (2012) simulated cumulative distributions of ALT under current and future climates. These were compared with field-based organic matter (C and N) profiles of three classes of permafrost soils in the American soil classification system. This provides an estimate of the amount of permafrost carbon that will thaw, within a given time period and a given greenhouse gas emission scenario. These estimates are for the uppermost 3 m, and should not be seen as the total amount of C that will be *released* due to thaw, only the amount of *thawed* permafrost carbon. The carbon release to the atmosphere depends on decomposition rates and quantity of C that is available for rapid decomposition, and is not taken into account in this study. For RCP 4.5, this yielded an estimate of 61–399 Pg of thawed permafrost carbon, for RCP8.5 108–706 Pg by 2100. Koven et al. (2015b) also used incubation data of Schädel et al. (2014) to further constrain permafrost emissions and obtain true carbon losses. With the RCP4.5 scenario, permafrost soil C losses of 12.2–33.4 Pg C are projected; under RCP8.5, 27.9–112.6 Pg C.

None of the land surface models consider abrupt thaw, the rapid degradation of ice-rich permafrost by thaw lake formation and erosion. A key parameter for abrupt thaw is the ground ice content of permafrost, which is poorly quantified. A solution is to reduce the complexity of the normal soil carbon model, and to include a sub-model that describes abrupt thaw. Schneider von Deimling et al. (2015) developed a two-dimensional soil carbon model, of which the initial carbon reservoirs are data-constrained, and contains a sub-model of abrupt thaw.

Simulations of soil carbon processes operate on latitudinal bands instead of grid cells in this model. The model contains a number of soil carbon pools, differentiated by soil environment and carbon amount and quality. The thaw lake fraction in each latitudinal band is assumed to depend on surface air temperature anomaly. It is limited by a maximum temperature, when drainage and terrestrialisation of lakes is assumed to take over lake expansion. The model is highly simplified, but that allows the use of a Monte Carlo approach. This entails performing a large number model runs with randomized parameters, to estimate confidence bands of the model outcomes. Figure 8.7 shows the trajectory of CO_2 and CH_4 release modelled for the high warming RCP 8.5 scenario. A large increase of emission of CO_2 and CH_4 from thawing deep permafrost carbon by thaw lake expansion is modelled, starting in this century, while the wetland ecosystem CH_4 emission does not increase before the mid of the next century. The emission of deeper CH_4 stores (previous chapter, Walter Anthony et al. 2012) is not included.

Chadburn et al. (2017a) evaluate three land surface models (JSBACH, JULES, ORCHIDEE) against observation data collected at five research stations in Arctic and High Arctic tundra research sites in Sweden, Svalbard, Greenland and Siberia. These models include soil carbon processes and soil temperature in a relatively

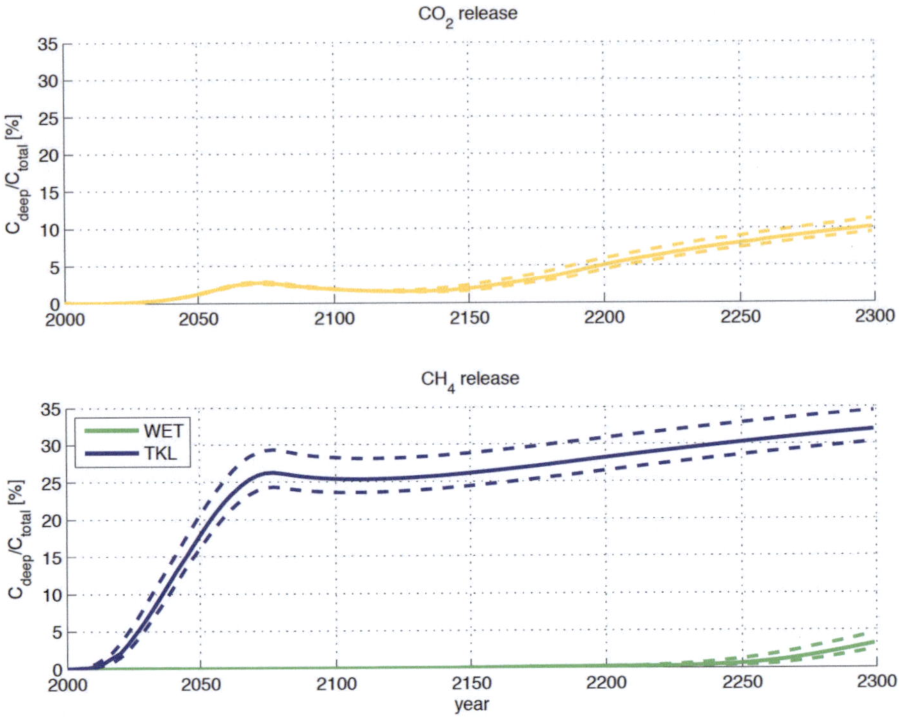

Fig. 8.7 Contribution of deep permafrost carbon deposits to total carbon fluxes under aerobic (upper panel) and anaerobic (lower panel) conditions. Shown is the contribution of cumulated CO_2 and CH_4 fluxes from deep deposits (3–15 m) to total circum-arctic carbon release (0–15 m) under strong warming (RCP8.5). Solid lines represent median values, dashed lines 68% ranges. The contribution of deep deposits to wetland-affected CH_4 release (green) and to thermokarst (thaw lake) affected CH_4 release (blue) is shown separately. (From: 'Observation-based modelling of permafrost carbon fluxes with accounting for deep carbon deposits and thermokarst activity' by Schneider von Deimling et al. (2015) in Biogoesciences 12(11) https://doi.org/10.5194/bg-12-3469-2015, Fig. 4; © The Authors; published under Creative Commons License 4.0 https://creativecommons.org/licenses/by/4.0/)

large detail. There are still differences in performance of the models in modelling of soil temperature, active layer, soil moisture, soil carbon quantity and carbon fluxes. Based on this model-data comparison, Chadburn et al. (2017a) propose several improvements to land surface models:

- Improvement of the vegetation phenology to simulate vegetation biomass and leaf area;
- Including nutrient limitation and dynamic vegetation;
- Include mosses for both photosynthesis and peat accumulation;
- Improvement of the soil carbon profile for organic soils (including peat accumulation);

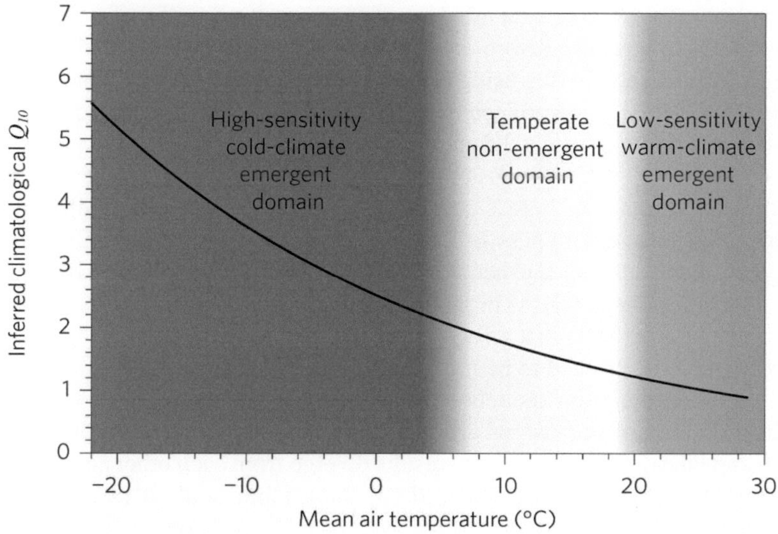

Fig. 8.8 Inferred 'climatological Q_{10}' as a function of temperature. Climatological Q_{10} is calculated from the derivative of a regression relationship between apparent turnover time and mean annual air temperature. Emergent domains are the temperature ranges where the climatological Q_{10} differs appreciably from short-term Q_{10} values derived from instantaneous decomposition data (that is, $Q_{10} > 2$ or $Q_{10} < 1.4$). (From Koven et al. (2017) 'Higher climatological temperature sensitivity of soil carbon in cold than warm climates', Nature Climate Change 7(11). © 2017 Springer Nature AG)

- Improvement of soil moisture and snow, with better representation of different tundra plant functional types and their interactions with snow.

In global scale earth surface models, a rather low temperature sensitivity of soil carbon decomposition is assumed, together with simplified schemes of the decomposition processes. As seen above, these models do not predict present permafrost soil carbon stocks very well; dedicated permafrost carbon models predict higher decomposition rates. The temperature and freeze-thaw processes in permafrost soils result in high apparent – or climatological – temperature sensitivity of decomposition (Koven et al. 2017). This climatological temperature sensitivity is derived by Koven et al. (2017) from a global database of soil carbon stocks, vegetation input and soil carbon turnover time estimates (Carvalhais et al. 2014). It can be expressed as a climate-related Q10 value which can be seen as an ecosystem property (Fig. 8.8). In cold climates, this climatological Q_{10} is appreciably higher that the traditionally measured Q_{10} that is based on short-term decomposition rates. Koven et al. (2017) suggest this relation as a benchmark for the representation of cold climate soil processes in land surface models.

The climate change sensitivity of the carbon cycle, γ, is the amount of carbon released from soils per degree of warming. For permafrost soils, γ is also higher than for temperate and tropical climates. Using the data-model combined approach, Burke et al. (2013) and Koven et al. (2015b) have given an estimate of γ on a

100 year time scale. Burke et al. (2013) arrive at an estimate of γ of 6–66 PgC °C^{-1} on a 100 year time scale; Koven et al. (2015b) at 14–19 PgC °C^{-1} (5–95% confidence interval; carbon losses using atmospheric sign convention). The estimate of Koven et al. (2015b) is constrained by both carbon stocks and measured decomposition rates, and derived from a different set of models.

As seen in Chap. 4, CH_4 emission processes differ in several respects from aerobic soil carbon turnover. There is consumption of CH_4 in soil and vegetation, while the vegetation is also a bypass to CH_4 oxidation. Several wetland CH_4 emission models have been developed to account for these processes (e.g. Walter and Heimann 2000). The emission of CH_4 is highly sensitive to the soil water table position and vegetation characteristics that affect emission. Wetland extent, hydrology and ecosystem characteristics need to be modelled correctly (Mi et al. 2014).

Wetland CH_4 emission models are increasingly used to forecast wetland CH_4 emissions. However, because of differences in model structure and uncertainty in wetland and lake distribution, model results diverge from each other and need model tuning on data (Petrescu et al. 2007, 2010; Budishchev et al. 2014). Melton et al. (2013) evaluated the performance of ten wetland CH_4 emission models with various resolution and model structure. Simulated wetland areas and observed CH_4 emissions over large subarctic wetland complexes (West Siberia, Hudson Bay lowlands) showed very large differences among models and deviations from data. Wetland extent was not always modelled correctly, some models missed the presence of large permafrost wetland complexes, e.g. the Yana-Indigirka-Kolyma lowlands. A CH_4 emission process that is missing in all models, is the high early wintertime CH_4 emissions from permafrost soils (Sect. 4.6.2). Improvements are evidently needed for modelling of the ecosystem CH_4 emission, in particular because of the higher temperature sensitivity of methanogenesis compared to aerobic decomposition of organic matter, and the increasing labile organic carbon source produced by vegetation biomass.

Concluding, improvement of large-scale carbon cycle modelling is possible by improvement of permafrost active layer and temperature modelling and coupling with climate-interactive vegetation models. The main unknown factors, which are more difficult to tackle, are the effects of abrupt thaw of soils with high amounts of ground ice, lateral transport of carbon by groundwater and streams, and the interaction of vegetation with nutrient cycle, CO_2 fertilisation and disturbance regimes. For ecosystem CH_4 emission modelling also improvements are necessary. N_2O emission modelling is still a blank spot, but improvements of modelling of the nutrient cycle and abrupt thaw may also result in improvements on this subject.

8.4 Geomorphology: Lake Formation and Erosion in Models

Abrupt permafrost degradation is characterized by radical changes of the ecosystem, relief, drainage, and displaces large amounts of soil and carbon in a short time from its original soil environment into another environment with other organic mat-

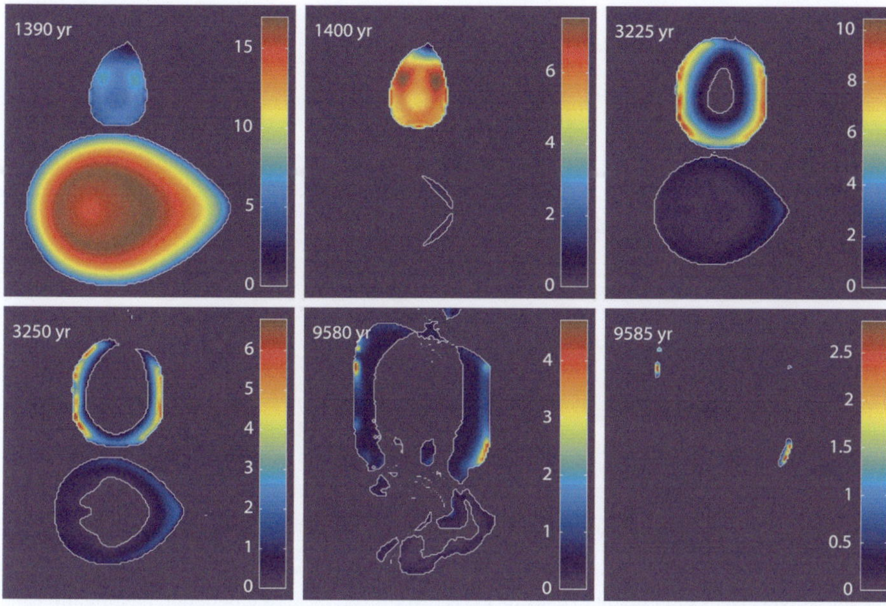

Fig. 8.9 Six snapshots of modelled lake water depth of two lakes spanning catastrophic (between 1390 and 1400) and partial drainage events (depth scale in meters) in a 10,000 year lake evolution simulation of the thaw lake model of Kessler et al. (2012). The ring-like lake shapes remaining after partial drainage are also seen in real thaw lakes (Fig. 1.24). (Republished with permission of John Wiley and Sons Inc., from Kessler et al. (2012): Simulating the decadal-to millennial-scale dynamics of morphology and sequestered carbon mobilization of two thermokarst lakes in NW Alaska. Journal of Geophysical Research: Biogeosciences 117(G2), doi:https://doi.org/10.1029/2011JG001796; permission conveyed through Copyright Clearance Center, Inc)

ter decomposition rates. Abrupt permafrost thaw is characterized by erosion, pond and lake formation. These processes are difficult to implement in large scale models in a deterministic way, because it needs small-scale information on ground ice and soil material.

An example in question is the modelling of thaw lakes. For thaw lakes, very interesting models have been developed that describe the geomorphological processes of thaw lakes in detail. The models of West and Plug (2008) and Plug and West (2009) describe the evolution of thaw lakes and taliks below the lakes in two dimensions. These models include talik formation, subsidence, bank erosion and redistribution of eroded bank material towards the deepest parts of the lake. Kessler et al. (2012) add the third dimension to it, providing a model with strong geomorphological realism (Fig. 8.9). The model can simulate thaw lake evolution over millennia, simulate lake drainage events, and generates thaw lake geomorphology that is observed in nature, such as the formation of a central bulge by thick sediment accumulation in the centre of the lakes. A CH_4 emission model that computes CH_4 production from old permafrost carbon in the talik is included.

Nevertheless, a thaw lake model application at this detail is difficult to implement in large scale land surface models or even in a regional model of the perma-

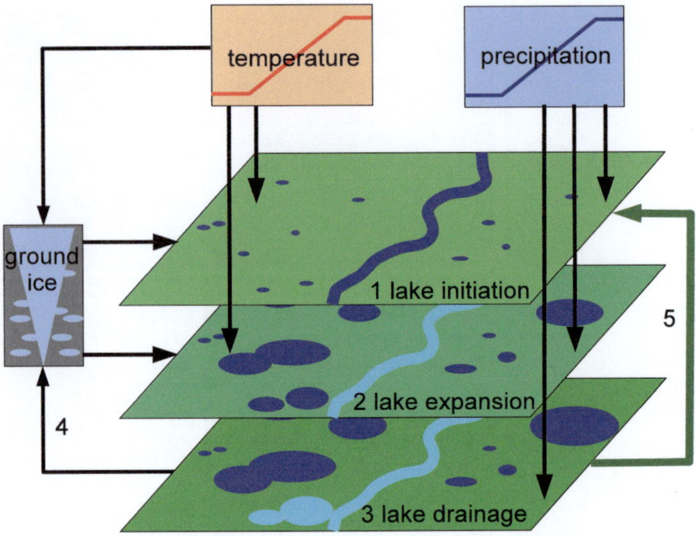

Fig. 8.10 Schematic of the thaw-lake cycle model by Van Huissteden et al. (2011). The rate of lake initiation (1) and subsequent expansion (2) is determined by air temperature, precipitation and ground ice content. Lake formation creates thawed permafrost underneath the lake. Subsequent drainage of lakes by river erosion (3) is determined by precipitation. After lake drainage, new ice-rich permafrost may be established (4), after which new lakes may form in the same area (5). From Van Huissteden et al. (2011) 'Methane emissions from permafrost thaw lakes limited by lake drainage', Nature Climate Change 1(2). © 2011 Springer Nature AG

frost carbon cycle, because of the required geological and ground ice detail and its computational demands. The alternative is to move to a large scale and simplify processes into a few bulk process equations. A model operating at a landscape scale of hundreds of km^2 with probabilistic relations of lake expansion with climate has been designed by Van Huissteden et al. (2011). This approach has been a starting point for other large-scale simulations (Tan and Zhuang 2015; Schneider von Deimling et al. 2015).

Lake initiation, expansion and surface drainage are simulated with relations to yearly averages of air temperature, precipitation and subsurface ice content (Fig. 8.10). The model is two-dimensional and does not include subsurface drainage or water balance. Model runs with future climate scenarios show a gradual expansion of lake area at first, followed by a decrease of lake area due to increased drainage of large lakes, which limits lake expansion. The model was tested, as a conceptual model only, on realistic but fictive data based on landscape characteristics in Northeastern Siberia.

Tan and Zhuang (2015) have used an adapted version of the model (coupled to their lake CH_4 emission model) with real topographical, landscape and ground ice data. They concluded that the model is robust in simulating the present-day and future abundance of thermokarst lakes in different pan-Arctic regions under complex environmental conditions, and at spatial scales much larger than the study of

Van Huissteden et al. (2011). The AThaw model by Schneider von Deimling et al. (2015) uses similar simple relations of thaw lake area with climate. The model scales thaw lake area expansion exponentially to air temperature anomaly, and assumes a maximum temperature for thaw lake expansion. It shows a similar pattern of lake growth, followed by a decrease of lake area as the model of Van Huissteden et al. (2011). The model does not include precipitation as a driver of thaw lake change.

The geomorphological mechanism of thaw lake drainage in a warming climate produces a surprising result. Using the AThaw thaw lake model of Schneider von Deimling et al. (2015), Walter Anthony et al. (2018) modelled CO_2 and CH_4 emissions by thaw lake expansion, and compared this to the carbon cycle effect of gradual thaw by active layer thickness growth (modelled by the CLM land surface model). This resulted in a higher radiative forcing by abrupt thaw greenhouse gas emission (compared to gradual thaw) for the RCP 4.5 scenario, than for the rapid warming RCP 8.5 scenario: respectively up to 162% and up to 130% in the second half of this century. Similarly, the increase of emission from thawed old carbon is respectively 190% and 127% (Fig. 8.11). Counter-intuively, a lower emission scenario with more modest warming would produce higher total emission from thaw lake expansion than a scenario that results in rapid and strong warming.

A similar effect was found with the thaw lake model of Van Huissteden et al. (2011) and to a lesser extent by Tan and Zhuang (2015) (Fig. 8.12). First, this can be understood by accounting for the effect of drainage of lakes after initial expansion of lakes. With rapid warming, drainage of thaw lakes occurs earlier than in a scenario with slower warming, in which lakes persist longer (Van Huissteden et al. 2011; Walter Anthony et al. 2018). Second, Walter Anthony et al. (2018) compare the abrupt thaw emissions with gradual thaw. For the RCP 8.5 scenario, the emission from gradual permafrost thaw is much higher; compared to this, the effect of thaw lake formation is relatively smaller. Third, the effect of thaw lake increase is an emission increase of CH_4, with a larger short term radiative effect than the CO_2 emission that dominates gradual thaw (Walter Anthony et al. 2018). It should be kept in mind that this result is based on models with simplified representation of the hydrological, geomorphological and biogeochemical processes that drive thaw lake development and greenhouse gas emission rates.

Despite a potential decrease in lake area by thaw lake drainage, the inclusion of modelled lake CH_4 emissions, and non-thaw lake emissions by Tan et al. (2015) results in a continued future rise of CH_4 emissions from pan-arctic lakes. The projected methane emissions by 2100 are 28.3 ± 4.5 Tg CH_4 year^{-1} under a low warming scenario (RCP 2.6) and 32.7 ± 5.2 Tg CH_4 year^{-1} under a high warming scenario (RCP 8.5) which are about 2.5 and 2.9 times the simulated present-day emissions. Figure 8.13 shows the pan-arctic lake emissions modelled by Tan et al. (2015). The areas which are susceptible to thaw lake expansion in Northern Siberia, Alaska and Canada are shown as high emission areas.

Modelling landslides in alpine areas with permafrost has attracted much attention (e.g Haeberli et al. 2017). However, the lower relief permafrost areas, with ice-rich and often carbon-rich unconsolidated material are under-represented in modelling literature. Leibman et al. (2003, 2014) and Lewkowicz and Harris (2005)

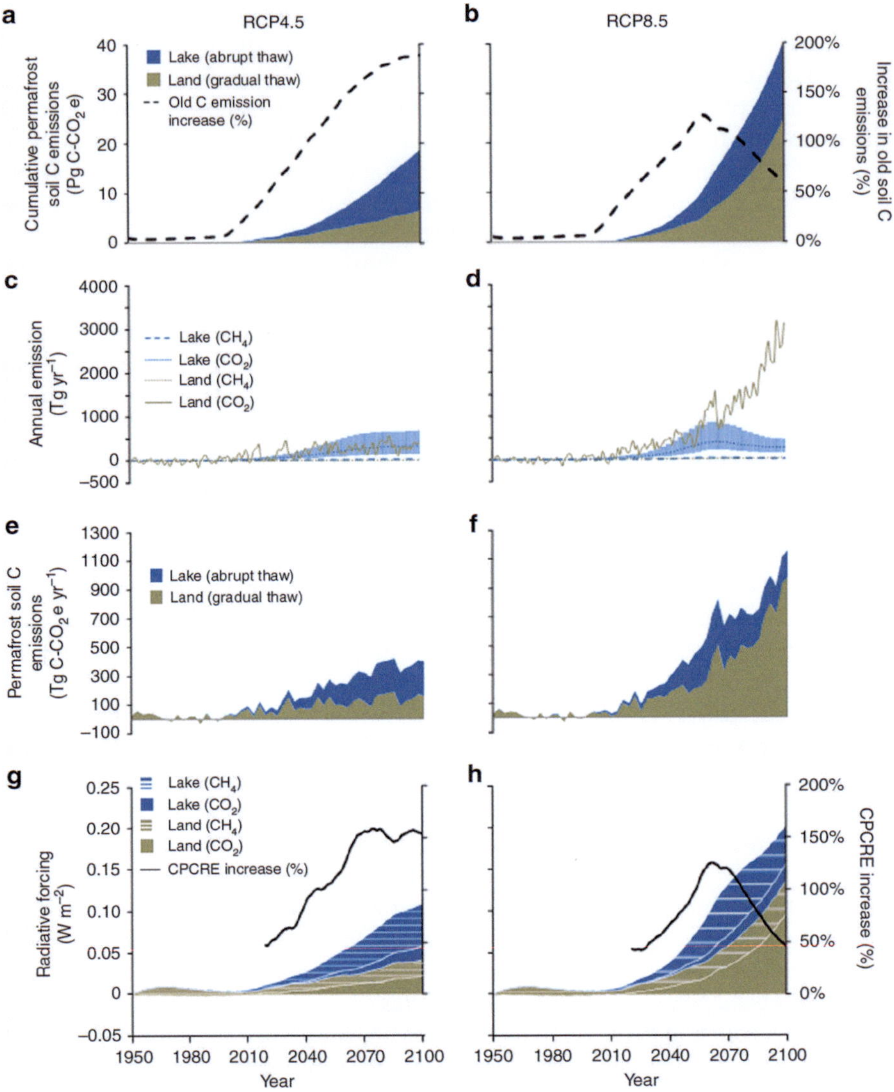

Fig. 8.11 Modelled circumpolar permafrost carbon emissions from gradual thaw (land) versus abrupt thaw (lakes) from 1950 to 2100; from Walter Anthony et al. (2018). Permafrost soil carbon emissions for RCP 4.5 and 8.5 scenarios, are distinguished for homogenous gradual thaw from the CLM4.5BGC model (brown) and heterogeneous abrupt thaw from AThaw modelled thaw lake formation (median values, blue). Cumulative emissions, including percent increase in emissions from old permafrost carbon by abrupt thaw (**a**, **b**). Annual CH_4 and CO_2 emissions expressed individually as Tg year^{-1} (**c**, **d**) and collectively as Tg C-CO_2 equivalent year^{-1} (**e**, **f**) based on a GWP100 of 28 (IPCC 2013). Error bars (**c**, **d**) surrounding the median lake emissions are the 68% uncertainty range (500 member AThaw model ensemble). Radiative forcing (**g**, **h**) associated with fluxes in **c** and **d**. In **g** and **h**, the increase in circumpolar permafrost-carbon radiative effect (CPCRE) attributed to abrupt thaw lakes is shown for the years 2018–2100. (From Walter Anthony et al. (2018): 21st-century modeled permafrost carbon emissions accelerated by abrupt thaw beneath lakes, Nature communications 9(1), DOI: https://doi.org/10.1038/s41467-018-05738-9; Creative Commons License 4.0, https://creativecommons.org/licenses/by/4.0/)

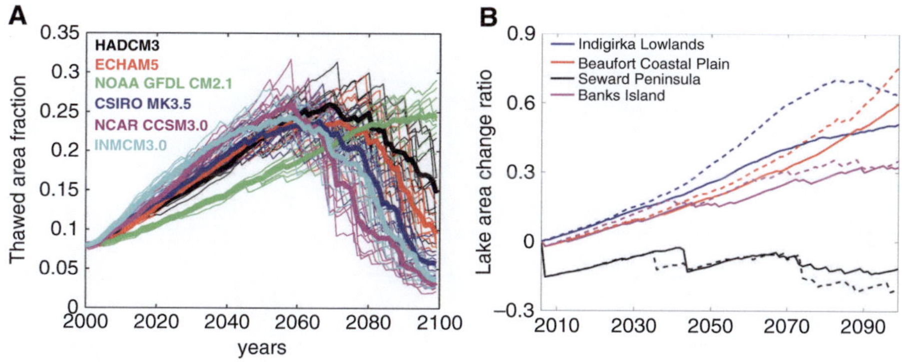

Fig. 8.12 (**a**) Ensemble of thaw-lake model runs for climate change expected in the twenty-first century, for randomly generated initial lake and ground ice distributions. The climate input is derived from six climate models for SRES A2 (rapid greenhouse gas emission increase) from the IPCC fourth assessment report (IPCC 2007). For every climate model, ten runs of the thaw-lake model have been used. The coloured lines indicate the development of the thawed area; colours indicate climate models. The thick lines represent the average thaw-lake area for each climate model. For further reference to the climate models see the original publication (Van Huissteden et al. 2011). (**b**) Change ratios of thaw lake areas in the four Arctic regions under two CMIP5 RCP climate scenarios, and with real topography and ground ice information. Solid lines represent the RCP 2.6 simulations, and dashed lines represent the RCP 8.5 simulations. The ratio is the relative increase or decrease of lake areas compared to the lake areas in 2006. (From Tan and Zhuang (2015). (**a**): From Van Huissteden et al. (2011) 'Methane emissions from permafrost thaw lakes limited by lake drainage', Nature Climate Change 1(2). © 2011 Springer Nature AG. (**b**): Republished with permission of John Wiley and Sons Inc., from Tan and Zhuang (2015): Methane emissions from pan-Arctic lakes during the 21st century: An analysis with process-based models of lake evolution and biogeochemistry. Journal of Geophysical Research: Biogeosciences 120(12), doi:https://doi.org/10.1002/2015JG003184; permission conveyed through Copyright Clearance Center, Inc)

have described cryogenic landslides and their relation to climate (Sect. 5.3.4). The large increase of cryogenic landslides in Northern Canada reported by Lewkowicz and Way (2019) and the relation of erosion with nitrous oxide emissions (Abbott and Jones 2015) show that inclusion of abrupt thaw erosion processes in permafrost carbon models is urgently required. Cryogenic landslide activity is strongly linked to weather extremes, and therefore carbon and nitrogen release from increased landslide activity should feed back on climate change. Application in models should be feasible, e.g. by application of empirical landslide susceptibility indices based on terrain parameters and climate (e.g. Blais-Stevens et al. 2015).

The inclusion of ice wedge thaw in large scale models will improve the water balance modelling of the surface. Liljedahl et al. (2016) modelled the effects of development of high centred polygons on the surface drainage. Low centred polygons on low-relief areas promote surface ponding of water and slow runoff during snowmelt and rain events, while high centred polygons promote rapid runoff, and drier surface conditions. A change from low to high centred polygons therefore affect the ratio of CO_2 to CH_4 emission and total carbon emission. For correct simu-

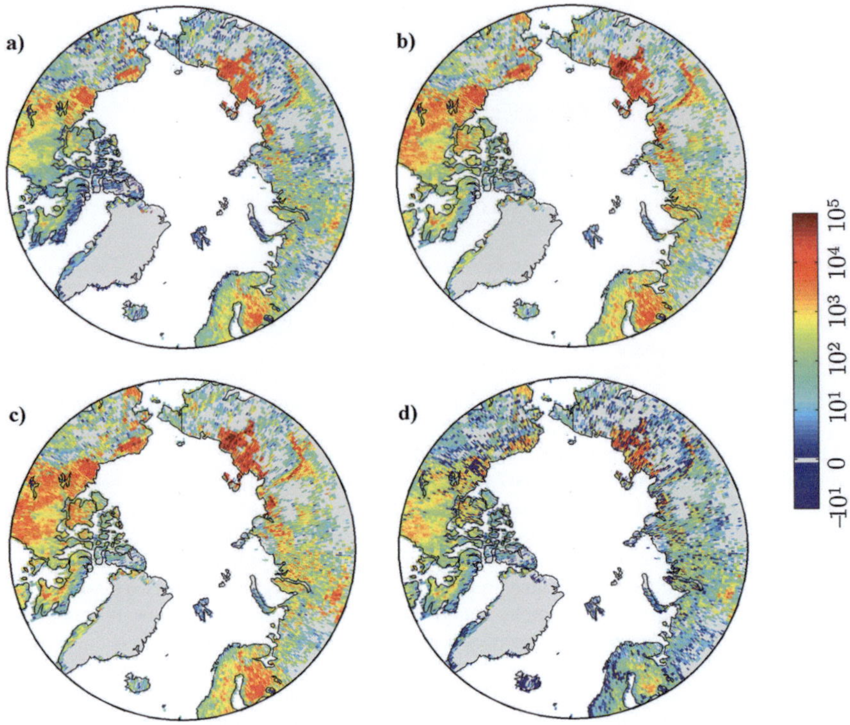

Fig. 8.13 The projected evolution of CH_4 emissions from pan-Arctic lakes under the future climate scenarios: (**a**) the CH_4 emissions in 2006, (**b**) the CH_4 emissions in 2099 under CMIP5 RCP 2.6, (**c**) the CH_4 emissions in 2099 under CMIP5 RCP 8.5, and (**d**) the difference of the CH_4 emissions in 2099 between c and b. (From Tan and Zhuang (2015). Republished with permission of John Wiley and Sons Inc., from Tan and Zhuang (2015): Methane emissions from pan-Arctic lakes during the 21st century: An analysis with process based models of lake evolution and biogeochemistry. Journal of Geophysical Research: Biogeosciences 120(12), doi:https://doi.org/10.1002/2015JG003184; permission conveyed through Copyright Clearance Center, Inc)

lation of this runoff enhancement, a realistic snow cover distribution, including snow accumulation in the ice wedge troughs, proves indispensable. Mapping of micro-relief types (Lara et al. 2015, 2018) will be helpful in providing the basic data improvement for modelling surface hydrology. Alternatively, mathematical approaches exist to model polygon networks (Cresto Aleina et al. 2013).

8.5 Outlook

The models discussed in Chap. 8 paint a bleak picture of future permafrost. The well underpinned sensitivity of permafrost area to global warming developed by Chadburn et al. (2017b) – 4 million km^2 loss of permafrost area with each degree of global warming – shows that 40% of the permafrost will thaw if global warming will stabilize at 2 °C according the Paris Agreement.

There is little doubt that the permafrost carbon feedback will increase climate warming, but there are more doubts on the projections of the strength of this carbon feedback. Even under moderate greenhouse gas emissions scenarios, a substantial decrease of the permafrost area at the surface is imminent this century. According to the IPCC 5th Assessment Report (IPCC 2013), 37–81% of the permafrost area will be lost (upper 3.5 m thawed) by the end of the twenty-first century under the RCP2.6, respectively RCP8.5 pathways. There is large uncertainty on the carbon emissions resulting from permafrost thaw: the release of CO_2 or CH_4 to the atmosphere from thawing permafrost carbon over the twenty-first century is estimated to be anywhere between 50 and 250 Gt C for RCP8.5 scenario. IPCC (2013) list as sources of uncertainty the physical thawing rates, the fraction of carbon that is released after thawing, the possible enhanced carbon uptake by ecosystems created by liberated nutrients, the spatial variability of thawing processes and their influence on the ratio of CO_2 to CH_4 release.

At the time of writing of the IPCC 5th Assessment Report, the complexity of permafrost processes was not included in the models used for the report (IPCC 2013). The discussion of models in Chap. 8 indicates that progress still can be made – and is made – in modelling physical processes of thawing and the modelling of spatial variability in thawing processes. However, the complexity of ecosystem processes that occur upon thawing still needs to find its ways into models. For instance, experimental data discussed in Chap. 6 show that there is considerably uncertainty on the beneficial effects of nutrients on the ecosystem carbon sink; carbon stores in biomass may increase but this does not necessarily increase the soil carbon.

Despite these uncertainties, the rate of change of permafrost is extremely high. Let us put current and future changes of permafrost in the perspective of past climate change, at the termination of the Last Glacial. At the Last Permafrost Maximum between 22 and 17 kyr ago, the permafrost area on the present land surface in Asia was likely about 21.60 million km^2, which was reduced to 14.26 million km^2 in pre-industrial times (Lindgren et al. 2016; the calculation is restricted to Asia, because during the Holocene new permafrost was established after the demise of the Laurentide and Scandinavian ice sheets on the North American and European continents; Fig. 1.22).

If we assume that LPM permafrost in Asia started to thaw after the Younger Dryas Stadia (11,700 years BP) and completed in the Early Holocene by 9000 years ago, this would result in a loss rate of 2737 km^2 $year^{-1}$. This time span is corroborated by datings of widespread permafrost thaw by Walter et al. (2007), which shows that most thaw occurred during this time span. A start of widespread LPM permafrost thaw after the Younger Dryas is a conservative estimate – the radiocarbon dates of Walter et al. (2007) and other dates from Last Glacial eolian sediment successions in Europe (e.g. Bateman and Huissteden 1999) indicate that thawing of LPM permafrost started well before the Younger Dryas. Based on the *lowest,* most conservative model estimates by McGuire et al. (2018), globally 3.2 million km^2

permafrost loss is expected in 2300. If downscaled to Asian permafrost area (about 60% of global permafrost, Zhang et al. 2008), the estimated permafrost loss rate is 6644 km^2 $year^{-1}$ for Asia. This future loss rate is *2.4 times faster* than during the termination of the Last Glacial.

This is a conservative minimum; using other future permafrost area estimates, such as the RCP 8.5 simulations of McGuire et al. (2018) and Slater and Lawrence (2013), or the climate sensitivity estimate of permafrost of Chadburn et al. (2017b), could make this comparative loss rate a few times higher. Although this is a coarse back-of-the-envelope estimate, it indicates that permafrost thaw will likely proceed at an unprecedented rate in Quaternary geological history. This will not only affect permafrost carbon, but also ecosystems on a scale and rate for which there is no past analogue.

At the time of writing of this chapter (July 2019), Arctic sea ice area is at a record low after decades of decrease in extent. As discussed previously (Parmentier et al. 2015), decrease of the sea ice surface also feeds back on the land carbon cycle; higher land temperatures and higher precipitation warm the terrestrial permafrost bordering the Arctic ocean and increase the CH_4 emission from wetlands. In particular the abrupt thaw in the shape of erosion and thaw lake expansion is highly disruptive for ecosystems, the permafrost carbon store and human infrastructure in the Arctic (Turetsky et al. 2019). The alarming rate of change is corroborated with recent observations of abrupt thaw occurring on a very large scale in sensitive areas (Liljedahl et al. 2016; Lewkowicz and Way 2019). Also, from my colleagues visiting our research site in Northeast Siberia I received accounts of increasing erosion of a Yedoma plateau, after a record warmth in June. Again, these processes are currently not included in in present climate model forecasts; only a few recent attempts have been made to capture these processes in models.

Increased carbon uptake in vegetation is simulated by the latest generation of land surface models, which is assumed to compensate a part of the permafrost carbon losses. However, this compensation is likely overestimated and will over time be overwhelmed by increased carbon emission from permafrost soils (Abbott et al. 2016). Some ecosystems will sequester part of the carbon that is lost (Schuur et al. 2015). In particular High Arctic ecosystems might increase carbon uptake – if present ecosystem water and nutrient limitations are relieved (Lupascu et al. 2014; Chaps. 5 and 6). Ecosystem resilience after disturbance by thaw also may compensate for part of the losses. However, ecosystem resilience is limited by the rate of ecosystem disturbance by abrupt thaw. This could stress ecosystems to their limits. A high rate of ecosystem disturbance may lead to abrupt, widespread ecosystem disruption with irreversible changes (Chap. 6).

Schuur et al. (2015) expect a limited temperature feedback to global climate from the permafrost carbon feedback by 2100 (0.15–0.3 °C). This is a substantial part of the limits to temperature rise agreed in Paris (1.5–2 °C). Although the current carbon emissions from permafrost are an order of magnitude lower than those of fossil fuel emissions (Schuur et al. 2015), the permafrost carbon feedback makes a reduction greenhouse gas emission an increasingly steep uphill struggle for our societies. Moreover, permafrost thaw will not stop after 2100, even if drastic reduction of greenhouse gas emissions is achieved. Permafrost thaw starting in this century will have impact in the centuries that follow. Schuur et al. (2015) argue that most of the permafrost carbon emission will be gradual and sustained, rather than sudden and disruptive. Nevertheless, they warn for surprises. In the previous chapters several potential surprises have been discussed, of which the current rate of abrupt thaw of ice-rich permafrost is the most import one. More poorly quantified potential surprises exists, such as the CH_4 and old carbon stores in deeper permafrost (Chap. 7).

Permafrost thaw and carbon release from Yedoma permafrost has been seen as a possible climate tipping point (Lenton 2012). A tipping point in a system is a threshold at which large, abrupt and possibly irreversible changes in the state of the system may occur, when that threshold is crossed by one of the system inputs. Crossing the threshold may be induced by a minor change – the proverbial drop that makes a bucket overflow. Rapid crossing of such a threshold in permafrost regions was not seen as imminent by Lenton (2012). However, at least one of the models discussed in Chap. 8 (Westermann et al. 2016) shows potential tipping point behaviour for thaw of large areas of ice-rich permafrost. Next, the rapid pace of abrupt thaw seen in some permafrost areas is alarming (Turetsky et al. 2019). This suggests that permafrost thaw as an Arctic climate tipping point is not unrealistic, if widespread high-latitude ecosystem damage is considered besides emitted carbon. Furthermore, other tipping points in the climate system could enhance permafrost thaw (Cai et al. 2016).

I risk here the accusation of 'alarmism', an incrimination that is all too often made when scientists report the effects of climate change. However, as seen in the previous chapters, permafrost science consist of a sound and huge body of interdisciplinary work by Earth scientists, biologists, physicists and model developers, in which there is a lively discussion on research results and their interpretation. This science is based on field data, gathered painstakingly under harsh field conditions, often with low budgets. Hopefully, this book contributes to a better understanding among a wider audience about the fascinating permafrost soils, ecosystems and landscapes. Let us hope also that the political inaction, the ignorance, or even downright suspicion against science will not last, and well-informed choices in society will be possible to reduce the loss of permafrost ecosystems and their carbon stores.

References

Abbott BW, Jones JB (2015) Permafrost collapse alters soil carbon stocks, respiration, CH_4, and N_2O in upland tundra. Glob Chang Biol 21(12):4570–4587

Abbott BW, Jones JB, Schuur EA, Chapin FS III, Bowden WB, Bret-Harte MS, Epstein HE, Flannigan MD, Harms TK, Hollingsworth TN (2016) Biomass offsets little or none of permafrost carbon release from soils, streams, and wildfire: an expert assessment. Environ Res Lett 11(3):034014. https://doi.org/10.1088/1748-9326/11/3/034014

AMAP (2017) Snow, water, ice and permafrost in the Arctic (SWIPA) 2017. Arctic Monitoring and Assessment Programme (AMAP), Oslo. 269 p

Anisimov O, Sherstiukov A (2016) Evaluating the effect of environmental factors on permafrost factors in Russia. Earth's Cryosphere: 90–99

Bateman MD, Huissteden JV (1999) The timing of last-glacial periglacial and aeolian events, Twente, eastern Netherlands. J Quat Sci 14(3):277–283

Beven K (2007) Environmental modelling: an uncertain future? CRC Press, 310 p

Blais-Stevens A, Kremer M, Bonnaventure PP, Smith SL, Lipovsky P, Lewkowicz AG (2015) Active layer detachment slides and retrogressive thaw slumps susceptibility mapping for current and future permafrost distribution, Yukon Alaska Highway Corridor. In: Engineering geology for society and territory – volume 1. Springer International Publishing, Cham, pp 449–453

Bonan G (2016) Ecological climatology: concepts and applications, 3rd edn. Cambridge University Press, New York. 754 p

Budishchev A, Mi Y, van Huissteden J, Belelli-Marchesini L, Schaepman-Strub G, Parmentier FJW, Fratini G, Gallagher A, Maximov TC, Dolman AJ (2014) Evaluation of a plot-scale methane emission model using eddy covariance observations and footprint modelling. Biogeosciences 11(17):4651–4664. https://doi.org/10.5194/bg-11-4651-2014

Burke EJ, Hartley IP, Jones CD (2012) Uncertainties in the global temperature change caused by carbon release from permafrost thawing. Cryosphere 6(5):1063–1076

Burke EJ, Jones CD, Koven CD (2013) Estimating the permafrost-carbon climate response in the CMIP5 climate models using a simplified approach. J Clim 26(14):4897–4909

Cai Y, Lenton TM, Lontzek TS (2016) Risk of multiple interacting tipping points should encourage rapid CO_2 emission reduction. Nat Clim Chang 6(5):520–525

Carvalhais N, Forkel M, Khomik M, Bellarby J, Jung M, Migliavacca M, Mu M, Saatchi S, Santoro M, Thurner M, Weber U, Ahrens B, Beer C, Cescatti A, Randerson JT, Reichstein M (2014) Global covariation of carbon turnover times with climate in terrestrial ecosystems. Nature 514:213. https://doi.org/10.1038/nature13731

Chadburn S, Burke E, Essery R, Boike J, Langer M, Heikenfeld M, Cox PM, Friedlingstein P (2015a) Impact of model developments on present and future simulations of permafrost in a global land-surface model. Cryosphere Discuss 9(2):1965–2012

Chadburn S, Burke E, Essery R, Boike J, Langer M, Heikenfeld M, Cox P, Friedlingstein P (2015b) An improved representation of physical permafrost dynamics in the JULES land-surface model. Geosci Model Dev 8(5):1493–1508

Chadburn SE, Krinner G, Porada P, Bartsch A, Beer C, Belelli Marchesini L, Boike J, Ekici A, Elberling B, Friborg T, Hugelius G, Johansson M, Kuhry P, Kutzbach L, Langer M, Lund M, Parmentier FJW, Peng S, Van Huissteden K, Wang T, Westermann S, Zhu D, Burke EJ (2017a) Carbon stocks and fluxes in the high latitudes: using site-level data to evaluate Earth system models. Biogeosciences 14(22):5143–5169. https://doi.org/10.5194/bg-14-5143-2017

Chadburn SE, Burke E, Cox P, Friedlingstein P, Hugelius G, Westermann S (2017b) An observation-based constraint on permafrost loss as a function of global warming. Nat Clim Chang 7(5):340

Cresto Aleina F, Brovkin V, Muster S, Boike J, Kutzbach L, Sachs T, Zuyev S (2013) A stochastic model for the polygonal tundra based on Poisson-Voronoi diagrams. Earth Syst Dynam 4(2):187–198

Euskirchen ES, McGuire AD, Chapin FS III, Yi S, Thompson CC (2009) Changes in vegetation in northern Alaska under scenarios of climate change, 2003–2100: implications for climate feedbacks. Ecol Appl 19(4):1022–1043

Gettelman A, Rood RB (2016) Demystifying climate models. A users guide to Earth system models. Springer Open, Berlin. 274 p

Haeberli W, Schaub Y, Huggel C (2017) Increasing risks related to landslides from degrading permafrost into new lakes in de-glaciating mountain ranges. Geomorphology 293:405–417

Harden JW, Koven CD, Ping CL, Hugelius G, McGuire AD, Camill P, Jorgenson T, Kuhry P, Michaelson GJ, O'Donnell JA (2012) Field information links permafrost carbon to physical vulnerabilities of thawing. Geophys Res Lett 39(15):L15704. https://doi.org/10.1029/2012GL051958

IPCC (2007) Climate change 2007: the physical science basis. Contribution of Working Group I to the Fourth Assessment Report of the Intergovernmental Panel on Climate Change. Cambridge University Press, Cambridge/New York

IPCC (2013) Climate change 2013: the physical science basis. Contribution of Working Group I to the Fifth Assessment Report of the Intergovernmental Panel on Climate Change. Cambridge University Press, Cambridge. https://doi.org/10.1017/CBO9781107415324

Jafarov EE, Marchenko SS, Romanovsky V (2012) Numerical modeling of permafrost dynamics in Alaska using a high spatial resolution dataset. Cryosphere 6(3):613–624

Kaplan D, Glass L (2012) Understanding nonlinear dynamics. Springer Science & Business Media, New York. 420 p

Kessler M, Plug LJ, Walter Anthony K (2012) Simulating the decadal-to millennial-scale dynamics of morphology and sequestered carbon mobilization of two thermokarst lakes in NW Alaska. J Geophys Res Biogeosci 117(G2):G00M06. https://doi.org/10.1029/2011JG001796

Koven C, Friedlingstein P, Ciais P, Khvorostyanov D, Krinner G, Tarnocai C (2009) On the formation of high-latitude soil carbon stocks: effects of cryoturbation and insulation by organic matter in a land surface model. Geophys Res Lett 36(21):L21501

Koven CD, Riley WJ, Stern A (2013) Analysis of permafrost thermal dynamics and response to climate change in the CMIP5 earth system models. J Clim 26(6):1877–1900

Koven CD, Lawrence DM, Riley WJ (2015a) Permafrost carbon– climate feedback is sensitive to deep soil carbon decomposability but not deep soil nitrogen dynamics. Proc Natl Acad Sci 112(12):3752–3757

Koven CD, Schuur E, Schädel C, Bohn T, Burke E, Chen G, Chen X, Ciais P, Grosse G, Harden JW (2015b) A simplified, data-constrained approach to estimate the permafrost carbon–climate feedback. Philos Trans R Soc A Math Phys Eng Sci 373(2054):20140423

Koven C, Hugelius G, Lawrence DM, Wieder WR (2017) Higher climatological temperature sensitivity of soil carbon in cold than warm climates. Nat Clim Chang 7:817–824. https://doi.org/10.1038/NCLIMATE3421

Langer M, Westermann S, Heikenfeld M, Dorn W, Boike J (2013) Satellite-based modeling of permafrost temperatures in a tundra lowland landscape. Remote Sens Environ 135:12–24

Lara MJ, McGuire AD, Euskirchen ES, Tweedie CE, Hinkel KM, Skurikhin AN, Romanovsky VE, Grosse G, Bolton WR, Genet H (2015) Polygonal tundra geomorphological change in response to warming alters future CO_2 and CH_4 flux on the Barrow Peninsula. Glob Chang Biol 21(4):1634–1651. https://doi.org/10.1111/gcb.12757

Lara MJ, Nitze I, Grosse G, Martin P, McGuire AD (2018) Reduced arctic tundra productivity linked with landform and climate change interactions. Sci Rep 8(1):2345

Leibman M, Kizakov A, Sulerzhitsky L, Zaretskaia N (2003) Dynamics of landslide slopes and their development on Yamal Peninsula. In: Permafrost. Proceedings of the 8th international conference on permafrost. Swets and Zeitlinger, Lisse, pp 651–656

Leibman M, Khomutov A, Kizyakov A (2014) Cryogenic landslides in the West-Siberian plain of Russia: classification, mechanisms, and landforms. In: Landslides in cold regions in the context of climate change. Springer, Cham, pp 143–162

Lenton TM (2012) Arctic climate tipping points. Ambio 41(1):10–22. https://doi.org/10.1007/s13280-011-0221-x

Lewkowicz AG, Harris C (2005) Frequency and magnitude of active-layer detachment failures in discontinuous and continuous permafrost, northern Canada. Permafr Periglac Process 16(1):115–130

Lewkowicz AG, Way RG (2019) Extremes of summer climate trigger thousands of thermokarst landslides in a high Arctic environment. Nat Commun 10(1):1329. https://doi.org/10.1038/s41467-019-09314-7

Liljedahl AK, Boike J, Daanen RP, Fedorov AN, Frost GV, Grosse G, Hinzman LD, Iijma Y, Jorgenson JC, Matveyeva N, Necsoiu M, Raynolds MK, Romanovsky VE, Schulla J, Tape KD, Walker DA, Wilson CJ, Yabuki H, Zona D (2016) Pan-Arctic ice-wedge degradation in warming permafrost and its influence on tundra hydrology. Nat Geosci 9(4):312–318. https://doi.org/10.1038/ngeo2674

Lindgren A, Hugelius G, Kuhry P, Christensen TR, Vandenberghe J (2016) GIS-based maps and area estimates of northern hemisphere permafrost extent during the last glacial maximum. Permafr Periglac Process 27(1):6–16

Lupascu M, Welker J, Seibt U, Maseyk K, Xu X, Czimczik C (2014) High Arctic wetting reduces permafrost carbon feedbacks to climate warming. Nat Clim Chang 4(1):51

McGuire AD, Koven C, Lawrence DM, Clein JS, Xia J, Beer C, Burke E, Chen G, Chen X, Delire C (2016) Variability in the sensitivity among model simulations of permafrost and carbon dynamics in the permafrost region between 1960 and 2009. Glob Biogeochem Cycles 30(7):1015–1037

McGuire AD, Lawrence DM, Koven C, Clein JS, Burke E, Chen G, Jafarov E, MacDougall AH, Marchenko S, Nicolsky D (2018) Dependence of the evolution of carbon dynamics in the northern permafrost region on the trajectory of climate change. Proc Natl Acad Sci 115(15):3882–3887. https://doi.org/10.1073/pnas.1719903115

Melton JR, Wania R, Hodson EL, Poulter B, Ringeval B, Spahni R, Bohn T, Avis CA, Beerling DJ, Chen G, Eliseev AV, Denisov SN, Hopcroft PO, Lettenmaier DP, Riley WJ, Singarayer JS, Subin ZM, Tian H, Zürcher S, Brovkin V, van Bodegom PM, Kleinen T, Yu ZC, Kaplan JO (2013) Present state of global wetland extent and wetland methane modelling: conclusions from a model inter-comparison project (WETCHIMP). Biogeosciences 10(2):753–788. https://doi.org/10.5194/bg-10-753-2013

Mi Y, van Huissteden J, Parmentier FJW, Gallagher A, Budishchev A, Berridge CT, Dolman AJ (2014) Improving a plot-scale methane emission model and its performance at a northeastern Siberian tundra site. Biogeosciences 11(14):3985–3999. https://doi.org/10.5194/bg-11-3985-2014

Mishra U, Jastrow J, Matamala R, Hugelius G, Koven C, Harden JW, Ping C, Michaelson G, Fan Z, Miller R (2013) Empirical estimates to reduce modeling uncertainties of soil organic carbon in permafrost regions: a review of recent progress and remaining challenges. Environ Res Lett 8(3):035020

Nicolsky DJ, Romanovsky V, Panda S, Marchenko S, Muskett R (2017) Applicability of the ecosystem type approach to model permafrost dynamics across the Alaska North Slope. J Geophys Res Earth Surf 122(1):50–75

Parmentier F, Van Der Molen M, Van Huissteden J, Karsanaev S, Kononov A, Suzdalov D, Maximov T, Dolman A (2011) Longer growing seasons do not increase net carbon uptake in the northeastern Siberian tundra. J Geophys Res Biogeosci 116(G4):G04013. https://doi.org/10.1029/2011JG001653

Parmentier FW, Zhang W, Mi Y, Zhu X, van Huissteden J, Hayes DJ, Zhuang Q, Christensen TR, McGuire AD (2015) Rising methane emissions from northern wetlands associated with sea ice decline. Geophys Res Lett 42(17):7214–7222. https://doi.org/10.1002/2015GL065013

Petrescu A, Van Huissteden J, Jackowicz-Korczynski M, Yurova A, Christensen T, Crill PM, Maximov T (2007) Modelling CH4 emissions from arctic wetlands: effects of hydrological parameterization. Biogeosciences 5:111–121. https://doi.org/10.5194/bg-5-111-2008

Petrescu A, Van Beek L, Van Huissteden J, Prigent C, Sachs T, Corradi C, Parmentier F, Dolman A (2010) Modeling regional to global CH_4 emissions of boreal and arctic wetlands. Glob Biogeochem Cycles 24(4):GB4009. https://doi.org/10.1029/2009GB003610

Plug LW, West JJ (2009) Thaw lake expansion in a two-dimensional coupled model of heat transfer, thaw subsidence, and mass movement. J Geophys Res Earth Surf 114(F1):F01002. https://doi.org/10.1029/2006JF000740

Schädel C, Schuur EA, Bracho R, Elberling B, Knoblauch C, Lee H, Luo Y, Shaver GR, Turetsky MR (2014) Circumpolar assessment of permafrost C quality and its vulnerability over time using long-term incubation data. Glob Chang Biol 20(2):641–652

Schneider von Deimling T, Grosse G, Strauss J, Schirrmeister L, Morgenstern A, Schaphoff S, Meinshausen M, Boike J (2015) Observation-based modelling of permafrost carbon fluxes with accounting for deep carbon deposits and thermokarst activity. Biogeosciences 12(11):3469–3488

Schuur EA, McGuire AD, Schadel C, Grosse G, Harden JW, Hayes DJ, Hugelius G, Koven CD, Kuhry P, Lawrence DM, Natali SM, Olefeldt D, Romanovsky VE, Schaefer K, Turetsky MR, Treat CC, Vonk JE (2015) Climate change and the permafrost carbon feedback. Nature 520(7546):171–179. https://doi.org/10.1038/nature14338

Slater AG, Lawrence DM (2013) Diagnosing present and future permafrost from climate models. J Clim 26(15):5608–5623

Tan Z, Zhuang Q (2015) Methane emissions from pan-Arctic lakes during the 21st century: an analysis with process-based models of lake evolution and biogeochemistry. J Geophys Res Biogeosci 120(12):2641–2653

Tan Z, Zhuang Q, Walter Anthony K (2015) Modeling methane emissions from arctic lakes: model development and site-level study. J Adv Model Earth Syst 7(2):459–483

Tarnocai C, Canadell JG, Schuur EAG, Kuhry P, Mazhitova G, Zimov S (2009) Soil organic carbon pools in the northern circumpolar permafrost region. Glob Biogeochem Cycles 23(2):GB2023. https://doi.org/10.1029/2008gb003327

Todd-Brown K, Randerson J, Post W, Hoffman F, Tarnocai C, Schuur E, Allison S (2013) Causes of variation in soil carbon simulations from CMIP5 Earth system models and comparison with observations. Biogeosciences 10(3):1717–1736. https://doi.org/10.5194/bg-10-1717-2013

Turetsky MR, Abbott BW, Jones MC, Anthony KW, Olefeldt D, Schuur EA, Koven C, McGuire AD, Grosse G, Kuhry P (2019) Permafrost collapse is accelerating carbon release. Nature 569:32–34. https://doi.org/10.1038/d41586-019-01313-4

Van Huissteden J, Dolman AJ (2012) Soil carbon in the Arctic and the permafrost carbon feedback. Curr Opin Environ Sustain 4(5):545–551. https://doi.org/10.1016/j.cosust.2012.09.008

Van Huissteden J, Berrittella C, Parmentier F, Mi Y, Maximov T, Dolman A (2011) Methane emissions from permafrost thaw lakes limited by lake drainage. Nat Clim Chang 1(2):119–123

Walker WE, Harremoës P, Rotmans J, Van Der Sluijs JP, Van Asselt MB, Janssen P, Krayer von Krauss MP (2003) Defining uncertainty: a conceptual basis for uncertainty management in model-based decision support. Integr Assess 4(1):5–17

Walter Anthony KM, Anthony P, Grosse G, Chanton J (2012) Geologic methane seeps along boundaries of Arctic permafrost thaw and melting glaciers. Nat Geosci 5(6):419–426. https://doi.org/10.1038/ngeo1480

Walter Anthony K, von Deimling TS, Nitze I, Frolking S, Emond A, Daanen R, Anthony P, Lindgren P, Jones B, Grosse G (2018) 21st-century modeled permafrost carbon emissions accelerated by abrupt thaw beneath lakes. Nat Commun 9(1):3262

Walter BP, Heimann M (2000) A process-based, climate-sensitive model to derive methane emissions from natural wetlands: application to five wetland sites, sensitivity to model parameters, and climate. Glob Biogeochem Cycles 14(3):745–765

Walter K, Edwards M, Grosse G, Zimov S, Chapin F (2007) Thermokarst lakes as a source of atmospheric CH_4 during the last deglaciation. Science 318(5850):633–636

West J, Plug L (2008) Time-dependent morphology of thaw lakes and taliks in deep and shallow ground ice. J Geophys Res Earth Surf 113(F1):F01009. https://doi.org/10.1029/2006JF000696

Westermann S, Langer M, Boike J, Heikenfeld M, Peter M, Etzelmüller B, Krinner G (2016) Simulating the thermal regime and thaw processes of ice-rich permafrost ground with the land-surface model CryoGrid 3. Geosci Model Dev 9(2):523–546. https://doi.org/10.5194/gmd-9-523-2016

Zhang Y, Wang X, Fraser R, Olthof I, Chen W, Mclennan D, Ponomarenko S, Wu W (2013) Modelling and mapping climate change impacts on permafrost at high spatial resolution for an Arctic region with complex terrain. Cryosphere 7(4):1121–1137

Glossary

Avulsion The abandonment of a river channel and the formation of a new river channel.

Bryophyte Non-vascular (without water transport vessels) land plants: mosses, liverworts and hornworts

Cryptogams Plants that reproduce by spores

Ebullition Gas transport through water or soil pores by bubbles.

Evapotranspiration The combined evaporation from plant and soil surfaces and the active release of water vapor by plants through their leaf stomata

Gibbs free energy In short, the energy associated with a chemical reaction at that can do work; see for more precise definitions textbooks on thermodynamics.

Glacio-isostatic rebound The slow rising of the land surface after unloading of the Earth's crust when ice caps melts.

Graminoids Plants with a grass-like morphology (e.g. besides grasses, also sedges, rushes etc).

Gyttja Organic lake sediment, consisting of small size plant fragments and plankton.

Marine isotope stages Stratigraphy (see below) based on oxygen isotopes in diatoms in deep sea sediment cores; the stable $\delta^{18}O/^{16}O$ ratio is indicative of the climate (glacials and interglacials).

Mycorrhiza Symbiotic association of fungi with plant roots.

Priming effect Addition of organic matter substances that influence the decomposition of all soil organic matter, either positively or negatively. Here only the positive priming effect is referred to.

Petagram (Pg) A mass of 10^{15} g, equivalent to Gigaton (10^9 metric ton); often used to quantify carbon fluxes on a global scale.

J. van Huissteden, *Thawing Permafrost*,
https://doi.org/10.1007/978-3-030-31379-1

Reanalysis Climate reanalysis is produced by combining models with observations; it gives a grid-based numerical description of the recent climate and weather for all locations on Earth, also where weather station data is missing.

Rhizosphere The region of soil that is influenced by plant roots and their excretions.

Stratigraphy The branch of geology that studies sequences of layered rocks; both the age and type of rock.

Index

J. van Huissteden, *Thawing Permafrost*,
https://doi.org/10.1007/978-3-030-31379-1

MIX
Papier aus verantwortungsvollen Quellen
Paper from responsible sources
FSC® C105338

If you have any concerns about our products,
you can contact us on
ProductSafety@springernature.com

In case Publisher is established outside the EU,
the EU authorized representative is:
Springer Nature Customer Service Center GmbH
Europaplatz 3, 69115 Heidelberg, Germany

Printed by Libri Plureos GmbH
in Hamburg, Germany